Collins
Advanced Science

Human
Biology

Mike Boyle & Kathryn Senior

The complete guide to Human Biology

William Collins's dream of knowledge for all began with the publication of his first book in 1819. A self-educated mill worker, he not only enriched millions of lives, but also founded a flourishing publishing house. Today, staying true to this spirit, Collins books are packed with Inspiration, innovation and practical expertise. They place you at the centre of a world of possibility and give you exactly what you need to explore it.

Collins. Freedom to teach.

Published by Collins
An imprint of HarperCollinsPublishers
77-85 Fulham Palace Road
Hammersmith
London
W6 8JB

Browse the complete Collins catalogue at
www.collinseducation.com

© HarperCollinsPublishers Limited 2008

10 9 8 7

ISBN-13 978-0-00-726751-4

Mike Boyle and Kathryn Senior assert their moral rights to be identified as the authors of this work.

British Library Cataloguing in Publication Data. A Catalogue record for this publication is available from the British Library.

Commissioned by Penny Fowler
Project management by Laura Deacon
Edited by Rosie Parrish
Proof read by Penelope Lyons and Susan Watt
Indexing by Jane Henley
Photo research by Alex Riley
Design by Newgen Imaging
Cover design by Angela English
Production by Arjen Jansen

Contents

1

What's in a cell?

1 WHAT'S IN A CELL?

You started life as a stem cell

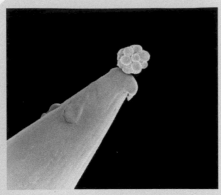

Scanning electron micrograph (SEM) of an
embryo on a pin point

The photo shows an embryo on the point of a pin. You were once that small, about nine months before you were born.

This embryo consists of just eight very special cells. They may not look like much, but they are stem cells – they have the potential to develop into any of the different cells that make up your body. Given time, these cells will divide and then differentiate into particular cell types, such as nerve, muscle, bone or blood.

Just how these cells divide and grow into a complete, healthy organism is the central puzzle of biology. We know that the key lies in the genes – tiny sections of DNA that code for making proteins – but we are still a long way from understanding how genes interact to control the growth and development and repair throughout an organism's life.

When we do unravel these mysteries, the potential benefits are enormous. Although there is a lot of research still to do, stem cells could be used to regenerate tissue that normally can't regenerate itself after injury or damage. Stem cells could be used to bridge spinal cord injuries to bring feeling and function back into limbs, or they could be used to regenerate heart muscle after a heart attack. There have already been some successful attempts to restore sight in people with damaged retinas. Transplanting stem cells in sheets over the damaged part of the retina can allow the eye to reclaim some of its function.

1 THE COMPLEXITY OF CELLS

The artist who drew the cell in Fig 1.1 attempted to show something of the detailed structure of a human cell. This contrasts sharply with the images of animal cells that you will have seen with a light microscope (Fig 1.2). The first time you saw these little bags with dots you were probably disappointed and it is difficult to believe that something that looks like this can be so complex.

But all of your body cells are complex and you have about 50 million million (or 50 trillion) of them. Many are actively dividing. Cells from the lining of the digestive system, the blood and the skin are dying constantly and need to be replaced. In this chapter, you start to look at the complex internal structure of basic cell types. You will also investigate the techniques that have helped us learn more about the relationship between the structure of cells and their function.

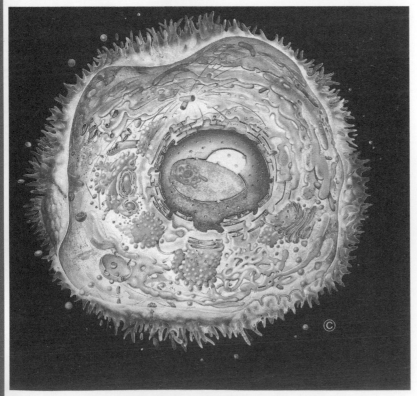

Fig 1.1 This painting illustrates something of the complexity of an animal cell. It was commissioned by the Science Museum on behalf of the Biochemical Society to provide the centrepiece of a permanent exhibition opened in 1987

2 THE CELL THEORY

Following the invention and refinement of the microscope came the **cell theory**, a general acceptance that all living things are made of cells. Modern cell theory has three central ideas:

- The cell is the smallest independent unit of life.
- The cell is the basic living unit of all organisms – all organisms are made up of one or more cells.
- Cells arise from other cells by cell division. They cannot arise spontaneously.

As you study human biology in more detail, you will discover that there are exceptions to every rule. In the case of cell theory, that exception is the **virus**. Viruses do not have a cellular structure or organisation, and whether they are actually living organisms is a subject of debate.

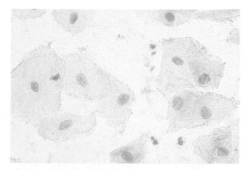

Fig 1.2 The light microscope can reveal the basic features of animal cells; this is what cheek cells look like

3 USING CELL STRUCTURE TO CLASSIFY ORGANISMS

Organisms can be classified on the basis of the internal organisation of their individual cells. With the exception of viruses, all organisms are either **prokaryotic** or **eukaryotic**. Prokaryotic cells are relatively simple, they have no separate nucleus and show little organisation. Bacteria are prokaryotes. In contrast, eukaryotic cells are larger and show much more internal organisation. Animals, plants, fungi and protoctists are all eukaryotes. The main differences between prokaryotes and eukaryotes are shown in Table 1.1.

Table 1.1 Differences between prokaryotes and eukaryotes

	Prokaryotes	Eukaryotes
organisms	bacteria	plants, animals, protoctists, fungi
diameter of cells	0.1–10 µm	10–100 µm
site of genetic material	DNA in cytoplasm	DNA inside distinct nucleus
organisation of genetic material	DNA is circular; no histone proteins; DNA does not condense at cell division	DNA is linear; attached to histone proteins; condenses into visible chromosomes before cell division
internal structure	few organelles	many organelles with complex membrane systems
cell walls	always present	present in plants and fungi and some protoctists; never in animals
flagella	have simple flagella	have modified cilia that consist of microtubules in a distinctive '9 + 2' arrangement

4 PROKARYOTES – THE FIRST ORGANISMS?

A few billion years ago, the first living organisms to evolve on Earth were probably prokaryotes. The term literally means 'before the nucleus' because the genetic material (DNA) of these organisms is not enclosed by a membrane, and therefore they do not have a true nucleus.

It is tempting to think of prokaryotes as inferior to eukaryotes, but in some ways they have achieved greater success. They have been on Earth more than twice as long as eukaryotes, they are present in greater numbers (there are more bacteria living on your skin than there are people on Earth) and they occupy an enormous number of different habitats. Some bacteria, for example, are able to live in volcanic springs at temperatures as high as 90 °C.

 REMEMBER THIS

No matter how far you take the study of biology, all you need to know about microscopic distances is these two simple facts:

- 1 mm = 1 000 µm
  1 millimetre = 1 000 micrometres
- 1 µm = 1 000 nm
  1 micrometre = 1 000 nanometres

How small is small?

Cells and the molecules they contain are very small. When we study them, we must think about measurements that are minute beyond our imagination. It is easy, for example, to develop a mental block when confronted by the statement 'The nanometre is 10^{-9} m.'

The two units commonly used to describe microscopic objects are the **micrometre** (µm) (commonly called the micron) and the **nanometre** (nm) (Fig 1.3). Starting with a familiar unit, the millimetre (mm), one thousandth of one millimetre is known as a micrometre (µm). The micrometre is used to describe cells and organelles. An average animal cell is 30 to 50 µm across;

the nucleus has a diameter of about 10 µm. Plant cells can reach 150 µm or more in length.

When describing small cellular components and molecules, the useful unit is the nanometre (nm). A nanometre is one thousandth of a micrometre. As a rough guide, the light microscope reveals structures that can be measured in micrometres, but you need an electron microscope to see objects measured in nanometres.

Fig 1.3 Units of measurement; how they are used and how they relate to each other

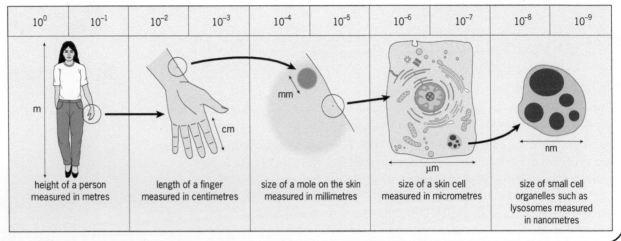

10^0	10^{-1}	10^{-2}	10^{-3}	10^{-4}	10^{-5}	10^{-6}	10^{-7}	10^{-8}	10^{-9}

height of a person measured in metres

length of a finger measured in centimetres

size of a mole on the skin measured in millimetres

size of a skin cell measured in micrometres

size of small cell organelles such as lysosomes measured in nanometres

? QUESTION 1

1 **a)** Estimate the diameter (in micrometres) of a full stop on this page.
b) How many nanometres are there in one millimetre?

BACTERIA

Most bacteria are spherical or rod-shaped cells, and several micrometres long (Fig 1.4) . Their rigid protective **cell wall** is made of **peptidoglycan**, a substance unique to bacteria. Beneath the cell wall is the **plasma membrane**, which is similar in structure to the membrane of eukaryotic cells. This completely encloses the contents of the cell. Some types of bacteria, such as *Neisseria meningitidis*, which can cause

(a)

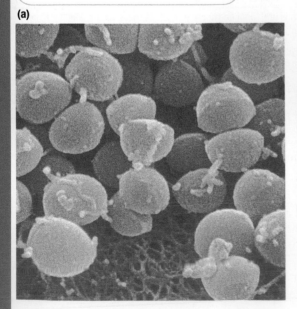

(b)

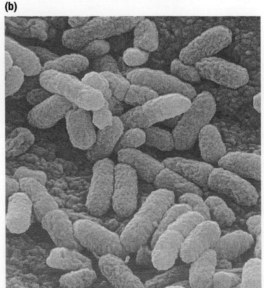

Fig 1.4a SEM showing *Staphylococcus aureus* cells. These spherical cells are called cocci (singular coccus), and are one of the two main types of bacterial cell (the others are bacilli – singular bacillus). *S. aureus* is commonly found on human skin where it usually does no harm. However, it can cause serious infection if it enters the body through a cut or wound

Fig 1.4b SEM showing *Salmonella enteritidis* cells. These sausage-shaped cells are called bacilli. This species of *Salmonella* can cause severe food poisoning

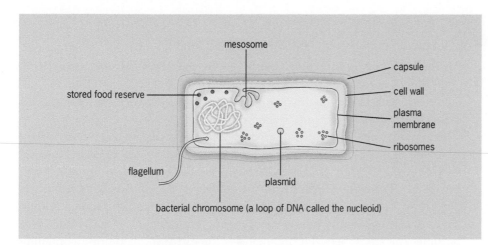

Fig 1.5 The internal rod-shaped structure of a bacterial cell – *Escherichia coli*

mesosome

capsule

cell wall

stored food reserve

plasma membrane

ribosomes

flagellum

plasmid

bacterial chromosome (a loop of DNA called the nucleoid)

meningitis, also have a **capsule**. This is a sticky coat outside the cell wall that prevents the bacterium from drying out, from being digested by host intestinal enzymes, or from being attacked by the host's immune system.

Fig 1.5 shows the internal structure of a rod-shaped bacterium, *Escherichia coli*.

Inside the plasma membrane, the bacterial cell is a single cytoplasmic compartment that contains DNA, RNA, proteins and small molecules. Bacteria have a circular piece of DNA in a region of the cytoplasm known as the **nucleoid**. There are smaller rings of DNA, known as **plasmids**, elsewhere in the cytoplasm. Plasmids are of great interest to biologists because they often contain genes that code for antibiotic resistance, and can be used to carry genes between cells in genetic engineering.

Bacteria feed by **extracellular digestion**. They release enzymes into the surrounding medium and absorb the resulting soluble products. **Glycogen granules** and **lipid droplets** often occur in the bacterial cytoplasm and provide a limited store of these polymers. Bacteria can synthesise a wide variety of enzymes, and some species are able to digest unlikely substances such as oil and plastic. Proteins are synthesised on **ribosomes** (page 13), and cell respiration occurs on **mesosomes**, inner extensions of the plasma membrane.

Some bacteria are **motile**: they can swim. They have thin fibres called **flagella** (singular flagellum) that are corkscrew-shaped and that rotate, propelling the bacteria in different directions. Other fibres, called **pili**, enable the bacteria to perform a primitive form of sex, as shown in Fig 1.6. During **conjugation**, the pili become joined and form a channel to allow plasmids to be copied and transferred to other individuals.

? QUESTION 2

2 Some bacteria can digest, or break down, oil and plastic. Suggest how this could be useful to people.

The role of plasmids in genetic engineering is discussed further in Chapter 27.

✔ REMEMBER THIS

Until the invention of the microscope, people had no idea that cells existed. They knew nothing of bacteria and had no idea about how infectious diseases spread.

The role of bacteria in disease and in the environment is discussed in Chapters 29 and 35 respectively

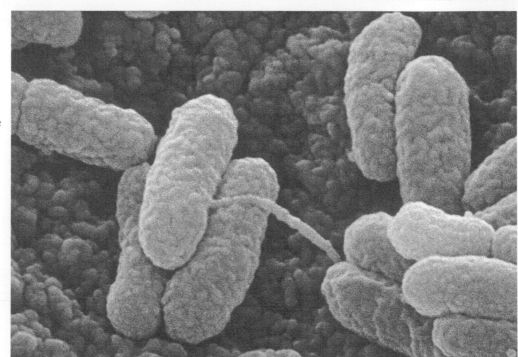

Fig 1.6 SEM showing conjugation in two bacterial cells

5 EUKARYOTES: ANIMAL CELLS

All species of animals, plants, **fungi** and **protoctists** are made of eukaryotic cells. The term **eukaryote** means 'true nucleus', because the DNA of eukaryotic cells is confined to a definite area inside the cell enclosed by a **nuclear envelope**.

Eukaryotic cells also have other **organelles** that form compartments. By being in a compartment, the chemicals involved in a particular process, such as respiration or photosynthesis, are kept separate from the rest of the cytoplasm. This allows the chemical reactions of the process to take place quickly and efficiently. This high degree of internal organisation is one of the reasons why eukaryotic cells are larger than prokaryotic cells. The fluid that occupies the space between organelles is the **cytosol**, a solution containing a complex mixture of enzymes, the products of digestion (amino acids, sugars, and so on) and waste materials.

INSIDE AN ANIMAL CELL

If you look at an animal cell under a **light microscope**, the right staining and illumination techniques and a good quality microscope should allow you to see the **nucleus**, **nucleolus**, the **chromosomes** in a dividing cell, and even the **Golgi body**, **mitochondria** and food storage particles. But to make sure you see the inside of a cell in more intricate detail, you really need an **electron microscope**.

The **electron micrograph** in Fig 1.7 (opposite) shows the internal structure of an animal cell. You can see far more of the cell's components – the **endoplasmic reticulum** (**ER**), the internal features of the mitochondria, the plasma membrane, **lysosomes, ribosomes** and **cytoskeleton** are all now visible.

In this chapter, we look at the detailed structure of each individual organelle and find out how each type contributes to the function of the cell as a whole.

Table 1.2 summarises the functions of some of the major organelles and structures in the eukaryotic cell.

Table 1.2 Summary of the functions of major eukaryotic cell organelles and structures

Organelle	Occurrence	Size	Function
nucleus	usually one per cell	10 µm	site of the nuclear material – the DNA
nucleolus	inside nucleus	1–2 µm	manufacture of ribosomes
mitochondrion	numerous in cytoplasm; up to 1 000 per cell	1–10 µm	aerobic respiration
rough endoplasmic reticulum	continuous throughout cytoplasm	extensive membrane network	isolation and transport of newly synthesised proteins
smooth endoplasmic reticulum	usually small patches in cytoplasm	variable	synthesis of some lipids and steroids
ribosome	free in cytoplasm or attached to rough ER	20 nm	site of protein synthesis
Golgi body	free in cytoplasm	variable	modification and synthesis of chemicals
lysosome	free in cytoplasm	100 nm	digestion of unwanted material
chloroplast	cytoplasm of some plant cells, e.g. mesophyll	4–10 µm	site of photosynthesis
vacuole	usually large, single fluid-filled space in plant; smaller and more numerous in animals	up to 90% of volume of whole plant cell	storage of salts, sugars and pigments; creates turgor pressure by interaction with cell wall
plasma membrane	encloses the cytoplasm of all cells	7–10 µm	exchange and transport of materials into and out of the cell
cell wall	surrounds all plant and fungal cells (but of different structure)	thickness varies	provides rigidity and strength

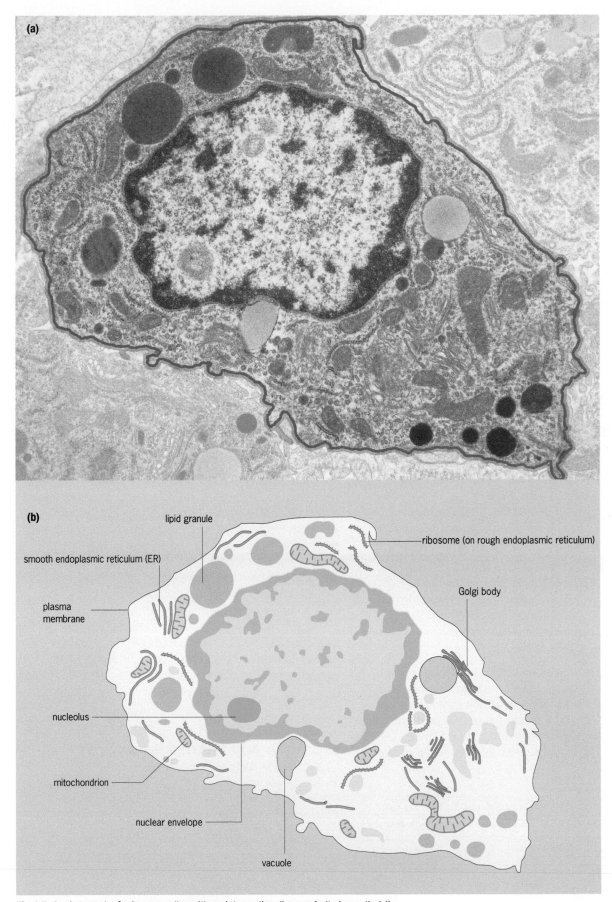

Fig 1.7 A micrograph of a human cell **a** with an interpretive diagram **b**. It shows that the cytoplasm of animal cells contains a complex system of membrane-bound organelles

Using microscopes
THE LIGHT MICROSCOPE

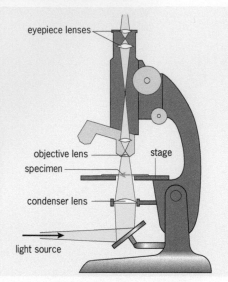

Fig 1.8 The standard compound microscope

Fig 1.8 shows the structure of a **light** or **optical microscope**. This instrument is also known as a **compound microscope** because two lenses, the **eyepiece lens** and the **objective lens**, are combined to produce a much greater **magnification** than is possible with a single lens. The total magnification is calculated by multiplying the magnification of the two lenses together. For example, if the eyepiece lens has a magnification of ×10 and the objective lens is ×50, the total magnification is ×500.

The light microscope has powers of magnification of up to × 1500, good enough to see cells, larger organelles and individual bacteria, but not powerful enough to reveal smaller structures such as plasma membranes, viruses or individual molecules. Table 1.3 shows organelles in animal and plant cells that are visible with a light microscope.

An important feature of a microscope is its **resolving power**, which should not be confused with the magnification. Two objects close together may appear as one single image when viewed under the light microscope. Increasing the magnification does not allow you to **resolve** the two objects into separate images; the objects just appear to be a larger single image. Resolving power is as important as magnification when investigating structural details.

The limitation of the light microscope is due to the nature of light itself. The wavelength of light determines the maximum effective magnification and the resolving power. The wavelength of visible light is around 500–650 nm and the resolving power – the resolution – of the light microscope is 200 nm (0.2 μm), so two objects separated by less than 200 nm appear as one object.

Table 1.3 Structures normally visible in animal and plant cells with the light microscope

Organelle	Function	Animal cells	Plant cells
nucleus	control of cell activities	✓	✓
vacuole	storage and support	✗	✓
chloroplasts	photosynthesis	✗	✓
cell wall	support	✗	✓

THE ELECTRON MICROSCOPE

The electron microscope (EM) (Fig 1.9) was invented in the 1930s. Today's modern transmission EMs can magnify up to 50 million times and can resolve two objects that are 78 picometres apart (one picometre is one trillionth of a metre). They can visualise individual atoms. The development of the EM has had a huge impact on biology. Organelles inside cells can be seen in great detail and new ones have been discovered.

While the light microscope uses lenses to focus a beam of light, the EM uses electromagnets to focus a beam of electrons. The wavelength of the electrons is much smaller than the wavelength of light, so the resolving power of the EM is much greater than the light microscope.

The main disadvantage of the EM is that the electron beam must travel in a vacuum because, being so small, electrons are scattered when they hit air molecules. Specimens for the EM must therefore be prepared (killed, dehydrated and fixed) so that they retain their structure inside a vacuum. Such harsh preparation methods can damage cells, and cause **artefacts** – features that do not exist in the living cell – to appear. For example, microsomes, tiny vesicles surrounded by ribosomes, were seen when animal cells were examined using the electron microscope. At first, cell biologists thought that these were organelles they had not noticed before, but they later realised that microsomes were fragments of endoplasmic reticulum (see Table 1.2, page 6) produced by the fixing process.

The main differences between electron microscopy and light microscopy are shown in Table 1.4, opposite.

SCANNING AND TRANSMISSION ELECTRON MICROSCOPES

There are now several types of EMs, including the transmission EM, the scanning EM, the reflective EM, the scanning transmission EM and the tunnelling EM.

In a transmission electron microscope (TEM), a beam of electrons is transmitted through the specimen. The specimen must be thin and it is stained using electron-dense substances such as heavy metal salts. These substances deflect electrons in the beam and the pattern that the remaining electrons produce as they pass through the specimen is converted into an image (Fig 1.7a).

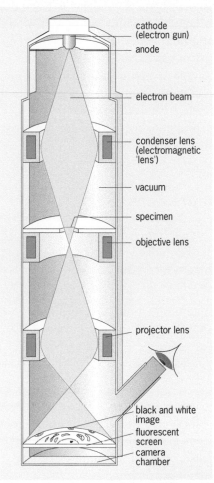

cathode (electron gun)

anode

electron beam

condenser lens (electromagnetic 'lens')

vacuum

specimen

objective lens

projector lens

black and white image

fluorescent screen

camera chamber

Fig 1.9 The basic features of the transmission electron microscope. The beam of electrons is focused by electromagnets before passing through the specimen, and the image appears on a fluorescent screen

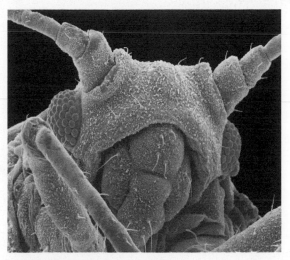

Fig 1.10 SEM of the head of an aphid

STM was invented in 1981 by Gerd Binnig and Heinrich Rohrer who worked for IBM at the time. They won the Nobel Prize for their development in 1986.

The STM is special because it allows scientists to 'see' individual molecules and even atoms. The three-dimensional images produced (Fig 1.11) can show the individual atoms within a crystal lattice, for example. This kind of EM can resolve to a distance of 0.2 nanometres and it is interactive. It can be used to manipulate atoms, control or trigger chemical reactions or add or remove electrons.

It is mainly used in the physical sciences but has been used to visualise the surface of cell membranes.

The scanning EM and the tunnelling EM are the most useful in biology. Scanning EMs record the electrons that bounce off the surface of an object, so they produce stunning three-dimensional images that can provide a wealth of information (Fig 1.10).

The **scanning tunnelling microscope** (STM) is completely different to a scanning electron microscope. It scans an electrical probe over a surface and picks up the weak electric current that flows between the probe and the surface. The

Table 1.4 Comparing light and electron microscopes

	Light microscope	Electron microscope
illumination	light	electrons
focused by	lenses	magnets
maximum magnification	×1 500	×50 million
resolving power	200 nm (0.2 μm)	78 picometres
specimens	living or dead	dead
preparation of specimens	often simple	more complex
cost of equipment	relatively cheap	very expensive
images in colour	yes	no (colour is added by computer)

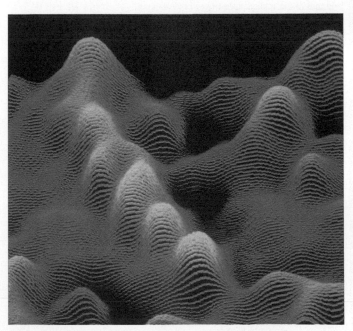

Fig 1.11 False-colour scanning tunnelling micrograph (STM) of double-stranded DNA

? QUESTION 3

3   If a light microscope had an eyepiece lens of ×25 and an objective lens of ×40, what would the total magnification be?

Chapter 4 covers the structure and function of the plasma membrane in detail.

Fig 1.12 The plasma membrane is a complex organelle but its structure can be represented by this simplified diagram

Fig 1.13 As the electron micrograph **a** and the interpretive diagram **b** show, the nucleus is bounded by the nuclear envelope, a double membrane which contains many pores, each one about 100 nm in diameter. The nuclear pores represent about 15 per cent of the total surface area of the nuclear envelope, indicating the heavy traffic of materials in and out of the nucleus. The nucleus contains the cell's genetic material that exists as chromatin, loosely packed DNA attached to proteins called histones. The nucleolus is clearly visible

THE PLASMA MEMBRANE

The **plasma or cell surface membrane** is the boundary between the cell and its environment. It has little mechanical strength but plays a vital role in controlling which materials pass in and out of the cell.

Although basically a double layer of **phospholipid molecules**, arranged tail to tail, the plasma membrane is a complex structure, studded with proteins. These can be embedded in the membrane or they can penetrate the **bilayer** forming **pores** (holes or channels – Fig 1.12 and see the section on the plasma membrane in Chapter 4) through which molecules can pass.

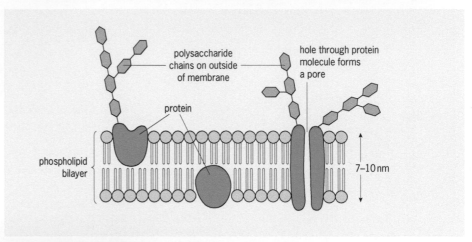

THE NUCLEUS

The **nucleus** is the largest and most prominent organelle in the cell. Almost all eukaryote cells have a nucleus – red blood cells in mammals and phloem cells in plants are exceptions. Every nucleus is surrounded by a **nuclear envelope**. As Fig 1.13 shows, this consists of two membranes that are separated by a gap of 20 to 40 nm.

The nucleus is usually spherical and about 10 μm in diameter. It contains the cell's DNA, which carries information that allows the cell to divide and carry out all its cellular processes. When the cell is not actively dividing (as in Fig 1.13), the DNA is spread throughout the nucleus as chromatin. Close examination of the chromatin reveals two different levels of density. Dark-staining chromatin, consisting of tightly packed DNA, is known as **heterochromatin**; the lighter, more loosely packed material is called **euchromatin**. Euchromatin contains the DNA that is being actively read

(a)

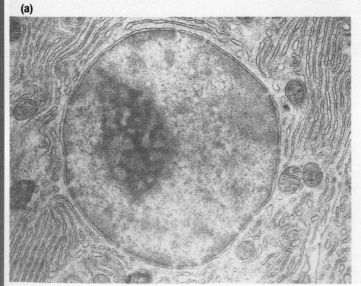

(b)

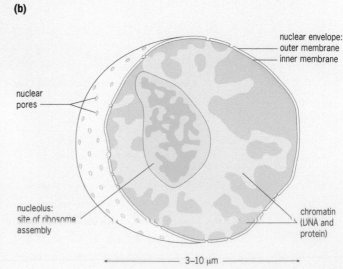

to produce proteins; in heterochromatin, the DNA is packed together, and is not being read.

Individual segments of DNA called genes contain the information necessary to make individual proteins, including the enzymes that control most of the cell's activities. In fact, a central concept in biology that is true for all cells, prokaryotes and eukaryotes, is that:

Many genes code for making enzymes that, in turn, control the activities of the cell.

When a cell is dividing, its chromosomes become visible, as Fig 1.14 shows.

Nuclei also have one or more nucleoli (Fig 1.15). These dark-staining, spherical structures are ribosome-producing centres: they synthesise ribosomal RNA and package it with ribosomal proteins to make ribosomes.

> Enzymes are discussed in Chapter 5 and protein synthesis in Chapter 24. Cell division is covered in detail in Chapter 23.

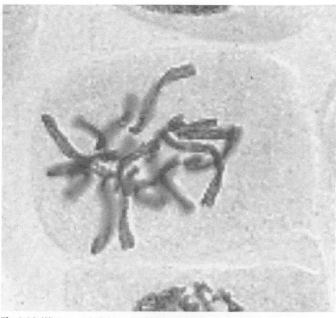

Fig 1.14 When a cell divides, its DNA condenses into visible chromosomes

(a)

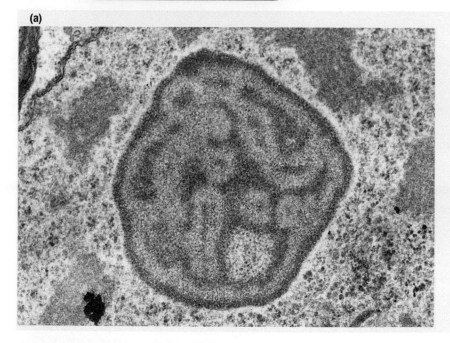

(b)

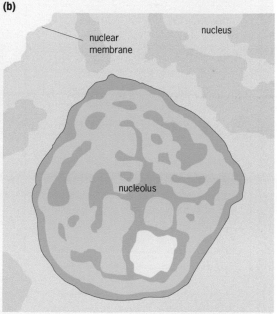

Fig 1.15 As the electron micrograph **a** and diagram **b** show, the nucleolus has a highly organised structure; its DNA codes for the RNA and proteins that are used to make ribosomes

ENDOPLASMIC RETICULUM

The nuclear envelope joins with the membrane of the **endoplasmic reticulum** (ER), a system of complex tunnels that are spread throughout the cell.

On much of the outside surface of the ER in a eukaryotic cell are the sites of attachment for ribosomes. This gives it a grainy appearance (Fig 1.16 overleaf) and its name, **rough ER**.

The main function of rough ER is to keep together and transport the proteins made on the ribosomes. Instead of simply diffusing away into the cytoplasm, newly made proteins are threaded through pores in the membrane and accumulate in the space called the **ER lumen** (Fig 1.17). Here, they are free to fold into their normal three-dimensional shape. Not surprisingly, a mature cell that makes and secretes large amounts of protein – such as one

(a)   **(b)**

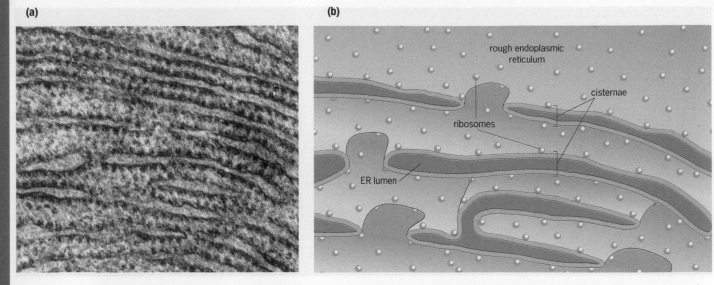

Fig 1.16 Electron micrograph (**a**) and interpretive diagram (**b**) showing rough ER. The ER is a large sheet of membrane that is folded over on itself many times, forming stacked layers called cisternae. The space inside the cisternae, the ER lumen, forms an extensive transport system throughout the cytoplasm

The smooth ER in a binge drinker

Liver cells contain large amounts of smooth ER. The enzymes on the inner surface of the smooth ER help the body cope with the sudden influx of large amounts of alcohol – as in binge drinking. The smooth ER can double its surface area and its enzymes within a few days to cope with the extra demand. It returns to normal when the alcohol has been dealt with. Drinking too much and too often can overwhelm this process and can cause permanent liver damage.

that makes digestive enzymes – has rough ER that occupies as much as 90 per cent of the total volume of the cytoplasm. The rough ER is also a storage unit for enzymes and other proteins.

Small vesicles containing newly synthesised proteins pinch off from the ends of the rough ER and either fuse with the Golgi body or pass directly to the plasma membrane.

ER with no ribosomes attached is known as **smooth ER** (Fig 1.18 opposite). Smooth ER tends to occur in small areas that are not continuous with the nuclear membrane. Smooth ER is not involved in protein synthesis but is the site of steroid (lipid hormone) production. It also contains enzymes that **detoxify**, or make harmless, a wide variety of organic molecules, and it acts as a storage site for calcium in skeletal muscle cells.

?   QUESTIONS 4–5

4   What function might the cells of the liver, gut and glands have in common? How could you recognise such cells from electron micrographs?

5   Why was ER not discovered until after the invention of the electron microscope?

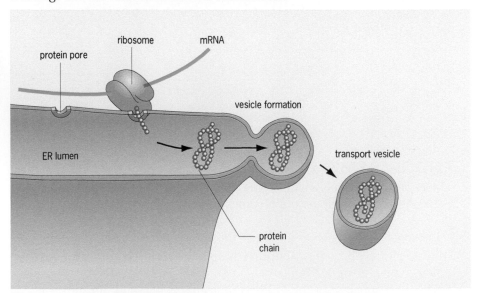

Fig 1.17 Accumulation of proteins in the ER lumen. Many proteins made on the ribosomes are threaded through pores in the ER membrane

RIBOSOMES

Ribosomes are small, dense organelles, about 20 nm in diameter, present in great numbers in the cell. Most are attached to the surface of rough ER but they can occur free in the cytoplasm, as in Fig 1.19. This artist's impression of protein synthesis shows the ribosome's distinctive shape. Ribosomes are made from a combination of ribosomal RNA and protein (65 per cent RNA : 35 per cent protein); they are involved in protein synthesis, as described in Chapter 24.

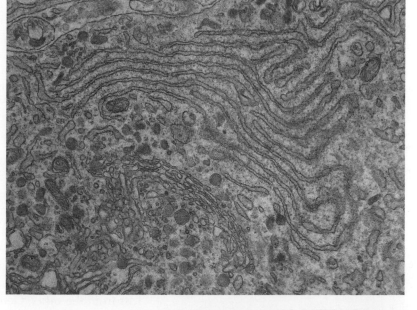

Fig 1.18 EM of smooth ER, which has no attached ribosomes. It is usually not as abundant as rough ER, although it is common in the cells of the liver, gut and some glands

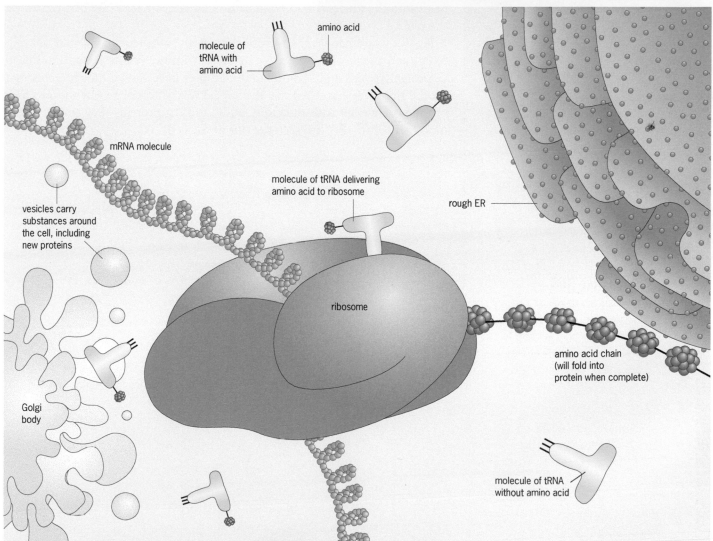

Fig 1.19 An artist's impression of protein synthesis. A ribosome can be thought of as a giant enzyme on which a protein is assembled. Transfer RNA molecules (tRNA) bring specific amino acids to the ribosome. Each one is added to the growing amino acid chain according to the code on the messenger RNA molecule

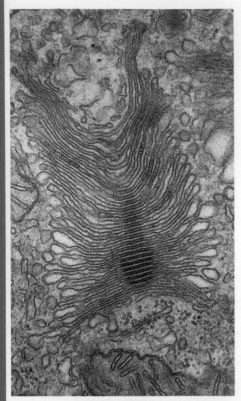

Fig 1.20 The Golgi body with a very large number of vesicles. The forming face is at the foot and secretory vesicles are at the top

Ribosomes assemble amino acids in the right order to produce new proteins. The ribosome uses the code on messenger RNA (mRNA) to put amino acids together in chains to form specific proteins. This is another central concept in biology that you will learn more about later on:

A gene is a piece of DNA that codes for a particular protein. A copy of the gene in the form of messenger RNA passes out of the nucleus and travels to the ribosome where it controls protein synthesis.

Generally, proteins that are to be used inside the cell are made on free ribosomes, while those that are to be secreted out of the cell are made on ribosomes that are bound to ER membranes.

THE GOLGI BODY

The **Golgi body**, also called the **Golgi apparatus**, is a tightly packed group of flattened cavities or vesicles (Fig 1.20). The whole organelle is a shifting, flexible structure; vesicles are constantly being added at one side and lost from the other. Generally, vesicles fuse with the forming face (the one nearest to the nucleus) and leave from the maturing face (the one nearest to the plasma membrane).

Fig 1.21 summarises the relationship between the rough ER and the Golgi body, which appears to be involved with the synthesis and modification of proteins, lipids and carbohydrates. Studies have shown that proteins made on the ribosomes attached to ER are packaged into vesicles by the ER. Some of the vesicles join with the Golgi body, and the proteins they contain are modified before they are secreted out of the cell.

Protein synthesis is covered in detail in Chapter 24.

 REMEMBER THIS

A vesicle is a small spherical organelle, bounded by a single membrane, which is used to store and transport material around the cell.

cell surface membrane

material passes out of cell

material passes into cell

lysosome

secretory vesicle

endocytotic vesicle

maturing face

Golgi body

forming face

rough ER

Fig 1.21 The various functions of the Golgi body are summarised in this diagram. Vesicles from the ER or from outside the cell bring material to the Golgi body. After processing, material passes out of the cell or enters lysosomes

LYSOSOMES

Lysosomes are small **vesicles** 0.2–0.5 µm in diameter (Fig 1.22) that contain a mixture of digestive enzymes called **lytic enzymes**. It is important that the membrane of lysosomes remains intact because if the enzymes leak out, they could digest vital molecules in the cell.

Why do cells need such potentially lethal structures? They have several uses:

- To supply the enzymes which destroy old or surplus organelles.
- To digest material taken into the cell. After a white cell has engulfed a bacterium, for example, lysosomes discharge enzymes into the vacuole (Fig 1.23) and digest the organism. This process is called **phagocytosis**.
- To destroy whole cells and tissues. Parts of tissues and organs often need to be removed after they have performed their function. The muscle of the uterus is reduced after giving birth and milk-producing tissue is destroyed after breast-feeding finishes. Bone is also constantly made and reabsorbed throughout life.

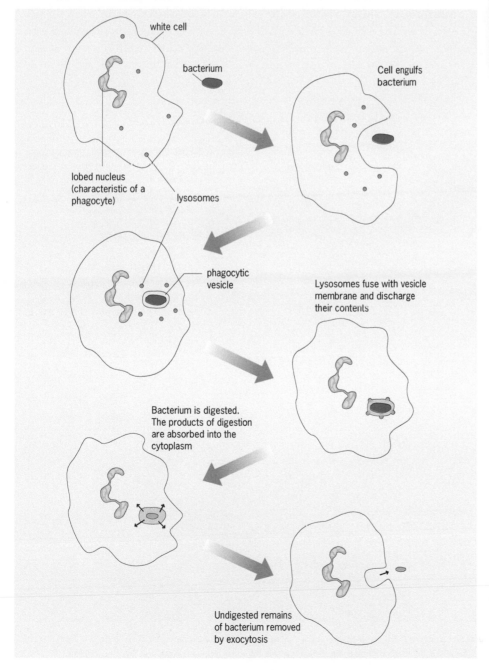

white cell

bacterium

lobed nucleus
(characteristic of a
phagocyte)

lysosomes

Cell engulfs
bacterium

phagocytic
vesicle

Lysosomes fuse with vesicle
membrane and discharge
their contents

Bacterium is digested.
The products of digestion
are absorbed into the
cytoplasm

Undigested remains
of bacterium removed
by exocytosis

? QUESTIONS 6–7

6 What would happen to an organism if it lost control of its lysosomes?

7 If we could control lysosome activity in specific tissues, how could this be used in the treatment of cancer?

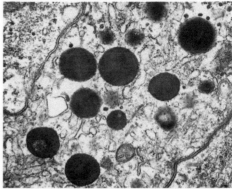

Fig 1.22 Lysosomes are simply bags of digestive enzymes. They can be distinguished from other vesicles in the cell only by using a stain specific for the chemicals inside the lysosomes

? QUESTION 8

8 Why are digestive enzymes synthesised in an inactive form?

Fig 1.23 One of the functions of lysosomes is to digest material taken into the cell by the process of phagocytosis (see also Chapter 4). Here, a white cell, known as a phagocyte, is ingesting a bacterium. Lysosomes discharge their enzymes into the temporary vacuole and the bacterium is digested

Lysosomes are similar in structure to many other vesicles in the cell and are thought to be made in the same way: inactive digestive enzymes from the rough ER pass through the Golgi body where they are activated and packaged into vesicles.

MITOCHONDRIA

Mitochondria are relatively large, individual organelles that occur in large numbers in most cells. They are usually spherical or elongated (sausage-shaped) and are 0.5–1.5 µm wide and 3–10 µm long. Their function is to make **ATP** by **aerobic respiration**. ATP is a molecule that diffuses around the cell and provides instant chemical energy to the processes that require it.

As Fig 1.24 shows, mitochondria have a double membrane; the outer membrane is smooth while the inner one is folded. This arrangement gives a large internal surface area on which the complex reactions of aerobic respiration can take place. Mitochondria are particularly abundant in metabolically active cells, tissues such as muscle and tissues involved in active transport.

Cell respiration is discussed in detail in Chapter 3.

The function of mitochondria is covered in more detail in Chapter 3.

Active transport is discussed in detail in Chapter 4.

? QUESTION 9

9 Why does a sperm need so many mitochondria?

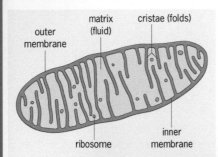

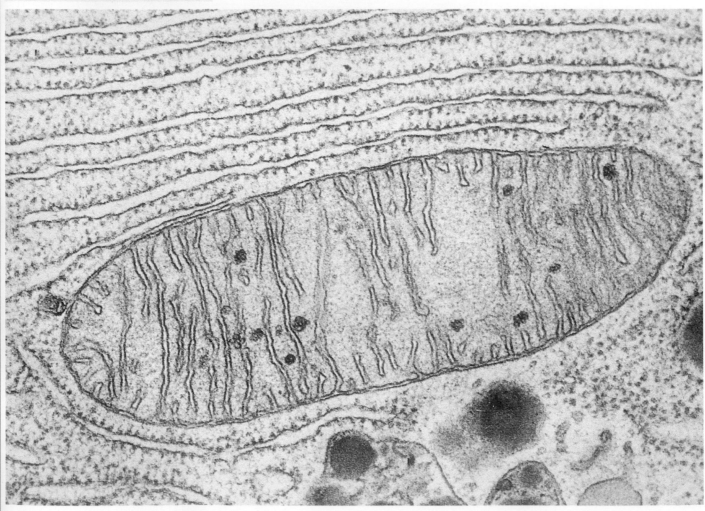

Fig 1.24 Mitochondria are the sites of aerobic respiration in the cell. The inner membrane is folded into cristae, which provide a large surface area that is occupied by many of the enzymes involved in cell respiration. Generally, the more metabolically active the cell, the more cristae there are in the mitochondria

CELL FRACTIONATION

In order to study the function of a particular organelle it is often helpful to isolate it from the rest of the cell. This can be done by cell fractionation, as outlined in Fig 1.25.

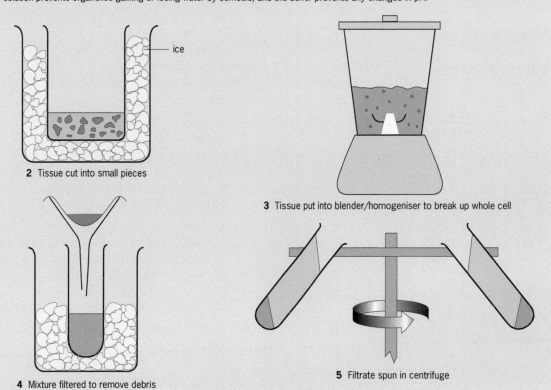

1 Liver tissue is placed in a fluid that is an **ice-cold isotonic buffer**. When the cell is broken up, membranes are ruptured and many compounds that do not normally mix are brought together. The low temperature minimises unwanted reactions, including self-digestion by digestive enzymes. The isotonic solution prevents organelles gaining or losing water by osmosis, and the buffer prevents any changes in pH.

— ice

2 Tissue cut into small pieces

3 Tissue put into blender/homogeniser to break up whole cell

4 Mixture filtered to remove debris

5 Filtrate spun in centrifuge

Principle: The organelles will separate out in a particular order, according to their density and shape. After a time, the sediment containing a particular type of organelle can be separated from the supernatant (the liquid that contains the remaining organelles). The exact times vary from tissue to tissue.

	Organelle		Centrifuge setting (g)	Time (minutes)
First to separate out	nuclei		800–1000	5–10
	mitochondria		10 000–20 000	15–20
	lysosomes			
	rough ER		50 000–80 000	30–50
	plasma membranes		80 000–100 000	60
	smooth ER			
Last to separate out	free ribosomes		150 000–300 000	> 60

Fig 1.25 Cell fractionation: how to isolate particular types of organelle from liver tissue

17

Size and scaling – getting a sense of proportion

An essential part of microscopy is to know the actual size of the objects being studied. Examiners often use this topic to test your basic numeracy.

There are only three bits of information to work with.
1 The actual size of the object
2 The magnification
3 The size of the image on the paper or screen
If you know two of these values, you can work out the third one. The formula is:

$$\text{magnification} = \frac{\text{observed size}}{\text{actual size}}$$

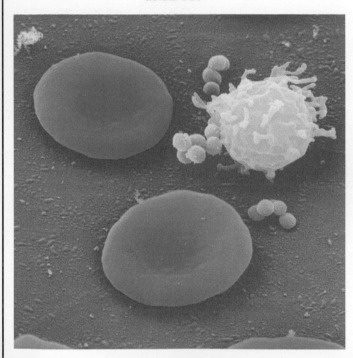

Fig 1.26 Coloured scanning electron micrograph of red blood cells, white blood cells and bacteria (a type of staphylococcus). Magnification is x7500. The bacteria are a lot smaller than the blood cells

This can be simplified further to $\triangle{\tiny O \atop M\ A}$. In exam questions, cover up the value you need and the calculation you need to do should be clear.

Often an exam question will show a micrograph or diagram, give you the magnification and ask you to work out the actual size. For example, in Fig 1.26 the magnification is X7500, i.e. the image is ×7500 times larger than the cell really is. Measuring the diameter of the red cell, we find it is 45 mm.

Converting mm into micrometres, 45 mm is 45 000 µm. The real cell must be 7500 times smaller than that.

So actual size is 45 000/1800 = 6 µm

Try working out the size of the stoma plus guard cells in Fig 1.27.

Recognise a silly answer when you see one

It's a great help if you know the sort of answer you are looking for. If you measure a bacterium and calculate it to be 3 mm long, there's something wrong somewhere.

Remember the golden rules

1 mm = 1000 µm (micrometres)

1 µm = 1000 nm (nanometres)

The following are useful guidelines

- Animal cells are usually 20 – 50 µm in diameter, while plants cells are usually larger, perhaps 100–150 µm

- Organelles such as nuclei and mitochondria are obviously smaller. An average nucleus may be 10 µm in diameter while mitochondria range from 1 to 10 µm.

- Most bacteria are much smaller than plant and animal cells. An average bacterium would be about the same size as a mitochondrion (there may be a good reason for this – see page 22 about the endosymbiont theory). The largest known bacteria are 250 µm long but the smallest are just 0.3 µm long.

- Small organelles and large molecules are measured in nm. The plasma membrane, for example, is only about 8 nm thick.

- Viruses are really, really tiny. The largest ones measure about 400 nm – slightly larger than the smallest known bacteria. The smallest viruses are just 20 nm in diameter.

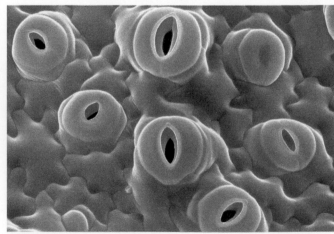

Fig 1.27 Micrograph of guard cells from the underside of a tobacco leaf, showing the stomata (pores). Magnification is ×450

Autoradiography

Electron microscopes are useful tools for investigating cell structure, but they don't reveal much about the processes going on inside a living cell. After World War II, work on the atomic bomb produced chemicals called **radioactive isotopes** as a by-product. These are very useful for studying metabolism because they behave in the same way as non-radioactive atoms and their movements through the cell can be traced (Fig 1.28).

After cells are exposed to a radioactive isotope such as **carbon-14**, they are washed to remove excess isotope and then fixed, embedded, sectioned and mounted on microscope slides. The slides are then taken to a darkroom and coated with photographic emulsion that has been heated to make it melt. After coating, the slides are cooled and kept in the dark for various time periods from a few days to several weeks. During this time, radioactive decay causes a chemical change to occur in the film of photographic emulsion. When this is developed using standard photographic techniques and the slide is examined under the light microscope, silver grains can be seen in the areas exposed by the radioactivity. The pattern of grains shows the structures of the cell that have incorporated the radioactive label.

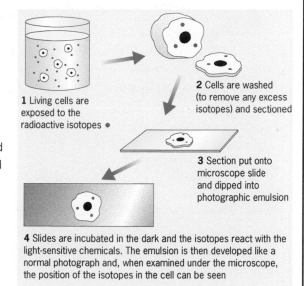

1 Living cells are exposed to the radioactive isotopes •

2 Cells are washed (to remove any excess isotopes) and sectioned

3 Section put onto microscope slide and dipped into photographic emulsion

4 Slides are incubated in the dark and the isotopes react with the light-sensitive chemicals. The emulsion is then developed like a normal photograph and, when examined under the microscope, the position of the isotopes in the cell can be seen

Fig 1.28 The basic method of autoradiography

6 EUKARYOTES: PLANT CELLS

Typical leaf cells from a plant, seen with the light microscope, are shown in Fig 1.29. These are **palisade mesophyll cells**.

Plant cells have several features that are not found in animal cells (Table 1.5). A plant cell is surrounded by a cell wall made of cellulose. A large

? QUESTION 10

10 How many of the following cells would fit into a line 1 cm long?
a) Animal cells of 20 μm
b) Plant cells of 50 μm
c) Bacterial cells of 2 μm

Table 1.5 Comparing plant cells and animal cells

Feature	Cell: Animal	Plant
nucleus	✓	✓
plasma membrane	✓	✓
mitochondria	✓	✓
rough ER	✓	✓
smooth ER	✓	✓
Golgi bodies	✓	✓
lysosomes	✓	✓
cell wall	✗	✓
plastids, eg chloroplasts	✗	✓
vacuoles	✓	✓
centrioles*	✓	✓*

*Centrioles are not present in the cells of higher plants

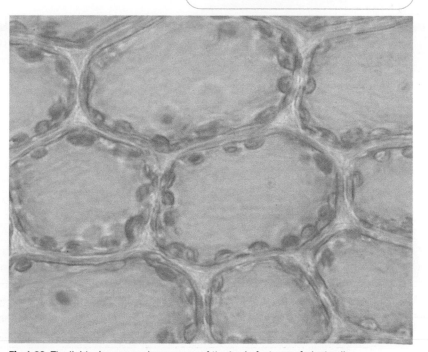

Fig 1.29 The light microscope shows some of the basic features of plant cells

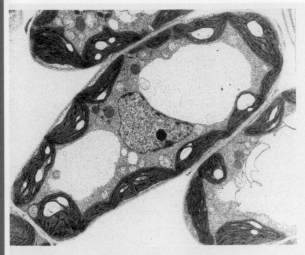

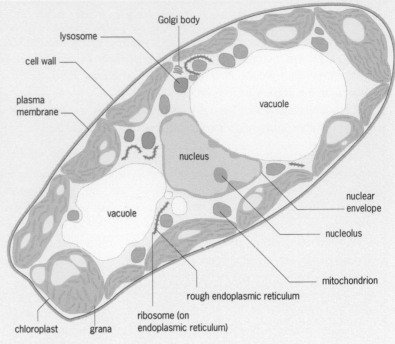

Fig 1.30 Transmission electron micrograph (above) and explanatory diagram (right) showing a palisade mesophyll cell from a soya bean leaf

? **QUESTION 11**

11 Which parts of a plant would be unlikely to contain chloroplasts, and why?

The structure of cellulose is discussed in Chapter 3.

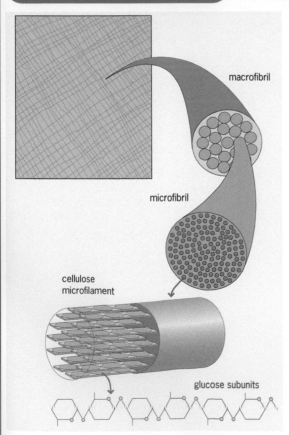

Fig 1.31 The cellulose cell wall is a composite material. Fibrils containing thousands of cellulose molecules are cemented together by a 'glue', a complex mixture of compounds

proportion of the inside of the cell is taken up with a fluid-filled compartment known as the **vacuole**. Together, the cell wall and vacuole maintain the shape of the whole cell.

As you can see from Fig 1.30, plant cells have specialised organelles, the **chloroplasts**, which enable them to make their own food by photosynthesis. Chloroplasts bring together all the chemicals involved in photosynthesis, and provide sites for the chlorophyll molecules so that they can absorb the maximum amount of light.

THE CELLULOSE CELL WALL

Unlike animal cells, plant cells are enclosed by a **cell wall** that consists of many **cellulose fibres** cemented together by a mixture of other organic substances.

Cellulose is a **polysaccharide** (a sugar polymer) which consists of long straight chains of glucose molecules bonded to adjacent molecules by hydrogen bonds. In the cell wall, around 2000 parallel cellulose molecules are packed together to form **microfibrils**. These in turn are bundled together to form **macrofibrils**, as you can see in Fig 1.31. Like fibreglass, the cell wall has great strength because of the many strong fibres and the 'glue' that holds them together.

The main functions of the cell wall are:
- to provide rigidity and strength to the cell. Cell walls need to be strong enough to resist expansion to allow the cell to become turgid,
- to force the cell to grow in a certain way. A particular arrangement of fibrils causes the cell to assume a particular shape – for example a long, thin tube.

VACUOLES

Plant cells have a large, fluid-filled cavity called a **vacuole**. In mature cells the vacuole can occupy over 80 per cent of the cell volume, and is filled with a fluid called **cell sap** (Fig 1.32). The sap consists of a complex mixture of sugars, salts, pigments and waste products in water.

Vacuoles have several functions:

- They absorb water by **osmosis** and therefore swell, pushing the cytoplasm against the cell wall. In this state, a plant cell is said to be turgid.
- They store food substances such as sugars and mineral salts.
- They store pigments that give colour to plant structures such as petals.
- They can accumulate waste products and by-products of metabolism. Sometimes these chemicals may be toxic or have an unpleasant taste and are used by the plant to make it less palatable to herbivores.

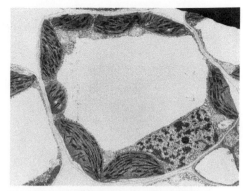

Fig 1.32 In a mature plant cell, the vacuole becomes so large that the cytoplasm is reduced to a thin layer around the cell

CHLOROPLASTS AND OTHER PLASTIDS

Chloroplasts are one of a group of plant cell organelles known as **plastids**. Chloroplasts are surrounded by a double membrane and contain an elaborate internal membrane system that houses the chemicals of photosynthesis (Fig 1.33).

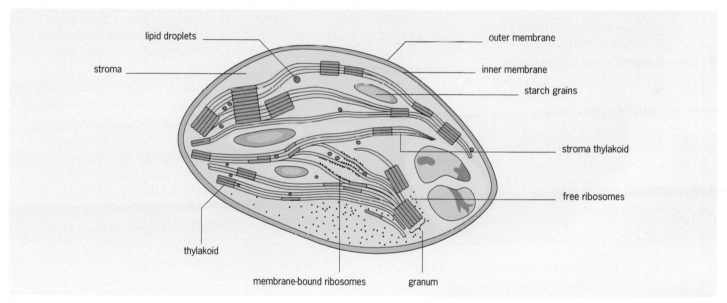

lipid droplets

stroma

outer membrane

inner membrane

starch grains

stroma thylakoid

free ribosomes

thylakoid

membrane-bound ribosomes

granum

Fig 1.33 EM of chloroplast (right) and interpretive diagram (above). Chloroplasts are organelles in which all the chemicals associated with photosynthesis are brought together. Chlorophyll molecules are housed on membranes to maximise light absorption. In a flowering plant, most chloroplasts are found in the palisade cells of the leaves

The structure and function of chloroplasts are covered in detail in Chapter 33.

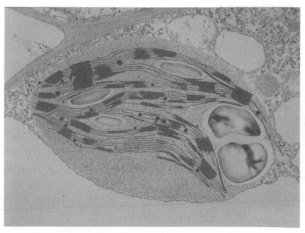

The origin of mitochondria and chloroplasts: the endosymbiont theory

Scientists continue to speculate whether or not mitochondria and chloroplasts originated from outside eukaryotic cells. Many researchers think that they were originally free-living prokaryotes, similar to bacteria and blue-green bacteria, which, at some point, became incorporated into larger cells.

Let's look at the evidence. Mitochondria and chloroplasts have several similarities to prokaryotes:

- They are similar in size to prokaryotes.
- They have their own DNA, as prokaryotes do, suggesting an ability to reproduce independently.
- The DNA of both organelles is circular, like that of bacteria.
- Both organelles have their own ribosomes, smaller than those in eukaryotic cells but identical to those in prokaryotes.
- All reproduce by binary fission.
- The biochemistry of mitochondria is closer to that of aerobic bacteria than that of the eukaryotic cell; chloroplasts and blue-green bacteria are even more similar.

This theory was first suggested by a team led by Lyn Margulis at Boston University, USA. Assuming the theory is true, how could these organisms have become incorporated into a larger cell without being destroyed by lysosomes?

Imagine a heterotrophic, anaerobic organism – similar to an amoeba in structure and movement – that feeds by phagocytosis and respires without oxygen, though oxygen is present in its environment. If such a cell were to ingest aerobic bacteria that it did not digest, the bacteria could respire aerobically, and in doing so could provide the host organism with far more ATP than the host could produce on its own. The host cell in turn would give the bacteria material to respire – food.

Similarly chloroplasts, which may originally have been free-living blue-green bacteria, could also have become incorporated into larger cells. If these were ingested – but not digested – they would continue to photosynthesise within the host and so provide it with organic food and oxygen – an obvious advantage. The host, in turn, could provide the blue-green bacterium with carbon dioxide and nitrogen.

REMEMBER THIS

An **autotroph** is an organism that can make its own food: a green plant is an autotroph that make its own food by photosynthesis. A **heterotroph** is an organism that must obtain ready-made food: a cow is a heterotroph that obtains its food from grass.

Associations in which an autotroph lives inside a heterotroph, or an aerobe lives inside an anaerobe, are both very convenient arrangements. One organism produces what the other needs. This is the key to the success of several other associations, for example, lichens (algae and fungi) and corals (algae inside animals). Both host and 'passenger' gain a survival advantage: the host gets food or energy and the smaller organism is protected from predators.

The idea of incorporation by ingestion could also explain why both organelles have an extra external membrane. The inner membrane would represent the original prokaryotic membrane while the outer one could represent the food vacuole into which the organism was taken.

SUMMARY

When you have studied this chapter, you should know and understand the following:

- All living organisms are composed of either **prokaryotic** or **eukaryotic** cells.
- **Bacteria** are prokaryotes; **plants**, **animals**, **protoctists** and **fungi** are eukaryotes.
- Prokaryotic cells are relatively small, with very little internal organisation and few internal membranes.
- Eukaryotic cells are relatively large, with a high degree of internal organisation.
- Eukaryotic cells have a true, membrane-bound **nucleus** and many **organelles** – membrane-bound compartments in which particular chemical processes take place.
- The nucleus, **ribosomes**, **rough endoplasmic reticulum** (ER), **vesicles** and **Golgi body** in eukaryotic cells all take part in different stages in the making and packaging of chemical products. Some chemicals are used within the cell.

Others – such as hormones or digestive enzymes – are released.

- **Mitochondria** are the site of production of most of the eukaryotic cell's **ATP**, which provides the energy for other cellular processes.
- Plant cells contain structures not found in animal cells: **chloroplasts**, a **cell wall**. The **vacuole** in most plant cells is much larger than the vacuoles found in an animal cells.
- The cell can be studied using a variety of experimental techniques: the **electron microscope** reveals the structure of the cell, **cell fractionation** isolates individual organelles and, with the use of radioactive isotopes in **autoradiography**, the passage of materials through the cell can be followed.

Practice questions and a How Science Works assignment for this chapter are available at www.collinseducation.co.uk/CAS

2

Tissues and organs

2 TISSUES AND ORGANS

Growing tissues and body parts

As we learn more about how tissues interact to form organs and organ systems, we come across new opportunities to create artificial tissues to use in medicine. Artificial heart valves and tubes to keep blood vessels open have been used for quite a while, but more recently a team of bioengineers in the USA have gone one better. They have developed a way of treating the lining from the small intestine, stomach, bladder and liver of pigs and other animals to make an organic mesh that can be used to repair internal organs such as vocal cords and damaged ligaments. This mesh is also proving useful for treating skin wounds and chronic sores.

The tissues are removed and sterilised, then processed into sheets, tubes and other shapes which are grafted into areas that need to be replaced or repaired. Because the material is obtained from living tissue, it not only contains collagen, the fibrous protein that has great strength and flexibility, but also growth factors, proteins that send messages to cells to tell them where and how to grow and differentiate.

Once implanted into the body, the growth factors send instructions to the surrounding cells. The resulting new tissue growth is essentially the same shape and performs the same functions as the damaged tissues.

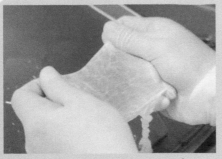

A senior research scientist holds a sheet of material harvested from the urinary bladder of a pig. The material holds promise as a scaffold for the repair or replacement of many different body parts, such as the trachea and oesophagus in human patients

1 THE COMPLEXITY OF LIVING ORGANISMS

Most human cells are microscopic. The largest cell in the human body is the egg cell, or **ovum**. This is just visible with the naked eye. It needs to be larger than other cells because it needs space to store food reserves (Fig 2.1).

Humans, like all large organisms, are **multicellular**. Our body systems consist of complex organs and tissues (Fig 2.2) that interact and co-operate to keep all the cells of the body in the best possible environment. Different groups of cells carry out different tasks and functions: they have **differentiated** and become **specialised**. Different types of cells develop to carry out specific functions that contribute to the success of the organism as a whole. This is known as **division of labour**.

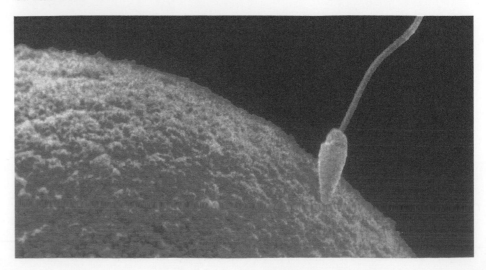

Fig 2.1 This photograph shows the huge difference in size between an ovum (a human egg cell) and the sperm that will fertilise it. The cytoplasm is packed with food reserves that allow the cell to divide several times before it implants in the uterus. The DNA in a fertilised egg carries all the information necessary to make a complete human baby in just nine months

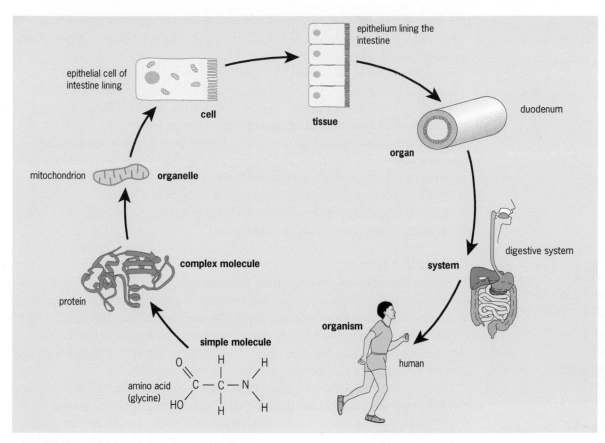

Fig 2.2 The body of a human being is made up from 50 million million (or 50 trillion) cells and its complexity is amazing. It is organised on many different levels

2 TISSUES

A **tissue** is a collection of similar specialised cells that work together to achieve a particular function. Different tissues combine to form more complex structures called **organs**. An organ usually has a specific function in the organism, for example, to detect light, to absorb food or to produce a hormone.

Studying the detailed structure of individual tissues is a lot more interesting if you look at how the structure of a tissue relates to its function in the body of the organism. Getting familiar with normal tissues is particularly important in medicine; recognising what is normal and abnormal in a particular type of tissue can mean the difference between life and death.

> ✓ **REMEMBER THIS**
> A **tissue** is a collection of similar specialised cells.
> **Histology** is the study of cells, tissues and organs, using microscope techniques.
> **Mitosis** is a type of cell division (see Chapter 23).

HUMAN TISSUES

In humans, as in all mammals, each organ of the body is made up from a combination of these four basic tissue types:

- **Epithelial tissue** forms thin sheets that line and cover body structures. Epithelial tissue lines the intestines, for example.
- **Connective tissue** is tough and fibrous and forms structures that hold the body together. Ligaments and tendons are mainly connective tissue.
- **Muscular tissue** can contract to produce movement.
- **Nervous tissue** has the ability to conduct impulses, allowing communication between different parts of the body.

Epithelial tissue

Epithelial tissues form continuous sheets that line or cover most structures and cavities in the body. Different of epithelial tissue types are classified according to the size and shape of their cells, and the number of cell layers they contain. For example, simple **squamous epithelium** contains only

? QUESTIONS 1

1   Smoking has been found to affect the lining of the respiratory system. What type of tissue is found here?

✓ REMEMBER THIS

Connective tissue consists of cells in a matrix that usually contains strengthening fibres.

The structure of proteins is discussed in Chapter 3.

✓ REMEMBER THIS

Fibrous cartilage is a good shock absorber because it is compressible (it can withstand pressure without being damaged) and it can absorb the impact of a rapid jolt which would damage a more brittle structure such as a bone.

✓ REMEMBER THIS

Tendons are made from fibrous connective tissue that contains many closely packed collagen fibres. In some situations, such as in the legs during vigorous exercise, the tendons withstand enormous forces.

one layer of cells and is therefore thin and ideally suited to be a membrane where substances are exchanged, such as at the surface of alveoli. **Stratified squamous epithelium** is several cells thick and can be replaced continuously, making it perfect as a protective covering for the skin.

Epithelial tissues have very little mechanical strength; they are supported and attached to underlying connective tissue by a **basement membrane**, a continuous sheet consisting of **collagen** and other proteins.

Epithelial tissues form barriers that keep different body systems separate. They also have many other functions. For instance, the epithelial cells which line the mammalian respiratory tract are **ciliated**: they form a 'carpet' of tiny hair-like processes which trap inhaled dust and sweep it away from the lungs and out towards the throat where it is swallowed. Other epithelial tissues, such as the lining of the mammalian intestine, secrete **mucus**.

Connective tissue

As their name suggests, **connective tissues** usually connect structures together. Connective tissue cells are not packed closely together but are separated by a non-cellular **matrix** (a mesh) which they secrete. (Notice the contrast to epithelial tissue where the cells are tightly packed.) The nature of the matrix accounts for the property of the tissue. In bone, for example, the matrix is a hard mineral strengthened by fibres of collagen, a tough protein.

The main types of connective tissue found in the human body are listed in Table 2.1. Cartilage, tendons, ligaments, blood and bone are all connective tissues. Each one is adapted to perform a particular function. All except blood have a structural role in the body. Blood is described as a fluid connective tissue because plasma, a fluid, separates and surrounds the cells. Because connective tissues are mainly structural, their metabolic demands (their need for food and oxygen) are relatively small.

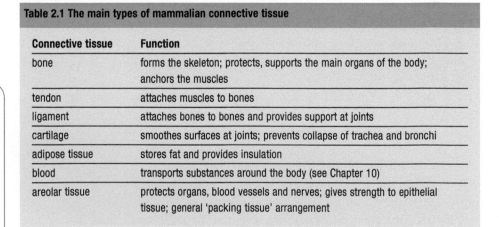

Table 2.1 The main types of mammalian connective tissue

Connective tissue	Function
bone	forms the skeleton; protects, supports the main organs of the body; anchors the muscles
tendon	attaches muscles to bones
ligament	attaches bones to bones and provides support at joints
cartilage	smoothes surfaces at joints; prevents collapse of trachea and bronchi
adipose tissue	stores fat and provides insulation
blood	transports substances around the body (see Chapter 10)
areolar tissue	protects organs, blood vessels and nerves; gives strength to epithelial tissue; general 'packing tissue' arrangement

Cartilage

Cartilage is a tough, smooth and flexible connective tissue. There are three types of cartilage:

● **Hyaline cartilage**. This is a compressible and elastic tissue found in the nose and trachea, and at joints where it covers the ends of bones to provide smooth surfaces.
● **White fibrous cartilage**. This is found between the vertebrae and acts as a shock absorber.
● **Yellow elastic cartilage**. This is a very elastic tissue found in the ears, pharynx (throat) and epiglottis (the flap that closes over the airway during swallowing).

Skin grown from stem cells for treating serious burns

A skin burn is classified into first degree, second degree or third degree, depending on what depth of skin is affected. Third-degree burns are severe, full thickness burns that completely destroy the skin and some of the underlying tissue (Fig 2.3a).

About 20 years ago, people with third-degree burns that covered over 75% of their body rarely survived. Since then, the critical care of burns patients has managed to treat the shock and fluid loss and many more people survive. Today, people with almost 100% burns can be treated successfully.

Another thing that has changed is the way skin is grafted to treat the burns themselves. Doctors used to cut away healthy pieces of skin and use them to cover the burned areas. In people with nearly 100% burns, this was not possible.

Cadaver skin, artificial skin and cultured skin have all been used with varying levels of success but the most promising treatment for serious burns involves skin made from stem cells.

Artificial skin is now produced commercially and is available in burns centres throughout the world (Fig 2.3b). When a patient arrives with large area, third-degree burns, surgeons have learned that the best treatment is to excise any damaged skin still left. In this operation the patient is 'peeled' with an instrument that does look quite like a potato peeler. The healthy underlying tissue is then covered as quickly as possible with artificial skin. This stops loss of fluid from the huge wound and prevents microorganisms getting in to cause infection. The artificial skin also acts as a matrix for the patient's own skin to regenerate into, and patients now suffer much less scarring than previously. The patient is kept sedated for long periods, so that he or she can recover with as little psychological trauma as possible.

One recent project has been led by the Singapore-based biotech company CellResearch Corporation in association with the National Hospital of Traditional Medicine and St Paul's Hospital Burns Centre in Vietnam. Researchers and doctors there have worked out a way to grow stem cells on synthetic scaffolds to make sheets of skin to put onto the wounds of their patients. These heal quickly and are not rejected. They are really useful when the person does not have enough healthy skin for any grafting to be attempted.

The stem cells used to make the skin are taken from the umbilical cord lining of newborn babies. This is less controversial than taking stem cells from embryos originally developed in *in vitro* fertilisation. In normal births the umbilical cords are collected from the placenta – this organ would normally be discarded. The stem cells that can be taken from it are versatile enough to turn into many of the body's cells.

Various skin cell types produced by the stem cells have also been used to produce 'cultured skin', which could one day be used as replacement skin for other types of injury, or as a tool to study the effects of different compounds on the skin by cosmetics companies.

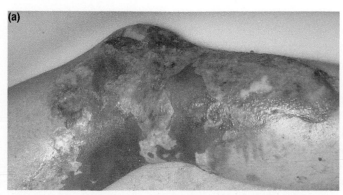

Fig 2.3a A third-degree burn, the deepest and most serious type of burn injury

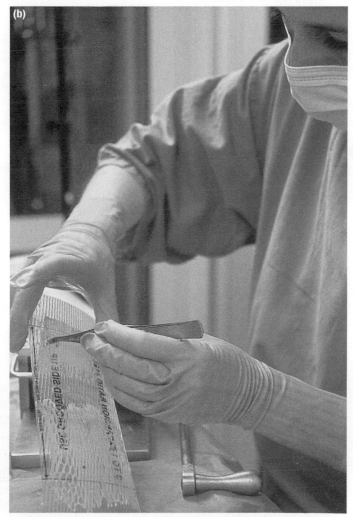

Fig 2.3b Cultured skin is sliced and stretched in preparation for burns' treatment

Muscle structure and function are discussed in Chapter 17.

Muscular tissue

Muscular tissues are contractile: they contain protein fibres, actin and myosin, which produce movement when they slide over each other.

As Fig 2.4 shows, there are three types of muscle: skeletal muscle, smooth muscle and cardiac muscle. The properties of the three types are summarised in Table 2.2.

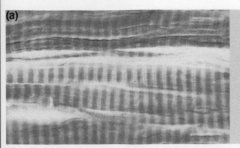
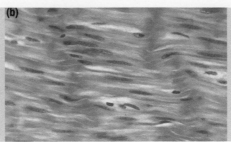

Fig 2.4 Micrographs of muscle: skeletal **a**; smooth **b**; cardiac **c**. See Table 2.2 for descriptions

? QUESTIONS 2–3

2 Which type of muscle would you expect to have the fewest mitochondria? Explain.

3 What features of skeletal muscle make it unsuitable to form the muscles of the heart?

Table 2.2 Comparing the three types of mammalian muscle

Muscle:	Skeletal	Smooth	Cardiac
other names	striped, striated, voluntary	unstriped, unstriated, involuntary, visceral	heart
site in the body	attached to skeleton	tubular organs; gut, reproductive system, glands, bronchioles	heart
function	movement and maintenance of posture	usually controls movement of substances along tubes	heartbeat
control	voluntary	involuntary (not under conscious control)	myogenic (self-generating)
speed of contraction	rapid	slow, usually known as peristalsis (rhythmic squeezing movements) (see Chapter 8)	rapid
speed of fatigue	rapid	slow	slow

The muscles of most vertebrates make up a large proportion of their body mass. As you might expect, active muscle tissue has a very high metabolic rate, and many mitochondria are packed between the muscle fibres.

Nervous tissue

Nervous tissue contains neurones, specialised elongated cells that transmit electrical impulses from one end to the other. Nervous tissue controls and co-ordinates the activities of the body. Changes in its internal and external environment act as stimuli that are detected by receptor cells in sense organs. From here, impulses travel along sensory neurones to the central nervous system (CNS). The CNS consists of the brain and spinal cord, both also made up mainly of neurones. The brain processes the information it receives and decides which response to make, often taking into account previous experience (memory). Impulses are sent out along motor neurones to effector organs: the muscles and glands.

Control and co-ordination in the nervous system are discussed in Chapters 15 and 16.

PLANT TISSUES

Like animals, plants are made up of different tissues that, together, form a whole functioning organism. Of course, there are basic differences between plants and animals, and plant tissues reflect these differences. Tissues of a multicellular plant are adapted for the following functions:

- photosynthesis,
- transport,
- support,
- storage,
- protection,
- reproduction,

> The process of photosynthesis is covered in Chapter 33.

3 ORGANS AND BODY SYSTEMS IN ANIMALS

An **organ** is a collection of tissues that work together to perform a particular function. For example, think about one of the most sophisticated of organs, the eye (Fig 2.5):

- The iris contains muscular tissue, as do the suspensory ligaments that control the lens.
- The retina is made from nervous tissue that detects light and sends information to the brain.
- Epithelial tissues make up blood vessels and other structures such as the capsule around the lens, and the conjunctiva that covers the front of the eye.
- The whole organ is held together and protected by connective tissue. The sclera, in particular, forms a tough, protective outer coating.

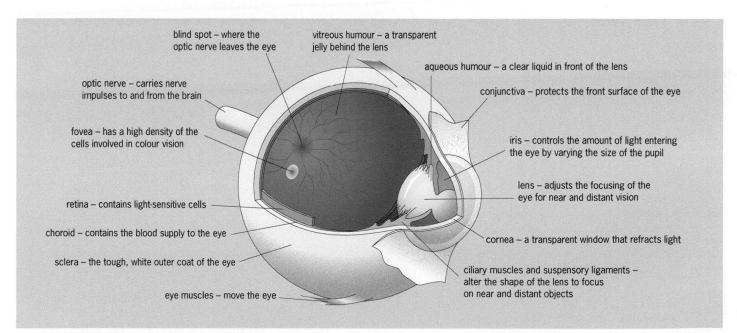

Fig 2.5 The human eye. The eye is a sphere, about 2.5 cm across. Its outer wall has three layers: an inner layer, the retina; a middle layer, the choroid; and a tough outer layer, the sclera

In a whole organism, organs form **systems** that carry out life processes such as digestion, excretion and reproduction. The combined systems of the body also need to create a stable internal environment to keep cells bathed in fluid that has the optimum temperature and composition.

? QUESTION 4

4 Which body system in the list on this page does not contribute directly to the maintenance of internal conditions in the body?

Here is a short summary of the main systems in the human body together with their primary functions:

- The **digestive system** extracts from food the simple molecules that cells need (Fig 2.6).
- The **excretory system** removes the cells' waste products from the body.
- The **respiratory system** provides oxygen and removes carbon dioxide.
- The **circulatory system** transports all necessary substances to the cells, and removes waste products.
- The **muscles**, **skeleton** and **nervous system** combine to ensure that the organism obtains food and avoids danger.
- The **reproductive system** enables the organism to produce offspring and to pass on its genes.

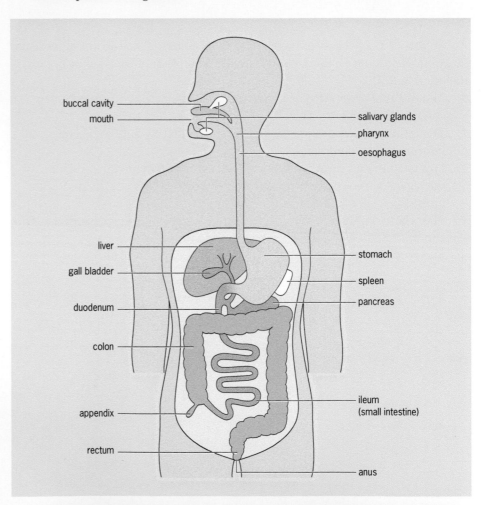

Fig 2.6 The digestive system. Many organs form the system that enables us to process our food

SUMMARY

When you have studied this chapter, you should know and understand the following:

- The cells that make up the human body are specialised for a variety of different functions. This is called **division of labour**.
- Specialised cells form **tissues**, tissues form **organs** and organs form **organ systems**.
- A tissue is a collection of similar specialised cells.
- The tissues of the human body can be classified

into four broad groups: **epithelial**, **connective**, **muscular** and **nervous**.

- The whole organism consists of all the systems working together to produce an individual that is able to control its internal conditions.

 Practice questions and a How Science Works assignment for this chapter are available at www.collinseducation.co.uk/CAS

3

The chemicals of life

3 THE CHEMICALS OF LIFE

Back in time to 'RNA world'

All new organisms come from other organisms – they don't just arise spontaneously. So how did life first come about? Could it have developed from non-living chemicals?

Some experiments, carried out in the USA in the 1950s by Stanley Miller, showed that this is a definite possibility. In his laboratory, Miller recreated the conditions thought to have been present on Earth 3500 million years ago. Then, the atmosphere contained no oxygen. Instead, it was probably a mixture of methane, ammonia, carbon dioxide, hydrogen sulfide and hydrogen. There were frequent electrical storms, and no ozone layer to protect the surface of the Earth from the Sun's ultraviolet rays.

When these conditions were recreated in the laboratory, the chemical reactions in the mixture produced several biological molecules: sugars, amino acids and bases. The findings were important because these molecules are the building blocks of nucleic acids and proteins – the chemicals on which all life on Earth is based. Since these groundbreaking experiments, further work over the past half century has suggested that one of the earliest self-replicating 'life-forms' may have been RNA. An 'RNA world' may have persisted for a long time, until conditions became milder and DNA was able to take over. However, we still have many jigsaw pieces to find before we work out how these complex molecules might have made the giant step to become life.

Was there ever life on Mars? In 2003, instruments on a Mars lander looked for minute traces of molecular fossils as some scientists believe that similar conditions to those on Earth may have once produced organic molecules. Finding amino acids on Mars could finally solve the mystery of whether life of any sort did once exist on Mars and it could also help us understand life's origins on our own planet

1 STUDYING THE CHEMICALS OF LIFE

The study of the chemicals and reactions that take place inside living organisms is a science in its own right – **biochemistry**. Biochemists aim to understand how living organisms work by studying how the molecules within their cells interact. In this chapter we look at **carbohydrates**, **lipids**, **proteins** and **nucleic acids**, and describe the characteristics that make them the chemicals of life.

SOME CENTRAL THEMES IN BIOCHEMISTRY

The chemical systems that make up living things seem incredibly complicated, but there are some simple underlying patterns:

- Living organisms contain a huge range of **macromolecules**, but these large molecules are built from a small number of simple molecules.
- The simple building-block molecules are very similar in all organisms, suggesting that all life had a common origin.
- The characteristics of an organism are determined by the information contained in its DNA.
- The DNA contains information that the cell can use to make proteins. Many proteins are **enzymes**, which control the physical and chemical activities of an organism.
- The chemical activities that go on inside an organism are given the general term **metabolism**.
- Metabolic reactions are divided into two general categories: **anabolic** and **catabolic**. Anabolic reactions build up large molecules from smaller ones, while catabolic reactions do the reverse, breaking down larger molecules.

Photosynthesis

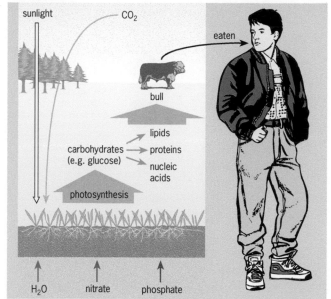

Respiration

respiration
food + $O_2 \rightarrow H_2O + CO_2$ + energy

Fig 3.1 Organic molecules are normally made in plants by photosynthesis. These molecules are then available to the rest of the food chain

Fig 3.2 All organisms, including plants, need energy. They obtain energy by breaking down organic molecules (food). This process is respiration

- Anabolic reactions usually involve **condensation reactions** in which building-block molecules are joined together and a water molecule is released.
- Catabolic reactions, such as those that occur during digestion, usually involve hydrolysis reactions in which larger molecules are split as they react with water.
- In photosynthesis (Fig 3.1), plants use the energy from sunlight to build up organic molecules such as sugars from simple ones such as carbon dioxide and water.
- All organisms need a supply of energy, which they obtain via respiration (Fig 3.2). In respiration, organic molecules are oxidised into simpler molecules, usually carbon dioxide and water. The resulting energy is used to fuel the many energy-requiring processes within the organism.

LIFE ON EARTH IS CARBON-BASED

Why is life on Earth based on carbon? It is because this element can bond with itself repeatedly, to produce an infinite variety of molecules. Living things are composed mainly of organic molecules: proteins, carbohydrates, lipids and other molecules that have 'skeletons' made from carbon.

The carbon atom

Carbon has an atomic number of 6: it has a nucleus containing six protons and six neutrons, orbited by six electrons, two in an inner shell, the other four in an outer shell (Fig 3.3, overleaf). Carbon acquires a full, and therefore stable, outer shell of eight electrons by sharing four more. So, carbon forms four covalent bonds – bonds with electrons shared between the atoms. We say it has a **valency** of 4.

Carbon can form covalent bonds with other carbon atoms to form stable chains or rings. C—C bonds are the most common; compounds that contain single carbon bonds are said to be **saturated**. C=C bonds are also frequent and C≡C are possible. Compounds with double or triple bonds are said to be **unsaturated**. This means that they are not saturated with hydrogen (more hydrogen atoms could be added at their multiple bonds). This is particularly relevant with respect to the fatty acids, as we see later in this chapter.

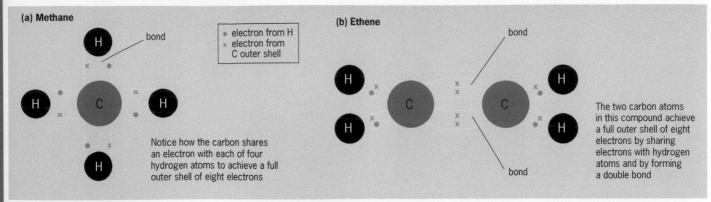

(a) Methane

bond

- electron from H
- × electron from C outer shell

Notice how the carbon shares an electron with each of four hydrogen atoms to achieve a full outer shell of eight electrons

(b) Ethene

bond

bond

The two carbon atoms in this compound achieve a full outer shell of eight electrons by sharing electrons with hydrogen atoms and by forming a double bond

Fig 3.3 Dot-and-cross diagrams showing two simple carbon compounds

Carbon atoms commonly form covalent bonds with hydrogen, nitrogen, oxygen, phosphorus and sulfur. These elements make up the vast majority of organic molecules.

What is an organic molecule?

The word 'organic' is much used today, but the use of the word (as in 'organic' vegetables) to mean 'no added chemicals' is not scientific. Originally, the word organic meant 'of living origin', in a time when scientists thought that the chemicals that made living things had a divine origin and could not be made by people. When scientists started to make organic molecules in the laboratory (urea was first) this idea was abandoned.

For our purposes, we can say that organic molecules are ones with carbon backbones – at least one carbon bonded to another one. Some organic molecules are enormous, with relative molecular masses in the millions. Table 3.1 shows some common examples of simple organic and inorganic compounds.

In this chapter, we look at four types of large organic molecules commonly found in living things: carbohydrates, lipids, proteins and nucleic acids.

Table 3.1 Some simple organic and inorganic compounds	
Inorganic	**Organic**
water (H_2O)	ethane (C_2H_6)
nitrogen gas (N_2)	ethanol (C_2H_5OH)
ammonia (NH_3)	ethanoic acid (CH_3COOH)
carbon dioxide (CO_2)	glucose ($C_6H_{12}O_6$)
hydrogen sulfide (H_2S)	amino acid ($CHRNH_2COOH$)
nitrate ion (NO_3^-)	

> ✔ **REMEMBER THIS**
> A **monomer** is a single molecule. A **dimer** is formed when two of the same molecules link together. A **polymer** is formed when many of the same molecules link together.

2 CARBOHYDRATES

Carbohydrates are organic molecules that contain three elements: carbon, hydrogen and oxygen. Carbohydrates include **sugars** and **starches** (Fig 3.4, below, and Table 3.2, opposite page). There are three basic types of carbohydrate molecule, named according to structure and size:

- **Monosaccharides** are single sugars.
- **Disaccharides** are double sugars (made from two monosaccharides).
- **Polysaccharides** are multiple sugars (polymers of many monosaccharides).

Carbohydrates are the first molecules made in photosynthesis. Lipids, proteins and nucleic acids are formed from carbohydrates.

Fig 3.4 The term 'carbohydrate' covers a range of chemicals which includes sugars, starches and cellulose. This picture shows food containing complex carbohydrates – these foods are staples of the human diet

MONOSACCHARIDES AND DISACCHARIDES

Monosaccharides and disaccharides are classed as sugars and usually have names ending in **-ose**, such as **sucrose** and **lactose**. In monosaccharides, the three elements carbon, hydrogen and oxygen are always present in the same ratio and they have the basic formula $(CH_2O)_n$. Monosaccharides are classified according to the number of carbon atoms they have: 3, 5 and 6 are the most usual (Table 3.3). In glucose, for example, n is 6, so its formula is $(CH_2O)_6$ or $C_6H_{12}O_6$.

Table 3.2 Some common carbohydrates

CHO type	Compound	Sub-units	Occurrence in living things
monosaccharide	glucose		widespread
	fructose		sweet fruits
	galactose		milk
disaccharide	maltose	$2 \times$ glucose	germinating seeds
	sucrose	glucose + fructose	fruit
	lactose	glucose + galactose	milk
polysaccharide	starch	glucose	plants (storage)
	glycogen	glucose	animals (storage)
	cellulose	glucose	plant cell walls

Table 3.3 Common monosaccharides

Number of carbon atoms	Type	Common examples
3	triose	glyceraldehyde (see Chapter 33)
5	pentose	ribose, deoxyribose (in RNA and DNA)
6	hexose	glucose, fructose, galactose

Glucose

It is useful to begin a study of carbohydrates with **glucose**: it is the main source of energy for many organisms – including humans – and most of the common polysaccharides are glucose polymers. Glucose is a hexose (6-carbon) sugar that has the formula $C_6H_{12}O_6$. All other hexose sugars, such as **fructose** and **galactose**, have the same formula (Fig 3.5).

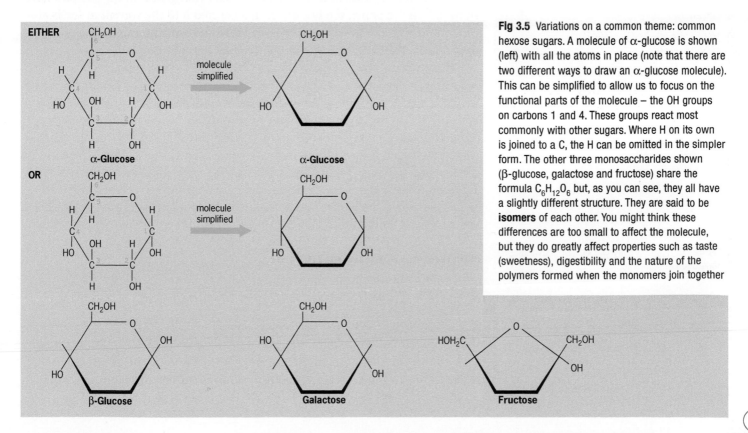

Fig 3.5 Variations on a common theme: common hexose sugars. A molecule of α-glucose is shown (left) with all the atoms in place (note that there are two different ways to draw an α-glucose molecule). This can be simplified to allow us to focus on the functional parts of the molecule – the OH groups on carbons 1 and 4. These groups react most commonly with other sugars. Where H on its own is joined to a C, the H can be omitted in the simpler form. The other three monosaccharides shown (β-glucose, galactose and fructose) share the formula $C_6H_{12}O_6$ but, as you can see, they all have a slightly different structure. They are said to be **isomers** of each other. You might think these differences are too small to affect the molecule, but they do greatly affect properties such as taste (sweetness), digestibility and the nature of the polymers formed when the monomers join together

BIOCHEMICAL TESTS

Foods are usually mixtures of carbohydrates, lipids and proteins; one way to find out what types of molecules a food contains is to do biochemical tests. Some of the common ones you will use at A level are shown in Table 3.4 – these are basically the food tests you learned at GCSE level.

Table 3.4 Some simple biochemical tests

Nutrient	Reagent used	How test is carried out	Positive result
reducing sugar	Benedict's solution	Add Benedict's solution to sample in a test tube. Heat in a water-bath	orange-red precipitate
non-reducing sugar	hydrochloric acid, and Benedict's solution	Once the reducing sugar test has proved negative, boil with dilute acid. Add sodium hydrogencarbonate to neutralise. Carry out reducing sugar test, as described above	orange-red precipitate
starch	iodine solution	Add a few drops of iodine solution to the sample	blue-black staining
lipid	ethanol	Shake the sample with ethanol in a test tube. Allow to settle. Pour clear liquid into water in another test tube	cloudy-white emulsion
protein	Biuret solution	Add biuret solution to sample in a test tube. Warm very gently	lilac/mauve colour

CHROMATOGRAPHY

This is useful when only tiny amounts of each substance is present and may not be detectable by one of the biochemical tests. It can also distinguish between substances that give the same result in biochemical tests. For example, the Benedict's test gives a positive result for both glucose and fructose because both are reducing sugars but chromatography can separate them very easily.

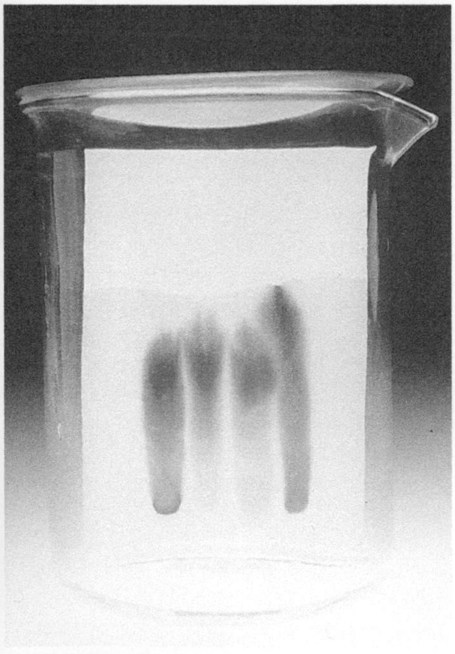

Doing chromatography

Fig 3.6 shows a basic **chromatography tank**. The paper has a pencil line at the bottom – the samples are dotted onto the paper along this line at regular intervals. Each sample is allowed to dry before adding the next, to avoid spreading and merging of the substances under test. When all the samples have been put on the paper, the bottom end is lowered into the solvent at the bottom of the tank. The lid is put on to prevent evaporation of the solvent, and the apparatus is then left for a few hours to allow the solvent to soak up the paper.

Different substances dissolve in a particular solvent to different degrees – a substance that dissolves very well will move further up the paper, perhaps to near the top. A substance that hardly dissolves at all remains near the original spot placed on the paper. If you are separating out pigments, for example, from a leaf extract, it is easy to see where the different substances are on the paper. However, if the test substance is colourless, you just end up with a piece of white paper – this must be treated with a stain to reveal the spots. For amino acid or protein mixtures, the paper is sprayed with **ninhydrin** which shows up the spots. This is highly toxic, so spraying is always carried out in a fume cupboard.

Fig 3.6 Chromatography in a beaker showing separation of pigments

Identifying your substance

The paper with the separated substances is called a chromatogram. Substances can be identified by working out their R_f value.

$$R_f = \frac{\text{distance moved by spot}}{\text{distance moved by solvent front}}$$

R_f values vary between 0 (didn't move at all) and 1 (went as far as the solvent front). A particular compound will always have the same R_f value in

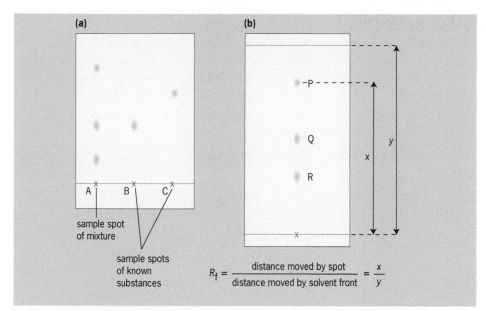

Fig 3.7 Chromatograms

$$R_f = \frac{\text{distance moved by spot}}{\text{distance moved by solvent front}} = \frac{x}{y}$$

a particular solvent. So, for instance, every time you run lysine in the solvent propanone, it will give you exactly the same R_f value. Fig 3.7b shows how an R_f value is calculated.

Chromatography: a worked example

Paper chromatography was used to find out which amino acids were present in polypeptide X (a polypeptide is a small section of a protein). First, enzymes were used to digest the polypeptide into its amino acids. Chromatography was then carried out using a specially coated plastic sheet (this has exactly the same role as blotting paper, but gives clearer results). After running with solvent Y, the dried chromatogram was sprayed with ninhydrin.

EXAMPLE

Q Fig 3.8 shows the chromatogram produced by polypeptide X; Table 3.5 gives R_f values of some amino acids. Which amino acids are present in polypeptide X?

A The R_f of an amino acid is:

$$\frac{\text{distance moved by the spot}}{\text{distance moved by the solvent front}}$$

Spot A has moved x mm, the solvent front has moved y mm. Therefore the R_f value is: $x/y = 48/50 = 0.96$. This amino acid is therefore probably proline.

Spot B has moved x mm, the solvent front has moved y mm. Therefore the R_f value is: $x/y = 45/50 = 0.90$. This amino acid is therefore probably leucine.

Spot C has moved x mm, the solvent front has moved y mm. Therefore the R_f value is: $x/y = 33/50 = 0.66$. This amino acid is therefore probably tyrosine.

Spot D has moved x mm, the solvent front has moved y mm. Therefore the R_f value is: $x/y = 25/50 = 0.50$. This amino acid is therefore probably glycine.

Spot E has moved x mm, the solvent front has moved y mm. Therefore the R_f value is: $x/y = 19/50 = 0.38$. This amino acid is therefore probably glutamic acid.

? **QUESTIONS 1–2**

1 Fig 3.7a shows that mixture A contains substance B. How many substances does mixture A contain? Does the mixture contain substance C? Give the reason for your answer.

2 In Fig 3.7b, the R_f value of substance P is 40/50, which is 0.80. Calculate the R_f values of substances Q and R. Which substance is least soluble in the solvent used?

✔ **REMEMBER THIS**
If you work out an R_f value to be more than 1, you've probably got the formula upside down.

Table 3.5 R_f values of some amino acids

Amino acid	R_f value
alanine	0.70
arginine	0.72
glutamic acid	0.38
glycine	0.50
tyrosine	0.66
leucine	0.91
proline	0.95

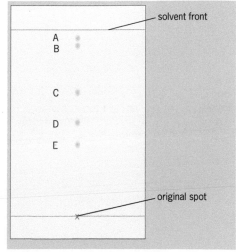

Fig 3.8 Amino acid chromatography

TWO-WAY CHROMATOGRAPHY

Sometimes, two substances will have the same R_f values in a particular solvent, so they won't be separated out. The problem is overcome by using two-way chromatograms. The first chromatogram is run in the usual way but then the paper or plastic sheet is turned through 90° and run again with a different solvent. As Fig 3.9 shows, this reveals extra spots not distinguishable in the first chromatogram.

Fig 3.9 Two-way chromatography

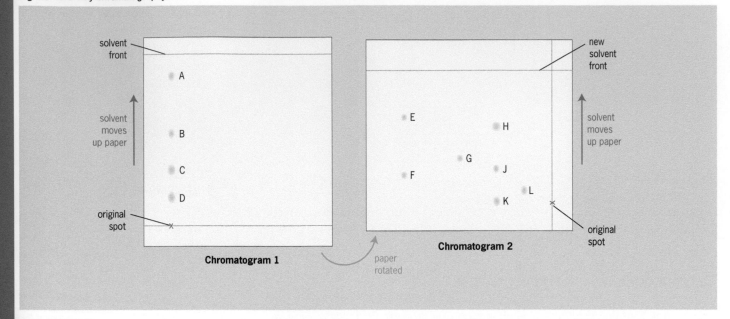

EXAMPLE

Q Fig 3.9 shows the results of two-way chromatography on a mixture of sugars. How many sugars were present in the mixture?

A The first chromatogram suggests only four, but the second chromatogram reveals that there were actually seven. Three of the sugars must have very similar solubility in the first solvent used.

Q Which spots in Chromatogram 1 contained more than one substance?

A Spots A and C. Spot A contains E and F. Spot C contains K, J and H.

Q Which substance was insoluble in the second solvent?

A Substance K because it has not moved towards the second solvent front.

WHAT IS A 'REDUCING SUGAR'?

For a fuller explanation of reduction–oxidation reactions (redox reactions), see Chapter 33.

The term 'reducing sugar' reflects the fact that some sugars can reduce other chemicals. This basically means that the sugar can donate electrons to other substances.

A standard test for a reducing sugar involves boiling the sample with Benedict's solution, a blue solution that contains copper sulfate. If a reducing sugar is present, the Cu(II) ions in copper sulfate are reduced to Cu(I) ions, resulting in an orange-red precipitate (Fig 3.10, opposite). Glucose, fructose, galactose, maltose and lactose are all reducing sugars, but sucrose is not. However, after sucrose is boiled with dilute acid to hydrolyse (split) it into its monosaccharides, it does produce a positive result.

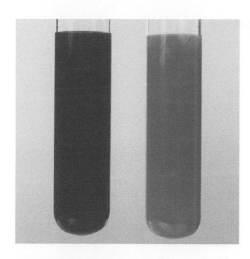

Fig 3.10 The Benedict's test can be used to estimate the amount of reducing sugars present in foods such as fruit juice. Samples with no reducing sugars remain blue, as left; those with a low concentration produce a green suspension; those with more produce yellow and orange suspensions; and juices, which are very rich in reducing sugars, produce an orange precipitate, as shown right

? **QUESTIONS 3–4**

3 In the Benedict's test, why does sucrose give a negative result before hydrolysis but a positive result after hydrolysis?

4 Fig 3.11 shows an α-1,4-glycosidic link. What do the numbers refer to?

Monosaccharides link by means of glycosidic bonds

When two monosaccharides join together in a condensation reaction, the bond between them, a **glycosidic** link, centres around a shared oxygen atom (Fig 3.11). Two α-glucose molecules join together to make one molecule of **maltose**. Sucrose, the familiar sugar we buy in bags, consists of one molecule of glucose and one of fructose. Lactose, the main sugar in milk, is a disaccharide that contains glucose and galactose. The structures of sucrose and lactose are shown in Fig 3.12.

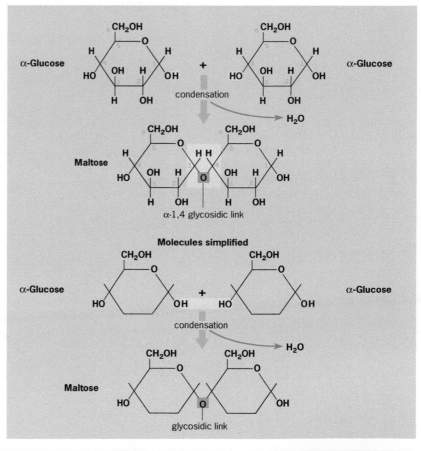

Fig 3.11 Two glucose molecules join to form maltose. Like many anabolic (building-up) reactions, this is a condensation reaction which involves the production of water

Fig 3.12 The structures of sucrose and lactose. Sucrose is made by the condensation of α-glucose and fructose. Lactose is made by the condensation of β-glucose and galactose

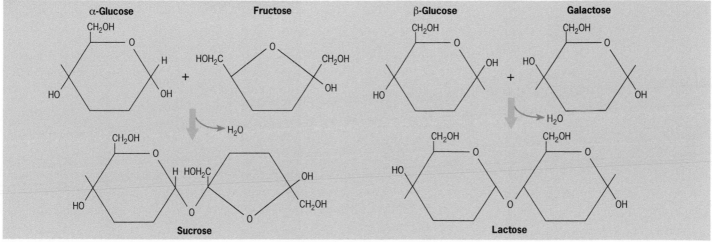

POLYSACCHARIDES

Starch

Starch is the most abundant storage chemical in plants (Fig 3.13) and it is the single largest provider of energy for most of the world's population.

Starch has the three properties that are necessary for a storage compound. Starch is:

- compact,
- insoluble,
- readily accessible when needed.

Starch is a mixture of two compounds, **amylose** and **amylopectin**. Amylose is an unbranched polymer in which glucose monomers are joined by α-1,4-glycosidic linkages. These bonds bring the monomers together at a slight angle and, when they are repeated many times, a spiral molecule is produced. In amylose, there are six glucose residues in a turn of the spiral.

The glucose chains of amylopectin have α-1,4-glycosidic linkages and α-1,6-glycosidic linkages. This allows branching (Fig 3.14).

QUESTIONS 5–6

5 If a storage compound was soluble, what would it do to the cytoplasm of the cells in which it was stored?

6 Which component of starch – amylose or amylopectin – would the plant be able to build up and break down the quickest? Explain your answer.

Fig 3.13 In many plants, starch is the main storage carbohydrate. Foods such as rice, pasta and potatoes all have a high starch content. When treated with an iodine/potassium iodide solution, a blue-black starch–iodide complex is formed. This reaction has been used here to demonstrate the presence of starch in a potato

Fig 3.14 Amylose and amylopectin, the two different polymers that make up starch. When a plant needs to break down starch to provide glucose for respiration, it removes the terminal (end) units of amylose and amylopectin to release glucose. Since the branched amylopectin molecule has more terminal glucose units that can be removed simultaneously, it can be broken down more quickly than amylose

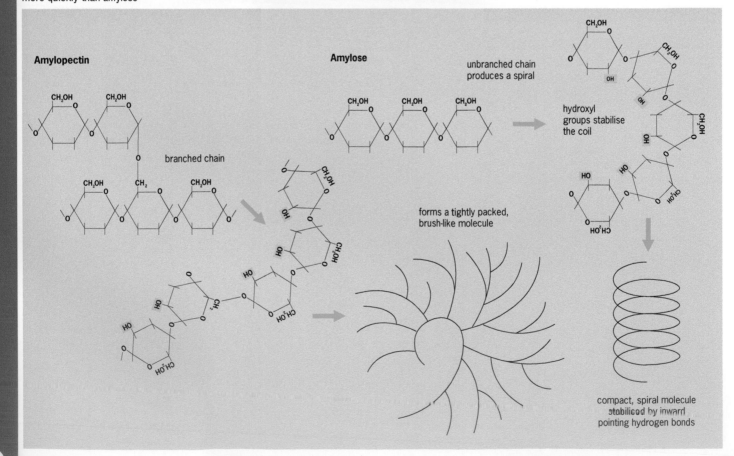

Amylopectin

CH₂OH

branched chain

CH₂OH

Amylose

CH₂OH

unbranched chain produces a spiral

hydroxyl groups stabilise the coil

forms a tightly packed, brush-like molecule

compact, spiral molecule stabilised by inward pointing hydrogen bonds

Glycogen

In humans, **glycogen** is the main storage carbohydrate. Its structure is similar to amylopectin, but it is even more frequently branched. The advantage of this is more 'ends' for enzymes to add or remove glucose. In this way glycogen can be built up and broken down quickly, allowing the animal to store energy efficiently and release it on demand. In humans, glycogen is stored in large amounts in the liver and the muscles. During prolonged exercise, when the immediate supply of glucose is used up, the body restores its supplies by breaking down glycogen. If an average person goes without food, his or her glycogen stores last for about a day, but prolonged exercise such as marathon running can use all of the body's glycogen in less than two hours. When glycogen runs out, the body turns to using its lipid stores. This is why eating less while taking more exercise is the quickest way to lose weight.

Cellulose

Cellulose is a structural polysaccharide: it gives strength and rigidity to plant cell walls. Individual cellulose molecules are long unbranched chains containing many β-1,4-glycosidic linkages (Fig 3.15). The molecules are straight and lie side by side, forming hydrogen bonds along their entire length. This results in microfilaments which are bundled into **microfibrils** and then into strong bundles of chains called **macrofibrils**.

Cellulose is probably the most abundant structural chemical on Earth, but few animals can digest it because they do not make the necessary enzyme, **cellulase**. Herbivorous animals, whose diet contains large amounts of cellulose, can deal with it because they have cellulase-producing microorganisms in their digestive system. Humans cannot digest cellulose, but we make good use of it in other ways – it provides the bulk necessary for peristalsis and its strength is used in products such as paper, cotton, Lycra and nail varnish.

> **? QUESTION 7**
>
> 7 Why do cellulose-based materials such as cotton take a long time to rot?

> **✔ REMEMBER THIS**
>
> Carbohydrates, proteins and nucleic acids are polymers; lipids are not.

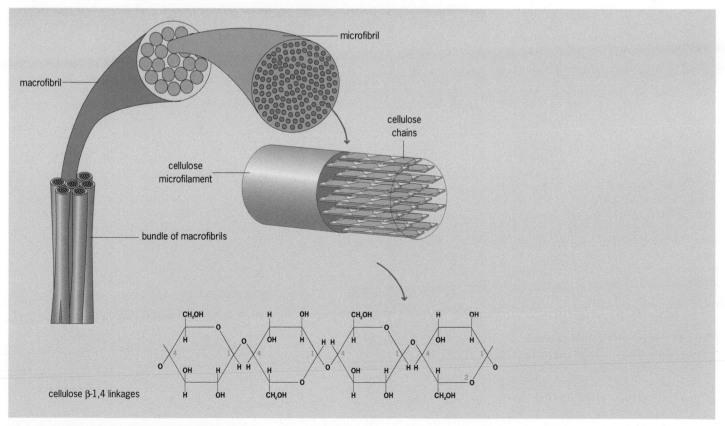

Fig 3.15 Cellulose and the structure of the plant cell wall

macrofibril

microfibril

cellulose chains

cellulose microfilament

bundle of macrofibrils

cellulose β-1,4 linkages

3 LIPIDS

Lipids are a varied group of compounds that include the familiar **fats** and **oils**. As they are non-polar molecules, most lipids are insoluble in water but soluble in non-polar solvents such as alcohol and ether. Important exceptions are phospholipids, which have polar heads. The **emulsion test** for lipids, shown in Fig 3.16, is based on the solubility of lipids in ethanol.

Lipids contain the elements carbon, hydrogen, oxygen and sometimes phosphorus and nitrogen. They are intermediate-sized molecules that do not achieve the giant sizes of the polysaccharides, proteins and nucleic acids.

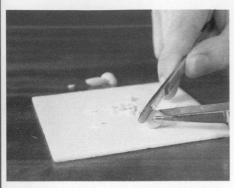 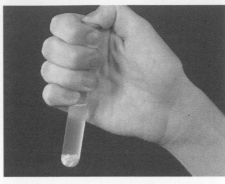 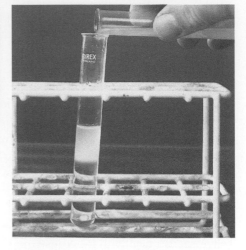

Fig 3.16 The emulsion test for lipids. The food to be tested is broken up into very small pieces, mixed with pure ethanol and shaken vigorously. Any lipid present dissolves in the alcohol. This top layer is poured off and mixed with water. If lipid is present, the mixture turns white as an emulsion, and a suspension of fine lipid droplets forms. If no lipid is present, the mixture remains clear

LIPID STRUCTURE AND FUNCTION

The **triglycerides**, which act mainly as energy stores in animals and plants, are a large important group of lipids.

Triglycerides consist of one molecule of **glycerol** and three **fatty acids**. Fig 3.17 shows how the four components join by condensation reactions to form an E-shaped molecule.

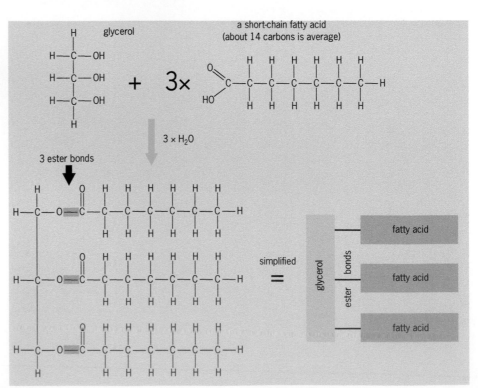

Fig 3.17 A triglyceride is formed from a condensation reaction between one molecule of glycerol and three fatty acids. The bonds formed are called ester bonds. The bottom right diagram shows the final shape of the triglyceride: glycerol is combined with three fatty acids to form an E-shaped molecule

The glycerol molecule is common to all triglycerides and so the properties of different triglycerides depend on the nature of the fatty acids. Fatty acids vary in the length of their chain and in the degree of **saturation** they show (Fig 3.18).

Table 3.6 shows the chain length of some common fatty acids. Chains of about 14 to 16 carbon atoms are the most usual, but they range from 4 to over 28. A **saturated fatty acid** has the maximum amount of hydrogen and therefore has no double bonds. **Monounsaturated fatty acids** possess one C=C bond and **polyunsaturates** contain more than one.

? QUESTIONS 8–9

8 Look at Table 3.6. Which fatty acids are unsaturated?

9 Of the fatty acids listed in Table 3.6, which are liquid at room temperature (20 °C)? What do they have in common?

Table 3.6 Some common fatty acids

Fatty acid	No. of carbon atoms	No. of double bonds	Abundant in	Melting point/°C
palmitic acid	16	0	palm oil	63.1
stearic acid	18	0	cocoa	69.6
lauric acid	12	0	coconut, palm oil	44.2
oleic acid	18	1	olive, rapeseed	13.4
linoleic acid	18	2	sunflower, maize	−5.0

Physical and chemical characteristics of different triglycerides

Triglycerides, which contain longer chain fatty acids and saturated fatty acids, generally form 'hard' fats such as lard and suet that are solid at room temperature. Animal tissues contain a higher proportion of saturated fats than plants. Plant lipids contain shorter chain, unsaturated or polyunsaturated fatty acids and so are 'light' oils that are liquid at room temperature. Unsaturation leads to a lowering of the melting point because double bonds produce kinks in the carbon chain. This increases the distance between molecules by stopping chains from lying parallel and so reducing weak intermolecular forces. This makes the lipid more 'fluid'.

Triglycerides as energy stores

Many animals store energy in the form of triglycerides: gram for gram, they yield more than twice as much energy as proteins or carbohydrates. Triglycerides are **highly reduced** compounds; they contain many C–H bonds which can yield energy during respiration. If you look at the formula for a lipid, such as $C_{14}H_{26}O_2$ you can see that there is a far greater ratio of hydrogen to oxygen than in a carbohydrate such as sucrose, which has the formula $C_{12}H_{22}O_{11}$.

Phospholipids

Phospholipids have a similar structure to triglycerides but one of the fatty acids is replaced by polar **phosphoric acid** (Fig 3.22). This gives the molecule a polar head and a non-polar tail. When placed in water, phospholipids arrange themselves with their hydrophobic ('water-hating') tails pointing inwards and their hydrophilic ('water-loving') heads pointing outwards. This is vitally important because it results in double layers called bilayers. Phospholipid bilayers form the basis of all biological membranes.

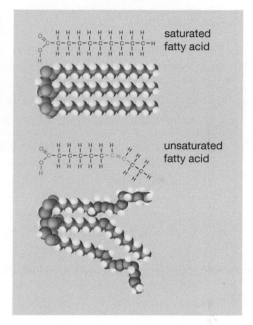

saturated fatty acid

unsaturated fatty acid

Fig 3.18 Saturated and unsaturated fatty acids. Unsaturated fatty acids have C=C bonds, which cause kinks in the chain. These bonds account for the difference in melting points. The saturated fatty acids lie tightly packed alongisde each other, forming many hydrogen bonds. These tend to be solids at room temperature, i.e. fats. The unsaturated ones cannot lie parallel, so there are fewer H bonds. These tend to have lower melting points so are usually liquids at room temperature, i.e. oils

Biological membranes are discussed in more detail in Chapter 4.

Fat cells have hidden depths

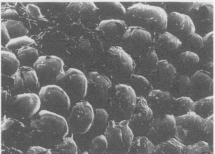

Fig 3.19 Fat cells

Fig 3.20 Leptin-deficient rat

People used to think that fat cells were just little blobs of fat surrounded by a membrane – which is certainly what they look like under the scanning electron microscope (Fig 3.19). However, we are now beginning to realise that fat cells are an important tissue in their own right, and that they help to regulate our metabolism.

Fat cells produce and secrete a wide range of proteins that act on other cells. Particularly interesting are the hormones such as leptin that act on centres in the brain that tell us to stop eating when we have had enough. (But how often do we listen?) The importance of leptin was demonstrated in the 1980s when rats that don't make any leptin were first bred. These animals look normal at birth, but they gain weight rapidly because they don't know when to stop eating. When they are a few months old, they become grossly fat, resembling spheres of fur with a head, four feet and a tail (Fig 3.20).

Other hormones produced by fat cells that also help control appetite and feeding behaviour have since been found. The interactions between them are so complex that we don't yet understand how they fit together. Even so, researchers are confidant that progress will be made and that we may have that elusive weight control pill in the next decade or so. The ideal weight for a woman has long been a matter of fashion rather than health (Fig 3.21).

Fig 3.21a Lipids are vital chemicals in all living organisms. In humans, triglycerides occur mainly in adipose (fat storage) tissue, which forms under the skin and around internal organs. As well as storing energy, adipose tissue insulates against the cold, protects organs against physical damage, and contributes to a person's overall shape. The female body shape is, to a large extent, determined by adipose tissue distribution. This painting by Reubens shows the fuller female figure that was fashionable in previous centuries

Fig 3.21b The ideal body has long been a source of debate; it seems to be dictated by fashion and culture rather than common sense. For a while, it was considered fashionable to be very thin – size zero models appeared on catwalks – but few women can achieve that shape and stay healthy

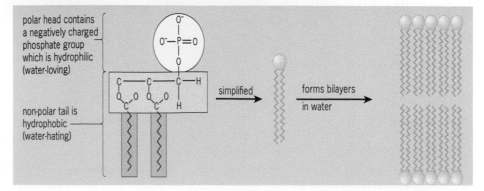

polar head contains a negatively charged phosphate group which is hydrophilic (water-loving)

non-polar tail is hydrophobic (water-hating)

simplified

forms bilayers in water

Fig 3.22 The structure of a phospholipid. The phosphoric acid gives the molecule a polar head and a non-polar tail. Plasma membranes – and other membranes in the cell – form automatically due to the behaviour of phospholipids. In water, the hydrophilic polar heads face outwards while the hydrophobic tails point inwards

Cholesterol

Many people associate **cholesterol** with heart disease, but this lipid is a perfectly normal constituent of every cell in our body (Fig 3.23). As well as eating food that contains cholesterol, we can also synthesise cholesterol in the liver. The more there is in the diet the less the liver needs to make. Vegans eat no animal products, yet they are easily able to make all the cholesterol they need.

Steroid hormones

Steroid hormones have a similar structure to the cholesterol from which they are made. They include testosterone, progesterone and the oestrogens.

Waxes

Waxes are lipids that are often used to waterproof surfaces, so preventing water loss. The cuticle of a leaf and the protective covering on an insect's body are both waxes. Waxes consist of a very long chain fatty acid joined to an alcohol molecule (not glycerol as in triglycerides). They have no nutritional value because they cannot be digested by lipases (lipid-digesting enzymes).

The role of cholesterol in cell membranes is discussed in Chapter 4.

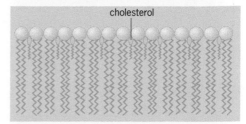

Fig 3.23 Cholesterol molecules are found between phospholipid molecules in membranes. Plasma membranes are particularly rich in cholesterol, while organelle membranes contain less cholesterol

The role of steroid hormones is discussed in Chapter 18.

HOW SCIENCE WORKS

Steroids in athletics

Anabolic steroids are synthetic compounds that are similar in structure to **testosterone**, the male sex hormone. They increase anabolic (building-up) reactions, which enhance muscle size.

Athletes, notably weightlifters, began injecting testosterone in the 1950s to gain extra strength.

When steroids were banned, athletes looked for other natural substances that would be undetectable in blood tests. **Human growth hormone** (HGH, or **somatotropin**) and also **erythropoietin** are other illegal but potentially performance-enhancing substances. HGH increases muscle size in adults and erythropoietin increases the oxygen-carrying capacity of the blood (see the Assignment for Chapter 10 on CAS website).

HOW ANABOLIC STEROIDS WORK

Steroid hormones are lipid soluble and so can pass through the plasma membrane. Studies have shown that anabolic steroids pass into the nucleus, where they force the cell to 'read' the genes that code for the muscle proteins actin and myosin. As well as increasing muscle size, athletes can train for longer, and the greater aggression that they tend to develop may also give them an added competitive edge (Fig 3.24).

The use of anabolic steroids is unfair and dangerous. To achieve a muscle-enhancing effect, athletes need to inject 10 to 40 times the normal dose, and this can have irreversible side effects. Symptoms of steroid abuse include acne, impotence, sterility, diabetes, heart disease and even liver cancer. Studies of male steroid abusers have shown that their testes may

Fig 3.24 Athletes such as Ben Johnson have risked the shame of being banned from their sport by taking steroids to enhance their performance

decrease in size by as much as 40 per cent. Their sperm count is correspondingly lowered. Female athletes who use steroids tend to have fewer periods and can develop masculine features such as excessive body hair and a deeper voice.

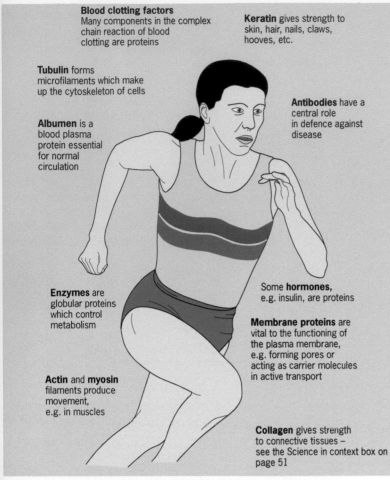

Blood clotting factors
Many components in the complex chain reaction of blood clotting are proteins

Keratin gives strength to skin, hair, nails, claws, hooves, etc.

Tubulin forms microfilaments which make up the cytoskeleton of cells

Antibodies have a central role in defence against disease

Albumen is a blood plasma protein essential for normal circulation

Enzymes are globular proteins which control metabolism

Some **hormones,** e.g. insulin, are proteins

Membrane proteins are vital to the functioning of the plasma membrane, e.g. forming pores or acting as carrier molecules in active transport

Actin and **myosin** filaments produce movement, e.g. in muscles

Collagen gives strength to connective tissues – see the Science in context box on page 51

Fig 3.25 After water, proteins are the major constituent of our bodies

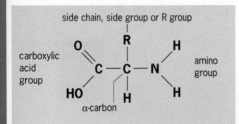

Fig 3.26 The basic structure of an amino acid

side chain, side group or R group

carboxylic acid group

amino group

α-carbon

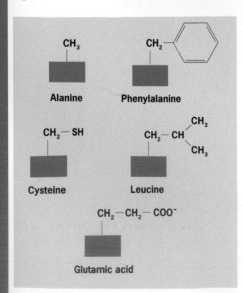

Alanine

Phenylalanine

Cysteine

Leucine

Glutamic acid

4 PROTEINS

Proteins play a central role in the structure and metabolism of all living organisms. Protein molecules have a huge variety of shapes and sizes (the structure of carbohydrates and lipids is relatively limited by comparison). This versatility of shape is the key to the role of proteins in the cell.

THE IMPORTANCE OF PROTEINS TO LIVING ORGANISMS

We can appreciate the extent to which organisms use proteins by looking at their roles in the human body (Fig 3.25). Consider some of the jobs that proteins need to do:

- Each cellular metabolic reaction must be catalysed by a different **enzyme**.
- Each substance that passes across a cell membrane requires a different **carrier molecule**.
- A different **antibody** is needed to combat the chemicals produced by the many (and also constantly changing) disease-causing organisms such as bacteria and viruses.

Enzymes, carrier molecules and antibodies are all proteins that are 'tailor-made' to fulfil all these requirements.

THE STRUCTURE OF PROTEINS

Proteins are large and complex molecules (Table 3.7). If a water molecule (relative molecular mass = 18) was the size of a brick, proteins would be whole buildings.

In addition to the elements carbon, hydrogen and oxygen, proteins always contain nitrogen and sometimes sulfur. The building blocks of proteins are called **amino acids**. Fig 3.26 shows the basic structure of an amino acid. As the name suggests, all amino acids contain an amino group ($–NH_2$) and a carboxylic acid group (–COOH). Both of these groups are attached to a central carbon atom, known as the α-carbon. The 'backbone' of an amino acid is made up from the three atoms C–C–N.

The R group varies between different amino acids. Fig 3.27 shows five of the different R groups in the 20 amino acids that make up all of the proteins in the human body.

Fig 3.27 Five of the 20 amino acids found in the human body. The fundamental and unchanging amino acid structure is shown as a red box; R groups are shown as chemical representations. A further 200 or more amino acid structures are possible, but these need to be synthesised (made artificially). They are never found in living systems

Table 3.7 The relative molecular masses of several proteins	
Protein	**Relative molecular mass**
insulin	5 700
haemoglobin	64 500
myoglobin	16 900
hexokinase	102 000
glycogen phosphorylase	370 000
glutamine synthetase	592 000

Amino acids in solution

Amino acids are readily soluble in water. When in solution they can resist a change in pH by mopping up or releasing hydrogen ions (H^+) and hydroxyl ions (OH^-), acting as buffers, that is, regulators of pH. In whole proteins, the buffering effect is largely due to the R groups.

Peptides, polypeptides or proteins?

It is easy to be confused about the precise meaning of the terms **peptide**, **polypeptide** and **protein**; there isn't a universally accepted rule to follow. This is how we use them in this book.

When amino acids are assembled into a short chain a peptide is formed. Longer chains are known as polypeptides. The term 'protein' is reserved for the finished, functional molecule. Some proteins consist of one polypeptide, others consist of two or more than two. Haemoglobin, for example, contains four polypeptides.

HOW AMINO ACIDS FORM PROTEINS

Amino acids join together in long chains to form proteins by means of **peptide bonds**. Fig 3.28 shows how two amino acids join to form a dipeptide. This is an example of a **condensation reaction**.

All proteins are complex molecules and biochemists look at their structure at four different levels: **primary**, **secondary**, **tertiary** and **quaternary**.

Primary structure

The primary structure of a protein refers to the sequence, or order, of amino acids in that protein (Fig 3.29).

A simple primary structure of a tiny protein could be shown as:

alanine–histidine–phenylalanine–glutamine–cysteine

Real proteins usually consist of a lot more than five amino acids. The hormone insulin, for example, a relatively small protein, has 51 amino acids.

The code for the primary structure of any protein is contained in the gene or genes that code for that protein. This code determines the precise order in which amino acids are assembled. This order then dictates the way they will twist and turn to produce the precise three-dimensional shape that allows the protein to carry out its specific function in the body. The first level of three-dimensional twisting is described as the secondary structure of the protein.

Secondary structure

When combinations of amino acids join together in a chain they tend to fold into particular shapes and patterns (such as spirals) in some places. These shapes form because the amino acids twist around to achieve the most stable arrangement of hydrogen bonds. The secondary structure of the protein refers to the patterns contained within the amino acid chain. Such patterns exist in different places in different proteins, producing an almost infinite variety of molecular shapes.

Maintaining a constant pH and other aspects of homeostasis are discussed in Chapters 11 to 14.

REMEMBER THIS

No matter how many amino acids are in the chain, if there is a —COOH group on one end, there will be an —NH_2 group on the other.

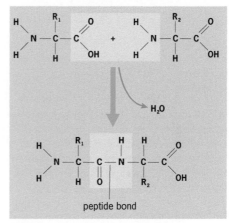

Fig 3.28 Two amino acid molecules react together in a condensation reaction (a reaction that releases water) to produce a dipeptide

Fig 3.29 When amino acids join, there is a repeating sequence of C—C—N—C—C—N— etc. This backbone runs throughout the length of the protein

N-terminal end H_2N ... C-terminal end

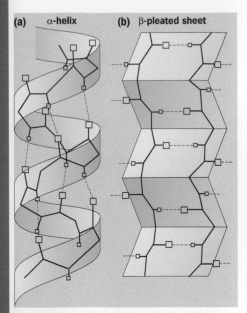

(a) α-helix **(b)** β-pleated sheet

Fig 3.30 In an α-helix, the polypeptide is coiled into a helix that is held in position by hydrogen bonds between different groups in the backbone. In a β-pleated sheet, the amino acid chains lie side by side, forming a sheet that is held together by hydrogen bonds between adjacent parts of the polypeptide

 REMEMBER THIS

A **residue** is the name given to the remains of a monomer that has become incorporated into a polymer.

The main types of secondary structure in proteins are:

- the α-**helix**, a spiral, and the most common type of secondary structure (Fig 3.30a). Amino acid **residues** in the spiral twist on their axis, each residue forming a hydrogen bond with another residue four units along. These hydrogen bonds stabilise the α-helix,
- the β-**pleated sheet**, a flat structure that consists of two or more amino acid chains running parallel to each other, linked by hydrogen bonds (Fig 3.30b).

The secondary structure of a protein depends on its amino acid sequence: some amino acids (or combinations of amino acids) tend to produce α-helices, others usually make β-sheets. A few amino acids tend to produce a sharp bend in the chain, a vital function that allows the chain to fold back on itself. Fig 3.31 shows the secondary structures within the enzyme lysozyme.

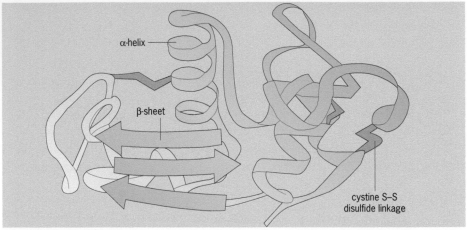

α-helix

β-sheet

cystine S–S disulfide linkage

Fig 3.31 The secondary structures in the enzyme lysozyme: the α-helices and β-pleated sheets are clearly visible. The α-helices are shown as spiral ribbons, the β-pleated sheets as broad, flat arrows. This enzyme is present in tears and sweat, where it catalyses the breakdown of some bacterial cell walls

Tertiary structure

The tertiary structure of a protein is its overall three-dimensional shape and is produced as a result of the following:

- the sequence of amino acids that produces α-helices, β-sheets and bends at particular places along the chain.
- the hydrophobic nature of many amino acid side chains. Globular proteins are surrounded by water and so the hydrophobic side chains tend to point inwards.

Tertiary structure is maintained by attractive forces that arise when the amino acid chain folds (Table 3.8) and also by **disulfide bridges** – covalent bonds that form when two **sulfur-containing cysteine residues** react. Disulfide bridges occur most often in structural proteins, where they contribute to strength.

The importance of tertiary structure for protein function cannot be over-emphasised. Functional proteins, such as enzymes and antibodies, must have an exact shape – and sometimes the ability to change shape – to fulfil their role in the organism. Many structural proteins depend on their tertiary structure for strength. The large number of disulfide bridges in keratin, for example, makes body structures such as hair and nails very tough.

Table 3.8 The different types of bond that maintain the secondary, tertiary and quaternary structure of a protein

Type of bond	Formed between	Relative strength
hydrogen bonds	H and an electronegative atom, usually O or N	weak, very common
ionic bonds	oppositely charged ions	weak
van der Waals forces	non-specific, between nearby atoms	weak
disulfide bridges	two SH-containing cysteine residues	strong, contribute to strength of fibrous proteins
hydrophobic interactions	R groups of some amino acids and surrounding water	weak, hyrophobic (non polar) R groups point inwards

If a protein consists of only one polypeptide, the tertiary structure is the final shape of the molecule. If, however, the protein has more than one, it has a further, quaternary level of structure.

Quaternary structure

Many proteins consist of more than one polypeptide chain and sometimes also have non-protein **prosthetic** groups, often vital to the function of the protein. The quaternary structure refers to the three-dimensional structure, or **conformation**, produced when all the sub-units combine to give the final, active molecule (Fig 3.32).

Fig 3.32 A computer-generated model of the protein molecules of collagen, the most abundant protein in the human body. The red, green and yellow chains represent three different polypeptide chains

Fibrous and globular proteins

The final three-dimensional structure of proteins results in two main classes of protein – **fibrous** and **globular**. Fibrous proteins contain polypeptides that bind together to form long fibres or sheets. They are physically tough and are insoluble in water (Fig 3.33).

Fig 3.33 This diagram shows how the fibrous protein keratin is arranged inside a human hair

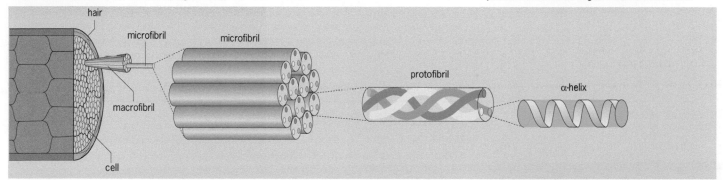

Globular proteins are usually individual molecules with complex tertiary and quaternary structures. They are spherical, or globular, in shape, hence the name. Most are soluble in water and they tend to have a biochemical rather than a structural function (Fig 3.34).

How stable are proteins?

As the final shape of globular proteins is maintained by relatively weak molecular interactions such as hydrogen bonds, proteins are very sensitive to temperature increases and other changes in their environment. As the temperature goes up beyond 40 °C, molecular vibration increases and bonds that are holding together the tertiary or quaternary structure break, changing the shape of the molecule. This is known as **denaturation**. Different proteins are denatured at different temperatures. Some begin to be denatured after about 40–45 °C or even below, but many are not totally denatured until 60 °C or even higher. It is an oversimplification to say 'organisms die at temperatures over 44 °C because their proteins become denatured'. In practice, organisms die because of a metabolic imbalance caused when enzymes work at different rates.

Proteins can also be denatured by adverse chemical conditions. Chemicals that affect weak bonds alter the overall structure; even a slight change in protein shape can mean loss of function. Some proteins are particularly sensitive to changes in pH.

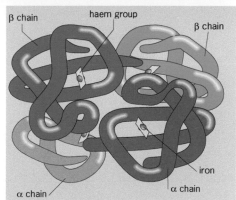

Fig 3.34 Haemoglobin is an example of a globular protein. This complex molecule picks up oxygen from the lungs and releases it to respiring tissues. Each molecule consists of four polypeptide chains held together by disulfide bonds. There is a prosthetic group, the haem group (which contains iron), at the centre of the each polypeptide chain. The function of haemoglobin is discussed more fully in Chapter 9

Enzymes are discussed in detail in Chapter 5.

When diabetes is not helped by insulin

Like many other protein molecules with a biochemical function, such as enzymes and hormones, the hormone insulin is a globular protein (Fig 3.35). Produced by the pancreas, it is a major regulator of metabolism. When blood sugar levels are high, insulin is secreted into the blood and it quickly travels all around the body. All cells of the body have insulin receptors on their surface and when insulin binds, this causes glucose channels to change shape slightly, particularly in muscle cells and fat cells. This then allows these cells to take up glucose from the blood and store it as glycogen or fat.

Type I diabetes, which we cover in more detail in Chapter 11, arises when the cells in the pancreas that produce insulin become damaged. It usually occurs in young people (under 20) and so is sometimes called juvenile-onset diabetes. It is treatable by insulin injections. However, there is also another type of diabetes – called Type II. This affects much older people – in their 40s and 50s and beyond – and often does not respond to insulin injections. The problem here is that muscle and fat cells become resistant to the effect of insulin – one reason is that they don't have enough insulin receptors.

One of the hormones produced by fat cells has recently been linked to this type of diabetes. Mice that don't make any adiponectin show the same symptoms as people with Type II diabetes. This hormone is now being considered as a future treatment as it is able to increase cells' sensitivity to insulin, possibly by increasing the number of insulin receptors made.

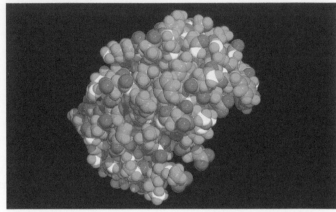

Fig 3.35 Although this model looks like a blob, each insulin molecule contains two sub-units. Each sub-unit has two chains and is bound to a single zinc atom. Atoms and molecules that bind to proteins are called ligands. Ligands are bound to the proteins by weak bonds and are usually more easily removed from the protein than are prosthetic groups. Proteins can have many different types of ligand and these are often important for the function of the protein

ANALYSING PROTEINS

We can find out if a sample contains protein using the **Biuret test** (Fig 3.36). More complex biochemical techniques can be used to tell us more about proteins in living systems. We can:

- find out which proteins are present in a mixture, such as plasma,
- work out the exact three-dimensional shape of a protein using techniques such as X-ray diffraction and nuclear magnetic resonance,
- find out which amino acids are present in a particular protein,
- analyse the exact amino acid sequence of a protein,
- use computers to predict the three-dimensional shape of a protein, using only its amino acid sequence.

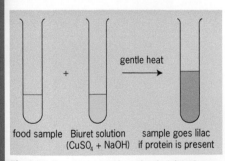

gentle heat

food sample | Biuret solution (CuSO$_4$ + NaOH) | sample goes lilac if protein is present

Fig 3.36 The Biuret test is a simple laboratory test that detects the presence of peptide bonds

For more on DNA fingerprinting please see the How Science Works assignment for this chapter at www.collinseducation.co.uk/CAS

Fig 3.37 Electrophoresis can be thought of as chromatography using electricity. Instead of separating chemicals according to their solubility, electrophoresis causes individual molecules to move to the positive or negative terminal at speeds that depend on their mass and overall charge. This is also the principle behind DNA fingerprinting

Identifying individual proteins

The proteins present in a body fluid such as blood plasma can be separated from each other by electrophoresis. Each protein carries a particular overall electrical charge. Electrophoresis uses this fact to separate individual proteins, as Fig 3.37 explains.

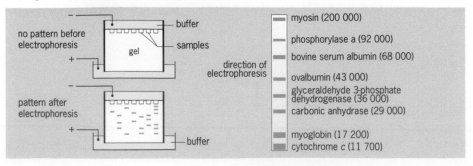

no pattern before electrophoresis — buffer — samples — gel

pattern after electrophoresis — buffer

direction of electrophoresis

myosin (200 000)
phosphorylase a (92 000)
bovine serum albumin (68 000)
ovalbumin (43 000)
glyceraldehyde 3-phosphate dehydrogenase (36 000)
carbonic anhydrase (29 000)
myoglobin (17 200)
cytochrome c (11 700)

Collagen

Collagen is probably the most widespread structural protein in the human body. It is a fibrous protein that gives strength to tissues such as tendons, ligaments, bone and skin. Fig 3.38 shows the structure of collagen. A single collagen molecule contains three polypeptides – each of about 1000 amino acids – intertwined to form a triple helix. This arrangement has great strength, mainly due to the large number of hydrogen bonds that occur along the length of the polypeptides.

Collagen is secreted in an unassembled form because the instant formation of large fibres would damage the cell that made it. Complete collagen is produced as enzymes act on the individual polypeptides, causing them to twist together to form very long fibres (sometimes several millimetres long). These have the tensile strength of steel and are used to strengthen bone in much the same way as metal rods reinforce concrete. Brittle bone disease (Fig 3.39) is a genetic disorder which results in a fault in the bonding between collagen and the mineral component of bone.

The primary structure of collagen is very regular. It consists of a repeating sequence of glycine and two other amino acids, often proline and hydroxyproline.

These amino acids do not cause the chain to gain the normal α-helix or β-sheet structure. Instead, they form long, separate chains that allow the collagen triple helix to form. This is a good example of how the primary structure is ultimately responsible for the shape and properties of the whole protein.

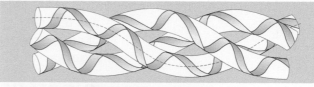

Fig 3.38 The structure of collagen

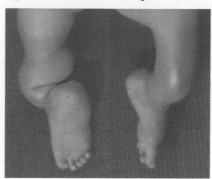

Fig 3.39 The legs of a small child suffering from 'brittle bone' disease – an unusually severe case. Frequent fractures of the leg result in the deformities

Working out the three-dimensional shape of a protein

X-ray diffraction is useful here. When a beam of X-rays is fired at a crystal of purified protein, the atoms in that protein diffract (bend) the X-rays, producing a specific pattern on a photographic plate. The shape of a protein is not immediately obvious from the information produced, but a trained scientist with a computer can produce an accurate three-dimensional model of the molecule.

Identifying which amino acids are present

Protease (protein-degrading) enzymes break up the protein and then the constituent amino acids are identified using chromatography or electrophoresis. These techniques do not reveal the sequence of amino acids.

Determining the amino acid sequence

The protein is digested into manageable chain lengths of amino acids and each length is analysed in turn. The polypeptide is treated with a chemical that binds to the N-terminal (N-end) amino acid but not to any of the others. The 'tagged' terminal amino acid is removed from the rest of the chain and identified by chromatography. Frederick Sanger, who won a Nobel Prize in the 1950s, was the first scientist to sequence a complete protein (insulin). Today, the process is fully automated and has become a standard technique.

Computer modelling

As biochemists accumulate knowledge about the structure of many individual proteins, they produce computer software to help work out the shape of other complete proteins using their amino acid sequences. This is incredibly useful in genetics. Sometimes a new gene is sequenced and no one knows its function. The DNA sequence can be given to a computer which then works out the sequence of amino acids that would be produced if the gene were expressed in a cell. It can then 'build' the final protein and compare it to other proteins with known functions.

5 NUCLEIC ACIDS

Nucleic acids are so called because they are slightly acidic molecules, and it was thought originally that they occurred only in the nucleus. The two types of nucleic acid, **DNA** and **RNA**, both contain carbon, hydrogen, oxygen, nitrogen and phosphorus.

THE STRUCTURE OF NUCLEIC ACIDS

Nucleotides are the building blocks of nucleic acids. A nucleotide consists of three units (Fig 3.40):
- a **sugar** (ribose or deoxyribose),
- a **phosphate** group,
- a nitrogen-containing **base**.

As the names imply, **deoxyribo**nucleic acid has nucleotides in which the sugar is **deoxyribose**, while **ribo**nucleic acid contains the sugar **ribose**.

DEOXYRIBONUCLEIC ACID

Deoxyribonucleic acid (DNA) is the macromolecule that carries the **genetic code**, the information for making the cell's proteins. Most of the DNA in a eukaryotic cell is in the nucleus (Fig 3.41).

The nucleotides in DNA can contain any one of four nitrogenous (nitrogen-containing) bases: **adenine**, **guanine**, **cytosine** or **thymine**. When DNA **replicates** (copies itself), it makes new strands by adding nucleotides. These are available as free molecules in the cytoplasm. Generally, cells can synthesise their own nucleotides.

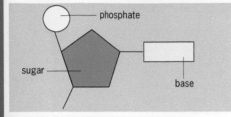

Fig 3.40 The basic structure of a nucleotide

More details about nucleic acids can be found in Chapter 24.

More details about nucleic acids can be found in Chapter 24.

? QUESTION 10

10 If there are 46 chromosomes in a human cell, each with an average of 5 cm uncoiled length, estimate the length of DNA contained within each cell.

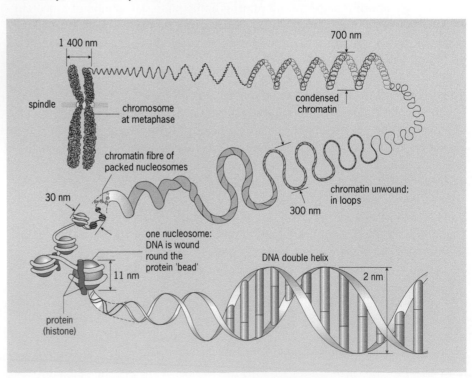

Fig 3.41 Studies of chromosomes have shown that each one is a giant, highly coiled DNA molecule. Its shape is achieved by supercoiling – coils within coils, condensing a huge amount of DNA into a tiny length. Each chromosome can contain 300 000 000 nucleotide units and, unravelled, the DNA within it would be between 5 and 10 centimetres long

How DNA carries the genetic code

DNA has two remarkable characteristics:
- It is a store of genetic information.
- It can copy itself exactly, time after time.

Looking at the structure of DNA helps us to understand how it does this.

How the bases pair

Adenine and guanine belong to a group of chemicals called **purines** while thymine and cytosine are **pyrimidines**.

Because of the shape of the two types of molecule, each purine always bonds with only one pyrimidine. So, in DNA, adenine always bonds with thymine, and cytosine with guanine. In RNA, cytosine bonds with guanine and adenine bonds with a fifth base, uracil:

DNA: A=T   RNA: A=U
    G≡C   G≡C

The base pairs are held together by hydrogen bonds. There are two H-bonds between A and T (or U) and three between C and G.

Fig 3.42 shows how the base pairs within DNA fit together to form a double-stranded helix. The sides are formed by alternating sugar–phosphate units, while the base pairs form the cross-bridges, like the rungs of a ladder. Each base pairing causes a twist in the helix and there is a complete 360° turn every ten base pairs.

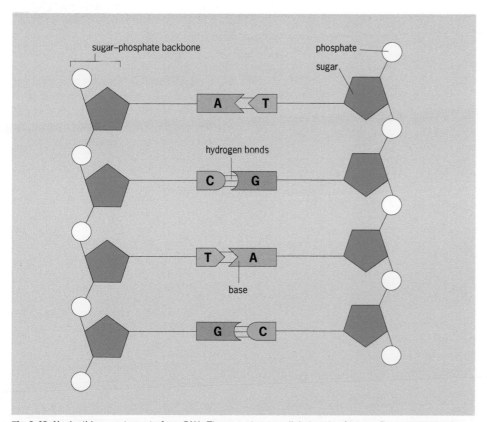

Fig 3.42 Nucleotides condense to form DNA. There are two parallel strands of bases. For any gene, only one strand of the DNA – known as the sense strand – is used to make proteins. The other side serves to stabilise the molecule. The sense strand can alternate between one gene and the next; it is not always on the same side of the molecule. Note that the pyrimidine bases are half the size of the purine bases

? QUESTION 12

12 List the four major differences between RNA and DNA.

RIBONUCLEIC ACID

Three of the bases in ribonucleic acid (RNA) – adenine, guanine and cytosine – are the same as those in DNA. The fourth is different: RNA contains uracil instead of thymine.

RNA molecules are much smaller than DNA molecules. DNA can consist of over 300 000 000 nucleotides; RNA usually consists of a few hundred. RNA is also less stable. DNA molecules are the permanent store for genetic information and last for many years. In contrast, RNA molecules have a short-term function and are easily replaced. There are three forms of ribonucleic acid (RNA) in the cell:

- **Messenger RNA** (mRNA) (Fig 3.43a) can be thought of as a mobile copy of a gene. Small lengths of mRNA are assembled in the nucleus using a single gene within the DNA as a template. When a complete copy of the gene has been produced, the mRNA moves out of the nucleus to the ribosome, where the protein is synthesised according to the code taken from the DNA.

- **Transfer RNA (tRNA)** (Fig 3.43b) is found in the cytoplasm and is a carrier molecule, bringing amino acids to the ribosomes for assembly into a new amino acid chain, according to the order specified on the mRNA code.

- **Ribosomal RNA (rRNA)** makes up part of the ribosome, a small organelle that brings together all the chemicals associated with protein synthesis.

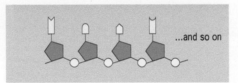

Fig 3.43a Messenger RNA is a relatively delicate, short-lived molecule composed of a single strand of nucleotides. It carries the base sequence from the gene to the ribosome and provides a template for protein synthesis

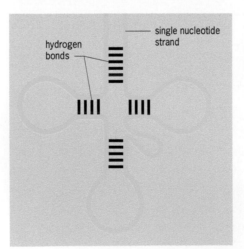

Fig 3.43b Transfer RNA is a single folded strand of nucleotides that is stabilised by hydrogen bonds, producing what is often described as a 'clover leaf' structure. The tRNA brings amino acids to the ribosome so that they can be added on to the growing amino acid chain

SUMMARY

When you have finished this chapter, you should know and understand the following.

CARBOHYDRATES

- **Carbohydrates** contain the elements carbon, hydrogen and oxygen. They are the first products made by plants in photosynthesis.
- The term **sugars** describes **monosaccharides** and **disaccharides**. Their names end in the suffix **-ose**.
- Monosaccharides include **glucose**, **fructose** and **galactose**. These sugars are **isomers**: they have the same formula but their atoms are arranged in different ways. Monosaccharides are linked by **glycosidic bonds**. These are formed by **condensation reactions**.
- Disaccharides, such as **maltose**, **sucrose** (cane sugar) and **lactose** (milk sugar), consist of two monosaccharides linked together.
- Most **polysaccharides** are **polymers** of glucose. **Starch** is used for storage in plants, **glycogen** is used for storage in animals, and **cellulose** gives strength to plant cell walls.

LIPIDS

- **Lipids** contain carbon, hydrogen, oxygen, often phosphorus and occasionally nitrogen. Most are **non-polar** chemicals and therefore insoluble in water.
- Lipids are used for energy storage, protection and insulation.
- In living things there are two main types of lipids: **triglycerides** and **phospholipids**. Triglycerides are the familiar fats and oils. Phospholipids form cell membranes.
- Fatty acids vary in chain length, and may be **saturated** (with hydrogen) or **unsaturated**. These factors determine the properties of the triglyceride, such as its melting point and viscosity.
- Phospholipids are similar to triglycerides, but phosphoric acid replaces one of the fatty acids. They have a polar head, allowing them to form **bilayers** (membranes) in water.
- **Cholesterol** is a normal constituent of cell membranes.

PROTEINS

- **Proteins** consist of chains of **amino acids** linked by **peptide bonds**.

- There are 20 different amino acids in living things. All have a **carboxylic acid** group and an **amino group** but differ in their **R group**.
- **Fibrous proteins** often join to form large fibrils whose function is to provide strength or produce movement. **Globular proteins** – including enzymes, antibodies and some hormones – are usually individual molecules with a chemical function.
- The **primary structure** of a protein is the sequence of its amino acids.
- The **secondary structure** refers to the patterns and shapes formed within the **polypeptide** chain, for example, an **α-helix**.
- The **tertiary structure** refers to the three-dimensional shape of a polypeptide chain, which results from the interactions as the chain folds back on itself. If the protein consists of one polypeptide chain, the tertiary structure refers to the overall shape of the molecule.
- The **quaternary structure** refers to the overall three-dimensional shape of a protein that consists of more than one polypeptide chain.

NUCLEIC ACIDS

- There are two types of **nucleic acid** – **deoxyribonucleic acid** (**DNA**) and **ribonucleic acid** (**RNA**). RNA itself has three forms, **messenger RNA**, **transfer RNA** and **ribosomal RNA**.
- DNA carries the genetic code. Its structure allows it to store information, pass information on to RNA so that proteins can be made, and to copy itself, allowing the genetic code to pass to new cells.
- Messenger RNA copies the genes, to allow them to be used as templates for protein synthesis. Transfer RNA brings amino acids to the ribosome during synthesis. Ribosomal RNA is a major structural component of the ribosome.
- DNA and RNA are composed of **nucleotides**, which themselves contain a **sugar**, a **phosphate group** and a **nitrogen-containing base**.
- DNA contains the bases A, C, T and G. RNA contains the bases A, C, U and G. A always pairs with T (or U) and C always pairs with G.

 Practice questions and a How Science Works assignment for this chapter are available at www.collinseducation.co.uk/CAS

4

Movement in and out of cells

4 MOVEMENT IN AND OUT OF CELLS

Cool kidneys

The list of people awaiting an organ transplant is growing. Improving surgery and reducing organ rejection rates has made transplants increasingly successful, leading to a greater demand for kidneys, hearts, livers and other body parts.

Surgeons face problems: an organ can become available in Dundee, but the most suitable recipient might be in Norwich. So, the organ usually has to be transported. But cells in tissues that are separated from their blood supply cannot carry out their normal metabolic functions because they are unable to exchange materials with blood. They run out of oxygen, accumulate waste and can die within minutes.

The solution is to cool the cells so that their metabolic demands are lowered. Organs packed in ice can survive a journey of several hours. During the transplant, the kidney is warmed to normal body temperature and connected without delay to the recipient's blood system.

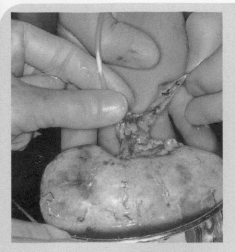

A kidney waiting to be transplanted. The pale pink colour changes to a much darker red when the new blood supply is connected

1 THE CELL AND ITS ENVIRONMENT

Each living cell is a dynamic system that can exchange a volume of fluid several times greater than its own volume every second! This happens only because cells are very small: if they were any larger the nutrients could not be taken up quickly enough to satisfy demand and waste could not be expelled efficiently enough to prevent poisoning of the cell. All organisms larger than about 1 mm have developed strategies to increase the exchange of materials, to meet the needs of all their cells.

Material passes into or out of cells by these basic processes:

- **diffusion** (and **facilitated diffusion**),
- **osmosis**,
- **active transport**,
- **endocytosis** and **ectocytosis**.

In this chapter, we look at each of these in detail but, since they all involve the plasma membrane, we first take a detailed look at this important structure.

2 THE PLASMA MEMBRANE

The **plasma membrane**, often called the **cell membrane** and sometimes known as the **cell surface membrane** or **plasmalemma**, is the boundary between a cell and its surroundings. It has little physical strength, but it plays a vital role in regulating the materials that pass in and out of the cell. Many of the organelles of the eukaryotic cell are also made up of membranes (Fig 4.1) with the same basic structure as the plasma membrane.

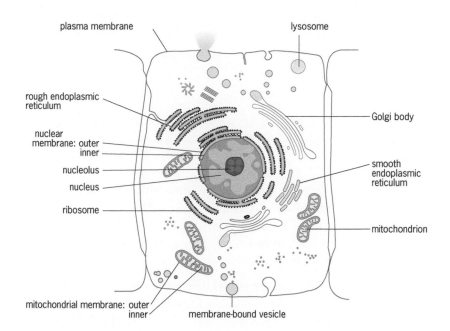

Fig 4.1 Whenever you draw a eukaryotic cell, nearly every line you draw represents a membrane. In this diagram of an animal cell, all the membranes are drawn as red lines

THE STRUCTURE OF THE PLASMA MEMBRANE

As we saw in Chapter 1, the plasma membrane is basically a double layer of **phospholipid** molecules about 7 to 10 nm thick (Fig 4.2). It cannot be seen with a light microscope and so its structure could not be studied directly until the electron microscope was developed. Instead, early cell biologists deduced its structure by investigating its properties.

Fig 4.2 The phospholipids in a membrane are arranged tail to tail, forming a bilayer

The fluid mosaic theory

In 1972, aided by electron microscope studies and evidence from other techniques, Singer and Nicholson put forward the **fluid mosaic theory** (Figs 4.3, below, and 4.4, overleaf). Today, most scientists accept it as the model that best represents the structure of living cell membranes.

Fig 4.3 The fluid mosaic model of the structure of a plasma membrane. The membrane has been described as 'protein icebergs in a lipid sea'

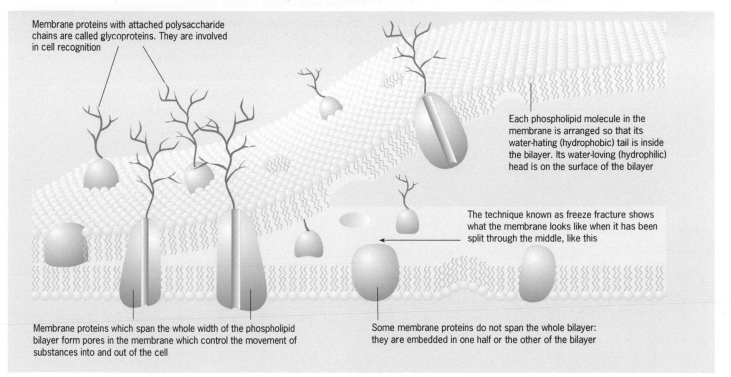

Membrane proteins with attached polysaccharide chains are called glycoproteins. They are involved in cell recognition

Each phospholipid molecule in the membrane is arranged so that its water-hating (hydrophobic) tail is inside the bilayer. Its water-loving (hydrophilic) head is on the surface of the bilayer

The technique known as freeze fracture shows what the membrane looks like when it has been split through the middle, like this

Membrane proteins which span the whole width of the phospholipid bilayer form pores in the membrane which control the movement of substances into and out of the cell

Some membrane proteins do not span the whole bilayer: they are embedded in one half or the other of the bilayer

Fig 4.4 A simplified diagram of a plasma membrane, showing the features described in the fluid mosaic theory. This level of detail should be enough to enable you to explain the essential properties and functions of the membrane

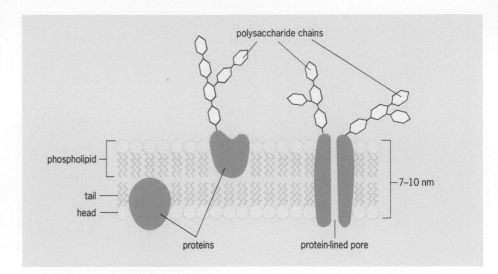

 REMEMBER THIS

Water is a **polar molecule** (it has regions of positive and negative charge) and is a solvent for **polar substances** such as sugars, charged ions (Na^+, Cl^-, Ca^{2+}, K^+), B and C vitamins and amino acids. Polar substances do not dissolve in lipid and so can cross a cell membrane only by going through pores.

Most fats, oils and lipids are **non-polar molecules** (they do not have charged regions) and do not dissolve in water. Other non-polar substances (such as vitamins A, D, E and K) can dissolve in lipids and so can cross cell membranes without going through pores.

According to the fluid mosaic model, the cell membrane consists of a double layer of phospholipid molecules, known as a **lipid bilayer**. This is studded with proteins and other molecules. The name **fluid mosaic** is used because the bilayer is a very fluid structure (the phospholipid molecules are in constant sideways motion – a bit like table tennis balls on the surface of a stretch of water) – and it contains a 'mosaic' of protein molecules.

THE CHEMICAL MAKE-UP OF CELL MEMBRANES

Cell membranes contain **cholesterol** (not shown in Fig. 4.4), **phospholipids**, **proteins** and **polysaccharides**.

Phospholipids are the major constituent of cell membranes. They naturally form membranes in water because they automatically arrange themselves into a bilayer that is virtually impermeable to water and to anything that is water-soluble (Fig 4.2). So the membrane keeps the cell contents in and it keeps everything else out – except it's not as simple as that. Some substances, mainly water-soluble chemicals, need to get in and out of cells all the time – so how does the cell manage this?

The answer is – **membrane proteins**. These protein molecules act as **hydrophilic pores**, water-filled channels that allow water-soluble chemicals to pass through (Fig 4.5). Pores are usually small and highly selective; they allow only specific molecules or ions through. Proteins in the membrane that form pores usually span the entire membrane, but other proteins with other functions can occur only in the top or only in the bottom layer of **lipids**.

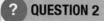 **QUESTION 2**

2 Explain the terms 'hydrophilic' and 'hydrophobic'. How do these terms relate to a phospholipid molecule?

Fig 4.5 Some of the proteins in the membrane form pores. These allow particles that cannot dissolve in lipids to enter and leave the cell. Membrane pores are not simply 'holes'; they can control what passes through them

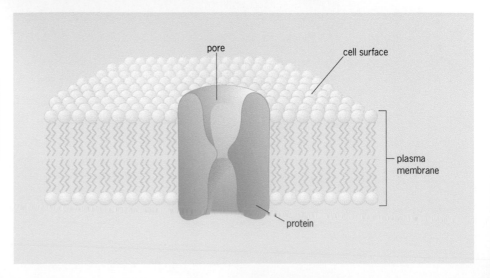

Membrane proteins have several functions. They:

- form pores through which water and water-soluble chemicals can pass,
- act as carriers in active transport,
- form receptor sites for hormones,
- are important in cell recognition.

Plasma membranes can also contain **cholesterol** molecules, which fit in between the phospholipids (Fig 4.16, page 67). Cholesterol is a vital constituent of animal cells, despite being notorious for its connection with heart disease (Fig 4.6).

Polysaccharides, branched polymers of simple sugars, stick out from the outer surface of some membranes like antennae (Figs 4.3 and 4.4, pages 59 and 60). They attach to lipids, forming **glycolipids,** or to proteins, forming **glycoproteins**. Glycolipids and glycoproteins help cells to recognise each other – allowing the immune system to tell the difference between body cells and invading bacteria, for example.

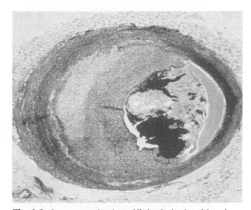

Fig 4.6 A narrowed artery. High cholesterol levels are associated with the development of coronary heart disease. The fatty deposit, atheroma, builds up under the endothelium of the blood vessel, narrowing the lumen

3 DIFFUSION

Diffusion is basically a 'mixing of molecules'. In a gas or liquid the molecules or ions move continuously, randomly bumping into each other and changing direction. In this way, particles tend to diffuse, or spread, so that they are evenly spaced rather than being concentrated in one place (Fig 4.7).

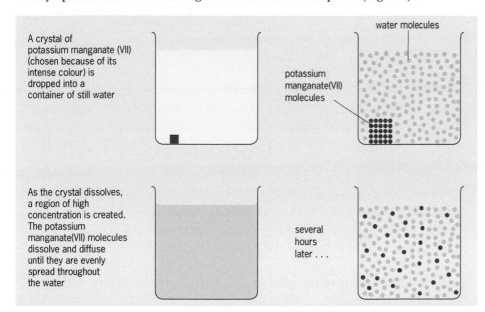

Fig 4.7 If you drop a crystal of potassium manganate(VII) (chosen for its visibility) into still water, it dissolves and the ions diffuse until evenly spread. This takes several hours (depending on temperature). Diffusion is a passive process which requires no input of energy: it does, however, depend on the kinetic energy (energy of movement) of the molecules in a gas or a liquid

? QUESTION 3

3   Why does diffusion apply to particles in liquids and gases but not solids?

A useful working definition of diffusion is:

Diffusion is the movement of particles within a gas or a liquid from a region of high concentration to a region of lower concentration.

Diffusion mainly moves substances over short distances. It is too slow to move substances efficiently over distances much greater than a fraction of a millimetre. In the mammalian lung, for example, oxygen diffuses through the thin epithelium of the alveolus and into the blood, a journey of usually less than a hundredth of a millimetre (10 µm). It is then carried away to other parts of the body by the circulatory system. The rapid movement of materials around an organism in a stream of fluid is called **mass flow**.

DIFFUSION AND ENERGY

The difference in concentration of a substance between two regions is called a **concentration gradient**. Particles that are free to move have **kinetic energy**. The region of a gas or liquid with the highest concentration of particles of a particular substance has the highest kinetic energy for that substance. Particles move down a concentration gradient by diffusion, until they are spread evenly.

The most important point to remember is that diffusion is a **passive process**: it requires no input of energy from the cell. It follows that movement of a substance against a concentration gradient – **active transport** (page 69) – requires energy.

FACTORS THAT AFFECT THE RATE OF DIFFUSION

The rate at which molecules of a substance diffuse from one region to another has a great influence on the organisation of cells, organs and whole organisms (Fig 4.8).

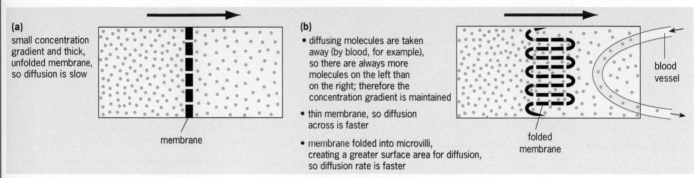

(a) small concentration gradient and thick, unfolded membrane, so diffusion is slow

membrane

(b)
- diffusing molecules are taken away (by blood, for example), so there are always more molecules on the left than on the right; therefore the concentration gradient is maintained
- thin membrane, so diffusion across is faster
- membrane folded into microvilli, creating a greater surface area for diffusion, so diffusion rate is faster

blood vessel

folded membrane

Fig 4.8 A schematic diagram to show some of the features which increase the rate of diffusion. An efficient exchange of materials requires a large surface area, maintenance of a diffusion gradient and a thin plasma membrane. Increasing the temperature would also speed up diffusion, but this is not an option in living systems

Several factors affect the rate of diffusion:
- the surface area between the two regions; the greater the surface area, the greater the rate of diffusion,
- the distance over which diffusion occurs; this is known as the **length of the diffusion pathway**,
- the concentration gradient – the relative concentration of the substance in the two areas. Diffusion is more efficient if the concentration gradient can be maintained. One way to achieve this is to use blood to transport the substance away from the immediate area into it has diffused. Another way is to combine the substance with another chemical to prevent it from diffusing back,
- the size and nature of the particles. Fat-soluble substances can diffuse through the lipid bilayer of the membrane. Water-soluble substances must pass through the protein pores, which tend to be small and selective. Generally, very large molecules cannot diffuse into or out of cells at all,
- the temperature at which the process takes place; at higher temperatures, molecules have more kinetic energy and so diffuse more quickly.

Calculating the rate of diffusion

The rate at which substances diffuse can be estimated from a simple formula that takes into account the factors that affect diffusion. Rate of diffusion is proportional to:

$$\frac{\text{surface area} \times \text{concentration difference}}{\text{length of diffusion pathway}}$$

This relationship is known as **Fick's law**. For efficient diffusion, the values on the top line of this equation should be as large as possible and the value on the bottom line should be as small as possible.

? QUESTION 4

4   Phosphorylated chemicals (those with a phosphate group added) cannot pass easily through cell membranes. Suggest why glucose molecules are phosphorylated as soon as they enter a cell.

Diffusion in the human body

Diffusion is not fast enough to be effective over distances greater than a few micrometres. The human body has many organs that are adapted for the exchange of materials – the lungs, intestines and kidneys are obvious examples. However, these organs are of little use without a circulatory system that transports the exchanged materials to all other areas of the body. The blood system is an example of what is known as a **mass flow system**.

The main function of this system is to bring materials to within diffusing distance of living cells, and to take away cell products. In the human body, no cell is more than a few micrometres away from a capillary.

4 FACILITATED DIFFUSION

Some substances enter and leave cells much faster than you would expect if only diffusion occurred. We now know that some membrane proteins assist, or **facilitate**, the diffusion of some substances across the cell membrane. Two types of protein are responsible for **facilitated diffusion**:

- Specific **carrier proteins** take particular substances from one side of the membrane to the other.
- **Ion channels** are proteins that open and close to control the passage of selected charged particles.

CARRIER PROTEINS

Until the 1970s, cell biologists thought carrier proteins worked by rotating within the membrane, like turnstiles. Newer research points to a different explanation (Fig 4.9). As soon as the diffusing molecule binds to the carrier protein, the protein undergoes a change in shape so that the diffusing molecule ends up at the other side of the membrane, where it is released.

Like diffusion, facilitated diffusion involves movement *down* a concentration gradient and requires no *input* of metabolic energy.

ION CHANNELS

Ion channels are proteins with a central 'hole' lined with polar groups (Fig 4.10). Ion channels facilitate the diffusion of charged particles such as Ca^{2+}, Na^+, K^+ and Cl^- ions. Many channels are **gated**, so can open or close. Cells use ion channels to control the movement of ionic substances between themselves and other cells, and to regulate the ionic composition of their cytoplasm.

QUESTION 5

5 Name some organs where exchange of materials is the main function.

Gas exchange in the human body is discussed in Chapter 9.

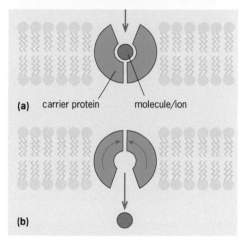

(a) carrier protein   molecule/ion

(b)

Fig 4.9 Facilitated diffusion using a carrier protein. The diffusing molecule interacts with the carrier protein, causing a change in shape which 'squeezes' the molecule through the channel

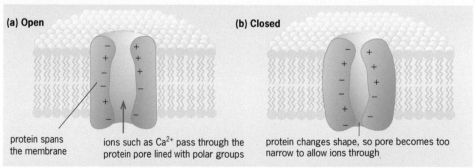

(a) Open

protein spans the membrane

ions such as Ca^{2+} pass through the protein pore lined with polar groups

(b) Closed

protein changes shape, so pore becomes too narrow to allow ions through

Fig 4.10 An ion channel. Because they are charged particles, ions cannot easily pass through the non-polar lipid bilayer. Specific membrane proteins form polar pores through which ions can pass. These channels are usually specific for one type of ion and can open and close according to the needs of the cell. When fully open, over one million ions per second can flow through a single channel

SCIENCE IN CONTEXT

Diabetes and carrier proteins

The concentration of our blood sugar is kept relatively constant by controlling how much glucose passes from the blood into cells. After a meal, when blood glucose levels are high, the hormone insulin is released. This hormone activates a mechanism in the cell membrane that facilitates the diffusion of glucose into the cell, so lowering the blood sugar concentration.

One form of diabetes in humans is caused by a gene mutation that alters the structure of glucose carrier proteins in plasma membranes. An affected person cannot get enough glucose into their cells. Unlike diabetes that results from the inability to make insulin (see Chapter 11), this form of the disease is managed by diet, because injecting insulin has little effect.

Transmission of nerve impulses is discussed in detail in Chapter 15. Gas exchange is discussed detail in Chapter 9.

A difference in concentration of ions may lead to a net positive or negative charge in a particular region. This type of concentration gradient is called an **electrochemical gradient**. Charged particles move towards regions of opposite charge. This is important in the process of transmission of a nerve impulse and in the process by which red blood cells exchange materials with the tissues.

5 OSMOSIS

In biology, we usually talk about the diffusion of substances that are dissolved in water. But what about the water molecules themselves? Does water diffuse down a diffusion gradient? The answer is yes, and the diffusion of water is known as **osmosis**.

A working definition of osmosis is:

> **Osmosis is the diffusion of water only. It is the net movement of water molecules from a region of their higher concentration to a region of their lower concentration, through a partially permeable membrane.**

Water, water everywhere, and not a drop to drink … The concept of osmosis may be easier to grasp if you think of a real life-and-death situation. Imagine a ship-wrecked survivor floating in a dinghy on the open sea, the sun burning down on him for 15 hours a day; no land or rescue in sight. After a few hours, he gets so thirsty, he gives in and drinks seawater. However, because seawater contains salt, its sodium chloride concentration is about three times that found in human blood. The survivor is already dehydrated, and the salt causes water to move out of his blood and into his stomach by osmosis, causing further dehydration of his body cells. Death from dehydration is more likely than if he had drunk nothing.

IMPORTANT FACTS ABOUT WATER

Understanding some of the properties of water is the key to understanding osmosis, so before going into detail about this process, we need to look at what happens when a substance dissolves in water.

As you can see from the information in the Stretch and Challenge box on page 66, water molecules are polar – they carry a tiny positive and a tiny negative charge. In pure water, all the water molecules present are able to move around freely, not associating with each other or with anything else – think of them as 'free' water molecules. However, when a substance is placed in the water that is soluble, things change. As each of the solute molecules dissolves, it becomes charged and attracts the charges on the water molecules. It soon becomes surrounded by a shell of water molecules. When this happens, the water molecules that form the 'shell' are no longer 'free' as they were before: they have been 'tied up' by the solute molecules. This reduces the concentration of 'free' water molecules in the solution.

AN OVERVIEW OF OSMOSIS

Now think about what happens if you have two solutions, each with different concentrations of solute, on either side of a partially permeable membrane.

Fig 4.11a shows what happens. There are more solute molecules in solution B (on the right of the membrane) than in the solution A (on the left). In solution B, more water molecules are 'tied up' in shells around the solute molecules, so there are fewer 'free' water molecules in solution B than in solution A. There is therefore

REMEMBER THIS

The following terms are useful when describing osmosis in human cells.

Hypertonic: a greater solute concentration.

Hypotonic: a lower solute concentration.

Isotonic: an equal solute concentration.

These are relative terms and so can only be used to compare solutions. For example, seawater is hypertonic to human blood plasma, distilled water is hypotonic to human blood plasma.

a concentration difference between the two solutions, both of solute molecules and of water molecules.

The pores in the membrane are too small to allow solute molecules through, so their concentration has to stay the same. What can change is the concentration of 'free' water molecules on either side of the membrane. These water molecules can pass through the pores and water can diffuse down its concentration gradient. Water molecules move from the left, where there is a higher concentration of water molecules, to the right, where there is a lower concentration of water molecules (Fig 4.11b). This is osmosis.

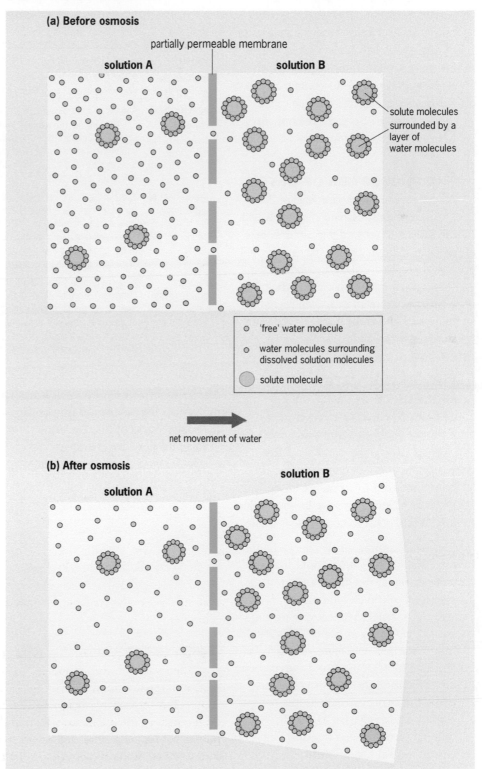

Fig 4.11a A 'weak' solution is one with a low concentration of solute molecules but a high concentration of water molecules. It has more 'free' water molecules than the 'concentrated' solution on the right. In osmosis, water molecules move from a region of their higher concentration to a region of their lower concentration – in this case from the solution on the left to the solution on the right

Fig 4.11b As osmosis occurs, the number of 'free' water molecules on either side of the membrane starts to equalise. Eventually, both solutions will have the same concentration of solute molecules relative to water. Because water has moved from left to right, the solution on the left will have a much lower volume than the one on the right

All about water

There is no life on Earth without water: life almost certainly originated in water. Most biochemical reactions take place in solution and all organisms that live on land can do so only because they have their own internal aquatic environment. Some organisms can survive dehydration as spores or seeds, but these structures simply allow a state of dormancy, which ends when water becomes available once again. The bodies of all organisms contain a high percentage of water. Even one of the most complex of organs, the human brain, is 85 per cent water.

THE PROPERTIES OF WATER

All cells are tiny compartments of fluid. Living cells must exchange materials with the surroundings, and this can only take place in solution. So, as well as being full of water, cells are also surrounded by it. Water has unique physical and chemical properties, many of them significant in biology. Some of the most important are summarised in Table 4.1.

WHY IS THE WATER MOLECULE SO STICKY?

The water molecule consists of two atoms of hydrogen and one atom of oxygen. This gives water a relative molecular mass of 18. Many compounds of this size are gases, but water is a liquid because its molecules are 'sticky' – they cling to each other.

Electrons are negatively charged. Since opposite charges attract and like charges repel, the lone pairs of electrons of the oxygen atom effectively repel the electrons of the hydrogen – oxygen bonds. This results in a 'v-shaped' molecule, as Fig 4.12 shows. The region around the oxygen atom has a slight negative charge, while the two hydrogen atoms have a slight positive charge. Molecules with different areas of positive and negative charge are said to be **polar**.

Property	Significance in living systems
Table 4.1 Some properties of water	
cohesion	Hydrogen bonding causes water molecules to stick to each other. This is important, for example, when water evaporates from the leaves of a tall tree: water is so cohesive that a continuous column can be sucked up from roots which might be over 100 metres below. This cohesion causes surface tension. Water has a very high surface tension, and several organisms, such as pond skaters and rat-tailed maggots, take advantage of this property
the polar nature of water	This makes water a very good solvent – it has been called the 'universal solvent'. Though obviously not true since it doesn't dissolve everything, it does act as a solvent for all ionic and polar compounds
melting and boiling point	Water freezes at 0 °C and boils at 100 °C and is therefore a liquid over most of the temperature range found on Earth
specific heat capacity	A gram of water absorbs a lot (4.2 joules) of heat before its temperature rises by 1 °C. Organisms have a high water content and so can absorb a lot of energy (or lose a lot) before their temperature changes significantly. This makes water a useful thermal buffer. Most organisms cannot tolerate great changes in temperature, and the water in their bodies protects them from rapid changes in external temperature
latent heat of vaporisation	The hydrogen bonding between water molecules makes it difficult for them to separate from each other and vaporise (evaporate) to form a gas. When water does evaporate, the escaping molecules take a lot of energy with them, and so vaporisation is a very efficient cooling process. Anyone who has ever stepped out of the shower and stood in a draught will know just how efficient! The processes of sweating, panting and transpiration all involve cooling by evaporation
thermal conductivity	A naked person sitting in a room at 25 °C would find it comfortable. However, if immersed in water at 25 °C, the same person would feel cold. This is because the thermal conductivity of water allows heat to be transferred away from the body more quickly through the water than through air. Materials with a low thermal conductivity are good insulators. Fur is a good insulator because it traps air, but the insulating properties are lost if the fur gets wet. The same is true of feathers

Attractive forces exist between water molecules: the positive charge on the hydrogen atoms of one molecule attracts the negative charge on the oxygen atoms of another molecule.

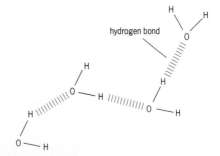

Fig 4.13 Weak attractive forces exist between the positive areas of one water molecule and the negative areas of adjacent ones. Many of the unique properties of water are due to this hydrogen bonding

These attractive forces are called hydrogen bonds (Fig 4.13). In effect, each water molecule acts as a 'mini magnet'. As well as 'sticking' to each other, water molecules also cling to other charged particles.

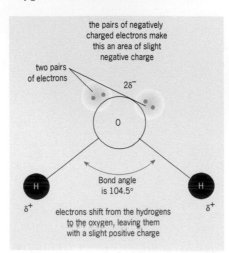

Fig 4.12 Water is a polar molecule: it has distinct areas of positive and negative charge

OSMOSIS IN LIVING CELLS

All cells contain cytoplasm, a complex solution separated from its surroundings by a partially permeable plasma membrane. So, all cells have the potential to gain or lose water by osmosis. The overall tendency for water to enter or leave a cell is determined by **water potential**.

Water potential

If no other factors are involved, a cell that is placed in a solution of the same concentration will neither gain nor lose water. Water molecules pass equally in either direction and there is no net change in the cell volume: Fig 4.14a.

A cell placed in a solution that contains less solute than the cytoplasm will gain water by osmosis: Fig 4.14b. If the surrounding solution contains more solute than the cytoplasm, the cell will lose water by osmosis: Fig 4.14c.

The water potential is the potential of any system to absorb water molecules by osmosis.

Fig 4.14 Factors that determine whether water enters or leaves a cell by osmosis
Fig 4.14a When the solution surrounding the cell contains the same amount of solute as the cytoplasm (i.e. is **isotonic**), the cell neither gains nor loses water
Fig 4.14b When the cell's cytoplasm contains more solute than the surroundings (i.e. is **hypertonic**), the cell gains water by osmosis
Fig 4.14c When the cell's cytoplasm contains less solute than the surroundings (i.e. is **hypotonic**), the cell loses water by osmosis

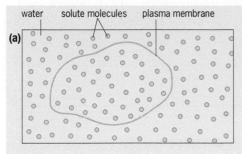
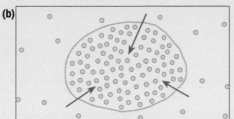
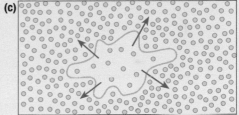

water   solute molecules   plasma membrane
(a)   (b)   (c)

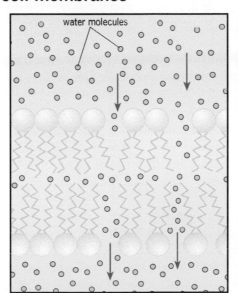
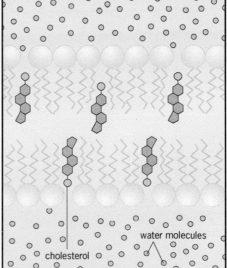

HOW SCIENCE WORKS

How water molecules cross cell membranes

We have already said that the plasma membrane is virtually impermeable to water. In fact, it is not 100 per cent impermeable. A tiny amount is able to pass through the lipid bilayer. But recent studies have shown that water passes through purified phospholipid bilayers, with all the proteins removed, at rates approximately 100 to 1000 times faster than you might expect.

The exact reason for this is not known, but scientists think the constant sideways motion of the phospholipids, coupled with the flexing of the fatty acid tails, creates temporary holes (Fig 4.15) through which the water molecules can slip. These holes appear in only one half of the membrane at a time, so the water molecules 'wait' at the halfway point until a hole appears in the other side, rather like crossing a busy road, one lane at a time.

Membranes with a high proportion of cholesterol are less permeable to water,

water molecules

Fig 4.15 A possible explanation for the fact that lipid bilayers are permeable to water. Temporary holes, created by the random sideways movement of the lipid molecules, act as channels through which water can pass

and to simple ions such as sodium and chloride. It could be that cholesterol has a stabilising effect on the phospholipids.

water molecules
cholesterol

Fig 4.16 Cholesterol is an important constituent of the membranes of many types of cell. The effect of cholesterol is to reduce leakage of water and ions through the lipid bilayer, probably by reducing the sideways motion of the lipid molecules

By slotting in between them, the cholesterol molecules minimise the creation of temporary holes (Fig 4.16).

Osmosis explains why AIDS doesn't spread by kissing

AIDS is a sexually transmitted disease caused by the human immunodeficiency virus (HIV). Semen, breast milk and vaginal secretions are known to contain sufficient HIV to set up an infection in another person. Scientists studying how the disease spreads could detect the virus in these body fluids but saliva seemed to be the odd one out. And there was no evidence that the virus could be passed to another person by kissing. The answer to this puzzle is that saliva is very low in salt compared to other body fluids. It is actually less concentrated than the inside of blood cells that naturally find their way into the mouth, because of wear and tear of the cheeks and tongue. In a person infected with HIV, these cells could carry a high load of virus – and if passed on through kissing or sneezing, could theoretically infect the person on the receiving end. Fortunately, because cells are hypertonic to saliva, water passes from the saliva into the cells by osmosis, causing them to lyse and die, giving no chance of transmission of HIV.

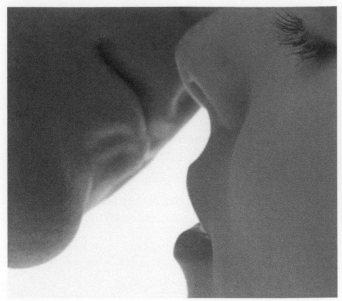

Fig 4.17 No risk of HIV transmission …

QUESTION 6

6 Preserves such as jam and marmalade have a high sugar content. Suggest why bacteria do not thrive when they land on these foods.

Water potential is measured in units of pressure, kilopascals (kPa), and is a negative scale. For example, a relatively weak solution of sucrose can have a water potential of –200 kPa while a more concentrated solution can have water potential of –500 kPa. Pure water has the highest possible water potential: zero.

If the two solutions described above were separated by a partially permeable membrane, water would move from the more dilute solution into the more concentrated one. So, if there are no other factors involved, such as physical pressure, water will move from a region of high water potential to a region of lower water potential. This is from the solution with the water potential of –200 kPa to the solution with the more negative value of –500 kPa. If you have a problem with negative scales, as many people do, compare it to temperature. Of two solutions, one at –40 °C and the other at –60 °C, it is obvious which is the colder. If the two solutions were placed in contact with each other, heat would pass from the warmer to the colder until they were both about equal. It's the same with the movement of water in osmosis.

Osmosis in animal cells

If you place a human cell, or any animal cell in distilled water, it absorbs water by osmosis and swells up. As plasma membranes have virtually no physical strength, the cell often bursts.

Fig 4.18 (opposite) shows what happens to red blood cells placed in **hypertonic**, **isotonic** and **hypotonic** solutions. When blood plasma becomes hypertonic to red blood cell cytoplasm, as it does during severe dehydration, water is lost from the red blood cells, which shrink and become crinkled, or **crenated**. At the other extreme, hypotonic plasma causes red blood cells to swell and burst, leaving 'ghosts' of plasma membrane. This process in known as **haemolysis**: literally 'blood splitting'.

In order to protect our cells from the effects of osmosis, we must maintain our body fluids at the same solute concentration as our cell cytoplasm. We do this by **osmoregulation**.

QUESTION 7

7 Why is it incorrect to say that 'seawater is a hypertonic solution'?

Osmoregulation is discussion in detail in Chapter 13.

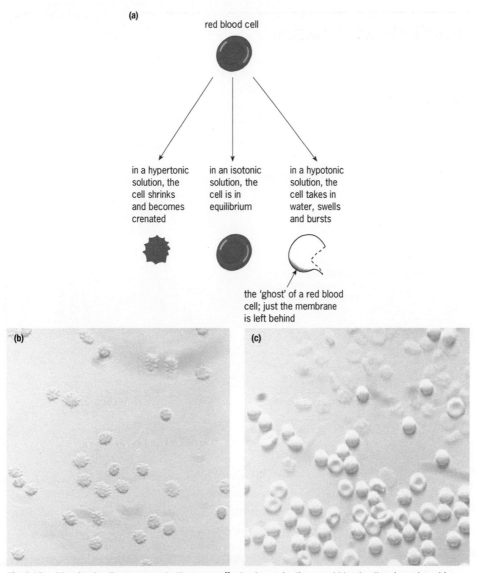

(a)

red blood cell

in a hypertonic solution, the cell shrinks and becomes crenated

in an isotonic solution, the cell is in equilibrium

in a hypotonic solution, the cell takes in water, swells and bursts

the 'ghost' of a red blood cell; just the membrane is left behind

(b)

(c)

Fig 4.18a All animal cells are prone to the same effects shown by these red blood cells when placed in solutions of different concentrations
Fig 4.18b Red blood cells become crenated in a hypertonic solution
Fig 4.18c Red blood cells take in water and burst in a hypotonic solution (some red cells have not yet burst)

6 ACTIVE TRANSPORT

We have seen that diffusion is movement of molecules down a diffusion gradient:

Active transport is movement of molecules against a concentration gradient.

Cell membranes have evolved specialised carrier proteins to transport particular ions or molecules across the membrane, often from a region of low concentration to a region of much higher concentration. Active transport requires an input of energy from cell respiration, usually provided by ATP. These processes all involve active transport:

- absorption of amino acids from the gut into the blood,
- reabsorption of glucose and other useful molecules in the kidney,
- pumping sodium ions out of cells,
- the exchange of sodium and potassium ions which allows conduction of nerve impulses.

Fig 4.19 An example of an active transport mechanism. The H⁺–ATPase pump removes hydrogen ions from a cell against a diffusion gradient. It is found, for example, in the membrane of the kidney tubule cells, where it contributes to the production of acidic urine

? QUESTION 8

8   Why do respiratory poisons such as cyanide prevent active transport?

✔ REMEMBER THIS

The membrane proteins responsible for facilitated diffusion and active transport are similar to enzymes. Carrier proteins have binding sites for specific chemicals, they can work only at a particular rate, becoming saturated at high concentrations of transported substance (Fig 4.20), and they can be inhibited. But they differ from enzymes in one important respect: they do not alter the substances they transport.

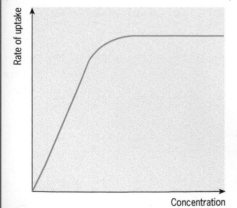

Fig 4.20 The effect of increasing concentration on rate of uptake. Note that the rate of active transport increases to the point at which all the carrier proteins are working at full capacity. If this graph was showing diffusion, it would not level off

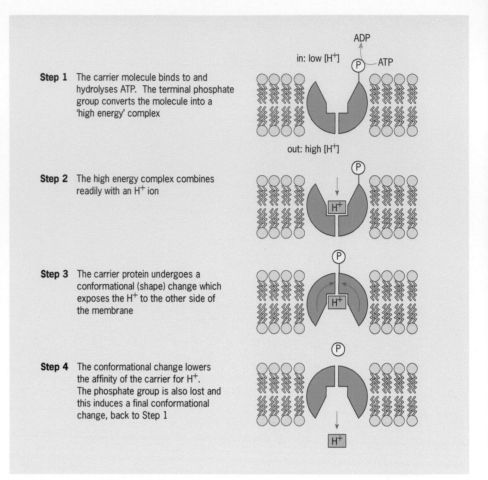

Step 1   The carrier molecule binds to and hydrolyses ATP. The terminal phosphate group converts the molecule into a 'high energy' complex

Step 2   The high energy complex combines readily with an H⁺ ion

Step 3   The carrier protein undergoes a conformational (shape) change which exposes the H⁺ to the other side of the membrane

Step 4   The conformational change lowers the affinity of the carrier for H⁺. The phosphate group is also lost and this induces a final conformational change, back to Step 1

The mechanism of active transport is similar to that of facilitated diffusion (Table 4.2). Cell membranes contain specific carrier proteins that combine with the substance to be transported, enabling it to pass rapidly across the membrane. The key difference is that, while facilitated diffusion is a passive process, the carrier protein complex in active transport is activated by an input of energy (Fig 4.19).

The rate of active transport is closely linked to the rate of respiration. We can assume that active transport is taking place if investigations show that:

● movement of particles takes place against a diffusion gradient,
● the rate of transport increases as the rate of respiration increases,
● the process is inhibited by respiratory poisons such as cyanide.

Table 4.2 Comparing diffusion, facilitated diffusion and active transport

	Simple diffusion	Facilitated diffusion	Active transport
type of membrane molecule involved	lipids	proteins	proteins
force driving the process	concentration gradient	concentration gradient	ATP hydrolysis
direction of transport	with concentration gradient	with concentration gradient	against concentration gradient
specificity	non-specific	specific	specific
saturation at high concentration of transported molecules	no	yes	yes

7 ENDOCYTOSIS AND EXOCYTOSIS

The transport processes covered so far involve the movement of individual molecules or ions across the cell membrane. There are other processes, **endocytosis** and **exocytosis**, which transport larger volumes of material (solids and liquids) into or out of a cell. Fig 4.21 shows the main features of these two processes and highlights the differences between them.

ENDOCYTOSIS – BRINGING MATERIAL INTO THE CELL

In endocytosis, the cell surrounds material with a plasma membrane, bringing it into the cell within a small vesicle. There are two main types of endocytosis:

- **Phagocytosis**. Cells **phagocytose** (take in) solid particles such as bacteria and old red blood cells in order to destroy them.
- **Pinocytosis**. A process identical to phagocytosis but which takes in fluid (Fig 4.21a).

Phagocytosis and pinocytosis

In the process of phagocytosis, a cell's cytoplasm flows round solid material. When this is completely surrounded, a segment of cell membrane is pinched off and a vesicle is formed within the cytoplasm. The membrane and the solid material inside the vesicle form a **food vacuole**. Lysosomes fuse with the vacuole membrane, allowing digestive enzymes to enter the vacuole and begin to digest the contents. Soluble products, such as amino acids and sugars, are absorbed into the cytoplasm, while undigested remains are discharged from the cell by exocytosis.

Examples of phagocytosis include the removal of foreign matter by white cells (see Fig 1.23, page 15) and the removal of old red blood cells from circulation by the Kupffer cells of the liver.

EXOCYTOSIS – RELEASE OF MATERIAL FROM THE CELL

In exocytosis (Fig 4.21b), materials enclosed in vesicles are expelled from the cell when the vesicle membrane fuses with the plasma membrane. When the process involves a liquid it is called reverse pinocytosis and this is how the cell secretes many of its synthesised products, such as digestive enzymes, mucus, hormones and the components of milk.

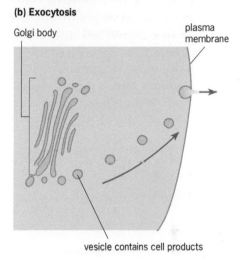

(a) Endocytosis (pinocytosis)

an invagination (pocket) in membrane fills with fluid

membrane closes round droplet, forming vesicle

(labelled) vesicles pass to various parts of cell according to the nature of their contents

(b) Exocytosis

Golgi body

plasma membrane

vesicle contains cell products

Fig 4.21 Endocytosis and exocytosis. These processes appear to be the reverse of each other, but endocytosis must be selective and the contents must be 'labelled' by the cell so that they can be processed in the correct way

Pus

When someone has a heavy bacterial infection, such as an infected cut or sweat gland, white blood cells gather in great numbers and go into what can be described as a 'phagocytosis frenzy'. They engulf so many bacteria and dead cells that they become literally 'full'. Toxins secreted by the bacteria then kill the white cells to form pus. The lysosomes of dead white cells tend to liquefy the material surrounding them, forming a smelly, runny fluid.

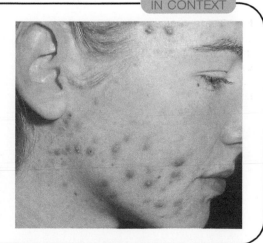

Fig 4.22 This person's acne is caused by an oversecretion of sebum, a greasy substance that blocks sebaceous glands and hair follicles, trapping bacteria underneath. Pus is formed when white cells do their best to deal with the problem

SUMMARY

When you have finished this chapter, you should know and understand the following:

- **Diffusion** is the movement of particles down a concentration gradient until they are evenly distributed.
- In order to maximise the process of diffusion, organisms have a large **surface area**, make the exchange surface as thin as possible and maintain a **diffusion gradient**.
- **Facilitated diffusion** is diffusion enhanced by specific proteins in cell membranes.
- **Osmosis** is the diffusion of water from a region of high water potential to a region of lower water potential access a diffrentially permeable membrane.
- **Water potential** is the tendency for water to move by osmosis. It has a negative scale. Pure water has a water potential of zero. More concentrated solutions have a more negative water potential.

- Diffusion, facilitated diffusion and osmosis are passive processes; they do not require energy.
- **Active transport** is the movement of particles across plasma membranes against a diffusion gradient. This process requires energy, usually in the form of ATP. It is achieved by carrier proteins in cell membranes.
- **Endocytosis** is the passage of droplets of fluid (**pinocytosis**) or solid particles (**phagocytosis**) into the cell by becoming enveloped in a plasma membrane.
- **Exocytosis** is the passage of material (usually a cellular secretion) out of the cell.

Practice questions and a How Science Works assignment for this chapter are available at www.collinseducation.co.uk/CAS

5

Enzymes and metabolism

5 ENZYMES AND METABOLISM

The secret of the runny filling in a Creme Egg

It is easy to see how confectioners make chocolates with hard centres: they simply pour melted chocolate over the centre and wait for it to set. But what about the soft centres? It isn't possible to pour liquid chocolate over a liquid centre and still keep the shape. So, how do they make the runny, yolklike inside to chocolate eggs?

Chocolate lovers everywhere may be surprised to learn that the answer is: use an enzyme. To start with, the centre is solid and contains an enzyme and a polysaccharide. After the chocolate coating has set, the enzyme breaks down the long polysaccharide chains, turning the hard centre into the familiar runny filling. Increasingly, enzymes are being used in medicine, industry and biotechnology.

1 WHAT ARE ENZYMES?

Enzymes are complex chemicals that control reactions in living cells. They are biochemical **catalysts**, speeding up reactions that would otherwise happen too slowly to be of any use to the organism. This definition is a bit of an oversimplification. An active enzyme may speed up a particular reaction, but living organisms do not need all reactions to be going at the maximum rate all of the time. The key word is control. It is more accurate to say that enzymes interact with other molecules to produce an ordered, stable reaction system in which the products of any reaction are made when they are needed, in the amount needed.

THE ROLE OF ENZYMES IN AN ORGANISM

Fig 5.1 is part of a metabolic pathway chart, showing some of the large number of different but interconnected chemical reactions in a living cell. Charts like this are too complex to learn but they illustrate several important points:

- Many of the complex chemicals that living organisms need cannot be made in a single reaction. Instead, a series of simpler reactions occurs, one after another, forming a **metabolic pathway**. A single pathway may have many steps in which each chemical is converted to the next. A specific enzyme controls each reaction.

- Each individual step in the chart represents one of the simple chemical reactions. At first glance, there seem to be a great variety of reactions, but look more closely and you will see that the same types of reaction occur again and again.

- Enzymes control cell metabolism by regulating how and when reactions occur. Using this very simple pathway as an example,

$$A \rightarrow B \rightarrow C \rightarrow D$$

the final product is substance D, the chemical needed by the living organism. The pathway needs three different enzymes and, when D is no longer needed, or if too much has been produced, one of the three enzymes is 'switched off'.

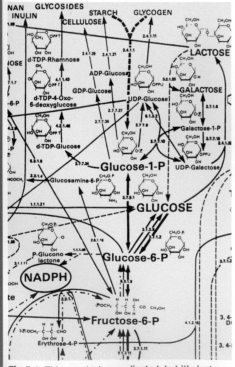

Fig 5.1 This may look complicated, but it's just a small part of a large chart that shows all the different reactions that go on inside a cell. Each arrow is a different reaction, and each number is a particular enzyme that catalyses that reaction

One enzyme, one reaction – but what a difference

When carbon dioxide dissolves in water, a small proportion of the carbon dioxide molecules combine with water to form carbonic acid.

$$CO_2 + H_2O \rightarrow H_2CO_3 \rightarrow H^+ + HCO_3^-$$

It is a rather slow reaction, but, in the presence of the enzyme **carbonic anhydrase**, it is speeded up about 10^7 (ten million) times. One molecule of carbonic anhydrase can convert 600 000 molecules of carbon dioxide into carbonic acid every second. In living cells, this reaction allows other processes to occur much faster:

- In red blood cells, the enzyme speeds up the production of acid, which in turn causes oxyhaemoglobin to give up its oxygen. Without the enzyme, delivery of oxygen to the tissues would be much slower.
- In certain cells in the stomach lining, the activity of carbonic anhydrase in a pathway allows hydrochloric acid to be secreted rapidly. This creates the acidic conditions necessary for digestive enzymes in the stomach to work properly.
- In the cells of the kidney tubule, carbonic anhydrase speeds up excretion of excess acid, and so helps to maintain the pH of the body at the correct level.

2 THE CHEMICAL NATURE OF ENZYMES

Enzymes are globular proteins. They have a very precise overall shape (tertiary structure) formed from one or more polypeptides. Usually, they are roughly spherical, hence the term **globular** protein (Fig 5.2). This overall three-dimensional shape of an enzyme molecule is very important: if it is altered, the enzyme cannot bind to its substrate and so cannot function. Enzyme shape is maintained by hydrogen bonds and ionic forces (Fig 5.3, overleaf) and enzyme function can be affected by changes in temperature and pH.

? QUESTION 1

1 By what process are enzymes made in the cell?

✓ REMEMBER THIS

The active site of an enzyme is part of the enzyme molecule and not part of the substrate.

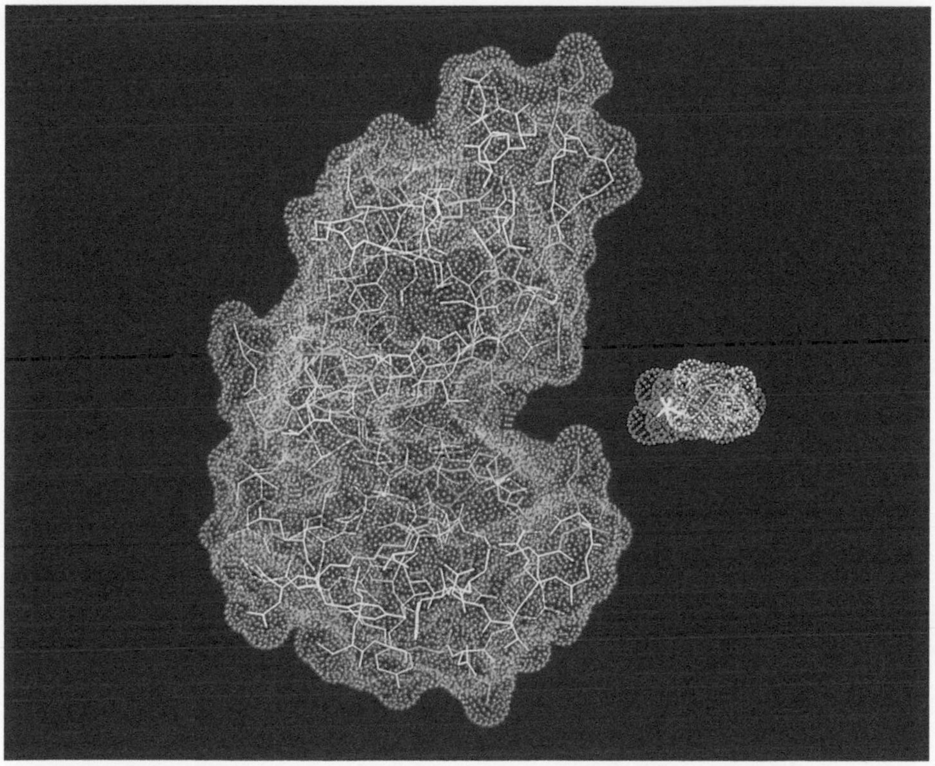

Fig 5.2 The three-dimensional shape of ribonuclease A, an enzyme that helps break up mRNA in the cytoplasm of bacteria. The **active site**, the 'pocket' into which the substrate fits, is clearly visible

REMEMBER THIS

An enzyme acts on a chemical known as its substrate. The name of an enzyme often comes from substituting or adding -ase in the name of the substrate, so, for example, lactose is the substrate of the enzyme lactase.

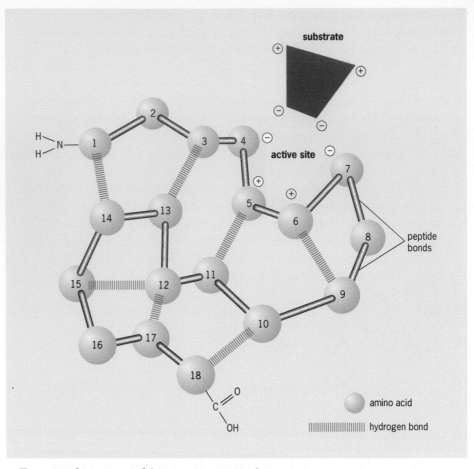

Fig 5.3 This schematic diagram shows the three-dimensional structure of an enzyme molecule. Hydrogen bonds are shown in red. The shape and the electrical charges on the substrate closely match those of the active site

Enzymes have several important properties:

- Enzymes are **specific**: each one catalyses only one reaction.
- Enzymes combine with their substrates to form temporary enzyme–substrate **complexes**.
- Enzymes are not altered or used up by the reactions they catalyse, so can be used again and again.
- Enzymes work very rapidly and each has its own **turnover number** (Table 5.1, page 79).
- Enzymes are sensitive to temperature and pH.
- Enzyme function can be slowed down or stopped by **inhibitors**.

3 THE SPECIFICITY OF ENZYMES

Taking digestive enzymes, you might find it difficult to believe that each enzyme catalyses only one reaction. Trypsin, for example, can begin the digestion of a wide variety of foods rich in protein: eggs, pork, chicken and soya, for example. But when you look at how trypsin works at the molecular level, you can see that this enzyme *is* **specific**. Trypsin cuts an amino acid chain at a point between two particular amino acids, arginine and lysine, and nowhere else. Most proteins have these two amino acids next to each other at some points in their polypeptide chain, and so can be partly digested by trypsin.

Several scientists in biochemistry have devised mechanisms – often called **models** – to explain how enzymes work. In coming up with ideas, they have had to take account of enzyme specificity. Two models that are used to explain how enzymes work are the **lock-and-key hypothesis** and the **induced-fit hypothesis**.

QUESTIONS 2–3

2 Predict substrates for the following digestive enzymes: protease, lipase, nucleotidase.

3 If the enzyme amylase is specific, how can it catalyse the digestion of bread, potatoes and rice?

The lock-and-key hypothesis

This idea assumes that enzyme function depends on an area on the molecule known as the **active site**, visible in Figs 5.2 and 5.3 (pages 75 and 76). The active site is a groove or pocket in the surface of the enzyme into which the substrate molecule fits. Typically, the active site is formed by 3 to 12 amino acids. The size, shape and chemical nature of the active site correspond closely with that of the substrate molecule, so they fit together like a key fits into a lock (Fig 5.4) or, perhaps more realistically, like two pieces in a three-dimensional jigsaw.

Although this model helps us to understand some of the properties of enzymes, it is now generally accepted that a modified version, known as the induced-fit hypothesis, better represents what happens when an enzyme catalyses a reaction.

> **REMEMBER THIS**
> The substrate and the active site of an enzyme are not the same shape: they are **complementary**.

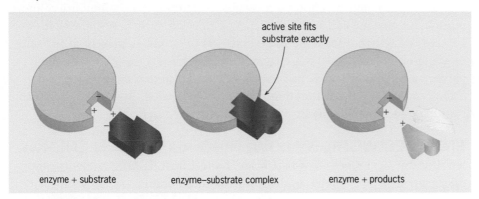

active site fits
substrate exactly

enzyme + substrate   enzyme–substrate complex   enzyme + products

Fig 5.4 The lock-and-key hypothesis. The active site is a particular shape (the lock) into which only one substrate (the key) will fit. The enzyme and substrate combine for an instant to form an enzyme–substrate complex. The formation of this complex brings about the desired chemical reaction, converting substrate into product(s), which are then released

The induced-fit hypothesis

Experimental evidence suggests that the active site in many enzymes is not exactly the same shape as the substrate, but moulds itself around the substrate as the enzyme–substrate complex is formed (Fig 5.5). Only when the substrate binds to the enzyme is the active site the correct shape to catalyse the reaction. As the products of the reaction form, they fit the active site less well and fall away from it. Without the substrate, the enzyme reverts to its 'relaxed' state, until the next substrate molecule comes along.

Both models show why enzymes are not altered by the reactions that they catalyse: they bind to a substrate momentarily, allowing a reaction to happen, but do not themselves undergo any chemical change.

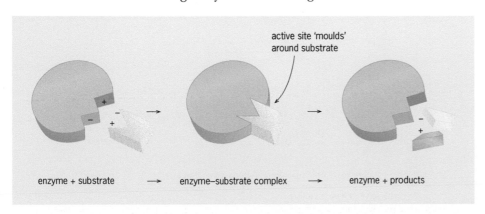

active site 'moulds'
around substrate

enzyme + substrate   →   enzyme–substrate complex   →   enzyme + products

Fig 5.5 The induced-fit hypothesis. Before substrate binding, the enzyme's active site is 'relaxed'. When the substrate binds, the active site is pulled into the correct shape by molecular interactions between the two molecules, and an enzyme–substrate complex forms. As the products fall away from the active site, the molecule becomes 'relaxed' again

Recycling of molecules is discussed in Chapter 34.

THE CHEMISTRY OF ENZYME ACTION

In biology, almost all reactions are **reversible**. If they were not, there would be no recycling of molecules because once large molecules had been built up they could not be broken down again (see Chapter 34).

A simple reversible reaction can be expressed as:

$$A + B \rightleftharpoons C + D$$
$$\text{reactants} \qquad \text{products}$$

In this reaction, the **reactants** (A and B) combine to give the **products** (C and D). If this reaction were to take place in a test tube, a proportion of the products would react to form A and B again. Eventually, an **equilibrium** would be reached in which the relative proportion of reactants and products would remain the same.

In theory, an enzyme allows a reversible reaction to reach equilibrium more quickly. An enzyme can speed the reaction in either direction; the way it does depends on whether there are more reactants or products present at the start. In living organisms, however, enzymes usually speed up reactions in one particular direction. This is because, in any organism, an enzyme acts only if the product of the reaction that it catalyses is needed. And, as the product is used as soon as it is made, equilibrium is rarely reached.

ENZYMES WORK BY LOWERING ACTIVATION ENERGY

For any chemical reaction to take place, bonds must be broken before new ones can form. The energy needed to break these bonds, and so set the reaction in motion, is the **activation energy** (Fig 5.6).

Many reactants need a large amount of energy to push them to a state where they can take part in a reaction, so many reactions take place only at high temperature. Many substances burn, for instance, but only after the initial activation energy has been supplied, perhaps by lighting a match.

Fig 5.6 In order for a reaction to happen, activation energy must be supplied. Then the rest of the reaction proceeds, just as the boulder rolls down the hill once the energy has been supplied to push it to the top. Catalysts such as enzymes work by lowering the activation energy

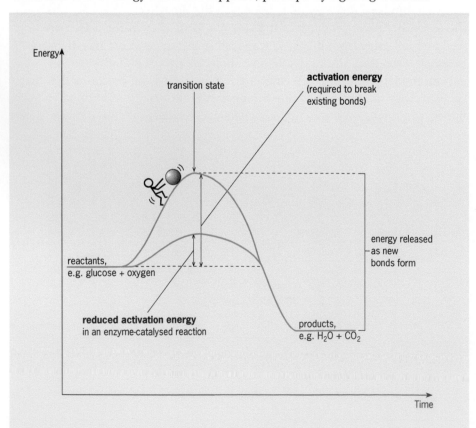

In the presence of enzymes, the activation energy is greatly lowered and this allows reactions to take place at the relatively low temperatures normally found in living organisms. The 'halfway' point in a reaction is called the **transition state**, and this is represented by the top of the curve in Fig 5.6. The transition state represents the stage when the old bonds have been broken in order to allow new ones to form. Enzymes lower the activation energy by making it easier to achieve the transition state.

Enzymes speed up reactions in a number of ways:

- They can hold the substrates close together at the correct angle – this would otherwise have to occur by chance collision.
- They can position any charged groups on the active site to help the reaction to occur.
- By acid or base catalysis – the active site of the enzyme can behave as an acid or a base, donating or accepting protons (hydrogen ions, H^+) from the substrate.
- Through structural flexibility – the flexible shape of the enzyme can change during catalysis. This ensures that the substrates are brought together in the correct sequence for the reaction.

Many metabolic reactions require energy in addition to the presence of the relevant enzyme. This energy is supplied by ATP made by cell respiration. Reactions that need energy are made to happen by coupling them with ATP breakdown, and so are called **coupled reactions**.

HOW FAST DO ENZYMES WORK?

The speed at which an enzyme works is expressed as its **turnover number**. This is usually defined as the number of substrate molecules turned into product in one minute by one molecule of enzyme. Values range from less than 100 to many millions. Some examples are given in Table 5.1.

NAMING AND CLASSIFYING ENZYMES

Older enzyme names such as **pepsin**, **catalase** and **trypsin** give no clues about the nature of the reaction they catalyse. To cope with the rapidly expanding number of new enzymes, the International Union of Biochemistry has developed a scheme for naming and classifying enzymes. Very generally, enzymes are named be adding the suffix -ase to the name of their substrate. The rest of the name attempts to indicate the nature of the reaction taking place. Alcohol dehydrogenase, for example, catalyses the removal of hydrogen from alcohol (ethanol). Further examples are given in Table 5.2.

> **? QUESTION 4**
>
> 4 The enzyme lysozyme helps to split bacterial cell walls. Where would you expect this enzyme to be present in the human body?

> Cell respiration is discussed in Chapter 33.

Table 5.1 Some turnover numbers

Enzyme	Turnover number
carbonic anhydrase	36 000 000
catalase	5 600 000
β-galactosidase	12 000
chymotrypsin	6 000
lysozyme	60

(Source: *Biochemical society guidance notes 3, Enzymes and their role in biotechnology*)

Table 5.2 Some common enzymes and their substrates

Enzyme	Substrate	Reaction catalysed
maltase	maltose	hydrolysis of maltose to glucose
amylase	starch	hydrolysis of starch to maltose
alcohol dehydrogenase	alcohol (ethanol)	removal of hydrogen from alcohol
DNA ligase	DNA	joining together two DNA strands
RNA polymerase	nucleotides that make RNA	synthesis of mRNA on a DNA molecule
glycogen synthetase	glucose	polymerisation of glucose into glycogen
ATPase	ATP	synthesis or splitting of ATP

QUESTION 5

5 People with a great deal of physical stamina have been found to have a lot of **glycogen synthetase** in their muscles. Suggest what this enzyme does.

Redox reactions and glycolysis are discussed in Chapter 33.

 REMEMBER THIS

Heating denatures enzymes; it does not destroy them.

 REMEMBER THIS

The factors that affect enzyme activity also affect the functions of the cell and, ultimately, the organism.

 QUESTIONS 6–7

6 Suggest which types of bonds are broken when an enzyme is denatured. (Look at Fig 5.3, page 76, for help.)

7 Egg albumen, when heated, goes solid and white. What do you think has happened to the proteins in the egg white, and why does the white not become runny again when the egg is cooled?

 QUESTION 8

8 Why is it an advantage for humans to have a constant body temperature of around 37 °C?

Types of enzymes

Although there are many different enzymes, each one can be put into one of six main categories according to the type of reaction they catalyse:

- **Oxidoreductases**. These catalyse oxidation and reduction (redox) reactions. In aerobic respiration, most of the cell's ATP is generated by redox reactions.
- **Transferases**. These catalyse the transfer of a chemical group from one compound to another, such as the transfer of an **amino group** from an amino acid to another organic acid in the process of **transamination**.
- **Hydrolases**. These catalyse **hydrolysis** (splitting by use of water) reactions. Most digestive enzymes are hydrolases.
- **Lyases**. These catalyse the breakdown of molecules by reactions that do not involved hydrolysis.
- **Isomerases**. These catalyse the transformation of one isomer into another, for instance the conversion of glucose 1,6 diphosphate into fructose 1,6 diphosphate. This is one of the first reactions in **glycolysis**, the first stage of respiration.
- **Ligases**. These catalyse the formation of bonds between compounds, often using the free energy made available from ATP hydrolysis. DNA ligase, for example, is involved in the synthesis of DNA.

4 FACTORS THAT AFFECT ENZYME ACTIVITY

Enzymes are proteins and their function is therefore affected by:

- Temperature
- Enzyme concentration
- pH
- Inhibitors
- Substrate concentration

Temperature

For a non-enzymic chemical reaction, the general rule is: the higher the temperature, the faster the reaction (Fig 5.7a). This same rule holds true for a reaction catalysed by an enzyme, but often only up to about 40 to 45 °C. Above this temperature, enzyme molecules begin to vibrate so violently that the delicate bonds that maintain tertiary and quaternary structure are broken, irreversibly changing the shape of the molecule. When this happens the enzyme can no longer function and we say it is **denatured**.

The effect of temperature on a reaction can be expressed by the temperature coefficient, commonly known as the Q_{10}.

Where t is the chosen temperature, the formula for the Q_{10} is:

$$\frac{\text{rate of reaction at } t + 10\,°C}{\text{rate of reaction at } t\,°C}$$

Fig 5.7b shows how to calculate the Q_{10}. To avoid denaturing the enzyme, the values for living organisms need to fall in the range 4 to 40 °C, so we have chosen t as 20 °C.

$$Q_{10} = \frac{\text{rate at } 30\,°C}{\text{rate at } 20\,°C} = \frac{4}{2} = 2$$

In practice, most enzymes have a Q_{10} between 2 and 3. A value of 2 means that the rate of reaction doubles with a 10 °C temperature rise, 3 means that it triples.

HOW SCIENCE WORKS

Stable enzymes for washing powders

Enzymes are unstable, particularly at high temperature, so their commercial usefulness is limited. Many industrial processes need to take place in 'unnatural' environments, at high temperature and extremes of pH. But there are organisms that can thrive at high temperatures. Heat-loving bacteria have thermostable enzymes that are also more resistant to extremes of pH and other unfavourable conditions, such as organic solvents.

Genes that code for some thermostable enzymes have been transferred into bacteria that are easy to grow in large quantities, using recombinant DNA technology (see Chapter 27). As the bacteria that have received a particular gene multiply, the gene is translated into protein and large amounts of the enzyme are produced for commercial use.

An example of a thermostable enzyme is the alkaline protease **subtilisin**, the famous stain digester in biological washing powders. This enzyme is produced on a grand scale by the bacterium *Bacillus subtilis*, and is active in alkaline environments, and so is compatible with the other ingredients in washing powder. It is active at temperatures up to 60 °C, allowing it to be used in a wide variety of wash programmes.

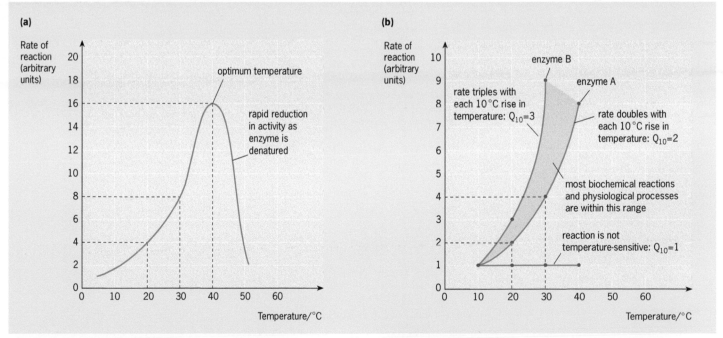

Fig 5.7a Up to about 40 °C, the rate of enzyme-controlled reactions increases with temperature. The optimum temperature for many enzymes is about 40 °C, although the activity of an individual enzyme may increase up to about 50 °C or beyond. However, as the temperature passes 43–44 °C, most enzymes lose their activity

Fig 5.7b Not all enzymes have the same Q_{10} values. This variation is important because enzymes, even those in the same pathway, can vary significantly in their sensitivity to temperature. Outside an organism's normal temperature range, the enzymes may begin to work at different rates. This causes a metabolic imbalance that may be lethal, and may explain why some organisms die at temperatures that would seem to be relatively mild. Some Antarctic fish, for example, live in water that remains very constant at around −2 °C, and die if placed in water above 6 °C

QUESTION 9

9 Chymotrypsin is found in the small intestine at pH 8. What would happen to this enzyme if it found its way into the stomach?

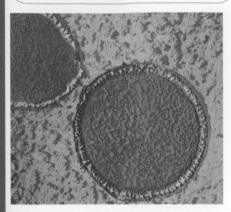

Fig 5.8 This bacterium *Staphylothermus marinus* lives near deep sea hot-water vents. It thrives at temperatures of up to 98 °C. Such heat-loving bacteria have thermostable enzymes (enzymes that are very resistant to heat damage). These have great potential in industry

Some organisms have enzymes that are less sensitive to heat than those found in mammals. For example, certain bacteria can survive in hot volcanic springs and deep sea hydrothermal vents (Fig 5.8) at temperatures of over 90 °C, so their enzymes must be active at these extreme temperatures.

pH

Like other proteins, enzymes are stable over a limited range of pH. Outside this range, enzymes are denatured. Free hydrogen ions (H^+) or hydroxyl ions (OH^-) affect the charges on amino acid residues, distorting the three-dimensional shape and causing an irreversible change in the protein's tertiary structure.

Enzymes are particularly sensitive to changes in pH because of the great sensitivity of their active site. Even if a slight change in pH is not enough to denature the molecule, it may upset the delicate chemical arrangement at the active site and so stop the enzyme working (Fig 5.9).

Most enzymes are **intracellular**: they work inside cells, and their optimum pH is around 7.3 to 7.4. Most organisms have buffer systems that resist changes in pH, and many are able to excrete excess acid or alkali.

Fig 5.10 shows some enzymes that have optimums of pH that are distinctly acid or alkaline. These include digestive enzymes that normally work in the stomach or inside lysosomes. We describe enzymes that work outside cells as **extracellular**.

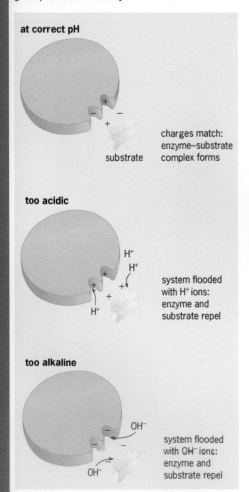

Fig 5.9 The effect of pH on enzyme activity

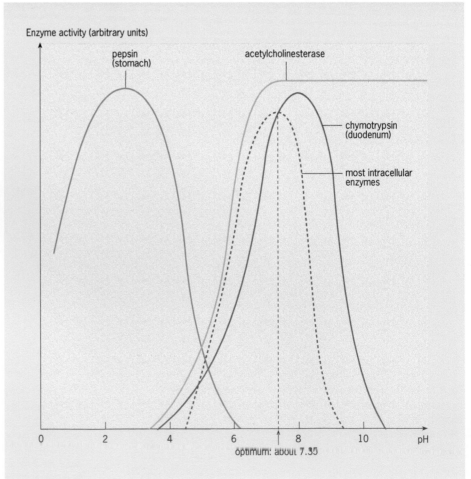

Fig 5.10 The optimum pH of some different enzymes

Substrate concentration

The rate of an enzyme-controlled reaction increases as the substrate concentration increases, until the enzyme is working at full capacity. At this point the enzyme molecules reach their turnover number and, assuming that all other conditions such as temperature are ideal, the only way to increase the speed of the reaction even more is to add more enzymes (Fig 5.11).

Enzyme concentration

In any reaction catalysed by an enzyme, the number of enzyme molecules present is much smaller than the number of substrate molecules. Look back at Table 5.1 (page 79), which shows the turnover numbers of some enzymes, and you can see that one molecule of an enzyme can convert millions of substrate molecules into products every minute.

When there is an abundant supply of substrate, the rate of reaction is limited by the number of enzyme molecules. In this situation, increasing the enzyme concentration increases the rate of reaction.

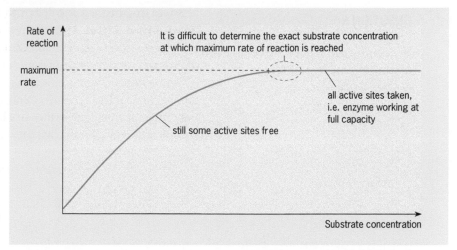

Fig **5.11** The effect of substrate concentration on the rate of enzyme action

5 INHIBITORS

Inhibitors slow down or stop enzyme action. Usually, enzyme inhibition is a natural process, a means of switching enzymes on or off when necessary. Inhibition tends to be **reversible**, and the enzyme returns to full activity once the inhibitor is removed. Many 'external' chemicals, such as drugs and poisons, can inhibit particular enzymes. This type of inhibition is often **non-reversible**. Reversible inhibitors are either **competitive** or **non-competitive**.

The different types of inhibitors are summarised in Fig 5.12.

Competitive inhibitors

Competitive inhibitors compete with normal substrate molecules to occupy the active site. The inhibitor molecules must be a similar size and shape to the substrate to fit the active site, but cannot be converted into the correct product. Effectively, competitive inhibitors 'get in the way' (Fig 5.13) and reduce the number of interactions that can happen between enzyme and substrate.

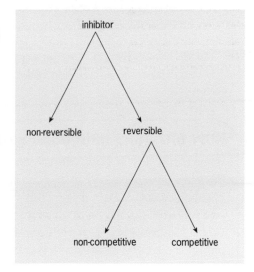

Fig **5.12** A summary of the different types of enzyme inhibitors

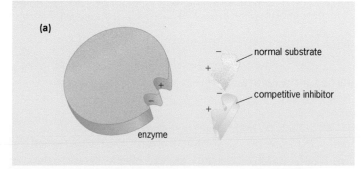

Fig 5.13a A competitive inhibitor fits into the active site of the enzyme, preventing the real substrate from gaining access. The inhibitor cannot be converted to the products of the reaction and so the overall rate of reaction is slowed down. If an inhibitor is present in equal concentrations to the substrate, and if both types of molecules bind to the active site equally well, the enzyme can only work at half its normal rate

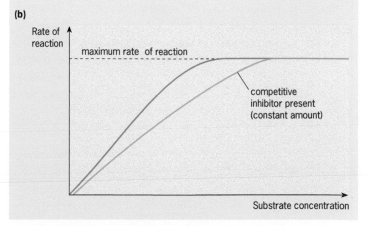

Fig 5.13b The effect of a competitive inhibitor on the rate of reaction. Note that the effect of the inhibitor can be overcome by adding more substrate

? QUESTION 10

10 What would the graph in Fig 5.14b look like if more non-competitive inhibitor was added?

Non-competitive inhibitors

Non-competitive inhibitors bind to the enzyme away from the active site but change the overall shape of the molecule, modifying the active site so that it can no longer turn substrate molecules into product (Fig 5.14a).

Non-competitive inhibition has this name because there is no competition for the active site. The presence of a non-competitive inhibitor has the same effect as lowering enzyme concentration: all inhibited molecules are taken out of action completely. Fig 5.14b shows the effect of a non-competitive inhibitor.

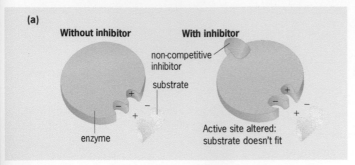

Fig 5.14a Non-competitive inhibitors attach to enzyme molecules and alter the overall shape, so that the active site cannot function. Although the substrate may still be able to bind, forming an enzyme–inhibitor substrate complex, the substrate cannot be turned into product. When the inhibitor molecule is removed, normal function is restored

Fig 5.14b Non-competitively inhibited enzymes molecules show no activity at all, but unaffected ones work normally. Thus, the maximum rate of reaction is lowered, as it would be if we used less enzyme

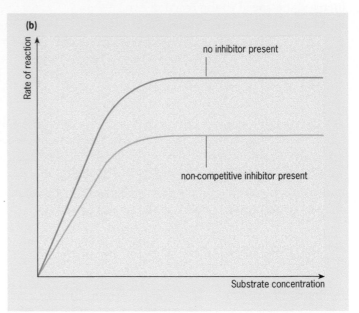

HOW SCIENCE WORKS

How enzymes help babies digest cow's milk

When babies are first born, their intestines don't behave at all like those of an adult. So that they can absorb and make use of important antibodies in their mother's breast milk, they have intestinal walls that have a structure that resembles a string vest. It has lots of large pores through which whole proteins can pass, without being digested into their component amino acids.

Everything is fine, as long as the baby is fed breast milk, but some mothers cannot breast-feed for very good reasons. What happens then? A hundred years ago, babies that couldn't feed from their mother usually fell behind in their growth, sickened and often died. Cow's milk does not contain the same proteins as human milk, and it also has a completely different balance of minerals such as sodium. Feeding tiny babies on untreated cow's milk leads to a generalised immune response against the cow proteins and can also damage the delicate kidneys, which are not able to cope with large amounts of sodium.

Today, many babies are fed on infant formula milk (Fig 5.15) and they thrive just as well as babies that are breast-fed (although the natural method is still considered the best). Enzyme technology has had a lot to do with this – enzymes have been used in the production of infant formula milk from cow's milk for over 50 years. Manufacturers used proteases to digest the milk

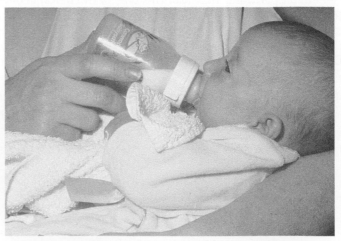

Fig 5.15 Babies can thrive on infant formula milk, which is produced using enzymes

proteins into short peptides and individual amino acids, so preventing whole cow proteins from entering the baby's system. This 'pre-digested' cow's milk is also adjusted for fat and mineral content, and today's formula milks are as close to human milk as possible. Whether biotechnology will ever manage to add all the antibodies, cells, human enzymes and other components of naturally produced human milk remains a question for the future.

Irreversible inhibitors

Irreversible inhibitors bind permanently to the enzyme, rendering it useless. For obvious reasons organisms rarely produce this type of inhibitor for their own enzymes. However, irreversible inhibitors are splendid weapons to use against other organisms. A wide variety of natural toxins are irreversible inhibitors, as are many pesticides.

How inhibitors help to control metabolism

Many metabolic pathways are self-controlling: when a substance is needed, a particular pathway is activated to produce it. When enough has been produced, the pathway is deactivated.

This happens because some enzymes in a metabolic pathway are inhibited by the end product. As Fig 5.16 shows, if too much product begins to accumulate, this inhibits one of the enzymes in the pathway. When the product is once more in short supply, the inhibition is lifted and the pathway becomes active again. This self-regulation is an example of **negative feedback**. This is a fundamental principle important in homeostasis.

REMEMBER THIS

Cyanide is an irreversible inhibitor of **cytochrome oxidase**, one of the enzymes involved in respiration. Organisms poisoned with cyanide die because they are deprived of ATP, their immediate energy source.

Homeostasis is discussed in Chapter 11.

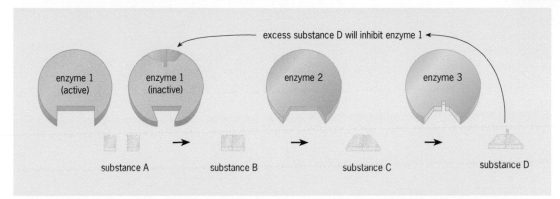

Fig 5.16 A metabolic pathway can be self-regulating if the end product acts as a non-competitive inhibitor on one of the enzymes in its production pathway

SUMMARY

After reading this chapter, you should know and understand the following:

- Enzymes are **globular** proteins with a precise, but delicate, three-dimensional shape maintained by **ionic forces** and **hydrogen bonds**.
- During a reaction, the substrate fits into a region on the enzyme surface called the **active site**.
- Enzymes are **specific**: each enzyme catalyses one particular reaction.
- Enzymes speed up reactions by lowering the **activation energy** needed to get the reaction started.
- Between 4 °C and 40 °C, the rate of an enzyme-controlled reaction increases between two- and threefold for every 10 °C rise in temperature. Increase in temperature beyond 40 °C usually **denatures** the enzyme. Activity is lost when its three-dimensional structure is destroyed.
- Most enzymes have an optimum pH. For **intracellular** enzymes this is usually about 7.35.

- **Extracellular** enzymes, such as those found in digestive juices, may have optimums of extreme pH values.
- The rate of an enzyme-controlled reaction is limited by the supply of substrate, enzyme or the enzyme cofactor.
- **Inhibitors** are substances that slow down or stop enzyme activity. They may be **reversible** or **non-reversible**. Reversible inhibitors may be **competitive** or **non-competitive**.
- Competitive inhibitors tend to be similar in structure to the substrate and compete for the active site. Non-competitive inhibitors do not bind to the active site itself but alter the shape of the enzyme so that the active site is no longer functional.

 Practice questions and a How Science Works assignment for this chapter are available at www.collinseducation.co.uk/CAS

6

Biotechnology

6 BIOTECHNOLOGY

Pollution-digesting bacteria

Scientists in the USA have identified new strains of bacteria that could revolutionise how we deal with pollution. One 'eats' benzene, a toxic chemical found in petroleum-based fuels. Fuel spills and leaking storage tanks mean that benzene is a common contaminant of groundwater supplies and, since long-term exposure to benzene can lead to cancer in humans, it must be dealt with before it can affect soil and drinking water.

Benzene breaks down when exposed to the atmosphere but, without oxygen, this degradation process slows to a crawl. Removing the compound from airless environments is therefore difficult. However, *Dechloromonas* strain RCB and *Dechloromonas* strain JJ apparently oxidise benzene to carbon dioxide without the help of oxygen. These organisms are creating a lot of interest because they can speed up the break-down of benzene in anaerobic conditions by up to 10 times. What once took 70 days now happens in 7 days.

Another bacterial species digests toluene, another common but toxic groundwater contaminant that comes from fuels, adhesives and household solvents. These bacteria also work in the absence of oxygen and research has already identified four genes that code for enzymes in the metabolic pathway responsible.

Using these organisms, or other bacteria given pollution-busting genes, to clean up contamination would have several advantages over current methods. Not only would they be cheaper, but also, the bacteria could be brought to the contaminated area, which would avoid the environmental scarring caused by the removal of tonnes of contaminated soil for dumping or cleaning.

Culture of oil-degrading bacteria. Genetically engineered microbes may become more widely used to extract oil from the ground and valuable metals from factory wastes. Toluene-digesting species of bacteria could be used to prevent pollution by treating industrial waste before it is released into the environment

1 WHAT IS BIOTECHNOLOGY?

A definition of biotechnology is quite difficult to pin down. Broadly speaking, it is the manipulation of biological organisms, systems or processes for the benefit of people, in areas such as agriculture, food production and medicine.

The oldest biotechnologies include the **fermentation process**, which has been used to make bread, beer, wine and cheese for centuries, and also **plant and animal breeding**. More recent biotechnology techniques include the **manipulation of genes** to make **transgenic** plants and animals, using **monoclonal antibodies** to target cancer cells, and using advanced chemistry to **immobilise** enzymes to control and improve manufacturing processes.

Elsewhere, you will also see biotechnology defined more narrowly – as the manipulation of DNA and genes to make organisms that are commercially useful. In this book, we consider both aspects of biotechnology. This chapter covers the wider aspects, with sections on the use of biotechnology in farming and food production, in industry and in medicine. We also consider some of the social and ethical implications of biotechnology. The manipulation of DNA and genes is also here, particularly in relation to farming and food production.

Gene technology, DNA technology, cloning, therapeutic cloning, genetic engineering and the Human Genome Project are covered in more detail in Chapters 22–27.

2 BIOTECHNOLOGY AND FARMING

MANIPULATING FARM ANIMALS

Some people might argue that farming animals is manipulating them too much already, but biotechnology is now used increasingly as a tool to develop more productive animals. Not surprisingly, many of the developments are controversial.

Using hormones to increase milk production

In order to produce milk, cows need to produce a calf each year – they are constantly either pregnant or lactating, or both. With careful timing of one pregnancy after another, a dairy cow can produce a steady supply of milk. Her calf is usually taken away and hand reared, or is used for food. The quality and quantity of this supply depend on the genetic make-up of the cow, how old she is, how much food she is given and the quality of that food. It will also depend on whether she is healthy and free from infection: the overuse of antibiotics in farm animals is a contentious issue (Fig 6.1).

For some years now, farmers in the USA have been enhancing the level **bovine somatotrophin (bST)** in dairy cows. Also known as **bovine growth hormone (BGH)**, this is a natural protein hormone produced by the **pituitary gland** of all cattle.

Biotechnology has enabled scientists to produce a **recombinant** form of this protein, called **rbST**. This means that they have isolated the gene for rbST from cows, and then inserted it into the genome of bacteria. These bacteria are then cultured and produce the hormone in vast quantities in industrial vats. The industrial production of recombinant growth hormone is summarised in Fig 6.2 (overleaf).

When this readily available hormone is injected into cows on a regular basis, it reliably improves their efficiency as milk producers. The mammary glands of such dairy cows take in more nutrients from the bloodstream and produce more milk.

It is very unlikely that bST, whether recombinant or natural, has any biological effect on humans. It already exists naturally in milk, so there is just a bit more of it in the milk from rbST-treated cows. However, the human digestive juices break down the protein into the amino acids that make it up, so it disappears. Babies cannot do this and should not be fed on cow's milk. The hormone cannot gain entry to the human body as a complete protein. And even if it did, the hormone is **species-specific**. The form found in cows does not fit into any receptor in the human body; it is a completely different shape from human growth hormone, for example.

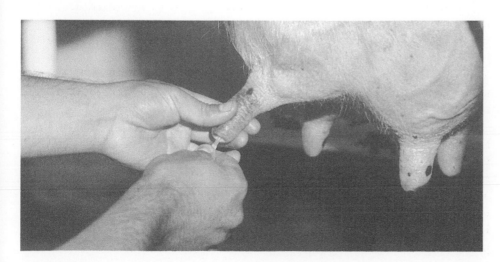

Fig 6.1 Cow's udder being an injected with antibiotic. Farmers in Canada have reported that cows given growth hormone suffer more udder infections than untreated cows and that the greater quantity of antibiotics used to treat them then gets into the human food supply. These antibiotics can increase the risk that bacteria that infect people then evolve to become antibiotic resistant

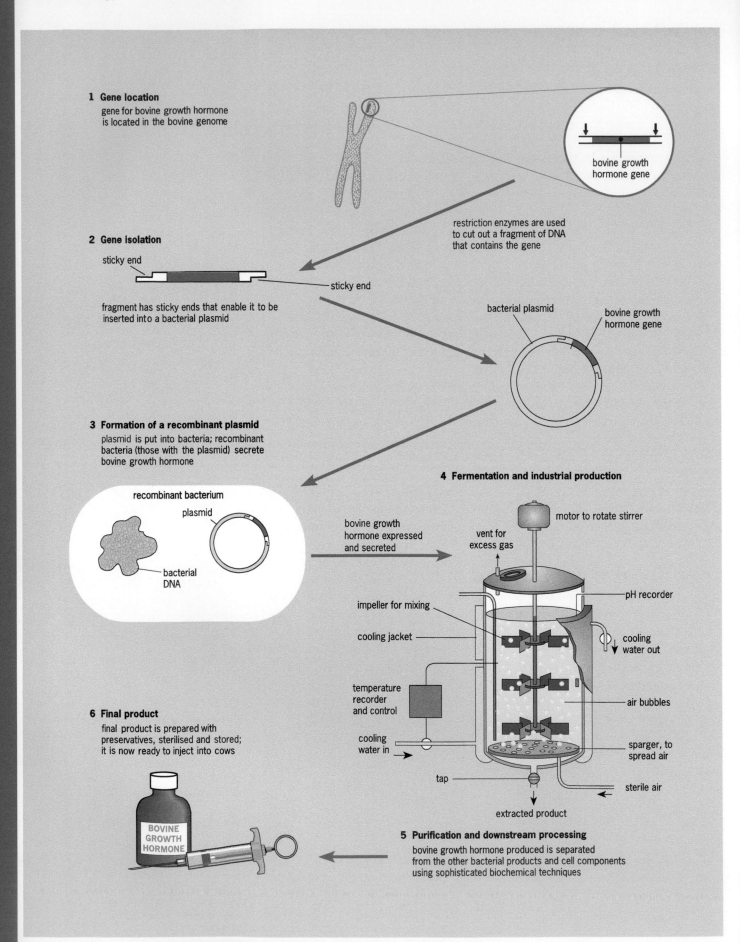

1 Gene location
gene for bovine growth hormone
is located in the bovine genome

bovine growth
hormone gene

restriction enzymes are used
to cut out a fragment of DNA
that contains the gene

2 Gene isolation

sticky end

sticky end

fragment has sticky ends that enable it to be
inserted into a bacterial plasmid

bacterial plasmid

bovine growth
hormone gene

3 Formation of a recombinant plasmid
plasmid is put into bacteria; recombinant
bacteria (those with the plasmid) secrete
bovine growth hormone

4 Fermentation and industrial production

recombinant bacterium

plasmid

bacterial
DNA

bovine growth
hormone expressed
and secreted

motor to rotate stirrer

vent for
excess gas

pH recorder

impeller for mixing

cooling jacket

cooling
water out

temperature
recorder
and control

air bubbles

6 Final product
final product is prepared with
preservatives, sterilised and stored;
it is now ready to inject into cows

cooling
water in

sparger, to
spread air

tap

sterile air

extracted product

BOVINE
GROWTH
HORMONE

5 Purification and downstream processing
bovine growth hormone produced is separated
from the other bacterial products and cell components
using sophisticated biochemical techniques

Fig 6.2 Steps in the production of recombinant bovine growth hormone. This diagram is not drawn to scale

Pros and cows of rbST

Cows that are given rbST increase their overall milk production between 5 and 10 per cent but they don't need to eat 5–10 per cent more food, so this is a cost-effective technique. Those in favour argue that it allows dairy farmers to be more efficient and that it enables more milk to be produced in areas where food is needed urgently. They propose that nations suffering from hunger and malnutrition could benefit from rbST.

Many farmers in the USA are still using injections of rbST to increase milk yields in cows but this product is banned in Canada, the European Union, Japan and Australia. People are worried that the treatment is not good for the cows, which get more infections, and that there are dangers to those who drink the milk. There is still concern that the milk does contain high levels of growth hormone and that this could increase the risk of breast and colon cancer in humans.

Those against growth hormone use say that milk produced by treated cows should be labelled so consumers can decide whether to drink it or not. This could lead to an unsold surplus of milk, which may need to be bought by governments to prevent farmers going out of business.

Embryo duplication

The technique of **artificial insemination** – the use of sperm from a bull or ram with characteristics that are particularly sought after in the farming world – has been used to breed 'better' animals for several years. In the last 15 years of the twentieth century reproductive technology became a lot more sophisticated and it became possible to make several female animals pregnant with identical embryos. When these are produced by mating a prime female with a prime male, the resulting embryos can be split, sometimes into two, sometimes more, and implanted into any healthy female of the same species (Fig 6.3). This means that a whole herd of prime stock animals can be produced from one breeding pair, whose embryos are carried by **surrogate mothers**.

Synchronising sheep breeding

In Britain, the changing day-length of the seasons dictates the breeding behaviour of sheep. Sheep synchronise their reproductive cycle to produce lambs in early spring (March/April) when new grass is plentiful. This also gives the young as much time as possible to mature before the next winter. The gestation period in sheep is 21 weeks, and ewes and rams must mate in October/November to get the timing just right.

Normally, left to their own devices, sheep rely on their hormones to do this. In the autumn, the stimulus of shortening days affects the **hypothalamus**, which responds by secreting a hormone called **gonadotrophin-releasing factor**. As its name suggests, this acts on the pituitary gland to cause the release of a gonadotrophin hormone – **follicle stimulating hormone (FSH)**. FSH stimulates the development of follicles in the ovaries; these secrete oestrogen, and this causes the ewe to come into season. As she begins to give off vaginal secretions that contain **pheromones**, nearby rams are aroused and mating occurs.

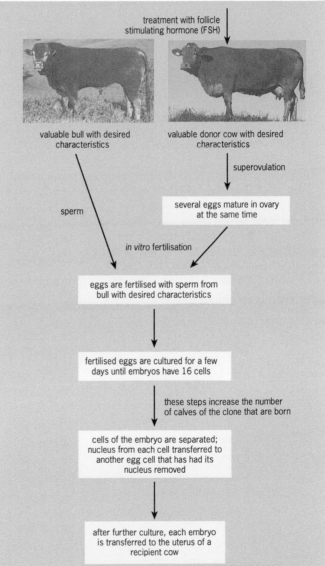

treatment with follicle stimulating hormone (FSH)

valuable bull with desired characteristics

valuable donor cow with desired characteristics

superovulation

sperm

several eggs mature in ovary at the same time

in vitro fertilisation

eggs are fertilised with sperm from bull with desired characteristics

fertilised eggs are cultured for a few days until embryos have 16 cells

these steps increase the number of calves of the clone that are born

cells of the embryo are separated; nucleus from each cell transferred to another egg cell that has had its nucleus removed

after further culture, each embryo is transferred to the uterus of a recipient cow

Fig 6.3 Embryo duplication allows up to 16 embryos to be obtained from one mating between a valuable female and a valuable male. The calves are carried by surrogate mothers. The cow that is the biological mother does not carry a calf; she is inseminated again soon afterwards to obtain another embryo for splitting between another set of surrogates. In this way, one pair of prime animals can be used to generate dozens of calves every year

REMEMBER THIS

Sheep have an **oestrous cycle** that is controlled by hormones. These hormones are affected by day-length, and so oestrus in sheep is seasonal. In humans, ovulation occurs once a month through the reproductive life of a woman and is not affected too much by external environmental factors. Pheromones (see Chapter 18) can modify the timing of ovulation slightly.

Fig 6.4 When ewes come into season, the farmers release rams into their field for a night. In the morning, all the females should have been impregnated. Just to make sure none has been missed, some farmers take the precaution of taping a block of dye to each ram's chest. This rubs off during mating. Any ewes that do not have dyed bottoms by morning have to stay with the rams for another night

? **QUESTION 1**

1 Why could a castrated ram not be used as a teaser when a farmer is trying to synchronise mating in his flock of ewes?

✔ **REMEMBER THIS**

Artificial insemination is the transfer of sperm collected from a prime ram or bull into the vagina of several female animals, ensuring that they all produce offspring from the same male.

The problem with this for farmers of today is that the mating process, and therefore the lambing, can take place over a period of about a month. This makes it very difficult to care for all the ewes and lambs, particularly if the flock is very large. To enable farmers to concentrate their effort and resources (such as vets on standby), they try to ensure that lambing actually takes place over a five-day period. To do this, they must ensure that most ewes mate on the same day. Achieving this takes some extraordinary measures:

- Ewes that are to be mated must be kept away from all sexually active males during the time they are weaning their previous lambs. Sexually active males include not only rams, but male lambs and billy goats, which can secrete some of the hormones that attract the female.
- This absence of males allows the next stage to have its maximum effect. Vasectomised but thoroughly randy rams are then put with the ewes four weeks before the intended mating day to bring them into season. These rams are termed 'teasers', for obvious reasons.
- Two weeks after the introduction of the teasers, the ewes are separated again and small sponges impregnated with the hormone **progesterone** are inserted just inside the vagina of each one. The progesterone prevents ovulation in the ewes until it is removed 14 days later, at which point all the ewes should be at the same stage of their oestrous cycle.
- Sometimes, another hormone, called **pregnant mare serum gonadotrophin** (**PMSG**), is injected into the ewes just after sponge removal. This slightly increases the litter size of the ewes, and may also make the timing more precise.
- The ewes come into heat 36 to 48 hours after sponge removal. It's time for action.
- The rams are introduced to the ewes and, after few hours of frenzied activity, the farmer achieves his or her objective (Fig 6.4). Alternatively, the farmers use **artificial insemination** – important if they want to use a particular ram to father as many lambs as possible.
- If all has gone well, the ewes should lamb 145 days later, with some lambs born two days or so either side of the expected date.

USING BIOTECHNOLOGY IN CROP DEVELOPMENT

Farmers have been practising a crude form of biotechnology for centuries by **cross-breeding** plants to introduce and maintain desirable characteristics. This is called **selective breeding**. More recently, plant technologists have bombarded plant cells or small plants with radiation that damages DNA, causing random changes that sometimes result in 'better' plants. However, the newest, truly biotechnological techniques involve **genetic engineering** and produce the **genetically modified** (**GM**) plants and crops that are so often in the news.

How are plants genetically modified?

Genetic modification usually involves the transfer of a single gene, together with an obvious marker so that the plant breeder knows which plants have taken up the gene. The gene chosen usually codes for a protein that gives the plant a higher rate of growth, or makes it resistant to drought, pests or weedkillers, or that makes part of the plant more useful for the food industry (see the How Science Works box on golden rice on page 96).

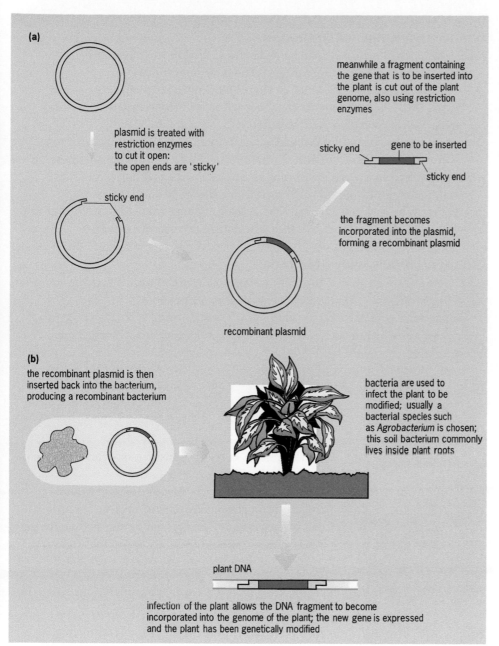

(a)

plasmid is treated with
restriction enzymes
to cut it open:
the open ends are 'sticky'

meanwhile a fragment containing
the gene that is to be inserted into
the plant is cut out of the plant
genome, also using restriction
enzymes

sticky end

gene to be inserted

sticky end

sticky end

sticky end

the fragment becomes
incorporated into the plasmid,
forming a recombinant plasmid

recombinant plasmid

(b)
the recombinant plasmid is then
inserted back into the bacterium,
producing a recombinant bacterium

bacteria are used to
infect the plant to be
modified; usually a
bacterial species such
as *Agrobacterium* is chosen;
this soil bacterium commonly
lives inside plant roots

plant DNA

infection of the plant allows the DNA fragment to become
incorporated into the genome of the plant; the new gene is expressed
and the plant has been genetically modified

Fig 6.5a Production of a recombinant bacterial
plasmid containing a gene for transfer into a plant
Fig 6.5b Bacteria such as *Agrobacterium* sp. are
forced to take up the recombinant plasmid and are
then introduced into the plant. The plasmid enters
plant cells where it opens up to release the gene.
This is then incorporated into the DNA of the plant.
As the plant genome is expressed, so is the new
gene. This diagram is not drawn to scale

Fig 6.5a shows how a gene is first incorporated into a **plasmid** (a small circular piece of DNA) and is then introduced into the cells of the plant that is being modified (Fig 6.5b). Sometimes bacteria can infect a whole plant; other techniques introduce the bacteria carrying the plasmid into individual plant cells. These are then cultured to produce plant tissue and then whole plants that carry the 'foreign' gene.

The expanding use of GM crops
The use of GM crops has been expanding steadily over the last 10 years. In 2006, 10.3 million farmers in 22 countries across the world planted a total of 252 million acres of land with them. Over half of all GM crops were grown in the USA, with a further 30% in South America; Canada and South Africa grew some GM varieties.

Indian and Chinese farmers represent about 7 per cent of the total. These farmers are the most enthusiastic farmers of GM crops because, in the developing world, such crops can mean the difference between making a good living and starving. The greatest increase in GM crops is expected to be in developing countries during the next decade.

REMEMBER THIS

Genetically modified plants could also be called **transgenic plants**, but this tends to be a term used more commonly for animals that have been bred to carry specific genes.

REMEMBER THIS

There is a lot of controversy about owning and patenting genes and genetically modified organisms. Naturally occurring organisms cannot be patented or owned, but genetically engineered plants and animals can. The issue of owning genes is even more complex, because they are naturally occurring. To be able to patent a gene, the discoverer must have found out its function and that it could be manipulated in some way for medical or scientific good.

Fig 6.6 The use of genetic engineering in food production has provoked strong feelings in many people

Some advantages of GM crops

There is a lot of controversy about GM crops. Like most things, they have some advantages and some problems. The advantages include:

- a better-tasting crop product, perhaps with more nutrients,
- crops that take less time to become mature and ripen or crops that stay fresh longer once they are harvested,
- crops that produce higher yields and are better able to withstand stress, such as cold or hot weather,
- plants that have better resistance to disease, pests and herbicides.

Overall, people in favour of GM crops argue that they allow more land to be used effectively, they make farmers in developing countries more self-sufficient, they result in cheaper food and they can reduce the use of pesticides.

Some problems with GM crops

The major problem that is preventing more widespread use of genetically modified crops and foods is public suspicion and hostility.

Opponents of GM foods are particularly vocal in the UK (Fig 6.6). In other countries in Europe, and in the USA, people have accepted them more readily.

Here are some of the arguments made by people opposed to GM foods:

- GM foods are bad for your health: swapping genes from one organism to another is unnatural, and therefore dangerous. What might be the effect of eating 'foreign' genes from bacteria?
- Genes from GM plants might escape and get into other plants, creating 'superweeds'. Bacteria might pick up the plasmids that have been used to carry genes into plant cells, and then infect other plants. The genes may spread outside crop plants, making weeds resistant to all known weedkillers.
- GM plants might breed with traditional varieties, producing hybrids that are of no use. Organic farmers might find that crops that have taken years to establish become contaminated with genes from GM crops.
- Growing GM foods resistant to weedkillers and pesticides may affect insects and other useful wildlife which may then be lost.
- The technology behind GM crops is only available to the multinational, large agrobusinesses, not the small, independent farmer. Organic farmers may find their produce outsold by GM foods, which can be produced at a much lower cost.

EXAMPLES OF GM FOODS AND CROPS

The major GM crops grown today are herbicide- and insect-resistant soybeans, corn, cotton (see the How Science Works box, opposite), canola and alfalfa. Others include:

- a sweet potato resistant to a virus capable of destroying most of the African harvest,
- golden rice – rice with increased iron and vitamins grown in Asian countries (see the How Science Works box on page 96),
- plants that are able to survive in extreme conditions, such as tomatoes that can grow well in salty water.

Growing GM cotton in India

India is a major exporter of cotton and the cotton crop is important to the Indian economy (Fig 6.7). The cotton plant is easy to grow in hot climates but it can be severely damaged by insect pests – bollworm, aphids, jassids and whiteflies. These not only harm the plants directly, they also introduce plant viruses, leading to disease.

Within the last decade, a genetically modified cotton plant, which has a gene from the bacterium *Bacillus thuringiensis*, has been developed. The *Bt* gene produces a protein that is toxic to bollworms, so a plant with this gene is resistant to attack by this pest. The first commercial cultivation was allowed in 2002. There was quite a lot of controversy over the results of the trials in the next two years, but a general conclusion emerged that the *Bt* cotton did have advantages over wild varieties.

Since then, more farmers in India have chosen to grow *Bt* cotton; 5 million acres were planted in 2005 and this went up to 8.1 million acres in 2006.

Farmer Eknath Shivram Pandit has grown corn and cotton for nearly 15 years and recently switched to *Bt* cotton. He has noticed many benefits on his farm and for his family. The cotton crop now suffers fewer insect infestations and so he needs to spray with insecticides only two or three times, instead of 15 to 20 times. The cotton crops he has produced have also had higher yields, and the family's standard of living has improved.

Pandit says: 'This technology is going to be very helpful. There has been a lot of benefit. This product has brought in money.

Fig 6.7 Growing GM cotton in India.

Fig 6.8 The money brought in from cotton growing means farmers can invest in new technology

With money, there can be education. I will buy more land for farming. Dig more wells. I can build a house. Save some money for my daughter's wedding and son's education.'

Since introducing *Bt* cotton to his farm, Pandit has been able to replace his ox cart with a tractor (Fig 6.8), which speeds up his visits to and from the local market. 'I am more enthusiastic about farming. It's worth it now,' he says.

Disease-resistant sweet potatoes

The sweet potato is an important crop in Africa. It is grown as a staple food source by the women and children of poor families who harvest tubers, as they need them, for daily meals. Larger harvests are done occasionally to trade for other products.

Sweet potatoes are very prone to a viral disease called sweet potato virus disease (SPVD). This is caused by the sweet potato feathery mottle virus and a strain of the sweet potato chlorotic stunt virus. Affected plants produce only tiny tubers. Some naturally resistant varieties do exist, but these produce low yields compared to the wild-type plant.

Genetically modified sweet potato varieties have been developed that are disease-resistant and also give good yields of tubers, and trials have proved successful. Farmers in Africa, from Uganda to Tanzania have adopted these new varieties and the production of sweet potatoes has increased.

Salt-loving tomatoes

Millions of hectares of land all over the world become impossible to farm every year because irrigation techniques cause an increase in the salt content of the water and soil. Plants generally dislike salty conditions; a few specialised species can survive but most crop plants cannot.

In the last few years, agricultural scientists have developed a type of tomato that thrives in salty conditions. This makes it more likely that other plants can also be modified to become salt tolerant.

The tomatoes take up the salt through their roots, but they then isolate it, so that only water passes to the rest of the plant – the tomatoes themselves are not salty. The GM tomatoes can grow in water that contains 50 times more salt than freshwater. Seawater is 150 times more salty than freshwater, but researchers hope that further modifications may produce plants capable of living in seawater in the future.

HOW SCIENCE WORKS

Golden rice

Rice is a staple food for over three billion people in South East Asia and Africa. Many people there have very little else to eat. Although rice is a good source of carbohydrate and protein, it does not contain very many of the essential vitamins and minerals that are needed to stay healthy. Those that are present are not easily taken up by the body during digestion of the rice.

Many people in the developing world therefore have deficiencies of iron, zinc and vitamins A, C and E. These deficiencies make people less able to work, and cause reduced mental capacity, stunted physical growth and blindness. Because the deficiencies are particularly severe in pregnant women, there is an increased chance that babies and young children die.

Providing supplements to large parts of the population of South East Asia is theoretically possible, but it would cost a lot, so is not really an option. General moves to extend the diet are difficult because of poverty. One answer is genetic modification of the rice to produce more of the required vitamins and minerals and this has now become a reality.

Golden rice is now widely grown and is a good example of biofortification (Figs 6.9 and 6.10). The rice plants normally cultivated have been genetically modified to produce β-Carotene in rice endosperm. β-Carotene is converted in the body to vitamin A.

The rice plants were given two new genes that express the proteins phytoene synthase and carotene desaturase, both of which allow the plant to make β-carotene and concentrate it in the endosperm. The first generation of GM plants produced in 2002 was able to make a small amount of β-carotene. Later generations were improved so that the rice plants can now make higher β-carotene levels.

The GM plants produce rice that is high in β-carotene and which looks golden because of this pigmented molecule in its endosperm. It looks attractive, so people enjoy eating it – if it had turned out bright blue, there could have been a problem.

Some studies have already shown that growing golden rice does help reduce vitamin deficiencies in the local people. In the future,

the plants may be modified again to enrich them with other vitamins and minerals. Over-expression of ferritin, an iron storage protein, has already been achieved in rice grains and the zinc levels in barley grains have been doubled by over-expressing a zinc-binding protein in the barley plant. Eventually, food scientists think it will be possible to combine all these genetic modifications into one type of rice plant, which would then provide all the necessary vitamins and minerals missing from the diet.

Fig. 6.9 Lush growth of golden rice in fields in South East Asia

Fig. 6.10 Golden rice 2 (left), golden rice 1 (top right) and normal white rice (bottom right). Golden rice 1 provided people with 15–20 per cent of the recommended daily allowance (RDA) of vitamin A. Golden rice 2, which was developed subsequently, provides 50 per cent of the RDA of vitamin A

3 BIOTECHNOLOGY AND FOOD

LACTIC ACID FERMENTATION

Making yoghurt

Yoghurt is basically milk (cow, goat or ewe) that is fermented by bacteria. Although it originated in Western Asia and Eastern Europe, it is now eaten all over the world. To make yoghurt, a starter culture of the lactic acid bacteria *Lactobacillus bulgaricus* and *Streptococcus thermophilus* is added to whole or skimmed milk. The bacteria multiply and ferment lactose, the disaccharide sugar in the milk, producing lactic acid. As you will know if you have ever added lemon juice to milk, acid causes the milk to curdle. Separation of the acidified milk into solid curds gives yoghurt its taste and texture.

Yoghurts can be pasteurised to destroy the fermentation bacteria but if this is not done, the result is 'live' yoghurt. Many supermarkets now sell yoghurts they describe as 'bio-yoghurts'. These have been produced in the usual way, left unpasteurised, and then extra bacteria added. Bio-yoghurts usually contain live *Lactobacillus acidophilus* and *Bifidobacterium bifidum*. Various health claims are made for these yoghurts, but whether they actually do keep you healthy or not is still an open question.

Cheese-making: one of the oldest biotechnologies

Cheese-making dates back at least 5000 years and the basic method used has changed very little over that time. Three main ingredients are needed:

- milk,
- a source of the protein-degrading enzyme **chymosin**, which causes the milk proteins to clump together. Traditionally, this was added in the form of calf **rennet** (chymosin can also be called **rennin**). Non-animal sources of this enzyme are now available and can be used in the manufacture of cheese suitable for vegetarians,
- a starter culture of lactic acid bacteria.

In many ways, cheese-making is similar to yoghurt-making. The lactic acid starter culture is added to the milk to convert the lactose to lactic acid, which helps to preserve the cheese. Chymosin causes the milk to set, producing hard cheese. Once bacterial fermentation stops, the maturation of the cheese then begins. As the bacteria die, they are digested by their own enzymes – a process called **autolysis**. This process produces more enzymes, including **peptidases**, that produce the strong flavours that we associate with a 'good' cheese. Some strains of bacteria produce carbon dioxide during maturation – this makes the holes in cheeses such as Gouda and Emmental.

A fungus of the Penicillin genus (*Penicillium roquefortii*) is used to inoculate fermented curd to produce blue cheeses such as Stilton and Danish Blue. This fungus grows through the cheese during the maturation process, producing the characteristic blue-green veins.

✔ **REMEMBER THIS**

Anaerobic bacteria that can ferment carbohydrates to lactic acid are widely used in the food industry. They lack the enzymes of the citric acid cycle and the respiratory chain, and so ferment carbohydrates into lactic acid – this process is called **fermentation**.

For more on yoghurt please see the How Science Works assignment for this chapter at www.collinseducation.co.uk/CAS

✔ **REMEMBER THIS**

Pasteurisation of milk involves heating it to 60 °C so that most of the bacteria in it are killled, but the taste is not unduly affected. Sterilised milk is heated to 100 °C to kill all the bacteria, so that it lasts much longer. However, at this temperature, some of the proteins in it are denatured and this changes the taste of the milk – many people dislike the new taste and sterilised milk has never been popular for drinking.

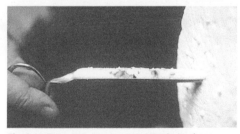

Fig 6.11 The photographs show the holes in Gruyere (below) and a blue cheese, Roquefort (above), being probed during its maturation to control the development of the blue veins

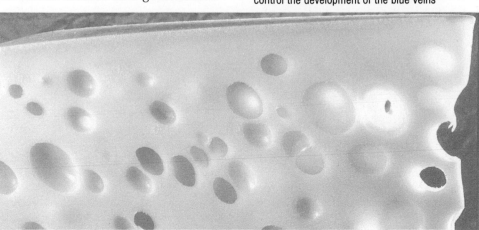

FERMENTATION USING YEAST

Yeast is an essential microorganism in the production of bread, wine and beer. Under the right conditions, yeast will ferment carbohydrates to release carbon dioxide and alcohol in a series of chemical reactions.

Bread-making

Bread is basically a mixture of flour, water and a little fat that is baked. Some breads, such as flatbread or matzos, are still produced flat like this, but yeast fermentation is usually used to cause the dough to rise, so that a lighter, less dense loaf can be made. As the flour, water and yeast are mixed, enzymes called amylases that are present in the flour break down the starch into the disaccharide, maltose. Other enzymes break this down into glucose, which is then fermented by the yeast to carbon dioxide and alcohol. In bread, the alcohol evaporates during baking, but the carbon dioxide bubbles become trapped in the sticky dough, causing the baked loaf to have lots of tiny pockets.

Wine and beer

In wine-making, the fermentation of grape juice is a complex process that involves yeasts, bacteria and filamentous fungi – but the yeasts have the starring role. During fermentation, yeasts use the sugars and other components of grape juice and convert them into ethanol, carbon dioxide and other end-products that contribute to the chemical composition and taste of wine. Fermentation stops when the yeast becomes poisoned by its own waste – the alcohol – which is why drinks made by fermentation only are no stronger than about 14 per cent alcohol by volume. Spirits such as whisky and gin, which have a much higher alcohol content, are **distilled** from a fermented mixture.

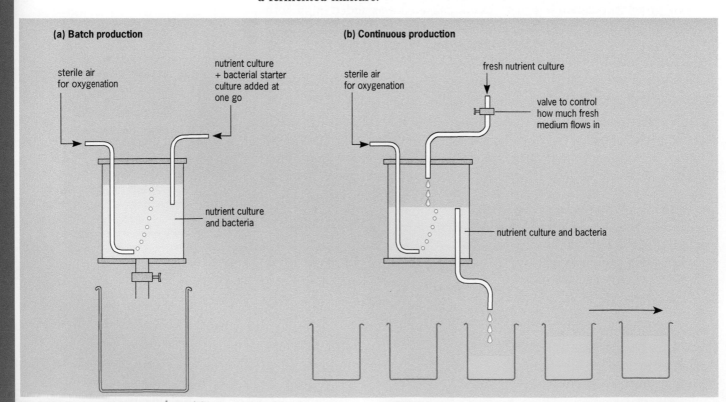

Fig 6.12a Batch production: when fermentation is complete, the entire contents of the tank are released for processing. The tank is cleaned and then the whole process is repeated. Each production is a different batch

Fig 6.12b Continuous production: fermentation is continuous and can carry on indefinitely as long as fresh medium is added, product is removed and the conditions for optimum growth are maintained. Continuous production runs a higher risk of contamination than batch production, so great care has to be taken to keep the incoming medium sterile, apart from the microorganism in the culture

Most wine-makers are more interested in the alcoholic by-product of yeast fermentation but a brewer producing beer needs both the alcohol and the carbon dioxide made by the yeast. Carbon dioxide gives the beer its characteristic fizziness.

The brewing industry uses barley as the source of the food that the yeast ferments to make the alcohol in beer. However, barley stores food in the form of starch – which yeast cannot use directly. In order to solve this, the brewer allows the barley grains (seeds) to germinate. During germination, enzymes in the barley convert the starch into maltose sugar, which the yeast can ferment. This process is called **malting**.

Beer production is made as efficient as possible by providing the best possible conditions for yeast to grow and ferment. This means that the temperature, oxygen supply and amount of glucose must be carefully controlled, and unwanted microorganisms must be kept out. The easiest way to make sure that this happens throughout the fermentation is to set up all the conditions at the start, together with the raw materials, and then to leave the whole system closed and untouched until the fermentation is complete. This has the disadvantage that production is only possible in batches rather than a continuous process. The main features of batch processing and continuous processing are shown in Fig 6.12.

USING OTHER MICROORGANISMS AS A FOOD SOURCE

Biotechnology has been used in the past few years to make foods that are as rich in protein as meat, but that are much cheaper. Pruteen™ is one example (Fig 6.13), mycoprotein (for example, Quorn™) is another. Both can be eaten by vegetarians. Mycoprotein has the following nutritional advantages over meat:

- it has no animal fat, little overall fat and no cholesterol,
- it has a high protein content (as high as that of skimmed milk),
- it is high in fibre,
- it contains useful amounts of trace elements and B vitamins.

Mycoprotein is produced by the fungus *Fusarium graminearum*, which is related to mushrooms and truffles. This fungus can use the carbohydrate present in cheap, readily available carbohydrates – wheat in the UK, potato in Ireland, and cassava, rice or sugar in tropical countries, for example. Mycoprotein contains 45 per cent protein and 13 per cent fat, a composition similar to that of grilled beef. It is also high in fibre and has a complete amino acid content.

? QUESTION 2

2 What is the difference between fermentation and glycolysis? (See Chapter 33)

✓ REMEMBER THIS
In exam questions on fermentation you will be expected to know about the basics of respiration, particularly glycolysis, and also be able to apply your knowledge of the properties of enzymes.

✓ REMEMBER THIS
Mycoprotein is made by a filamentous fungus – one that produces lots of thin strands called **hyphae**.

✓ REMEMBER THIS
Pruteen™ is a microbial protein produced by bacteria that can break down methanol.

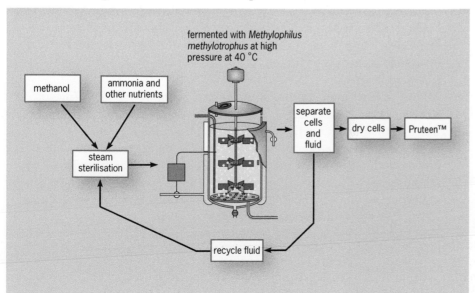

Fig 6.13 Pruteen™ production. Pruteen™ is the trade name for microbial protein produced by growing *Methylophilus methylotrophus* bacteria on methanol, which is derived from methane or natural gas. Since the product is a low-value commodity, large quantities have to be manufactured to make profits. Usually very large (3000 m³) airlift fermenters are used with a diameter of 7 m and a height of 60 m

The industrial fermenters currently being used to manufacture mycoprotein are 40 metres high (similar in height to Nelson's Column). They run continuously for six weeks, after which there is a two-week period for cleaning and preparing the fermenter for the next run. After production, the mycoprotein looks like pastry, but it is then mixed with binding agents and colours to form a product that looks like different kinds of meat.

OTHER FERMENTED FOOD PRODUCTS

Sauerkraut

Sauerkraut is a German delicacy that is now popular in other parts of the world, particularly the USA. It is made by fermenting cabbage that has been salted. The main bacteria responsible for the fermentation are lactic acid bacteria. *Leuconostoc mesenteroides* ferments the cabbage juices to produce lactic acid, acetic acid, carbon dioxide and ethanol. Two others, *Lactobacillus plantarum*, which produces lactic acid only, and *Lactobacillus brevis*, are used to complete the process.

Soya products

Biotechnology is also used to process soya beans to give the following food products:

- **Textured vegetable protein** (**TVP**) – defatted soya flour that has been processed and dried to give a substance with a sponge-like texture that resembles meat. Soya beans are **dehulled** (their skins are taken off) and their oil is extracted before they are ground into flour. This flour is then mixed with water to remove soluble carbohydrate and the residue is textured by either **spinning** or **extrusion**. Extrusion involves passing heated soya residue from a high-pressure area to a reduced pressure area through a nozzle – this causes the soya protein to expand. The soya protein is then dehydrated and may be either cut into small chunks or ground into granules.
- **Tofu** – soya bean curd made from coagulated soya milk. Soya beans are soaked, crushed and heated to produce soya milk to which a coagulating agent such as calcium sulfate or calcium chloride is added. The resulting soya curd is then pressed to give tofu (Fig 6.14). Tofu is sometimes known as soya cheese, and is sold as blocks packaged in water.
- **Tempeh** – a fermented soya bean paste made by inoculating cooked soya beans with the fungus *Rhizopus oligosporous*. This forms a mycelium holding the soya beans together and is responsible for the black specks in tempeh. Tempeh has a chewy texture and distinctive flavour and can be used as a meat substitute in recipes. It may be deep-fried, shallow-fried, baked or steamed.
- **Miso** – a fermented condiment made from soya beans, grain (rice or barley), salt and water. Miso production involves steaming polished rice, which is then inoculated with the fungus *Aspergillus oryzae* and left to ferment to give an end-product called **koji**. Koji is then mixed with soya beans that have been heated and extruded to form strands, together with salt and water. This is then left to ferment in large vats. Miso varies widely in flavour, colour, texture and aroma. It is used to give flavour to soups, stews, casseroles and sauces.
- **Soy sauce** – true soy sauce, called **shoyu**, is made by fermenting soya beans with cracked roasted wheat, salt and water. **Tamari** is similar but slightly stronger and made without wheat (and so is gluten free)

Fig 6.14 Soya products: dried soya beans, soya milk (blue bowl) and two types of tofu or bean curd – dried tofu (upper right) and fresh tofu (brown bowl)

Fermentation for shoyu and tamari takes about one year. Much of the soy sauce available in supermarkets is not true soy sauce but is made by chemical hydrolysis from defatted soya flour, caramel colouring and corn syrup without any fermentation process.

4 BIOTECHNOLOGY IN INDUSTRY

Enzymes were first used on a large scale by the textile industry. Thread used for weaving is protected by a coating of starch paste – useful during the weaving process to make the thread more 'slippy' so that weaving was easier, but totally useless afterwards. The starch used to be removed by acids, alkalis or oxidising agents but these often discoloured the cloth. In the early part of the twentieth century, the industry started using enzyme extracts from malt, and later from an animal pancreas. A thermostable bacterial amylase was discovered in 1917, but its use in the mass production of cloth took a few years to perfect. At around the same time, the leather industry also started to experiment with enzymes to reduce costs. Protease enzymes began to replace calcium hydroxide and sodium sulfide for removing animal hair from skins.

These early attempts at using enzymes seem crude by today's standards and it took until after 1945 for large-scale production and use of enzymes to be commonplace in industry.

ENZYMES IN THE INDUSTRIES OF TODAY

The ability of enzymes to catalyse specific chemical reactions at body temperature makes them very useful tools in the commercial world. Table 6.1 outlines some of industrial applications of enzymes. This is a rapidly changing field, and new applications of enzyme technology appear all the time.

 REMEMBER THIS

Thermostable enzymes, those that work outside the normal physiological range of the human body, are particularly useful in industry. Many have been discovered in bacteria and other organisms that live in the extreme environments of deep ocean vents. The enzymes in these organisms are not denatured at high temperatures – and so can be used to catalyse industrial processes at temperatures approaching 100 °C. An application of thermostable enzymes in washing powders is discussed in Chapter 5.

Table 6.1 Some applications of enzymes (Source: *Biochemical Society Guidance Notes 3, Enzymes and their role in biotechnology*)

Enzyme	Reaction	Source of enzyme	Application
Industrial applications			
α-amylase	breaks down starch	bacteria	converts starch to glucose in the food industry
glucose isomerase	converts glucose to fructose	fungi	production of high-fructose syrups
proteases	digest protein	bacteria	washing powder
rennin	clots milk protein	animal stomach linings; bacteria	cheese-making
catalase	splits hydrogen peroxide into $H_2O + O_2$	bacteria; animal livers	turns latex into foam rubber by producing gas
β-galactosidase	hydrolyses lactose	fungi	in dairy industry, hydrolyses lactose in milk or whey
Medical applications			
L-asparaginase	removes L-asparagine from tissues – this nutrient is needed for tumour growth	bacteria (*E. coli*)	cancer chemotherapy – particularly leukaemia
urokinase	breaks down blood clots	human urine	removes blood clots, e.g. in heart disease patients
Analytical applications			
glucose oxidase	oxidises glucose	fungi	used to test for blood glucose, e.g. in Clinistix™, used by people with diabetes
luciferase	produces light	marine bacteria; fireflies	binds to particular chemicals indicating their presence, e.g. used to detect bacterial contamination of food
Manipulative applications			
lysozyme	breaks 1–4 glycosidic bonds	hen egg white	disrupts bacterial cell walls
endonucleases	break DNA into fragments	bacteria	used in genetic manipulation techniques, e.g. gene transfer, DNA fingerprinting

Commercial production of enzymes

Table 6.2 shows some important industrial enzymes and their sources. As you can see, the majority of enzymes used in industry are **extracellular**. This type of enzyme is easier to isolate, as it is secreted into the growth medium by the source organism, and can be extracted relatively easily. **Intracellular** enzymes are more tricky, as their removal requires the cultured cells to be broken open, and then the enzyme must be separated from all the other proteins and other molecules. Fig 6.15 provides a flow chart to show how a protease for use in a stain remover is produced on a large scale from a bacterial culture.

Table 6.2 Some important industrial enzymes and their sources

	Enzyme	Source	Intracellular/ extracellular	Industry/ industrial use
Animal enzymes	catalase	liver	I	food
	chymotrypsin	pancreas	E	leather
	lipase	pancreas	E	food
	rennet	abomasum	E	cheese
	trypsin	pancreas	E	leather
Plant enzymes	actinidin	kiwi fruit	E	food
	α-amylase	malted barley	E	brewing
	β-amylase	malted barley	E	brewing
	bromelain	pineapple latex	E	brewing
	β-glucanase	malted barley	E	brewing
	ficin	fig latex	E	food
	lipoxygenase	soya beans	I	food
	papain	pawpaw latex	E	meat
Bacterial enzymes	α-amylase	*Bacillus*	E	starch
	β-amylase	*Bacillus*	E	starch
	asparaginase	*Escherichia coli*	I	health
	glucose isomerase	*Bacillus*	I	fructose syrup
	penicillin amidase	*Bacillus*	I	pharmaceutical
	protease	*Bacillus*	E	detergent
	pullulanase	*Klebsiella*	E	starch
Fungal enzymes	α-amylase	*Aspergillus*	E	baking
	aminoacylase	*Aspergillus*	I	pharmaceutical
	glucoamylase	*Aspergillus*	E	starch
	catalase	*Aspergillus*	I	food
	cellulase	*Trichoderma*	E	waste
	dextranase	*Penicillium*	E	food
	glucose oxidase	*Aspergillus*	I	food
	lactase	*Aspergillus*	E	dairy
	lipase	*Rhizopus*	E	food
	rennet	*Mucor miehei*	E	cheese
	pectinase	*Aspergillus*	E	drinks
	pectin lyase	*Aspergillus*	E	drinks
	protease	*Aspergillus*	E	baking
	raffinase	*Mortierella*	I	food
Yeast enzymes	invertase	*Saccharomyces*	I/E	confectionery
	lactase	*Kluyveromyces*	I/E	dairy
	lipase	*Candida*	E	food
	raffinase	*Saccharomyces*	I	food

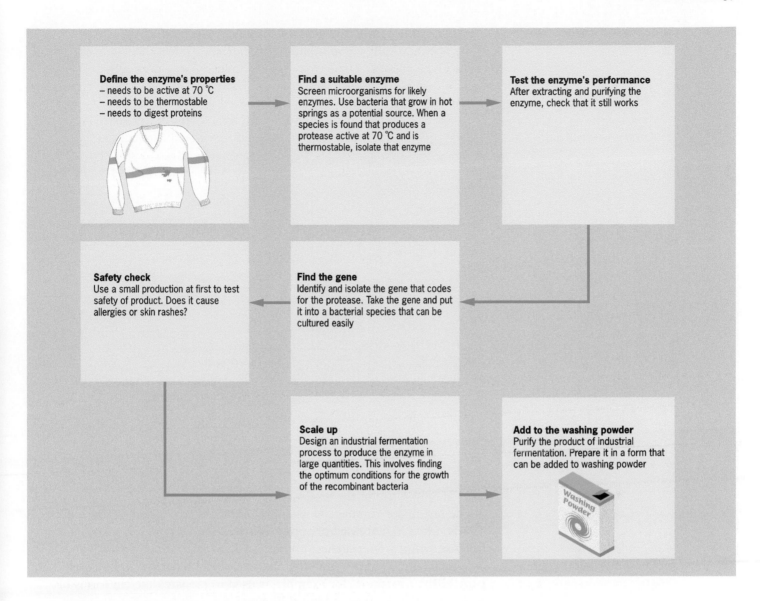

Define the enzyme's properties
– needs to be active at 70 °C
– needs to be thermostable
– needs to digest proteins

Find a suitable enzyme
Screen microorganisms for likely enzymes. Use bacteria that grow in hot springs as a potential source. When a species is found that produces a protease active at 70 °C and is thermostable, isolate that enzyme

Test the enzyme's performance
After extracting and purifying the enzyme, check that it still works

Safety check
Use a small production at first to test safety of product. Does it cause allergies or skin rashes?

Find the gene
Identify and isolate the gene that codes for the protease. Take the gene and put it into a bacterial species that can be cultured easily

Scale up
Design an industrial fermentation process to produce the enzyme in large quantities. This involves finding the optimum conditions for the growth of the recombinant bacteria

Add to the washing powder
Purify the product of industrial fermentation. Prepare it in a form that can be added to washing powder

Fig 6.15 Production of protease from *Bacillus* bacteria on a large scale for use as a stain remover

Immobilised enzymes

In an industrial process controlled by an enzyme, the enzyme is usually the component that costs the most. Like all catalysts, enzymes can be used several times, so it is a shame to waste an enzyme. In a process in which the enzyme is simply added to the substances that you want to react, a lot of enzyme is wasted because not all of it can be removed from the products.

To solve this, biotechnologists have developed methods to **immobilise** enzymes (see also Fig 6.16 overleaf):

- **Cross-linkage**. The enzyme molecules can be linked together to make much larger molecular structures that are easy to extract from the products. The enzyme can also be chemically attached to a supporting material.
- **Entrapment**. The enzyme can be mixed with ingredients that form a gel. When the gelling reaction is complete, the enzyme becomes trapped. Gaps in the gel are large enough to let the substrate in, but not let the enzyme out. Enzymes can also be trapped in a compartment behind a selectively permeable membrane that, again, lets reactants in, but doesn't allow the enzyme molecules out.
- **Adsorption**. The enzyme can be stuck temporarily to various surfaces – usually this method is only used for experimental studies, not full-scale industrial manufacture.

Fig 6.16 Immobilising enzymes

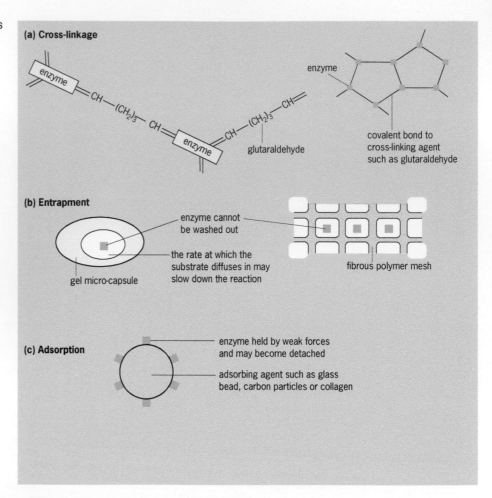

The first major commercial use of immobilised enzymes took place in the 1970s, when immobilised **glucose isomerase** was used to make high-fructose syrups from starches. Immobilised enzymes are now widely used in the food industry and the pharmaceutical industry. Immobilised **penicillin amidase**, for example, is used to prepare different forms of penicillin from the type produced by fungi.

Enzyme immobilisation is also important in **biosensors** (see page 106). These often require immobilisation of an enzyme into an electrode tip or into a paper strip (see the Clinistix™ test in the Science in Context box opposite).

BIOFUELS

Biotechnology is also used industrially to produce fuel. A biofuel is defined as a gaseous, liquid or solid fuel that contains energy that has come from a biological source. An example of a biofuel is rapeseed oil, which can be used in place of diesel fuel in modified engines. The methyl ester of this oil, **rapeseed methyl ester** (**RME**), can be used in unmodified diesel engines and is sometimes known as **biodiesel**. Other biofuels include **biogas** and **gasohol**.

Gasohol

Gasohol is produced by fermentation of sugar cane by the yeast *Saccharomyces cerevisiae* (commonly known as 'baker's yeast'). It has been produced on a large scale in Brazil, where it is manufactured in large distilleries and provides an alternative source of energy for motor vehicles (Fig 6.17). In Brazil all cars are adapted so as to be able to use gasohol or a mixture of gasohol and petrol.

Fig 6.17 Filling up with biofuel made from sugar

Fig 6.18 A biogas digester in India

Biogas

Biogas is a mixture of methane and carbon dioxide that is produced by the anaerobic decomposition of such waste materials as domestic, industrial and agricultural sewage. The decomposition is carried out by **methanogenic bacteria**, anaerobes that respire to produce methane, the main component of biogas. Methane is collected and used as an energy source for domestic processes, such as heating, cooking and lighting. The production of biogas is carried out in special **digesters**, which are widely used in China and India (Fig 6.18).

 REMEMBER THIS

Biofuels, such as wood, straw, vegetable oil, alcohol and methane are carbon neutral. But what does this mean? Put simply, it means that the amount of carbon dioxide released into the atmosphere when biofuels burn is balanced by the amount of carbon dioxide removed from the atmosphere during the growing of the biofuel crop.

For every gram of carbon dioxide released by burning a biofuel, a gram was removed from the atmosphere by photosynthesis just a few months before. This perfect balance is why biofuels are carbon neutral. It is important to remember that biofuels cease to be carbon neutral if fossils fuels are used to transport them from where they are made to where they used.

SCIENCE IN CONTEXT

The Clinistix™ test for glucose in urine

Enzymes are both specific and sensitive: this makes them ideal for use in analysis, which often involves very small samples. One such application is in analysis of glucose – an important technique in both medicine and industry. You may have heard of Clinistix™ (Fig 6.19) – the sticks used to test for the presence of glucose in urine. Clinistix™ contain two enzymes: glucose oxidase and peroxidase.

Glucose oxidase catalyses the following reaction:

glucose + oxygen　　gluconic acid + H_2O_2 (hydrogen peroxide)

In a simple, visible test, the production of peroxide is coupled to the production of a coloured dye or chromagen.

The second enzyme, peroxidase, catalyses the following reaction:

$$DH_2 + H_2O_2 \quad 2H_2O + D$$

D stands for the chromagen, a colourless hydrogen donor. When it loses its hydrogen it becomes coloured. The intensity of colour indicates the amount of glucose present.

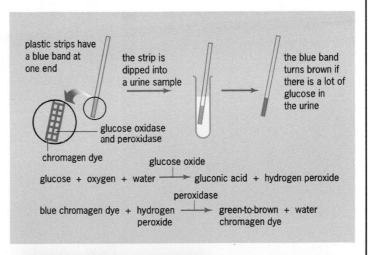

plastic strips have a blue band at one end

the strip is dipped into a urine sample

the blue band turns brown if there is a lot of glucose in the urine

glucose oxidase and peroxidase

chromagen dye

glucose oxide
glucose + oxygen + water ⟶ gluconic acid + hydrogen peroxide

peroxidase
blue chromagen dye + hydrogen peroxide ⟶ green-to-brown + water chromagen dye

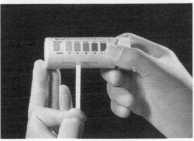

Fig 6.19 The urine glucose test for diabetes. Glucose oxidase, peroxidase and DH_2 are fixed on a cellulose fibre pad. When this is dipped into a sample of urine, the colour reaction gives a quantitative measure of glucose present in the urine

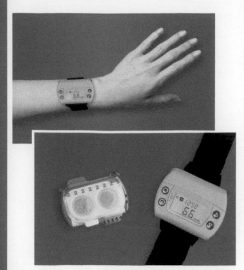

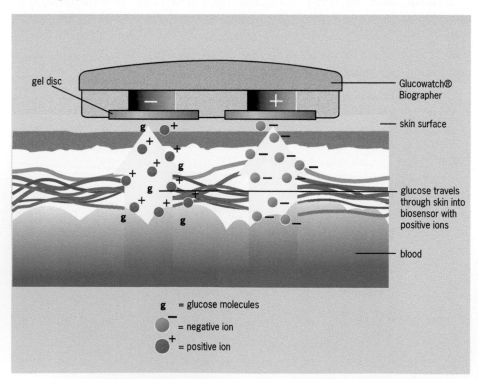

5 BIOTECHNOLOGY IN MEDICINE

BIOSENSORS

Biosensors are being developed for many applications in medicine. One such device that is already available is the Glucowatch® Biographer (Fig 6.20), which measures blood glucose levels without puncturing the skin to take a blood sample.

The device looks like an ordinary watch; it is worn on the wrist and it has an LCD dial. however, people with diabetes use it to measure their blood glucose level, not the time.

How the glucose biosensor works

The back of the Glucowatch® Biographer that comes into contact with the skin has two circular flat plates, as shown in Fig 6.21. The two plates generate a low electric current that is directed into the skin. This causes positive and negative ions to move out of the skin, pulling glucose molecules from the blood with them. The glucose collects in gel discs that contain glucose oxidase. This enzyme catalyses a reaction that generates a small electrical signal. The size of the signal is proportional to the glucose level in the blood. The Glucowatch® Biographer uses this signal to calculate a reading, which is then displayed.

Fig 6.20 The Glucowatch® Biographer

Fig 6.21 How the Glucowatch® Biographer detects the level of glucose in the blood

gel disc

Glucowatch®
Biographer

skin surface

glucose travels
through skin into
biosensor with
positive ions

blood

g = glucose molecules

= negative ion

= positive ion

How is the biosensor used?

Usually, people with diabetes who control their condition by insulin injections measure their blood glucose using the standard finger-prick method. They do this three or four times a day; before each meal and last thing at night. The disadvantage of this is shown in Fig 6.22. The curve shows how the blood glucose level actually changes during a 12-hour period. This can be detected by the Glucowatch® Biographer because it takes readings every 20 minutes, rather than by much less frequent finger-prick tests. These show the glucose levels at points A, B and C, lulling the person into thinking that their blood glucose is stable.

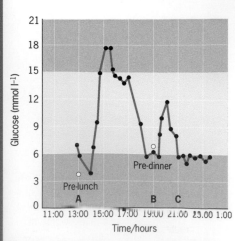

Fig 6.22 Daily variations in blood glucose level in a person with diabetes

MONOCLONAL ANTIBODIES

Monoclonal antibodies are antibodies that can be produced in large amounts and that bind only to one, very specific target. This is usually part of a molecule, not even a whole molecule.

How are monoclonal antibodies produced?

The method used to create monoclonal antibodies is called **hybridoma technology**. It was first devised in the 1980s by Cesar Milstein, working in laboratories in Cambridge, UK. It basically involves fusing a cell that produces antibodies with one that is **immortal**. Fig 6.23 summarises the process of making a monoclonal antibody.

The first step in the production of a mouse monoclonal antibody is to immunise a mouse with the molecule against which we want the monoclonal antibody to react. White blood cells in the mouse produce a range of different antibodies, directed against different parts of the molecule.

Cells called B lymphocytes that are responsible for making these antibodies are isolated from the mouse's spleen. These cells are then fused with tumour cells that divide continuously and rapidly. This creates a single, large cell called a **hybridoma**. These hybridomas show properties of both 'parent' cells; they produce antibodies and they divide continuously.

One hybridoma can divide to produce a whole clone of identical hybridomas, all secreting the same antibody molecule. Once a clone has been established and is growing well in culture, the antibody produced by this collection of cells is called a **monoclonal antibody**.

Monoclonal antibodies used in diagnosis

Monoclonal antibodies can be linked to a radioactive isotope and, in this form, they have been used for diagnosing and monitoring disease since the early 1980s. Some forms of cancer can be diagnosed using monoclonal antibodies directed against molecules found only on the surface of cancer cells. When injected into patients, such a monoclonal antibody will carry its radioactive marker straight to the tumour, where it can be detected using **CT** (**computerised tomography**) scanning (Fig 6.24).

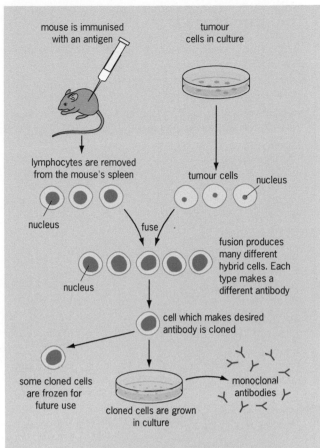

Fig 6.23 How a monoclonal antibody is made. Tumour cells, which divide without control, are fused with lymphocytes that produce an antibody. The hybrid cell produced shares the properties of the two cells that formed it. It divides endlessly (it is immortal) and it produces an antibody. When one cell is then used to form a whole culture, all the cells are clones. They all produce the same antibody, a monoclonal antibody

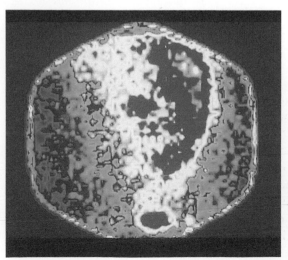

Fig 6.24 A cancer of the colon. Radioactivity from the increased concentration of monoclonal antibody in the colonic cancer cells is shown by the red, pink and white areas at the centre-right

REMEMBER THIS
Monoclonal antibodies also form the basis of modern pregnancy testing kits (see the How Science Works box overleaf).

The pregnancy test

By the time a pregnant woman would have been due for her next period, the embryo will have implanted and will be starting to secrete human chorionic gonadotrophin (hCG). Enough of this hormone passes into the urine for it to be detected by a modern pregnancy testing kit (Fig 6.25). This kit makes use of monoclonal antibodies that are specific for hCG and which are attached to a coloured chemical.

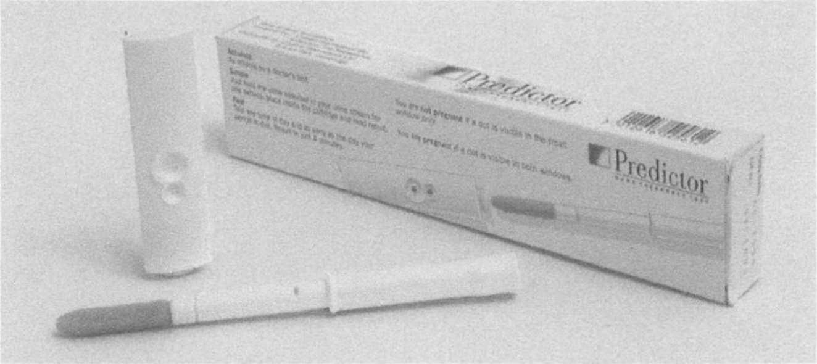

Fig 6.25 A pregnancy testing kit

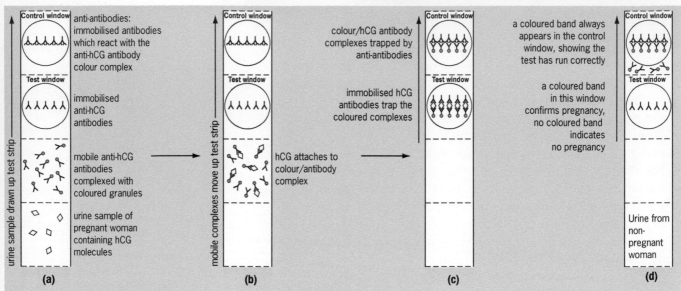

Fig 6.26 How a home pregnancy test kit works for a pregnant woman (a, b and c) and for a woman who is not pregnant (d)

Fig 6.26a Urine is drawn up the test strip by capillary action

Fig 6.26b hCG attached to the colour/antibody complex moves up the test strip

Fig 6.26c HCG attached to the colour/antibody complexes moves up to dock with the immobilised (fixed) hCG antibodies in the test window. All the complexes are stopped at one place, forming a visible coloured band and confirming pregnancy. Colour/antibody complexes without hCG attached move on to the control window

Fig 6.26d With the urine of a non-pregnant woman, the colour/antibody complexes have no attached hCG, so all the complexes move past the immobilised hCG antibodies in the test window and attach to the anti-antibodies in the control window, showing that the test has worked

Treating disease with monoclonal antibodies

The use of monoclonal antibodies in the treatment of different diseases has really expanded during the last 15 years. Most monoclonal antibodies are used for cancer treatment.

Normally, antibodies are produced in the body by B cells when they come into contact with an antigen – such as a surface molecule on a bacterium. Researchers have developed techniques to take one of these B cells, which is making an antibody they are interested in, and fuse that B cell with an immortal cancer cell. The resulting cells divide again and again but never die and they all make exactly the same sort of antibody. These become manufacturing centres for that type of antibody – which is then called a monoclonal antibody.

Monoclonal antibodies can be made to directly attack specific molecules on cancer cells – they kill only the cancer cells and no other body cells. They do a lot of damage to tumours in this way. Monoclonal antibodies can also be attached to radioactive molecules or to toxins and can act as 'guided missiles' to target specific cancer cells.

Developing new drugs and treatments

Developing any new therapy is a time-consuming and expensive process and developing new drugs and monoclonal antibody therapies takes several years.

The process starts in a laboratory with researchers who are doing fundamental research into how the body works or what goes wrong in a particular disease. When they identify the mechanism of a disease at the molecular level, this often suggests how pathways can be altered or blocked – by a potential treatment for that disease.

Potential treatments are often based on a naturally occurring molecule, perhaps a substrate for an enzyme in a biochemical pathway. Modifying this molecule using chemical techniques can then create a library of thousands of potential compounds that could one day become drugs. These molecules are screened using automated high-throughput methods, and the ones that seem to have the desired activity are chosen for further testing. These are usually called **lead compounds**.

Lead compounds are then tested in cell systems in culture to see if they show activity in living cells. If they show promise, they can be further modified to try to create a molecule of higher activity. This then goes on to the next round of testing – in small animals like mice and rats. There is a lot of opposition to animal testing, but there is no other way to see how a potential treatment behaves in a cell in a living system. Although mice are not the same as people, testing how a drug affects a mouse can provide a great deal of useful information that helps to refine the drug. This can take several years. Eventually, when researchers are confident that a drug may be ready for use, it is tested in people. These tests are called **clinical trials.** There are three main phases of a clinical trial and a drug must pass through all of them before it can be approved for use.

Phase I trials

Phase I studies are mainly concerned with the drug's safety and are generally done in a small number of people, usually less than 100. Researchers look for possible side-effects, how well the drug is tolerated and how the drug passes through the body.

A Phase I clinical trial of Trabio, a fully human monoclonal antibody, was designed to prevent scarring in patients having surgery for glaucoma. The trial was a double-blinded randomised trial, using Trabio versus placebo (an injection of saline).

As this was a Phase I trial, its main aim was to establish if Trabio was safe and if it caused swelling or other problems when injected at/near the site of operation. Trabio was shown to be safe and well tolerated in this group of patients: there were no serious local injection site reactions and no serious adverse events were reported. There were also some signs that scarring was reduced but it was difficult to see any significant difference in the small number of patients in the trial (24). A follow up Phase II study was then planned.

Phase II trials

Phase II studies are designed to test out how well the drug can treat a particular disease. The second phase of testing may last from several months to a few years and involve up to several hundred patients. Most Phase II studies are placebo-controlled, double-blinded randomised trials.

Tefibazumab, a humanised monoclonal antibody, was tested for its ability to treat *Staphylococcus aureus* infections. This monoclonal antibody binds to a protein that is expressed in the body in infected cells, called **adhesion protein clumping factor A**.

A total of 60 patients took part in the trial; 30 were given the monoclonal antibody and the other 30 a placebo. People were randomly assigned to the two groups. The results showed that there was no difference in the two groups with respect to adverse reactions to the treatment, which told the researchers that the monoclonal antibody was safe. Fewer patients in the tefibazumab group died and fewer got worse, so the study concluded that this monoclonal antibody would be worth investigating further as a treatment for this type of bacterial infection.

Phase III clinical studies

In a Phase III study, an experimental drug is tested in several hundred to several thousand patients with the disease. Most Phase III studies are also randomised and double blinded. Phase III studies can last years. Once a drug or treatment has passed through Phase III trials for one type of disease, it may well then go on to other Phase III trials for other, related diseases – such as a different type of cancer.

Panitumumab is a fully human monoclonal antibody directed against the epidermal growth factor receptor (EGFR). In a Phase III trial, researchers compared two groups of people with advanced colon cancer, who had not responded to chemotherapy; 232 of them were given supportive care and the other 231 were given supportive care and panitumumab. Researchers then measured the time that each patient remained at the same level of health without their symptoms getting any worse.

Patients who received the monoclonal antibody remained the same for 14 weeks but patients in the other group started to get worse after only eight weeks. Analysing the data showed that this result was statistically significant. Although eight weeks doesn't seem a long time, for these seriously ill patients it did

 REMEMBER THIS

A placebo is something that looks sufficiently like a drug to 'fool' people into thinking they are taking a real drug. It has long been known that many people who feel ill start to feel better if they are given a sugar pill. This has no physiological effect but they feel better because of the psychological effect of believing that the pill will do them good.

In clinical trials, experimental drugs are tested in comparison with placebos in order to assess the actual impact of the drug, and to make sure researchers are not measuring the psychological 'placebo effect'.

Testing a drug against a placebo nullifies this effect and makes the trial results more reliable.

Developing new drugs and treatments (Cont.)

make a big difference and pantiumumab therapy has since been approved for use in advanced colon cancer.

Control group

A control group is a group of people who are identical in every way to an experimental group in a clinical trial. The two groups contain the same numbers of people and are usually matched for age, sex, ethnic grouping and lifestyle (all non-smokers, etc.). This is done to reduce the number of variables in the experiment.

Double-blind

A double-blind test is a control group test where neither the person running the trial nor the people in the trial know who is receiving the experimental drug and who is receiving the placebo. The drug and the placebo appear to be identical and they are prepared and labelled by someone who doesn't interact with the people in the trial.

Randomised

A randomised trial is one in which people are randomly assigned to either the control or the experimental group. Neither the people in the trial nor the people running the trial choose who takes part in which group.

> **✔ REMEMBER THIS**
>
> The point of control groups, double-blinding and randomised testing is to reduce error, self-deception and bias on the part of those running and taking part in the trial. For example, although they might not mean to, if someone knew they were taking the drug, they might exaggerate how much better they felt, which would distort the results.

SUMMARY

- Biotechnology is defined generally as the **manipulation** of biological organisms, systems or processes for the benefit of people, in areas such as agriculture, food production and medicine.
- Biotechnology is widely used in agriculture. For example, bovine growth hormone injections are used to increase milk production in cows; artificial insemination of a prime cow using semen from a prime bull is used to generate multiple embryos carried by surrogate mothers (ordinary cows); hormones are used to control oestrus and mating in sheep to synchronise lambing in a large herd.
- The technology that creates **transgenic animals** is advancing rapidly. Some animals are bred with genes that make them more productive for the food industry (fish with an extra growth hormone gene); other animals are given genes that allow them to produce human proteins used in medicine.
- Biotechnology is also used to **modify plants** used in food production. Genetically modified foods offer both advantages and disadvantages. Some people and pressure groups in the UK are very suspicious of the development and use of GM foods. Others are in favour but don't know anything about them. Knowledge, evaluation and caution are important.
- Biotechnology is not new. The process of **fermentation** has been used to make yoghurt, cheese, beer, wine and bread for thousands of years.
- New biotechnology techniques are used to make edible protein from fungus (Quorn™) and bacteria (Pruteen™).

- Other **fermented food products** include sauerkraut, and products made from soya beans (e.g. tempeh and soy sauce).
- **Enzymes** are used in industry, particularly in the food and manufacturing sectors. Enzymes allow chemical processes to occur quickly and at relatively low temperatures, so are useful for keeping costs down.
- Most of the enzymes used in industry are extracellular. These are secreted by bacteria into the growth medium and are easy to separate and purify. Some intracellular enzymes are used but extracting them involves breaking up the bacterial cells first.
- Enzymes are often immobilised so they can be reclaimed and reused. The main methods of immobilising enzymes are **cross-linkages**, **entrapment** and **adsorption**.
- Immobilised enzymes are important in **biosensors** and in **diagnostic tests**.
- Monoclonal anibodies are used mainly for cancer treatment.
- Developing new drugs is time-consuming and expensive. The process starts with identification of a potentially useful molecule and progresses through testing in cell cultures and animal testing. Next comes three phases of testing in humans: Phase I trials, Phase II trials and Phase III clinical studies.

Practice questions and a How Science Works assignment for this chapter are available at www.collinseducation.co.uk/CAS

7

Nutrition

7 NUTRITION

Is it will-power or liposuction?

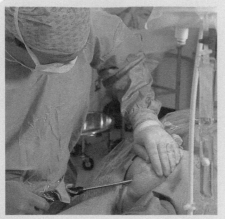

In the increasingly popular cosmetic treatment of liposuction, the fatty adipose tissue is literally hoovered out of 'problem' areas

In the affluent countries of the western world a lot of people eat too much. The number of overweight people is on the increase, and at the same time, others have made dieting an obsession.

We know that to lose weight we have to eat less food and exercise more, but the trick is to find the self-discipline to do it. One glance at the newsagent's shelves shows that there are many magazines devoted to slimming. Many of the articles are basically the same – diets, exercise regimes and inspirational stories such as: *I lost ten stones and so can you*.

If will-power fails, there are more drastic methods. Liposuction removes excess fatty tissue, usually from the buttocks and thighs, while 'beer bellies' can be removed in the same way. To cut down on intake, desperate people have even had their jaws wired up or part of their stomach surgically stapled so that it cannot hold much food.

> ### ✔ REMEMBER THIS
>
> **Inorganic substances** are generally made from simple molecules. Water, mineral ions (sodium chloride, etc.) and gases such as carbon dioxide are all inorganic.
>
> **Organic substances** – those based on carbon – include carbohydrates, lipids, proteins and nucleic acids; see Chapter 3.

> Photosynthesis is described in Chapter 33. Digestion is discussed in Chapter 8.

1 WHY DO WE NEED TO EAT?

All living things need food to obtain energy, to grow and to repair damaged cells and tissues. This chapter is about nutrition – the process of obtaining food. There are two types of nutrition, **autotrophic nutrition** and **heterotrophic nutrition**. For most human biology courses, you need only be aware of a broad definition of each.

Autotrophic nutrition involves taking in simple **inorganic** molecules and using them to form complex substances. Nutrition in plants is **autotrophic**. Their method of nutrition is more commonly known as **photosynthesis**, a process in which they build up large complex molecules from carbon dioxide and water.

Heterotrophic nutrition involves taking in complex organic molecules, and then breaking them down into simpler molecules that are absorbed into the body. This process of breaking down food is called **digestion**.

Humans are heterotrophic organisms, or **heterotrophs**. All the food we eat comes ready-made. Heterotrophs can be divided into three groups: **herbivores**, **omnivores** and **carnivores**. Herbivores, such as cows and sheep, obtain all their energy and nutrients directly from plants; carnivores, such as tigers and polar bears, eat only other animals. Humans are omnivores; they eat both plants and other animals.

The old cliché 'we are what we eat' is perfectly true. As developing babies, we were built when molecules from the food our mothers ate were assembled according to the genes we inherited, and ever since we have been taking in the food we need to build and power our bodies.

In Chapter 3 we looked at the chemicals of life: carbohydrates, proteins, lipids and nucleic acids. This chapter looks at how we provide our bodies with these and other essential substances.

2 WHAT DO WE NEED TO EAT?

You will have heard much talk of a **balanced diet**. Many assume that to achieve a balanced diet, all you need to do is eat a little bit of everything. However, as we shall see, like many things in life, it's just not that simple.

THE CONCEPT OF A BALANCED DIET

A diet must be balanced in terms of energy and in terms of its chemical content.

Whether a person is fat, thin or 'just right' depends on the balance of their energy intake versus their energy expenditure. Put simply, this means that if you take in more energy in the food you eat than you use up in your everyday activities, you will start to put on weight. Someone who is **obese** eats more than they need over long periods of time (Fig 7.1a). If you take in less energy that you use up, you lose weight. Someone who is **starving** cannot meet their energy demands, and they start to digest their own body fabric (Fig 7.1b). A diet that is balanced in terms of energy, enables us to maintain a constant, healthy weight.

It is also possible to eat a balanced diet in terms of energy, but then suffer from **malnutrition**. This is an imbalance in the chemical composition of the food we eat. If, for example, you ate only chocolate, you could easily maintain your body weight but your body would suffer from a range of deficiencies after a time – chocolate contains lots of sugar and fat, but few vitamins, minerals, complex carbohydrates and no fibre.

The energy in food

As you sit here reading this book, you might think that you are not using much energy, but even at rest, your body has a steady energy demand. Think about some of the processes going on:
- Your heart is beating.
- You are breathing.
- Food is being pushed along your intestines.
- Food is being absorbed from your intestines into your blood.
- Urine is moving from your kidneys to your bladder.
- Nerves are taking information from your sense organs to your brain, keeping you informed of changes in your environment.
- You are thinking: your brain is processing information coming in from the sense organs, sorting out what is important and what can be ignored.
- Unless you are somewhere very hot, you are probably losing heat to your surroundings. You need to replace the heat you lose in order to maintain a constant internal body temperature.
- Some of the molecules from digested food, such as amino acids, are being built up into molecules that will help to make new cells and tissues.

All these processes (and many others) require energy that is released from food molecules by respiration, a process that goes on constantly in all of our cells.

The amount of energy a person needs depends on:
- their age;
- their level of activity;
- their size;
- their genetic background (some people inherit a very high or a very low metabolic rate).

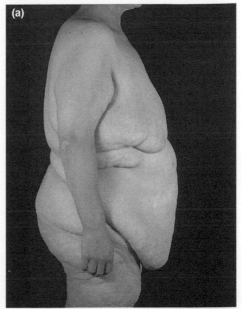

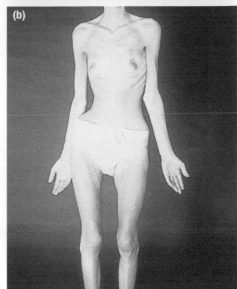

Fig 7.1a When the body takes in more energy from food than it needs, the excess is stored as fat under the skin and around the internal organs. The distribution of body fat is determined by the sex hormones; males tend to accumulate fat around the waist, while females accumulate fat more on the buttocks and thighs

Fig 7.1b When the body does not get enough energy from food, it has to respire stored fat and then protein. The actual fabric of the body – such as the muscles – starts disappearing

? QUESTION 1

1 An adult man, weighing 80 kg, needs to eat about 12 000 kJ per day to maintain his body weight. However, a toddler, weighing just 20 kg, still needs about 6000 kJ per day. Explain why, kg for kg, the toddler needs more energy.

The energy requirements of people at different times of life are covered in detail on page 122.

Ideally, we need to eat enough food to maintain an energy balance so that:

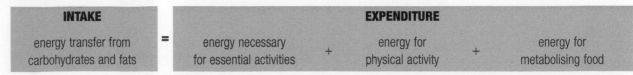

INTAKE	=	EXPENDITURE		
energy transfer from carbohydrates and fats		energy necessary for essential activities	+ energy for physical activity	+ energy for metabolising food

The energy required for the body's essential activities, that is, for the processes that take place when we are at rest, is called the **basal metabolic rate** (BMR).

HOW SCIENCE WORKS

Measuring the energy in food

The SI unit of energy is the joule (J) and the amount of energy in food is measured in joules. When we talk of the energy requirement of people, or of the energy content of food, we talk about kilojoules.

$$1000 \text{ J} = 1 \text{ kJ}$$

The energy value of food can be estimated by burning a known amount of the food in oxygen and measuring the energy released as heat. This process is called calorimetry. The information on energy content given on food packaging has been calculated in this way.

Fig 7.2a shows one very simple method of measuring the energy contained in a food sample. The sample is weighed and set alight. The energy released as the sample burns is used to heat a known volume of water. We can then calculate the total amount of energy contained in the food sample since we know that 4.18 joules of

energy will raise the temperature of 1 cm³ of water by 1 °C. Unfortunately, although this method is simple, it is not very reliable:

● A lot of the energy escapes and does not go into heating the water. In particular, heat is lost to the surroundings.

● Food samples are very difficult to burn. There is usually a lot of unburned material left at the end of the experiment.

The apparatus shown in Fig 7.2b is much more accurate.

The calorie is an older unit for energy in food, but is still widely used (Fig 7.2c). In popular nutrition guides, the calories referred to are actually kilocalories, which should be abbreviated Cal, with a capital C.

$$1000 \text{ cal} = 1 \text{ kcal} = 1 \text{ Cal}$$

A calorie is defined as the energy needed to raise the temperature of 1 cm³ of water by 1 °C.

$$1 \text{ calorie} = 4.18 \text{ joules}$$
$$1 \text{ Calorie} = 4.18 \text{ kilojoules}$$

Fig 7.2c Packaging for muesli shows the energy given in kilojoules and kilocalories

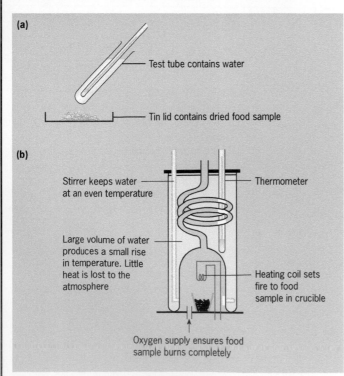

(a)

Test tube contains water

Tin lid contains dried food sample

(b)

Stirrer keeps water at an even temperature

Thermometer

Large volume of water produces a small rise in temperature. Little heat is lost to the atmosphere

Heating coil sets fire to food sample in crucible

Oxygen supply ensures food sample burns completely

Fig 7.2a A simple calorimeter. **Fig 7.2b** A more accurate calorimeter

NUTRITION INFORMATION		Typical value per 100g	Per 40g Serving with 125ml of Semi-Skimmed Milk
ENERGY	kJ	1350	800*
	kcal	310	180
PROTEIN	g	8	8
CARBOHYDRATE	g	66	33
(of which sugars)	g	(31)	(19)
(starch)	g	(35)	(14)
FAT	g	1.5	2.5 *
(of which saturates)	g	(0.3)	(1.5)
FIBRE	g	13	5
SODIUM	g	0.7	0.4

Anorexia and bulimia

In Chapter 4, we saw how someone adrift in the sea can die from dehydration even though surrounded by water. Some people in our society can also starve, even though they have access to plenty of food. Unlike the shipwreck victim, their problem is psychological rather than physiological (Fig 7.3a) – they have the eating disorder **anorexia nervosa**.

Anorexia usually affects teenage girls, but the number of boys affected is rising. A diet usually triggers the condition. Restricting food to lose a bit of weight gets out of control and the affected teenager becomes obsessed with food. They lose confidence as well as weight, and they often try to control the eating patterns of those around them. Some anorexics will, for example, make large meals for others, and encourage them to overeat, but eat virtually nothing themselves. Treatment does help, but an unlucky few do die from anorexia. Once the body weight of a teenage girl drops below about 7 stones (44.5 kg), her periods stop. Below 6 stones, she is unable to carry out ordinary activities like going to school, and once her weight has dropped to below 5 stones, she is in real danger of sudden death because of heart or other organ failure.

A related eating disorder, **bulimia nervosa**, also tends to affect young adults, mainly girls, and can also have devastating effects. Bulimia can also be triggered by dieting, but instead of under-eating, a bulimic will binge on food, eating vast quantities until overfull. Often, she will then make herself vomit to get rid of the uncomfortable feeling, and because she feels shame and self-disgust (Fig 7.3b). The vomiting leads to further negative feelings, and is then followed by another bingeing session. This cycle can continue for hours, or may only happen now and again. The body weight of a bulimic does not change drastically, and if the bingeing and vomiting are carried out in secret, friends and family may not know about the problem.

Fig 7.3a People with eating disorders don't see themeselves as they are

Fig 7.3b Vomiting to get rid of the feeling of fullness

BALANCING THE DIET

What constitutes a balanced diet varies from animal to animal; in a big cat like a tiger, a balanced diet is a large meal of antelope every two days. Humans, because they are omnivores, need to eat a wider range of foods and the healthiest diets seem to be those based mainly on carbohydrates, with relatively low amounts of fat and salt, some protein, and plenty of fresh fruits, vegetables and other plant products such as cereals and grains. The typical 'western' diet tends to have too much fat and protein, and too few fresh fruits and vegetables. Taking in a high proportion of fat in the diet leads to obesity because fat contains almost twice as much energy per unit mass than either carbohydrate or protein:

- 1 g carbohydrate produces in the body approximately 17 kJ;
- 1 g protein produces approximately 18 kJ;
- 1 g fat produces about 39 kJ.

A balanced diet for most people provides more than 55 per cent of its energy in the form of carbohydrates. Any diet should also include non-energy rich components – vitamins, minerals and fibre (see below). These compounds are needed only in tiny amounts, but they are vital components of the enzymes and other molecules that are involved in life processes.

? **QUESTION 2**

2 How do you test for the presence of a simple sugar in a food? (Hint – you may need to revise Chapter 3, page 36.)

REMEMBER THIS

Foods containing carbohydrates are put in three groups: those with a high **glycaemic index** (GI), those with a medium GI and those with a low GI. Foods in the high GI group release glucose into the blood much faster than those in the other two groups.

To cut the risk of developing diabetes in later life, the advice is to eat more foods from the medium and low GI group and eat smaller and fewer portions of foods in the high GI group. This helps keep your blood sugar more constant, so you need to produce less insulin to keep it in balance.

Carbohydrates

Carbohydrate is the most important energy provider among the food types, accounting for between 40 and 80 per cent of total energy intake of most human diets. It is also necessary for the metabolism of proteins and fats. Carbohydrates can be divided into three groups:

- **Sugars** such as sucrose and glucose – sweet foods and drinks contain large amounts. Sugars can be thought of as 'instant' fuels. The fastest way to get energy into our bodies is to have a glucose drink: the glucose does not have to be digested and so passes quickly into the bloodstream where it can be carried to the working muscles.
- **Starches**, abundant in many staple foods such as rice, potatoes, pasta and bread.
- **Non-starch polysaccharides** (NSPs) such as cellulose which collectively make up much of the fibre in our diet.

Although almost all types of carbohydrates provide similar amounts of energy, the nature of the carbohydrate can have a significant effect on our health. For instance, sugars can be classified in the following way:

- **Intrinsic sugars** are those contained within the cell walls of food, e.g. sugars within the cells of a banana.
- **Extrinsic sugars** that are not contained within cells, such as those found in processed foods such as sweets and soft drinks.
- **Milk sugars** occur in milk and milk products. Lactose is the primary milk sugar.

A major health problem caused by extrinsic sugars is tooth decay, or dental cavities, probably the most widespread disease in the western world. In the UK, about 13 per cent of the total dietary energy comes from extrinsic sugars – efforts are being made to reduce this to below 10 per cent to prevent tooth decay.

Generally, nutritionists disapprove of eating too many simple carbohydrates, advising instead to base your diet on starchy foods. In planning a meal such as breakfast, for example, you should eat a wholegrain, unsweetened cereal, or wholemeal bread with fruit and maybe low-fat yoghurt instead of a doughnut or Danish pastry. The energy from complex carbohydrates is released slowly over a few hours and this stops you being hungry until the next meal, so you can avoid 'snacking' on more high-sugar foods.

This makes sense because it keeps your blood glucose level from rising and falling rapidly. A major factor in hunger is blood glucose level – when it is low we feel hungry. Foods containing a lot of extrinsic sugars, or processed foods that allow the rapid digestion of starch, both enable glucose to be released rapidly, and it enters the blood. This triggers the body's insulin-based control mechanism, which reduces the glucose level again, causing hunger to return with a vengeance.

The importance of fibre

Fibre, the indigestible component of carbohydrate-rich foods, provides bulk to the digestive tract, stretching the gut wall and stimulating a faster through-put of food. Generally there is very little fibre in animal products; the vast bulk of indigestible material comes from plant tissues. If we eat a highly processed diet (one based on white bread, sweet foods, high-fat foods with little fresh fruit, vegetables or whole grains) we do not take in enough fibre. In the short term, this means chronic constipation and it increases our risk of developing gallstones and cancer of the colon later in life.

What makes for a healthy colon?

Many large studies involving thousands of people have shown that a diet low in fibre and high in red meat and fat makes it more likely that someone will develop cancer of the colon or rectum later in life. It is one of the clearest cases of a lack of a food type being associated with a particular cancer.

One large study, the European Prospective Investigation into Cancer and Nutrition (EPIC), published its latest findings in 2004. This study started in 1993, and monitored 521 000 people across nine different countries to investigate the links between diet and cancer. It concluded that eating a diet high in fibre and fish and low in red and processed meats can cut the risk of developing colon cancer. It is also important to drink alcohol very moderately and to keep to a healthy weight and exercise regularly.

The study has shown that breast and prostate cancer risk is not affected by how much fibre or fruit and vegetables you eat – these cancers are affected by other risk factors.

Proteins

The human body is made largely from protein; it accounts for more than half of the dry mass of our bodies. The proteins in our diet are digested to amino acids that then travel to our cells to be reassembled in a different order to make the proteins of the body. Proteins are needed for growth and repair of tissues, and not surprisingly are an important part of the diet of bodybuilders (Fig 7.4). Table 7.1 shows the protein content of some familiar foods.

Although there are many different types of protein, there are only 20 different amino acids. These can be divided into two dietary groups: the **non-essential amino acids** are those that the body can make, the **essential amino acids** are those that the body cannot make (Table 7.2). These must be obtained from food. We make non-essential amino acids in the liver.

As a general rule, proteins of animal origin contain all the essential amino acids, while different plants contain only some of them. If you are a vegetarian, you need to combine proteins from different plants in the same meal so that you take in the full range of amino acids. This is not as difficult

Table 7.1 The protein content of some familiar foods

Food	Protein content/ g per 100 g
milk (whole, cow's)	3.2
Cheddar cheese	25.5
beef	20.2
chicken	17.6
turkey	20.6
cod	17.4
baked beans	5.2
red kidney beans (dry)	22.1
peanuts	25.5
eggs	12.3
apples	0.4
cornflakes	7.9
potatoes	2.1

Source: MAFF reference book 342, *Manual of nutrition*

Fig 7.4 In order to make the extra muscle tissue, bodybuilders must have a high-protein diet

Fig 7.5 Each one of these eggs provides enough daily protein for an adult human. When proteins are heated they are denatured and often coagulate. This accounts for the change in colour and texture when an egg is cooked. However, the amino acid content is usually unchanged

Table 7.2 Essential and non-essential amino acids

Essential	Non-essential
isoleucine	alanine
leucine	arginine
lysine	aspartic acid
methionine	asparagine
phenylalanine	cysteine
threonine	glutamic acid
tryptophan	glutamine
valine	glycine
	proline
	serine
	tyrosine
	histidine (essential in infants)

REMEMBER THIS

Plasma proteins are found in the blood, and one of their vital functions is to draw water back into the blood from tissue fluid. A lack of blood protein leads to the swollen abdomen seen in cases of starvation (Fig 7.6).

QUESTION 3

3 How do you test for the presence of protein in a food? (Hint – you might need to revise Chapter 3, page 36.)

Non-essential amino acids and the liver are discussed in Chapter 11.

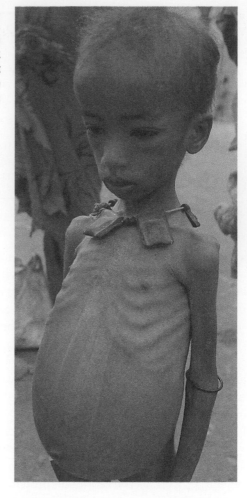

Fig 7.6 This child is suffering from kwashiorkor, a Swahili word which means 'displaced child', referring to the fact that the condition often develops after breast-feeding has stopped. A lack of protein in the blood means that water cannot be drawn back from the tissues, resulting in the swollen abdomen

as it sounds: a simple meal of beans on wholemeal toast (pulses plus grains) contains all the essential amino acids.

It may come as a surprise to know that a healthy adult human needs only about 60 g of protein per day, about the size of a large egg (Fig 7.5, page 117). The human body cannot store protein. If we eat too much, the excess amino acids that are produced are **deaminated** in the liver. The resulting ammonia is combined with carbon dioxide to form urea, which is excreted in the urine. Although we need only a small amount of protein, this tends to be the most expensive part of the diet and is therefore often lacking (Fig 7.6).

Fats

What we usually think of as 'fat' is mainly triglyceride, but the term includes edible oils, fats, waxes and related compounds. Approximately 95 per cent of the lipids in the human diet are triglycerides (three fatty acids attached to a glycerol molecule, see Fig 3.17). The remaining 5 per cent comprise mainly cholesterol and the phospholipids. The body can synthesise many of its fatty acids from carbohydrates and proteins: only a few, the essential fatty acids, must come from food.

The fat content of most vegetables is low (seeds are an exception). In general, animal products have a higher fat content as animals store fat in their bodies – even lean meat is quite high in fat. Animal sources include mammals (beef, pork, lamb, etc.) and oily fish such as herring and pilchard. Milk and milk products, such as cheese, are also rich in fats: milk contains 4.5 per cent fat by mass, and cheese is 30 per cent fat.

Vitamins

Vitamins are essential organic compounds needed by the body in small amounts (Table 7.3). For example, the recommended daily amount of vitamin A for a young adult male is 1000 µg (only one thousandth of a gram). This means that the average human needs to take in only about 25 g of vitamin A (a mouthful) during an entire lifetime.

Obviously, we get most of our vitamins from our food, but there are other sources. Gut bacteria synthesise vitamin K (essential for blood clotting). In addition, humans can make vitamin A from carotene, a substance found in most plant foods, particularly carrots. Ultraviolet radiation from sunlight acts on a compound in our skin called **ergosterol**, converting it to vitamin D.

 REMEMBER THIS

Vitamins help to regulate cell chemical processes, often through interaction with important enzyme systems. Lack of a particular vitamin in the diet leads to a **deficiency disease**.

Table 7.3 Vitamins important in the human diet

Vitamin	Name	Minimum daily requirement	Rich food source	Function	Deficiency disease
Fat-soluble vitamins					
A	retinol	1000 µg	fish liver oils, dairy products, liver; most leafy vegetables and carrots contain carotene that can be converted into retinol	needed for healthy epithelial cells and for regeneration of rhodopsin in rod cells of the eye	dry (keratinised) skin; night blindness
D	calciferol	10 µg	fish oils, egg yolk, butter; it can be made by the action of sunlight on skin	promotes absorption of calcium from intestines; necessary for formation of normal bone and reabsorption of phosphate from urine	rickets in children ('soft' bones, bend easily); painful bones in adults
E	tocopherol	10 µg	vegetable oils, cereal products; found in many foods	in rats, formation of red blood cells, affects muscles and reproductive system; unclear in humans; because the vitamin is widely found in foods, deficiency is rare in humans	mild anaemia and sterility in rats
K	phylloquinone	70 µg	fresh, dark green vegetables; gut bacteria	formation of prothrombin (involved in blood clotting)	delayed clotting time; may occur in newborn babies before the establishment of gut bacteria
Water-soluble vitamins					
The B vitamins form a vitamin complex. They are chemically unrelated, but tend to occur together and have similar functions					
B_1	thiamine	1.5 mg	all plant and animal tissues contain thiamine but rich food sources are yeast, cereals, nuts, seeds, pork	coenzyme in cell respiration, necessary for complete release of energy from carbohydrates	accumulation of lactic acid and pyruvate; beriberi (severe muscle wastage, stunted growth, nerve degeneration)
B_2	riboflavin	1.8 mg	liver, milk, eggs, green vegetables	coenzyme in cell respiration; precursor of FAD	cracked skin, blurred vision
B_3	niacin	20 mg	liver, yeast, whole cereals, beans	coenzyme in cell respiration; precursor of NAD/NADP	pellagra (dermatitis, diarrhoea, dementia)
B_6	pyridoxine	2 mg	meat, fish, eggs, cereal bran, some vegetables	interconversion of amino acids – transamination	dermatitis, nerve disorders
B_{12}	cyano-cobalamin	2 µg	liver, milk, fish, yeast; none in plant foods	maturation of red blood cells in bone marrow; maintenance of myelin sheath	pernicious anaemia; nerve disorders; anaemia, especially during pregnancy
	folic acid	200 µg	liver, raw green vegetables, yeast, gut bacteria	formation of nucleic acids and red blood cells	
C	ascorbic acid	60 mg	blackcurrants, potato, rose hips, citrus fruits	formation of collagen and intercellular cement	scurvy (Fig 7.8 overleaf); poor wound healing

Vitamin K injection and new-born babies

Vitamin K is made in the body by helpful bacteria in the large intestine. Everybody has them, so no-one should be short of vitamin K. However, one very special group of people are at risk of vitamin K deficiency. All babies are born with sterile guts – not a bacterium in sight – and it takes about a week for the normal collection of different bugs to build up. During that first 7 days of life, this makes new babies prone to a disease called **haemorrhagic disease of the new-born**. Vitamin K is vital for blood clotting, and a lack of vitamin K can lead to bleeding all over the body. This can occur if the skin is broken, at the navel, but also in the gut, the bowel and the brain. Brain bleeding is rare but is highly dangerous. Minor bleeding seems to happen in about one baby in 400, while more serious bleeding is very rare, about one baby in 10 000 suffers.

Fortunately, haemorrhagic disease of the new-born can be prevented very simply by giving vitamin K either by injection or by mouth on the first day of life (Fig 7.7). Recently, there has been some controversy about doing this because some studies suggested that babies given vitamin K by injection were more prone to leukaemia. Other studies have shown no such link, and most babies in the developed world now receive the vitamin K supplement.

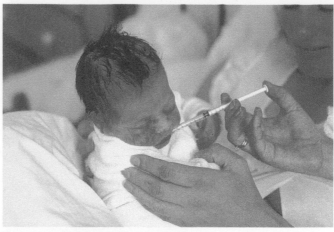

Fig 7.7 A new-born baby being given vitamin K

? QUESTION 4

4   Early Arctic explorers learned from the resident Eskimos not to eat the liver of polar bears. It causes headache, drowsiness, vomiting and extensive peeling of the skin. Since polar bear liver may contain up to 600 000 μg vitamin A per 100 g, what is the likely explanation for this observation?

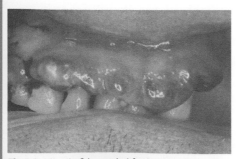

Fig 7.8 Vitamin C is needed for healthy gums; the inflammation seen in this photograph is caused by vitamin C deficiency – otherwise known as scurvy

Vitamins can be divided into two groups: fat-soluble vitamins (A, D, E and K), which contain only carbon, hydrogen and oxygen, and water-soluble vitamins (vitamin C and the B complex of vitamins), most of which also contain nitrogen. Fat-soluble vitamins can be stored in the body (especially in the liver) but water-soluble ones cannot, and are excreted in the urine. Water-soluble vitamins need to be taken into the body every day, and it is virtually impossible to 'overdose' on them. Taking too much of fat-soluble vitamins can cause problems: they can build up in fatty tissues until they reach toxic levels.

Minerals

The minerals essential to human health are listed in Table 7.4. Minerals have four functions in the body:

- They are raw materials for the construction of body tissues.
- They have a homeostatic role, providing the correct chemical environment for cells.
- They are metabolites for various cell processes.
- They are essential partners for enzymes and vitamins controlling cell activity.

Generally, the minerals in the first two categories are needed in relatively large amounts and are known as **macronutrients**. Minerals in the last two categories are needed in specific chemical reactions and in tiny amounts. For this reason they are known as **micronutrients**.

The presence of other vitamins and minerals can affect the actual amount of a mineral that the body takes up. For example, calcium, the most abundant mineral in the body, is taken up more quickly in the presence of vitamins C and D, and iron is better absorbed in the presence of vitamin C.

Table 7.4 Minerals essential in the human diet

Mineral	Major food source	Function	Total body content/g
Major minerals (macronutrients)			
calcium	milk, cheese, bread, watercress	needed for muscle contraction; nerve action; blood clotting and the formation of bone	1000
phosphorus	cheese, eggs, roast peanuts, most foods	bone and tooth formation; energy transfer from foods; DNA, RNA and ATP formation	780
sulfur	dairy produce, meat, eggs, broccoli	formation of thiamin, keratin and coenzymes	140
potassium	potatoes, meat, chocolate	muscle contraction; nerve action; active transport	140
sodium	any salted food, meat, eggs, milk	muscle contraction; nerve action; active transport	100
chlorine	salted foods, seafood	anion/cation balance; gastric acid formation	95
magnesium	meat, green vegetables	formation of bone; form coenzymes in cell respiration	19
Trace elements (micronutrients)			
iron	liver, kidney, cocoa powder, watercress	formation of haemoglobin, myoglobin and cytochromes	4.2
fluorine	water supplies, tea, seafood	resistance to tooth decay	2.6
zinc	meat, liver, legumes	enzyme activator; carbon dioxide transport	2.3
copper	liver, meat, fish	enzyme, melanin and haemoglobin formation	0.072
iodine	seafood, iodised salt, fish	constituent of thyroxine	0.013
manganese	tea, nuts and spices, cereals	bone development; enzyme activator	0.012
chromium	meats, cereals	uptake of glucose	>0.002
cobalt	meat, yeast, comfrey	component of vitamin B_{12}; formation of red blood cells	0.0015

Water

Just under 70 per cent of the human body is water: two-thirds of this is found within the cells and one-third outside the cells in tissue fluid and blood plasma. It is essential in the diet and can come from three main sources:

- as a drink,
- as food (lettuce, for example, contains 94 per cent water),
- as metabolic water: water is released from cellular chemical reactions, especially cell respiration.

An individual must take in enough water to balance what is lost in urine, sweating, breathing out and in faeces. Otherwise, they lose water and can become dehydrated. Severe dehydration leads to an increase in heart rate and blood urea concentration and, eventually, death.

The chemical properties of the water molecule are discussed in Chapter 4; the role of the kidneys in water balance is covered in Chapter 13.

QUESTION 5

5 To what extent does a meal of brown bread, butter and cheese satisfy dietary requirements for an adult man? What does the meal provide, and what does it lack?

QUESTION 6

6 Why do experts recommend that a combination of dieting and increasing the amount of exercise is the most effective way to lose weight if you are overweight? Why not just diet or do more exercise?

The nutritional requirements of a human change according to age, sex, level of physical activity and state of health. Within groups of people there are still variations, but it is possible to say broadly what the general food and energy requirements of the groups are.

Fig 7.9 shows the energy requirements for different people at different stages of life. The Tour de France cyclist is included as a contrast to the 'ordinary' lifestyles of people who are very active. This is one of the most demanding sporting events in the world, requiring peak output for several hours at a time, often over several days.

Fig 7.9 The approximate energy needs of different people

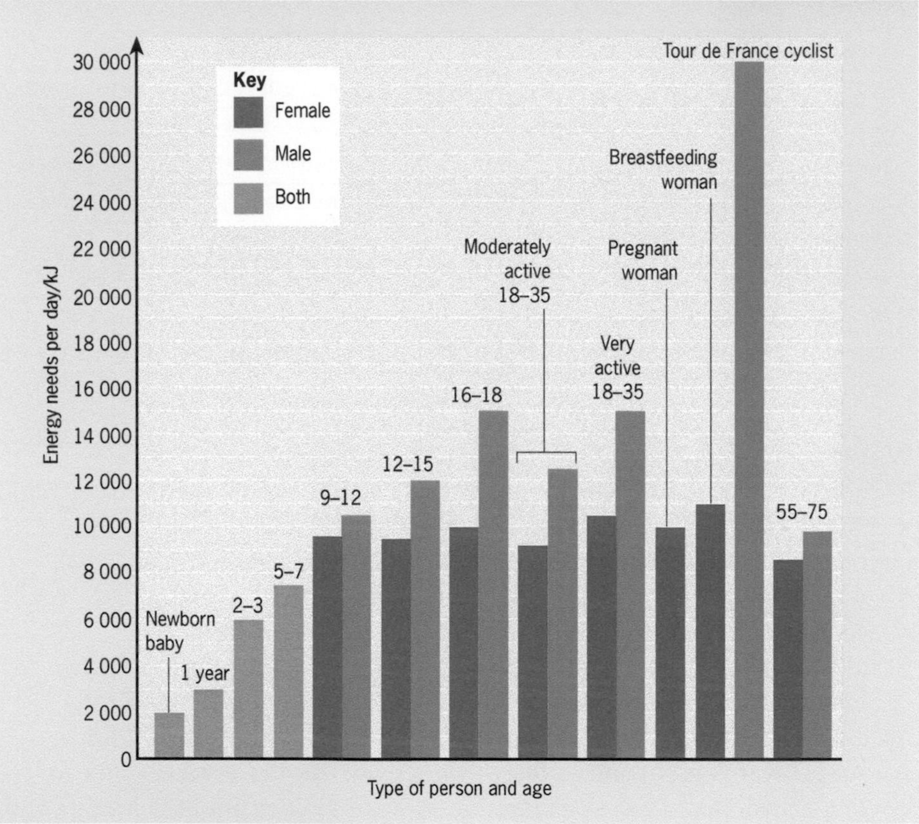

THE FIRST YEAR OF LIFE

When a baby is first born, its mother's breasts produce a fluid called **colostrum**. This is rich in antibodies, which pass whole into the blood of the baby through large pores in its intestine. Generally, the colostrum also has about three times more protein than the breast milk produced later. There is a gradual transition between colostrum and early breast milk over the first week. As the baby suckles, it first takes in **foremilk** – a more dilute breast milk that quenches its thirst – before richer, fattier **hindmilk** is produced towards the end of the feed.

Breast milk produced at about the third month of lactation has 75 calories per 100 ml; its main components are shown in Table 7.5.

During the first 4–6 months of life growth and development are rapid. Breast milk (or infant formula) contains all the nutrients required during this period; babies don't really need any solid food before the age of 4 months. As the baby grows, the fat content of breast milk decreases. After 4–6 months, milk is no longer enough and the baby starts taking in other foods – this is called **weaning** (Fig 7.10). Iron requirements increase rapidly and it is important that the diet given during weaning contains enough iron.

Table 7.5 Composition of human milk

Component	Amount present per 100 ml
Carbohydrates	
lactose	7.3 g
oligosaccharides	1.2 g
Proteins	
caseins	0.2 g
lactalbumin	0.2 g
lactoferrin	0.2 g
secretory IgA	0.2 g
β-Lactoglobulin	None
Milk lipids	
triglycerides	4.0%
phospholipids	0.04%
Minerals	(milli moles)
sodium	5.0 mM
potassium	15.0 mM
chloride	15.0 mM
calcium	8.0 mM
magnesium	1.4 mM

Requirements for protein, thiamin, niacin, vitamin B_6, vitamin B_{12}, magnesium, zinc, sodium and chloride also increase between 6 and 12 months.

TODDLERS

Once babies start to toddle, walk and then run, they need a lot more energy from food. Growth and development are very rapid at this stage and, although protein requirements do not increase much, children between the ages of 1 and 3 years need more of all the vitamins, except vitamin D (this is synthesised in the skin in sunlight) and all the minerals, except zinc. Slightly lower amounts of calcium, phosphorus and iron are needed.

Children at this stage of life need energy-dense diets. The usual rules of what makes a healthy diet for an adult just do not apply. They should drink whole milk, not skimmed or semi-skimmed, and they shouldn't eat too much fibre. If the diet is too bulky, there is a possibility that a child cannot absorb enough energy from the food he or she eats. After the age of 2 years, children can switch to semi-skimmed milk, as long as the rest of the diet contains enough energy. Children shouldn't switch to completely skimmed milk until they are at least 5 years old.

CHILDHOOD

From the age of 5 until about 11, growth is still rapid and children of this age are very active (Fig 7.11):

- 5–7 years: children need more energy, protein, and more of all the vitamins (except C and D) and all the minerals.
- 7–10 years: there is a marked increase in requirements for energy and protein. Children of this age need no more thiamin, vitamin C or vitamin A than they did at the age of 5 but their need for the other vitamins and minerals increases.
- 11–14 years: the energy needed continues to increase and young teenagers need 50 per cent more protein than they did at the age of 10. By the age of 11, the vitamin and mineral requirements for boys and girls start to differ. Boys of 11–14 need more of all the vitamins and minerals. Girls need no more thiamin, niacin, vitamin B_6, but they need more of all the minerals. When periods start, girls need more iron than boys to make up the iron lost in menstrual blood.
- 15–18 years: boys need more energy and protein and more thiamin, riboflavin, niacin, vitamins B_6, B_{12}, C and A, and more magnesium, potassium, zinc, copper, selenium and iodine. In girls, the need for energy,

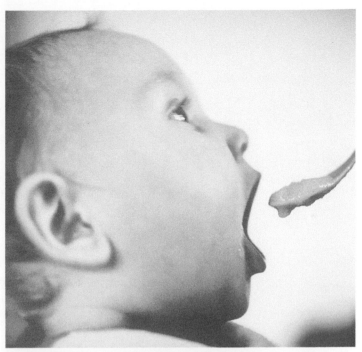

Fig 7.10 He won't stay this clean for long. Modern baby foods are well balanced and nutritious, giving a great start in life. It's not until he reaches the age of the children shown below that he'll start nagging for high-fat, high-salt or high-sugar foods. Broccoli? No chance

Fig 7.11 Young children are growing rapidly and are very active, so they require lots of energy from their food

protein, thiamin, niacin, vitamins B_6, B_{12} and C, phosphorus, magnesium, potassium, copper, selenium and iodine all increase, but girls continue to need slightly less energy than boys of the same age. Boys and girls have the same requirement for vitamin B_{12}, folate, vitamin C, magnesium, sodium, potassium, chloride and copper. Girls still need more iron.

Fig 7.12 Tour de France cyclist being pushed to the limit

ADULTHOOD

As young adults stop growing, their need for energy and some vitamins and minerals stabilises, and usually decreases. The energy requirements of adults depend mainly on their lifestyle – a very active person will need more energy (see Figs 7.9 and 7.12).

- 19–50 years: adult men and women both need less energy in their diet than when they were teenagers and their requirements for calcium and phosphorus fall as they are no longer making large amounts of new bone. There is also a reduced requirement in women for magnesium, and in men for iron. An adult's need for protein and most of the vitamins and minerals remain virtually unchanged from the teenage years.

- 50+ years: energy requirements decrease gradually after the age of 50 in women and age 60 in men (Fig 7.13). Protein requirements decrease for men but increase slightly in women. Vitamin and mineral intake should continue at the same level as younger adults, for both men and women. The only exception is that once women have gone through the menopause and their periods have stopped, they need the same amount of iron as men of the same age.

Fig 7.13 As we age, our metabolic rate slows and we tend to become less active. This can lead to an increase in weight but elderly people often lose interest in food and eating, and can become very thin and frail

PREGNANCY AND BREAST-FEEDING

Women are advised to take folic acid supplements when planning to get pregnant and in the first three months of pregnancy. A pregnant woman requires no other additional food components or energy until three months before the birth (Fig 7.14).

Fig 7.14 Eating for two? A developing baby is fed via the placenta, an organ which takes the molecules from food directly out of the mother's blood. When her baby is born, the mother needs even more food energy to produce milk. The actual amount of extra energy needed depends on how much milk is produced, the size of the fat stores that have accumulated during pregnancy and how long the woman chooses to breast-feed

Dietary reference values

In the UK, estimated requirements for particular groups of the population are based on advice given by the Committee on Medical Aspects of Food and Nutrition Policy (COMA). The COMA panel reviews scientific evidence and makes proposals that are used by the government. In 1991, the COMA report 'Dietary Reference Values for Food Energy and Nutrients for the United Kingdom' was published. This estimated the nutritional requirements of the age groups shown in Table 7.6.

Dietary reference values (DRVs) for energy intake exist for all different age groups, but dietary reference values for specific food groups (carbohydrates and vitamins, for example) only apply to people over 5 years old.

Source: Department of Health. Dietary Reference Values for Food Energy and Nutrients for the United Kingdom. London: HMSO, 1991.

Table 7.6 How energy requirements change during life

| | Estimated average requirements for energy | | | |
| | Megajoules per day | | Kilocalories per day | |
Age	Males	Females	Males	Females
0–3 months	2.28	2.16	545	515
4–6 months	2.89	2.69	690	645
7–9 months	3.44	3.20	825	765
10–12 months	3.85	3.61	920	865
1–3 years	5.15	4.86	1230	1165
4–6 years	7.16	6.46	1715	1545
7–10 years	8.24	7.28	1970	1740
11–14 years	9.27	7.72	2220	1845
15–18 years	11.51	8.83	2755	2110
19–50 years	10.60	8.10	2550	1940
51–59 years	10.60	8.00	2380	1900
60–64 years	9.93	7.99	2380	1900
65–74 years	9.71	7.96	2330	1900
74+ years	8.77	7.61	2100	1810
Women only pregnancy		8.9		2140

OBESITY AND BODY MASS INDEX

Obesity is a term that means 'grossly overweight' and, despite more people understanding the problems of the condition, the proportion of overweight and obese people is on the increase (Table 7.7).

Body mass index

In 1869, a Belgian astronomer Quetelet worked out that people's body mass varied not in proportion to their height, but in proportion to the square of their height. So the formula for body mass index (BMI) is:

$$\frac{\text{mass in kg}}{\text{height}^2 \text{ in m}^2}$$

For example, a man who is 1.83 m tall (6 feet) and weighing 82 kg would have a BMI of

$$\frac{82}{(1.83)^2} = \frac{82}{3.35} = 24.47$$

The BMI is now universally used to define underweight, overweight and obesity, as shown in Table 7.8.

The problems of obesity

There was a time when fat people were looked upon just as happy, jolly folk. Nowadays, we know the problems associated with obesity, and obese people have to deal with the intolerance of others towards their size and body image.

Table 7.7 The increasing incidence of obesity

Country	Weight level	Age range	Year	Incidence/% men	women
England	obese (BMI >30)	16–64	1980	6	8
			1994	13	16
England	overweight (BMI 25–30)	16–64	1980	35	24
			1991	40	26
Germany	obese (BMI >30)	25–69	1985	15	17
			1990	17	19
USA	obese* (white people)	20–74	1978	24	24
			1988–91	34	34
USA	obese* (black people)	20–74	1978	26	45
			1988–91	32	49

*In the USA survey, the threshold for obesity was BMI >27.8 men, >27.5 for women

Sources: Obesity Research Information Centre, London, and Dept of Health Report 46 (1994) *Nutritional Aspects of Cardiovascular Disease*, HMSO.

Table 7.8

BMI/kg m⁻²	Description
less than 20	underweight
20–24.9	normal
25–29.9	overweight
30–40	obese
over 40	severely obese

Fig 7.15 This is Daniel Lambert (1770–1809), once thought to be the heaviest man in England. He weighed 336 kg (52 stone 11 lb) and was 180 cm (6 ft) tall

One look at the population of an old people's home will support the blunt statement that 'old people aren't fat and fat people aren't old'. Obesity puts a great strain on the body that more often than not leads to severe medical problems and a premature death (Fig 7.15). Consider some of the problems:

- **Coronary heart disease**. Obesity increases the levels of triglyceride and cholesterol in the blood (see Chapter 31).
- **Type 11 Diabetes**. This is the inability to control blood sugar levels. In an obese person, insulin becomes less effective than normal, a problem known as insulin resistance.
- **High blood pressure**. The heart has to create enough pressure to force blood around all that extra tissue. High blood pressure leads to a variety of problems, not least of which is the increased risk of a stroke.
- **Joint problems**. Excess weight puts a lot of strain on some joints.
- **Accidents.** Obese people tend to be clumsier than less heavy people, and unable to move quickly.
- **Depression and suicide**. These could be responses to a poor body image.

The causes of obesity

Many obese people claim that eating too much is not the problem; it's in their genes. Recent research suggests that there may be some truth in this, but we can only inherit a tendency towards obesity or 'leanness'. We must still eat the food which allows us to fulfil our potential, and studies suggest that social and environmental factors play a huge part. Children born to overweight parents tend to be given larger portions and generally follow the influence of their parents.

TACKLING OBESITY

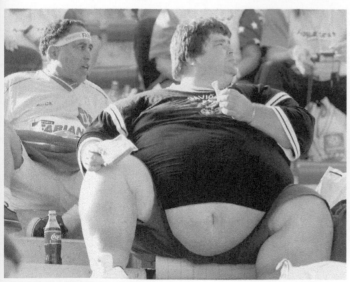

Fig 7.16 The USA has the highest proportion of obese people in the world. However, with the amount of processed, high-fat and high sugar foods that people are eating, the UK and other countries in Europe are seeing increasing numbers of people who are very obese

Lifestyle

People are generally eating more and exercising less, using the car for even the shortest journeys and taking part in leisure activities with minimal energy needs (Fig 7.16).

Drugs

Research has shown that adipose (fat storage) tissue gives off a hormone, leptin (Greek, *Ieptos* = slender/thin) which suppresses the appetite centre of the brain. There are artificial appetite suppressants available, but their effect is only temporary. Clearly, there would be a huge market for an effective drug.

Surgery

Adipose tissue can be surgically removed; the process of liposuction is one of the cosmetic surgeons' biggest moneyspinners. However, all surgery carries some risk and, without a long-term change in eating and exercise habits, the problem will return.

SUMMARY

When you have read and studied this chapter, you should know and understand the following:

- Food supplies us with the energy and raw materials we need for repair, growth and development of body tissues.
- Humans need a **balanced diet** which contains carbohydrate, protein, fat, **water**, **fibre**, vitamins and minerals.
- **Carbohydrate** foods (sugars and starches) should be the main source of energy in the diet, although **lipids** provide more than twice as much energy per gram.
- When intake of 'energy' foods exceeds requirements, the body will store the excess as glycogen and fat. When intake is too low, the body will turn to the lipid stores and increasingly use **protein** when the other stores run low.
- **Vitamins** are either water soluble (A, D, E and K) or fat soluble (B complex and C). Lack of particular vitamins or **minerals** in the diet can lead to **deficiency diseases**.
- Body mass index (BMI) is measured as weight (kg) divided by height^2 (m^2). BMI is now used universally to define underweight, normal weight, overweight and obese.

 Practice questions and a How Science Works assignment for this chapter are available at www.collinseducation.co.uk/CAS

8

Digestion

8 DIGESTION

Fructose is present in many fruits

Irritable bowels don't tolerate fructose

Humans are omnivores; they have a digestive system that is adapted so they can eat a range of foods, from both animals and plants. But that doesn't mean that all foods suit everybody. Some people suffer food allergies or are intolerant to specific foods. Lactose intolerance is well known (page 137) but researchers are uncovering a problem with sugar digestion – an intolerance to fructose.

Everyone experiences indigestion or stomach upsets now and again, but some people have chronic digestive problems – this is known as irritable bowel syndrome (or IBS), and is extremely common. Its main symptoms are abdominal discomfort, a bloated feeling and experiencing diarrhoea and then constipation for no apparent reason. In the autumn of 2001, researchers in the USA revealed that about 10 per cent of IBS cases could be explained by an intolerance to fructose. Fructose is a simple sugar found in honey and many fruits.

When people with IBS symptoms were given fructose, tests showed that over three-quarters of them produced breath that contained hydrogen and methane. This indicated that they were not metabolising fructose normally, and these abnormal by-products were being produced in the gut.

1 THE HUMAN DIGESTIVE SYSTEM

The overall function of the **digestive system** is to break down the larger molecules in food, such as protein and starch, into their monomers – amino acids and sugars – so that they can be absorbed easily into the body.

The **alimentary canal**, or **gut**, can be thought of as a long coiled tube that runs through the body. The food we eat is subjected to 'conveyor-belt' food processing as it passes through the different regions of the gut. A common misconception is that the food we eat is 'inside' us, when really it travels through a cavity in the middle of the body. If we cannot digest the food, it cannot be absorbed into the body and passes straight through.

The alimentary canal, or 'gut', is a muscular tube that leads from the mouth to the anus. The overall process of nutrition can be divided into several stages:

- **Ingestion**: this is taking in food. In humans, food is put into the mouth and chewed. **Swallowing** takes it down through the **oesophagus** into the stomach. Food is propelled through the alimentary canal by **peristalsis**, rhythmic contractions of the gut wall.
- **Mechanical breakdown**: the breaking up of food material into smaller pieces, mainly by the action of chewing or the churning action of the stomach.
- **Digestion**: the chemical breakdown of complex food molecules such as carbohydrates, proteins and fats into simpler ones.
- **Absorption**: the passage into the bloodstream of simple food molecules such as amino acids, sugars and fatty acids, vitamins, minerals and water.
- **Egestion**: the elimination of undigested food material from the body.

 REMEMBER THIS

Humans are **heterotrophs**; we cannot make our own food and so must obtain it in the form of large organic molecules from other organisms. All these molecules were originally made by green plants.

2 STRUCTURE OF THE HUMAN DIGESTIVE SYSTEM

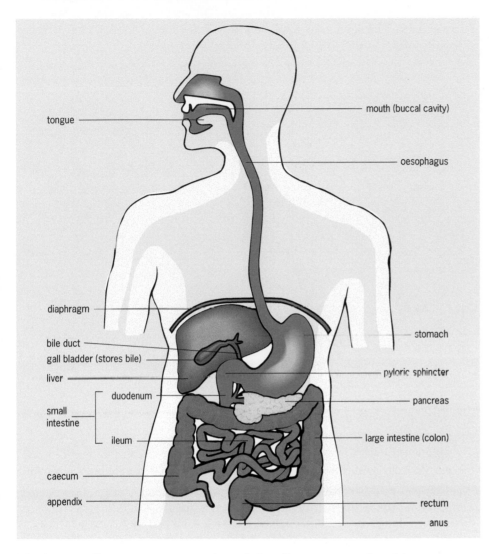

tongue

mouth (buccal cavity)

oesophagus

diaphragm

bile duct

gall bladder (stores bile)

liver

stomach

pyloric sphincter

pancreas

duodenum

small intestine

ileum

large intestine (colon)

caecum

appendix

rectum

anus

Fig 8.1 The human digestive system

The human digestive system consists of the alimentary canal and its associated glands, the salivary glands, the liver and the pancreas (Fig 8.1).

The alimentary canal begins at the mouth and ends at the anus. Between the two openings is a long convoluted tube organised into several distinct regions. The **oesophagus** carries food from the mouth to the **stomach**, a muscular bag or sac that stores food and is the first significant site of digestion.

Beyond the stomach is the **small intestine**. This has two main parts, the **duodenum** and the **ileum**. These are the main sites of digestion and absorption. The large intestine includes the **appendix**, the **colon**, whose main function is to absorb water, and the **rectum**. It ends at the **anus**.

BASIC FEATURES OF THE GUT WALL

The wall of the gut is not the same all the way along the alimentary canal. Each region has specific features, which can be related to the function of different parts of the gut.

The gut wall is divided into three main layers:

- an outer **muscle layer**, protected by a thin coating of fibres;
- a middle layer, the **submucosa**;
- an inner layer, the **mucosa**.

The oesophagus

The main function of the oesophagus is to push food from the mouth to the stomach. The structure of the wall of the oesophagus is shown in Fig 8.2. The endoscope image shows that the inside surface has large ridges, but it is fairly smooth. The squamous epithelial layer is several cells thick and protects the underlying tissues from damage when swallowing.

Fig 8.2 The oesophagus

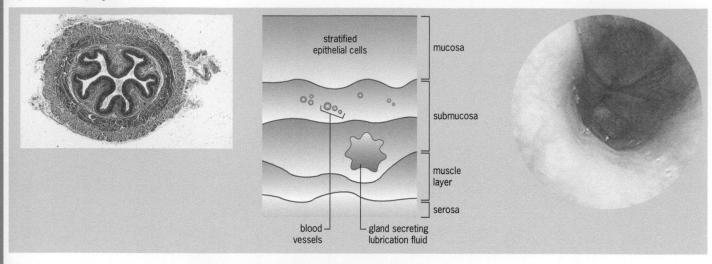

The stomach

The stomach has deep ridges called **rugae** that help with the mechanical breakdown of food (Fig 8.3).

Overall, the stomach:

- mixes food with gastric juice by muscular action;
- retains food, giving enzymes time to act;
- digests proteins through the action of the enzyme pepsin;
- curdles milk with the enzyme rennin, absorbs some simple chemicals such as water, salts as ions, and alcohol.

The muscle layer is thick; three different layers of muscle run in different directions. This gives the stomach the power to contract and relax to churn food up. Glands in the **gastric pits** produce a strong acid that reduces the pH of the stomach contents. The stomach also produces large quantities of **mucus** to protect its own cells from damage from the acid.

 REMEMBER THIS

Note that all regions of the gut have two layers of muscle, apart from the stomach, which has three. The extra *oblique* layer runs at 45° to the other two, and helps churn the food.

Fig 8.3 The stomach

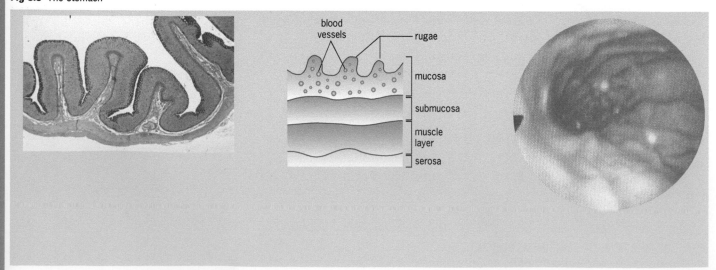

The duodenum and ileum

The small intestine is about 5 metres long and is made up of two main parts, the duodenum and the ileum (Figs 8.4 and 8.5). The duodenum is the first 25 cm of the small intestine, the ileum forms the rest of it. While the duodenum is the main site of digestion, most absorption takes place in the ileum.

The small intestine:

- moves food from the stomach to the large intestine by peristalsis – food moves through the intestine at about 1 cm per minute;
- secretes the large amounts of water necessary for most of the chemical reactions involved in digestion;
- completes the digestion of carbohydrates, proteins and fats;
- absorbs most of the small soluble food molecules produced by digestion.

The pancreatic duct and bile duct join with the duodenum, releasing digestive enzymes and bile into the lumen. This neutralises stomach acid, and creates a pH that is optimal for digestion. The wall of the small intestine looks less folded than that of the stomach, but looking at it under a light microscope reveals the tiny microscopic infoldings that increase the surface area. The tiny folds are called villi; there are 4–5 million of them in the ileum. They allow intimate contact between the wall of the intestine and the products of digestion inside this part of the gut, ensuring that the absorption of digestion products is as efficient as possible.

Fig 8.4 The duodenum

Fig 8.5 The ileum

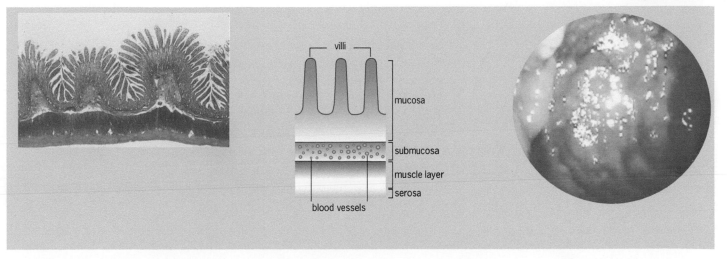

The colon

The ileum joins the large intestine at the caecum, near the appendix (look back to Fig 8.1). The **colon**, **rectum** and **anus** make up the final section of the digestive tract. Sodium ions leave the lumen of the colon and re-enter the blood by active transport; water follows by osmosis. The colon therefore absorbs water from the undigested food that reaches it. What is left in the colon is called **faeces**. As well as undigested food, faeces contain dead cells, bacteria and bile pigments. These wastes are then **egested** from the anus during **defecation**.

Fig 8.6 The colon

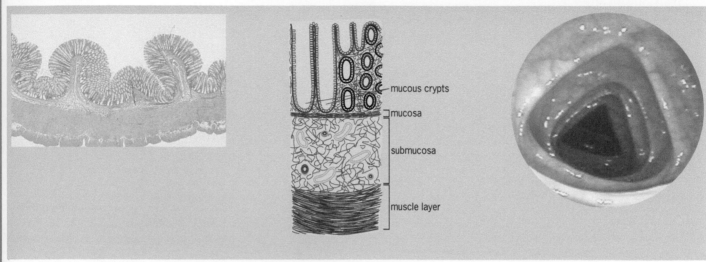

mucous crypts

mucosa

submucosa

muscle layer

3 INGESTION AND MECHANICAL BREAKDOWN OF FOOD

Whatever your table manners are like, you ingest food by getting it into your mouth. Your lips and tongue then mix the food with saliva and the chewing action of grinding the food between your teeth breaks up the food into small pieces. The technical term for this is **mastication**.

Adding saliva to food softens it, making it easier to swallow. Salivary glands (Fig 8.7) produce saliva at the rate of about 1 to 1.5 litres per day. It is produced constantly, but more is released when we see, smell, taste or even just think about food.

Saliva is mainly water (99.5 per cent) with some dissolved substances (0.5 per cent) including:

- **mineral salts** such as phosphates and hydrogencarbonates;
- **salivary amylase**, a starch-digesting enzyme that breaks molecules of starch into maltose;
- **mucin**, a slimy glycoprotein lubricant;
- **lysozyme**, an enzyme that kills bacteria.

Salivary amylase, the starch-splitting enzyme in saliva, begins the process of chemical breakdown. However, the speed at which most people chew and swallow their food means that salivary amylase has little chance to act, which is probably why people who lack the gene to make this enzyme are rarely aware of their 'genetic deficiency'.

REMEMBER THIS

The mouth is lined by stratified epithelium (epithelial cells stacked on top of each other). This protects the deeper tissues of the mouth from friction damage and has a very high turnover: we replace the lining of our mouth – and the rest of the gut – every 24 hours or so.

SWALLOWING

Swallowing is not a voluntary action, but it is under some voluntary control because you can decide to swallow if you want to. When you are eating, the act of moving food to the back of your throat is the voluntary action but swallowing is a reflex response to food or liquid touching the back of your throat.

Theoretically, when the tongue forces food or liquid to the back of the throat, the food is able to travel in one of four directions:

- back out of the mouth;
- into the nasal cavity;
- into the trachea, or windpipe;
- into the oesophagus (where it should go).

Normally, when we swallow, the first three options are closed, and food or liquid is forced into the oesophagus. If you have ever been laughing, eating crisps and drinking a fizzy drink, you may have experienced fizzy liquid going 'up your nose' or crisps going 'down the wrong way'.

The oesophagus is about 25 cm long and 2 cm in diameter. When you swallow food, smooth muscles in the oesophageal wall contract rhythmically to propel it, by **peristalsis**, from the **pharynx** to the stomach (Fig 8.8). Elastic tissue in the walls enables the oesophagus to expand and, as in the mouth, stratified epithelium protects against friction damage.

Mechanical breakdown continues in the stomach, as the churning action breaks the food up into yet smaller pieces.

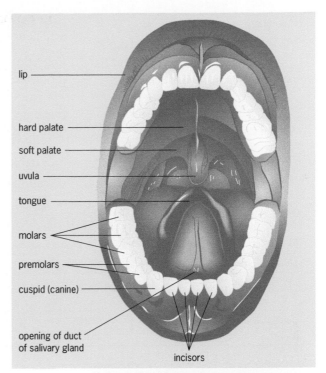

Fig 8.7 There are three pairs of salivary glands in the mouth and one pair can easily be seen. If you look under the tongue and dry the area, you will be able to see saliva oozing out

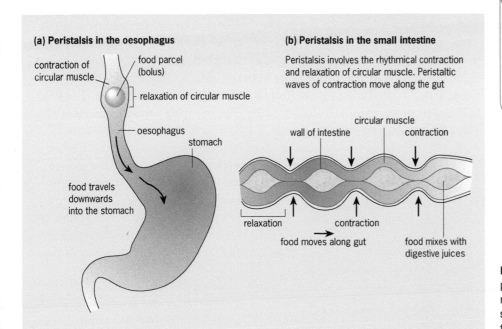

(a) Peristalsis in the oesophagus

contraction of circular muscle

food parcel (bolus)

relaxation of circular muscle

oesophagus

stomach

food travels downwards into the stomach

(b) Peristalsis in the small intestine

Peristalsis involves the rhythmical contraction and relaxation of circular muscle. Peristaltic waves of contraction move along the gut

wall of intestine

circular muscle

contraction

relaxation

contraction

food moves along gut

food mixes with digestive juices

> ✔ **REMEMBER THIS**
> The intestine, along with many other tubular organs in the body (ureters, vas deferens, uterus), is made from smooth muscle whose main function is the slow rhythmic contraction known as peristalsis.

Fig 8.8 Food is propelled through the gut by peristalsis. The squeezing action of peristalsis moves semi-solid food from the mouth to the stomach in about 4 to 8 seconds. Food moves through the intestines at a rate of about 1 cm/min

REMEMBER THIS

Organisms can only digest foods for which they have enzymes. The reason why humans can digest starch and not cellulose is simply because we can make amylase but not cellulase.

4 DIGESTION

Digestion is usually thought of as the chemical breakdown of complex food molecules into simpler ones that can be absorbed and used by the body. Enzymes are key players in this process; they control all of these breakdown reactions. Table 8.1 gives an overview of the main enzymes involved in human digestion.

In the first part of this section, we look at where the different enzymes act, and what they do; in the following section we see how the release of these secretions is controlled.

Table 8.1 A summary of the main human digestive enzymes

Secretion	Enzymes produced	Site of production	Site of activity	pH	Substrate	Products
saliva	salivary amylase	salivary glands	mouth	6.5–7.5	starch	maltose
gastric juice	pepsin	stomach	stomach	2.0	proteins, polypeptides	
	rennin				caseinogen (milk protein)	
pancreatic juice	trypsin	secretory cells of the pancreas (acini)	duodenum	7.0	proteins, polypeptides	short polypeptides
	chymotrypsin				proteins	polypeptides
	carboxypeptidase				polypeptides	dipeptides, amino acids
	pancreatic amylase				starch	maltose
	maltase				maltose	glucose
	sucrase				sucrose	glucose, fructose
	lactase				lactose	glucose, galactose
	lipase				fats and oils	fatty acids, glycerol

? QUESTION 1

1 A person who has developed stomach cancer may have to have their stomach removed. Suggest what effect this will have on their eating habits.

DIGESTION IN THE STOMACH

The stomach is a muscular sac under the diaphragm. When empty, it is the same size as a large sausage, but it can stretch to the size of a melon. The stomach is the first area of any significant digestion.

Food remains in the stomach for between 30 minutes and 4 hours depending on the type of meal (fatty meals stay there the longest). The food is digested mechanically and chemically. The resulting semi-liquid material, **chyme**, passes into the duodenum, the first part of the small intestine.

Specialised groups of cells inside **gastric pits** in the mucosa secrete **gastric juice** (Fig 8.9). Each type of cell produces a specific secretion:

- **Oxyntic cells** secrete a solution of hydrochloric acid which brings the pH of gastric fluid down to between pH 2.0 and 3.0.
- **Zymogen cells** (peptic cells) secrete the enzyme pepsinogen which is later converted into the protein-splitting enzyme, pepsin.
- **Mucous cells** secrete the mucus that protects the stomach lining from the digestive action of its own secretions.

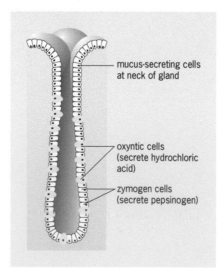

mucus-secreting cells at neck of gland

oxyntic cells (secrete hydrochloric acid)

zymogen cells (secrete pepsinogen)

Fig 8.9 A gastric pit from the mucosa of the stomach showing the specialised cells that secrete gastric juice

The importance of stomach acid

The hydrochloric acid in gastric fluid:

- provides the optimum pH for **pepsin** and **rennin**;
- denatures proteins and helps to soften tough connective tissue in meat;
- is a strong **bactericide** (it kills bacteria) and so protects the body from some of the harmful microbes that might enter the body in food.

Protein digestion in the stomach

One of the main functions of the stomach is to begin to digest proteins. **Pepsin**, is a powerful **endopeptidase enzyme**, digests proteins. This enzyme breaks peptide bonds in the middle of the protein chain, turning protein molecules into polypeptides. Protein digestion is completed when **exopeptidase** enzymes remove amino acids from the ends of the polypeptides.

The stomach avoids digesting its own tissue by secreting pepsin in an inactive form, **pepsinogen**. This is converted to pepsin in the lumen of the stomach only after contact with hydrochloric acid. The hydrogen ions in the acid cause the pepsinogen to unfold and become pepsin, the active form of the enzyme.

Milk digestion

Caseinogens, the proteins in milk, are water soluble. They are valuable nutrients (milk is the sole source of food for young mammals; see the Science in Context box below) but, if they remained in their soluble form, they would leave the stomach before protein digestion had finished. To avoid this, the stomach produces **rennin**. This curdles milk, converting soluble caseinogen into insoluble **casein**. Like pepsin, rennin is also secreted in an inactive form, **prorennin**. Like pepsinogen, prorennin is converted to its active form by contact with stomach acid.

? QUESTIONS 2–3

2 At what pH would you expect the human form of the enzyme pepsin to display optimum activity?

3 What is the difference between an endopeptidase and an exopeptidase enzyme?

Proteins, polypeptides and amino acids are discussed in Chapter 3.

? QUESTION 4

4 **a)** Would you expect rennin production to increase or decrease with age in mammals? Give a reason for your answer.
b) Would you expect rennin to be found in animals other than mammals?

SCIENCE
IN CONTEXT

Lactose

Like all mammals, humans feed their young on milk (Fig 8.10). The main carbohydrate in breast milk is the disaccharide lactose. But why should a mother go to the trouble of combining two monosaccharides into lactose when the baby will simply have to break it down again to use it as an energy source?

The answer lies in the size of the lactose molecule. In the mother, milk is made in the mammary gland and is stored there between feeds. Since it is a disaccharide, lactose stays in the milk, rather than diffusing into the surrounding tissue as smaller monosaccharides would tend to do. In the baby, lactose remains in the gut for the same reason and is broken down only gradually to form glucose. So there is a steady

absorption of glucose into the infant's blood, rather than a sudden surge at feeding time.

Several medical conditions can arise because of a failure to digest lactose. In adult life, some people stop making the enzyme lactase and so cannot break down lactose. Undigested lactose cannot be absorbed and it accumulates in the gut, encouraging the growth of bacteria that produce large amounts of carbon dioxide and lactic acid. The result is diarrhoea and wind. Affected people are said to be **lactose intolerant**.

A type of lactose intolerance that results in far more serious problems is found in people with an inherited condition called **galactosaemia**. Affected people can break down

lactose into glucose and galactose but their liver cannot convert galactose to glucose. Galactose builds up to dangerous levels and the condition can be fatal.

Fig 8.10 Milk contains all of the food chemicals that a new baby needs. The main carbohydrate in milk is lactose

DIGESTION IN THE SMALL INTESTINE

The small intestine is the major site of digestion. Two secretions, **pancreatic juice** from the pancreas and **bile** from the liver, carry on the digestive processes that began in the mouth and stomach. In addition, the digestive process is finished off by enzymes in the intestinal lining itself.

Pancreatic juice

Pancreatic juice enters the duodenum via the pancreatic duct. Over a litre of alkaline pancreatic juice is secreted every day. The fluid contains several enzymes that are produced by secretory cells in the acini of the pancreas:

- **Endopeptidase enzymes** (including trypsin and chymotrypsin) carry on the protein digestion, which started in the stomach, by cutting long chains of amino acids into shorter ones.
- **Exopeptidase enzymes** complete protein digestion by removing the terminal (end) amino acids from the chains.

Note: these two types of enzyme work together. Endopeptidase enzymes make more 'ends' for exopeptidase enzymes to work on.

- Amylase enzymes hydrolyse starch into maltose;
- Lipase enzymes hydrolyse triglycerides into fatty acids and glycerol.

Hydrogencarbonate ions make pancreatic juice slightly alkaline (pH = 7.1 to 8.2), allowing it to neutralise the stomach acid and helping to create optimum conditions for the intestinal digestive enzymes.

The pancreas and liver both have other functions that are not connected directly with digestion. The pancreas produces the hormones insulin and glucagon and the liver has many regulatory – homeostatic – functions.

Bile

Liver cells called **hepatocytes** produce thick yellow-brown or olive green bile. This contains:

- water;
- **bile salts** (sodium glycocholate and sodium taurocholate);
- **bile pigments** (breakdown products of red blood cells, for example bilirubin) and some mucus;
- cholesterol.

Liver cells produce around 0.8 to 1.0 litre of bile daily. Secretions from individual cells pass into tiny canals called **bile cannaliculi**. These lead to the gall bladder, a small sac-like organ, which stores the bile until it is needed. Bile is released into the duodenum when the muscle wall of the gall bladder contracts. Bile reaches the duodenum through the bile duct. Bile:

- **emulsifies** fats (breaks large fat or oil droplets into an emulsion of microscopic droplets) – this process massively increases the surface area available for fat-digesting enzymes;
- neutralises the (acidic) chyme from the stomach and creates the ideal pH for intestinal enzymes;
- stimulates peristalsis in the duodenum and ileum;
- allows the excretion of cholesterol, fats and bile pigments.

The structure of the pancreas is discussed in Chapter 11.

 REMEMBER THIS

We know that enzymes are very specific, so how can just a few enzymes digest such a wide variety of foods? The answer is that the same types of bond are found in many foods. For instance, the enzyme trypsin will hydrolyse the peptide bond between the amino acids argenine and lysine. These two amino acids are found together in many protein food such as pork, chicken, beef, egg and soya.

The functions of the pancreas and the liver are discussed in Chapter 11.

? QUESTION 5

5   Why do the protein-splitting enzymes trypsin and chymotrypsin need to be produced in an inactive form?

CARBOHYDRATE DIGESTION

Carbohydrate digestion is completed by enzymes fixed in the membranes of the epithelial cells. The folded membrane in the epithelium is known as a brush border. The three vital **brush border hydrolase enzymes** are maltase, lactase and sucrase.

- Maltase hydrolyses maltose into two molecules of glucose.
- Lactase hydrolyses lactose into a glucose molecule and a galactose molecule.
- Sucrase hydrolyses sucrose into a glucose molecule and a fructose molecule.

CONTROL OF DIGESTION

Since we do not eat food continuously, it is important that we produce digestive enzymes only when food is in the gut. If large quantities of the acids and protein-attacking enzymes were released into an empty stomach, there would be a real danger that you would digest yourself.

Control of gastric secretions

Nerves and hormones control the secretion of gastric juice in a process that has three distinct phases (Fig 8.11):

- **Nervous phase.** The sight, smell or taste of food initiates a nerve reflex in which impulses from the brain trigger gastric glands to release their secretions.
- **Gastric phase** (hormonal). Food in the stomach stimulates the lining to secrete the hormone **gastrin**. Gastrin increases gastric juice secretion through direct action on the gastric glands.
- **Intestinal phase** (hormonal). The duodenal lining, stimulated by partially digested food, produces a second hormone, **enteric gastrin**. This hormone also acts on the stomach's gastric glands, producing further small amounts of gastric juice.

Pressure sensors and chemical sensors in the stomach detect stomach stretching and the presence of chyme. The resulting nerve impulses, together with the action of gastrin, direct the stomach to start emptying its contents into the duodenum.

Control of pancreatic secretions

Pancreatic secretion, like gastric secretion, is controlled by both nervous and hormonal mechanisms. Impulses travel from the brain down the **vagus nerve** to trigger the secretion of pancreatic juice.

The lining of the duodenum secretes two hormones in response to acidic chyme arriving in the lumen of the duodenum. **Secretin** stimulates the pancreas and liver to secrete pancreatic juice rich in hydrogencarbonate ions. **CCK-PZ** (**cholecystokinin-pancreozymin**) stimulates the gall bladder to release bile and the pancreas to release digestive enzymes.

Control of bile secretion

The hormone secretin acts together with nervous stimulation by the vagus nerve to increase the rate of bile secretion. The acidity of chyme in the duodenum and the hormone CCK-PZ stimulate the gall bladder to contract.

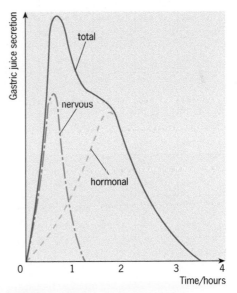

Fig 8.11 Secretion of gastric juice is partly controlled by a nerve (the vagus nerve) and partly by a hormone (gastrin). Gastrin stimulates gastric glands, causing them to release their secretions. It increases movement in the tract and relaxes the sphincters that control the movement of food into and out of the stomach

5 ABSORPTION OF THE PRODUCTS OF DIGESTION

Food absorption takes place mainly in the ileum of the small intestine. The small soluble molecules produced by digestion in the stomach and duodenum can be easily absorbed through the gut lining. The epithelial lining has a much greater surface area than that of an equivalent smooth-sided tube; an average 70 kg man has approximately 100 square metres of absorbing surface in his small intestine. This is possible because of three important structural features:

- folds in the inner surface of the intestinal wall;
- movable projections (0.5–1.0 mm in length) called **villi** (Fig 8.12) that are present on the folded surface of the wall;
- microscopic projections called **microvilli** on the cell surface membranes of epithelial cells that line the villi.

The epithelial cells absorb amino acids, monosaccharides, fatty acids and glycerol, water and other substances including vitamins, nucleic acids, ions and trace elements. The type of transport process used varies depending on the substance, but **diffusion**, **facilitated diffusion** and **active transport** are all involved.

> Transport processes for moving molecules into and out of cells are discussed in Chapter 4.

ABSORPTION OF CARBOHYDRATES

Glucose is actively transported into the cells of the small intestine and across the walls of blood capillaries. Active transport is necessary because diffusion alone would be too slow to supply the body's needs. Also, there would be diffusion out of the epithelial cells if the concentration of food in the intestines were very low. Energy for the transport is obtained from the hydrolysis of ATP.

ABSORPTION OF PROTEINS AND FAT

Amino acids and small peptide molecules (di- and tripeptides) are actively transported into epithelial cells of the small intestine.

Bile salts and lipase enzymes break up complex lipid molecules into monoglycerides and free fatty acids. The monoglycerides and some of the fatty acids combine with bile salts to form microscopic droplets called **micelles**. Micelles are soluble in water and diffuse easily into epithelial cells, along with the remaining fatty acids. Inside the cells, the fatty acids and glycerol recombine, forming triglycerides, which acquire a protein coat (that stops them sticking together) and form particles called **chylomicrons**. It is in this state that lipids leave epithelial cells and enter, not a blood capillary this time, but a branch of the lymphatic system called the **lacteal**.

Fig 8.12 Epithelial cell and villus

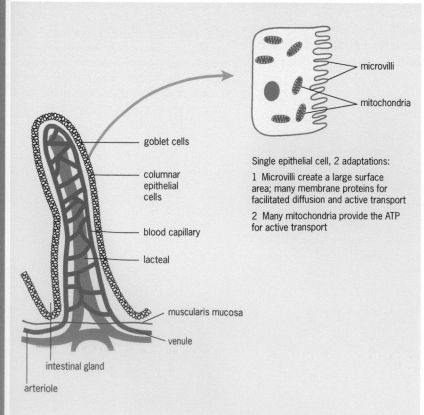

goblet cells

columnar epithelial cells

blood capillary

lacteal

muscularis mucosa

venule

intestinal gland

arteriole

microvilli

mitochondria

Single epithelial cell, 2 adaptations:

1 Microvilli create a large surface area; many membrane proteins for facilitated diffusion and active transport

2 Many mitochondria provide the ATP for active transport

TRANSPORT OF ABSORBED FOOD PRODUCTS

Amino acids, dipeptides and simple sugars diffuse out of epithelial cells in the small intestine, directly into blood capillaries within the villi. From here they pass to the liver via the **hepatic portal** vein. The liver processes absorbed food, converting some into storage products. Glycogen, copper, iron and vitamins A, D, E and K can all be stored. The liver breaks down other food products, including excess amino acids which are **deaminated** – their amine groups are removed. The breakdown products are then passed to the kidneys for excretion.

The lymphatic system plays a major role in the transport of absorbed lipids. Chylomicrons that have entered the lacteals remain in suspension, giving the lymph fluid a milky-white appearance. From here, the chylomicrons move into larger lymphatic vessels that eventually drain into a large duct that empties into the blood. In the blood, these complexes are broken down into fatty acids and glycerol and enter cells to be used in the synthesis of complex lipids.

Liver function is discussed in Chapter 11. Kidney function and excretion are discussed in Chapter 13.

EXAMPLE

Q Explain how the structure of the ileum relates to its function of absorbing the products of digestion.

A Relating structure to function is one of the basic skills a biologist must master. The structure of the small intestine is a classic example.

When faced with this question, many students will simply state a well-learned GCSE list of 'large surface area, good blood supply, thin walls, etc.', and expect full marks. At A-level a little more depth is required. Why is a large surface area vital? What does a good blood supply do?

Large surface area

The basic idea behind a large surface area is that a lot of intestine is in contact with a lot of digested food molecules. These are absorbed by diffusion or active transport and so the more membrane there is, the more pores and carrier proteins there are available. The fact that the intestine is long is the most obvious way to increase the surface area, but there are also the villi and the microvilli.

Good blood supply

Like most organs which are adapted for the exchange of materials, there is also a way of maintaining the diffusion gradient so that there is always more digested food on one side of the absorbing membrane than the other, thus ensuring that absorption is continuous and as rapid as possible. As soon as the digested food molecules are absorbed into the blood, they are taken away and replaced with blood containing fewer of these molecules.

Thin walls

Fick's law (Chapter 4) states that the speed of diffusion is inversely proportional to the diffusing distance. Put simply, the thinner the membrane, the faster the diffusion, and so the barrier between food and blood is as thin as possible, being as little as a few micrometres – the thickness of the epithelial cell and the cell lining the capillary.

The small intestine also has another feature of note: the epithelial cells contain many mitochondria which make the ATP needed to power the active transport mechanisms.

6 EGESTION

Egestion of undigested food from the body occurs after it has been through the large intestine. This is about 1.5 m long and extends from the ileum to the anus. It has four sections, the **caecum**, the **colon**, the **rectum** and the **anus**. The large intestine:

- absorbs water;
- is the site of manufacture of certain vitamins (microorganisms within the colon of some animals produce vitamin K and folic acid);
- forms and expels undigested food residue as faeces.

The caecum receives material from the ileum. In humans, the bottom of the caecum is attached to a small blind tube, the **appendix**. This tube is twisted and coiled and is about 8 cm long. The human appendix plays no part in digestion, but can become inflamed, causing appendicitis.

Much of the water that is poured on to food during its passage through the digestive tract (e.g. as saliva, gastric juice, pancreatic juice) is reabsorbed. This reabsorption is vital: the digestive system pours up to 8 litres of fluid into the gut each day. If we lost all this fluid we could dehydrate. Most water absorption occurs in the small intestine, but of the litre or so that enters the large intestine, all but 100 cm^3 is reabsorbed by the colon. Minerals (as ions) can also diffuse or be actively transported into the bloodstream from this segment of the large intestine.

Diseases such as cholera and dysentery that cause severe diarrhoea can result in potentially lethal dehydration within a very short time because the toxins produced by the bacteria that cause these diseases reverse the reabsorption of water. To combat this, oral rehydration therapy can be given, a mixture of glucose and salts in water – see the Science Feature box on page 203.

The final sections of the large intestine, the rectum and anus, are concerned with the compaction of faeces and the act of **defecation**.

? QUESTION 6

6   Why are the nutrients produced by colon bacteria (in the large intestine) of less use to us than nutrients provided by bacteria in the small intestine?

✔ REMEMBER THIS

Faeces contain some water, mineral salts, undigested food, bile pigments, products of bacterial decomposition and epithelial cells that have become detached from the digestive tract.

SUMMARY

After reading this chapter, you should be able to read and understand the following:

- The human **alimentary canal** is a tube that leads from the mouth to the anus. It is associated with glands such as the **salivary glands**, **pancreas** and **liver**.
- The alimentary canal is also called the **gut** or **digestive system**. It is responsible for the **ingestion**, **digestion**, **absorption** and **egestion** of food.
- Food is broken down **mechanically** by the chewing action of teeth and by the muscular churning of the alimentary canal.
- Digestive juices are poured on to food as it travels down the alimentary canal. These soften and lubricate the food and also contain **digestive enzymes** that break the food down **chemically**.
- The muscles of the alimentary canal 'squeeze' the food – a process called **peristalsis** – and push it along from one end to the other.
- Large food molecules are broken down to their building blocks: carbohydrates are digested to form simple sugars, proteins are broken into amino acids, and fats are converted to fatty acids and glycerol.
- Digestion of food in the small intestine is due to the actions of bile and pancreatic juice, and the process is completed on the surface of the epithelial cells themselves.
- Digested food products are absorbed into the body across the wall of the alimentary canal. Most enter the bloodstream, but the products of fat breakdown enter the **lymphatic system**.
- Most digestion occurs in the **duodenum**; most absorption occurs in the **ileum**.
- A sophisticated control system ensures that digestive enzymes are released in the right place at the right time.
- Water is absorbed from undigested food that enters the colon, turning it into **faeces** that are egested from the body through the **anus**.

Practice questions and a How Science Works assignment for this chapter are available at www.collinseducation.co.uk/CAS

9

Gas exchange

9 GAS EXCHANGE

The endless pathways in the human lung

Diseases of the lung such as asthma, bronchitis and emphysema can be investigated at first hand thanks to the endoscope, a camera on the end of a flexible tube. This technology allows doctors to investigate parts of the body – such as the lungs, the stomach and the intestines – without the need for surgery. Not only does it relay enlarged pictures to a TV screen, but tools on the end of the device can take samples of tissue (such as tumours), remove blockages and even perform small operations.

Having inserted an endoscope into a person's airway, the doctor's first decision is whether to turn into the right or left lung. Here, he or she could look for obstructions, tissue damage or tumours. In practice, an endoscope can be used only for the larger branches, the trachea and the bronchi. If the camera were small enough to go right into a terminal alveolus, the endoscope operator would face another 23 decisions of left or right, such is the extent of the branching of the airways.

In total, there are about 2^{23} terminal bronchioles that lead from the trachea to the alveoli of the human lung. The number of alveoli is even larger, about 300 million. The combined surface area for gas exchange is huge, about the size of a badminton court.

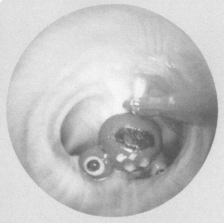

An endoscope picture that shows a toy frog stuck at the top of a child's airway

---

Organic molecules like glucose are discussed in Chapter 3.

✔ REMEMBER THIS

A summary of the overall process of respiration:

glucose + oxygen

↓

carbon dioxide + water + energy
(in ATP)

This is why we need to breathe.

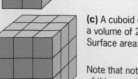
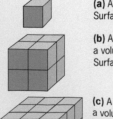

(a) A 1 cm³ cuboid organism. Surface area: volume ratio = 6:1

(b) A cuboid organism with a volume of 8 cm³. Surface area: volume ratio = 3:1

(c) A cuboid organism with a volume of 27 cm³. Surface area: volume ratio = 2:1

Note that not all the cells of this organism are on the surface

Fig 9.1 These simple shapes illustrate the problems faced by an organism as it gets larger. Think of each block as a cell and each complete cube as an organism. You can see that increasing the size of the organism lowers the surface area available to each cell. If the third 'organism' could not evolve a system to transport materials to the cell in the middle, it would not have much chance of survival

1 GAS EXCHANGE

Humans, like all living organisms, need energy to carry out the processes of life. In particular, we need energy for movement, growth and repair and to keep our body temperature stable. We obtain this energy from the oxidation of organic molecules such as glucose in the process of **respiration**. Respiration uses glucose and oxygen and produces carbon dioxide, water and energy in the form of adenosine triphosphate (ATP). This goes on constantly in every living cell of our bodies and, to keep the process going, we must obtain a constant supply of oxygen and expel the carbon dioxide as waste. We humans are also large warm-blooded organisms, and so we need a massive amount of oxygen to supply our metabolic demands.

Compared to water, air contains a lot of oxygen and is not very dense, so it is relatively easy for us to move several litres in and out of our bodies in a single breath. Our highly developed lungs are adapted to bring enough air sufficiently close to the blood to allow rapid diffusion to take place.

In this chapter, we look at the structure of the lungs, the mechanism of ventilation, the process of gas exchange and, finally, the way in which our breathing rate changes according to the varying demands of the body.

The surface area and volume problem

Although the amount of material an organism *needs* to exchange is proportional to its *volume*, the amount of material it *can* exchange is proportional to its *surface area*. Fig 9.1 shows that as organisms get larger, their volume increases at a faster rate than their surface area and so their need to exchange materials could quickly outstrip their ability to do so.

Organisms with a diameter larger than a few millimetres can survive only if they evolve strategies to increase their surface area:volume ratio. Many species of worm have evolved a flat or cylindrical body. However, this sort of adaptation allows only a limited size increase. Much larger organisms have developed specialised organs to increase the surface area that is available for exchanging materials. Lungs, gills, intestines and kidneys are all examples of such organs.

The evolution of specialist exchange organs is linked to the development of a **transport system**. Without it, exchanged materials cannot reach other parts of the body quickly enough.

2 STRUCTURE OF THE LUNGS

The intestines, kidneys and lungs, organs that are adapted for exchange of materials, all show several features that maximise the exchange process. The **lungs** are responsible for exchanging gases between the blood and the air outside the body (Fig 9.2). Oxygen enters the body by diffusion, and carbon dioxide leaves in the same way. The lungs increase the efficiency of diffusion by having:

- **A large surface area**. The huge surface area for gas exchange is the result of many little air pockets called alveoli, which are surrounded by many blood capillaries.
- **Thin, permeable walls**. The gases have only to pass through two thin cell layers, thus ensuring that the diffusing distance is as short as possible.
- **A good blood supply**. The lungs receive as much blood per minute as the rest of the body put together. Blood transports incoming oxygen

REMEMBER THIS
Gas exchange always occurs by diffusion alone, not osmosis, active transport or facilitated diffusion.

REMEMBER THIS
It is not correct to list 'moist' as an adaptation for gas exchange. Gas exchange surfaces must be permeable to small molecules, including water. Being moist is a side effect, not an adaptation.

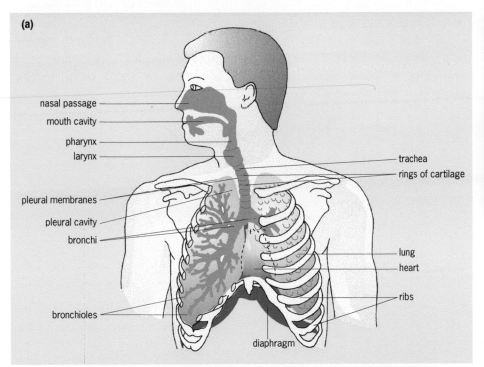

(a)

nasal passage
mouth cavity
pharynx
larynx
trachea
rings of cartilage
pleural membranes
pleural cavity
bronchi
lung
heart
ribs
bronchioles
diaphragm

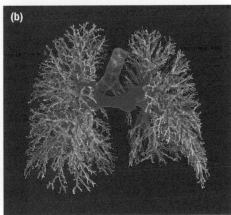

(b)

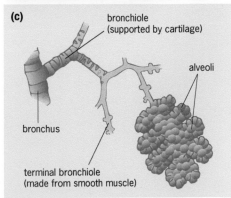

(c)

bronchiole (supported by cartilage)
alveoli
bronchus
terminal bronchiole (made from smooth muscle)

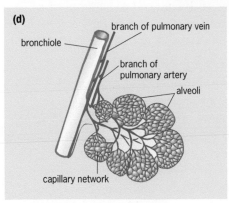

(d)

branch of pulmonary vein
bronchiole
branch of pulmonary artery
alveoli
capillary network

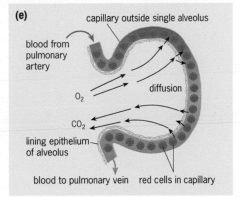

(e)

capillary outside single alveolus
blood from pulmonary artery
O_2
diffusion
CO_2
lining epithelium of alveolus
blood to pulmonary vein   red cells in capillary

Fig 9.2 The structure of the human gas exchange system
Fig 9.2a The overall structure
Fig 9.2b A resin cast of human lungs
Fig 9.2c, d and **e** The fine structure of the alveolus and the mechanism by which gas is exchanged

145

> ✔ **REMEMBER THIS**
> Alveoli don't collapse when we breathe out because their surface is covered by an anti-sticking substance called surfactant (see the How Science Works box on page 149).

Fig 9.3 The mechanism of breathing

Fig 9.3a Inspiration (breathing in) happens when the external intercostal muscles contract and pull the ribcage upwards and outwards, away from the spinal column. At the same time, the diaphragm contracts and flattens, pushing down on the abdominal organs such as the liver. These movements increase the volume and therefore lower the pressure in the thorax. As the pressure in the thorax falls below that of the atmosphere, air is forced into the lungs to equalise the pressure

Fig 9.3b Expiration (breathing out) is normally a passive process: it uses no energy. When inspiration is over, the muscles in the thorax relax and breathing out follows due to a combination of gravity, the elastic recoil of the connective tissues of the lungs and the pressure exerted by the abdominal organs. We can consciously speed up expiration by forcing air out of our lungs using our internal intercostal muscles. This happens, for instance, when we blow up a balloon or play a wind instrument

Fig 9.3c The pressure changes associated with breathing

away quickly to a different part of the body. This ensures that a diffusion gradient is always present: there is always less oxygen in the blood that flows through the lungs than there is in the air.

- **Efficient ventilation**. This is a physical pumping mechanism, usually called breathing, that continually brings fresh air to the gas exchange surfaces in the alveoli and keeps the diffusion gradient as large as possible.

3 MECHANISM OF VENTILATION

Lungs have no muscle – they cannot move on their own – so how do we breathe? Lungs **inflate** and **deflate** because they are *expanded* and *compressed* by movements of the ribcage and diaphragm. This is possible because two **pleural membranes** hold the lungs to the ribcage and diaphragm. The outer membrane lines the **thoracic cavity**: the inner membrane encloses the lungs. Between the two is a very narrow space, the **pleural cavity**, filled with **pleural fluid**. This allows the pleural membranes to slide smoothly over each other during breathing, and prevents them from separating. Fig 9.3 shows how we breathe in and out.

To understand why two pleural membranes continue to stick together during breathing, imagine two wet pieces of glass pressed together. They can easily slide over each other, but it is virtually impossible to pull them apart without introducing air into the middle. If air is introduced into the pleural cavity, after a stab wound for example, the lung collapses, a situation known as a **pneumothorax**. The pleural cavity is filled with a foamy mixture of air and blood, while the elastic lungs shrink back from the thorax wall. The ribs can still move (though it is painful) but they don't inflate the lungs.

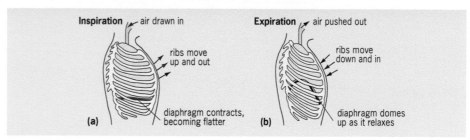

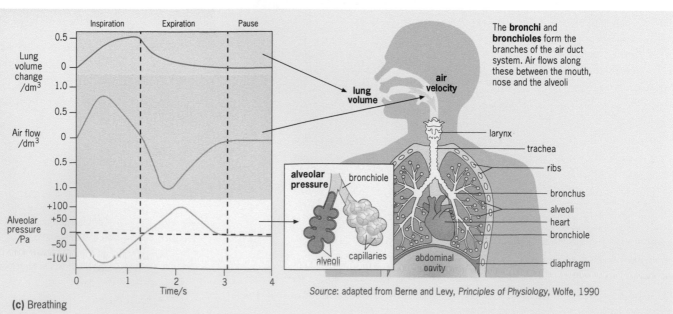

(c) Breathing

Source: adapted from Berne and Levy, *Principles of Physiology*, Wolfe, 1990

USING A SPIROMETER TO DETERMINE LUNG VOLUME

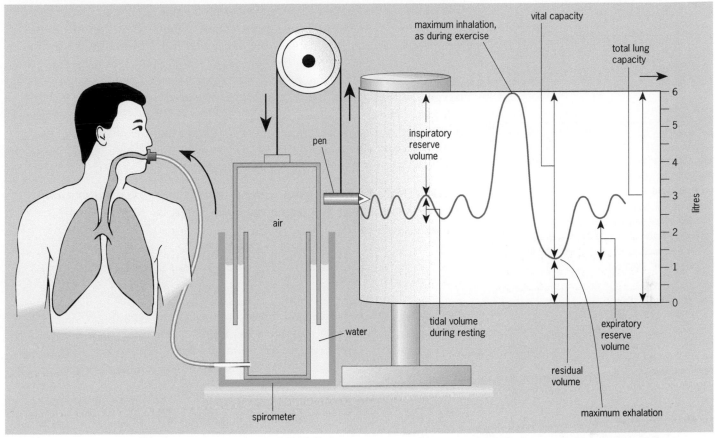

Fig 9.4 A spirometer and the trace it produces

Fig 9.4 shows a **spirometer**, a device for measuring the volumes of air that a person breathes in and out. The trace that this apparatus produces tells us a lot about lung volume.

The **total capacity** of the lungs is the amount of air that they can hold during the deepest possible breath. It is not possible to exhale all of it, however, because there must be some left in the airways (which are held open by rings of cartilage) and the alveoli can never collapse completely. The volume of air that remains in the lungs after breathing out is called the **residual volume**. The maximum usable lung volume (the total lung capacity minus the residual volume) is called the **vital capacity**. So if you breathe in as far as you can, then exhale as much as you can, that's your vital capacity. The average vital capacity is 4.5 to 5 litres for men and 3.5 to 4 litres for women.

Secondly, during normal breathing, the volume of air that moves in and out of the lungs in each breath is called the **tidal volume**. In a normal adult at rest, this is about 0.5 litres.

Thirdly, the trace shows that after breathing in at rest, the person could inhale an extra 3 litres, a value known as the **inspiratory reserve volume (IRV)**. He or she could also breathe out another litre or so: the **expiratory reserve volume (ERV)**. These are the extra volumes of air that we breathe in and out during exercise.

Finally, we can use the trace to work out the **ventilation rate**: the volume of air taken into the lungs in one minute. To do this, multiply the number of breaths taken in a minute by the tidal volume.

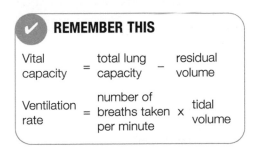

✔ REMEMBER THIS

$$\text{Vital capacity} = \text{total lung capacity} - \text{residual volume}$$

$$\text{Ventilation rate} = \text{number of breaths taken per minute} \times \text{tidal volume}$$

? QUESTION 1

1 If the average breathing rate is 15 breaths per minute, and the tidal volume is 0.5 litre, calculate the ventilation rate.

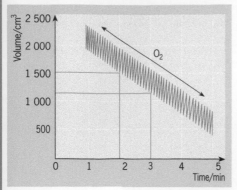

Fig 9.5 In addition to lung volume, the spirometer can also be used to measure the volume of oxygen used. If the carbon dioxide is absorbed as soon as it is exhaled, the total volume of air falls, showing how much oxygen is being used. We can use the measurements to determine how much oxygen is used at rest, and to see how this changes with exercise

? **QUESTION 2**

2 The alveolar membrane is 0.2 μm thick. What fraction of a millimetre is that?

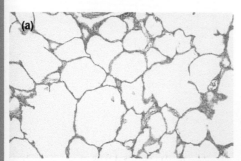

Fig 9.6a A micrograph of healthy mammalian lung tissue, showing the thin alveolar epithelium and large surface area for gas exchange

Fig 9.6b Gas exchange at the alveolar surface. In the human lung, the alveolar endothelium, the membrane that lines the air-sacs, is only 0.2 μm thick. This minimises the distance oxygen has to diffuse to enter the blood. Breathing maintains a fresh supply of air and the circulatory system takes away oxygenated blood so that constant gas exchange is possible

Haemoglobin and red blood cells are discussed in Chapter 10.

USING A SPIROMETER TO DETERMINE OXYGEN CONSUMPTION

We can use a spirometer to estimate a person's rate of oxygen consumption. If the person re-breathes the air in the closed system of the spirometer, the composition of that air changes: the oxygen level decreases and the carbon dioxide level increases. Soda lime placed in the apparatus absorbs the carbon dioxide and so the volume of air in the spirometer decreases in volume as the oxygen is used up. Fig 9.5 shows a trace produced in this way.

To calculate the amount of oxygen used, we measure the volume decrease in a given time. In our example, the trace shows that the air volume fell from the 1600 cm³ mark to the 1300 cm³ mark in one minute. So, this person used up 300 cm³ of oxygen in one minute.

WARNING! Re-breathing your own air can be dangerous. You should never do investigations such as this without close supervision.

4 GAS EXCHANGE AT THE ALVEOLI

Gas exchange between air and blood occurs at the alveoli (Fig 9.6). These tiny air-sacs create a huge surface area: 1 cm³ of frog lung tissue has a surface area of about 20 cm², but the corresponding figure for a mouse lung is over 800 cm². Other mammals have a similar value. The total surface area of one human lung is about 70 to 100 m².

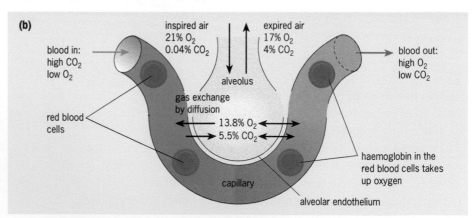

As we breathe in, fresh air enters the lungs and passes into individual alveoli. Oxygen diffuses rapidly through the walls of the alveoli and into the blood. Here, most of it combines with haemoglobin in red blood cells. At the same time, carbon dioxide diffuses out of the blood and into the alveoli. It is breathed out during the next few expirations.

Table 9.1 shows the composition of atmospheric, alveolar and exhaled air. Each value for exhaled air is an average of the values for inhaled and alveolar air because exhaled air is a mixture of the two.

Table 9.1 The composition of inhaled, alveolar and exhaled air

| | Percentage of total volume | | | | |
	O_2	CO_2	N_2 + inert gases	H_2O vapour	Temperature/°C
atmospheric	21	0.04	79	variable	variable
alveolar	13.8	5.5	80.7	saturated	37
exhaled	17	4	79.6	saturated	37

The composition of exhaled air varies during the course of a single expiration. The first air to emerge has a very similar composition to atmospheric air because it has been nowhere near the alveoli – it has simply filled the **dead space** in the trachea and bronchi. As the exhalation continues, air that has been deep inside the alveoli is breathed out. The composition of this air has been altered by gas exchange: it contains more carbon dioxide and less oxygen than atmospheric air (Fig 9.7).

Fig 9.7 The air we exhale is also saturated with water vapour. When you wake up on a cold day to see condensation on the windows, it is likely that a significant amount of that water came from your lungs

? **QUESTION 3**

3 Why would it be impractical to breathe under water using a 6 metre snorkel?

HOW SCIENCE WORKS

The lungs of premature babies

Alveoli are minute bubble-like air-sacs lined with moisture, and are liable to collapse because of surface tension. To prevent this, the alveolar epithelium secretes a surfactant, a mixture of phospholipids, which greatly reduces the surface tension and keeps the alveoli open.

Without surfactant, the lungs cannot function effectively and severe breathing problems can develop. An unborn baby does not start to secrete surfactant until about the 22nd week of pregnancy and the lungs have not accumulated enough surfactant to cope with breathing until about the 34th week.

Any babies born before this have immature lungs and suffer from a condition called **respiratory distress syndrome**. The effort needed to inhale and inflate the collapsed alveoli becomes too great and, without medical help, the baby can die from exhaustion and suffocation. Surfactant can now be made artificially. It is introduced into the lungs of premature babies to help them to begin breathing. This major breakthrough

means that babies as young as 23 weeks (17 weeks premature) now have more of a chance of survival (Fig 9.8).

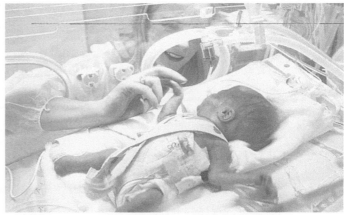

Fig 9.8 This baby was born prematurely, but survived owing to the treatment of her lungs with surfactant. Only a tiny amount (0.5 cm³) of surfactant is needed, enough to line the alveoli with a layer one or two molecules thick

THE EFFECT OF SMOKING ON GAS EXCHANGE

Smoking is the largest cause of preventable death in the western world, and is responsible for about 100 000 deaths per year in the UK. In addition to lung cancer and coronary heart disease, smoking causes bronchitis (Fig 9.9) and generally reduces the gas exchange capacity of the lungs.

Research has shown that there are two distinct types of smoke. **Mainstream** smoke passes into the smoker through the filter, **sidestream** smoke comes out of the burning end of the cigarette. In a room of smokers, about 85 per cent of the smoke is sidestream, which means that passive smokers don't get the benefit of the filter. Cigarette smoke contains three substances that have a direct effect on the lungs: nicotine, carbon monoxide and tar.

Nicotine is not the most harmful chemical in smoke, but it is highly addictive, and has a powerful effect on the nervous system (see page 237). It is thought to increase the stickiness of blood platelets, increasing the risk of blood clots forming inside arteries and veins. If these clots break away, they can cause a heart attack or a stroke.

Fig 9.9 The lungs have an efficient cleaning system: goblet cells (orange) secrete mucus that traps dust, while a carpet of cilia (tiny hairs, green) waft the mucus to the throat where it is swallowed (or to the nose, from where it is blown). Smoking disrupts this cleaning mechanism, and often leads to chronic bronchitis

Climbing Mount Everest – an explanation of partial pressures

As you study the workings of the lungs and blood in more detail, you will come across the term partial pressure. To explain this, think about someone climbing Mount Everest (height 8848 metres) (Fig 9.10a).

At sea level, there is a lot of air pushing down on us, and this atmospheric pressure has a value of about 100 000 pascals (Pa) or 100 kPa. We can therefore say that the barometric pressure is 100 kPa. Dry air is 20.9 per cent oxygen at sea level so the partial pressure of oxygen (P_{O_2}) is 20.9 per cent of 100, which is about 20.9 kPa.

As our mountaineer progresses up the mountain, the atmospheric pressure becomes less because there is less air pushing down on him. The partial pressure of oxygen decreases accordingly (Fig 9.10b).

Oxygen passes into the lung tissues because of the different concentration in the alveolar air and the blood, and at higher altitudes the difference is smaller, making it difficult to take in enough oxygen to meet demands. At 5300 metres, the P_{O_2} is only half of that at sea level, and

by the time the summit is reached it is only about 8 kPa – just over one-third that at sea level. This is why most mountaineers who attempt Everest do so with pressurised oxygen containers, and to conquer the mountain without the help of additional oxygen – as has been done – is a remarkable feat.

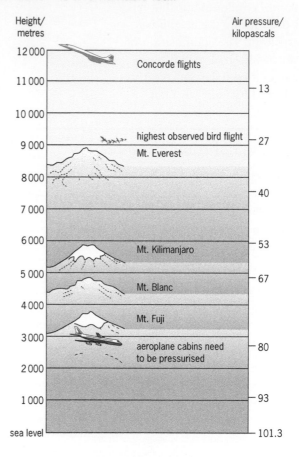

Fig 9.10a
Mountaineers usually use additional oxygen at very high attitudes

Fig 9.10b The effect of altitude on the partial pressure of oxygen. Aircraft cabins must be pressurised and mountaineers usually need the help of pressurised oxygen canisters. The human body is able to acclimatise to lower partial pressures of oxygen by increasing the amount of haemoglobin in the blood

Lifestyle diseases are discussed in Chapter 31.

Carbon monoxide binds to haemoglobin molecules, preventing them from carrying oxygen. Smokers therefore have a lower oxygen-carrying capacity than non-smokers.

Tar is a complex mixture of chemicals that paralyses the cilia but stimulates the goblet cells (Fig 9.9), resulting in a build-up of mucus that traps dust and pathogens. This results in 'smoker's cough' and increases the smoker's chance of developing chronic bronchitis.

In the long term, the epithelia become damaged and replaced by scar tissue, cilia are destroyed, and some of the smaller airways become blocked by mucus. Chronic bronchitis develops with its persistent cough that produces large quantities of phlegm (mucus, bacteria and white cells). Tar also contains carcinogens and heavy smokers have a greatly increased chance of developing lung cancer.

5 THE CONTROL OF BREATHING

Control of breathing is **involuntary**: we don't have to think continually about breathing in and out, and our breathing rate is automatically matched to our needs. Breathing rate changes as our bodies detect the physical and chemical variations that occur when we carry out different activities.

SETTING A REGULAR PATTERN

Breathing is controlled by the **respiratory centre**. This is located in the brain in an area called the **medulla oblongata** (Fig 9.11).

Regular nerve impulses travel down **efferent** (outgoing) nerves that pass from the respiratory centre out to both the external intercostal muscles and the diaphragm. These muscles then contract, initiating inhalation. As air enters the lungs, stretch receptors in the airways start firing and feed information to the brain about how the inflation of the lungs is progressing. The more the lungs inflate, the faster the stretch receptors feed back impulses. When the lungs are sufficiently inflated, signals from the respiratory centre stop for a short time and exhalation follows automatically.

CHANGING BREATHING RATE TO MEET DEMAND

Ensuring that the body has a constant supply of oxygen is an important aspect of homeostasis. But, surprisingly, the body is relatively insensitive to falling oxygen levels. It is much more sensitive to an increase in carbon dioxide and so this is a much more reliable indicator of the need for oxygen. The levels of oxygen in the arterial blood vary little, even during exercise, but the carbon dioxide levels vary in direct proportion to the level of exertion. The heavier the exercise, the greater the carbon dioxide concentration of the blood. Lactate levels also increase during exercise. Any increase in carbon dioxide or lactate concentration in the blood lowers its pH.

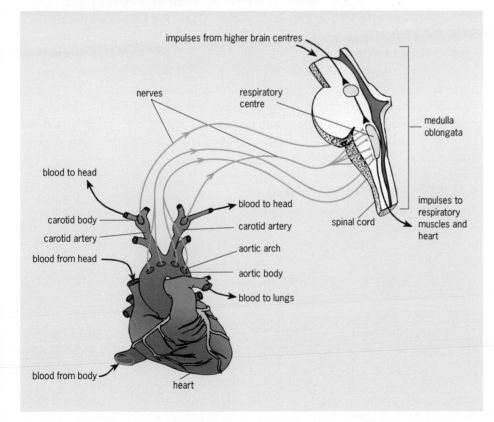

> ✓ **REMEMBER THIS**
>
> Together, the carotid and aortic bodies are known as the peripheral chemoreceptors. These cells sense changes in carbon dioxide and pH levels and, to a lesser extent, they are also sensitive to changes in oxygen levels.

Fig 9.11 Regulation of the levels of oxygen and carbon dioxide in the blood. Chemoreceptors (carbon dioxide/pH sensitive cells) occur in both the carotid and aortic bodies, as well as in the respiratory centre itself

Chemoreceptors (see Fig 9.11) are cells that are extremely sensitive to the composition of the blood that flows past them, and can detect very small changes in pH. They occur at three sites:

- **Central receptors** in the medulla oblongata. These are sensitive to the carbon dioxide concentration in the blood that flows through this region of the brain.
- The **carotid bodies** in the wall of the **carotid artery**.
- The **aortic bodies** situated on the **aortic arch**, just above the heart.

When chemoreceptors register a change in carbon dioxide level or pH, they send nerve impulses to the respiratory centre in the brain. This responds by sending more frequent impulses to the external intercostal muscles and diaphragm. When this happens, our ventilation rate increases: you breathe harder and faster. Heart rate also increases and so the body automatically increases oxygen delivery at the same time as removing the extra carbon dioxide.

The control of breathing rate is very similar to the control of heart rate (see page 162) with one important difference: we can control our breathing rate by thinking about it. This suggests that the higher, 'conscious' centres of the brain are more closely linked to the respiratory centre than they are to the cardiovascular centre, the part of the brain that controls heart rate. Also, research shows that pulse and ventilation rates change dramatically during exercise, even before the concentration of blood gases has a chance to change. It is as if the body predicts what is about to happen. How this works is not understood and is an active area of research.

THE EFFECTS OF OXYGEN DEPRIVATION

In some situations, at high altitudes for example, oxygen levels can fall without carbon dioxide levels increasing. In a rarefied or artificial atmosphere, normal breathing flushes carbon dioxide out of the lungs but there may not be enough oxygen to replace it. When this happens, the chemoreceptors often fail to register a problem, and the brain can become starved of oxygen.

The first symptoms of oxygen starvation are feeling ridiculously happy, having impaired senses and lacking judgement. When mountaineers, fighter pilots or deep-sea divers start giggling and making stupid mistakes, it is a sure sign that they are not getting enough oxygen. It is also a signal for their colleagues to act fast and, if possible, provide emergency oxygen. If they don't get help quickly, they soon lapse into unconsciousness, and brain damage and death follow (Fig 9.12).

ATHLETES AND V_{O_2}(MAX)

Many physical activities, such as jogging, swimming and team sports, rely on energy released by the aerobic pathway of cell respiration. The level of performance an athlete can achieve is largely governed by how fast oxygen gets to the muscles.

The rate at which a person uses oxygen is called the V_{O_2} and is measured in terms of the volume of oxygen consumed (cm^3), per kilogram of body weight, per minute.

The V_{O_2}(max) is the maximum rate at which oxygen is consumed and is the amount of oxygen that can be delivered to the tissues when the lungs and heart are working as hard as possible. Athletes use a knowledge of V_{O_2}(max) in their training, as a measure of how hard they are working. A training schedule, for example, might require an athlete to work at 65 per cent of their V_{O_2}(max) for a set length of time.

Heart rate is discussed in Chapter 10.

Cell respiration is discussed in Chapter 33.

Fig 9.12 In 1875, three French physiologists decided to investigate the effects of low oxygen levels on the human body. The easiest way to get oxygen-poor air was to go up in a hot-air balloon to observe the effects of the rarefied atmosphere on each other. At first there were no obvious effects and they happily continued to throw out ballast, going up to 8000 metres. At this point they all fainted. The balloon eventually came down on its own, and one of the scientists woke up to find the other two dead

The Example shows how to calculate V_{O_2} and V_{O_2}(max) for an average person. (You can work out your own, if you are able to measure your tidal volume, the number of breaths you take per minute and your weight in kg.) You can see that the amount of oxygen consumed during exercise increases tenfold or 1000 per cent from 4.28 to 51.42 cm^3 O_2 kg^{-1} min^{-1}.

EXAMPLE

Q What is the rate of oxygen consumption of a normal 70 kg adult at rest? Assume that the tidal volume of a normal adult at rest is 0.5 litres and that he or she takes 15 breaths per minute.

A We can work out the ventilation rate from the formula:

$$\text{Ventilation rate} = \frac{\text{number of breaths taken per minute}} {} \times \text{tidal volume}$$

Ventilation rate $= 15 \times 0.5$ litres $min^{-1} = 7.5$ litres min^{-1}

Atmospheric air is one-fifth oxygen and from Table 9.1 we know that only one-fifth of the available oxygen is absorbed. So:

amount of oxygen used per minute

$= 7.5 \times 0.2 \times 0.2$ litres min^{-1}

$= 0.3$ litres min^{-1}

And the V_{O_2} for this person is:

$\frac{0.3}{70} = 0.004\,28$ litres O_2 kg^{-1} min^{-1},

or: $\quad$ 4.28 cm^3 O_2 kg^{-1} min^{-1}

Q What is the V_{O_2}(max) for the same adult? Assume that the volume of air taken in during each breath during strenuous exercise is 3 litres and that the number of breaths per minute increases to 30.

A Knowing that the tidal volume of the same adult during exercise is 3 litres and that he or she takes 30 breaths per minute, we can work out the ventilation rate:

$$\text{Ventilation rate} = \frac{\text{number of breaths taken per minute}} {} \times \text{tidal volume}$$

Ventilation rate $= 30 \times 3$ litres $min^{-1} = 90$ litres min^{-1}

Atmospheric air is one-fifth oxygen and from Table 9.1 we know that only one-fifth of the available oxygen is absorbed. So:

amount of oxygen used per minute

$= 90 \times 0.2 \times 0.2$ litres min^{-1}

$= 3.6$ litres min^{-1}

And the V_{O_2}(max) for this person is:

$\frac{3.6}{70} = 0.051\,42$ litres O_2 kg^{-1} min^{-1},

or: $\quad$ 51.42 cm^3 O_2 kg^{-1} min^{-1}

SUMMARY

After reading this chapter, you should know and understand the following:

- We must **exchange gases** to provide oxygen for cell respiration, and to take away the waste carbon dioxide that this process produces.
- Lungs are **respiratory organs** that increase the surface area for gas exchange. These organs have a large surface area and thin membranes. Ventilation and a good blood supply help to maintain a high diffusion gradient.
- Our breathing movements serve to ventilate air-sacs called **alveoli**, bringing in a continuous supply of fresh oxygen and flushing out waste carbon dioxide.
- The lungs cannot move on their own: they inflate when pulled outwards by the **external intercostal muscles** and the **diaphragm**. The lungs are attached to the inside of the ribcage by attraction between the two pleural membranes and the pleural fluid between them.

- At rest, our lungs inflate mainly by movement of the diaphragm. The amount of air moved in and out at rest is called the **tidal volume** and the **ventilation rate** can be worked out by multiplying the tidal volume by the number of breaths taken per minute. The total amount of air that we can inhale, after exhaling as much as possible, is called the **vital capacity**.
- The rate of breathing is controlled by the **respiratory centre** in the **medulla oblongata** of the brain. This sets a regular breathing pattern that is modified according to information received from **chemoreceptors**. These detect the rise in carbon dioxide in the blood and the fall in pH that indicate the body's need for oxygen.

 Practice questions and a How Science Works assignment for this chapter are available at www.collinseducation.co.uk/CAS

10

The circulatory system

10 THE CIRCULATORY SYSTEM

During open-heart surgery, the job of the heart and lungs is performed by a heart–lung machine. This sophisticated device also has the ability to lower the core temperature of the body by about 10 °C. This reduces the body's demand for oxygen, giving the surgeons approximately twice as long to perform the operation

The miracle of open heart surgery

When the blood vessels supplying the heart muscle become narrowed by fatty deposits, the heart can become starved of oxygen. The patient will suffer heart pains on exercise – called angina – and is very susceptible to heart attacks. Often, the only effective remedy is surgery.

One of the available operations – sometimes known as a 'cabbage' (CABG = coronary artery bypass graft) – takes a team of two or three surgeons about five hours to complete. The overall aim is to take pieces of vein from the patient's leg and graft them from the aorta, over the blocked artery and into the heart muscle itself. But how do you sew small blood vessels onto an organ that won't keep still?

The answer is that you don't. The heart has to be stopped and this is only possible with a heart–lung machine – a device that can take blood from the patient, oxygenate it and return it to the body at the right temperature and pressure. Once this machine has been connected, the heart can be stopped; this is done simply by pouring iced water over it. Once the heart is still the surgeon can sew on the new pieces of blood vessel. When complete, the heart is re-started by means of two electrodes placed on the heart itself. This standard operation can greatly improve the quality of life for patients, as well as prolonging their life expectancy.

1 WHY DO WE NEED A CIRCULATORY SYSTEM?

 REMEMBER THIS

The human circulatory system is an example of a **mass flow system**, in which large volumes of fluid are carried to all parts of the organism.

The human body contains specialised organs such as intestines, lungs and kidneys whose function is the exchange of materials such as digested food, oxygen and waste. However, these organs could not function without a **circulatory system**: a network of tubes filled with fluid that can deliver vital materials to all the cells of the body, and then take away their waste.

The circulatory system is so extensive that every living cell is only a few micrometres from a capillary. The blood that passes nearby provides the cell with the materials it needs, takes away its waste and generally bathes it in favourable conditions.

The human circulatory system has three components:

- A fluid, **blood**, which flows in the system, carrying materials around the body.
- A system of tubes or vessels – **arteries**, **veins**, **capillaries** – which carry the fluid.
- A pump, **the heart**, which keeps the fluid moving through the vessels.

In this chapter, we look at these three components in detail.

A DOUBLE CIRCULATION

Fig 10.1a shows that humans, like other mammals, have a double circulation: a **pulmonary circulation** that carries blood between the heart and the lungs and a **systemic circulation** that carries blood between the heart and the rest of the body.

To understand the advantage of a double circulation, you might find it helpful to consider the situation in fish (Fig 10.1b). Fish have a single circulation and a two-chambered heart. Blood first travels to the gills where it passes through thin systemic capillaries, picking up oxygen but losing pressure. Blood continues to travel around the body of the fish, back to the heart, but more slowly because it is at low pressure. This situation works for fish but is no use in warm-blooded animals such as ourselves; the slow single circulation would not be able to supply all our cells with sufficient amounts of the freshly oxygenated blood they need.

With a double circulation, the right side of the heart receives deoxygenated blood from the body and pumps it to the lungs. Here, the blood gains oxygen and loses pressure. However, the blood returns to the heart, which gives it a boost so that it can reach all the body parts quickly. The volume of blood going around each circulation in any given time is the same, but the systemic circulation is pumped under greater pressure by the heart because the blood has to reach all the extremities of the body.

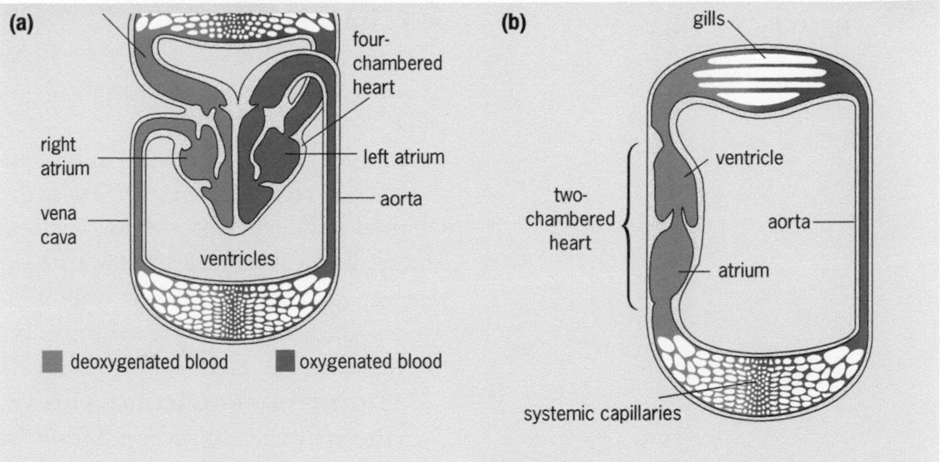

Fig 10.1a Humans have a double circulation and a four-chambered heart. Deoxygenated blood, coloured blue in the diagram, passes into the right side of the heart and is pumped to the lungs where it picks up oxygen and releases carbon dioxide. Oxygenated blood, coloured red, returns to the heart to be pumped to all parts of the body except the lungs

Fig 10.1b The single circulation of the fish is a simple circuit in which blood loses pressure as it passes through the gills and so moves slowly around the rest of the body

2 THE HUMAN CIRCULATORY SYSTEM

The human heart beats over 100 000 times a day, creating the pressure to force blood through more than 80 000 kilometres of arteries, veins and capillaries. Fig 10.2 illustrates the main vessels that make up the human circulatory system.

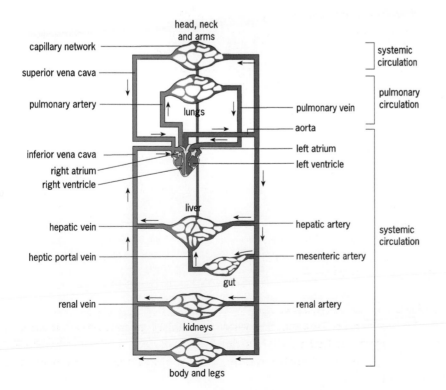

Fig 10.2 An overview of human circulation. Deoxygenated blood (coloured blue) passes into the right side of the heart through the superior vena cava and the inferior vena cava. The right ventricle pumps it into the pulmonary circulation and it picks up oxygen and releases carbon dioxide as it passes through the lungs.
Now oxygenated (coloured red), the blood returns to the heart. The left ventricle then pumps it around the systemic circulation and it travels to all parts of the body except the lungs

STRUCTURE OF THE HEART

The heart is composed mainly of **cardiac muscle**, a specialised tissue that contracts automatically, powerfully and without fatigue, throughout our lives (Fig 10.3).

The thickness of the walls in the different heart chambers reflects their function. The atria are thinly muscled: they pump blood the short distance to the ventricles directly below them. The right ventricle is more heavily muscled than either of the atria: it has to force blood a much further distance to the lungs. The left ventricle has the thickest wall: it has to push blood all around the body.

It is important that blood flows through the heart in one direction only. Two sets of valves close to prevent backflow:

- The **atrio-ventricular valves** (**AV valves**) lie between the atria and the ventricles to prevent blood from returning to the atria when the

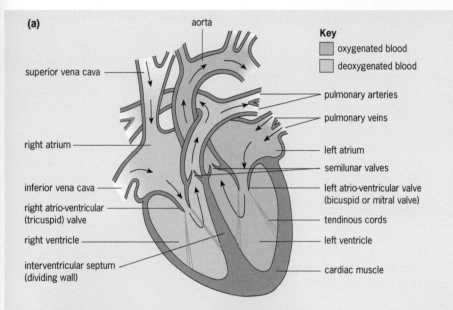

(a)

aorta

superior vena cava

right atrium

inferior vena cava

right atrio-ventricular (tricuspid) valve

right ventricle

interventricular septum (dividing wall)

Key
- oxygenated blood
- deoxygenated blood

pulmonary arteries

pulmonary veins

left atrium

semilunar valves

left atrio-ventricular valve (bicuspid or mitral valve)

tendinous cords

left ventricle

cardiac muscle

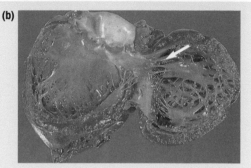

(b)

Fig 10.3a A longitudinal section of the heart

Fig 10.3b The heart is a slightly twisted, asymmetrical organ: the best way to appreciate its three-dimensional structure is to dissect a pig's or sheep's heart or to study a model

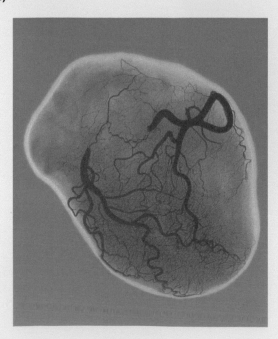

(c)

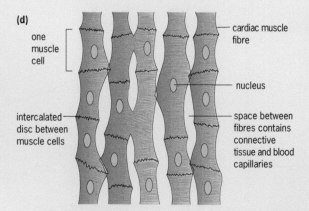

(d)

one muscle cell

intercalated disc between muscle cells

cardiac muscle fibre

nucleus

space between fibres contains connective tissue and blood capillaries

Fig 10.3c The coronary circulation contains all the blood vessels that supply the heart muscle. In the photograph, these blood vessels are easy to see because they have been filled with a dye that is opaque to X-rays

Fig 10.3d The fine structure of cardiac muscle. The rapid spread of impulses through cardiac muscle is possible because individual cells are connected by specialised junctions called intercalated discs. This system of communication ensures that a wave of contraction passes through the cardiac muscle, so that the chambers contract at the right time

ventricles contract. The **tricuspid valve** on the right side has three **cusps**, or flaps. The **bicuspid valve**, or **mitral valve**, on the left has two cusps. Both atrio-ventricular valves are subjected to great pressure and ultra-tough tendinous cords, arrowed in Fig 10.3b, prevent them turning inside-out.

- The **semilunar** valves guard the openings to the pulmonary artery and the aorta and prevent backflow of blood into the ventricles. Both sets of valves have three semilunar ('half moon'-shaped) cusps.

HOW THE HEART BEATS

The sequence of events in a single heartbeat is known as the **cardiac cycle**. The cycle has four overlapping stages:

- **Atrial systole**. Both atria contract, forcing blood into both of the ventricles. This stage lasts 0.1 seconds.
- **Ventricular systole**. Both ventricles contract, forcing blood through the pulmonary artery to the lungs and through the aorta to the rest of the body. This takes 0.3 seconds.
- **Atrial diastole**. The atria relax, although the ventricles are still contracted. Blood enters the atria from the large veins coming from the body. This takes about 0.7 seconds.
- **Ventricular diastole**. The ventricles relax, and become ready to fill with blood from the atria as the next cycle begins. This takes about 0.5 seconds.

Given an average heart rate of 75 beats per minute, each cycle takes 0.8 seconds.

> ✓ **REMEMBER THIS**
> The QRS complex (Fig 10.4b) is associated with the electrical impulse passing through the AVN, the bundle of His and the Purkyne fibres (see overleaf).

STRETCH AND CHALLENGE

The electrocardiogram (ECG)

The electrical events that control the cardiac cycle are recorded by placing electrodes on certain parts of the body (Fig 10.4a). A normal, healthy heartbeat produces a distinctive trace (Fig 10.4b). Certain heart defects produce a modified trace and this makes the ECG a useful diagnostic tool (Fig 10.4b).

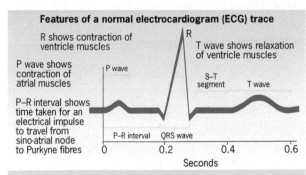

Features of a normal electrocardiogram (ECG) trace

R shows contraction of ventricle muscles

T wave shows relaxation of ventricle muscles

P wave shows contraction of atrial muscles

P–R interval shows time taken for an electrical impulse to travel from sino-atrial node to Purkyne fibres

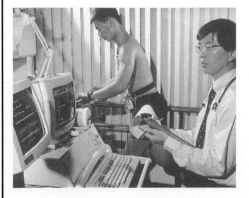

Fig 10.4a Electrodes taped to the chest pick up the electrical events of the cardiac cycle as they pass through the body. People with heart problems can now transmit their ECG by telephone to a specialist who can detect problems at a distance

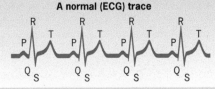

A normal (ECG) trace

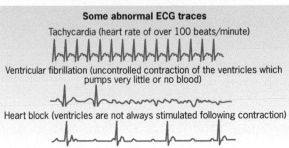

Some abnormal ECG traces

Tachycardia (heart rate of over 100 beats/minute)

Ventricular fibrillation (uncontrolled contraction of the ventricles which pumps very little or no blood)

Heart block (ventricles are not always stimulated following contraction)

Fig 10.4b A healthy ECG trace traces showing common heart defects that can be diagnosed using the ECG

Control of the cardiac cycle

The heartbeat must be carefully controlled so that each chamber contracts only when full of blood. To achieve this, the events of the cardiac cycle are carefully co-ordinated (Fig 10.5).

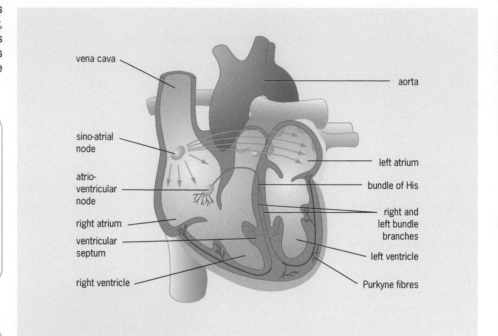

Fig 10.5 The conduction system of the human heart initiates and controls a heartbeat. Cells in the sino-atrial node act as a pacemaker, initiating impulses which spread through the walls of the atria and, after a delay, through the walls of the ventricles via the atrio-ventricular node

REMEMBER THIS

Valves are simply strong flaps of tissue: they cannot move on their own. A common mistake is to credit valves with the active control of blood flow. As blood begins to flow back, the valve is forced shut and this prevents any further backflow.

REMEMBER THIS

The heart continues to beat when removed from the body. Individual heart muscle cells grown in culture beat on their own! Because of this, we say that the heart is **myogenic**: the stimulus which drives it to beat originates in the muscle itself. (Some animals have hearts that are **neurogenic**: they beat only when stimulated by external nerves.)

A single heartbeat starts with an electrical signal from a region of specialised tissue called the **sino-atrial node** (**SAN**), on the wall of the right atrium. This is the 'pacemaker', or heartbeat regulator. The electrical signal spreads out over the walls of the atria, causing them to contract.

From there, the signal does not pass directly to the ventricles. If it did, the ventricles would begin to contract before they had filled with blood. Instead, the impulse is delayed slightly. A second node, the **atrio-ventricular node** (**AVN**) picks up the signal and channels it down the middle of the **ventricular septum** through a collection of specialised cardiac muscle fibres called the **bundle of His**. From here, the signal spreads throughout the wall of the ventricles, through the **Purkyne** (or **Purkinje**) fibres, and the ventricles contract after they have filled with blood.

Electrical events and volume/pressure changes in the cardiac cycle

We have looked at what happens to the blood that passes through the heart during the cardiac cycle, and at the electrical events that co-ordinate the contraction and relaxation of the atria and ventricles. Fig 10.6 shows how these events correspond to pressure and volume changes in the heart and blood vessels, and also to the sounds that you hear when you listen to someone's heart beating. You will need to refer to this figure as you read on.

During **atrial systole**, the atria fill with blood from the vena cava and the pulmonary vein. Some of the blood that enters the atria flows straight into the ventricles, without any need for contraction. Atrial systole is initiated when the SAN sends out an electrical signal that spreads out over both the atria, causing contraction and forcing the remainder of the blood into the ventricles. There are no heart sounds. Atrial systole is very short, little more than a

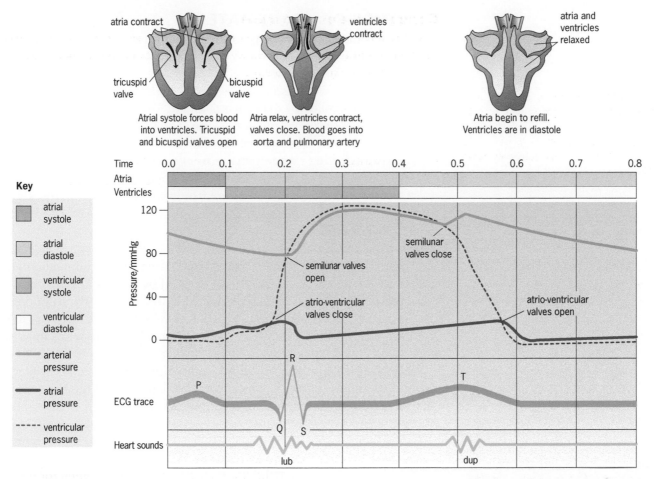

Fig 10.6 The cardiac cycle. See Fig 10.4b for an explanation of the ECG trace. Many exam questions use this diagram, without labels, to test understanding of the events of one heartbeat. It is important to remember what is cause and what is effect:
- Pacemaker cells initiate systole (contraction)
- The squeezing of the muscle walls reduces the volume and so increases the pressure in the chambers, forcing blood in a particular direction
- The direction of blood flow causes the valves to open or close. The valves ensure that blood flow through the heart is in one direction only

'twitch'. Atrial relaxation, or **atrial diastole**, lasts for the remainder of the cycle. All the rest of the 'action' is in the ventricles.

Ventricular systole begins when the ventricles have filled with blood. The AVN picks up the signal from the SAN and then conducts impulses down through the bundle of His and on through the Purkyne fibres in the walls of the ventricles. This stimulates the ventricles to contract. Blood is forced upwards, forcing the semilunar valves open and the AV valves shut. The closing of the AV valves causes the 'lub' of the 'lub-dup' heart sound. Pressure in the arteries rises sharply as blood is forced into them.

During **ventricular diastole**, the ventricle walls relax, arterial pressure falls and blood begins to flow back into the ventricles. This reversal of the flow causes the semilunar valves to shut, causing the second, 'dup', heart sound. Meanwhile, the atria have been filling with blood and, as the ventricles relax, blood flows from the atria into the ventricles, forcing the AV valves open again.

Stroke volume and cardiac output

The volume of blood pumped by the heart during one cardiac cycle is the **stroke volume**. A typical stroke volume in an adult is about 80 cm^3: every time the heart beats, 80 cm^3 of blood is forced through the pulmonary artery to the lungs and 80 cm^3 is forced into the aorta to the body. So, the heart pumps over 8500 litres of blood per day. Stroke volume increases during exercise, and regular exercise results in a permanent resting increase to 110 cm^3, or more.

The volume of blood pumped in one minute is called the **cardiac output**. It is calculated by multiplying the stroke volume by the heart rate and is expressed in litres of blood per minute.

 REMEMBER THIS

Cardiac output =
stroke volume × heart rate
(measured in litres:
1000 cm^3 = 1 litre)

? QUESTION 1

1 Calculate the cardiac output:
   a) at rest when the stroke volume is 80 cm^3 and the heart rate is 75 beats per minute;
   b) during vigorous exercise when the stroke volume is 100 cm^3 and the heart rate is 150 beats per minute.

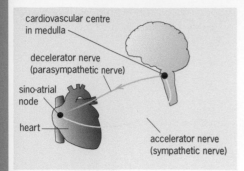

Fig 10.7 The nerves that connect the cardiovascular centre in the brain to the heart. The accelerator and decelerator nerves are part of the autonomic nervous system

> Breathing control and gas exchange are discussed in more detail in Chapter 9.

CONTROL OF HEART RATE

Heart rate is modified according to the needs of the body. It increases during physical exercise to deliver extra oxygen to the tissues and to take away excess carbon dioxide. At rest, a normal adult human heart beats about 75 times per minute: during very strenuous exercise it might beat 200 times per minute. Heart rate is controlled by the sino-atrial node. The rate goes up or down when the SAN receives information via two autonomic nerves which link the sino-atrial node with the **cardiovascular centre** in the brain (Fig 10.7):

- A **sympathetic** or **accelerator nerve** speeds up the heart. The synapses at the end of this nerve secrete noradrenaline.
- A **parasympathetic** or **decelerator nerve**, a branch of the vagus nerve, slows down the heart. The synapses at the end of this nerve secrete **acetylcholine**.

A **negative feedback system** controls the level of carbon dioxide and, indirectly, the level of oxygen in the blood. During exercise, the blood level of carbon dioxide starts to rise. This is detected by **chemoreceptors** (cells sensitive to chemical change) situated in three places: the carotid artery, the aorta and the medulla. Nerve impulses travel from these receptors to the cardiovascular centre. In response, the cardiovascular centre sends impulses down the sympathetic nerve to increase the heart rate.

When the carbon dioxide level drops, the cardiovascular centre responds by sending impulses down the parasympathetic nerve, and the heart rate returns to normal. The control of heart rate is closely related to the control of breathing.

Several factors affect heart rate:

- Secretion of **adrenaline** in response to stress, excitement or other emotions. Adrenaline is the hormone that prepares the body for action, and one of its effects is to increase heart rate.
- Movement of the limbs, as in exercise. It is thought that stretch receptors in the muscles and tendons relay information to the brain, telling the cardiovascular centre that oxygen levels will soon fall and that carbon dioxide will soon build up. This initiates signals that speed up heart rate and breathing rate.
- The level of respiratory gases in the blood, as described above.
- Blood pressure. When this gets too high, a fail-safe mechanism prevents any further increase in heartbeat.

3 BLOOD VESSELS

There are three types of blood vessels: **arteries**, **veins** and **capillaries**. The structure of each is closely related to its function (Fig 10.8 and Table 10.1).

Arteries carry blood away from the heart towards other organs of the body. Arteries branch into smaller **arterioles**, which branch into tiny capillaries. Capillaries are **permeable** ('leaky') vessels whose walls are one cell thick, allowing exchange of materials between blood and nearby cells. Blood flows from capillaries into venules, which drain into larger veins.

As Fig 10.8 and Table 10.1 show, the walls of arteries and veins have the same three layers: the **tunica interna**, the **tunica media** and the **tunica externa**. The relative thickness and composition of these layers vary according to the function of the vessel. Arteries, which have to withstand pulses of high pressure, have a thick tunica media containing smooth muscle and elastic fibres. Veins have a thinner tunica media and a larger lumen and carry slower flowing blood at low pressure. Valves prevent blood flowing backwards.

 REMEMBER THIS

Arterial blood is under high pressure and surges occur every time the ventricles contract.

The walls of the arteries are able to absorb and smooth out these pulse waves so that by the time blood reaches the arterioles, it is flowing steadily.

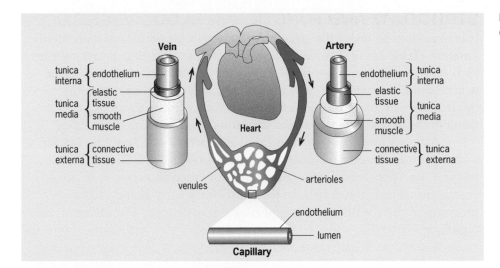

CAPILLARY CIRCULATION

The circulatory system keeps all cells bathed in tissue fluid. The composition of tissue fluid stays reasonably constant because permeable capillaries allow exchange of materials between tissue fluid and the blood.

How tissue fluid is formed

Fig 10.9 shows what happens in living tissue. Blood from the arterioles is under high **hydrostatic pressure**. When blood enters the capillary, substances begin to leak out through the permeable capillary wall. The capillary walls act as filters and a proportion of all chemicals below a particular size is squeezed out, forming tissue fluid. The composition of tissue fluid is very similar to that of plasma, but tissue fluid lacks most of the large proteins that cannot pass through the capillary wall. Plasma proteins remain in the blood where they play an important part in the drainage of tissue fluid.

> ✔ **REMEMBER THIS**
> All living cells are surrounded by tissue fluid. This is also known as **interstitial** or **intercellular fluid**.

> ✔ **REMEMBER THIS**
> Hydrostatic pressure is physical pressure. The contraction of the ventricles of the heart creates hydrostatic pressure in the blood vessels.

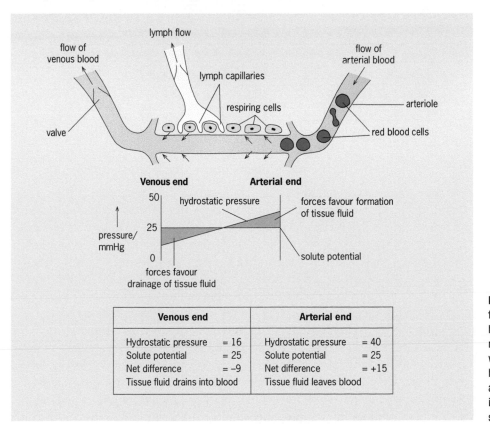

Venous end	Arterial end
Hydrostatic pressure = 16	Hydrostatic pressure = 40
Solute potential = 25	Solute potential = 25
Net difference = –9	Net difference = +15
Tissue fluid drains into blood	Tissue fluid leaves blood

Fig 10.9 The drainage and formation of tissue fluid. At the arterial end of the capillary, the high hydrostatic pressure forces water and small molecules out of the blood, providing the tissues with a fluid that contains nutrients and oxygen. The hydrostatic pressure decreases as blood flows along the capillary. When it falls, water drains back into the blood by osmosis, taking with it wastes such as urea and carbon dioxide

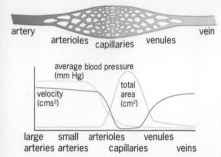

artery · arterioles · capillaries · venules · vein

average blood pressure (mm Hg)

velocity (cms²) · total area (cm²)

large arteries · small arteries · arterioles capillaries · venules veins

Fig 10.10 The capillaries have a very large combined cross-sectional area, much more than the arteries and veins. Consequently, as blood flows into capillaries pressure drops and blood flows more slowly, allowing more efficient exchange of materials across capillary walls

STRUCTURE AND FUNCTION IN BLOOD VESSELS

It is important to relate the structure of different blood vessels with their function (Table 10.1). Arteries have tough elastic walls that withstand the high pressure of blood that flows through them as it is forced out of the heart. Blood pressure is at its highest in the large arteries supplied by the left ventricle (Fig 10.11). The pressure then reduces as blood passes through the smaller arterioles and then into the capillaries. The rate of blood flow is slowest through the capillaries, which also have the largest surface area of the entire blood vessel system (Fig 10.10). This is because this is where gas exchange occurs. Blood flow then increases as blood enters the veins to travel back to the heart. Veins carry blood at low pressure, so their walls are thin and have a large lumen. The valves prevent backflow.

Table 10.1 Relating structure to function in blood vessel walls.

Vessel	Diagram	Structure of wall	Function
Artery		rich in elastic fibres; very tough elastic walls with some muscle fibres	recoil; withstands surges in pressure and maintains diastolic blood pressure
Arteriole		higher proportion of muscle to elastic fibres compared to artery	can constrict or dilate to control blood supply to particular areas
Capillary		thin, permeable	allows rapid diffusion and exchange
Vein		relatively thin; contain valves; have larger lumen than similar sized artery	returns blood to heart under low pressure, valves prevent backflow

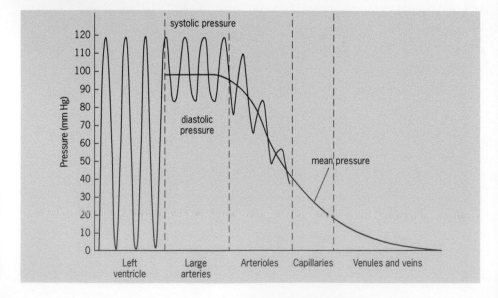

Fig 10.11 Graph showing the changes in pressure as blood flows away from the heart. Note the huge surges in the large arteries, whose elastic walls absorb the pressure. By the time blood leaves the arterioles it is flowing smoothly. By the time blood leaves the capillaries, most pressure has been lost

How tissue fluid returns to the blood

Two main forces act on the blood as is passes along the capillary:

- The hydrostatic pressure of the blood: this tends to force water and solutes *out* of the capillary.
- The water potential of the plasma; this tends to draw water *into* the capillary by osmosis.

As blood flows along a capillary, fluid passes to the tissues, and the volume and hydrostatic pressure of the blood in the capillary decrease (Fig 10.9). As the water potential created by the large plasma proteins is relatively constant, water begins to drain back into the blood when hydrostatic pressure falls below the water potential. Cell waste, such as urea, and substances that have been secreted into the tissue fluid, diffuse into the capillary (Fig 10.12), contributing to the composition of venous blood.

BLOOD FLOW IN VEINS

Blood that drains into the venules is deoxygenated, under low pressure and contains many waste products and cell secretions. Several features allow blood to return to the heart:

- Valves in veins close and prevent backflow.
- Working muscles surrounding the veins squeeze blood along as they contract. The action of muscles, especially those in the legs, is so important that it has been called the 'secondary heart' or the 'venous pump' (Fig 10.13). This venous pump is very important to the circulation, and regularly standing still for any length of time can lead to problems. Millions of people suffer from varicose veins or from haemorrhoids (piles), damaged veins whose walls have been stretched by pools of accumulated blood.
- Gravity helps blood to flow from organs 'above' the heart.
- The negative pressure created in the thorax during **inspiration** (breathing in) draws blood from surrounding veins.
- The action of the heart. After systole (contraction), the elastic walls of the chambers recoil, drawing blood in from the veins.

CONTROL OF BLOOD FLOW

The body often needs to alter blood flow to different areas, according to circumstances. For example, when we are hot, we lose excess heat because more blood flows to the surface of the skin. Blood flow can be modified by **sphincters** and **shunt vessels**, or by altering the diameter of the vessel itself.

A sphincter is a ring of muscle around a blood vessel which can contract, reducing the lumen size of the vessel and so reducing or preventing blood flow to a particular area. Sphincters can redirect blood into shunt vessels and these bypass a particular area. For instance, when we are cold, sphincters reduce blood flow to the skin surface, redirecting it along shunt vessels that keep the blood deeper in the body.

Many blood vessels, particularly arterioles, contain smooth muscle fibres that can contract to reduce blood flow through the vessel; this is **vasoconstriction**. Blood flow increases when the muscle fibres relax; this is **vasodilation**. Blood pressure can be regulated by altering the degree of vasoconstriction or vasodilation. Peripheral vasoconstriction means constriction of the blood vessels in the skin (periphery = outer region).

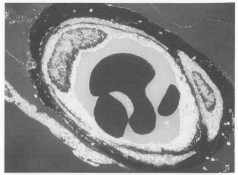

Fig 10.12 Capillary walls are about 1 μm thick. Most capillaries have a lumen that is a little larger than the diameter of a red blood cell. Note that cells in photos often seem to be strange shapes. There are two reasons for this. First, they are very flexible and will distort to squeeze along capillaries without causing blockages. Second, microscope slides take thin two-dimensional sections through three-dimensional objects, so they often give a false impression of shape

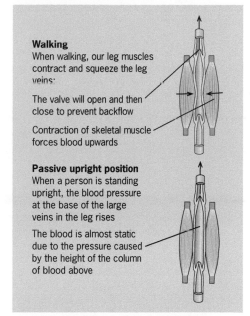

Walking
When walking, our leg muscles contract and squeeze the leg veins:

The valve will open and then close to prevent backflow

Contraction of skeletal muscle forces blood upwards

Passive upright position
When a person is standing upright, the blood pressure at the base of the large veins in the leg rises

The blood is almost static due to the pressure caused by the height of the column of blood above

Fig 10.13 When leg muscles contract, this squeezes blood along veins and back to the heart. The valves in the veins make sure that blood flows in one direction only

> Temperature regulation is discussed in more detail in Chapter 12.

? QUESTION 2

2 What effect do the following have on blood pressure?

a) vasoconstriction

b) vasodilation

4 THE LYMPHATIC SYSTEM

Fig 10.14 The major lymphatic vessels in the human body. The lymph system drains into the blood system in the upper thorax. A large lymph vessel, the thoracic duct, connects with the subclavian vein (sub-clavicle = under collarbone). Here lymph mixes with blood before entering the vena cava on its way to the heart

> **The immune system is discussed in Chapter 30.**

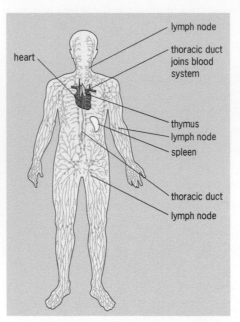

heart

lymph node

thoracic duct joins blood system

thymus
lymph node
spleen

thoracic duct

lymph node

The **lymphatic system** is part of the immune system. It is also part of the circulatory system: it returns to the heart the small amount of tissue fluid that cannot be returned by the veins.

We can think of the lymph system as an extra set of veins (Fig 10.14). Flow through lymphatic vessels is slow, but important. Just how important is obvious when you see someone with the disease elephantiasis (Fig 10.15).

Lymphatic flow begins in the **capillary beds**, where small amounts of tissue fluid drain into tiny **lymphatic capillaries** (Fig 10.16). The walls of these vessels are more permeable than blood capillaries to lipids and large molecules such as proteins, and so lymph contains a high proportion of these substances. Many cells secrete substances that are too large to enter the blood directly, and so can only pass into the general circulation via the lymphatics. The lymph capillaries drain into larger lymph vessels that look like thin, transparent veins. These vessels have valves to prevent backflow. Lymph contains no red blood cells, and so is pale and clear. All lymph vessels flow towards the upper chest, passing through numerous lymph nodes, which filter out bacteria and cell debris.

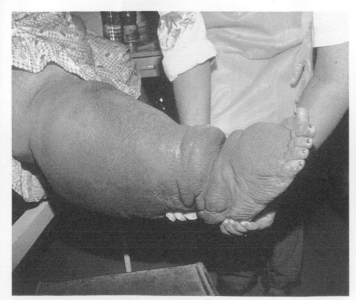

Fig 10.15 Lymphatic drainage takes about 120 cm³ of fluid out of the tissues every hour. When a lymphatic vessel is blocked, the limb quickly swells until the pressure interferes with normal blood flow. This photograph shows someone suffering from elephantiasis, a condition caused by a parasitic worm, *Wuchereria bancrofti*, which blocks lymphatic vessels

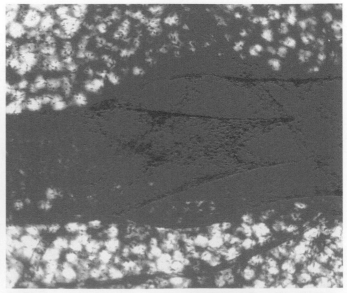

Fig 10.16 A longitudinal section through a lymphatic capillary. The cells that form these capillary walls overlap, forming tiny valves that allow flow of lymph in one direction only. Lymph capillaries in the villi of the intestines are called lacteals, and are important in the drainage of lipids and large molecules

5 BLOOD PRESSURE

The term **blood pressure** refers to the physical pressure of blood flowing through the arteries. This depends on the force created by the pumping of the heart, the volume of circulating blood and the size of the blood vessels. The body must keep blood pressure inside fairly strict limits. It must be high enough to force blood through all the capillaries so that all cells of the body are well nourished, but not too high because this would make the heart work unnecessarily hard and risk damage to blood vessels. High blood pressure is a risk factor for **atherosclerosis** (hardening of the arteries), and it causes a variety of other health problems. Fig 10.17 shows how we measure blood pressure.

The **diastolic pressure** is used as an indicator for medical problems: a normal reading is between 60 and 80 mmHg, with anything over 90 regarded as high. Elsewhere in the book we talk about pressure in terms of kPa, but blood pressure is still measured in the old mmHg units.

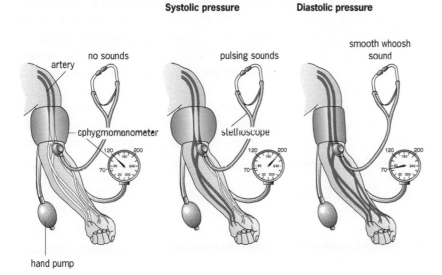

 REMEMBER THIS

Have you ever stood up very quickly, and then felt faint or dizzy? This happens because the rapid change in posture has altered the body's blood distribution, causing a temporary lack of blood to the brain. Fortunately, body mechanisms quickly bring the blood flow back to normal.

Lifestyle diseases such as atherosclerosis are discussed in Chapter 31.

Fig 10.17 Blood pressure is usually measured using a sphygmomanometer. An inflatable cuff is placed around the upper arm and inflated until all blood flow, in or out of the arm, stops. The blood flow in the brachial artery (at the elbow) is monitored using a stethoscope.

After inflation, there is no sound, but, as air escapes from the cuff, the pressure decreases until it falls just below that created by the heart as it contracts. At this point, blood is heard spurting through the constriction in the artery. This pressure is the **systolic value**.

Pressure in the cuff continues to drop until blood can be heard flowing constantly. This is the **diastolic value** and represents the pressure to which arterial blood falls between beats

CONTROL OF BLOOD PRESSURE

A negative feedback system keeps blood pressure inside safe limits. The pressure of the blood is detected by the **carotid sinus** (Fig 10.18), a small swelling in the carotid artery. When blood pressure gets too high, the walls of the sinus expand and **stretch receptors** in the artery wall inform the cardiovascular centre in the medulla. Signals from here then lower heart rate and vasodilation, so lowering blood pressure.

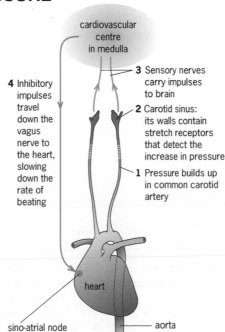

Fig 10.18 Regulation of the heart rate is influenced by the carotid sinus which detects blood pressure

cardiovascular centre in the medulla

4 Inhibitory impulses travel down the vagus nerve to the heart, slowing down the rate of beating

3 Sensory nerves carry impulses to brain

2 Carotid sinus: its walls contain stretch receptors that detect the increase in pressure

1 Pressure builds up in common carotid artery

heart

sino-atrial node

aorta

 REMEMBER THIS

Many mechanisms work together to keep blood pressure constant. The size of blood vessels, the heart rate and the volume of circulating blood can all change to increase or decrease blood pressure, according to the needs of the body. The kidneys are particularly important because they control how much fluid we lose. A drop in blood volume and pressure causes more fluid to be retained in the blood and less to be lost in the urine (see Chapter 13).

? **QUESTION 3**

3 'Blood pressure one twenty over seventy', shouts the doctor. What does this mean?

Temperature regulation is discussed in Chapter 12. Water balance is discussed in Chapter 13. Defence against infection is discussed in Chapter 30.

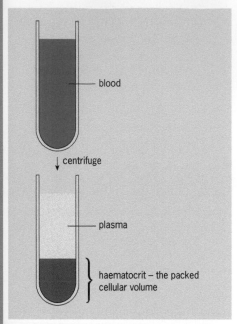

Fig 10.19 The percentage of blood volume taken up by cells is known as the haematocrit. A haematocrit of 39 would indicate that 39 per cent of the blood is composed of cells, mostly red cells. The average value for men is about 42 while women average 38. These values are affected by factors such as anaemia or high altitude

? QUESTION 4

4 Suggest how **a)** high altitude and **b)** anaemia will affect the haematocrit.

6 THE BLOOD

Blood is the fluid that flows through blood vessels, and the sight of blood is a sure sign that a vessel has ruptured. Losing large amounts of blood has serious consequences: death can result if the ruptured vessels are not sealed and blood volume is not rapidly returned to normal. In an emergency, when there is no time to find out the patient's blood group, accident victims are given fluids called **plasma expanders**. These are isotonic fluids that contain no cells but still increase blood volume.

Blood is a complex mixture containing cells, cell fragments and a range of dissolved molecules. It does the following:

● Transports materials. Blood transports digested food and oxygen to respiring tissues. It also takes carbon dioxide and waste products away from respiring cells to the various organs that remove them. It carries hormones from endocrine glands to target organs.

● Distributes heat. The blood helps to keep body temperature stable by distributing heat from metabolically active organs, such as the working muscles, to the rest of the body.

● Provides pressure. Many organs of the body depend on the physical pressure of the blood to carry out their function. For example, filtration in the kidney, formation of tissue fluid and erection of the penis all depend on blood pressure.

● Acts as a buffer. The blood contains many proteins and ions which act as buffers, keeping the pH constant by 'mopping up' any excess acid or alkali. Haemoglobin in red blood cells is also an important buffer.

● Defends the body against infection.

WHAT'S IN BLOOD?

Fig 10.19 shows that centrifuged blood separates out into two distinct layers: a **cellular portion** called the **haematocrit**, and the **plasma**. The cellular portion contains red cells, white cells and platelets.

Plasma

Plasma is the fluid part of blood. Its main constituents are shown in Table 10.2. The exact composition of the plasma varies greatly. For instance, in the hepatic portal vein, which leads from the intestines to the liver, the plasma is much richer in dissolved foods such as sugars, vitamins and amino acids than the plasma in any other vein.

Table 10.2 The main constituents of plasma

	Components	Function
Proteins	albumin	osmotic balance
	antibodies	immunity
	fibrinogen	blood clotting
Salts	sodium, potassium, chloride, hydrogencarbonate, calcium	osmotic balance, conduction of nerve impulses, carriage of carbon dioxide, buffering, blood clotting
Products of digestion	glucose, amino acids, fatty acids, glycerol, vitamins	nourishment of cells
Hormones	protein, e.g. insulin; lipid (steroids), e.g. testosterone	communication
Heat		distributed around body to maintain constant body temperature
Oxygen		vital in aerobic cell respiration
Waste products	urea, carbon dioxide	none: they must be removed by excretion (see Chapters 11 and 13)

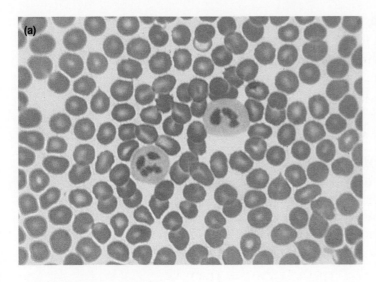

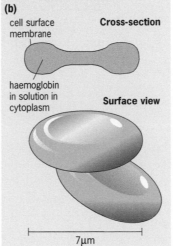

Fig 10.20a If you look at a drop of blood under the microscope, the most obvious feature is the mass of red blood cells. There are also two white blood cells in the centre of this picture
Fig 10.20b Red blood cells are biconcave discs filled with haemoglobin and other chemicals which help the loading and unloading of oxygen

Red blood cells

In this chapter we study red blood cells only (Fig 10.20): white cells and platelets are covered in Chapter 30.

Also known as **erythrocytes**, red blood cells are by far the most numerous in the blood. A single cubic millimetre of blood contains around 5 million, sometimes more. Red blood cells have one function: they carry the respiratory gases. The **haemoglobin (Hb)** they contain picks up oxygen in the lungs and swaps it for carbon dioxide in the respiring tissues.

Red cells have no nucleus: they are packed with Hb and the enzymes and chemicals that allow the Hb to carry oxygen effectively. Having Hb inside red blood cells rather than in solution in the plasma gives several advantages:

- A much greater volume of Hb can be carried in cells than could be dissolved in plasma.
- Hb can be kept in a favourable chemical environment to allow faster loading and unloading of respiratory gases.
- Hb molecules of a particular age are kept together and can be easily replaced when old.
- Hb in cells does not affect the osmotic properties of the blood (free Hb would).
- Hb in cells cannot be lost by excretion.

Red blood cells have a regular shape, described as a biconcave disc. This shape is maintained by the **cytoskeleton**, a complex but flexible internal scaffolding made from protein fibres. When you think about what red blood cells do it is easy to see why they have this shape. They must have a large enough volume to carry useful amounts of oxygen but they also need a large enough surface area to load and unload it quickly. The biconcave shape is a good compromise between the maximum volume of a sphere and the maximum surface area of a flat disc.

Red blood cells are distorted when they have to squeeze through capillaries, which often have lumens slightly smaller than the diameter of the red cell. Red cells rarely cause blockages because they are flexible and have a smooth surface.

Red blood cells are made in the bone marrow of the vertebrae, ribs and pelvis by specialised **stem cells**. The turnover of red cells is very rapid: we make millions of red cells every second. Red blood cells circulate for about 100 to 120 days before they are destroyed in the liver by the phagocytes called **Kupffer cells** and also in the spleen.

QUESTION 5

5   Why do you think that haemoglobin has been described as the 'Robin Hood' molecule?

REMEMBER THIS

Conjugated means 'joined'. A conjugated protein, such as haemoglobin, consists of a protein attached to a **prosthetic group** (a non-protein). The prosthetic group in haemoglobin contains iron, so we need a regular supply of iron in our diet to make this vital chemical.

REMEMBER THIS

The amount of oxygen in a mixture of gases such as air is often described in terms of its **partial pressure**, or **tension**, rather than its concentration – see Chapter 9. The greater the concentration of dissolved oxygen, the higher the partial pressure.

BLOOD AS A TRANSPORT MEDIUM

Oxygen is carried in the blood in two ways: 98 per cent travels as **oxyhaemoglobin**, the oxygen–haemoglobin complex. The remaining 2 per cent is dissolved in the plasma. Carbon dioxide is carried in three ways: as HCO^{3-} ions in the plasma (70 per cent), combined with haemoglobin as a **carbamino** compound (23 per cent) and in simple solution in the plasma (7 per cent).

The structure of haemoglobin

The Hb molecule is a large conjugated protein with a molecular mass of about 64 500 kilodaltons. Each Hb molecule consists of four **globin** sub-units consisting of a polypeptide chain and a prosthetic group called **haem**. At the centre of each haem is an iron ion (Fe^{2+}) which combines with oxygen. There are four haem groups, so the overall equation for the reaction is:

$$Hb + 4O_2 \rightarrow HbO_8$$
haemoglobin + oxygen → oxyhaemoglobin

The polypeptide chains hold the haem groups in place (see Fig 3.34) and help to load oxygen. When the first haem group combines with an oxygen molecule, the Hb molecule alters its shape so that the next haem group is exposed, making the loading of the second oxygen easier.

Haemoglobin in action

There are many substances that react readily with oxygen, but Hb is one of the few that can combine with oxygen where it is abundant and then release it when the concentration falls. This property is illustrated in a graph called an **oxygen dissociation curve** (Fig 10.21). This is plotted by analysing the percentage of Hb saturated with oxygen at different concentrations of oxygen. The graph shows that Hb becomes fully saturated with oxygen at the concentrations found in the lungs, and gives up a relatively large proportion of its oxygen in the lower oxygen concentrations that occur in the tissues.

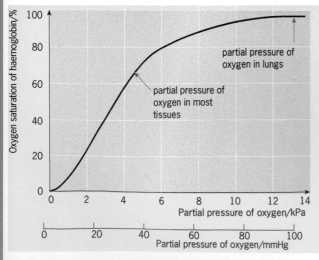

Fig 10.21 The oxygen dissociation curve. At the partial pressure found in the lungs, haemoglobin becomes 97–99 per cent saturated with oxygen. Surprisingly, haemoglobin releases only about 23 per cent of its oxygen in the respiring tissues, so the blood returning in veins is still about 75 per cent saturated. This suggests that three out of the four haem groups are still bound to oxygen. This allows great flexibility: if a tissue such as a working muscle becomes particularly oxygen starved, the blood can release large amounts of extra oxygen

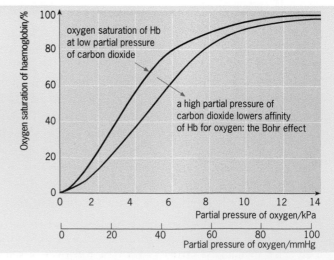

Fig 10.22 If the oxygen dissociation curve is plotted at higher carbon dioxide concentrations, it moves to the right, showing that haemoglobin has a reduced affinity for oxygen: this is called the Bohr effect

The dissociation curve in Fig 10.21 was plotted using mixtures of gases in which the concentration of carbon dioxide was constant. Fig 10.22 shows that, at higher levels of carbon dioxide, the curve moves to the right.

This is a vital point: it shows that carbon dioxide lowers the **affinity** of Hb for oxygen. This means that when carbon dioxide concentration is higher, haemoglobin does not hold on to its oxygen quite as well: Hb therefore tends to give up oxygen in areas of high carbon dioxide – such as in the respiring tissues which need it most. This lowering of affinity by carbon dioxide is called the **Bohr effect**, or the **Bohr shift**.

Fetal haemoglobin

Before birth, a fetus must obtain oxygen from its mother via the placenta. If fetal Hb had the same affinity for oxygen as the mother's Hb, no transfer of oxygen would be possible. But, as Fig 10.23 shows, fetal Hb has a higher affinity for oxygen than adult Hb. It can therefore pick up oxygen in the same conditions that cause the maternal blood to release it. After birth, the baby's body makes adult Hb which gradually replaces the fetal version.

UNLOADING OXYGEN AT THE TISSUES

The series of events that result in haemoglobin unloading its oxygen to supply respiring tissues is shown in Fig 10.24.

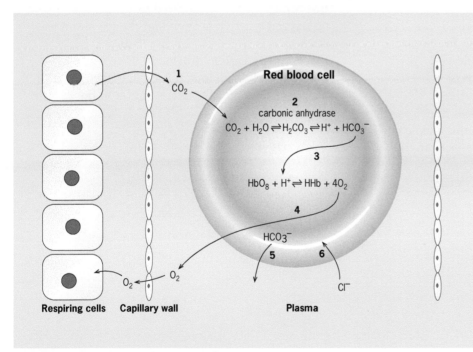

Fig 10.24 Unloading oxygen at the respiring tissues

Step 1 Carbon dioxide diffuses from the respiring tissue, through the capillary wall and the plasma, into the red blood cell

Step 2 Inside the red cell the enzyme carbonic anhydrase catalyses the conversion of carbon dioxide into carbonic acid (H_2CO_3)

Step 3 The hydrogen ions released by the carbonic acid makes the haemoglobin molecules less stable, causing them to release the four oxygen molecules

Step 4 The oxygen is free to diffuse into the respiring tissues

Step 5 The accumulating hydrogencarbonate ions diffuse out into the plasma, leaving the inside of the red cell with a net positive charge

Step 6 To maintain a neutral charge, chloride ions (the commonest negative ions in plasma) diffuse into the cell; this is called the chloride shift

Fig 10.23 The dissociation curve (above) for fetal haemoglobin is to the left of the adult version. So, at the oxygen concentrations found at the placenta there is an efficient transfer of oxygen from mother to fetus (below)

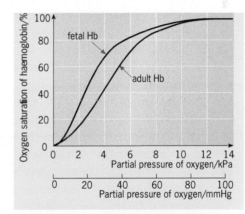

> ✔ **REMEMBER THIS**
>
> A human baby is known as an embryo up to about eight weeks of development. After this, it is known as a fetus for the rest of the pregnancy.

Myoglobin and muscles

Mammalian muscle contains the respiratory pigment, myoglobin (myo = muscle). The myoglobin molecule can be thought of as one-quarter of a haemoglobin molecule: it consists of one polypeptide chain and one haem, and combines with one oxygen molecule. Myoglobin has a higher affinity for oxygen than haemoglobin and so it can pick up oxygen from haemoglobin and store it until the muscle becomes short of oxygen. When this happens (during exercise, for example), the oxygen tension in the muscle drops. Haemoglobin has no more oxygen to deliver and myoglobin then releases its oxygen. See the Assignment for this chapter on the CAS website.

SUMMARY

When you have read this chapter, you should know and understand the following:

- Humans have a double circulation:
  the **pulmonary circulation** to the lungs and the **systemic circulation** to the rest of the body.
- The human **heart** is a four-chambered muscular pump made of **cardiac muscle**. This can contract powerfully and without fatigue.
- A single heartbeat, the **cardiac cycle**, consists of **atrial systole** (contraction) followed by **ventricular systole**, then **atrial** and **ventricular diastole** (relaxation).
- The heart is **myogenic**: the electrical impulses that control the cardiac cycle arise from the muscle itself.
- From the heart, blood passes into vessels in the following order: **arteries**, then **arterioles**, then **capillaries**, then **venules**, then **veins**.

- The **lymphatic system** acts as an extra set of veins, helping to drain fluid from the tissues back into the bloodstream.
- Blood consists of **red cells**, **white cells**, **platelets** and **plasma**.
- Red blood cells, or **erythrocytes**, contain **haemoglobin**, a conjugated protein that can pick up oxygen where it is abundant (lungs) and release it where it is needed (respiring tissues).
- Blood carries carbon dioxide in three ways: as hydrogencarbonate ions (HCO_3^-), combined with haemoglobin and as simple solution in the plasma.

 Practice questions and a How Science Works assignment for this chapter are available at www.collinseducation.co.uk/CAS

11

Homeostasis

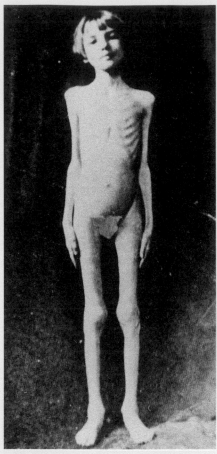

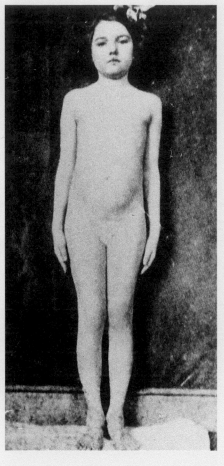

For Catherine Hayes, the discovery of insulin
came just in time. Daily injections of insulin
allowed the cells of her body to absorb glucose
effectively, so nourishing her tissues and
lowering her blood sugar

Insulin: a life-saver

Millions of people worldwide are unable
to control their blood glucose level.
They have a condition known as
diabetes mellitus. Early symptoms
include a raging thirst, extreme fatigue
and excessive urination.
If sufferers are not treated, they lose
weight and eventually die.

Until 1922, doctors regarded the
symptoms of diabetes mellitus as
a death sentence, especially when they
occurred in children. Physicians
confirmed a diagnosis by tasting
the patient's urine to find out if it
contained sugar (today, there are better
methods). One noted,
'In children, the disease is rapidly
progressive, and may prove fatal within
a few days... As a general
rule, the older the patient at the onset,
the slower the course.'

In 1923, a 16-year-old Canadian
girl, Catherine Hayes, wrote, 'Only a
year ago I was a human skeleton. Can
you imagine being 5 feet 4 inches and
weighing only 55 pounds [less than 4 stone; 24 kg]? My doctor had placed
me on a starvation diet, the only available treatment at the time.
I became too weak to engage in any physical activity, even walking, and
I eventually lost most of my muscle tissue. My skin became so dry that
it flaked and peeled. It was not only painful, but embarrassing, and
I wondered how my school friends could bear to look at me. Many
of them could not, and in the last year, I spent much of my time alone.

'However, what was even harder to bear was the knowledge that my life
could end at any time. It was very difficult to live with the thought that if I took
a turn for the worse I could lapse into a coma and die.

'Then came the news about the discovery of insulin...'

Temperature regulation is discussed
in Chapter 12, how the kidney
controls excretion and water balance
is discussed in Chapter 13, and
exercise physiology is discussed in
Chapter 14.

1 THE CONCEPT OF HOMEOSTASIS

In this part of the book, we focus on the basic concepts and mechanisms
of homeostasis, and then look at some examples including blood
sugar, temperature and the workings of the liver and kidney. Finally, we
investigate exercise physiology – the short-term and long-term changes that

happen to our bodies when we exercise. The fact that our muscles can work 20 times as hard and yet our body can still maintain acceptable internal conditions is one of the most impressive feats of homeostasis.

The word **homeostasis** means 'steady state'. Homeostasis describes how the body regulates its processes to keep its internal conditions as stable as possible. Homeostasis is necessary because cells, especially those of humans and other higher animals, are efficient but very demanding. To function properly they need to be bathed in tissue fluid that can provide the optimum conditions. Nutrients and oxygen must be delivered and waste needs to be removed. The concentration, temperature and pH of the fluid between cells must also be kept at levels that guarantee efficient cell functioning.

However, the phrase 'steady state' is a bit misleading. The conditions inside our bodies are not constant, but are kept within a narrow range. Some factors, such as core temperature and blood pH, fluctuate only slightly, while others, such as blood glucose, vary considerably throughout a normal day without producing any harmful effects.

In this chapter we look at the control of blood glucose and at the role of the liver. Other aspects of homeostasis are covered elsewhere in this book (Table 11.1).

THE MECHANISM OF HOMEOSTASIS

When you start to study how the body controls a physiological factor such as temperature, blood glucose or blood pressure, it is important to organise your thoughts by asking the following questions:

- What conditions bring about a change in the factor being considered?
- What detects the change?
- How is the change reversed?

You will soon notice a pattern. Whenever a physiological factor changes, the body detects the change and then, by using nervous or hormonal signals, or both, it reverses the change. The extent of the correction is monitored by a system called **negative feedback** (Fig 11.1). This makes sure that, as levels return to normal, corrective mechanisms are scaled down.

Table 11.1 Homeostatic mechanisms covered in this book

Homeostatic mechanism	Covered in
Temperature	Chapter 12
Blood pressure	Chapter 10
Solute concentration of blood	Chapter 13
Blood pH	Chapter 13
Blood volume	Chapter 13
Blood hormone levels	Chapter 18

? QUESTION 1

1 Why are homeostatic mechanisms often described as detection–correction systems?

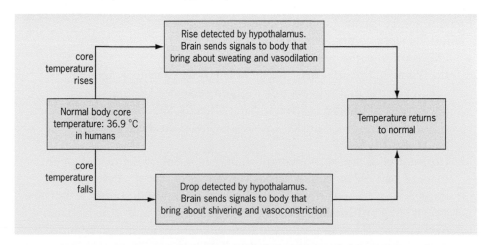

Fig 11.1 The mechanism of homeostasis as illustrated by temperature control. All other examples follow the same general pattern

Control of body temperature illustrates this mechanism well. When we are in a very hot environment or when we have been doing strenuous exercise, our body temperature rises. The brain detects this and sends signals to the body to bring the temperature down, using various corrective mechanisms such as sweating and increased blood flow to the skin. As the body cools, the drop in temperature is monitored by the brain, which begins to send out fewer signals. Sweating then decreases.

Temperature regulation is discussed more detail in Chapter 12.

Blood clotting is discussed in more detail in Chapter 30.

The opposite of negative feedback is **positive feedback**. In this situation, a change is amplified rather than returned to normal. Positive feedback in living systems is rare, but there are a few examples. Damage to a blood vessel causes a cascade reaction that brings about blood clotting: a few molecules of a substance become activated and each one then activates many more. The result is a tangled mesh of protein fibres that plugs the hole in the blood vessel.

Positive feedbacks also occur in abnormal situations and when normal homeostatic mechanisms get out of control. Elderly people whose sensory systems have deteriorated can suffer from hypothermia (see the Assignment for Chapter 12 on the CAS website). When they start to become cold, their body systems fail to respond, and the drop in body temperature goes uncorrected. As they grow colder, their metabolic rate decreases still further, they produce less heat and so in cold conditions they continue to cool down at an ever-increasing rate. Death occurs when their core temperature falls to about 25 °C.

Homeostatic mechanisms at the molecular level

We can also look at homeostasis at the level of molecules. In Chapter 5, we saw that enzymes control metabolic pathways, and that particular products are made, step by step, in a series of carefully controlled reactions. In a metabolic pathway, any product is made only as fast as it is needed. Excess product often prevents further quantities being made by inhibiting one of the enzymes in the pathway. When the excess has been used up, the inhibition is lifted and production continues. The process is self-regulating and so we describe it as a **homeostatic mechanism**.

2 THE CONTROL OF BLOOD GLUCOSE

In normal circumstances, we obtain most of our energy by respiring glucose. Cells therefore need a regular supply of this simple sugar. Some vital organs, notably the brain, cannot do without it for even a short time: lack of glucose can cause brain damage. Other cells and tissues, such as muscles, can respire lipids or even proteins for a short time if glucose is unavailable.

Blood glucose comes from:

- Digestion of carbohydrates in the diet.
- Breakdown of **glycogen**. This storage polysaccharide is made from excess glucose in a process called **glycogenesis**. Glycogen is particularly abundant in liver and muscle cells. When needed, glycogen can be broken down quickly to release glucose in the process of **glycogenolysis**.
- Conversion of non-carbohydrate compounds. Following deamination, the acid part of the amino acid is converted to glucose. Pyruvate and lactate can also be converted to glucose. The conversion process in either case is called **gluconeogenesis**, which means literally 'the generation of new glucose'. During prolonged fasting, blood glucose is maintained by conversion of the body's protein and lipid stores.

The blood of a healthy person contains between 80 and 90 mg of glucose per 100 cm³. This normal value is maintained even during prolonged fasting. But the value rises to around 120 to 140 mg per 100 cm³ shortly after a meal, when carbohydrate digestion is in full swing. Feedback mechanisms bring the levels back to normal in about two hours. Under normal conditions, the kidney is able to reabsorb all of the blood glucose passing through it, preventing any from being lost in the urine.

REMEMBER THIS

Glucose is one of the most abundant substances in our diet. Plant material contains starch and cellulose, and meat contains glycogen. All three are glucose polymers. During digestion, both starch and glycogen are broken down into glucose, which then passes into the blood in large amounts. Humans cannot digest cellulose, and this constituent of plant material forms much of our dietary fibre.

REMEMBER THIS

The terms used to describe glucose metabolism – glycogenesis, glycogenolysis and gluconeogenesis – can be confusing. Try remembering the origins of the words that make them up:

glyco, gluco = sugar
lysis = splitting
neo = new
genesis = generation, or creation.

THE MECHANISM OF BLOOD GLUCOSE CONTROL

The control of blood glucose level is a good example of homeostasis. A negative feedback mechanism operates to detect and correct the level of blood glucose, maintaining it within 'safe' limits. The pancreas plays a central role here: its **endocrine role** is to produce the hormones **insulin** and **glucagon** to control blood glucose (look ahead to Fig 11.3).

The pancreas itself detects any change in the level of blood glucose. If blood glucose becomes too high, β **cells** in the **islets of Langerhans** respond by releasing insulin. This hormone travels to all parts of the body in the blood, but exerts an effect mainly on cells in muscles, liver and adipose (fat storage) tissue. Insulin lowers blood glucose by making cell surface membranes more permeable to glucose. It activates transport proteins in the membranes, allowing glucose to pass into cells. Insulin also activates enzymes inside the cells. Some of these enzymes convert the glucose to glycogen, others increase protein and fat synthesis.

If the levels of blood glucose get too low, α **cells** in the islets of Langerhans secrete glucagon. This hormone fits into receptor sites on cell surface membranes, and activates the enzymes inside the cells that convert glycogen to glucose. The glucose then passes out of the cells and into the blood, raising blood glucose levels.

WHEN CONTROL OF BLOOD GLUCOSE FAILS

People with the disease **diabetes mellitus** are unable to control the level of glucose in their blood. This produces a range of symptoms. Blood glucose levels can get too high because the affected person produces little or no insulin. Without insulin, glucose cannot pass into cells and remains in the blood. The solute concentration of the blood increases, interfering with effective circulation and making the individual very thirsty. The cells are starved of their main fuel and are forced to respire lipids and proteins, leading to weight loss and eventual starvation. Glucose appears in the urine because blood sugar levels are so high that the kidney cannot reabsorb it all.

Diabetes has a variety of causes and varying degrees of severity. In the UK, about 25 people in every thousand suffer from diabetes in one form or another. That means that there are over a million sufferers. About a third of these have Type I diabetes, roughly two-thirds have Type II (see below).

Types of diabetes mellitus

There are two main types of diabetes: **Type I** and **Type II**.

Type I diabetes, also known as 'early onset diabetes' or 'insulin-dependent diabetes', occurs when the body cannot make insulin. This is often caused by an auto-immune reaction which attacks and destroys cells in the islets of Langerhans. This form of the disease usually appears before the age of 20, and the onset is sudden. Sufferers have this condition for the rest of their lives, but they can be treated by regular injections of insulin (Fig 11.2) matched to their glucose intake (in diet) and expenditure (in exercise). Before insulin was available, untreated Type I diabetes was usually fatal within a year of diagnosis. Most people with diabetes today need insulin injections, but they lead normal lives.

Type II diabetes, also called 'late onset diabetes', is more common than Type I, accounting for about 70 per cent of the cases in the UK. It tends to begin during middle age and is more common in those who are overweight. This form of diabetes is due to a decline in the efficiency of islet cells, or to a failure of the cell surface membranes to respond to insulin. Fortunately, in many cases it can be controlled by regulating the diet.

> The metabolism of glycogen and glucose is discussed in more detail in Chapter 3. The digestive functions of the pancreas are covered in Chapter 8.

> **REMEMBER THIS**
> Diabetes mellitus is an inability to control blood glucose levels. It is not the same as diabetes insipidus, which is caused by a lack of anti-diuretic hormone in the body.

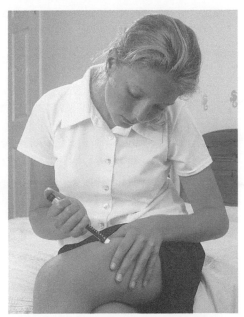

Fig 11.2 This girl has Type I diabetes and has to inject insulin every day. For many years, insulin for human treatment was derived from cows or pigs, but both are slightly different from human insulin and can cause an immune reaction, reducing their effectiveness. Today, diabetics use human insulin produced by genetically engineered bacteria (see Chapter 6)

The discovery of insulin

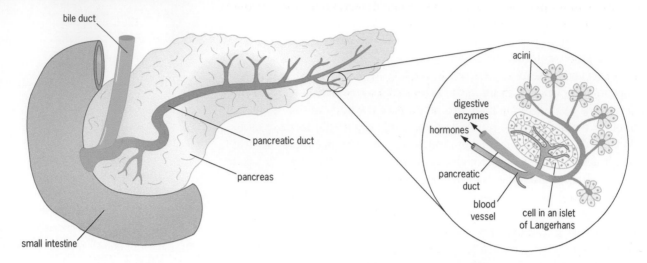

In the early days of science, one of the most direct ways to find out an organ's function was to remove it surgically and then observe the effects of its loss on the organism. If the symptoms could be relieved by injecting a ground-up extract of the same gland, then the organ in question was an endocrine gland – a gland that releases hormones into the bloodstream.

By the early years of this century, this effective but less-than-subtle approach had been used to clarify the role of several organs, including the thyroid gland. But, although researchers suspected that the pancreas was an endocrine gland, they ran into problems when they tried to use this method to demonstrate that the pancreas made a hormone responsible for controlling glucose metabolism (Fig 11.3). When the pancreas of a dog was removed, the animal developed the symptoms of diabetes. But injecting an extract of ground-up pancreas failed to relieve the symptoms, a result that recurred in several experiments.

Frederick Banting, a Canadian doctor, first had the idea that the digestive enzymes also made by the pancreas could be destroying any other active substances that were being made there. He persuaded the head of the physiology department at the University of Toronto, John J R Macleod, to let him have a laboratory, ten dogs and an assistant – Charles Best (Fig 11.4).

The dogs were subjected to one of two treatments. The pancreatic ducts of the dogs in group 1 were tied so that the animals could not produce pancreatic juice. Over the course of several weeks the acini, the cells that make pancreatic juice, degenerated, leaving just the islets of Langerhans functional. These dogs did not develop diabetes. The pancreases of the group 2 dogs were removed, and these animals did become diabetic.

Banting then made a pancreatic extract from the group 1 dogs and injected it into the diabetic dogs of group 2. The result was dramatic – an instant reduction of blood sugar that could be

Fig 11.3 The major part of the pancreas makes a juice containing digestive enzymes, but small patches of cells, called islets of Langerhans, produce the hormones insulin and glucagon

Fig 11.4 Banting and Best in the early 1920s. Their experiments on dogs that lead to the development of insulin did cause the suffering and death of some animals. However, we need to balance that against the millions of people and also the many dogs and other animals with diabetes that have benefited from insulin

achieved repeatedly in several different animals. There was great excitement, and then disappointment as the extracts were tried in humans and found to be too impure (they produced fever). Banting and Best enlisted the help of the biochemist James Collip. He went on to make a purer extract than had previously been possible, which could be used for people.

Some people feel that animal experiments are a bad idea, but the few dogs used in these early experiments enabled Banting's team to change a progressive and fatal illness into a chronic, manageable condition. As a direct result of these experiments, insulin became available to the world's diabetics. Today there are over 300 000 insulin-dependent diabetics in the UK, and as many as 30 million worldwide. Without insulin, they would not be alive.

LIVING WITH DIABETES

Diabetics must become experts in the management of their own condition, matching up dietary intake with insulin doses and exercise regimes in order to keep their blood glucose within definite limits.

Type I diabetes

In **Type 1** diabetes, little or no insulin is released into the blood after a meal. Without sufficient insulin, most of the glucose that enters the blood from the intestines after digestion cannot get into cells. Glucose accumulates in the blood, and the cells, which are starved of their main fuel, turn to alternatives – lipid and protein.

Five of the major symptoms of Type I diabetes are:

- excessive thirst – the excess glucose in the blood lowers the water potential and triggers the sensation of thirst;
- excessive urinating – a consequence of excessive drinking;
- weight loss – the cells respire stores of lipid;
- glucose in the urine – there is so much in the blood that the kidney cannot re-absorb it all;
- the fruity smell of ketones on the breath (a by-product of lipid metabolism).

All of the above are caused by **hyperglycaemia** – high blood sugar. Other symptoms include blurred vision, poor wound healing and frequent skin infections, such as thrush.

With Type I diabetes, the aim is to keep blood glucose levels within the range of 4 to 11 millimoles per litre. A millimole is one thousandth of a mole. Non-diabetics maintain their blood glucose to within about 4 to 9 millimoles per litre.

When they wake up in the morning, Type 1 diabetics usually test their blood sugar. This is done by pricking a finger and putting a drop of blood on a test strip, though there are sophisticated 'watches' than can take readings straight through the skin. Blood glucose levels will be low, following a night with no food. Any blood glucose value of less than 4 is regarded as **hypoglycaemic,** although some diabetics get no symptoms until the value approaches 1.

After taking the reading, the next step is usually to inject some insulin (Fig 11.5); it can't be taken orally, insulin is a protein and would be digested. Having taken insulin they have to eat, otherwise there won't be any glucose for the insulin to act on.

Two types of insulin are now available; fast-acting and slow-acting (Fig 11.6). Diabetics get used to judging the dose they need, for example two units of fast-acting and 14 of slow-acting might be enough to last until lunchtime. Fast-acting is normal, soluble insulin that works straight away. Slow-acting insulin is attached to a retarding agent that releases the insulin slowly over the next few hours. This allows patients to have a relatively normal day without having to constantly check their blood glucose and inject themselves. Diabetics get used to managing their condition and can match their moods and feelings to their blood glucose level. Many diabetics carry their insulin, testing kit and a few sweets or biscuits around with them.

Until relatively recently the insulin was obtained from animal pancreas. This was a problem because the non-human insulin is slightly different from human insulin and slowly brought about an immune reaction. Eventually, the 'foreign' insulin had little effect. With modern genetic engineering, however, it is now a relatively simple task to produce human insulin on a large scale.

Fig 11.5 For diabetics, there is now an alternative to syringes. Special pen-like injection devices such as this Novopen are widely available. They are very accurate and discreet to use

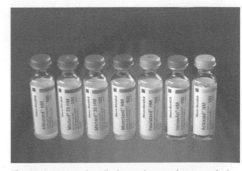

Fig 11.6 Human insulin is made on a large scale by genetic engineering and is supplied in different doses. A number of insulins are available that act at different speeds. The 'Actrapid' works straight away while the 'Ultratard' is very slow acting

REMEMBER THIS

Although control of blood sugar is a 'whole-body' process, the liver plays a central role. It is the first organ to receive the blood from the intestines that contains high levels of food molecules such as sugars following a meal. Liver cells are acutely sensitive to insulin and are also particularly rich in the enzymes involved in glucose metabolism.

? QUESTION 2

2 During their experiments, Banting and Best noticed that flies congregated around the urine of the diabetic dogs. Suggest an explanation for this.

Going 'hypo'

Hypoglycemia is low blood sugar, usually caused by too much insulin, not enough food, or vigorous exercise. Early symptoms may include confusion, dizziness, feeling shaky, headaches, irritability, pale skin and sweating. Without treatment, more severe symptoms may develop, including headache, passing out and coma.

Type II diabetes

Many diabetics have **Type II** diabetes, also known as 'late onset diabetes'. The treatment depends on the underlying cause. In most Type II cases, the condition is due to a combination of one or more of the following:

- muscle and fat cells do not respond to insulin;
- the body can't produce enough insulin to meet demand;
- liver cells release too much glucose from their stores.

The incidence of Type II diabetes has increased dramatically in the developed world. It is also predicted that many people in eastern countries such as India and China will follow suit, as they adopt a more western diet, i.e. one with higher levels of sugar and fat, and less fibre.

The underlying cause of most cases of Type II diabetes is obesity. Most new cases used to occur in the over 45s, but younger people are increasingly being diagnosed.

Diagnosis

Diagnosis is often made by an oral glucose tolerance test (OGTT). The patient first has a fasting glucose test (having not eaten for 8 hours) to measure a 'background' level. They then drink a set amount of glucose and their blood glucose levels are measured at intervals to see how the body deals with the new intake. The body should be able to detect the increased level, secrete insulin and absorb the new glucose into various cells around the body, lowering the blood glucose levels again. In diabetics, levels remain higher for longer. Table 11.2 shows how the results are interpreted after 2 hours.

Table 11.2 Interpretation of results of an OCTT (2 hours after a 75 g glucose drink)

Blood glucose level	Diagnosis
Less than 140 mg/100 ml (7.8 mmol/l)	Normal glucose tolerance
From 140 to 200 mg/100 ml (7.8 to 11.1 mmol/l)	Impaired glucose tolerance (pre-diabetes)
Over 200 mg/100 ml (11.1 mmol/l) on more than one testing occasion	Diabetic

Prognosis

There is no simple cure for Type II diabetes, but the condition can be managed and the complications minimised by a combination of diet and exercise to control weight.

Common complications include blindness (diabetic retinopathy), circulatory problems and kidney failure. For these reasons the eyesight, blood pressure and kidney function of diabetics is closely monitored.

3 THE LIVER PLAYS A CENTRAL ROLE IN HOMEOSTASIS

The liver receives blood from two sources: from the intestines via the hepatic portal vein, and from the hepatic artery (Fig 11.7b). The composition of blood that flows into the liver can fluctuate greatly, depending on factors such as the timing and nature of the last meal, but the content of blood leaving the liver is remarkably constant.

THE STRUCTURE OF THE LIVER

Before we go on to look at the functions of the liver in more detail, it is important to understand its structure (see Fig 11.7).

Blood is brought into the liver by two blood vessels. The 30 per cent that arrives in the **hepatic artery** is oxygenated blood, while the 70 per cent delivered by the **hepatic portal vein** contains relatively little oxygen but is rich in nutrients that have been absorbed from the intestines. The **hepatic vein** removes blood from the liver.

The liver consists of hundreds of thousands of **lobules**, each about 1 mm in diameter, which surround branches of the hepatic vein. Channels called **sinusoids** radiate out from each central vein. These are surrounded by rows of liver cells called **hepatocytes**. These apparently unspecialised cells perform the majority of the liver's functions, and the composition of the blood changes as it flows along each sinusoid towards the central vein.

Dotted along the sinusoids are numerous white cells called **Kupffer cells** (Fig 11.7c). These are phagocytes that engulf bacteria and debris. Parallel to the sinusoids are fine channels called **bile canaliculi** (singular: canaliculus). Hepatocytes secrete the constituents of bile into the canaliculi, and this drains into the gall bladder.

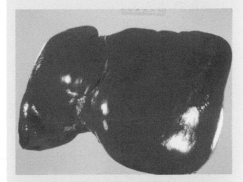

Fig 11.7a A liver prepared for organ transplant sugery. This liver has been cut to fit the recipient

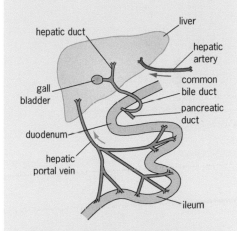

Fig 11.7b The blood supply to the liver and associated organs. The liver weighs about 1.5 kg in a normal adult

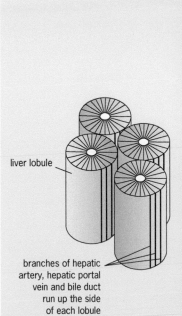

Fig 11.7c The liver consists of about a half to one million cylindrical blocks of cells called lobules

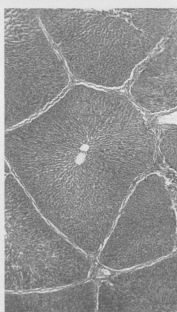

Fig 11.7d A micrograph of liver lobules. Radiating sinusoids drain into the central veins

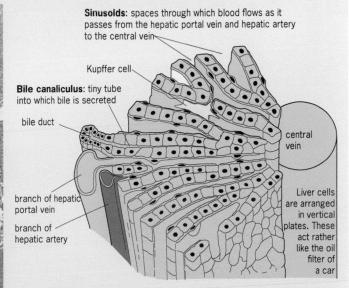

Sinusoids: spaces through which blood flows as it passes from the hepatic portal vein and hepatic artery to the central vein

Kupffer cell

Bile canaliculus: tiny tube into which bile is secreted

bile duct

branch of hepatic portal vein

branch of hepatic artery

central vein

Liver cells are arranged in vertical plates. These act rather like the oil filter of a car

Fig 11.7e The fine structure of the liver. Blood is delivered to the liver cells in branches of the hepatic artery and the hepatic portal vein. As blood flows along the sinusoids, some chemicals are removed while others are added by secretion. The liver secretes bile into the canaliculi. These drain into small branches of the bile duct

? QUESTION 3

3 Hepatocytes have many mitochondria and microvilli. What does this suggest about their function?

Phagocytes are discussed in Chapter 30. Bile is discussed in Chapter 8.

? QUESTION 4

4 Why can't we take in a week's supply of protein in a single meal by eating one large steak or omelette?

Excretion of urea is discussed in Chapter 13.

THE MAIN FUNCTIONS OF THE LIVER

The thousands of different chemical functions performed by the liver contribute greatly to the overall composition of the blood. They can be grouped under a few basic headings:

- Control of blood glucose levels
- Control of amino acid levels
- Synthesis of plasma proteins
- Synthesis of fetal red blood cells
- Destruction of red blood cells
- Detoxification
- Production of bile
- Control of lipid levels
- Storage of vitamins
- Cholesterol formation.

The control of amino acid levels

The human body cannot store proteins. Every day an adult needs a minimum amount (40 to 60 grams, about the weight of one egg) to provide the amino acids the body needs to repair and grow new cells. Most people take in more than this in their diet, and the liver breaks down any amino acids that are not used. Obviously, growing children, pregnant women and breast-feeding mothers need more protein.

Amino acids, like many other digested foods, reach the liver via the hepatic portal vein. They can be:

- **Deaminated**. This process removes the amino group (NH_2) of an amino acid and forms ammonia (NH_3): Fig 11.8a. The organic acid residue is usually respired, while the toxic ammonia is quickly converted into a more harmless substance, urea, via the ornithine cycle: Fig 11.8b. The kidneys remove urea from the body.
- **Transaminated**. There are 8 essential amino acids (10 in children) that must be present in the diet. The remaining 12 are termed non-essential amino acids because they can be made in the liver by transamination. This process involves the transfer of an amino group from an amino acid to an acid (derived from carbohydrate metabolism), thereby making a new amino acid.

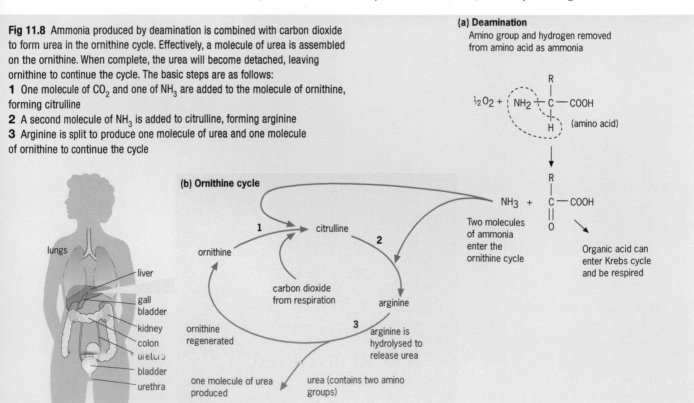

Fig 11.8 Ammonia produced by deamination is combined with carbon dioxide to form urea in the ornithine cycle. Effectively, a molecule of urea is assembled on the ornithine. When complete, the urea will become detached, leaving ornithine to continue the cycle. The basic steps are as follows:
1 One molecule of CO_2 and one of NH_3 are added to the molecule of ornithine, forming citrulline
2 A second molecule of NH_3 is added to citrulline, forming arginine
3 Arginine is split to produce one molecule of urea and one molecule of ornithine to continue the cycle

(a) Deamination
Amino group and hydrogen removed from amino acid as ammonia

(b) Ornithine cycle

lungs
liver
gall bladder
kidney
colon
ureters
bladder
urethra

1
citrulline
2
ornithine
carbon dioxide from respiration
arginine
3
ornithine regenerated
arginine is hydrolysed to release urea
one molecule of urea produced
urea (contains two amino groups)

NH_3 +
Two molecules of ammonia enter the ornithine cycle
Organic acid can enter Krebs cycle and be respired

- Used to synthesise plasma proteins such as fibrinogen.
- Released unchanged into the general circulation (most cells need a supply of amino acids for protein synthesis).

Detoxification and hormone breakdown

The liver concentrates and **detoxifies** (breaks down) many harmful chemicals. Some, such as hydrogen peroxide, are produced by the body itself. Others, alcohol and food additives for example, come from outside. Drinking large amounts of alcohol over a long period of time can cause the death of liver cells, followed by replacement with connective tissue ('scar' tissue). This is known as **cirrhosis** of the liver (Fig 11.9).

The liver also breaks down many circulating hormones. Removal of hormones from the blood is an important aspect of the control process as it ensures that hormones act only for as long as they are needed. Insulin, for instance, is rapidly metabolised by the liver and has a half-life of about 10 to 15 minutes.

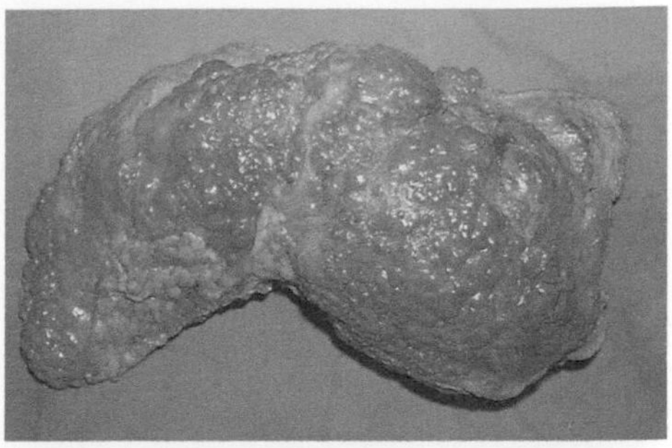

Fig 11.9 The liver of an alcoholic who developed cirrhosis of the liver. In the liver, alcohol (ethanol) is converted into ethanal by the enzyme alcohol dehydrogenase. Ethanal is toxic and seems to be responsible for much of the long-term alcoholic liver damage, although the exact mechanisms involved are unclear

Storage

The liver stores relatively large amounts of vitamins A, D and B_{12}, enough to supply the body for several months. Other vitamins, such as most of the B complex, and the minerals copper and iron are also stored but in smaller amounts. This high vitamin and mineral storage capacity explains why eating liver is good for you, even though some people don't find it a pleasant experience.

Manufacture of blood proteins and blood cells

The liver makes many important blood proteins, including the most abundant plasma protein, **albumin**, fibrinogen and other substances involved in blood clotting. Given a supply of vitamin K, the liver can make prothrombin and other blood clotting factors.

The liver of a fetus manufactures red blood cells, but this complex process is taken over by the bone marrow after birth.

Blood clotting is dicussed in detail in Chapter 30.

SUMMARY

When you have studied this chapter, you should know the following:

- The word **homeostasis** means 'steady state'. Homeostatic processes keep conditions in the body within narrow limits.
- Homeostasis is usually maintained by **negative feedback mechanisms**. The body detects a change in a particular internal factor, such as core body temperature, and then activates a corrective mechanism to reinstate the normal level.
- Blood glucose levels are monitored by the **islets of Langerhans** in the pancreas. If levels get too high, β cells in the islets secrete **insulin**. If blood glucose gets too low, α cells secrete **glucagon**.
- **Insulin**, a peptide hormone, increases the permeability of cell surface membranes to glucose. It also appears to activate intracellular

enzyme systems that convert glucose to glycogen, fat and protein.
- **Glucagon** promotes the breakdown of **glycogen**, releasing more free glucose into the blood.
- Fred Banting and Charles Best, two scientists working in Canada in the early twentieth century, discovered insulin. We may now see their experiments on dogs as controversial but the hormone treatment their work led to is now responsible for allowing many of the millions of diabetics alive today to lead normal lives.
- The liver is a large organ that plays a major role in controlling the composition of the blood.

Practice questions and a How Science Works assignment for this chapter are available at www.collinseducation.co.uk/CAS

12

Temperature regulation

SCIENCE
IN CONTEXT

Why it is easier to drown in warm water

There are many recorded cases of 'near drowning', incidents in which people have been plucked from water and resuscitated, and subsequently restored to full health. Rescuing the victim quickly – within minutes – lessens the chance of brain damage as a result of oxygen starvation.

Things are different, however, when the water is very cold. People have been pulled from freezing water – having fallen through thin ice, for example – and have later recovered completely, despite being under water for as long as one hour. How is this possible?

It seems that two factors contribute. First, a remarkable response known as the diving reflex occurs in people (especially infants) who are suddenly immersed in cold water. Heartbeat and breathing decrease and blood is redirected so that the organs at the core of the body – the heart, lungs and brain – receive much more blood than the limbs. In this way the vital organs get what little oxygen is available.

Secondly, the rapid cooling effect of the water, known as immersion hypothermia, reduces the oxygen demand of the brain and other organs. This means that people can survive for several minutes, perhaps even as long as an hour after the heart and breathing have stopped.

The same principle is used in surgery. Heart operations often involve stopping the heart and then connecting the patient to a heart–lung machine to supply the body with oxygenated blood. The heart itself, however, receives no blood and so to prolong the time available to operate, the core temperature of the patient's body is lowered to 27 °C. This reduces the oxygen demand sufficiently to double the time the surgeon has before the heart needs to be reconnected.

Babies are at home underwater – they spend nine months there. At this age the diving reflex is well developed.

1 THERMOREGULATION: SOME BACKGROUND

Different animals control their body temperature in different ways. Many animals (fish, insects, amphibians, etc.) have a body temperature that is more or less the same as their surroundings. Humans, other mammals and birds usually maintain a constant body temperature, despite the external temperature.

These two categories of animal are commonly described as **cold-blooded** and **warm-blooded.** But these terms are not really satisfactory: a cold-blooded animal, for example, is often not cold. In bright sunshine on a hot day, the core temperature of a crocodile can be several degrees higher than that of a mammal (Fig 12.1). Nor are the terms **poikilothermic** (cold-blooded) and **homoiothermic** (warm-blooded) strictly accurate. The word *poikilos* means changeable, while *homeo* means constant. However, many cold-blooded animals, especially aquatic ones, have a remarkably constant body temperature, due to the stability of their surroundings. Hibernating mammals, on the other hand, show a dramatic drop in their core temperature during their dormancy.

To overcome these problems, we now use two other terms: **ectotherm** ('heat from outside') and **endotherm** ('heat from within'):

- **Ectothermic organisms** can control their body temperature only by changing their behaviour. All animals except mammals and birds are ectotherms.
- **Endothermic organisms** maintain a stable core body temperature using physiological and behavioural means. Mammals and birds are the only endotherms. Humans are therefore endotherms.

HUMANS AND HEAT

Some mammals, such as polar bears, are obviously adapted to withstand the cold. In contrast, humans almost certainly evolved in Africa, and, as a species, we are adapted to a warm climate. We have a thin covering of insulating fat and our body hair is sparse. Most hairs are tiny and no use for insulation. We are one of the few animals to be covered in sweat glands, and our skin can make melanin, a dark pigment that blocks out harmful ultraviolet light.

It seems that we were able to spread to the colder areas of the planet only because of our ability to control our environment. We could build shelters, make clothes and control fire. These skills more than made up for the fact that we had few physical features to allow us to cope with the cold.

2 THERMOREGULATION: THE BASICS

Humans, like most mammals, can maintain a constant core body temperature, despite changes in the surroundings.

The core temperature of the human body remains reasonably constant at around 37 °C. It can fluctuate by a degree or so (more during fever) but it is generally very stable. Fig 12.2 is a thermal image of an adult male, showing the definition of the body core; this includes the trunk, the head and the upper part of the arms and legs. Body temperature is taken by placing a clinical thermometer in sites that detect the temperature of the core (Fig 12.3).

Heat is produced inside the body as a by-product of metabolic reactions. Heat production occurs throughout the body, and is especially high in working muscles. It is often stated that the liver is a particular source of heat, but tests have shown that the temperature of blood in the hepatic vein is no higher than in other vessels, suggesting that heat production occurs more generally in the organs of the body. Since body heat is a by-product of metabolism, the amount of heat produced depends on the metabolic rate.

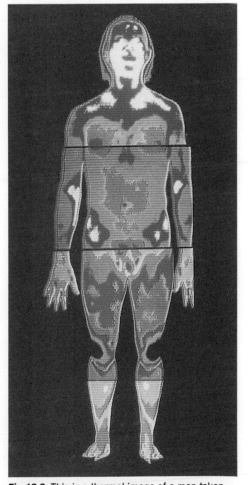

Fig 12.2 This is a thermal image of a man taken with a camera sensitive to infrared radiation. Areas of higher skin temperature look red or orange, cooler areas look blue or purple. The temperature of the skin can vary greatly, but the core temperature remains more or less stable

Hormones are discussed in Chapter 18.

This can be increased by doing more exercise and by the secretion of hormones such as thyroxine or adrenaline.

However, the principal way we control our body temperature is by increasing or decreasing heat loss to the environment, according to our needs. In humans, the basic mechanism that underlies temperature control, or **thermoregulation**, involves part of the brain called the **hypothalamus**. This acts as a thermostat. It can detect the temperature of the blood that passes through it and, if the temperature of the blood increases or decreases even slightly, the hypothalamus initiates corrective responses such as sweating or shivering.

When we encounter a particularly warm or cold environment, temperature receptors (or **thermoreceptors**) in the skin inform the hypothalamus. They also stimulate the higher, voluntary, centres of the brain. This means that we 'feel' hot or cold and decide to do something about it such as changing position (Fig 12.4), changing our clothing or turning the heating up or down. Often, this behavioural response corrects the situation without the need for any physiological response.

Before we can look at temperature control in detail, we need to understand a little about the processes of heat transfer, and to study the structure of the human skin.

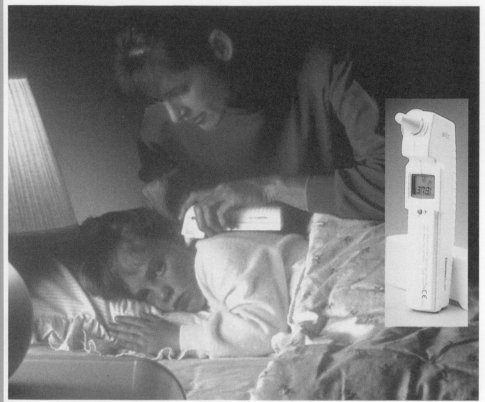

Fig 12.3 Measuring core body temperature. A clinical thermometer provides a good estimate of core temperature when it is placed in the mouth, armpit or rectum. This more recent thermal probe allows an instant reading of the temperature of the blood near to the brain, without damaging the ear

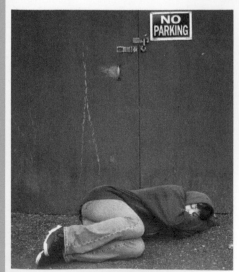

Fig 12.4 When it's cold, we may automatically assume a fetal position in bed to keep warm; when the arms and legs are drawn into the body this reduces the surface area that can lose heat. When it's hot we do the opposite, spreading out as much as possible

3 HEAT: WHERE DOES IT COME FROM? WHERE DOES IT GO?

For any organism to maintain a stable body temperature, its heat loss must equal its heat gain. **Endotherms** such as ourselves, which live in a temperate climate, produce their own internal body heat to keep warm but they also control the amount of heat they lose to the environment.

Heat can be gained or lost in four different ways:

- conduction
- evaporation
- convection
- radiation.

CONDUCTION

Conduction involves the transfer of heat between two objects that are in contact with each other. Heat is always conducted from a region of higher temperature to a region of lower temperature. When you sit on a cold seat, for example, heat is conducted from your body into the seat, until both are approximately the same temperature.

Fig 12.5 When the *Titanic* sank in the North Atlantic, the water temperature was similar to that of the air – a little above freezing. However, as we now know, those who remained dry in the lifeboats had a far greater chance of survival. The high thermal conductivity of the water drew the heat out of the unfortunate victims in the sea, most of whom died from hypothermia

The efficiency with which a material conducts heat is called its thermal conductivity. Different materials have different conductivities. Air has a low thermal conductivity, water has a much higher one (Figs 12.5 and 12.6).

A clothed human walking in air at 15 °C can maintain body temperature comfortably. If immersed in water at that temperature, it would not be long before their core temperature dropped and hypothermia began to set in. The heat loss into the water would be much greater – water can 'draw out' heat approximately 25 to 30 times faster than air at the same temperature.

Materials with a low thermal conductivity are very good insulators. Animals with fur keep warm largely by trapping air between the hairs. Humans rely more on fatty (adipose) tissue for insulation. This has a lower thermal conductivity than other body tissues and so conducts heat more slowly to the surroundings.

Fig 12.6 Most divers wear wet suits, even in tropical waters, because the body loses heat rapidly to any water that is below 29 °C. The wet suit traps a layer of water between the skin and the rubber. This warms up quickly as a result of conduction from the skin. The warm water layer cannot escape and so further heat loss is slow. For very cold water, dry suits are available. These trap a layer of air next to the skin. This provides even better thermal insulation

CONVECTION

Convection is heat transfer due to currents of air or water. A person immersed in cold, absolutely still water would be able to heat up the water immediately next to the skin, and could reduce their heat loss to some extent by not moving.

However, this situation does not happen in real life. For a person who has fallen into the sea or a river, there are usually strong currents that continually move the water over the skin, causing heat to be lost quickly by convection. Similarly, fast-moving air causes greater heat loss by convection. You have probably heard of the **wind chill factor**. A cold windy day feels a lot colder than a cold still day, even when the air temperature is the same.

In air, our clothes reduce heat lost by convection by trapping a layer of air next to the skin. In water, divers reduce the risk of heat loss by convection by wearing wet suits or dry suits (see Fig 12.6).

EVAPORATION

Evaporation is the change in state from a liquid to a gas. The evaporation of water uses up a large amount of energy, and this is known as the **latent heat of evaporation.** What this means for everyday life is that evaporation of water from a surface has a great cooling effect. Anyone who has ever stepped out of the bath and stood in a draught will have felt the power of this cooling. It is also why sweating is usually an effective way of losing heat when we get too hot. Even hot air – such as from a hair dryer – can have a cooling effect provided that the skin is wet so that evaporation can take place.

RADIATION

Radiation is the loss of infrared heat into the surroundings. A person sitting in a room at 20 °C radiates heat into the surrounding air, particularly from the exposed skin on the head, neck and hands. Under normal situations, at rest at comfortable room temperature, most of our heat loss is due to radiation. The heat that we radiate is in the infrared range. This is why infrared cameras can be used to search out humans and other warm-blooded animals from their colder surroundings.

In conduction, convection and evaporation, heat transfer occurs as a result of the movement of molecules. Radiation is fundamentally different: it does not depend on molecular movement. This explains why radiated heat can pass through a vacuum.

4 THE PROBLEMS OF TEMPERATURE EXTREMES

Humans can maintain a stable body temperature over a wide range of external temperatures. But what happens when we are no longer able to thermoregulate effectively, and our internal temperature changes? What happens to people stranded in the desert under a baking sun, or to those trapped in freezing water?

Experiments have been carried out in which a naked man is asked to sit in a room while the temperature around him is gradually raised or lowered. Fig 12.7 shows a generalised graph that plots what happens.

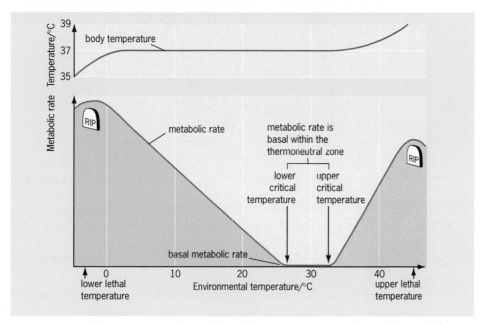

Fig 12.7 The effect of temperature on metabolic rate

Throughout the range 15 to 30 °C the naked man is able to maintain a constant core temperature. A point is reached, however, when his sweating and vasodilation are no longer able to keep him cool and he is said to have reached the **upper critical temperature**. After this, his metabolic rate starts to rise, but the experiment is always stopped at this point. If it were not, a positive feedback would begin. The man's enzymes would work faster and consequently make even more heat, raising his temperature further **(hyperthermia)**. At about 42 to 44 °C, the **upper lethal temperature** would be reached and he would die.

Why do humans die when the upper lethal temperature is reached? A common theory says that vital enzymes become denatured, but this cannot be the whole story. It is more likely that death from excess heat is due to a metabolic imbalance, caused by enzymes working at different rates. As the temperature increases, some enzymes work faster than others: some intermediate chemicals build up while some vital end-products become scarce. The end result is that cells, notably those in the brain, become irreversibly damaged.

If the same type of experiment is done, but temperature is lowered, the room eventually becomes so cold that the body is no longer able to maintain a constant core temperature: the **lower critical temperature** has been reached. At this point the metabolic rate begins to rise, and this helps the situation by generating more heat. Again, the experiment is always halted at this stage but, if heat loss were allowed to continue, a **lower lethal temperature** would be reached and the subject would die. Find out more about this in the Assignment for this chapter on the CAS website.

REMEMBER THIS

In Chapter 5 we saw that enzymes, the chemicals that control metabolism, are temperature sensitive. Enzyme activity increases with temperature and this increases our metabolic rate. The Q_{10} is a value that describes the effect of a 10 °C rise in temperature on the rate of reaction. Most enzymes have Q_{10} values of between 2 and 3. This means that they double or triple their activity with every 10 °C rise in temperature, up to the temperature when the enzymes are denatured by heat.

? QUESTION 1

1 Why does the upper critical temperature vary with humidity?

5 THE STRUCTURE OF HUMAN SKIN

The skin is in direct contact with the external environment and so plays a central role in maximising or minimising heat loss. Fig 12.8 shows the structure of mammalian skin.

Fig 12.8 The structure of human skin. Human skin differs from that of most other mammals by being quite bare. We do not necessarily have fewer hairs, but most are tiny and no use for protection or insulation. We are also covered in sweat glands, and can lose heat by evaporation over most of our body surface. In contrast, most mammals have far fewer sweat glands, because evaporation from wet fur is not effective as a cooling mechanism

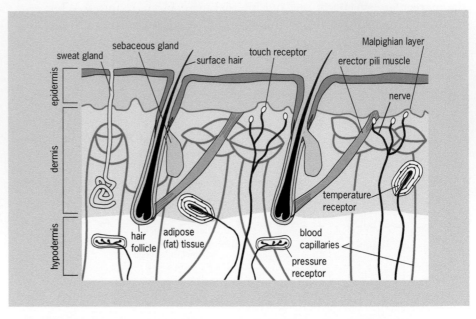

Structurally, the skin is divided into two layers, the outer **epidermis** and the **dermis** underneath. Forming the boundary between the two is the **Malpighian layer**, the function of which is to produce new epidermal cells by mitosis. These cells make the protein keratin, and once they are pushed up into the epidermis they flatten, die and dry up because they have no blood supply. We are constantly losing epidermal cells in a process called **desquamation**. House dust is mainly desquamated skin cells, and a significant proportion of the mass of our mattresses and pillows is due to accumulated skin cells (one estimate says that skin adds 1 per cent to the weight of your mattress every year).

The dermis is much thicker than the epidermis, and contains many different structures such as nerve endings, hair follicles and blood vessels, held together with elastic connective tissue. Beneath the dermis is the **hypodermis** that usually contains at least some **subcutaneous fat**. This fat storage tissue, called **adipose tissue**, determines human body shape to a large extent.

Human skin has several functions:

- It detects stimuli using cells that are sensitive to heat, cold, touch, pressure and pain.
- It prevents excessive water loss or gain.
- It plays a role in thermoregulation by adjusting heat loss according to the circumstances.
- It prevents entry of microorganisms.
- It secretes hair, fingernails and toenails.
- It allows subtle forms of communication. We secrete natural scents, **pheromones**, from modified sweat glands.

Pheromones are discussed in more detail in Chapter 18.

? **QUESTION 2**

2 Are tattoos placed above or below the Malpighian layer? Explain.

6 THERMOREGULATION

PHYSIOLOGICAL RESPONSES TO COLD

There are four main physiological responses when the hypothalamus detects a drop in blood temperature:

- We **shiver** when muscles contract and relax rapidly. Shivering muscles give out four or five times as much heat as resting muscles.
- **Vasoconstriction** occurs when the arterioles that lead to the capillaries in the surface layers of the skin **constrict** (narrow), so reducing blood flow to the skin surface (Fig 12.9a). This cuts down the amount of heat lost through the skin. Vasoconstriction is controlled by sympathetic nerves that pass from the **vasomotor centre** in the brain. This centre, in turn, receives information from the hypothalamus.
- **Piloerection** means literally 'erection of hairs' and involves a reflex. In most mammals, piloerection makes the fur 'thicker', so that it traps more air to provide extra insulation. In humans, the **erector pili** muscles in the skin (Fig 12.8) pull our tiny hairs upright, but only succeed in creating goose pimples.
- **Increased metabolic rate**. The body secretes the hormone adrenaline in response to cold. This raises metabolic rate and therefore increases heat production. People who live in cold conditions for a period of several weeks or months show a more permanent increase in metabolic rate due to the secretion of thyroxine.

PHYSIOLOGICAL RESPONSES TO HEAT

There are two main physiological responses to heat:

- **Vasodilation** occurs when arterioles that lead to skin capillaries dilate and shunt vessels are closed off, resulting in a greatly increased blood flow to the skin (Fig 12.9b). As a result, more heat can be lost to the environment. Coupled with sweating, vasoconstriction is a very effective cooling mechanism.
- **Sweating** is the production of a salty solution made by sweat glands. Evaporation of sweat from the skin's surface leads to cooling. The efficiency of sweating depends on the humidity. In dry air, humans can tolerate temperatures of 65 °C for several hours. In humid air, however, when the sweat cannot evaporate, temperatures of only 35 °C (lower than the core body temperature) cause overheating.

? **QUESTION 3**

3 Which of the responses to cold will be the least effective in humans? Explain your answer.

✔ **REMEMBER THIS**

A common misconception is that blood vessels move to the surface of the skin. Blood vessels are fixed. It is the blood flow through them that can be changed.

Thyroxine and other hormones are discussed is more detail in Chapter 18.

? **QUESTION 4**

4 Why do you think that people in the tropics are recommended to take salt tablets regularly?

Fig 12.9 Blood flow to the skin can be altered by controlling the flow of blood in the arterioles leading to the surface capillaries
(a) When tiny sphincter muscles in the arteriole walls contract to prevent blood flow to the skin surface, blood is forced through shunt vessels and away from the surface, so less heat is lost
(b) When the sphincter muscles relax, the arteriole walls relax, allowing more blood to flow into the dermis to increase heat loss

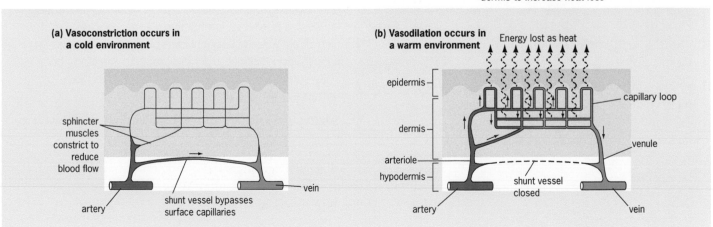

Hypothermia: no one is dead until warm and dead

The definition of hypothermia is a drop in core temperature to 35 °C or below. This condition is very common, and is regularly seen in elderly people (Fig 12.10) and in people who have been immersed in cold water – such as those caught in shipwrecks or people who fall through ice.

People can be revived from the most extreme states of hypothermia, even when all the vital signs point to death.

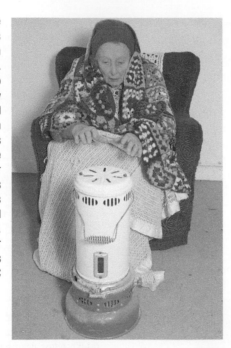

Fig 12.10 Elderly people whose sensory systems have deteriorated can suffer from hypothermia. When they start to become cold, their body systems fail to respond and the drop in temperature goes uncorrected. As the temperature drops, their metabolic rate decreases and they produce less heat, and so they cool down even more quickly. Death occurs when their core temperature reaches about 25 °C

Fig 12.11 A mountain rescue team uses an insulated bag to restore the core temperature of an accident victim

People with severe hypothermia, who are cold and blue, showing no heartbeat or breathing and with fixed and dilated pupils, have been revived and have later recovered.

There are several stages of hypothermia. When someone's core body temperature drops to 36 °C, they are in a state of **impending hypothermia** They start to feel uncomfortable and their skin gets pale, numb and waxy. Their muscles tense, they shiver and they feel tired and weak. This progresses to **mild hypothermia.** Core temperature drops below 35 °C. Uncontrolled, intense shivering begins. Victims are alert and aware of the situation and so may still be able to help themselves but their movements become clumsy and the cold causes pain and discomfort.

If there is no escape from the cold and core temperature drops below 33 °C, **moderate hypothermia** is reached. Shivering slows or stops, muscles begin to stiffen and the victim becomes confused and apathetic. Speech becomes slow and difficult to understand. Drowsiness and eccentric behaviour can follow. Finally, in **severe hypothermia,** the core temperature falls below 31°C. The person's skin is cold and often bluish-grey and their eyes are dilated. They often seem drunk and deny that there is a problem. There is a gradual loss of consciousness. In extreme cases the victim can appear dead, being rigid and not breathing.

The basic goals of hypothermia treatment are to stabilise the core temperature and to prevent cardiac arrest. One of the body's first responses to cold is peripheral vasoconstriction and wrapping the victim in a blanket and allowing them to warm up gently while preventing excessive heat loss is effective (Fig 12.11). Heating the skin directly is not a good idea because it causes vasodilation that returns cold blood to the body core. This leads to a sudden further drop in core body temperature that can stop the heart.

See the Assignment for this chapter on the CAS website.

Freezing humans

Can we keep people alive by freezing them? Can people suffering from incurable diseases be put into deep-freeze until such time as a cure is found, and then revived? The simple answer is no.

Interestingly, human life can survive sub-zero temperatures in a metabolically inactive dormant state. Sperm, eggs and even embryos kept in liquid nitrogen at −196 °C can be thawed out and used successfully in infertility treatment (see Chapter 20). However, it is not possible to freeze people and bring them back to life. One of the main problems is that the formation of ice crystals destroys the cells, but a critical factor in maintaining life seems to be the timescale involved: ice crystals take time to develop. When very small samples are immersed in liquid nitrogen, the freezing process happens rapidly: all the molecules simply 'lock' in position and ice crystals do not form. The cells and organelles are not damaged and, when thawed, function normally. This technique, however, only works with tiny samples of tissue, which freeze instantly and entirely.

Fig 12.12 If we developed a way of instantaneously freezing all the cells of a fully developed human, could they be thawed out and live to tell the tale? Only time will tell. In the meantime, this corpse is in cold storage.

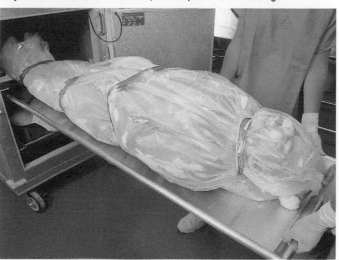

7 SKIN COLOUR AND SUNLIGHT

The colour of human skin is determined genetically. This is an example of continuous variation, which suggests that there are several genes operating that give rise to a whole range of skin colour from the palest to the darkest. The darkness of skin is due to the amount of **melanin** present (Fig 12.13). This pigment is able to absorb ultraviolet light, thus protecting the DNA in skin cells from damage (Fig 12.14).

> Continuous variation is discussed in more detail in Chapter 22.

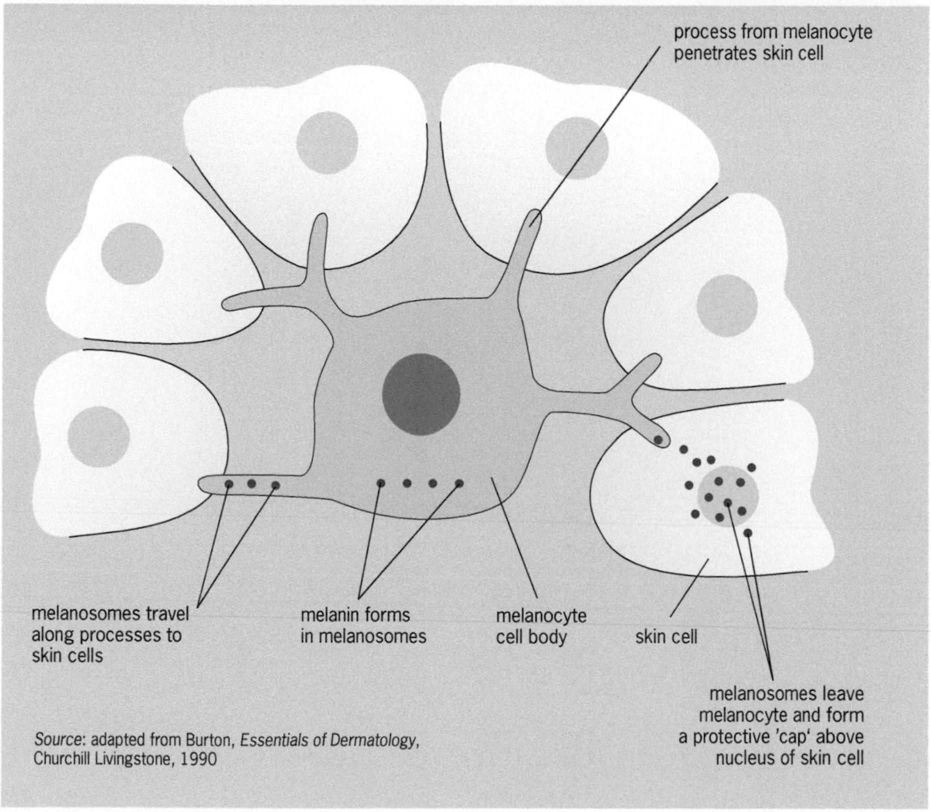

Fig 12.13 Skin contains cells called melanocytes that produce structures called melanosomes, which contain the melanin. Dark skin contains no more melanocytes than pale skin, but the melanosomes are larger and more evenly spread. Each melanocyte can protect the nuclei of several neighbouring cells. All but the darkest skins will respond to sunlight by producing more melanin

process from melanocyte penetrates skin cell

melanosomes travel along processes to skin cells

melanin forms in melanosomes

melanocyte cell body

skin cell

melanosomes leave melanocyte and form a protective 'cap' above nucleus of skin cell

Source: adapted from Burton, *Essentials of Dermatology,* Churchill Livingstone, 1990

Fig 12.14a Black skin colour is due to the pigment melanin, a substance that protects skin from the damage by ultraviolet light. Skin cancer is rare in dark-skinned races. Fair-skinned people are not adapted to cope with strong sunlight

Fig 12.14b This is not a tan, but a first degree burn! When the redness dies down the skin begins to make melanin, giving the desired tan. However, the tan wears away in a couple of weeks. People with pale skin lack the quantities of melanin that provide protection against the genetic damage that ultraviolet light can cause and so run a high risk of developing skin cancer if they sunbathe excessively

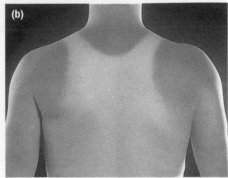

Humans originated in Africa, and needed deeply pigmented skin to protect them from the fierce tropical sun. Under such circumstances, those with darker skin would have had an advantage over paler skinned individuals. It is more difficult, however, to see why white-skinned races evolved. Scientists think that the reason for this may be to do with the ability to synthesise vitamin D.

At the Equator, the sunlight was strong enough to penetrate dark skin and enable the synthesis of enough vitamin D for health. At lower light intensities, however, dark-skinned individuals could not make enough vitamin D. In these circumstances, the paler skinned individuals had a selective advantage and so natural selection brought about the development of pale skin in northern or southern latitudes (Fig 12.15).

Fig 12.15 As a general rule, the further you go from the Equator, the paler the skin colour. The map shows the relationship between skin colour and ultraviolet light intensity. (Adapted from Jolly and Plug, *Physical Anthropology and Archaeology*, 2nd edn, Knopf 1979, and Goldsby, *Race and Races*, Macmillan 1971)

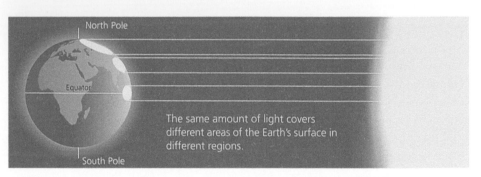

The same amount of light covers different areas of the Earth's surface in different regions.

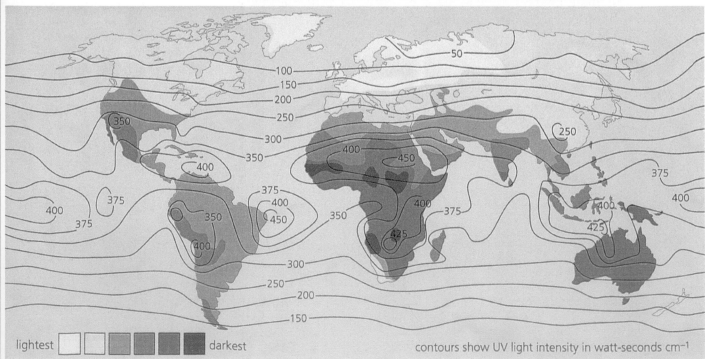

lightest ▢▢▢▢▢ darkest

contours show UV light intensity in watt-seconds cm⁻¹

HOW SCIENCE WORKS

The effect of sunlight on skin

The Sun produces electromagnetic radiation of many different wavelengths. Some ultraviolet radiation is filtered out by the ozone layer, but some reaches the Earth's surface. Ultraviolet light is a potential problem because it is mutagenic – it damages DNA.

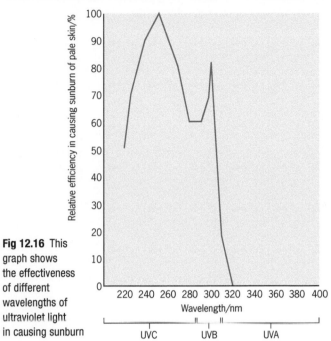

Fig 12.16 This graph shows the effectiveness of different wavelengths of ultraviolet light in causing sunburn

Ultraviolet radiation (Fig 12.16) is split into three categories:

- UVA (long wavelength, 315–400 nm) is the least likely to cause sunburn, and it cannot stimulate melanin production. However, it can damage DNA and lead to gene mutations that can give rise to skin cancer (Fig 12.17).
- UVB (medium wavelength, 295–314 nm) causes sunburn and stimulates melanin production.
- UVC (short wavelength, 200–294 nm) is very damaging to skin but fortunately much of it is filtered out by the ozone layer.

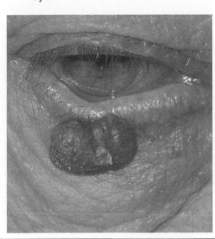

Fig 12.17 Skin cancer: this is a malignant melanoma, a common cancer especially in light-skinned people who are exposed to strong sunlight. If the cancer can be removed before it spreads to other parts of the body, the chances of survival are over 90 per cent

SUMMARY

After studying this chapter, you should know and understand the following:

- Animals are divided into two groups according to their ability to thermoregulate: **endotherms** ('heat from within') maintain a stable core temperature by behavioural and physiological means; **ectotherms** maintain a relatively steady body temperature using behaviour only.
- Humans, like all mammals, are ectotherms. We also say that mammals are warm-blooded, or **homoiothermic**. This means that they can maintain a stable core body temperature despite changes in the external temperature.
- To maintain a stable body temperature, heat loss must equal heat gain.
- Unless the climate is very hot, we regulate body temperature by producing heat as a result of metabolic reactions, and by controlling how much of it is lost to the environment. When we are too hot, we maximise heat loss; when we are cold, we try to lose as little as possible.
- Heat is transferred in four ways: **conduction**, **convection**, **evaporation** and **radiation**.

- Temperature receptors in skin provide an early warning system for temperature change, allowing us to thermoregulate by our behaviour (altering position, clothing, etc.). This often prevents the need for a physiological response.
- The **hypothalamus** can detect the temperature of the blood in the core of the body. If the blood temperature drops, the hypothalamus initiates corrective responses such as **vasoconstriction** and **shivering**. It the blood temperature increases, we respond by **vasodilation** and **sweating**. These are physiological responses.
- Extremes of both heat and cold can cause death. Death from cold occurs as a result of **hypothermia**. This is common in people who have fallen into freezing water.
- The pigment **melanin** protects the skin from ultraviolet radiation. This type of radiation is **mutagenic** and exposure to some types of UV light can cause skin cancer.

 Practice questions and a How Science Works assignment for this chapter are available at www.collinseducation.co.uk/CAS

13

Excretion and water balance

The inconveniences of kidney disease

If one of your kidneys were to stop working because of injury or disease, you could probably still lead a normal life. However, people whose kidneys both fail are faced with a crisis: water, urea and potassium build up rapidly in their bodies. They may continue to pass some urine, but they cannot get rid of all the waste produced by normal cell processes.

Most people suffering from kidney failure hope for a transplant, but there is a shortage of donors. Until a suitable organ becomes available, patients have to rely on dialysis to filter their blood and balance their fluid intake. As the photograph shows, dialysis is uncomfortable and inconvenient.

To reduce the time they need to spend in dialysis, kidney patients must stick to the strict Giovanetti diet. This comes as something of a shock. They must limit their fluid intake to only half a litre a day, about a quarter of the amount a human adult would normally drink. They must also control their protein intake to about 30 to 40 grams per day, the amount of protein in a small egg.

But perhaps the biggest problem is the need to regulate potassium. This ion is a normal constituent of the body, but large amounts cause serious problems, including heart failure. Potassium-rich foods include citrus fruits, bananas, instant coffee, peanuts, treacle and – a big blow for many people – chocolate. In this strange diet, carbohydrates are not a priority. Few patients feel like eating anyway, and this new restricted diet makes the task of finding appetising food even more difficult.

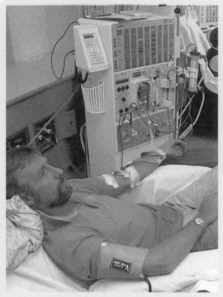

This patient is undergoing dialysis. For several hours, blood flows from the patient's forearm into the machine. Here, much of the waste, together with excess salt and water, is removed by filtration. Find out more about this process in the How Science Works box on page 208.

1 WASTE AND WATER CONTROL

The human body consists mainly of water. A person weighing 65 kg contains about 40 litres of water, of which 28 litres is **intracellular** (inside cells). The rest, the **extracellular** fluid, is made up from about 9 to 10 litres of tissue fluid and 2 to 3 litres of blood plasma.

The kidneys play a major role in regulating the volume and composition of these body fluids. They excrete or conserve water and salt so that the volume and composition of the blood and body fluids remain more or less constant (Fig 13.1). This is a vital aspect of homeostasis.

The kidneys also ensure that waste products do not build up by filtering the blood. Waste is allowed to pass through and then out of the body, while important substances such as glucose are reabsorbed and conserved. The removal of metabolic waste from the blood is called **excretion**.

Urine is the end-product of all processes that occur in the kidney.

Fig 13.1 When we drink a large amount of fluid, our kidneys have the job of getting rid of the excess water, to prevent body fluids from becoming too dilute. The alcohol in the beer also affects kidney function – more of that later

? **QUESTION 1**

1 If a person has 3 litres of blood plasma, and their haematocrit (the volume occupied by cells in the blood) is 40 per cent, what is their total blood volume?

Most excreted waste leaves the body in the urine, but some waste is also lost in **sweat** and in air that we breathe out. Sweat contains mainly salt and water, while **exhaled air** contains carbon dioxide and water vapour.

In this chapter, we concentrate on **nitrogenous excretion**, the removal of waste compounds that contain nitrogen. The main nitrogenous compound excreted by humans is **urea**. This is made in the liver following the breakdown of excess amino acids (Fig 13.2). Enzymes in the liver remove the amine (NH_2) group from amino acids in the process of **deamination**. The ammonia formed is a highly toxic compound that must not be allowed to build up. It immediately enters a series of reactions called the **ornithine cycle**, which produce the relatively harmless compound urea:

$$2NH_3 \quad + \quad CO_2 \quad \rightarrow \quad CO(NH_2)_2 \quad + \quad H_2O$$
ammonia + carbon dioxide → urea + water

Urea is carried in the blood from the liver to the kidneys to be excreted.

In this chapter, we look at the role of the kidneys in homeostasis. The kidneys selectively eliminate water and solutes, such as sodium, potassium and chloride ions, so that the water and solute balance of the body is kept at the correct level. The need to balance the solute concentration of body fluids is called **osmotic regulation**, or **osmoregulation**.

Osmoregulation is the maintenance of a constant solute concentration within the body. In Chapter 4, we saw why this is important: if we place animal cells in a *hypertonic* solution, they lose water by osmosis and shrivel. If we put them in a *hypotonic* solution, they gain water and may burst. Both situations are harmful to cells, and so it is important that we maintain the solute concentration of our body fluids within narrow limits.

This chapter also looks at what happens when something goes wrong. There are two types of kidney failure, also called renal failure. In many cases of **chronic kidney failure** there is a gradual decline in kidney function. **Acute kidney failure** is more of a crisis in which all kidney function stops much more suddenly. There are several causes of **acute kidney failure:**

- sudden loss of large amounts of fluid (blood or tissue fluid);
- inadequate blood flow to the kidneys;
- bacterial infection in the kidneys;
- effect of toxins;
- a blockage in the urinary tract caused, for example, by damage to the ureter.

Later on we look at the three main treatments for kidney failure – haemodialysis, continuous ambulatory peritoneal dialysis (CAPD) and kidney transplants (page 208).

 The ornithine cycle is discussed in Chapter 11.

 REMEMBER THIS
Excretion is the removal from the body of chemical waste produced by the metabolic processes within cells and that would be toxic if allowed to accumulate. You should not confuse excretion with **egestion** or **defecation**, the removal of undigested food and other debris from the intestine.

REMEMBER THIS
Fluids are often compared with respect to their solute concentrations. A solution that has a higher solute concentration than another is said to be **hypertonic**. A solution that has a lower solute concentration than another is **hypotonic**. Solutions that have exactly the same solute concentration are **isotonic**. Thus seawater is hypertonic to human blood plasma, because seawater has a much higher concentration of dissolved salts.

 REMEMBER THIS
The urinary system is also known as the **renal system**.

Fig 13.2 Adult humans need only 40 to 50 grams of protein per day. The body cannot store protein, so daily excess is broken down in the liver. The resulting urea is excreted in the urine

The circulatory system is discussed in Chapter 10.

2 A person has 5 litres of blood, and the kidneys filter 1.2 litres per minute. How many times on average does the total volume of blood in the body pass though the kidneys every hour?

2 AN OVERVIEW OF THE HUMAN URINARY SYSTEM

The human urinary system is shown in Fig 13.3. The kidneys lie at the back of the abdominal cavity, just below waist level, where they are protected to some extent by the spine and the lower part of the ribcage. If you place your hands on your hips, your thumbs show the position of your kidneys. Usually, the left kidney is slightly above the right.

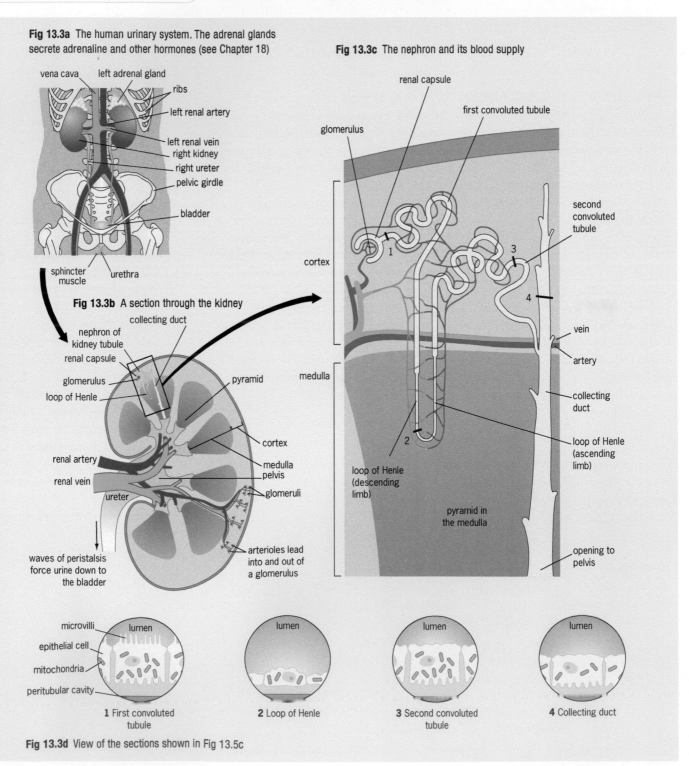

Fig 13.3a The human urinary system. The adrenal glands secrete adrenaline and other hormones (see Chapter 18)

vena cava
left adrenal gland
ribs
left renal artery
left renal vein
right kidney
right ureter
pelvic girdle
bladder
sphincter muscle
urethra

Fig 13.3b A section through the kidney

collecting duct
nephron of kidney tubule
renal capsule
glomerulus
loop of Henle
pyramid
renal artery
cortex
renal vein
medulla
pelvis
ureter
glomeruli
waves of peristalsis force urine down to the bladder
arterioles lead into and out of a glomerulus

Fig 13.3c The nephron and its blood supply

renal capsule
glomerulus
first convoluted tubule
second convoluted tubule
1
3
4
cortex
vein
artery
medulla
2
collecting duct
loop of Henle (ascending limb)
loop of Henle (descending limb)
pyramid in the medulla
opening to pelvis

microvilli
lumen
epithelial cell
mitochondria
peritubular cavity

1 First convoluted tubule

lumen

2 Loop of Henle

lumen

3 Second convoluted tubule

lumen

4 Collecting duct

Fig 13.3d View of the sections shown in Fig 13.5c

Oral rehydration therapy

Most people will have experienced the unpleasant symptoms that accompany a bout of mild food poisoning due to a dodgy late-night kebab, an undercooked chicken leg at a barbecue, or those prawns that lurked in the fridge just a little too long.

Vomiting is caused by bacterial toxins that irritate the gut lining. In the intestine, the frequency of peristalsis increases and the contents move along the gut rather more rapidly than usual. This doesn't give the large intestine enough time to absorb water from the waste and the result is diarrhoea. Fortunately, for most of us, the symptoms are short lived.

However, both vomiting and diarrhoea can be deadly if severe and prolonged. Dysentery and cholera can rapidly lead to death by dehydration as they cause the body to lose fluid faster than it can be replaced (Fig 13.4).

Every day the average person consumes about 2 to 3 litres of fluid in one form or another, and we also pour a huge volume of fluid (over 8 litres) into our intestines in the form of digestive juices. Diarrhoea does not allow for the efficient reabsorption of these fluids. Also, vital ions such a sodium, potassium and chloride, collectively known as electrolytes, are lost. If untreated, this loss can lead to muscle spasms, cramps, coma and heart failure.

Oral rehydration therapy (**ORT**) can be used to treat dehydration (Fig 13.5). This does not involve expensive drugs, simply a mixture of glucose and salt in water. In cases where the patient cannot keep anything down, the rehydration solution can be given directly into the bloodstream via a drip. This simple treatment has saved millions of lives in places where dysentery and cholera are very common.

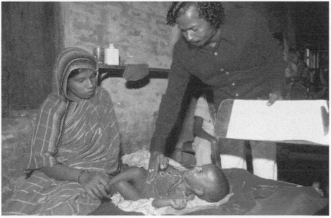

Fig 13.4 The baby in this picture has bacterial dysentery that he developed after drinking water contaminated with faeces. He risks death from dehydration if he does not receive rehydration therapy. An oral rehydration mixture is easy to make up: one level tablespoonful each of glucose and salt dissolved in a pint of boiled and cooled water

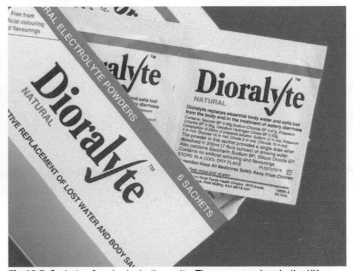

Fig 13.5 Sachets of oral rehydration salts. Those we can buy in the UK contain more than the essential ingredients

The kidneys receive blood from the two **renal arteries** that branch off the aorta. The kidneys receive the largest blood supply of any organ, per gram of tissue. About 1200 cm³ of blood flows to each of them every minute. Incoming blood must be at high pressure to ensure proper kidney function: they can't filter blood effectively if the pressure drops. Blood leaves the kidneys in the **renal veins**.

Urine made by the kidneys is pushed down muscular tubes, the **ureters**, by peristalsis (rhythmic muscular contractions). The ureters empty into the **bladder**, a muscular bag that stores urine until it is convenient to release it. The capacity of the human bladder varies from 400 to 700 cm³ or more. When it begins to get full, stretch receptors in the walls inform the brain of the urgency of the situation. **Urination**, or **micturition**, happens when the **sphincter muscles** relax, allowing urine to pass out of the body through the **urethra**.

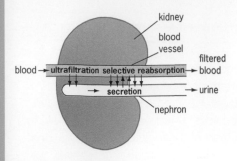

Fig 13.6 A simple summary of kidney function. At the near end of the nephron, blood is filtered to produce a fluid that is virtually identical to tissue fluid. The filtrate is then modified by active transport. This involves active secretion of substances into the filtrate and active reabsorption of substances into the blood. These processes are possible only because of the close association between the nephron and blood system (Fig 13.3c)

3 STRUCTURE AND FUNCTION OF THE MAMMALIAN KIDNEY

The structure of the kidney is shown in Fig 13.3b. The functional unit of the kidney is the **nephron** (or **kidney tubule**). It is important to know the position of the nephrons in relation to the overall plan of the kidney. The outer **cortex** of the kidney contains the **renal capsules** (also called **Bowman's capsules**) and the **first convoluted** and **second convoluted tubules** (also called the **proximal** and **distal tubules**), while the **medulla** houses the **loops of Henle** and the **collecting ducts**. Bundles of collecting ducts form **pyramids** that deliver urine into an open space called the **pelvis**. From here, urine flows down the ureters to the bladder.

Kidney function involves two processes: **ultrafiltration** and **active transport** (Fig 13.6). Ultrafiltration is filtration under pressure: blood is 'squeezed' to form a fluid called the **glomerular filtrate** (usually just the 'filtrate'). Active transport then modifies the filtrate, secreting some substances into it and reabsorbing others from it, according to the needs of the body. The end result is that blood flows back into the body without much of its harmful waste. This waste, a solution containing urea, salts and various other chemicals, is the urine.

Let's now look at how individual parts of the kidney contribute to this overall process.

THE NEPHRON

Each human kidney contains about a million nephrons, together with a maze of blood vessels and some connective tissue (Fig 13.7). In this section, we deal with the function of each region of the nephron in sequence, but you must remember that the nephron functions as a whole: the activities of one region are essential to the effectiveness of others.

Ultrafiltration in the renal capsule

Fig 13.8 shows the **renal capsule** in detail. It is shaped rather like a wineglass, with a central knot of blood vessels called the **glomerulus**. This area of the nephron filters blood by ultrafiltration (filtration under pressure). Obviously, this requires two things: a means of creating pressure and a filter.

Fig 13.7a Under the microscope, five renal capsules (containing glomeruli) can be seen, surrounded by sections of tubules. An underwater dissection allows some of the fine nephrons to be teased out and viewed individually without the aid of a microscope. Each nephron, although only 60 m in diameter, can be over 14 cm long when uncoiled

Fig 13.7b A scanning electron micrograph of a glomerulus with part of the torn renal capsule (whitish) round it

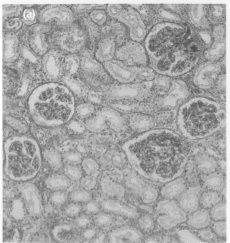

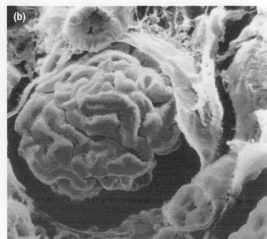

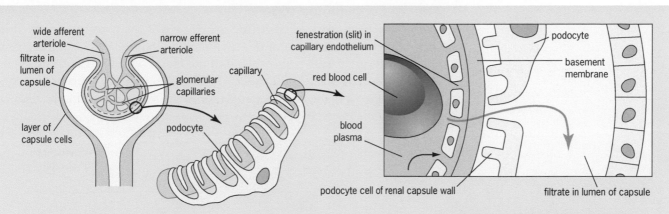

The walls of capillaries in the renal capsule are much more permeable than those of normal capillaries: the cells do not fit tightly together, but have thin slits in between (**fenestrations**), through which all the constituents of the plasma can pass

The renal capsule is lined with unique cells called **podocytes** ('foot-cells'). These cells, like those of the capillary, do not fit tightly together, but form a network of slits that fit over the capillary

Between these two relatively coarse filters is the continuous basement membrane. This finer filter prevents the passage of all molecules with a relative molecular mass greater than about 68 000 kilodaltons, so the larger molecules (mainly proteins) remain in the blood

The kidneys receive blood from the first branch off the aorta, so the blood is already under pressure when it reaches the nephron. This pressure is maintained and enhanced because the **afferent arteriole**, the blood vessel that takes blood into the glomerulus, is short and has a larger diameter than the longer **efferent arteriole** that takes blood away. This physical or **hydrostatic pressure** forces blood against a filter that consists of three layers:

- the lining or **endothelium** of the glomerular blood vessels,
- the **basement membrane**,
- the cells of the renal capsule itself.

Fig 13.8 shows how these membranes are arranged. The middle basement membrane acts as a fine filter and is therefore mostly responsible for the chemical composition of the filtrate. At this stage the filtrate is identical to tissue fluid.

The rate of filtrate production is high: about 125 cm³ per minute. Obviously, we don't produce anything like this volume of urine or we would be constantly in the toilet, and would dehydrate rapidly. On average we produce about 1 cm³ of urine per minute. The rest, over 99 per cent of the filtrate, is reabsorbed. In fact, after the renal capsule, the rest of the nephron is concerned with adjusting the volume and composition of the filtrate. Necessary substances are reabsorbed; toxic compounds and excess solutes and water are removed.

Several forces act on the fluids in the renal capsule, opposing or encouraging the filtration process. The high hydrostatic pressure of blood is the dominant force but it is opposed by the hydrostatic pressure and solute concentration of the filtrate (Table 13.1).

Fig 13.8 The fine structure of the renal capsule, a region of the kidney adapted for ultrafiltration of the blood. Note the difference in size between the afferent and efferent arteriole.

> **Tissue fluid is discussed in Chapter 10.**

> ✔ **REMEMBER THIS**
> Examiners like you to talk in terms of **water potential** when explaining osmosis and water movement. This is a negative scale – pure water has a water potential of zero, and the higher the solute concentration the lower the water potential. See Chapter 4 for details.

> ❓ **QUESTION 3**
> 3 If an individual lost a large amount of blood in an accident, what effect would this have on kidney function?

Table 13.1 Summary of the forces acting in the renal capsule

Force acting	Opposes or encourages filtrate formation	Approximate value/kPa
hydrostatic pressure of blood	encourages	8.0
hydrostatic pressure of filtrate in capsule	opposes	−2.4
solute concentration of blood	opposes	−4.3
overall filtration pressure	encourages	1.3

The first convoluted tubule

In the **first convoluted tubule** (also known as the **proximal convoluted tubule or proximal tubule**) (Fig 13.9), many solutes such as glucose and amino acids are totally reabsorbed into the blood by active transport. Normal urine should not contain glucose. Only when blood glucose becomes very high, because of diabetes for example, does the reabsorption mechanism fail (Fig 13.10).

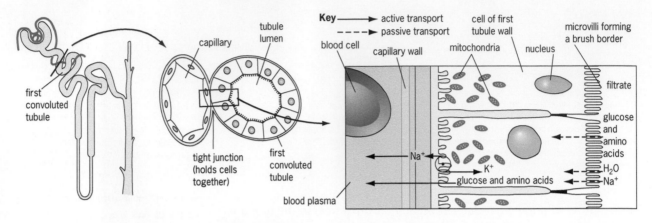

Fig 13.9 Most of the filtrate formed in the renal capsule is reabsorbed from the first convoluted tubule. Amino acids, glucose and sodium are removed from the tubule by active transport, and water follows passively by osmosis

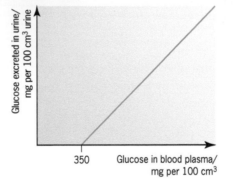

Fig 13.10 Normally the kidneys are able to reabsorb all of the glucose into the blood. Only when blood sugar exceeds a threshold of about 350 mg per 100 cm³ does glucose begin to appear in the urine

In addition, this part of the tubule is very close to blood vessels that carry blood away from the glomerulus. Blood in these vessels has a low hydrostatic pressure but a relatively high solute potential due to the plasma proteins that remain there because they could not pass through the filter. This allows the blood to reabsorb a large percentage of the water from the first convoluted tubule by osmosis.

As you can see from Fig 13.9, the cells that line the first convoluted tubule show all the classic adaptations to active transport: a large surface area, provided by microvilli, and many mitochondria to provide ATP to power the process.

The loop of Henle

The **loop of Henle** is a long U-shaped region of the nephron that descends deep into the medulla and then returns to the cortex (Fig 13.11). The loop creates a region of high solute concentration (low water potential) in the medulla. The collecting ducts pass through this region, and the osmotic gradient between the inside of the collecting duct and the outside draws water out of the duct by osmosis. Consequently, the urine becomes more and more concentrated (compared with body fluids) as it passes down the duct. So the loop is a vital adaptation that benefits humans and other land-living organisms; it allows us to get rid of waste without losing too much water.

? QUESTION 4

4 The loop of Henle is sometimes described as a **hairpin countercurrent multiplier**. Explain this description.

Fig 13.11 How the loop of Henle works. The numbers refer to the solute concentration in mg per 100 cm³

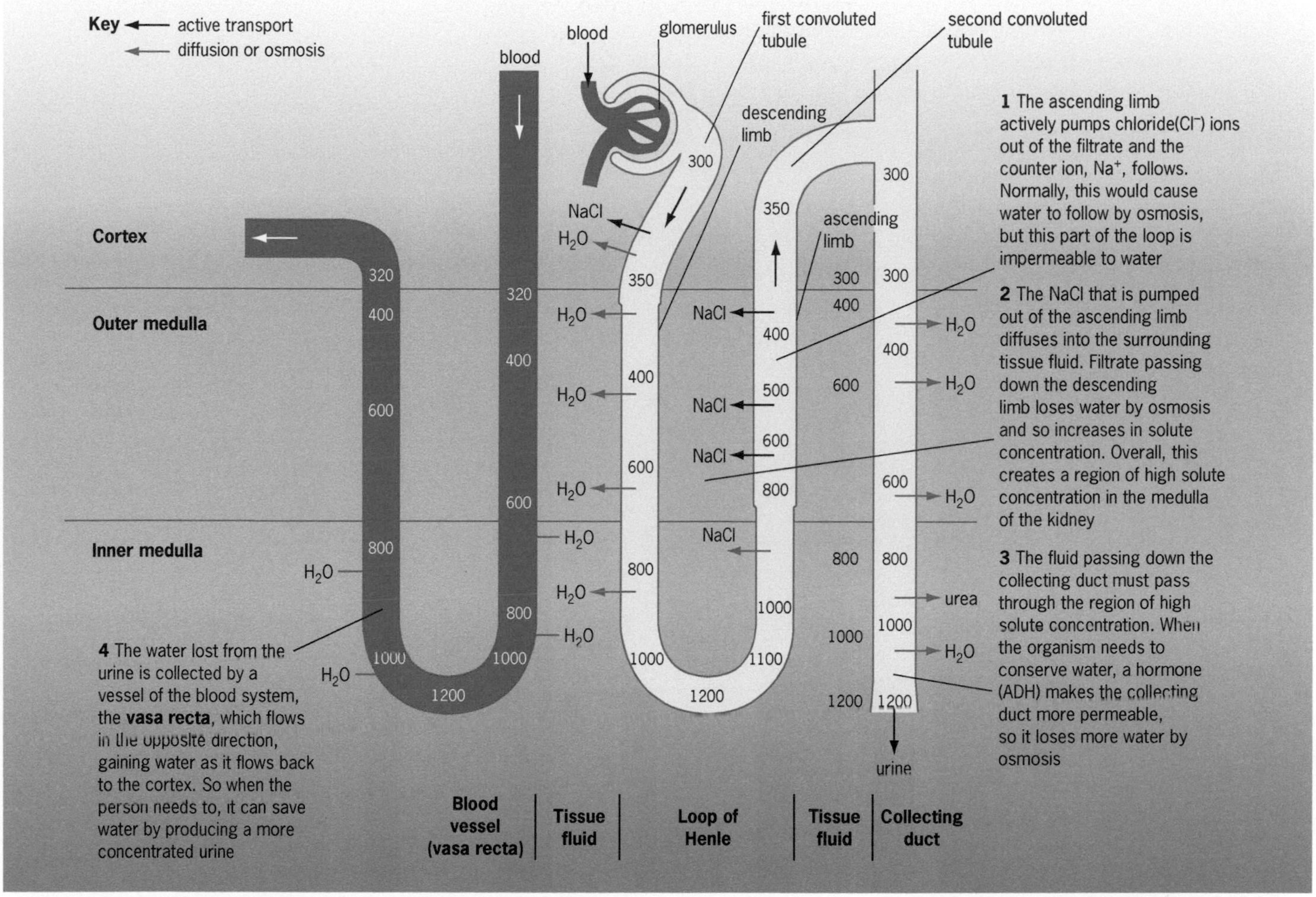

Key ← active transport
← diffusion or osmosis

blood · blood · glomerulus · first convoluted tubule · second convoluted tubule

descending limb

Cortex

Outer medulla

Inner medulla

NaCl
H₂O

ascending limb

1 The ascending limb actively pumps chloride(Cl⁻) ions out of the filtrate and the counter ion, Na⁺, follows. Normally, this would cause water to follow by osmosis, but this part of the loop is impermeable to water

2 The NaCl that is pumped out of the ascending limb diffuses into the surrounding tissue fluid. Filtrate passing down the descending limb loses water by osmosis and so increases in solute concentration. Overall, this creates a region of high solute concentration in the medulla of the kidney

3 The fluid passing down the collecting duct must pass through the region of high solute concentration. When the organism needs to conserve water, a hormone (ADH) makes the collecting duct more permeable, so it loses more water by osmosis

4 The water lost from the urine is collected by a vessel of the blood system, the **vasa recta**, which flows in the opposite direction, gaining water as it flows back to the cortex. So when the person needs to, it can save water by producing a more concentrated urine

Blood vessel (vasa recta) | Tissue fluid | Loop of Henle | Tissue fluid | Collecting duct

urine

Fig 13.11 outlines how the loop of Henle works. Fluid in the two limbs of the loop flows in opposite directions. We describe this sort of arrangement as a **countercurrent system**. As fluid travels up the ascending limb, sodium chloride (NaCl) is transported actively out of the limb into the surrounding area. This causes water to pass out of the descending limb by osmosis. The net result is that the solute concentration at any one level of the loop is slightly lower in the ascending limb than in the descending limb. The longer the loop, the more chance there is for this mechanism to build up a high sodium chloride concentration. If the loop in Fig 13.11 were only half the length shown, sodium chloride would accumulate to only about 600 units.

The second convoluted tubule

While the first convoluted tubule is reabsorbing most of the filtrate, the **second convoluted tubule** (also known as the **distal convoluted tubule**) 'fine tunes' the remaining fluid, according to the immediate needs of the body. This tubule plays an important role in the regulation of pH, salt and water balance.

? QUESTION 5

5 People on survival courses are taught to assess their level of dehydration by looking at the colour of their urine. Explain how they would do this.

Dialysis and kidney machines, kidney transplants

Dialysis is a method of separating small molecules from larger ones using a partially permeable membrane. Blood dialysis, or **haemodialysis**, separates the smaller constituents of plasma, such as urea and solutes, from the larger ones, such as proteins.

Blood is taken from the patient, usually from a vein in the forearm, and passed into the machine, where it runs through minute artificial capillaries Fig 13.12. These are made from a partially permeable plastic that filters the blood. While blood flows inside the artificial capillaries, a special fluid, the **dialysate**, flows round the outside in the opposite direction.

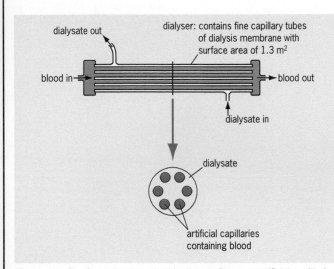

Fig 13.12a The fine tubes seen in this dialysis filter are artificial capillaries. Blood flows through the middle of these tubes, while the dialysing fluid flows along the outside in the opposite direction. Each filter is an expensive piece of precision engineering, but it can be used for only a few dialysis sessions and must then be discarded

In dialysis, molecules are exchanged between the blood and the dialysate. The composition of the dialysate is carefully controlled so that there is a net movement of urea, water and salts out of the blood.

Another type of dialysis is **CAPD**, which stands for **continuous ambulatory peritoneal dialysis**. People with kidney failure use one of their own membranes, the **peritoneum**, as a dialysing membrane. Over 3000 people in the UK use this technique because of the advantages it offers over conventional dialysis.

The basic principle is simple: the person who needs dialysis has a hole, or **stoma**, made in their abdomen wall near the navel, through which a large volume of dialysing fluid is introduced through a tube or **catheter**. He or she is free to walk around (hence the ambulatory part of the name) while dialysis occurs across the

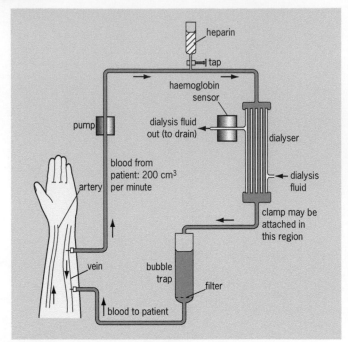

Fig 13.12b The circuit taken by the blood as it passes through the dialysis machine

peritoneum between the blood and the dialysate. The dialysate is replaced every six hours by a relatively simple exchange procedure that people can carry out themselves after some basic training.

When a donor kidney becomes available, it is a relatively simple operation to transplant it into another body. Surprisingly, the old kidneys are left in place: they are rather inaccessible and so are difficult to remove, but they do no harm. The new kidney is placed in the lower abdomen. Surgeons choose this site because the new kidney can be attached easily to a large artery (the **femoral artery**, supplying the leg) and is usefully right next to the bladder.

Finding a suitable donor for an organ transplant is difficult. Although several hundred thousand people die each year in the UK, only a tiny fraction can provide organs for transplant. For example, accident victims can become organ donors only if their injuries have not affected the organ itself. Owing to road safety improvements, the number of serious accidents is decreasing. This is good news, but it means that the number of organs available for transplant is getting ever smaller.

A further complication is that only a minority of people carry donor cards, and permission to use body parts from recently deceased people has to be given by distressed relatives, who often say no. The problems of transplants are covered in more detail in Chapter 30.

4 THE ROLE OF THE KIDNEY IN HOMEOSTASIS

The kidney contributes to several vital homeostatic mechanisms. Among the most important are regulation of water content and blood volume.

WATER BALANCE IN HUMANS

Table 13.2 shows a typical water balance sheet for an average person, assuming normal activity and a comfortable external temperature. On average, we get almost two-thirds of our water from drinks and a third from food. We obtain a small but important proportion of water from metabolic reactions, notably cell respiration.

Some of the water loss shown in Table 13.2 is unavoidable. Metabolic waste must be removed in solution, and so some water loss in urine is inevitable. Similarly, water is always lost from the lungs as we breathe out. A significant amount of water is also lost by diffusion through our skin (this is not the same as sweating).

Table 13.2 The water balance sheet for a 24-hour period

Water gain	Volume/cm^3	Water loss	Volume/cm^3
food and drink	2100	through skin	350
metabolic water	200	sweat	100
		in breath	350
		urine	1400
		faeces	100
Total	2300	Total	2300

The mechanism of water balance

Like most homeostatic mechanisms, maintenance of water balance involves a negative feedback loop that consists of a **detector** and a **correction mechanism**.

A part of the brain, the **hypothalamus**, contains **osmoreceptor cells** that are sensitive to the solute concentration of the blood. When the solute concentration rises, indicating that water loss has exceeded intake, the hypothalamus responds in two ways:

- It stimulates the thirst centre in the brain.
- It stimulates the pituitary gland to release **anti-diuretic hormone (ADH)**.

Fig 13.13 summarises the mechanism of ADH action. ADH acts on the kidney to reduce the volume of urine produced. It achieves this by increasing the permeability of the second convoluted tubule and the collecting duct to water. The action of ADH causes more water to leave the tubule and re-enter the blood. Much more concentrated urine is produced and vital water is conserved.

The exact mode of action of ADH is now becoming clear due to important new research. ADH is a peptide hormone (a short chain of amino acids) which fits into specific receptor proteins on the cell surface membranes of the second convoluted tubule and collecting duct. Once in place, they activate a second messenger molecule – cyclic AMP – which in turn activates protein molecules called aquaporins. These proteins move into the cell surface membrane where, as their name suggests, they act as water channels (aqua-pores), making the membrane more permeable to water.

? QUESTIONS 6–8

6 How would the water balance sheet change if the individual was suffering from diarrhoea?

7 Which areas of the nephron carry out active transport?

8 In desperation, a castaway drinks seawater. Why is this not a good idea?

Negative feedback is discussed in Chapter 11.

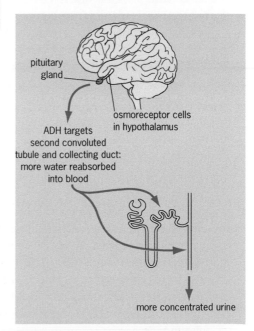

Fig 13.13 When the solute concentration of blood rises, the osmoreceptor cells in the hypothalamus stimulate secretion of ADH from specialised nerve cells in the posterior lobe of the pituitary gland. ADH makes the second convoluted tubule and collecting duct more permeable to water, and more water passes from the filtrate into the blood. The blood becomes more dilute and blood volume increases

Conversely, when fluid intake exceeds loss, the blood becomes more dilute. When the hypothalamus detects this, it reduces ADH production. The action of ADH on the kidneys lessens, resulting in less water reabsorption and the production of larger volumes of dilute, or **insipid**, urine.

People with the disease **diabetes insipidus** cannot produce ADH because they have a faulty pituitary gland. This condition results in the constant production of dilute urine, leaving the sufferer permanently thirsty and unable to venture very far from the toilet. Today, it can be treated by giving extracted or synthesised ADH.

CONTROL OF BLOOD VOLUME AND PRESSURE

Since it regulates water reabsorption, ADH also regulates blood volume. A drop in blood volume leads to a drop in blood pressure that is detected by stretch receptors in the walls of the aorta and carotid arteries. Impulses from these detectors pass to the hypothalamus, which then triggers the secretion of more ADH. This acts on the kidneys and causes them to retain more water, so increasing blood pressure.

Hangovers!

Sooner or later, most people experience the unpleasant 'morning after' feeling which tends to follow a bout of drinking too much alcohol. Many hangover symptoms are due to dehydration, rather than to the toxic effects of the alcohol or other ingredients. Research has shown that alcohol inhibits the production of ADH, causing water that the body needs to be lost in the urine. Many of the symptoms disappear when the body is rehydrated.

Fig 13.14 Had he known about the dehydrating effects of alcohol, this man could have minimised his headache by having a long drink of water before he went to bed (assuming, of course, that he could find the tap)

SUMMARY

After reading this chapter, you should know and understand the following:

- The kidneys remove metabolic waste and control water and solute levels in the body. As a result of these functions, the kidney also plays a vital role in the control of blood volume and pressure.
- Each kidney is made from around one million **nephrons**, narrow tubules closely entwined with blood vessels.
- At one end of the nephron, the **renal capsule**, the kidney filtrate is formed by ultrafiltration (pressure filtration) of the blood.
- The first filtrate has the same composition as tissue fluid. As it passes along the nephron, the composition of filtrate is altered by various active transport mechanisms that reabsorb some substances while allowing others to pass into the urine.
- A large amount of filtrate is formed, but over 99 per cent of it is reabsorbed in the **first convoluted tubule**, mainly by active transport

mechanisms and osmosis. Usually, all glucose and amino acids are reabsorbed into the blood.
- The movement of solutes from the **loop of Henle** creates a region of high solute concentration in the medulla, through which the collecting ducts must pass. As filtrate (now urine) flows along the collecting ducts, water leaves it by osmosis. The resulting urine is **hypertonic** (more concentrated than body fluids).
- The **second convoluted tubule** is involved in several homeostatic mechanisms including the regulation of salt, water and pH levels.
- When the water potential of the blood drops (e.g. when we dehydrate) it is detected by the **hypothalamus** which stimulates **ADH** release from the **pituitary**. ADH makes the second convoluted tubule and the collecting duct more permeable to water, so water leaves the filtrate and enters the blood.

 Practice questions and a How Science Works assignment for this chapter are available at www.collinseducation.co.uk/CAS

14

Exercise physiology

14 EXERCISE PHYSIOLOGY

Record breakers

At an athletics meeting in Oxford in 1954, Roger Bannister became the first man to run a mile in under 4 minutes. In 1999, the Moroccan Hithcem El Guerrouj ran the same distance in 3 minutes 43 seconds and he still holds that record in 2008. Genetically, humans today are no different from their predecessors. Yet year upon year, games upon games, athletes get better and world records keep tumbling. How is this being achieved?

First, there are advances in science and technology. A substantial support team now surrounds top athletes: sports physicians, sports physiotherapists, sports dieticians, etc. Training can now be tailored to an individual's needs. Their muscle types can be analysed so that they do just the right balance of different training methods and intensities. Training machines can isolate a particular muscle group and/or reproduce movement patterns such as a javelin throw or tennis serve. Movement can be filmed and analysed with a computer so that a biomechanics expert can target areas for improvement.

Diet can also play a big part both before and after the event. Endurance athletes prepare with a mainly carbohydrate diet so that their glycogen stores are topped up. Following an event, athletes need to rehydrate and replace lost fuels so that they can recover and begin training again as soon as possible.

Another important aspect in preparing a world-class athlete is the prevention of injury. Modern sports physiotherapists have an array of tools at their disposal to diagnose potential weaknesses. If an injury does occur, there is a variety of techniques available to speed the healing process (see photo).

Lastly, and crucially, there has been a change from amateur to professional status. Roger Bannister was a doctor who fitted in his athletics around his medical school training. Many of today's young athletes see their talent as a route to fame and fortune. They can devote all of their energies to excelling at their particular event or sport, knowing that success can set them up for life.

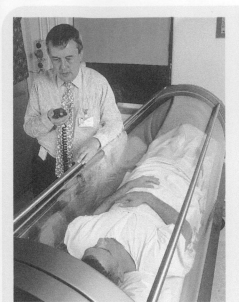

A variety of new techniques can speed up the healing process. In this controversial example, a patient is being treated in a hyperbaric chamber that supplies oxygen at a pressure 2.2 to 2.4 times greater than atmospheric. This causes rapid uptake of oxygen into the blood and body tissues, speeding up the healing process in a variety of injuries. Studies suggest that this can speed up the healing process by as much as a third.

WHAT IS FITNESS?

What do we mean by **fitness**? Overall, it refers to a person's ability to perform well in their chosen activity. But there are several different aspects to the concept of 'fitness'.

There are strength and speed. Both aspects depend largely on the power in the muscles. Some men can lift 250 kilograms – about the weight of three large men – above their heads, and there are top sprinters who can cover 100 metres in less than 10 seconds – by running at a speed of about 36 kilometres per hour. A person's strength depends to some extent on the cross-sectional size of their muscles, but other factors such as the shape of their skeleton, good technique and motivation, are all important. So is skill. Some people are naturally gifted; their brain is able to control their muscles in just the right way to perform a particular activity. Of course, skill levels are further improved through coaching, training and practice.

Yet another factor is suppleness, the flexibility of the body. Often an underestimated aspect of fitness, most sports require a degree of suppleness, and performance can be improved by incorporating suppleness work into a training programme.

In this chapter we mainly focus on a fourth element of fitness: **stamina** – the body's 'staying power'. We look at the way in which the body gets its energy, and the different types of 'energy currency' that are important in different sports.

1 GEARING UP FOR EXERCISE

As you sit and read this book, your body is ticking over nicely. Oxygen demand is low, and can be met comfortably by relatively shallow breathing and low pulse rate. Blood is delivering oxygen and glucose to your cells and waste products are being taken away. The levels of these chemicals remain relatively constant, and homeostasis, that all-important *steady state*, is being maintained with relatively little fuss. There is, literally, 'no sweat'.

All of this is rudely interrupted when you begin to exercise (Fig 14.1). The metabolic rate of the muscles increases by up to 20 times, or by 2000 per cent. To fuel this frantic activity, and to maintain some sort of stability, your body must adapt. The overall response of the body to exercise is an excellent example of how the different systems work together to carry on maintaining homeostasis.

Fig 14.1 This is Naoka Takahashi who can run a marathon in a time of 2 h 23 min 14 s. Throughout the 26 miles of this race, her body is able to keep her muscles supplied with oxygen and fuel (glucose and lipid), while ensuring that the waste products are eliminated before they build up. This is a magnificent feat of homeostasis

Homeostasis is discussed in Chapter 11.

REMEMBER THIS
The word **homeostasis**, meaning 'steady state', refers to the fact that conditions within the body need to be kept within certain limits. Homeostatic mechanisms include the control of blood gases (oxygen and carbon dioxide), blood pH, body temperature and blood glucose levels.

HOMEOSTASIS AND EXERCISE PHYSIOLOGY

Exercise physiology is the study of the responses of the body to exercise. The effects are familiar: in the short term we sweat, pant and go red, while in the long term we 'get fit', improving our muscle tone, strength, stamina and general well-being. This chapter builds on your knowledge of the human body, pulling together many different areas to show you that homeostasis is a whole-body process involving many different systems. Table 14.1 summarises the relevant topics that are covered elsewhere in the book.

Before we look at the ways in which the body responds to exercise, we need to set the scene by looking at some basic principles:

- The process of **cell respiration** releases the energy in organic molecules such as glucose and lipids, and transfers the energy to a chemical called **adenosine triphosphate**, ATP (Fig 14.2).
- ATP provides the energy for muscular contraction; it allows the fibres to slide over each other.
- When ATP splits by **hydrolysis** into ADP and P_i (phosphate), energy is released (Fig 14.3).
- The purpose of cell respiration is therefore to resynthesise ATP from ADP and phosphate.

Cell respiration is discussed in more detail in Chapter 33.

Muscle contraction is discussed in more detail in Chapter 17.

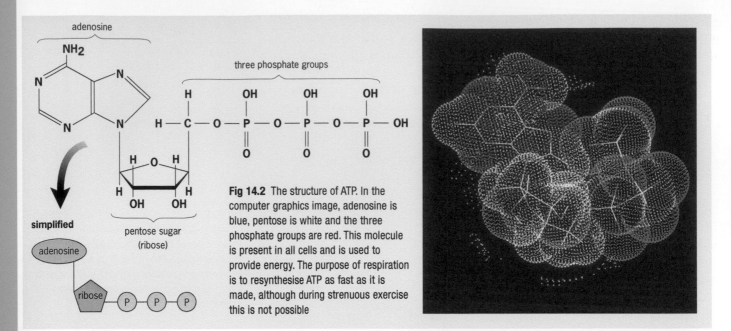

Fig 14.2 The structure of ATP. In the computer graphics image, adenosine is blue, pentose is white and the three phosphate groups are red. This molecule is present in all cells and is used to provide energy. The purpose of respiration is to resynthesise ATP as fast as it is made, although during strenuous exercise this is not possible

 REMEMBER THIS

ATP is a relatively small, simple molecule that is generated inside the mitochondria of cells. It can diffuse rapidly in and out of the cells and tissues, and moves to places where it is needed, such as muscle fibres. ATP delivers instant energy in small, usable amounts.

● ATP splitting is a **coupled** reaction. When we exercise, ATP hydrolysis is coupled with muscular contraction.

● When ATP is split, some of the energy is used to power the muscle, but as no energy transfer is 100 per cent efficient, some is always lost as heat. This is why vigorous exercise produces large amounts of heat that must escape from the body.

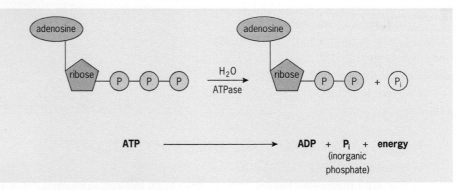

Fig 14.3 When the ATP molecule is split (hydrolysed), the terminal phosphate is removed and combined with water. This reaction releases energy, which is used to drive energy-requiring reactions such as muscular contraction

$$ATP \longrightarrow ADP + P_i + energy$$
(inorganic phosphate)

Table 14.1 Aspects of exercise physiology in this book

Topic	Found in
homeostasis and negative feedback systems	Chapter 11
the cardiac cycle and its control	Chapter 10
breathing and its control	Chapter 9
respiration – the production of ATP	Chapter 33
temperature control	Chapter 12
movement of muscle/muscle types	Chapter 17
electrocardiograms	Chapter 10
artificial pacemakers	Chapter 31
heart defects and their repair	Chapter 31
blood transfusions	Chapter 28
dehydration and water balance	Chapters 8 and 13
diet and energy content of food	Chapter 7

ENERGY AND EXERCISE

If the movement of muscles requires ATP, it follows that the ability of an athlete to move his or her muscles for any length of time requires a continued supply of this essential chemical. Muscles have three sources of ATP:

- **The ATP already present**. This provides instant energy and allows us movement on demand. When we contract our muscles as hard and as fast as possible, we use ATP far faster than it can possibly be made. So we rely on the ATP that has accumulated during periods of relative rest. During maximum effort there is only enough ATP for about three seconds, but there is a back-up chemical, **creatine phosphate**, CP. The energy in CP can be used to instantly resynthesise more ATP, allowing maximal exercise to continue for up to 10 seconds. This is called the **ATP/CP** system, or the **alactic anaerobic system**. Alactic means there is no build-up of **lactic acid**, or more accurately, **lactate ions**, and anaerobic means that this system does not require oxygen. All events that require explosive bursts of energy, such as weightlifting or short sprints, use the alactic anaerobic system.

- **The ATP provided by glycolysis, the first phase of respiration**. Glycolysis provides two ATP molecules per glucose molecule. This might not seem a lot compared with the 36 or so available from complete respiration, but it has two big advantages: it is relatively quick and it does not need oxygen. Thanks to this system, exercise can continue at near-maximum levels for up to one minute. However, there is a price to pay – the accumulation of lactate ions. Lactate lowers the pH in the muscles, causing fatigue and interfering with enzymes in muscle cells. The body can tolerate only limited levels of lactate. This system is known as the **glycolytic** or **lactic anaerobic system**, and is the main energy source for events that last between 10 and 60 seconds, such as the 400 metres hurdles (Fig 14.4).

- **The ATP provided by aerobic respiration**. This is the complete breakdown of glucose, when each molecule yields about 36 molecules of ATP. The problem is that this system takes time to provide energy, and then it still has its limits. However, provided that the level of activity stays within those limits, the aerobic system can fuel exercise for a couple of hours of more. This is the main energy system for many sports: all those that last for longer than one minute. Prolonged exercise that increases heart rate and breathing rate, but that is sustainable for hours rather than minutes, is commonly known as aerobics (Fig 14.5).

REMEMBER THIS

The stores of ATP and CP are known collectively as the **phosphate battery**, so called because they need to be recharged after a bout of maximum exertion. They are also called the **phosphagen system**.

Respiration and glycolysis are discussed in more detail in Chapter 33.

Fig 14.4 Sprinters and hurdlers rely on strength and are usually heavily muscled. To power their muscles during short races they rely on the ATP/CP system for up to about 10 seconds, and for the rest of the race the glycolytic system provides the energy

Fig 14.5 A familiar sight all over the western world, sustained aerobic exercise can strengthen the heart, lungs and circulation. As an added bonus, it also uses lipid (fat) as its main fuel, so aerobics are a vital part of many weight-loss programmes

The energy continuum

From what we have learned so far, it might appear that there are three separate energy systems. But, as we all know, we don't have to stop exercising after 10 seconds and then again after 1 minute to wait for the next energy system to cut in and give us some ATP. All three energy systems blend smoothly together, one taking over from the other. This phenomenon is known as the **energy continuum** (Fig 14.6).

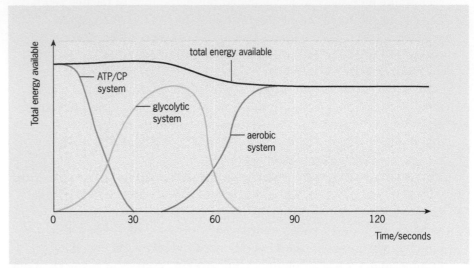

Fig 14.6 The energy continuum. The graph shows that one system gradually takes over from another, allowing us to move our muscles continuously

FUELS FOR EXERCISE

Respiration is the release of the energy contained in organic molecules such as glucose. All foods – carbohydrates, lipids or proteins – can be respired when the need arises. As a rough guide, the body uses the following fuels:

- **Glucose**. This supplies the normal, everyday energy needs when we are eating regularly (i.e. not dieting or starving). Anaerobic exercise always depends on glucose, because lipids cannot be used to fuel glycolysis.
- **Lipids**. Most lipid is stored in fat-storage, or **adipose** tissue. It takes a while to mobilise the lipid stores, but once the process has begun, this reservoir can fuel the body for as long as it lasts (for some of us almost indefinitely). Lipids cannot be used for short-term exercise, but they can be used to fuel aerobic exercise. This is why doing an aerobics class regularly can help to get rid of stored fat, if the exercise is done at the right intensity.
- **Protein**. The body usually gets around 5 per cent of its energy from protein. The proportion increases significantly only during starvation, when there is no other fuel available. Recent research suggests that protein can be used to fuel endurance events, but to what extent remains uncertain.

Overall, the main fuel used during exercise depends on the intensity of the activity. If you take running as an example, the main fuel for a long, gentle jog is lipid. But, as the intensity increases, so does the proportion of energy that comes from glucose. A summary of the energy systems used in exercise is shown in Table 14.2.

Table 14.2 A summary of the energy systems used in exercise

	Energy system		
	Anaerobic alactic	**Anaerobic lactic**	**Aerobic alactic**
other names	ATP/CP; phosphate battery	glycolytic; lactic system	aerobic system
energy comes from	ATP/CP already in muscles	glycolysis	complete respiration; glycolysis, Krebs cycle, electron transport chain
nature of energy supply	allows maximum strength but is short lived	allows exercise to continue at near-maximum for up to 1 minute	allows long-term exercise at lower intensity
timescale	up to 10 seconds	10 seconds to 1 or 2 minutes	anything longer than 1 to 2 minutes
by-product	no lactate	lactate	no lactate; carbon dioxide
activity	explosive 'full strength'	longer sprint	endurance events such as cycling and marathons
examples	sprints up to 100 metres; weightlifting and throwing events	200 to 400 metre sprints	800 metres, marathons
training needed	to improve speed or strength	to improve lactate tolerance	to increase endurance capacity and to strengthen cardiovascular system

Diet for athletes

It has long been known that the correct diet can improve an athlete's performance, and in recent years the diets of sports stars such as footballers have been increasingly in the spotlight (Fig 14.7).

The best approach is a **holistic**, or *whole-body*, one. The most dedicated athletes pay close attention to their diet, taking in the right amount and types of food at the right time. In addition, they ensure that they have the right amount of sleep and minimise their alcohol intake, especially before important events.

How do you plan a diet to maximise performance? A widely used technique is known as **carbo-loading** or **glycogen loading.** About six days before an event, the athlete goes on to a low-carbohydrate, high-protein diet. This depletes the body's glycogen reserves. Then, for the three days preceding the event the athlete goes on a high-carbohydrate diet, eating 8 to 18 grams of carbohydrate per kilogram of body weight per day. This significantly increases the amount of glycogen stored in the muscles, and this obviously helps during endurance events. It is of little help, however, if the event lasts for less than 90 minutes, and the effect wears off with repetition. Generally, athletes should aim for a diet in which 60 to 70 per cent of their energy comes from carbohydrate, increasing this figure in the day or two before an important event.

Fig 14.7 Nutritionists recommend that for a few days preceding an event, athletes should be on an easily digested high-carbohydrate diet, such as pasta, along with a little low-fat protein such as chicken or fish. This will leave them feeling neither hungry nor bloated, and will top up their glycogen reserves

Dehydration and isotonic drinks

Long-term exercise leads to prolonged sweating, and this can easily dehydrate an athlete. By 'dehydrate' we mean that the blood and body fluids become too concentrated. In the correct biological jargon we say that *the water potential is lowered*. The practical consequence of this is that the blood becomes more viscous and cannot flow as easily. In addition, sweating loses vital ions such as sodium, potassium and chloride, collectively known as electrolytes. A loss of electrolytes can lead to muscular cramps and a greatly reduced performance.

But it is not too serious; all the athlete needs to do is to replace the lost water, electrolytes and glucose. This is easy enough if the athlete happens to be at home, but dehydration can strike when you are three-quarters of the way through a sporting event. Getting a drink that can do the trick takes a bit of planning.

The fastest way to combat dehydration is to drink pure water. This way, water is absorbed as fast as possible by osmosis. However, if you add glucose and salts to the drink you lower the water potential and therefore slow down the water uptake. In fact, if the drink becomes too concentrated

it actually makes matters worse by drawing water out of the blood, in the same way as drinking seawater would.

The best compromise is to take an isotonic drink, a mixture that has the same water potential as body fluids. This way, all three components are absorbed as rapidly as possible. So do you have to spend money on expensive isotonic drinks? The simple answer is no. It is easy to make up an isotonic mixture on the same principle as the **oral rehydration therapy** given to people suffering from diarrhoeal diseases (see page 203).

For events lasting less than 90 minutes it is debatable whether isotonic drinks are any benefit at all. The fluid can be replaced by that cheapest of drinks, water, and the electrolytes and glucose are replaced later by any sensible meal.

Fig 14.8 Isotonic drinks contain a solution of electrolytes and glucose or short-chain polysaccharides, which can be easily digested and absorbed into the blood

2 SHORT-TERM RESPONSES TO EXERCISE

Exercise places great demands on the body. Think about the changes that occur when we exercise:

- oxygen levels fall,
- carbon dioxide and lactate levels increase,
- body temperature increases,
- blood glucose and glycogen levels fall,
- fluid and electrolytes (salts) are lost as we sweat.

All this presents a great challenge to the homeostatic mechanisms of the body. In an attempt to maintain some sort of stability, the body responds by:

- **Increasing heartbeat and breathing rate**. The increased respiration of the muscles raises the carbon dioxide level in the blood, and consequently lowers the pH (carbon dioxide is an acidic gas). These changes are detected by sensitive cells – **chemoreceptors** – situated in the **carotid** and **aortic bodies** in major blood vessels. In turn, these receptors inform the cardiovascular and respiratory centre in the brain, which responds by increasing the heart rate and the ventilation rate. The result is that gas exchange increases, as does the delivery of oxygen to the tissues and the removal of carbon dioxide.
- **Increased blood flow to the skin**. Muscle movement creates heat that must be lost. By increasing the diameter of the blood vessels that carry blood to the skin surface – a process called **peripheral vasodilation** – the blood can push heat to the body's surface where it can more easily be lost to the environment. This is why we 'go red' during and after exercise.
- **Increased sweating**. Sweat glands secrete a salty solution that evaporates from the skin, taking heat with it. Coupled with vasodilation, sweating is a very efficient means of removing excess heat. The problem

REMEMBER THIS

Ventilation rate is the amount of air taken in by our lungs in one minute. It can be calculated as depth of each breath (the tidal volume) × number of breaths per minute. At rest, this could be about 0.5 litres × 12 breaths = 6 litres per minute.

Gas exchange is discussed in more detail in Chapter 9.

Heat transfer is discussed in more detail in Chapter 12.

with excess sweating, of course, is that we also lose water and salts. If these are not replaced, the body becomes dehydrated and athletic performance can be impaired.

- **Increased mobilisation of glycogen.** In order to keep the muscles supplied with glucose, the glycogen stores in the muscles and the liver start to be mobilised. Glycogen is a highly branched polymer of glucose, which breaks down rapidly to release glucose. After prolonged exercise, glycogen stores are replenished from carbohydrate in the diet.

> The structures of glycogen and glucose are discussed in more detail in Chapter 3.

THE RECOVERY PROCESS

After exercise, following the short-term changes listed above, the body does not immediately return to normal. Each system gradually returns to resting levels. Generally, the fitter the individual, more quickly the resting state is achieved.

Fig 14.9 shows the pulse rate of an athlete before, during and after a session on an exercise bike. Following exercise, pulse rate follows a classic pattern; a rapid initial fall followed by a slower return to normal. The shape of the curve results from two processes: the recharging of the ATP/CP system – this is known as the **alactacid component** of recovery – and the removal of lactate, called the **lactacid component** of recovery. Both occur because the body was not able to deliver enough oxygen to keep up with demand. This shortfall is known as the **oxygen debt** and is repaid after exercise stops. On a longer term basis, the body must also restore its glycogen levels. After a particularly gruelling exercise session, restoring the glycogen can take up to 48 hours and is not associated with raised heart or breathing rate.

We shall look at each of these components in turn.

The alactacid oxygen debt component

This is the recharging of the ATP/CP system and takes one to two minutes to complete. Even fit people can't sprint continually, but after a short 'burst' it

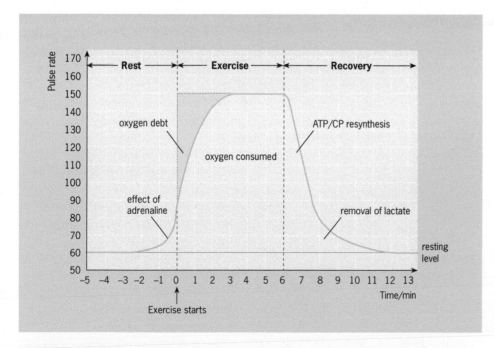

Fig 14.9 The effect of pulse rate on exercise. Note that the resting pulse is about 60, normal for a trained athlete. The increase in heart rate just before exercise starts is due to anticipation, caused by the release of adrenaline: the more important the event, the more noticeable is the effect of adrenaline. When exercise starts, oxygen demand is greater than supply, and an oxygen debt builds up. This is paid off after the activity has finished

The dangers of over-exercising

Some athletes train too much, and do not allow their bodies to recover properly between sessions. 'Over-training' can cause serious medical problems. In addition to poor performance and chronic fatigue, the immune system can be suppressed, leading to frequent infections, commonly sore throats and flu-like symptoms. At the end of the season, some professional footballers have been found to have a very low white cell count. Research showed a decrease in levels of natural killer cells, phagocytes, B cells and helper T cells (see Chapter 30). It is thought that psychological stress, which leads to over-secretion of the hormones adrenaline and cortisol, also contributes to immune suppression.

Joints are damaged by exercise. The knees are particularly vulnerable, because they take most of the weight of the body and can move in only one plane. Sudden twisting and/or heavy impacts can result in ligament damage. Long-term use can result in the articular cartilage wearing away, so that bone scrapes on bone, causing pain and inflammation – one type of arthritis.

Over-training in young athletes is particularly dangerous because their skeleton and joints are still developing.

Fig 14.10 This person has decided to improve his fitness by a little gentle running. He is comfortable at walking pace, but as soon as he starts to jog, he passes his anaerobic threshold, meaning that his muscles do not get oxygen quickly enough and lactate begins to accumulate, making his muscles ache. He will have to slow down until the lactate levels fall, then he will be able to run again. However, if he maintains regular exercise, he can raise his anaerobic threshold, allowing him to run for longer and/or faster without discomfort. In order to improve aerobic fitness, it is recommended that we take a minimum of 20 minutes exercise every day, or about 45 minutes every other day. The aim is to keep the heart elevated to about 65% of its safe maximum. That's around 130 for a healthy 20 year-old.

takes only a matter of seconds before there is enough ATP/CP to allow them to sprint again. In sports such as football, rugby, tennis or hockey, there are bursts of intense activity followed by periods of relative rest, during which the ATP/CP stores are restored by normal aerobic respiration.

The lactacid oxygen debt component

This is the removal of the lactate ions that accumulated during the exercise. This acidic chemical lowers the pH of the blood, interfering with the enzymes in aerobic respiration and so reducing ATP supply to muscles. A significant build-up is painful, and there is only so much muscle fatigue we can take before we simply have to stop (Fig 14.10). Interestingly, one of the main effects of training is to increase our tolerance to lactate, so that we can continue to exercise despite higher lactate levels.

Lactate builds up in the muscles and diffuses into the blood. It then has five possible fates:

- 65 per cent is oxidised to carbon dioxide and water,
- 20 per cent is converted to glycogen,
- 10 per cent is converted to protein,
- 5 per cent is converted to glucose,
- a trace is excreted in urine and sweat.

The list clearly shows that the anaerobic system is rather wasteful in terms of energy. Only two ATP molecules are made per glucose molecule, instead of a potential 36 (as in aerobic respiration). The production of two lactate molecules therefore represents a great waste of potentially useful energy: when each pair of lactate molecules are broken down in the liver, 34 molecules of ATP fail to be made available to the muscles.

The removal of lactate is speeded up by gentle exercise following the main activity. The **warm down** has the effect of reducing muscle soreness by keeping the capillaries dilated and therefore flushing oxygenated blood through the muscles.

Table 14.3 shows the recovery times that are recommended after exercise. This information is used by trainers to work out the optimum intervals between training sessions, and between training and events. The aim is to avoid chronic (long-term) fatigue that can have a drastic effect on an athlete's performance.

> ## ? QUESTION 1
>
> 1 Why does lactate accumulate when oxygen is not supplied quickly enough?

Table 14.3 Recommended recovery times after exhaustive exercise

Aspect of recovery	Recommended recovery time/min	
	minimum	maximum
restoration of muscle ATP and CP	2 minutes	3 minutes
repayment of alactacid oxygen debt	3 minutes	5 minutes
restoration of oxygen myoglobin	1 minute	2 minutes
restoration of muscle glycogen (after prolonged exercise)	10 hours	48 hours
removal of lactic acid from muscle and blood	1 hour *	2 hours*
repayment of lactacid oxygen debt	30 minutes	1 hour

* = speeded up by a warm down, i.e. if the muscles are kept working gently, the lactate is removed more quickly

3 LONG-TERM RESPONSES TO EXERCISE

If we exercise regularly, our bodies adapt and we 'get fit'. We feel better, look better and are able to cope easily with exercise that a few months earlier would have had us gasping in a heap on the floor. A remarkable feature of the human body is its ability to respond to exercise, making us more able to cope with our chosen activity.

SO WHAT IS ACTUALLY HAPPENING?

Observable long-term responses to exercise include changes to the heart, lungs and muscles, although the extent of the effects depends on the type of exercise done. For training to have an observable benefit, it must be above a certain intensity. Muscles must be **overloaded** before they begin to **adapt**.

For instance, if you were going to train for a rugby team, a brisk walk would be totally useless, because it would not overload your muscles. Rugby, like many sports, uses a combination of all the energy systems, and the training should reflect this balance. You would need to do endurance (cardiovascular) work as well as short and long sprints, along with training (e.g. with weights) that would exercise all of the relevant muscle groups. As the muscles adapt, the intensity of training must be increased continually, to ensure muscles are still overloaded. This is the concept of progressive **resistance** and ensures that the body continues to adapt and improve.

Changes to the heart

The heart responds to exercise like any other muscle; it enlarges, although the nature of the enlargement depends on the type of exercise done. Generally, there is an increase in the size of the **myocardium** (all the heart muscle) and therefore an increase in the size of the chambers. Consequently the **stroke volume** – the volume of blood pumped with each beat – increases. As a general guide, an untrained person has a stroke volume of about 90 cm^3, increasing to over 120 cm^3 after training.

When the heart can pump more blood per beat, it does not have to beat as often when the body is at rest. This is why getting fitter causes a decrease in resting pulse. This can go down from about 70 in the average untrained person to less than 50. Some of the world's top endurance athletes have a resting pulse of about 35 beats per minute.

Research also shows that exercise actually increases the strength of the blood vessels, allowing them to withstand higher pressures and reducing the risk of atherosclerosis (hardening of the arteries) later in life.

Changes to the lungs

The strength of the respiratory muscles (internal and external intercostals and the diaphragm) is increased, allowing a greater volume of air to be forced in and out. This means that **ventilation rate** improves. The total lung volume (the vital capacity) and surface area of the alveoli also increases, resulting in a greatly improved rate of gas exchange. Overall, the improvements in the circulation and gas exchange systems results in an increased V_{O_2}**(max)**.

Changes to the muscles

The way in which the muscles adapt to exercise depends on the type of muscle fibres involved and the nature of the exercise. Here are some of the improvements that can occur:

- As a general rule, repetitive exercise with moderate resistance (such as aerobics, or light weights) improves muscle tone and stamina. Training

Gas exchange is discussed in detail in Chapter 9, where you can refresh your memory of important terminology.

221

with greater resistance (such as heavy weights) brings about an increase in muscle size, a process known as **hypertrophy**. Muscular hypertrophy involves an increase in the cross-sectional size of existing muscles. This is due to an increase in the number of myofibrils, the sarcoplasmic volume, and in the amount of connective tissue (tendons and ligaments).

- A general improvement in performance occurs because of an improved coordination of the motor units. A significant aspect of improving skill involves the antagonistic muscles. Studies have shown that as we practise particular movements, such as a tennis serve, we get smoother and more efficient. Not only do we train the active muscle to work better, we also learn to completely suppress the antagonistic muscle, preventing it from interfering with the action.

- With training, muscles increase their **oxidative capacity** – their capacity for respiration. This is achieved by an increase in the number of mitochondria in the muscle cells, an increased supply of ATP and CP and a rise in the quantity of the enzymes involved in respiration. The ability of the muscles to store glycogen, the amount of myoglobin and the ability to use lipid as an energy store are also increased.

- The density of the capillaries running through muscles increases in response to exercise, a process called **capillarisation**. This allows a more efficient exchange between working muscles and the blood.

SUMMARY

By the end of this chapter you should know and understand the following:

- Exercise places great metabolic demands on the body. Homeostatic mechanisms must cope with a much greater oxygen demand as well as a build-up of carbon dioxide and heat. In the long term, there may be a significant loss of water and salts (electrolytes) through sweating.

- Movement of muscles requires **adenosine triphosphate**, **ATP**, the chemical that supplies the muscles with the energy released in respiration.

- There are three sources of ATP: these comprise three energy systems. The most immediate is the **ATP/CP**, or **alactic anaerobic** system: ATP already present in the muscles is backed up by creatine phosphate, which is used to make more ATP instantly. The second system that kicks into action is the **glycolytic** or **lactic anaerobic** system: ATP is supplied by glycolysis and this system results in a build-up of lactate ions. The third is the **aerobic** system in which energy comes from the complete oxidation of glucose or lipid.

- The duration of an event determines the type of energy system used: up to 10 seconds, the ATP/CP system; from 10 to 60 seconds, the glycolytic system; and more than 60 seconds, the aerobic system.

- In the short term, the body responds to exercise by maintaining homeostasis. There is an increase in heartbeat and ventilation rate, coupled with vasodilation and sweating.

- During exercise we can build up an **oxygen debt**. This is paid off after exercise has finished, or during rest periods within the activity. This is why pulse and ventilation rates remain higher than normal even when activity has stopped. Extra oxygen is needed to replace the ATP/CP stores and to oxidise accumulated lactate.

- The long-term responses of the body to exercise are generally those that improve the body's fitness. These include strengthening of the cardiovascular system (heart and blood vessels) and making the lungs and muscles more efficient. The overall result is a body that is better able to cope with the chosen activity.

Practice questions and a How Science Works assignment for this chapter are available at www.collinseducation.co.uk/CAS

15

Nerves and impulses

15 NERVES AND IMPULSES

Neurotoxins on the arrow tip

Some Amazonian Indians have a very effective way of hunting monkeys and other animals. They tip their blowpipe arrows in a preparation made from the bark of one of the local trees. The substance, known as curare, is a powerful neurotoxin.

For many centuries, the exact nature of curare was a mystery. In 1814, in an attempt to confirm his suspicions, the explorer Charles Waterton injected a donkey with curare. Within a few minutes, the donkey appeared dead. Waterton cut a small hole in her throat and inflated her lungs with a pair of bellows. According to his account, the donkey regained consciousness and looked around. This crude method of artificial respiration was continued for a couple of hours until the effects of the curare had worn off.

Analysis of curare has since shown that it prevents nerve impulses getting through to the muscles, causing paralysis. In addition to affecting the movement of skeletal muscles, curare interferes with heartbeat and breathing, although the exact effects depend on the dosage and whether the curare is taken orally or by injection. If heartbeat and breathing are severely affected, the result is usually fatal.

In 1939, the active ingredient in curare was isolated, and in 1943 doctors started to use it as an anaesthetic. It was later used as a muscle relaxant and is now a vital tool in surgery, relaxing muscles and generally making the surgeon's job a lot easier. It is also useful for treating conditions in which the muscles go into spasm – polio, tetanus, epilepsy and cholera. Today, synthetic analogues of the active ingredient in curare, such as d-tubocurarine, are used widely in medicine.

The Amazonian Indians, not being analytical chemists, assess the strength of their preparation by how long it takes a monkey to fall out of the tree. 'One-tree curare' is the most powerful; the monkey only escapes to the next tree before the paralysing toxin brings it tumbling to the forest floor. 'Two-tree curare' is obviously weaker and 'three-tree curare' is the weakest preparation that is useful. After that, the monkey travels too far and, although it still dies from the action of the poison, it's usually impossible to find.

 REMEMBER THIS

The entire human body consists of four basic types of tissue (see Chapter 2). Epithelial tissue forms coverings and linings, connective tissue holds other tissues together, muscle tissue can contract and nervous tissue is excitable – it can transmit impulses. All of our organs are formed from these four basic tissue types.

1 IT'S ALL ABOUT COMMUNICATION

The body of a human is incredibly complex. To ensure that the different parts work together effectively, there needs to be communication to register changes in internal and external conditions. In this block of chapters we look at the role of hormonal and nervous communication.

The nervous system is made up of specialised cells that allow the different parts of the body to communicate. These cells are called neurones (Fig 15.1). They carry information from one part of the body to another. We concentrate on the structure and function of these remarkable cells in this chapter.

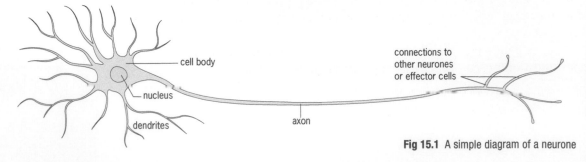

Fig 15.1 A simple diagram of a neurone

OVERVIEW OF THE NERVOUS SYSTEM

The nervous system as a whole ensures that the body responds appropriately to the external conditions at any given time. It means we can do the following:

- Gather information. Sense organs called **receptors** detect **stimuli** from the internal and the external environment.
- Transmit sensory information to the **central nervous system** by means of the **sensory nerves**.
- Co-ordinate information. Incoming information travels to the brain via the spinal cord. The brain then decides what to do. Decisions are often based on memory, the result of our past experience.
- Transmit the information to the **effectors**: the muscles and glands. Impulses pass from the central nervous system to the effectors via **motor nerves** (Fig 15.2).

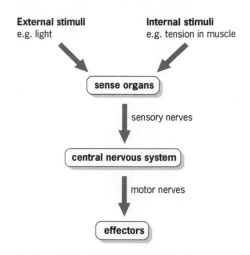

Fig 15.2 Flow diagram summarising nervous system function

HOW INFORMATION TRAVELS AROUND THE BODY

Information passes along **neurones** in the form of electrical signals called **nerve impulses**. A nerve impulse, known as an **action potential**, is not a message, nor is it an electrical current (a flow of electrons). It is more a change in ion balance in the nerve cell, which spreads rapidly from one end to the other, like the fire travelling along a burning fuse. It is little more than an electrical 'blip' – but the brain can make sense of the blips because they vary in frequency and arrive down specific nerves.

So what happens when the nerve impulse reaches the end of the neurone? It connects with other neurones at junctions called **synapses**. A nerve impulse crosses a synapse usually by means of a **chemical transmitter**. The whole of the nervous system therefore communicates by a mixture of electrical and chemical signals. This allows information to travel around an organism with far greater speed and precision than if only chemical signals were used.

The structure of the human nervous system is covered in detail in Chapter 16. The structure of muscles and their role in movement are covered in Chapter 17. Hormonal communication is the topic of Chapter 18. Senses and behaviour are discussed in Chapter 19.

Nerves

Neurones rarely act alone. They are bundled together into larger, visible structures called **nerves** (Fig 15.3). Nerves form a complex network throughout the body. Sensory nerves take information received from receptors, or sense organs, on the outer parts of the body – the eyes, ears, tongue and nose and touch receptors in the skin – into the central nervous system. Processing of that information happens in the brain, and then motor nerves take the information from the brain to effectors such as muscles and glands to cause the body to take some action. Think what happens if someone puts, say, a glass of water in front of you. Sensory nerves take the information from the eyes to the brain. If you are thirsty, the brain may then send information down motor nerves to your arm and hand and you would reach for the glass, pick it up and drink. If you aren't thirsty, information might travel down motor nerves to the muscles in your face and throat for you to say 'No thanks'.

In the reflex arc, the sensory and motor nerves are linked by a neurone in the spinal cord, rather than by neurones in the brain.

? QUESTION 1

1 What is the difference between a neurone and a nerve?

The reflex arc is discussed in Chapter 16.

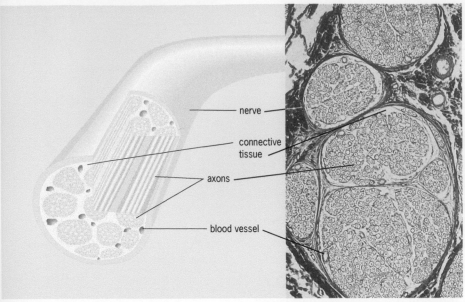

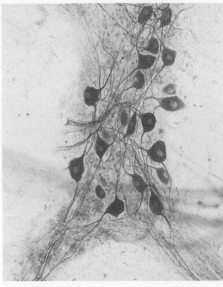

Fig 15.3a A nerve is a bundle of axons, together with connective tissue and blood vessels. The micrograph shows a cross-section of a nerve

Fig 15.3b The brown threads (axons) and their cell bodies form bundles that make up nerves

2 HOW DO NEURONES CARRY INFORMATION?

Neurones have two properties that enable them to carry information. They are **excitable** – that is, they can detect and respond to stimuli – and they are **conductive** – that is, they can transmit a signal from one end to the other. Before we look at how a neurone transmits information, let's find out what is going on in the neurone before information arrives.

THE NEURONE AT REST

At any given moment a neurone needs to be ready to conduct impulses. This state of readiness is called the **resting potential**. At this point, the axon membrane is **polarised**: the fluid on the inside is negatively charged with respect to the outside. This difference in charge, about –70 mV, results from an unequal distribution of ions known as an **electrochemical gradient**.

THE RESTING POTENTIAL

The resting potential results from an unequal distribution of ions, brought about by two processes; **active transport** and **facilitated diffusion**:

- **Active transport**. All animal cell membranes contain a protein pump called **Na+K+ATPase**. This splits ATP to gain the energy to pump ions. Three sodium ions move out of the cell at the same time as two potassium ions move in. It is an unequal exchange: more positive ions leave than enter the cell.
- **Facilitated diffusion**. There are also sodium and potassium **ion channels** in the membrane. These channels are normally closed, but they 'leak', allowing sodium ions to diffuse into the cell and potassium ions to leak out, down their respective concentration gradients. Generally, the potassium channels are more leaky than the sodium channels, so more potassium diffuses out. The potassium ions join the sodium ions that have been pumped out of the cell by the Na+K+ATPase pump.

The overall effect of the two processes is to cause an imbalance of sodium and potassium ions across the membrane: there are more positive ions outside the axon than inside, so the inside of the cell is negatively charged with respect

 QUESTION 2

2 **a)** What are the essential features of an active transport mechanism?
b) How is active transport different from diffusion?

✓ REMEMBER THIS

The Na+K+ATPase protein is thought to have evolved as an osmoregulator to keep the internal water potential high to prevent water entering animal cells and bursting them. Plant cells don't need this as they have strong cell walls that stop them bursting.

Working out how neurones carry information

It is difficult to study mammalian nerves because they are so small. Hodgkin, Huxley and Eccles did much of the early pioneering work on the nature of the nerve impulse in the 1940s and 1950s using the giant axons found in the squid. These axons are 1 mm across, making it far easier to insert electrodes into them (Fig 15.4).

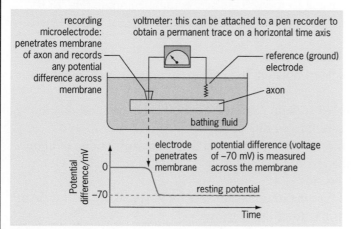

Fig 15.4 Inserting a microelectrode into a resting axon to measure its membrane potential

When the concentration of different ions was measured both in and out of the axon, they found the results in Table 15.1.

Table 15.1 Approximate concentrations of ions inside and outside the axon membrane (m mol kg^{-1})

Ion	Concentration outside the axon	Concentration inside the axon
Na$^+$	460	50
K$^+$	30	400
Cl$^-$	560	100
positively charged amino acids/proteins too large to cross the axon membrane	0	370

There are two opposing forces acting across the axon membrane:

1   Ions will diffuse down their **concentration gradient** as long as the membrane is permeable to them.

Potassium ions diffuse out because the membrane is more permeable to them than to sodium ions.

2   This movement is opposed by the electrochemical gradient – the balance of positive and negative ions either side of the membrane. Generally, positive ions will move towards areas of negative charge, and vice versa, until things are neutral. There are already more positive ions outside the membrane, and lots of negative ions inside the membrane, so the electrochemical gradient **opposes** the movement of potassium.

At –70mV these **two processes are balanced**, hence the value of the resting potential.

The experiments on squid axons also demonstrated why the trace of an action potential has its distinctive shape.

In one experiment, two electrodes were placed inside the large squid axon, one at each end (Fig 15.5). A short pulse of electrical current was applied through the stimulating electrode to mimic a nerve impulse. The recording electrode at the other end of the axon recorded a change of about 90 mV from –70 mV (the resting potential) to +20 mV. This was only very short-lived – the reading very quickly returned to –70 mV.

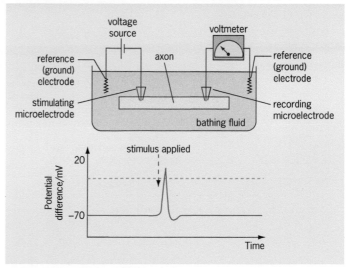

Fig 15.5 Recording the membrane potential in a squid axon when an electrical stimulus is applied

to the outside. This results in a potential difference across all animal cell membranes, called the **resting potential** or the **membrane potential**. The value of this potential varies from –20 to –200 mV in different cells and species, but is typically about –70mV.

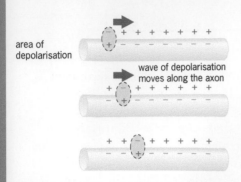

Fig 15.6 The basic concept of the nerve impulse – a wave of depolarisation which spreads along the axon. The active transport mechanism immediately re-establishes the resting potential as soon as the action potential has passed

 REMEMBER THIS

Once the action potential starts, the wave of electrical activity travels, or is **propagated**, along the axon at great speed.

 REMEMBER THIS

A nerve impulse is an **all or nothing** signal: if the threshold is not reached, depolarisation does not occur and no signal can travel along the axon.

THE ACTION POTENTIAL

An action potential is generated when a nerve is stimulated. The stimulus may come from a receptor cell (e.g. in a sense organ) or another neurone.

The action potential is brought about by a quick reversal in the permeability of the axon membrane that spreads rapidly down the axon as a wave of depolarisation (Figs 15.6 and 15.8). This allows sodium ions to flow suddenly into the axon as a wave of depolarisation, making the inside positive with respect to the outside (Fig 15.7). The sodium and potassium channels are **voltage gated**, which means that they can change their shape to let more or fewer ions pass, according to the voltage across the membrane (Fig 15.8).

The action potential has two stages: **depolarisation** and **repolarisation**.

Depolarisation step by step

1 When a neurone is stimulated, the voltage across the axon membrane changes.

2 A few voltage gated sodium channels detect this change and open to allow some sodium ions to diffuse in.

3 If the stimulus is large enough to reach the **threshold** value of about –50 mV, then the rest of the voltage gated sodium channels open for about 0.5 milliseconds.

4 This causes sodium ions to diffuse in very rapidly, making the inside of the cell more positive than the outside.

This is an example of a **positive feedback** ('change creating more change'). The more sodium ions there are, the more the voltage changes, so the more ion channels open, so the more sodium ions diffuse in.

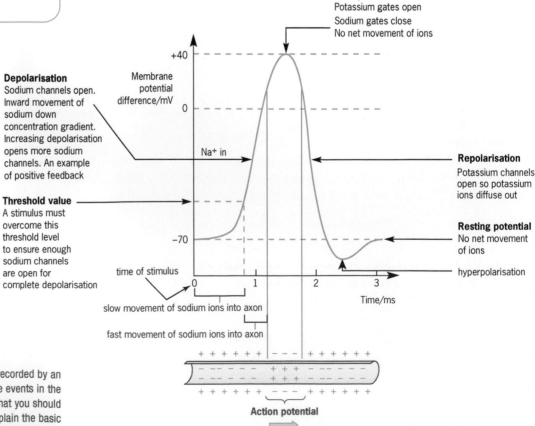

Fig 15.7 An action potential as recorded by an oscilloscope, and how it relates to the events in the axon. This is a key exam diagram that you should use when you are aiming to explain the basic sections of the trace

Repolarisation

When the membrane potential reaches zero, the potassium channels open for 0.5 milliseconds, causing potassium ions to rush out, making the inside of the cell more negative again. Since this restores the original polarity, it is called repolarisation.

Re-establishing the resting potential

The potassium channels remain open until after the resting potential value of −70 mV has been reached. This causes hyperpolarisation (the 'undershoot' visible in Fig 15.7) when the potential difference reaches about −80 mV. The potassium channels then close and the resting potential is established once again.

The refractory period

Nerves conduct messages by 'firing' repeated action potentials along the nerve fibres. The time delay between one action potential and the next is called the **refractory period**. This has two phases:

- The **absolute refractory period**. During this time, immediately after the sodium channels close, no further impulse can be conducted.
- The **relative refractory period**. During this time, the membrane begins to recover and becomes increasingly responsive. It is possible to initiate another action potential provided that the stimulus is greater than normal.

The refractory period imposes a limit on the frequency of nerve firing. Large nerve fibres recover in 1 millisecond and could theoretically propagate 1000 impulses per second. Small fibres take longer to recover – about 4 milliseconds – and so could propagate about 250 impulses per second. The refractory period ensures that each action potential is separated from the next, with no overlapping of signals. We can think of the information the signal conveys as coded information. The refractory period also ensures that a nerve impulse flows in one direction only: the wave of depolarisation can only move away from the refractory region, towards the axon terminal, and therefore onwards to the next neurone in the pathway.

The speed of an action potential

The speed at which a nerve impulse or action potential travels is k[...] its **conduction velocity**. In human nerve fibres, values range fro[...] 3 metres per second in unmyelinated fibres, and between 3 and 120 [...] per second in myelinated fibres (see below for what myelin is and wh[...] does). In general, conduction velocity depends on the following facto[...]

- **Axon diameter**. The larger the axon, the faster it conducts.
- **Myelination of the neurone**. A nerve impulse travels faster in [...] myelinated nerve than in an unmyelinated nerve.
- **Number of synapses involved**. Communication between neuro[...] across the tiny gaps at the synapses involves chemical release and a b[...] time delay. The greater the number of synapses in a series of neurones[...] slower the conduction velocity.

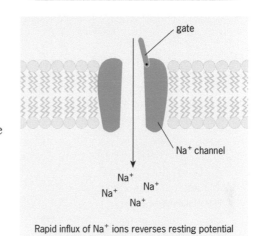

Rapid influx of Na+ ions reverses resting potential

Fig 15.8 A diagrammatic model of a voltage gated sodium channel in the axon membrane. When the gate opens, the rapid diffusion of sodium ions causes the action potential by reversing the resting potential

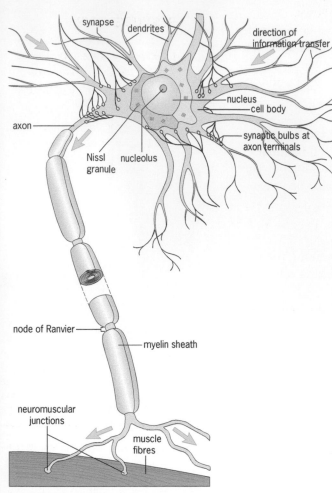

Fig 15.9 The structure of a generalised motor neurone

MYELINATED NEURONES AND SALTATORY CONDUCTION

Some nerves have exposed axon membranes but others are covered in a sheath of fatty material called myelin. As we have just seen, myelinated nerves conduct nerve impulses faster than unmyelinated nerves. This is an advantage as it allows information to travel around the body more quickly – this is probably why higher animals with more complex nervous systems – such as mammals – have a lot of myelinated nerves. The detailed structure of a generalised myelinated motor neuron is shown in Fig 15.9.

The importance of myelin

But what is myelin and how does it form around neurones? Specialised cells called **Schwann cells** wrap themselves round the axons of some neurones as they develop in a growing embryo (Fig 15.10a) The Schwann cells form a thick, lipid-rich insulating layer called the myelin sheath (15.10b). This insulates the axon electrically, rather like the plastic layer round a copper wire in an electrical flex. Neurones with myelin sheaths are said to be myelinated.

Saltatory conduction

Saltatory conduction occurs in myelinated nerves, when the action potential 'jumps' from node to node. This greatly increases the speed of signal transmission.

The myelin sheath insulates the axon, and so ion exchange can only occur at the nodes of Ranvier in between the Schwann cells, where the axon membrane is exposed.

Fig 15.10a Schwann cell growing round an axon

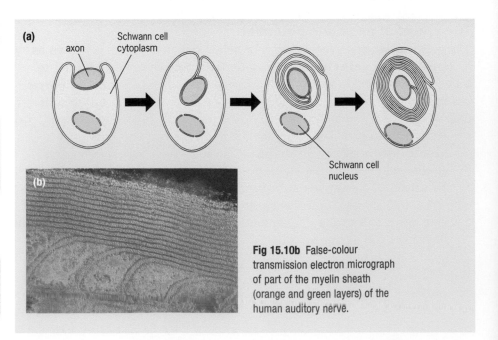

Fig 15.10b False-colour transmission electron micrograph of part of the myelin sheath (orange and green layers) of the human auditory nerve.

REMEMBER THIS

The pale, creamy colour of myelinated nerves is due to the fatty (lipid) nature of the myelin that surrounds them.

 QUESTION 4

4 What effect does a myelin sheath have on conduction velocity?

The mechanism of saltatory conduction appears to be as follows:

- When the action potential is present at one node, the influx of sodium ions causes the displacement of potassium ions down the axon (like charges repel).
- This diffusion of potassium down the axon makes the next node more positive and depolarises it until the threshold is reached.
- The impulse quickly jumps by this mechanism from node to node at speeds of over 100 metres per second, ten times faster than the best sprinters. Saltatory conduction, from '*saltare*' meaning to jump, refers to jumping conduction.

As well as being faster than non-myelinated conduction, saltatory conduction is very energy efficient in terms of ATP usage. Only a small part of the axon is involved in the exchange of ions, so fewer ions need to be pumped back after the action potential has passed.

REMEMBER THIS

In myelinated nerves, impulses can travel at up to 100 metres per second – the equivalent of a person running the length of a football pitch in one second, ten times quicker than the best sprinters can run.

STRETCH
AND CHALLENGE

Multiple sclerosis: demyelinisation of the central nervous system

Multiple sclerosis (MS) is a chronic, often disabling disease of the central nervous system that is caused by a loss of myelin around neurones. Symptoms may be mild, such as numbness in the limbs, or severe – paralysis or loss of vision. Most people with MS are diagnosed between the ages of 20 and 40 but the unpredictable physical and emotional effects can be lifelong.

MS is classified as an autoimmune disease: it results from the immune system attacking the body's own cells, mistaking them for foreign invading microbes. MS seems to be caused by a chemical found naturally in the body called interferon gamma. This molecule usually helps to activate the immune system to attack viruses and bacteria and other infectious agents. In people with MS, interferon gamma causes the immune system to wrongly identify nerve cells as foreign invaders. As a result, the myelin sheath coating nerves in the brain and spinal cord is destroyed by mistake (Fig 15.11). Transmission between nerve cells slows down and also becomes erratic.

At the moment, there is no cure for MS but there are drugs that can reduce the symptoms and minimise the number of relapses and attacks that people with the condition suffer.

In the UK, four drugs are approved for the treatment of MS. All work in much the same way – they block the damaging action of white cells that have been activated by the body's own inteferon gamma, so reducing the attacks on myelin. These drugs can help reduce the symptoms, but they cannot reverse the changes that have already occurred.

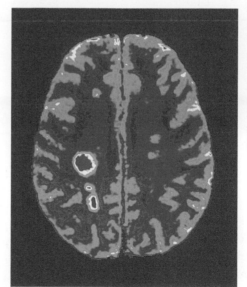

Fig 15.11 A coloured magnetic resonance image (MRI) scan of an axial section through the brain of a person with MS. The lesions due to MS are shown in red and yellow at the left and lower left

3 COMMUNICATION BETWEEN NEURONES: THE SYNAPSE

When an action potential reaches the end of an axon, it is passed on to the next neurone, or on to an effector cell such as a muscle or gland. The axon of one neurone does not usually make direct contact with the cell body of the next; the two cells are separated by a gap called a synapse (Figs 15.12 and 15.13).

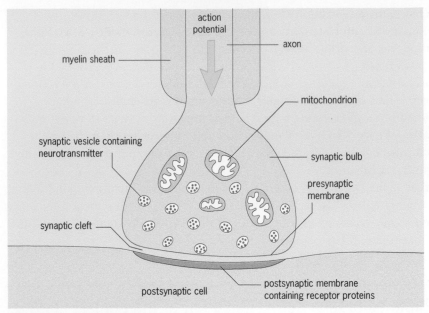

Fig 15.12 The basic structure of a synapse

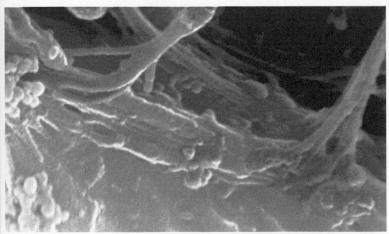

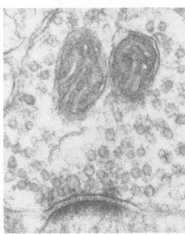

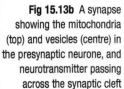

Fig 15.13a False-colour scanning electron micrograph of the junction sites (synapses) between nerve fibres (purple) and a neurone cell body (yellow)

Fig 15.13b A synapse showing the mitochondria (top) and vesicles (centre) in the presynaptic neurone, and neurotransmitter passing across the synaptic cleft

 REMEMBER THIS

The cell that carries a signal towards a synapse is a **presynaptic cell**; the cell carrying the signal away from the synapse is a **postsynaptic cell**. Presynaptic cells are always neurones but postsynaptic cells can be either neurones or effector cells.

 REMEMBER THIS

Many drugs and poisons exert their effect because they interfere with the functioning of synapses – see the Assignment for this chapter on the CAS website.

THE MAIN FEATURES OF A CHEMICAL SYNAPSE

The axon terminal of a presynaptic neurone is swollen, and is often called the **synaptic bulb** or **synaptic knob** (see Fig 15.12). It meets the cell body of the next axon, leaving a gap or **synaptic cleft** of about 20 nm. This is small – you would have to split a hair 250 times to get it to fit sideways into this gap. Even so, synapses have a high electrical resistance, and this gap is too big to allow the action potential to simply jump from one neurone to the next.

How do synapses work?

So how does the action potential get across? It is relayed by chemicals that diffuse across the gap and initiate an action potential in the neurone at the other side. The synaptic bulb contains many mitochondria, which provide energy for the manufacture of chemicals called **neurotransmitters**. Neurotransmitters are small molecules and they can diffuse easily across the synaptic cleft. Synaptic vesicles are temporary vacuoles (membrane-bound spheres) that store neurotransmitter chemicals, the most common being **acetylcholine**. Synapses that have acetylcholine as their transmitter are called **cholinergic synapses**.

Types of neurotransmitter

Hundreds of neurotransmitters have been identified and there are certainly more to find. There are four main groups:

- **Acetylcholine** (neurones that release acetylcholine are described as **cholinergic**).
- **Amino acids** such as **gamma-aminobutyric acid** (**GABA**), **glycine** and **glutamate**.
- **Monoamines** such as **noradrenaline**, **dopamine** and **serotonin** (neurones that release noradrenaline are described as **adrenergic**). Synapses that use noradrenaline affect heart rate, breathing rate and brain activity. This is similar to the effect of the hormone adrenaline, which prepares the body for emergencies.
- **Neuropeptides** (chains of amino acids) such as **endorphins**.

Acetylcholine also acts throughout the brain, modifying the activity of other neurotransmitters. Nerve pathways in which acetylcholine is a neurotransmitter seem to be involved in motivation and memory. A very fast-acting enzyme called **acetylcholinesterase** breaks down acetylcholine into ethanoic acid (acetic acid) and choline. These substances are reabsorbed through the presynaptic membrane. ATP energy from mitochondria is used to resynthesise acetylcholine, which is then returned to the vesicles. The chemicals in some 'nerve gases' work by inhibiting acetylcholinesterase.

> **✔ REMEMBER THIS**
>
> In some people, release of too much noradrenaline causes the heart to race. One way of treating this is to use drugs known as beta-blockers. These drugs have molecular shapes similar to noradrenaline.

Transmission at a synapse – a detailed explanation

As you read the next section, follow the stages of chemical transmission at a synapse numbered in Fig 15.14.

Fig 15.14 The sequence of events in chemical transmission at a synapse

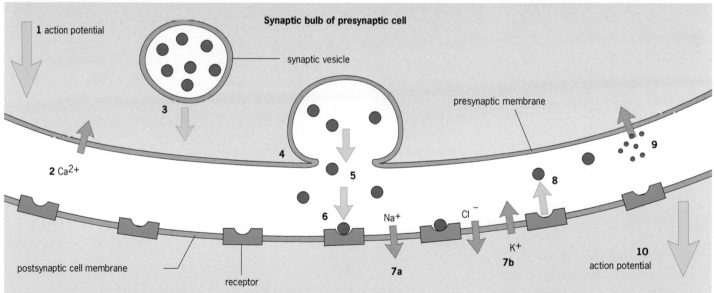

1. An action potential arrives at the synaptic bulb.
2. Calcium channels open in the **presynaptic membrane**. As the calcium ion concentration inside the bulb is lower than outside, calcium ions rush in.
3. As the calcium ion concentration increases, **synaptic vesicles** move towards the membrane.
4. The vesicles fuse with the membrane, releasing neurotransmitter into the synaptic cleft.

> **? QUESTIONS 5–6**
>
> 5 The synaptic cleft has a high electrical resistance but is very narrow. Suggest reasons for these two observations.
>
> 6 Why can transmission occur only one way across synapses?

REMEMBER THIS

Receptor binding can also lead to the formation of a second messenger (a transmitter substance) such as cyclic AMP (cAMP). This also changes the ionic permeability of the membrane, but it has a longer-lasting metabolic effect on the ion channels. Such long-term changes to brain neurones are thought to underlie memory.

5 The short journey across the synapse takes about a millisecond, longer than an electrical signal takes to travel the same distance. This time is therefore called the **synaptic delay**.

6 At the **postsynaptic cell**, the neurotransmitter binds to receptors on the postsynaptic cell surface membrane.

7a Some neurotransmitters open sodium channels in the postsynaptic membrane, causing an inflow of sodium ions. This creates an **excitatory postsynaptic potential** (EPSP) in the membrane. This potential lasts for only a few milliseconds and can travel only a short distance, but it makes the membrane more receptive to other incoming signals.

7b Alternatively, some neurotransmitters open potassium and chlorine channels (see inhibitory synapses, below).

8 Once the neurotransmitter has acted on the postsynaptic membrane, it is immediately released. If the transmitter remained, it could continue to stimulate the neurone, even without new impulses coming from the presynaptic cell.

9 At cholinergic synapses, the enzyme acetylcholinesterase splits acetylcholine into choline and ethanoic acid. These components then diffuse back into the presynaptic membrane, when they are resynthesised to acetylcholine using the ATP from the mitochondria.

10 At an excitatory synapse, an action potential is set up in the postsynaptic cell.

Inhibitory synapses

Does the arrival of an impulse at a synapse mean that an action potential is always generated on the postsynaptic neurone? The answer is no, because it would lead to chaos; all neurones would automatically pass on the signal to others. The significance of synapses is that they allow us to select particular pathways. Thus, at any one time, many more synapses need inhibiting than need stimulating. For this reason there are **inhibitory synapses**. Impulses arriving at these synapses make it more difficult for an action potential to be generated.

The neurotransmitters made by inhibitory synapses open potassium and chloride channels rather than sodium channels, and the resulting ion movement causes an **IPSP – inhibitory postsynaptic potential** – in which the membranes are hyperpolarised (to about –90mV) rather than depolarised (see step 7b above). Usually, whether or not an impulse is generated in a particular nerve depends on the balance of inhibition and excitation that neurones receive at any one moment.

? **QUESTION 7**

7 If the EPSPs generated in an neurone are cancelled out by the IPSPs, what will be the response of the neurone?

SYNAPSES IN ACTION: FACILITATION AND SUMMATION

Synapses have a vital role in **information processing**. Transmission of information across synapses is **graded**. They can amplify or damp down the information they receive. In many cases, they will not transmit it at all.

A neurone can be fed information by both excitatory synapses that produce EPSPs and inhibitory synapses that produce IPSPs (Fig 15.15). Whether the cell develops an action potential is determined by the sum of all the excitatory and inhibitory synapses at any particular moment. Put simply, impulses arriving at some synapses will 'excite' the cell, while others will 'calm it down'. Whether or not a neurone generates an action potential depends on the balance of the two types.

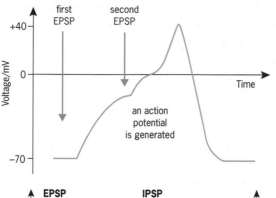

Fig 15.15a Summation of EPSPs. If many EPSPs are set in rapid succession in the same area of the postsynaptic membrane, the threshold potential may be reached and an action potential may result

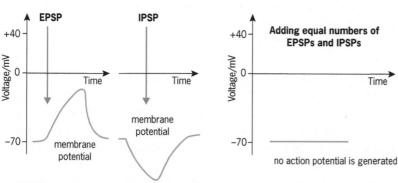

Fig 15.15b The effect of inhibitory synapses. If the same number of EPSPs as IPSPs are set up at the same time in the same area of the postsynaptic membrane, the two potentials cancel each other out

Imagine a synapse discharging its transmitter on to a postsynaptic neurone. This will set up an EPSP, but if it is not big enough to reach the threshold, no action potential is generated. However, if other synapses discharge their transmitter at the same time, or shortly after, the EPSPs will add up, or **summate**, until an action potential is generated. Generally, there are two types of summation:

- **Temporal summation** – summation of two or more impulses that arrive rapidly one after the other down the same neurone (Fig 15.15a).
- **Spatial summation** – summation of two or more impulses arriving down different neurones at the same time (spatial = related in space). One neurone can receive information from many others – this is synaptic convergence. It follows that the arrival of one excitatory impulse will leave the neurone more responsive to another one. This is known as **facilitation**, and results from the summation of two or more synapses discharging their transmitter substance at the same time.

As a simple example of this idea, imagine the touch receptors from one area of skin feeding into one sensory neurone. An action impulse from just one receptor is almost certainly an insignificant stimulus, and can be ignored. It will not create an EPSP large enough to generate an action potential in the sensory nerve. However, if several touch receptors are stimulated at the same time, they will summate and produce a sensory impulse.

 REMEMBER THIS
Temporal = related in time.
Spatial = related in space.

 REMEMBER THIS
Facilitation is result of spatial summation. It is not a result of temporal summation, which is simply the accumulation of EPSPs arriving before the preceding EPSP has died down.

WHY HAVE SYNAPSES?

Synapses are important because they allow the transfer of information in nerve networks to be controlled. Synapses:

- allow information to pass from one neurone to another;
- help ensure that a nerve impulse travels in one direction only;
- allow the next neurone to be excited or inhibited;
- can amplify a signal (make it stronger);
- protect nerve networks by not firing when over-stimulated. When this happens the synapse is said to be fatigued. Over-stimulation might damage muscle or gland tissue;
- can filter out low-level stimuli. For example, you fail to notice the sound of a clock ticking because synapses are 'filtering out' the signal of sound;
- aid information-processing by the action of summation (adding together the effect of all impulses received, see page 235);
- are modifiable and can form a physical basis for memory.

Overall, the significance of synapses cannot be over-emphasised. They allow us to select particular neural pathways. The process of learning is largely one of educating the synapses. People can play the violin or piano, or play tennis, because their synapses allow their brains to co-ordinate their senses and muscles in the right way. Your memories, too, have a basis in synapses choosing specific pathways. If you are asked, 'What's the capital of France?' your synapses will (we hope) select a pathway of neurones in your brain which will lead you to the answer 'Paris'.

> ✔ **REMEMBER THIS**
> Muscular contraction is discussed in more detail in Chapter 17.

A SPECIAL SORT OF SYNAPSE: THE NEUROMUSCULAR JUNCTION

When a motor neurone terminates on a muscle, it branches into many specialised synapses called **neuromuscular junctions**. Fig 15.16 shows a typical neuromuscular junction. These structures are wider than ordinary synapses and come into close contact with the surface membrane of the muscle, the sarcolemma. The area of sarcolemma in contact with the synapse is called the motor end-plate, and contains acetylcholine receptor sites. When an action potential arrives at a neuromuscular junction, vesicles of acetylcholine are released in the usual way. The transmitter changes the permeability of the motor end-plate to sodium ions and potassium ions, creating an end-plate potential (EPP) which results in an action potential passing along the sarcolemma. This impulse brings about the contraction of muscle fibres in that area. See the Assignment for this chapter on the CAS website.

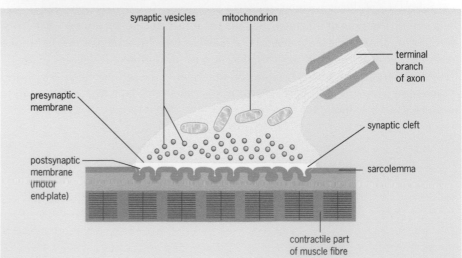

Fig 15.16 Studies have shown that each synaptic vesicle contains about 10 000 acetylcholine molecules, and that about 100 vesicles need to be released before an action potential (leading to a muscle twitch) can be generated

4 DRUGS AND SYNAPSES

Transmitters are released in tiny amounts: only 500–1000 molecules from each synaptic knob are required to transmit an impulse to a postsynaptic neurone. So, drugs that affect transmitters or their binding sites can have powerful effects when given in fairly small doses. Some chemicals, many of them from plants, have a dramatic effect on the nervous system.

NICOTINE AND ATROPINE

Nicotine is a substance found in tobacco. The nicotine molecule is a similar shape molecule to acetylcholine and so competes with acetylcholine to bind with its receptors. Once the nicotine has bound to the receptor, it opens the sodium channel that forms part of the receptor, causing nerve impulses to be generated (Fig. 15.17). See the How Science Works box (overleaf) for a discussion of the effects of nicotine.

Atropine is a drug, found in the plant deadly nightshade amongst others, which relaxes the muscles and glands controlled by the parasympathetic nervous system. Atropine eye drops, for example, will dilate the pupils. Atropine also binds to acetylcholine receptors, but does not open the sodium channels. It blocks the receptors, and prevents acetylcholine binding to them (Fig 15.17). When this happens in motor neurones, it causes muscle paralysis.

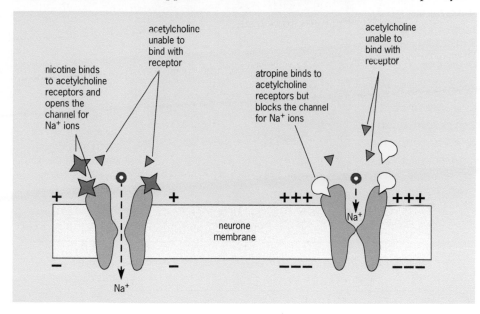

Fig 15.17 Drugs and transmitter binding sites

DRUGS AND NORADRENALINE

Beta-blockers are a class of drugs that stop the heart from being over-stimulated. They are given to patients suffering from a variety of heart disorders. Beta-blockers have a similar shape to noradrenaline and so compete with it for receptors on the postsynaptic membrane. Amphetamines and cocaine also affect noradrenaline synapses, but they work in a different way. They prevent the reabsorption of noradrenaline from the synaptic gap. So, noradrenaline remains in the gap and the neurone keeps on firing. One of the effects of amphetamines is to make a person feel energetic and carefree, which is why they are often known as speed. Before the harmful effects were known, amphetamines were given to pupils with poor attention spans, to help them concentrate.

Amphetamines are psychologically addictive. Users become dependent on the drug to avoid the 'down' feeling they often experience when the effect of the drug wears off. This dependence can lead a user to turn to stronger stimulants such as cocaine, or to larger doses of amphetamines to maintain a 'high'.

People who abruptly stop using amphetamines often experience the physical signs of addiction, such as fatigue, long periods of sleep, irritability and depression. How severe and prolonged these withdrawal symptoms are depend on the degree of abuse.

HOW SCIENCE WORKS

Nicotine – the drug that mimics acetylcholine

Do you wonder why people continue to smoke, even though they know the increased risk of lung cancer and heart disease associated with their habit? The answer lies partly with one of the components of tobacco: nicotine.

Nicotine is very addictive. It affects the brains of smokers, making them feel less stressed, better able to concentrate and less likely to eat sweet foods. Smokers become tolerant to nicotine over time, needing to smoke more to achieve the same effects. But how does nicotine cause addiction?

Studies carried out in the early 1980s using nicotine labelled with a radioactive tracer showed that it is taken into the brain very rapidly. Once there, it binds tightly to acetylcholine receptors, fooling postsynaptic cells into 'thinking' they are being stimulated. It also binds to other receptors which normally accept another neurotransmitter, dopamine.

The action of nicotine on both types of receptor in specific areas of the brain causes long-lasting changes to cell connections and may explain why it is addictive. We know that dopamine receptors, in particular, are involved in addictions to other substances such as amphetamines and cocaine.

Because nicotine is addictive but not carcinogenic (it is the chemicals in tobacco tar that have been shown to cause cancer), smokers keen to kick the habit can get help. They can buy skin patches and gum that deliver nicotine to the brain, but not tar to the lungs (Fig 15.18).

Although patches and gum do help smokers to cut down or stop smoking altogether, they are only a partial solution. Nicotine affects the acetylcholine receptors in the parasympathetic nervous system that are involved with the constriction of blood vessels. Over time, circulatory problems and heart disease can result, and it is best to avoid these effects altogether.

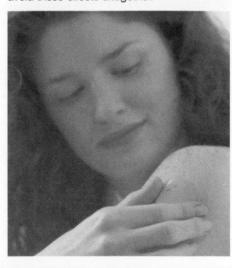

Fig 15.18
Transdermal patches can be used almost anywhere on the skin. They slowly release substances such as nicotine into the bloodstream

SCIENCE IN CONTEXT

Prozac and tranquillisers

Serotonin is a neurotransmitter normally active in the brain. Some forms of depression are caused by a reduced concentration of serotonin in the brain. The antidepressant drug Prozac is known as a serotonin re-uptake inhibitor. Prozac alleviates depression because it competes with serotonin for the 'active' sites on the proteins that reabsorb serotonin, leading to a higher concentration of serotonin at synapses in the brain. It also competes for the active sites on the enzymes that break down serotonin at synapses.

Tranquillisers are drugs that reduce tension. Benzodiazepine tranquillisers, such as Valium, work by increasing the binding of inhibitory transmitters in the brain. Inhibitory transmitters hyperpolarise rather than depolarise the membrane of the next neurone. This makes the next neurone less excitable. Valium reduces stress and anxiety, but it can be addictive.

SUMMARY

When you have read this chapter you should know and understand the following:

- The basic unit of the nervous system is the nerve cell, or neurone. This is a specialised cell with many **dendrites** that take impulses into the cell body, and a single, greatly elongated, axon that takes impulses away from the cell body.

- The axon membrane is able to use an active transport mechanism to establish a **resting potential**. This is an electrical charge across the membrane caused by an unequal distribution of ions.

- The nerve impulse itself, called the **action potential**, is a momentary reversal in the resting potential, caused by a sudden rush of sodium ions into the axon. The action potential spreads rapidly along the axon.

- The action potential lasts for only a millisecond or so, after which the resting potential is re-established. When an action potential has passed, there is a brief period of time – the **refractory period** – during which it is impossible to generate another action potential.

- The nerve impulse passes from one neurone to another (or from a neurone to a muscle cell) by means of **synapses**.

- Synapses are vital in selecting some neural pathways and not others. As such, synapses play a vital role in memory, skill and co-ordination of the body's activities.

- Transmission across synapses occurs when a chemical, the neurotransmitter, is released by the **presynaptic membrane**. This chemical diffuses across the gap and changes the permeability of the **postsynaptic membrane**, generating an **excitatory postsynaptic potential** (**EPSP**). If the EPSPs are sufficiently large, an action potential is generated in the next neurone.

- **Summation** means that the effect of several action potentials can add up to produce transmission at a particular synapse.

- Many drugs and poisons work by affecting synaptic transmission.

Practice questions and a How Science Works assignment for this chapter are available at www.collinseducation.co.uk/CAS

16

The human nervous system

What brain damage can tell us about brain function

The human brain is an organ of great mystery. In addition to being unbelievably complex, it is very difficult to investigate experimentally. We now know that different areas of the brain have particular functions, but how did we find out? Some of the earliest clues came from accidents in which people suffered damage to a particular part of the brain.

Consider the tale of Phineas Gage, a US railroad worker. He was a popular and reliable man, polite and responsible, and he had been made a foreman. In September 1848, he was jamming a stick of dynamite into a hole using a tamping iron. A spark from the iron ignited the dynamite, and the metal rod came out of the hole a lot faster than it went in.

The rod entered Gage's face below his left cheekbone, passing through the eyeball and through the top of his skull before landing several metres away. Gage fell back in a heap, as you would expect, but remarkably he did not die. He did not even pass out. He was driven by oxcart to a local physician, John Harlow. As the doctor stuck his fingers into the holes in Gage's head until the tips met, Gage asked when he would be ready to return to work. Within a couple of months he had recovered physically, but was no longer himself. Instead of being a gentle, honest, conscientious worker, Gage became 'a foul-mouthed and ill-mannered liar'.

He lived on as something of a celebrity for another 13 years before dying from an epileptic fit. On hearing of the death, Harlow managed to get Gage's body donated to medical research. He believed that the changes in Gage's behaviour had been caused by the damage to the frontal lobes of the brain. 'The equilibrium ... between his intellectual faculties and animal propensities seems to have been destroyed', Harlow wrote.

Some 130 years on, scientists were able to use computer modelling to trace the damage done to Gage's brain. The metal rod missed the areas associated with language and motor function, but destroyed the ventromedial region on the left side of his frontal lobe. This is what made Gage so antisocial. People who have had tumours in this area of the brain have undergone the same sort of transformation.

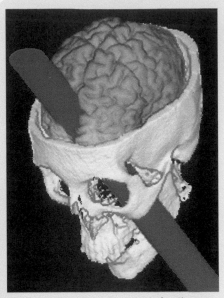

Ouch! Early knowledge about brain function was pieced together with the help of accidents such as that suffered by Phineas Gage. The metal rod missed the vital blood vessels and the crucial areas of the brain that keep us alive. It did, however, damage the frontal lobes, and the subsequent changes in Gage gave us important clues about the function of this part of the brain

1 GENERAL ORGANISATION

The human nervous system can be divided into two parts: the **central nervous system** (CNS) and the **peripheral nervous system**. The CNS consists of the **spinal cord**, which is protected by the vertebral column, and the **brain**, which is enclosed within the bony shell of the **cranium**. The peripheral nervous system brings information from the sense organs into the CNS, and then relays information out to the structures that bring about responses: the muscles and glands. The peripheral nervous system is divided into the **somatic** and **autonomic** nervous systems. The somatic system is

? QUESTION 1

1 How would you define intelligence?

responsible mainly for controlling voluntary actions, such as walking and talking, while the autonomic system keeps the involuntary body functions such as heartbeat, blood pressure and breathing ticking over. The autonomic system is further divided into the **sympathetic** system and the **parasympathetic** system.

REMEMBER THIS

Neurone – a nerve cell (see Chapter 15).

Stimulus – a change in an organism's environment (internal or external) that can be detected by receptor cells.

Receptor – a specialised cell that detects a stimulus and initiates a nerve impulse.

Sensory neurone – a nerve cell that carries impulses from the receptor into the central nervous system.

Central nervous system (CNS) – the brain and spinal cord. The CNS processes the incoming information and decides on a response – often based on previous experience.

Motor neurones – nerve cells that carry impulses from the CNS to the effectors.

Effectors – the organs that bring about the response, usually muscles or glands.

REMEMBER THIS

What is grey matter? Different areas of nerve tissue in the central nervous system are different colours. Grey matter contains nerve cell bodies; their nuclei are responsible for the grey colour. White matter consists largely of myelinated fibres (their fatty sheaths are creamy-white)

2 THE SPINAL CORD

The spinal cord starts at the base of the brain and ends at the first **lumbar vertebra** (this is roughly at waist level). The spinal cord is enclosed within the **vertebral column** and has a diameter of about 5 millimetres (Fig 16.1). Further protection is provided by three layers of tough membranes called **spinal meninges**, and by the **cerebrospinal fluid** that cushions the cord, acting as a shock-absorber. Each vertebra has an opening on its right and left sides to let spinal nerves pass through. These nerves extend into the body, forming the **peripheral nervous system**.

REMEMBER THIS

Meningitis is a potentially fatal disease characterised by inflammation of the meninges – the protective membranes that surround the brain and spine.

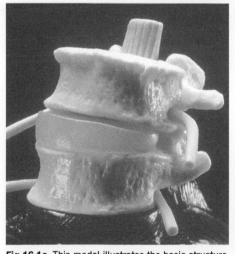

? QUESTION 2

2   How is the human spinal cord protected from damage?

Fig 16.1a This model illustrates the basic structure of the spinal cord. The delicate nerve cord itself runs down the middle of a channel in the centre of the vertebra. Between each bony vertebral disc is a shock-absorbing intervertebral disc made of cartilage. When the outer layer of the disc breaks or ruptures, part of the cartilage can push against the nerve cord, causing pain, numbness and – in severe cases – partial paralysis

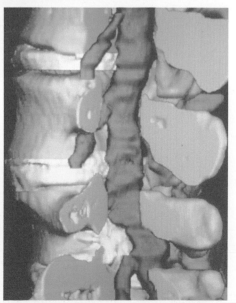

Fig 16.1b A coloured 3D computed tomography (CT) scan image of the same problem. The slipped disc (yellow, lower centre) can be seen pushing against the nerve cord (dark blue). Compare this with the two normal discs above

3 THE REFLEX ARC

A reflex arc is the simplest example of co-ordination, so is a good place to start a study of how the nervous system works. The key feature of a reflex is that a particular stimulus leads to a fixed response – this is very rapid and can't be controlled because it does not pass through the conscious parts of the brain. An important feature of reflex arcs is that they contain as few synapses as possible. This speeds up the response and in many cases – such as the blinking reflex or the reaction to heat and pain– avoids danger or minimises damage (Fig 16.2a). Other reflex arcs are mainly to do with posture.

THE KNEE-JERK REFLEX

Fig 16.2b shows a generalised representation of the knee jerk reflex – a classic example of a reflex arc. This is a postural reflex, one of the many mechanisms we have to maintain our position without having to constantly think about fine adjustments.

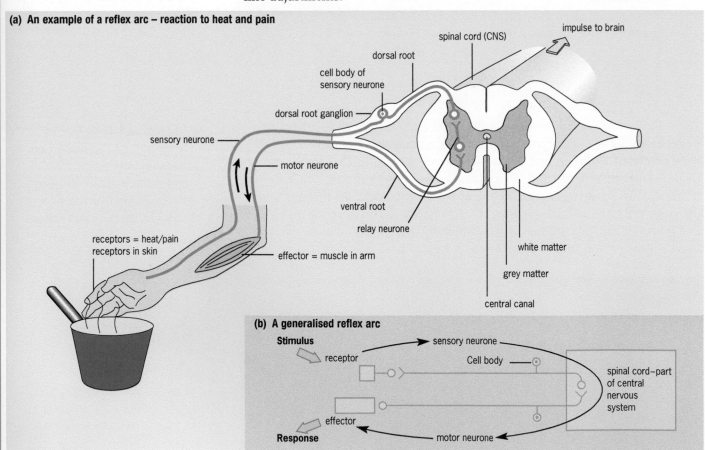

(a) An example of a reflex arc – reaction to heat and pain

(b) A generalised reflex arc

Fig 16.2 Reflex arcs

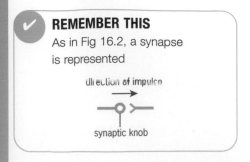

REMEMBER THIS

As in Fig 16.2, a synapse is represented

direction of impulse

synaptic knob

The knee-jerk reflex is initiated by the stretch receptor in the patellar tendon – a tap on this tendon just below the knee has the same effect as the knee bending. Nerve impulses pass up the sensory neurone and into the spine. Here, the sensory neurones synapse with motor neurones and the nerve impulses pass straight out of the spine in the motor nerve, where they pass to the thigh muscle (the quadriceps). Contraction of the quadriceps straightens the leg. Impulses will also pass from the sensory nerves up the spine to the brain, but we are conscious of the stimulus only after the response has been initiated.

OTHER REFLEXES

Most reflexes contain more synapses than the knee-jerk example. Blinking when a foreign object enters the eye and the withdrawal reflex, or pulling your hand away from a hot pan (Fig 16.2a), both involve a circuit containing sensory receptors, sensory neurones, spinal relay neurones (interneurones), motor neurones and effector muscles. The principle is the same: the heat of the pan stimulates receptors, and impulses pass along sensory neurones towards the spinal cord. Here, instead of making synapses with motor neurones, they pass their signals on to relay neurones. These connect with motor neurones that cause muscles to contract, moving your hand away from the pan.

You know that you have touched the hot pan because some impulses do travel to the brain, but the movement that is part of the reflex is involuntary, not under your conscious control. It is more difficult to persuade people that the swearing that accompanies such an event is also involuntary.

4 THE BRAIN

The human brain weighs about 1.5 kilograms, is 85 per cent water and has the consistency of thick blancmange (Fig 16.3). It is, however, the most complex material known. It copes with a huge amount of information from the various senses, deciding what is important and what can be ignored. It stores thousands and thousands of memories for decades, and can sort them into chronological order. It allows us to control the complex functions of the body, while at the same time allowing us to maintain posture, read, write and talk. Can we make a computer to do all of this? Not a chance.

The brain is the organ that makes humans human. We owe our success to this remarkable mass of nerves which takes over 20 years to mature, and which allows us to make tools and use language to an extent that far surpasses our nearest relative, the chimpanzee.

The brain of a human makes up about one-fiftieth of the body's mass. Its delicate tissues are protected by the skull or cranium and by the **cranial meninges**, membranes that are continuous with the spinal meninges. Cerebrospinal fluid bathes the outside of the brain and fills the chambers – the **ventricles**. Twelve pairs of cranial nerves innervate (supply nerves to) various regions of the head. The human brain is thought to contain ten thousand million (10^{10}) neurones. Each neurone may be in contact with a thousand other cells, providing an immense number of different communication routes.

THE AREAS OF THE BRAIN

The brain receives a vast number of impulses from receptors both inside and outside the body. It maintains basic involuntary body 'housekeeping' functions such as heart rate, breathing rate and temperature control. It also co-ordinates the semi-automatic muscular actions of the body such as swallowing, and it initiates and controls voluntary activities such as walking and running. The human brain is the site of higher mental functions such as reasoning, emotion and personality.

Like all vertebrate brains, the human brain consists of three parts, a hindbrain, midbrain and forebrain. However, our brains have evolved and enlarged to such an extent that the basic three-part pattern is only noticeable in the early stages of development of the embryo.

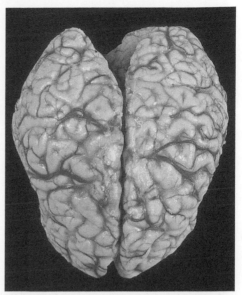

Fig 16.3a The most obvious feature of the human brain (seen here from above) is the two huge cerebral hemispheres. These are the site of higher conscious functions such as memory, language and emotion

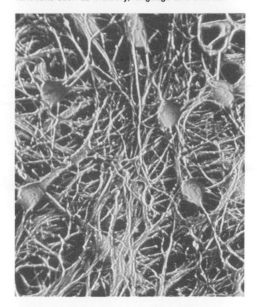

Fig 16.3b False-colour scanning electron micrograph (SEM) of neurones (nerve cells) from the cerebral cortex – the outer, heavily folded, grey matter of the brain. Have you ever thought about thinking? It's a strange idea, but our thoughts are pathways of nervous activity between the neurones of the brain. This is what makes us what we are, to the extent that there will never be a brain transplant, it would be a body transplant

Temperature regulation is discussed in more detail in Chapter 12, and homeostasis in Chapter 11. The endocrine system is discussed in Chapter 18.

The **hindbrain** has three distinct structures. The **medulla oblongata** (Fig 16.4) is a swollen portion at the bottom of the brain stem that houses vital centres controlling heart rate, breathing and blood supply. The **cerebellum** (Fig 16.4) controls body movement and maintains balance, and the **reticular activating system** (a collection of neurones in the centre of the brain stem) filters incoming stimuli and controls wakefulness and sleep.

As in other animals, the **midbrain** in humans links the forebrain to the hindbrain. Our emotions, which are located in the forebrain, can affect basic functions of the hindbrain such as control of blood vessel diameter, heart rate and sweating. When we are worried about something – exam results for example – the forebrain interprets this as stress and brings about the release of adrenaline, a hormone that prepares the body for action.

The **forebrain** has two main parts, the **cerebrum** and a region containing the **thalamus** and the **hypothalamus**.

Hypothalamus

The hypothalamus (Fig 16.4) is a key area. It receives a huge amount of internal and external sensory information and acts as a co-ordinating centre between the nervous system and the endocrine (hormonal) system. The hypothalamus is responsible for sensations such as hunger and thirst. It also helps control the autonomic nervous system since it regulates body temperature and the balance of water and salts in the blood. It is linked directly to the pituitary gland by means of blood vessels and nerves. Generally, the hypothalamus controls the release of hormones from the pituitary, including the antidiuretic hormone (which controls water reabsorption in the kidneys), growth hormones and some reproductive hormones. The thalamus (Fig 16.4) directs sensory information from the sense organs to the correct part of the cerebral cortex.

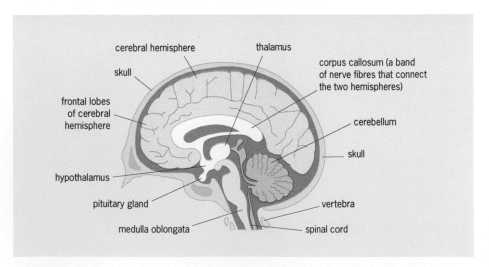

Fig 16.4 Main regions of the brain

Cerebrum

The cerebrum is made up of two large **cerebral hemispheres** (Fig 16.4). They have a thin outer layer, the cortex, which is thrown into many folds with fissures (grooves) between. The cortex is the surface layer of the hemispheres (Fig 16.5), and most of our conscious thought takes place here. The cerebral cortex carries out the 'higher' mental activities of reasoning, and is regarded as the site of personality and emotion and a sense of 'self'. This is what sets humans apart from other species.

Different areas of the cerebral hemispheres are associated with different sensory and motor functions – these are mapped out in Figs 16.6 and 16.7.

The cerebrum of mammals is very large compared with the forebrains of other vertebrates and is certainly the dominant feature of the human brain. Folds in the surface of the cerebrum increase the surface area for centres of control, where incoming nerve impulses are interpreted, or integrated, in the light of information already stored in the brain. Overall, the function of the cerebral cortex can be summarised in three stages:

- The **sensory areas** receive sensory information.
- The information is processed and interpreted in **association areas,** which decide on an appropriate response.
- Responses are initiated in the **motor areas** and modified by the **cerebellum**.

In this section we look at each of these areas in turn.

Vertical section

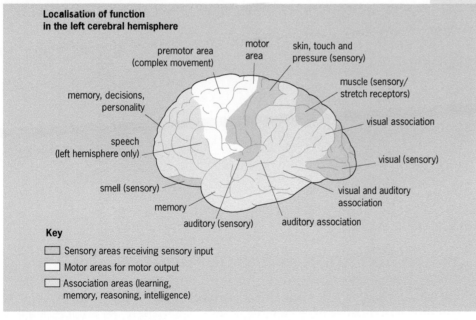

Fig 16.5 The cerebral cortex

Fig 16.6 The key areas of the human cerebral cortex

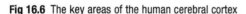

Fig 16.7 The cerebral hemispheres are divided into four main lobes
The **frontal lobe** is concerned with higher brain functions such as consciousness, reasoning, decision making and emotion. It includes the primary motor cortex that controls movement
The **temporal lobe** is concerned with processing auditory information including hearing, sound recognition, memory and speech
The **parietal lobe** is concerned with orientation, calculation, movement sensation, balance, and some aspects of memory
The **occipital lobe** is mainly concerned with sight; processing and interpreting information from the eyes. Includes; depth perception, colour, detail, perspective and recognition

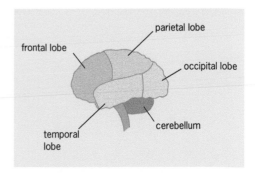

THE SENSORY AREAS OF THE CEREBRUM

The sensory areas of the brain receive incoming impulses from the vast number of receptors that provide us with information about the world outside. The external receptors might tell us that it is cold, dark, raining and that the skin is getting wet. Your brain has to decide what to do.

The model shown in Fig 16.8 shows how our brains 'see' our body. There are many receptors in the skin but as you will know, some parts are more sensitive than others. The lips, tongue and fingertips have more receptors per unit area of skin than anywhere else. Fig 16.8 reflects this; the largest parts are the most sensitive.

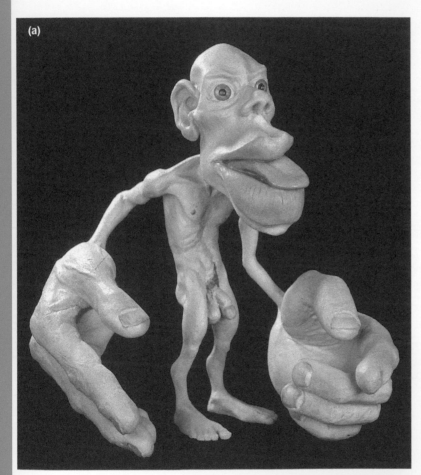
(a)

Fig 16.8a No, you won't meet him on a dark night. Some parts of the body contain more receptors than others, and consequently a larger area of brain is devoted to them. In this distorted model, anatomy is proportional to the area in the brain that deals with the sensory information. Thus the hands, mouth and tongue are huge – they have many receptors per square centimetre – while other areas such as the torso and legs are relatively small. The eyes, however, are not in proportion in this model. These take up more of the human brain than the rest of the body put together

Fig 16.8b The different areas of the body are also mapped out on the motor area of the cortex

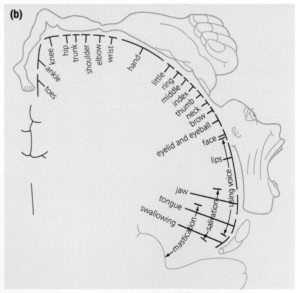
(b)

THE ASSOCIATION AREAS OF THE CEREBRUM

The association areas *interpret* the sensory information and make decisions in the light of previous experience, i.e. our **memory**. If you are cold and wet, you may see a friend's house and your visual association areas will compare it with the images of houses stored away. Once you have recognised it, impulses pass to your frontal lobes, which then make a conscious decision to knock at the door.

The way information from the eyes is processed gives us a particularly impressive example of brain function. The millions of rods and cones in the retinas of both eyes generate a stream of impulses that pass down the optic nerves to the visual cortex at the back of the brain. The different information from the two eyes is processed in the **visual association area** into a single image in which we have a perception of depth as well as colour and detail.

Speech and language

We take it for granted that we can listen to someone speaking our own language and understand what they say but have you ever wondered how this is possible?

When we hear speech, the **organ of Corti** in the ear converts the sounds to nerve impulses that travel to the **primary auditory cortex** in the cerebellum of both hemispheres of the brain. The two sides recognise different information in the signals. Both then send signals to an area called **Wernicke's area** in the left hemisphere. This is responsible for processing the sounds of speech. Wernicke's area is therefore the speech association area.

People who have damage in this area of the brain cannot understand someone speaking to them in their own language, even though they can hear perfectly. When they speak, it sounds OK in terms of rhythm but the sounds don't make any sense. This is a condition known as **receptive aphasia**.

In the normal brain, after the signals have been processed in Wernicke's area, the information is then passed to another area called **Broca's area** through a bundle of nerves called the **arcuate fasciculus**. Broca's area further processes the signals and transmits processed information to nearby motor-related areas. These connect with the parts of the face, jaw and throat and are responsible for speech production. So, Broca's area is the speech motor centre (Fig 16.9).

People with damage to Broca's area can understand what others say to them, but they can't actually speak. This condition is known as **expressive aphasia**. Broca did a lot of his work with a patient with a damaged Broca's area who could only say the word 'Tan'.

So, in summary, Wernicke's area lets you understand what someone else says, Broca's area lets you work out your reply.

THE MOTOR AREAS OF THE CEREBRUM

The motor areas generate the impulses that cause the muscles to contract. This simple statement hides the fact that the process is incredibly complex and requires such a degree of computation that we are unlikely to be able to make lifelike 'android' robots for a long time to come.

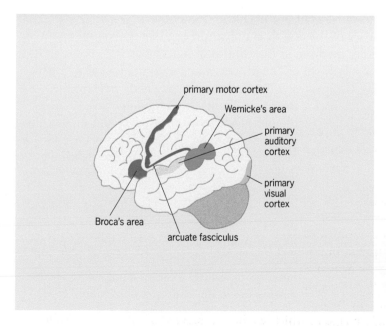

primary motor cortex
Wernicke's area
primary auditory cortex
primary visual cortex
Broca's area
arcuate fasciculus

Fig 16.9 The key areas of the brain associated with speech and language are found in the left hemisphere. The primary auditory cortex receives impulses from both ears. Impulses then pass to Wernicke's area, which is the speech association area, recognising the words heard. To respond, impulses pass along the arcuate fasciculus to Broca's area, the speech motor area which formulates the response. Impulses then pass to the primary motor cortex, which generates the motor impulses that makes us talk.

? QUESTION 3

3 Explain why it is earlier to stand on one leg with your eyes open than it is with your eyes shut.

The role of the inner ear in proprioception and balance is discussed in Chapter 19.

Muscular contraction is not a matter of all or nothing. If it were, muscles would be either relaxed or fully contracted. Movement, if possible at all, would be very jerky. However, we can move our muscles with a great deal of precision, creating just the right tension for the action we are performing.

This is made possible by the process of **proprioception**, in which the brain receives constant feedback about the degree of contraction of the muscles and the position of the joints in relation to each other. Vital information also comes from the inner ear about our position with respect to gravity, direction of movement, acceleration and deceleration. The process of movement control and postural maintenance is a matter of countless fine adjustments to alter the tension in the muscles. Interestingly, the left side of the body is controlled by the right motor area, and vice versa (Fig 16.10).

Fig 16.10 Two computer images of the left side of the human brain. These pictures were created by a process called magnetoencephalography, a technique that measures magnetic fields generated by neurone activity in the brain

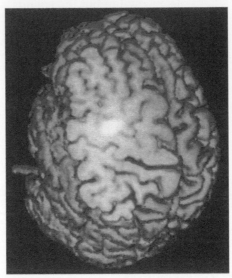 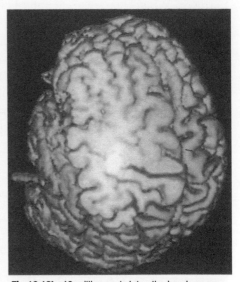

Fig 16.10a Just before the subject moves her right index finger, neurones in the motor area can be seen to 'fire'

Fig 16.10b 40 milliseconds later, the hand sensory area processes feedback from the muscles in the finger, so that ongoing modifications can be made

There are three key components to the process of moving:

- selection of a **motor programme**;
- initiation of a motor programme – the muscles contract;
- modification of movement in the light of feedback.

Given the complexity of the task, it is not surprising that several different areas of the brain are involved in the control of movement.

The process starts with the **basal ganglia**; a collection of structures lying deep in the forebrain. These are of fundamental importance because they select and initiate the motor programme. Fig 16.11 shows the pathway taken by motor impulses before they reach the muscles. From the basal ganglia impulses pass to the motor cortex.

The motor cortex contains two major regions, the **primary motor cortex** and the **supplementary motor cortex**. The primary motor complex contains neurones that send impulses to skeletal muscles along nerve fibres passing down the brain stem and spinal cord.

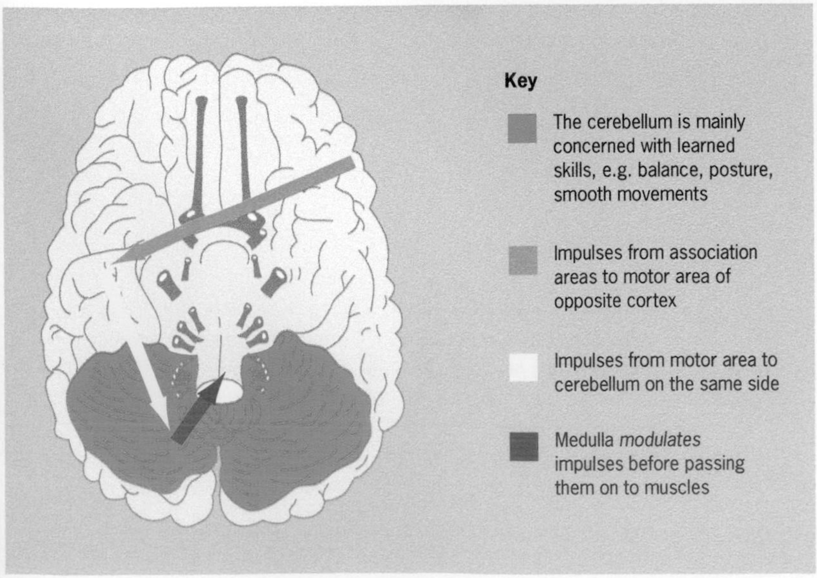

Key

The cerebellum is mainly concerned with learned skills, e.g. balance, posture, smooth movements

Impulses from association areas to motor area of opposite cortex

Impulses from motor area to cerebellum on the same side

Medulla *modulates* impulses before passing them on to muscles

Fig 16.11 Motor pathways and the cerebral cortex

Critical windows in brain development

We know that the way we develop is largely due to our genes, but it is becoming increasingly apparent that our nervous system does not develop properly without the right sensory stimulation at the right time.

There are **critical windows** or **critical periods** when a young animal must have sensory or motor input if 'normal' development is to take place. During a critical window, neural connections are made between sense organs and vital areas of the brain. These critical periods exist in virtually all species, from fruit flies to humans. If a skill is not acquired during a critical period, the acquisition of that skill in later life will be much harder, if not impossible.

For example, kittens are born with their eyes shut and if they do not open at the right time, permanent blindness results. There is nothing wrong with the eyes themselves, but the visual cortex where the incoming information is processed fails to develop properly.

Key work in this area came from the experiments of **David Hubel** and **Torsten Nils Wiesel**, who won the 1981 Nobel Prize in Physiology for their work. They found that if a kitten is deprived of light in one eye during the critical period, there is only partial development of the visual cortex. They can only see with one eye, and although they can perceive detail well, there is very poor judgement of distance. Electrical studies of the visual cortex showed that only about one-seventh of the visual cortex develops, compared with full development following input from two eyes.

To further refine their research, kittens were placed in striped tubes (where the stripes were horizontal for some kittens and vertical for others) or were fitted with goggles that presented vertical stripes to one eye and horizontal stripes to the other.

When tested a few months later, after removal of the striped tubes and goggles, the kittens seemed to be blind to stripes with the opposite orientation that they saw during rearing. Most of the cortical cells of the cats reared with horizontal stimuli subsequently responded strongly to horizontal stimuli and hardly at all to vertical stimuli. The opposite was true of the vertical stimuli-reared cats.

This has implications for children who have injured an eye, and have it covered with a bandage or patch. If the injury occurs during the critical window, the visual cortex will not develop properly and vision will be impaired.

The ethics of animal research

Few subjects in science cause so much fierce debate as that of **animal rights**. Was it right for Hubel and Wiesel to experiment on kittens (Fig 16.2) and make them partially sighted for life? Was it right for Banting and Best to experiment on dogs (page 178), giving them diabetes and ultimately killing them?

Fig 16.12 Kittens open their eyes about five days after birth – if they don't, their visual cortex fails to develop properly and they are blind

Most people believe that humans have certain rights, but it is very difficult to apply these ideals to animals. We can't do medical research on people without their consent, but it is obviously not possible for animals to give their consent. It makes more sense to consider animal welfare rather than animal rights. Most people think that animals should have clean water, food, exercise/stimulation and access to veterinary care.

Some people think that to use animals in experiments is being **speciesist** (in the same sense as racist or sexist). They argue that by experimenting we are automatically assuming that human life is more valuable than that of any other species. Others argue that it is very difficult to draw the line, and that if the argument is taken to its logical conclusion, we shouldn't ever kill pests like rats or even locusts.

How much do animals suffer?

There is a general consensus amongst scientists that humane treatment should be given to those organisms whose nervous system is advanced enough to appreciate it. The argument is

HOW SCIENCE WORKS

that pulling the legs off a daddy-long-legs, or boiling a lobster (which, for all its size, is no more sophisticated than an insect) is not inflicting pain because these animals do not have a nervous system advanced enough to perceive pain and suffering.

No country in the European Union is allowed to experiment on vertebrates in medical experiments if non-vertebrate alternatives exist. If there is no alternative – as would be the case with the kittens – animals can be used provided that strict guidelines are followed. Many drugs – sleeping pills, for example – can only be tested using an intact, conscious mammal.

Scientists commonly take a **utilitarian**, or 'greater good' approach, meaning that the right course of action is the one that leads to the greatest overall benefit i.e. the least suffering and loss of life in the long term. A utilitarian framework is in place today, allowing a certain amount of animal experimentation provided that the overall benefits are expected to be significantly greater than the expected suffering and or loss of life.

There are arguments for and against animal testing - see http://www.bbc.co.uk/science/hottopics/animalexperiments/ for more information.

? QUESTION 4

4 A woman has a stroke and as a result is unable to move the left side of her body properly. Where did the stroke probably occur?

The cerebellum is a complex structure lying on the posterior surface of the brain stem. Its function is to co-ordinate ongoing movements, thus producing smooth flowing movements. The cerebellum uses information from proprioceptors to decide whether the body's actions are going to plan or need modifying.

Parkinson's disease

Parkinson's disease is a disease in which the death of a small number of cells in the basal ganglia leads to an inability to select and initiate patterns of movement. The exact symptoms vary, but generally Parkinson's patients have an increased rigidity in their muscles, usually accompanied by a tremor. This leads to slowness of movement, poor balance and speech problems. This shows the importance of the basal ganglia – without them, the body is incapable of normal movement even though most of the motor system is intact.

Quite what triggers the onset of Parkinson's is still not clear. It appears to be due to a genetic predisposition and an as yet unidentified environmental trigger. What is known is that the underlying cause is an inability to produce **dopamine**, a neurotransmitter that has a number of functions, including enabling us to move smoothly and normally.

Some relief from Parkinson's has been achieved by treatment with **levadopa,** a precursor that is transformed into active dopamine in the brain. Dopamine reduces the symptoms – often spectacularly so – but is not a cure. Trials involving transplants of dopamine-producing cells from fetal or animal sources also show promise, but it is still early days.

Electroencephalograms and brain stem death

In 1929, the German scientist Berger discovered that the brain produces electrical activity that can be measured by electrodes placed on the scalp (Fig 16.13). The machine that measures brainwaves is called an **electroencephalogram**, or EEG. It was discovered that the patterns of EEG traces change in different situations, particularly with levels of consciousness.

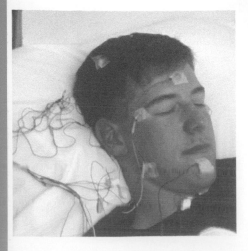

Fig 16.13 Person wired up to EEG machine

Epilepsy and the two sides of the brain

Epilepsy is a common disorder of the brain. Symptoms range in severity from a mild loss of concentration, known as an absence or petit mal, to full-blown convulsive fits (grand mal) in which the person blacks out and falls to the floor. These fits can be dangerous if the sufferer lashes out – injuring himself and others – or bites his own tongue.

The underlying cause of epilepsy is random, uncontrolled activity of some cells in the brain. This chaotic activity in both sensory and motor nerves causes patients to see and hear a variety of strange things, such as flashing lights and bells, while muscles jerk uncontrollably.

A diagnosis of epilepsy can be confirmed and studied using an EEG machine (Fig 16.13). Fig 16.14 shows a trace from a person during an epileptic seizure: you can compare it with the normal readings in Fig 16.16.

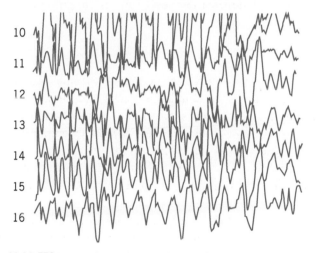

Fig 16.14 EEG traces taken during an epileptic seizure

Epilepsy can often be controlled successfully by drugs. However, in extreme cases, the condition is treated by brain surgery, and one such operation has given us a fascinating insight into the workings of the brain. The cerebral hemispheres have been described as two separate brains, and in order to work effectively as a whole, the two halves must communicate.
The bridge between the two halves is known as the **corpus callosum** (Fig 16.15). Neuroscientists discovered that the corpus callosum was involved in the spread of epileptic seizures.

Fig 16.15 The corpus callosum is a broad, thick mass of nerve fibres connecting the right and left cerebral hemispheres. The left eye supplies information to the right hemisphere, and vice versa. The corpus callosum allows the two sides of the brain to communicate

In a seemingly drastic operation, surgeons sever most of the corpus callosum. This often causes the seizures to be less intense and dangerous. However, there are other amazing consequences. Initially, subjects appear to be perfectly normal: they can talk and read, and have no problems in recognising the world around them. However, if they close their right eye, and are given a familiar object such as a comb, they cannot put a name to it. Open the other eye, however, and, 'Ah, it's a comb!'

The same happens with words. If a word such as 'TIGER' is looked at it with the left eye only, the patient can't read it. If they open the right eye, they can read the word immediately. This is because the left eye supplies information to the right side of the brain, and that is not where the language centre is situated. The right eye supplies information to the left side of the brain, to the language-processing neurones.

From studies of split-brain patients, and other studies, it appears that different sides of the brain have different functions. The left hemisphere contains the language centre, and the three Rs – reading, writing and 'rithmotic. The right side, in contrast, is responsible for our imagination and sense of humour. It can also appreciate form, geometry and music. It cannot, however, put words to things. If the right hemisphere needs a word, it has to put in a call to the left side, via the corpus callosum.

Split-brain patients do not experience the symptoms forever. Within a few months the right hemisphere develops more language skills and can function on its own. It has even been suggested that split-brain patients could read two books simultaneously, one with each eye!

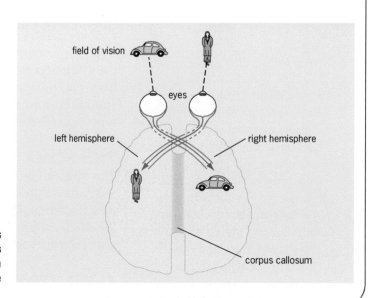

The four basic types of brainwaves, alpha, beta, theta and delta, are shown in Fig 16.16. The absence of any electrical activity from the brain of a patient indicates brain stem death. The absence of heartbeat, breathing movements *and* electrical brain activity is the clinical definition of death.

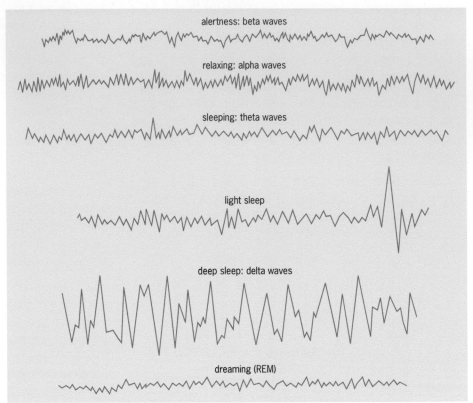

alertness: beta waves

relaxing: alpha waves

sleeping: theta waves

light sleep

deep sleep: delta waves

dreaming (REM)

Fig 16.16 EEG traces of the different brain waves

EEG traces tell us little about how the brain works, but they are useful for locating and diagnosing the various types of epilepsy, sleep disorders and brain tumours.

5 THE PERIPHERAL NERVOUS SYSTEM

The nerves of the peripheral nervous system behave like major road systems, carrying traffic in and out of the central nervous system. **Afferent nerves**, also called **sensory nerves**, carry information from sensory receptors into the CNS. **Efferent nerves**, also called **motor nerves**, carry information from the CNS out to effector organs. The efferent system can be further subdivided into the **somatic** and **autonomic** systems. These differ in their function, rather than their structure or position in the body.

THE SOMATIC NERVOUS SYSTEM

The somatic nervous system contains both afferent and efferent nerves. It receives and processes information from receptors in the skin, voluntary muscles, tendons, joints, eyes, tongue, nose and ears, giving an organism the sensations of touch, pain, heat, cold, balance, sight, taste, smell and sound. It also controls voluntary actions such as the movement of arms and legs.

THE AUTONOMIC NERVOUS SYSTEM

The autonomic nervous systems (ANS) consists of two sets of involuntary nerves that generally act antagonistically – these are the sympathetic and

 REMEMBER THIS

Afferent means 'incoming' while efferent means 'outgoing'. You can refer to 'afferent nerves' or 'efferent blood vessels', for example.

REMEMBER THIS

There is a parallel between the nervous and muscular systems of the human body. In the same way as we have voluntary and involuntary components of our nervous system, we also have skeletal (striped) muscle to deal with voluntary muscle contraction, and visceral (smooth) muscle to deal with involuntary muscle contraction. The different muscle types are described in Chapter 17.

parasympathetic systems (Fig 16.17) and they have opposing effects. The system is entirely motor, made up of efferent nerves only. It does not carry sensory information: feedback from muscles and glands travels via the somatic system.

The ANS controls basic 'housekeeping' functions such as heart rate, breathing, digestion and blood flow. Heart rate, for example, can increase or decrease. So, while the sympathetic system increases heart rate, the parasympathetic system lowers it.

Normally, the activity of both systems is balanced. But if the body is stressed, then the 'fight or flight' reactions of the sympathetic nervous system take over, causing an increase in heart rate, faster breathing, an increase in blood pressure and an increase in blood sugar level. This makes the body ready for sudden strenuous activity. When the emergency is over, the parasympathetic system takes over. It decreases the heart and breathing rates and diverts blood supply back to 'housekeeping' activities such as digestion and food absorption. The actions of the parasympathetic nervous system have been described as 'feed or breed' because parasympathetic stimulation leads to increased blood flow and peristalsis in the intestines, and sexual responses such as gaining an erection.

REMEMBER THIS

Generally, the sympathetic system has a stimulatory effect, and prepares the body for action, while the parasympathetic system returns body functions to normal.

Fig 16.17 The autonomic nervous system, showing the antagonistic (opposing) effects of the sympathetic and parasympathetic division. Autonomic neurones, like others in the nervous system, release chemical transmitter substances where they communicate with other cells. Autonomic neurones can synapse with other neurones within the ANS or with effectors such as muscles and glands. Neurones are classified as cholinergic or adrenergic on the basis of the transmitter they release

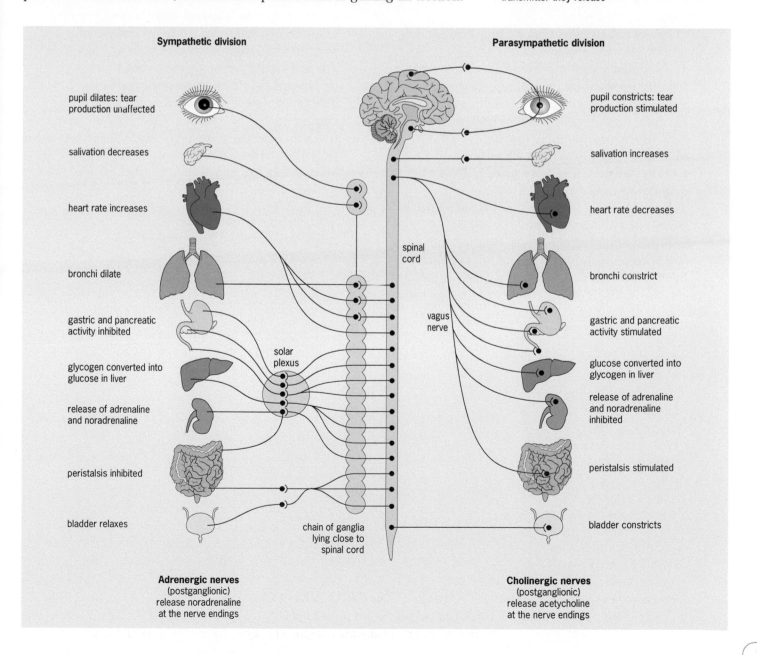

The autonomic system was originally thought to be independent of the rest of the nervous system, hence the term autonomic, meaning 'on its own' or 'self-governing'. Now we appreciate that it is not autonomous, but is regulated by areas within the central nervous system, including the hypothalamus, cerebral cortex and the medulla oblongata.

Do we have any conscious control over the autonomic nervous system?

The answer is yes, in some cases. As children we became potty trained when we learned conscious control over the muscular valves (sphincters) in our bladder and rectum. There are people who have learnt conscious control over some other autonomic functions; those who are adept at advanced meditation and yoga techniques can voluntarily lower their heartbeat.

SUMMARY

After reading this chapter, you should know and understand the following:

- The **central nervous system** (CNS) consists of the brain and spinal cord.
- The simplest co-ordinated response is the **reflex arc**. In the knee-jerk reaction, the pathway consists of the receptor in the tendon, the sensory nerve that connects directly to the motor nerve in the spine, and the effector, the thigh muscle.
- More complex reflex arcs such as blinking are **polysynaptic**. These have more intermediate connections inside the CNS.
- The human brain is divided into **hindbrain**, **midbrain** and **forebrain**. Generally, the midbrain and hindbrain are concerned with 'housekeeping' functions, while the **cerebral hemispheres**, which make up much of the forebrain, are responsible for the conscious functions such as language and memory.
- The **peripheral** nervous system consists of the **somatic** and **autonomic** nervous systems.
- The autonomic nervous system consists of the **sympathetic** and **parasympathetic** nervous systems. The two sets of nerves act antagonistically to control a variety of functions that are not under conscious control. Generally, sympathetic stimulation prepares the body for action while the parasympathetic system returns it to normal.

 Practice questions and a How Science Works assignment for this chapter are available at www.collinseducation.co.uk/CAS

17

Support and movement

17 SUPPORT AND MOVEMENT

Weightlessness is not good for the skeleton

Astronauts must often live and work in zero gravity. Weightlessness is not a restful state and it can put great stress on body systems, particularly the skeletal system.

Many changes occur in zero gravity. With less work to do, body muscles begin to break down and the bones start to lose calcium at an increased rate. The number of red blood cells in the body falls and there are dramatic shifts in fluid distribution: the face becomes puffy and the legs become thinner (astronauts call this condition 'bird's legs'). The lack of gravity also affects the spine. Without the constant downward force, the spine lengthens by as much as 8 millimetres. This can lead to blocked nerves and back pain. Often, astronauts also lose their touch sensitivity.

NASA scientists are busy looking at forms of treatment and exercise that might prevent bone breakdown. Such information might also have practical benefits closer to home. By studying and finding ways to slow the accelerated changes produced in space, it may be possible to develop new treatments for bone diseases such as osteoporosis, which occurs when bones become brittle due to a loss of calcium.

It might look fun, but in zero gravity the body is under a lot of stress

 REMEMBER THIS

Movement is a feature of all living things and occurs at all levels. Atoms move within molecules and cell contents move inside cytoplasm. Plants move when they tilt their leaves towards the Sun and when their flowers open and close. The human rib cage moves up and down as we breathe.

? QUESTIONS 1–2

1 Look at Fig 17.1. Why are the bones of the legs thicker than the corresponding bones in the arms?
2 Which structure is protected by the vertebral column or backbone?

1 THE HUMAN SKELETON

The human skeleton has several functions:
- It acts as a framework that supports soft tissues.
- It allows free movement through the action of muscles across joints.
- It protects delicate organs and structures such as brain and lungs.
- It forms red and white blood cells in the bone marrow.
- It stores and releases minerals from bone tissue.

Fig 17.1 shows the structure and features of the human skeleton.

THE STRUCTURE AND PROPERTIES OF HUMAN BONE

Bone is one of the hardest tissues in the human body and is second only to cartilage in its ability to absorb stress. It is made up of bone cells called **osteocytes**, which sit in a bone **matrix**. The matrix consists of inorganic matter (mainly compounds of calcium and phosphorus) interwoven with **collagen fibres** (see page 51).

Bone needs to be tough and resilient. These properties are provided by two types of bone tissue: **compact bone** and **spongy bone** (Fig 17.2). **Compact bone** is deposited in sheets called **lamellae** arranged as cylinders inside cylinders. Nerves and blood vessels run in a central canal. The compacted structure of the lamellae gives immense strength. The lamellae of **spongy bone** are arranged in a criss-cross pattern, forming a spongy honeycomb. This structure has excellent shock-absorbing properties.

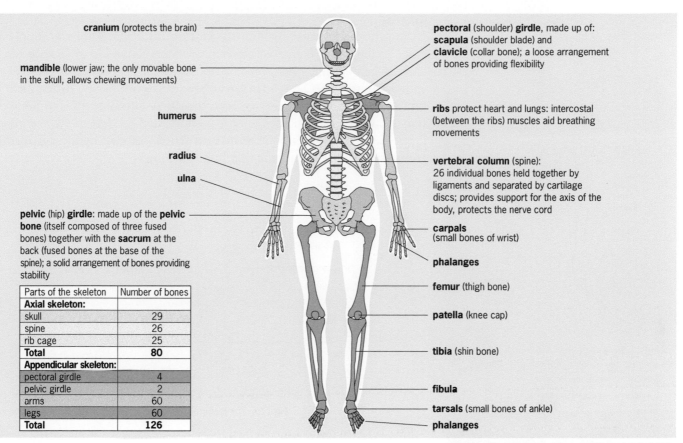

cranium (protects the brain)

mandible (lower jaw; the only movable bone in the skull, allows chewing movements)

humerus

radius

ulna

pelvic (hip) girdle: made up of the pelvic bone (itself composed of three fused bones) together with the sacrum at the back (fused bones at the base of the spine); a solid arrangement of bones providing stability

pectoral (shoulder) girdle, made up of: scapula (shoulder blade) and clavicle (collar bone); a loose arrangement of bones providing flexibility

ribs protect heart and lungs: intercostal (between the ribs) muscles aid breathing movements

vertebral column (spine): 26 individual bones held together by ligaments and separated by cartilage discs; provides support for the axis of the body, protects the nerve cord

carpals (small bones of wrist)

phalanges

femur (thigh bone)

patella (knee cap)

tibia (shin bone)

fibula

tarsals (small bones of ankle)

phalanges

Parts of the skeleton	Number of bones
Axial skeleton:	
skull	29
spine	26
rib cage	25
Total	**80**
Appendicular skeleton:	
pectoral girdle	4
pelvic girdle	2
arms	60
legs	60
Total	**126**

Fig 17.1 The human skeleton contains a total of 206 bones. It is divided into two parts: an axial skeleton that comprises the skull and vertebral column and an appendicular skeleton that is made up of the limbs and limb girdles.

All vertebrates show skeletal modifications related to their lifestyle. Since humans are bipedal (they walk on two legs), the hips and lower spine take most of the weight of the body and so the pelvic girdle (the hip) is larger and less flexible than the pectoral girdle (the shoulder)

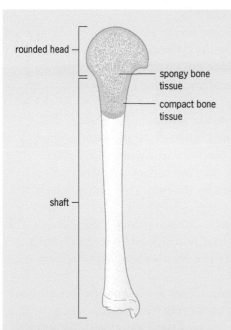

rounded head

spongy bone tissue

compact bone tissue

shaft

Fig 17.2a A human long bone. The part in section shows the distribution of compact and spongy bone

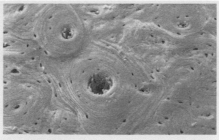

Fig 17.2b Compact bone makes up the shaft of the long bone where strength and rigidity are important

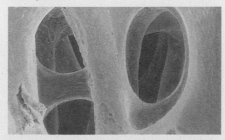

Fig 17.2c Spongy bone forms the rounded head of long bones which absorbs the shocks and jolts of movement

> ✔ **REMEMBER THIS**
>
> The human body also makes use of the supporting properties of water. The **amniotic fluid**, the fluid inside the uterus, surrounds and supports the developing fetus; and the **vitreous humour**, the jelly-like material inside the eyeball, supports the structures inside the eye.

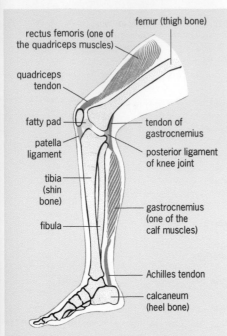

rectus femoris (one of
the quadriceps muscles)

femur (thigh bone)

quadriceps
tendon

fatty pad

patella
ligament

tendon of
gastrocnemius

posterior ligament
of knee joint

tibia
(shin
bone)

fibula

gastrocnemius
(one of the
calf muscles)

Achilles tendon

calcaneum
(heel bone)

Fig 17.3a A diagram of the human leg
showing the position of the main ligaments
(green) and tendons (blue)

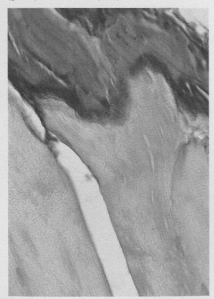

Fig 17.3b This micrograph shows a slice
of tendon (yellow) connected to a piece of
muscle (red). Tendons and ligaments look
similar but their composition is different:
tendons contain mainly **collagen fibres**
(see page 51); ligaments are made up of
fibres formed from the protein **elastin**

? **QUESTIONS 3–4**

3 Why are ligaments elastic?

4 Why does the tendon need to be
inelastic?

Joints

As in other vertebrates, the human skeleton is jointed. A joint is simply a
place in the body where two bones meet. Most joints are **movable** but some,
such as between the bones making up the **skull**, the **sacrum** (base of the
spine) and the **pelvis**, are **immovable**, or **fused**. Elastic **ligaments** bind
bones together, while tough inelastic **tendons** attach muscles to bone
(Fig 17.3). The human knee joint is seen in Fig 17.4. Internally, the joint is
lubricated by a viscous fluid, **synovial fluid**, secreted by the **synovial
membrane**, the membrane that lines the joint.

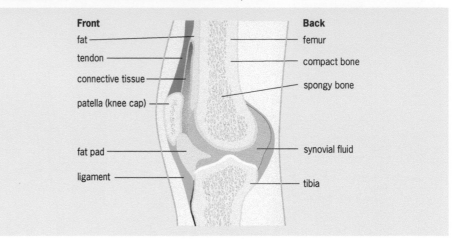

Fig 17.4 The features of a typical joint can be seen clearly in the knee joint

The ligaments and the synovial membrane together form a **joint capsule**
which surrounds the end of the bones. Different types of joint allow for
different kinds of movement (Fig 17.5).

HOW THE HUMAN BODY MOVES

A simple action, such as tapping your finger on the table, involves the skeletal
system, the muscular system and the nervous system. This voluntary action
begins as a stimulus in the cerebrum of the brain. Motor nerves carry
impulses to muscles and cause them to contract, pulling on bones through the
tough inelastic tendons.

Muscles can only pull, or contract; they cannot push. This means that
muscles rarely act alone: most of the time they work in groups. Contraction of
a muscle moves a bone at a joint, but a second (**antagonistic**) muscle
returns the bone to its original position. Take movement in the arm, for
example. The biceps muscle bends or **flexes** the arm; the triceps muscle
straightens or **extends** the arm.

We all know that machines can help us to lift heavy loads using levers. The
bones in our bodies also act as **levers**. The principle of leverage allows a
muscle to overcome the **resistance** supplied by a weight. This happens in
three ways:

● An arm bends because a force is applied between the **pivot** (the elbow)
and the resistance (the weight to be lifted) (Fig 17.6a).

● We stand on tiptoes by pivoting the toes on the ground and using our calf
muscles to raise the weight at the ankles. This is the wheelbarrow
principle (Fig 17.6b).

● When standing or sitting, we can raise our head because muscles at the
back of the neck tip the skull at the pivot at the top of the spine. This is
the seesaw principle (Fig 17.6c).

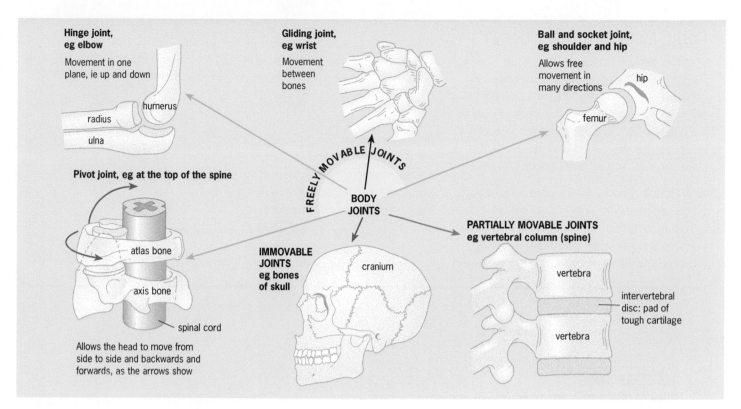

Hinge joint, eg elbow

Movement in one plane, ie up and down

humerus
radius
ulna

Gliding joint, eg wrist

Movement between bones

Ball and socket joint, eg shoulder and hip

Allows free movement in many directions

hip
femur

FREELY MOVABLE JOINTS

BODY JOINTS

Pivot joint, eg at the top of the spine

atlas bone
axis bone
spinal cord

Allows the head to move from side to side and backwards and forwards, as the arrows show

IMMOVABLE JOINTS eg bones of skull

cranium

PARTIALLY MOVABLE JOINTS eg vertebral column (spine)

vertebra
intervertebral disc: pad of tough cartilage
vertebra

Fig 17.5 There are many different types of joint in the vertebrate skeleton. They are classified both by their shape and by their mobility

The 'stuck finger' test demonstrates the effect of muscles working in groups (Fig 17.7). Each of your fingers has a separate tendon connecting it to muscles in the forearm. Try it. Curl up your middle finger and place the other four fingers on a hard surface as shown. Now lift up the other fingers one by one. You will find that you can lift all of your fingers except the ring finger. This is because these two fingers share the same tendon connection.

Fig 17.6 Muscles act across a joint using the principle of leverage

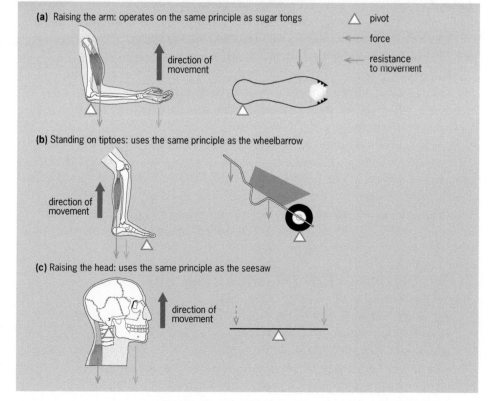

(a) Raising the arm: operates on the same principle as sugar tongs

direction of movement

△ pivot
← force
← resistance to movement

(b) Standing on tiptoes: uses the same principle as the wheelbarrow

direction of movement

(c) Raising the head: uses the same principle as the seesaw

direction of movement

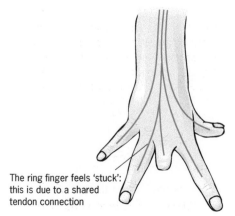

The ring finger feels 'stuck': this is due to a shared tendon connection

Fig 17.7 The stuck finger investigation

Coral: a substitute for bone?

Mending damaged bones is a major part of a surgeon's work, yet acceptable substitutes for bone are hard to find (Fig 17.8). It is possible to use bone from another part of the patient's body but only small amounts can be used. Artificial substitutes run the risk of being rejected by the body's immune system (see Chapter 30).

A few species of coral have a porous structure similar to that of bone. When grafted into the body, the honeycomb texture of coral provides the conditions necessary for new blood vessels to grow into it, and this promotes new bone growth.

In addition, coral is tough, carries no risk of infection and is unlikely to be rejected by the body.

'Liquid bone' can also be made from coral. This is based on a calcium-rich solution and a phosphate-rich solution. The surgeon mixes the compounds in a little acid before applying it, rather like toothpaste, to the site of a fracture. Within 12 minutes it has solidified, and after one hour the 'new' bone is as hard as real bone.

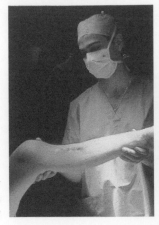

Fig 17.8 Coral grafts are used to repair shattered limbs, backbones and jaws. This broken leg is being examined before surgery

---

 REMEMBER THIS

There are three types of muscle.
Skeletal muscles are responsible for whole-body movement.
Smooth muscle is responsible for automatic movements such as those involved in peristalsis.
Cardiac muscle is found only in the heart. The properties of the different muscle types are described in Chapter 2 and cardiac muscle is covered in detail in Chapter 10.

? QUESTION 5

5 What type of muscle fibre (skeletal, cardiac or smooth) is found in the following structures?
a) stomach
b) aorta
c) biceps muscle
d) left ventricle of heart
e) face

2 MUSCLES AND MOVEMENT

The bones of the human skeleton provide a basic system of levers and joints that makes the skeleton potentially movable, but neither levers nor joints can move without muscles. Skeletal muscle (also called striped or striated muscle) provides the main source of power for human locomotion. In this section, we look at the structure and function of skeletal muscle.

THE PROPERTIES OF SKELETAL MUSCLE

Like other sorts of muscle, skeletal muscle has three basic properties:

- **Excitability**: it can receive and respond to a stimulus.
- **Extensibility**: it can be stretched and it can contract.
- **Elasticity**: it can return to its original shape after it has contracted.

Skeletal muscle is also described as **voluntary muscle** because we can contract it when we want to. Smooth muscle and cardiac muscle are **involuntary muscle**: we cannot consciously control their contraction.

Skeletal muscles contract and relax, moving bones at joints in the skeleton. This allows for movements such as walking, running and waving, and fine movements such as those needed to use tools. The action of skeletal muscles also helps to maintain the position of the body when we are sitting or standing, even though there is no obvious movement. The contractions of skeletal muscle also produce heat, and much of this is used to keep the body temperature at a normal level.

THE CONTRACTION OF SKELETAL MUSCLE

Setting the scene:

- An individual muscle is made up from hundreds of cylindrical **muscle fibres**, about 50 μm in diameter and ranging in length from a few millimetres to several centimetres (Fig 17.9).
- The muscle fibres are not composed of individual cells – the cell membranes have broken down and so each fibre has many nuclei.
- Each fibre is surrounded by a modified cell membrane called the **sarcolemma**, the cytoplasm of which is the **sarcoplasm**.

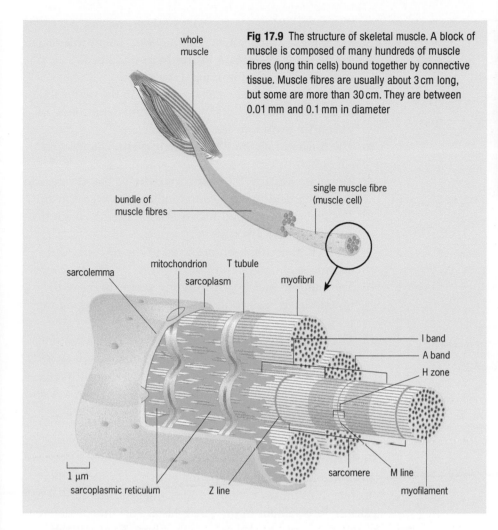

whole muscle

Fig 17.9 The structure of skeletal muscle. A block of muscle is composed of many hundreds of muscle fibres (long thin cells) bound together by connective tissue. Muscle fibres are usually about 3 cm long, but some are more than 30 cm. They are between 0.01 mm and 0.1 mm in diameter

bundle of muscle fibres

single muscle fibre (muscle cell)

sarcolemma   mitochondrion   T tubule

sarcoplasm   myofibril

I band
A band
H zone

1 μm

sarcoplasmic reticulum   Z line

sarcomere   M line

myofilament

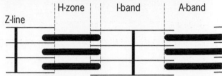

Z-line   H-zone   I-band   A-band

Fig 17.10 The ultrastructure of the myofibril. Each myofibril contains myosin and actin strands. Where the two overlap, the myofibril looks dark: this region is called the A band. In the regions where only actin is present, the myofibril looks lighter: this region is called the I band. At the centre of each I band is the dark Z line. At the centre of each A band is the dark M line. A bands and I bands alternate and so give skeletal muscle its striped appearance.

The sarcomere, the basic unit of muscle contraction, is the section of muscle fibre between one Z line and the next

- Each muscle fibre is composed many long cylindrical **myofibrils**, consisting of a repeating arrangement of proteins which causes the banding pattern (Fig 17.10).
- Each repeated pattern of proteins is called a **sarcomere**.
- The two main proteins involved in muscular contraction are **actin** (thin) and **myosin** (thick) filaments. Muscles contract when the actin fibres are pulled over the myosin fibres.
- Two other proteins involved are **tropomyosin** and **troponin**.
- **Tropomyosin** winds around the actin filament, preventing it from binding to myosin.
- **Troponin** (a globular protein) moves the tropomyosin out of the way, allowing actin to bind to myosin, thus initiating muscular contraction.

Within each sarcomere there are two light bands; the I zone consisting of just actin filaments and the H zone that consists of just myosin. Between them are darker areas where these proteins overlap. When muscles contract, the actin and myosin filaments are pulled over each other so that the light bands get smaller. Note that the width of the dark band corresponds to the width of the myosin molecules, so it doesn't get any narrower – molecules don't shrink, they just slide over each other.

The steps in muscular contraction are:

1   An impulse arrives down a motor nerve and terminates at the **neuromuscular junction** (a modified synapse).

2   The synapse secretes **acetylcholine**.

3   Acetylcholine fits into receptor sites on the **motor end plate**.

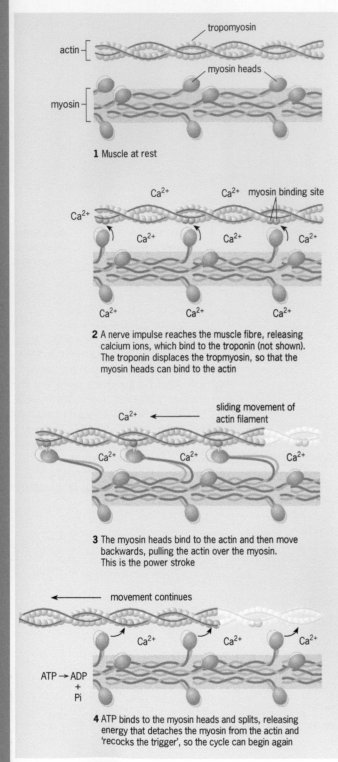

1 Muscle at rest

2 A nerve impulse reaches the muscle fibre, releasing calcium ions, which bind to the troponin (not shown). The troponin displaces the tropmyosin, so that the myosin heads can bind to the actin

3 The myosin heads bind to the actin and then move backwards, pulling the actin over the myosin. This is the power stroke

4 ATP binds to the myosin heads and splits, releasing energy that detaches the myosin from the actin and 'recocks the trigger', so the cycle can begin again

Fig 17.11 The sliding filament theory to explain muscle contraction

4 This initiates an action potential in the muscle cell membrane, which spreads throughout the myofilaments via the **sarcoplasmic reticulum** (modified endoplasmic reticulum) and **T tubules**.

5 The action potential causes a change in the permeability of the sarcoplasmic reticulum, resulting in an influx of **calcium ions** into the myofilament.

6 The calcium ions bind to the **troponin**, changing its shape.

7 Troponin displaces the **tropomyosin**, so that the myosin heads can bind to the myosin binding sites on the actin.

8 The myosin heads pull backwards, so that the actin is pulled over the myosin.

9 An ATP molecule becomes fixed to the myosin head, causing it to detach from the myosin.

10 The splitting of ATP provides the energy to move the myosin head back to its original position, 'cocking the trigger' again.

11 The myosin head becomes reattached to the actin once again, but further along. In this way, the actin is quickly pulled over the myosin in a rachet motion, shortening the sarcomere, the whole filament and the muscle itself.

The sliding filament theory of muscle action is shown in Fig 17.11.

Fast and slow muscle fibres

A nerve impulse is the trigger for muscle contraction, but the length of time a contraction lasts depends on how long calcium ions remain in the sarcoplasm. This time is different in different types of skeletal muscle fibres. **Fast twitch fibres** and **slow twitch fibres** are classified on the basis of their contraction times.

There are three important differences between fast and slow fibres:

- Slow twitch fibres have less sarcoplasmic reticulum than fast twitch fibres. This means that calcium ions remain in their sarcoplasm longer.
- Slow twitch fibres have more mitochondria which provide ATP for *sustained* contraction.
- Slow twitch fibres have significantly more myoglobin than fast twitch fibres. Myoglobin has a higher affinity for oxygen than the haemoglobin in blood and so is particularly efficient at extracting oxygen from the blood.

A central principle in physiology is that structure is closely related to function and this is illustrated clearly by the distribution of the muscle types in the body. For instance, the postural muscles – such as those in the neck and back – have a high proportion of slow twitch fibres. In contrast, the muscles that bring about fast, explosive movements such as running, jumping and throwing are packed with fast twitch fibres.

Different people have different ratios of fast twitch to slow twitch fibres in the muscles – this is determined by the type

Table 17.1 Analysis of the average proportions muscle fibre types in selected muscles in different male athletes

Type of athlete	Approx % fast twitch	Approx % slow twitch
marathon runner	18	82
swimmer	25	75
cyclist	40	60
800 metre runner	52	48
untrained person	55	45
sprinter and jumper	62	38

of muscle fibres they inherit and also by the amount and type of training they have done. Marathon runners have more slow twitch muscle fibres; sprinters have more fast twitch muscle fibres (Table 17.1). See the Assignment for this chapter on the CAS website.

Fast twitch muscle fibres can be further subdivided, thus giving us three different muscle types:

Slow twitch oxidative (type I) fibres

These fibres contain large amounts of myoglobin, which is why they appear red. They also have many mitochondria and a dense capillary network. These fibres release ATP slowly by **aerobic respiration**. The contractions they produce may not be very powerful, but vitally there is no build-up of lactate and so the muscles can go on working for long periods.

Fast twitch oxidative (type IIA) fibres

These show similar features to slow twitch fibres: a lot of myoglobin, many mitochondria and blood capillaries; but they are able to hydrolyse ATP far more quickly and therefore to contract rapidly. They are relatively resistant to fatigue, but not as resistant as type I fibres.

Fast twitch glycolytic (type IIB) fibres

These fibres have a relatively low myoglobin content, few mitochondria and few capillaries. They contain large amounts of glycogen, which provides fuel for anaerobic ATP production via the process of glycolysis. They contract rapidly but fatigue quickly, owing to the build-up of lactic acid. Type IIB fibres appear whiter than the other two types.

HOW SCIENCE WORKS

Time of death

When a body is discovered, a pathologist investigates and tries to estimate how long the person has been dead (Fig 17.12). This information can be important in a murder enquiry. The pathologist assesses the time of death, partly by taking the internal temperature, to see how much the body has cooled from the normal body temperature of 37 °C, and partly by looking at the state of the muscles.

When death occurs, ATP is no longer made. It is a short-lived chemical and so it runs out fairly quickly. This causes the muscles to lock into position as cross-bridges that formed between actin and myosin filaments before death can no longer be broken. This condition, **rigor mortis**, happens in all body muscles. It appears about four hours after death and lasts about 24 hours. After this time, muscle proteins are destroyed by enzymes within the cells and so rigor mortis disappears.

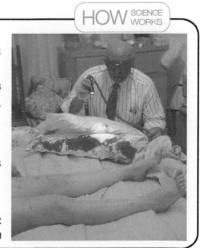

Fig 17.12 All suspicious deaths are investigated by a police pathologist: one of the first things he or she does is estimate the time of death

SUMMARY

After reading this chapter, you should know and understand the following:

- The skeleton has three main functions: **support**, **protection** and **movement**. The human skeleton is movable because different bones meet at **joints**, across which **muscles** are attached. Because muscles can pull but not push, muscles work in **antagonistic pairs**.

- Muscles are **excitable**, **extensible** and **elastic**. **Skeletal muscle**, also called **striped**, **striated** or **voluntary muscle**, is responsible for human locomotion, maintaining body posture and generating body heat.

- Skeletal muscle is made of **muscle fibres** which contain bunches of **myofibrils** which in turn contain **myofilaments**. Thick **myosin** filaments lie alongside thin **actin** filaments.

- The **sliding filament hypothesis** explains muscle contraction at the molecular level.

- The events of muscular contraction can be summarised as follows. A nerve impulse arrives at the **neuromuscular junction**. This releases acetylcholine, which changes the ionic permeability of the **sarcolemma**. Calcium ions are released into the **myofilaments**, where they bind to the tropomyosin–troponin–actin complex, and activate ATPase. Cross-bridges form between **actin** and **myosin**, the fibres slide over each other and the muscle contracts.

Practice questions and a How Science Works assignment for this chapter are available at www.collinseducation.co.uk/CAS

18

Hormones

18 HORMONES

The tallest man who ever lived

Robert Wadlow made it into the record books as the tallest man who has ever lived. There might have been people who have grown taller – stories exist of men over 9 feet tall – but their existence, unlike Wadlow's, cannot be proved. Today, such extremes in size are rare because conditions can be diagnosed early and treated effectively

The tallest man who ever lived was Robert Wadlow, who stood a touch under 9 feet tall (2.7 metres) and had size 36 feet (US sizes). Born in Alton, Illinois, USA, Wadlow reached the height of over 7 feet at the age of 13 and grew throughout his short life of 22 years. He might have continued to grow, but he died prematurely following an infected foot wound.

Wadlow's condition was caused by an over-productive pituitary gland that made too much human growth hormone, or somatotropin. Today people rarely suffer such extremes because the condition is diagnosed early. Blood tests done on a particularly large (or small) child can reveal hormone imbalances. Under-production can be remedied by injections of the hormone – either the real thing or a synthetic analogue (a chemical of similar structure that has the same effect). Conversely, over-production can be treated either by reducing the size of the gland, or by administering drugs that block the action of the hormone.

Many women find the symptoms of the menopause eased by hormone replacement therapy, and some women have even been able to conceive in their 60s if they have hormone treatment. Add this to the artificial hormones taken every day by women on the pill, and you can understand why making artificial hormones is big business.

1 HORMONES ARE CHEMICAL SIGNALS

The human body communicates in two fundamentally different ways: by nerves and by chemicals. As Fig 18.1 shows, there are three main types of chemical signal:

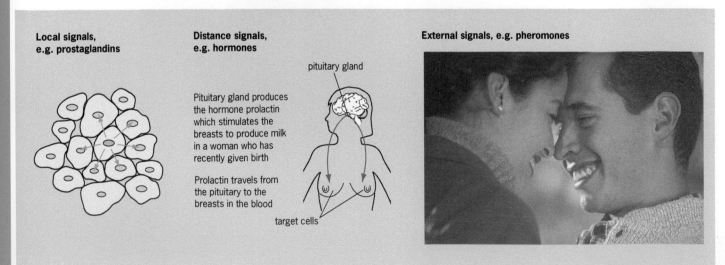

Local signals, e.g. prostaglandins

Distance signals, e.g. hormones

pituitary gland

Pituitary gland produces the hormone prolactin which stimulates the breasts to produce milk in a woman who has recently given birth

Prolactin travels from the pituitary to the breasts in the blood

target cells

External signals, e.g. pheromones

Fig 18.1 A summary of the different types of chemical signalling that exist within and between organisms. Local signals include neurotransmitters such as acetylcholine (see Chapter 15) and local messengers such as prostaglandins, endorphins and histamines. Hormones allow communication between different organs, via the blood. Externally produced messengers allow communication between organisms

- Locally acting chemicals such as **endorphins**, **prostaglandins** and **histamines** (Fig 18.2). These act on cells that are close to their site of production.
- **Hormones**. These are produced by special tissues called **endocrine glands**. They pass into the blood and act away from their site of production. Hormones affect **target cells** and **target organs** in the body.
- **Pheromones** are chemical messengers that allow external communication: simple exchanges of information between people. Generally, pheromones are not an important means of communication in humans, despite the claims of some newspaper reports and advertisements.

In this chapter, we focus mainly on hormones. The definition of a hormone in mammals is:

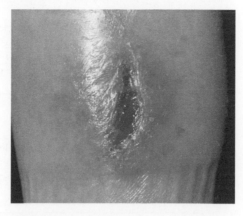

Fig 18.2 Histamine is a local messenger which is released by injured cells. It acts on the cells around an injury like the deep cut shown above, causing blood vessels near to the injury to dilate, bringing more blood into the area. This is an important part of the inflammation response (see Chapter 30)

A chemical messenger that is produced by cells or tissues of the endocrine system and that travels through the blood to act on target cells or target organs, before being broken down.

The main part of the chapter describes the human endocrine system, and we finish with a brief overview of prostaglandins and endorphins.

2 THE HUMAN ENDOCRINE SYSTEM

In humans, more than a dozen tissues and organs produce hormones. Some, including the **pituitary gland**, the **thyroid gland**, the **parathyroid glands** and the **adrenal glands**, are endocrine specialists: their major function is to secrete one or more hormones. Others, such as the pancreas, ovaries and testes, secrete hormones in addition to their other functions. Together, these glands make up the **endocrine system** (Fig 18.3, page 271).

The endocrine system has four main functions:

- It maintains **homeostasis**, the balance of the body, by making sure the concentration of many different substances in body fluids are kept at the correct level. The control of blood sugar level, blood pH and water balance provides examples of homeostasis.
- It works with the nervous system to help the body respond to stress. The release of adrenaline in the **fight or flight** response is an example of this (see the Assignment for this chapter on the CAS website).
- It controls the body's rate of growth.
- It controls sexual development and reproduction.

We know of more than 50 human hormones, and we divide them into two groups according to their origin: those made from *fatty acids* and those made from *amino acids*. The first group, which includes **steroids** such as oestrogen and progesterone, are *lipid-soluble*. Hormones based on **amino acids** are **modified amino acids**, **peptides** or **glycoproteins**: they are *water-soluble* and include insulin and adrenaline. Table 18.1 (overleaf) lists most of the hormones produced by the endocrine system and shows where they act and what they do. You might find it useful as a revision aid.

Examples of homeostasis are discussed in Chapters 11–14. Reproduction is discussed in Chapter 20.

Table 18.1 Human endocrine glands and their principal secretions. You do not need to learn all of this, but you should find it useful for reference as you read the rest of the chapter

Gland	Hormone	Chemical structure	Effect
Pituitary 1 posterior lobe	anti-diuretic hormone (ADH)	peptide	Reduces amount of water lost in urine. Raises blood pressure by constricting arterioles
	oxytocin	peptide	Contraction of smooth muscle during childbirth. Stimulates secretion of milk from mammary glands
2 anterior lobe	adrenocorticotrophic hormone (ACTH)	peptide	Stimulates production and release of hormones from adrenal cortex
	follicle stimulating hormone (FSH) also called ICSH in males	glycoprotein	Controls the development of follicles in the ovary, and sperm cells in the testis
	growth hormone	protein	Promotes growth (especially of skeleton and muscles). Affects body metabolism
	luteinising hormone (LH)	glycoprotein	Stimulates ovulation and formation of the corpus luteum (stimulates testosterone production in males)
	prolactin	protein	Stimulates milk production and release
	thyroid stimulating hormone (TSH)	glycoprotein	Stimulates growth of thyroid gland; synthesis and production of thyroid hormones
Thyroid	thyroxine + triiodothyronine	iodine-containing amino acids	Thyroxine increases rate of cell metabolism, controls aspects of growth and development and controls basal metabolic rate (BMR)
Parathyroid	parathormone	peptide	Raises blood calcium levels by stimulating release of calcium from bone
Pancreas (islets of Langerhans)	insulin (produced by the β cells)	protein	Lowers blood glucose levels by making cell membranes more permeable to glucose, increases glycogen storage in liver
	glucagon (produced by the α cells)	peptide	Raises blood sugar levels by stimulating glycogen breakdown in the liver
Adrenal 1 cortex	glucocorticoids, e.g. cortisol	steroids	In response to stress, raises blood glucose
	mineralocorticoids, e.g. aldosterone	steroids	Concerned with water retention. Increases reabsorption of sodium chloride in kidneys, so important in control of blood volume and pressure
2 medulla	adrenaline	modified amino acid	'Fear, flight and fight' reactions. Prepares the body for heightened activity. Mimics effects of the autonomic nervous system
	noradrenaline	modified amino acid	As adrenaline
Gonads 1 ovary (follicle)	oestrogens	steroids	Female sex characteristics, rebuilding of uterus lining after menstruation, inhibits FSH
2 ovary (corpus)	progesterone	steroids	Stimulates maturation of uterus lining, maintains pregnancy, inhibits FSH
3 testis	androgens, e.g. testosterone	steroids	Support sperm production. Important in the development of male secondary sexual characteristics
Placenta	human chorionic gonadotrophin (hCG)	steroid	Causes corpus luteum to secrete progesterone, thus maintaining pregnancy

REMEMBER THIS

There are two basic types of gland in the body: **exocrine glands** (e.g. salivary gland), whose secretions are released through a tube or **duct**; and **endocrine glands** (e.g. thyroid gland), which have no ducts, so they shed their secretions directly into the bloodstream for transport to their eventual destination.

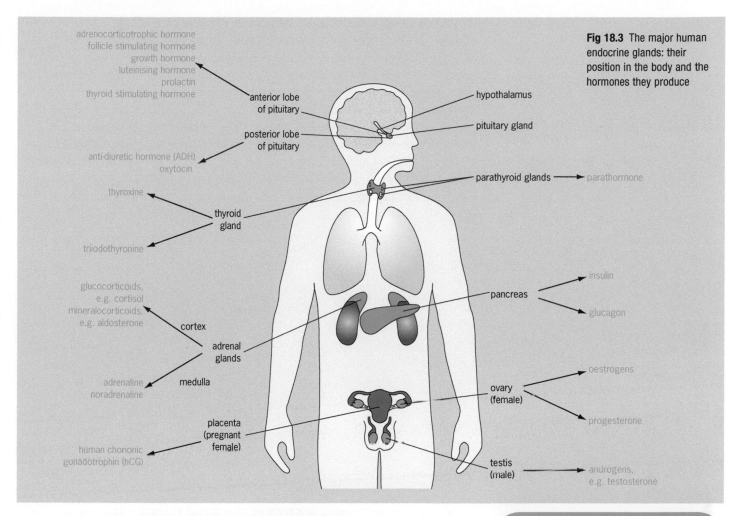

Fig 18.3 The major human endocrine glands: their position in the body and the hormones they produce

adrenocorticotrophic hormone
follicle stimulating hormone
growth hormone
luteinising hormone
prolactin
thyroid stimulating hormone

anterior lobe of pituitary

hypothalamus

posterior lobe of pituitary

pituitary gland

anti-diuretic hormone (ADH)
oxytocin

parathyroid glands → parathormone

thyroxine

thyroid gland

triiodothyronine

insulin

glucocorticoids, e.g. cortisol
mineralocorticoids, e.g. aldosterone

pancreas

glucagon

cortex

adrenal glands

adrenaline
noradrenaline

medulla

oestrogens

ovary (female)

progesterone

placenta (pregnant female)

human chorionic gonadotrophin (hCG)

testis (male) → androgens, e.g. testosterone

MAKING AND RELEASING HORMONES

Endocrine glands are tissues and organs that produce and release hormones. These glands have no ducts. The hormone is simply secreted into the blood and then travels to target organs or cells elsewhere in the body. A good example of an endocrine gland is the **adrenal gland**. Its structure is illustrated in Fig 18.4.

Hormones are formed inside endocrine cells. Some amino acid-based hormones, such as **insulin** and **glucagon**, are produced directly from a copy of a gene by the processes of transcription and translation. Most hormones, including the **steroids**, are made as an end-product of a series of chemical reactions. Each reaction makes minor changes to **precursor molecules**, until, eventually, the active hormone is produced. For the steroid hormones progesterone, oestrogen and testosterone, the precursor is cholesterol.

All endocrine glands contain some of the hormone they produce, at any given time. Although the hormone thyroxine is released continuously, most are not. **Negative feedback loops** control the release of hormones from many endocrine glands so that homeostasis is maintained.

 REMEMBER THIS

Fig 18.3 provides a useful reference point for the rest of this chapter, and also for the hormones described in Chapters 11, 13 and 20.

Transcription and translation are discussed in Chapter 24. Cholesterol is discussed in Chapter 3. Homeostasis is discussed in Chapter 11.

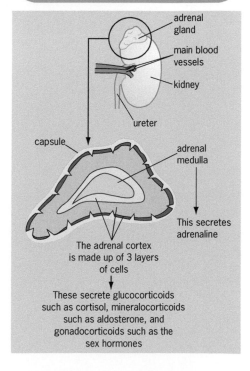

adrenal gland

main blood vessels

kidney

ureter

capsule

adrenal medulla

This secretes adrenaline

The adrenal cortex is made up of 3 layers of cells

These secrete glucocorticoids such as cortisol, mineralocorticoids such as aldosterone, and gonadocorticoids such as the sex hormones

Fig 18.4 The adrenal glands sit on top of the kidneys. They consist of the adrenal medulla, the tissue in the centre of the gland which secretes adrenaline, and an outer layer called the adrenal cortex

Thyroxine: an example of negative feedback

Metabolic rate is the rate at which all cells in the body carry out their biochemical reactions. It is a vital whole-body function that must be controlled within very strict limits. The hypothalamus in the brain detects even a small decrease in metabolic rate and responds by releasing more **thyrotropin releasing hormone** (Fig 18.5). This acts on the pituitary gland, causing it to release more **thyroid stimulating hormone**. This passes to the thyroid gland, which responds by secreting more **thyroxine**, the hormone that acts on individual cells to increase metabolic rate. As soon as metabolic rate gets back to normal levels, the hypothalamus responds by releasing less thyrotropin releasing hormone and, in a healthy person, homeostasis is maintained.

Fig 18.5 Control of metabolic rate involves a negative feedback loop

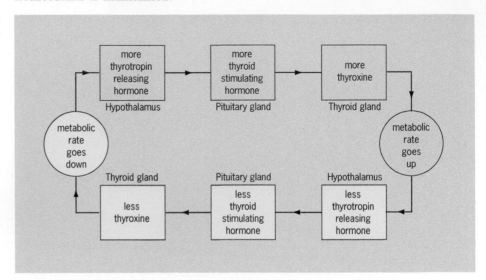

HOW HORMONES AFFECT TARGET CELLS

Hormones exert their effects in two different ways. Peptide hormones travel in the blood to all parts of the body. They do not, however, affect all cells in the body. Target cells or organs have specific proteins on their surface that act as receptor sites. The hormone fits into these sites as a key fits a lock, or like an enzyme fits its substrate. Once in place, the hormone–receptor complex brings about changes inside the cell, although the hormone itself never enters the cytoplasm. In contrast, steroid hormones pass easily into cells because they can get through the lipid bilayer. They bind to a specific receptor molecule inside the cytoplasm: this complex then causes biochemical changes inside the cell.

> **Enzymes are discussed in more detail in Chapter 5.**

Peptide hormones and second messengers

Hormones based on amino acids are water-soluble and vary in size from **thyroxine** (2 modified amino acids) to **growth hormone**, a protein made up of 190 amino acids. Because they are not lipid-soluble, they cannot get into cells through the lipid cell surface membrane. Instead, they act as first messengers, binding to target receptors on the outside of the cell surface membrane.

As Fig 18.6a shows, this binding event activates an enzyme, **adenylate cyclase**, which is located on the inside surface of the membrane. The activated enzyme converts ATP into a substance called **cyclic AMP**. This is the **second messenger** that moves about inside the cell, causing biochemical changes. The final effect on cell function depends on the type of hormone and the type of target cell involved.

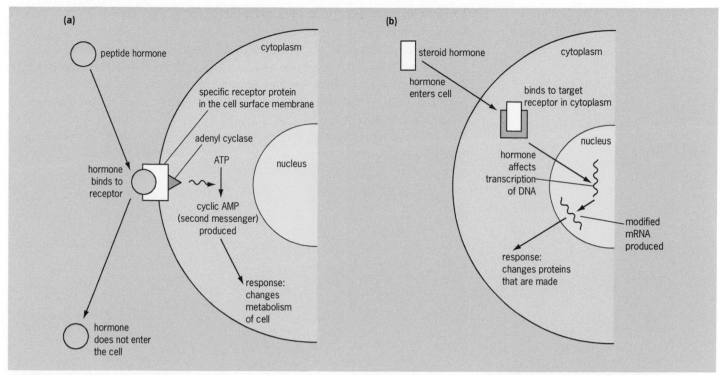

Fig 18.6a Water-soluble hormones work via a second messenger which is formed inside the cell when they bind to receptors on the cell surface

Fig 18.6b When steroid hormones bind to a receptor inside the cytoplasm, they form a complex similar to an enzyme–substrate complex. This enters the nucleus of the cell and binds directly to the DNA, interfering with the cell's ability to read some of its genes. A steroid hormone can either switch on or switch off protein synthesis of particular genes. The overall effect of hormone action is different for each steroid hormone (see Table 18.1)

Steroid hormones

Steroids are small lipid-soluble molecules that can get through cell membranes easily. They enter cells and bind with **target receptors** inside the cytoplasm (Fig 18.6b). Cortisol, progesterone, oestrogen and testosterone all act in this way.

3 COMPARING THE NERVOUS AND ENDOCRINE SYSTEMS

Hormones do not act independently. They work together with other hormones and also with the nervous system. The nervous system and the endocrine system both control and co-ordinate the function of different parts of the body and they both rely on chemical messengers, but they have obvious differences (Table 18.2).

Table 18.2 Differences between the nervous system and the endocrine system in humans

Property	System	
	Nervous system	**Endocrine system**
nature of signal	nerve impulses are **electrical** signals, transfer of information across synapses, is chemical (see Chapter 15)	all hormones are **chemical** signals
size of signal	**frequency modulated** – determined by the frequency of nerve impulses sent along a nerve fibre, and the number of nerve fibres being stimulated	**amplitude modulated** – determined by the concentration of hormone
speed of signal	**rapid** – human nerves conduct nerve impulses at speeds ranging between 0.7 metres per second and 120 metres per second	**usually slower**, by comparison – the release of insulin from the pancreas in response to a rise in blood sugar level takes several minutes
effect in the body	**localised** effect – each individual neurone links with only one or a few cells	more **general** effect – hormones can influence cells in many different parts of the body
capacity for modification	**can be modified** by learning from previous experience	**cannot be modified**: no learning from previous experience

THE ENDOCRINE AND NERVOUS SYSTEMS ARE LINKED

Despite these differences, we now realise that the link between the endocrine system and the nervous system in humans is more definite than once thought. The main physical link between the two systems is between the **hypothalamus**, part of the base of the brain, and the **pituitary gland**, an endocrine gland just beneath it (Fig 18.7).

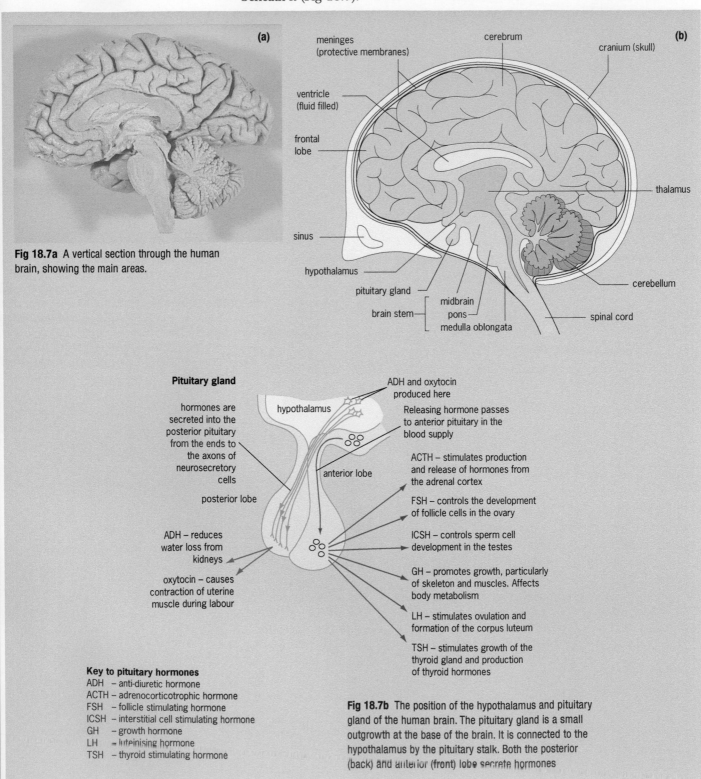

(a)

Fig 18.7a A vertical section through the human brain, showing the main areas.

(b)

meninges (protective membranes)

cerebrum

cranium (skull)

ventricle (fluid filled)

frontal lobe

thalamus

sinus

hypothalamus

cerebellum

pituitary gland

midbrain

brain stem — pons

spinal cord

medulla oblongata

Pituitary gland

hormones are secreted into the posterior pituitary from the ends to the axons of neurosecretory cells

hypothalamus

posterior lobe

anterior lobe

ADH and oxytocin produced here

Releasing hormone passes to anterior pituitary in the blood supply

ACTH – stimulates production and release of hormones from the adrenal cortex

FSH – controls the development of follicle cells in the ovary

ICSH – controls sperm cell development in the testes

GH – promotes growth, particularly of skeleton and muscles. Affects body metabolism

LH – stimulates ovulation and formation of the corpus luteum

TSH – stimulates growth of the thyroid gland and production of thyroid hormones

ADH – reduces water loss from kidneys

oxytocin – causes contraction of uterine muscle during labour

Key to pituitary hormones
ADH – anti-diuretic hormone
ACTH – adrenocorticotrophic hormone
FSH – follicle stimulating hormone
ICSH – interstitial cell stimulating hormone
GH – growth hormone
LH – luteinising hormone
TSH – thyroid stimulating hormone

Fig 18.7b The position of the hypothalamus and pituitary gland of the human brain. The pituitary gland is a small outgrowth at the base of the brain. It is connected to the hypothalamus by the pituitary stalk. Both the posterior (back) and anterior (front) lobe secrete hormones

THE HYPOTHALAMUS AND PITUITARY: PARTNERS IN COMMUNICATION AND CONTROL

The pituitary has been called the 'master gland' because it controls the activity of most of the other endocrine glands, but it is itself actually under the control of the hypothalamus. The range of hormones produced by the pituitary and the way the hypothalamus controls pituitary function are shown in Fig 18.7b.

The human pituitary gland has a **posterior lobe** and an **anterior lobe**. The **pituitary stalk** connects the posterior lobe to the brain. This direct physical link allows nervous communication between the pituitary and the hypothalamus. A rich network of blood vessels also links the hypothalamus with the anterior lobe, allowing hormonal signals to pass from the brain to the pituitary.

Neurosecretory cells and the posterior pituitary

A **neurosecretory cell** is a modified neurone that synthesises a hormone. The hormone is produced in the cell body and is then transported down the axon and released into the blood through the synapse, which terminates on a capillary. The neurosecretory cells that arise in the hypothalamus and end in the posterior pituitary make two hormones: **anti-diuretic hormone** (**ADH**) and **oxytocin** (look back to Fig 18.3 and Table 18.1). When the hypothalamus is stimulated by appropriate nerve impulses, it releases these hormones into the posterior pituitary. From here they pass into the blood and then travel around the body. ADH controls the reabsorption of water by the kidneys; oxytocin causes contraction of the uterus during childbirth.

Hormonal control of the anterior pituitary

Blood vessels that connect the hypothalamus with the anterior pituitary carry hormones and chemical messengers. These either stimulate or inhibit the release of pituitary hormones, including growth hormone, prolactin, the gonadotrophins, adrenocorticotrophic hormone and thyroid stimulating hormone (Fig 18.3).

NERVOUS AND HORMONAL CONTROL OF DIGESTIVE SECRETIONS

The body uses a combination of nerve communication and hormones to control the secretion of digestive juice to coincide with the presence of food in particular areas of the gut. Below is a brief summary of how this happens.

The taste or smell of food encourages the brain to signal the salivary glands and stomach to release saliva and stomach secretions. Food leaving the stomach stimulates the **vagus nerve**, which then triggers the release of bile and pancreatic juice. Digestive secretions can also be controlled by hormones. The hormone **gastrin**, produced by the stomach wall, travels in the bloodstream but exerts its effect locally, stimulating the production of both pepsinogen and hydrochloric acid.

Secretin and **cholecystokinin** control pancreatic and liver secretions. Both are formed by cells in the duodenal wall. Secretin causes the release of sodium hydrogencarbonate in the pancreas. In the liver, it increases the rate of bile formation. Cholecystokinin triggers the release of pancreatic enzymes such as lipase and trypsinogen.

REMEMBER THIS

Anterior means 'towards the front'.
Posterior means 'towards the back'.

Kidney function is discussed in Chapter 13. Childbirth is discussed in Chapter 20.

Control of the secretion of digestive juice is discussed in more detail in Chapter 8.

? QUESTION 1

1 What would happen if the production of digestive juices was not timed to coincide with the presence of food in the gut?

Acromegaly

Robert Wadlow, the tallest man who ever lived, had a condition called gigantism; he probably had a tumour that produced large amounts of growth hormone. The growth plates in his long bones didn't fuse as they normally do in the late teens, and he was still growing when he died at the age of 22. But what happens if someone develops a tumour that produces growth hormone later in life, after their growth plates have closed over? Tumours of the pituitary gland that cause excess production of growth hormone are rare, but they do happen. They cause a condition called acromegaly. The name acromegaly comes from the Greek words for 'extremities' and 'enlargement' and reflects one of its most common symptoms, the abnormal growth of the hands and feet. Soft tissue swelling of the hands and feet is often an early feature, with patients noticing a change in ring or shoe size. Gradually, bony changes alter the patient's facial features: the brow and lower jaw protrude, the nasal bone enlarges and spacing of the teeth increases. So, growth still occurs, but this does not lead to a change in height.

The disease does cause problems and is fatal if not treated. Overgrowth of bone and cartilage often leads to arthritis. When tissue thickens, it may trap nerves, causing numbness and weakness of the hands. Other symptoms of acromegaly include thick, coarse, oily, smelly skin; skin tags; enlarged lips, nose and tongue; deepening of the voice due to enlarged sinuses and vocal cords; snoring due to upper airway obstruction; excessive sweating, fatigue and weakness; headaches;

impaired vision; abnormalities of the menstrual cycle and sometimes breast discharge in women; and impotence in men. Body organs, including the liver, spleen, kidneys and heart, can also enlarge. This is bad enough, but in the longer term, the most serious health consequences are diabetes mellitus, high blood pressure and increased risk of cardiovascular disease. Patients with acromegaly are also at increased risk for polyps of the colon that can develop into cancer.

Acromegaly is difficult to diagnose as the changes it brings about happen very slowly and can be mistaken for changes that result from the normal ageing process. However, once it is diagnosed, it can be treated and people with acromegaly can avoid the changes in their facial features that were common until about 25 years ago, when treatment became available. The first treatment is surgery to remove the pituitary tumour – this reduces the production of growth hormone significantly, but perhaps not down to normal levels. The slight excess must then be 'mopped up' using drug treatment. The main drugs in use are called octreotide and lanreotide. These are analogues of the hormone somatostatin – this dampens down the production of growth hormone by the pituitary. Early forms of the drug had to be injected under the skin every 8 hours but longer-acting preparations that last a month are now available. After surgery, a patient is given the short-acting form to judge which dose is most suitable for them, and then they move on to the once-a-month injections, which need to continue for the rest of their lives.

REMEMBER THIS

The process of amplification seen in the enzyme cascade, which results in the fear, fight or flight response, allows one molecule of adrenaline to stimulate the production of millions of glucose molecules.

THE ACTION OF ADRENALINE

As the nervous system and the endocrine system are closely linked, we should not be surprised that hormones can affect the way that we feel emotionally. The classic example of this is the **fear, fight or flight response** (see the Assignment for this chapter on the CAS website).

At the molecular level, adrenaline binds to the outside of a target cell and activates the enzyme **adenyl cyclase**. This converts ATP to cyclic AMP. Cyclic AMP, is a second messenger which sets off a **cascade**, a chain of reactions that convert glycogen to glucose.

As adrenaline binds to receptor molecules on the surface of a cell, changes in the membrane cause the production of **G proteins**. For every one molecule of adrenaline that binds, 10 molecules of G protein are produced. Each of those 10 molecules of G protein catalyses the production of 10 molecules of adenyl cyclase. Each molecule of adenyl cyclase stimulates the production of 10 molecules of cyclic AMP. This **amplification** effect continues along a chain of reactions that eventually breaks down glycogen into glucose.

4 PROSTAGLANDINS

Prostaglandins are a group of hormone-like compounds. They were originally thought to be produced by the prostate gland – hence the term prostaglandins. They are not true hormones as they are not produced by endocrine glands. Most mammalian cells synthesise prostaglandins, which act locally on surrounding cells.

Prostaglandins control cell metabolism, probably by modifying levels of cyclic AMP inside the cell, and are involved in a wide range of activities including blood clotting, inflammation and smooth muscle activity. Prostaglandins are extremely powerful: one billionth of a gram produces measurable effects. The prostaglandins in human semen cause muscles of the uterus and oviduct to contract during female orgasm, helping the sperm on their journey towards the ovum. Synthetic prostaglandins can be used to stimulate labour in a pregnant woman whose baby is overdue.

5 ENDORPHINS

Endorphins are one of several morphine-like chemical messengers produced in the brain (endorphin = endogenous morphine). These polypeptide molecules mimic the effects of drugs such as morphine and heroin: like prostaglandins, they are involved in a wide range of activities, but their most important role seems to be in the management of pain. It is thought that endorphins bind to pain receptors and so block the sensation of pain. Long-distance runners who run until they collapse do so because abnormally high levels of endorphin block out the acute pain and discomfort.

Endorphins also seem to affect the 'pleasure centres' of the brain. Stimulation of these areas provides the intense feelings associated with orgasm. Scientists studying the chemistry of pleasure have shown that pleasurable sensations begin when the hypothalamus releases serotonin. This stimulates the release of endorphins, which turns on the supply of dopamine and simultaneously turns off the supply of GABA (an amino acid that suppresses dopamine). Dopamine stimulates the pleasure centre directly.

Drugs such as heroin work by mimicking endorphins. The brain of an addict stops making endorphins and so withdrawal from heroin results in a sudden lack of endorphin and a build-up of GABA, producing unpleasant withdrawal symptoms.

REMEMBER THIS

Prostoglandin pessaries can be used to set off labour in a pregnant woman whose baby is overdue. However, semen is rich in prostaglandins, and some experts recommend to expectant couples the natural alternative of having frequent sex. There is a sound physiological rationale behind this, but, from a practical viewpoint, it is rather difficult advice to follow.

6 PHEROMONES AND BEHAVIOUR

Fig 18.8 Sex attractants such as the pheromone bombykol are released into the air in minute quantities by female moths. The male is extremely sensitive to these pheromones, because his antennae are large and crammed with many specialist receptors. This male has about 17 000 receptors in each antenna that respond only to bombykol

Pheromones, sometimes called **ecto-hormones**, are substances that organisms release into the environment to communicate with members of the same species. Pheromones are small, volatile molecules that spread easily in the environment. They are active in very small amounts: the pheromone of the female gypsy moth causes a response in the antennae of the male moth at concentrations as low as one in a thousand million million molecules (Fig 18.8).

Pheromones are usually classified according to the type of response they produce:

- **Alarm pheromones** are produced by bees and ants when attacked. They excite other insects of the same species to swarm around the attacker.
- **Sex attractants** are released by moths, rats and possibly humans to attract members of the opposite sex. Humans do not have a particularly good sense of smell but there is some evidence that very young babies can recognise their mother by her characteristic smell. Some people also think that sexual partners can also recognise each other using their sense of smell, but there is less evidence to back this up.
- **Trail substances** are produced by ants to show other ants where to find sources of food.
- The **queen substance** is produced by the queen bee within a hive to suppress the production of other queens.

SUMMARY

After reading this chapter, you should know and understand the following:

- There are three basic types of chemical signal: **local signals** such as histamine, prostaglandins and endorphins; **hormones** such as insulin and adrenaline; external signals or **pheromones**.
- The human **endocrine system** has four main functions. It controls growth, sexual development and **fear, flight and fight reactions**. It maintains body homeostasis.
- Endocrine glands lack a duct: their secretions are delivered directly into the bloodstream.
- Hormone levels are usually controlled by **negative feedback loops**.
- There are two basic types of hormone: **peptide hormones** which are derived from amino acids (e.g. insulin) and **steroid hormones** which are made from fatty acids (e.g. oestrogen).
- Peptide hormones bind to receptors on the outer membrane of target cells and achieve their effect via a **second messenger** such as cyclic AMP. Steroid hormones pass straight through the cell membranes and bind to target receptors inside the cytoplasm.
- There is a close link between the nervous and endocrine systems, shown by the way in which, in the brain, the **hypothalamus** interacts with the **pituitary gland**.
- Adrenaline, the hormone that causes the classic fear, flight and fight response, acts at the molecular level to bring about the production of millions of glucose molecules.
- **Prostaglandins** and **endorphins** are local messengers which affect many different types of cell. Prostaglandins are best known for their effects on the female reproductive system: endorphins are chemicals that influence our perception of pleasure and pain.

 Practice questions and a How Science Works assignment for this chapter are available at www.collinseducation.co.uk/CAS

19

Senses and behaviour

19 SENSES AND BEHAVIOUR

A phantom limb – all in the mind

Prosthetic legs enable people who have lost both lower limbs to lead a fairly normal life. New technology is now making use of the 'phantom limb' sensations that amputees experience, by developing limbs with electrical connections. These give the person more control over the movement of their artificial limbs

Losing a limb is a surprisingly common event. Amputation can follow the horrific injuries suffered in terrorist bomb attacks or in serious industrial or road accidents. However, the amputee wards in Britain's hospitals are more likely to be full of smokers whose circulation is no longer effective enough to keep their extremities alive.

When a person is unfortunate enough to lose an arm or a leg, you might expect that they would lose all the sensations that come from that part of the body. However, amputees can still feel pain or itching in the removed limb. This is an added misery for a person already seriously traumatised by their injury.

The phenomenon of the phantom limb arises because of the nature of the nervous system. Our brain interprets the environment by knowing where all the incoming impulses from the nerves have originated. If someone stamps on your left foot, you know which foot hurts because impulses arrive via the left foot nerves. When a limb is removed, many of the nerves that connect it with the brain remain intact, and any stimulation of these nerves is interpreted by the brain as sensation in the missing limb.

1 RESPONDING TO THE ENVIRONMENT

We make sense of our surroundings by using our sense organs to gather information and relay it to the central nervous system. Table 19.1 shows the huge amount of sensory information that reaches the brain. Fortunately, we deal with most of it at a subconscious level, which is just as well, since otherwise we would have no time for thoughts. Overall, the brain copes with a huge range of sensory information and uses it to decide on appropriate courses of action.

In this chapter, we present an overview of how the sense organs work and then we examine the Pacinian corpuscle, the eye and the ear in more detail. Touch and the skin are covered in Chapter 12. We conclude the chapter with a brief look at behaviour.

Table 19.1 Summary of the sensory information that goes to the brain

External stimuli	Internal stimuli
sound: volume, pitch, harmonics	blood pressure
vision: colour, shape, intensity, movement	solute concentration of the blood: do we need a drink?
touch: light or heavy pressure?	internal pain: ulcer, wind, headache?
texture: sandpaper or velvet?	tension in muscles: posture?
temperature: hot or cold?	fatigue in muscles
pain: mild or severe?	tension in bladder or rectum
movement: direction, acceleration or deceleration	
direction of gravity	
smell: which one of thousands of different chemicals?	
taste: salty, sweet, sour or bitter?	

SOME IMPORTANT TERMS

Before going on to look at sensory systems in more detail, you will find it useful to become familiar with some important terms:

- A **stimulus** is a physical event, usually some form of energy change, that we can detect with our receptors. Examples include **mechanical stimuli** such as pressure, touch and movement of air, **thermal stimuli** such as heat and cold, **light** and **chemical stimuli** such as taste, smell, and the concentration of carbon dioxide or oxygen in the blood. There are many factors in our environment that are not stimuli – we cannot detect them because we do not have the appropriate receptors. We do not sense radio waves or high-pitched dog whistles, for example.

- A **sensation** refers to a general state of awareness of a stimulus. Aristotle first identified the five senses of hearing, sight, taste, smell and touch, but we now know that there are actually many more senses than this. The skin can experience the sensations of light touch or deep pressure, heat, cold and pain. There are also the general senses of balance and body movement. Internally, the body can sense changes in blood pressure and levels of chemicals in the blood; and we experience hunger and thirst when we need food or fluids.

- **Perception** is the interpretation of stimuli by the brain. Perception of a stimulus is different from just sensing that it is there – it allows us to assess the significance of the stimulus. Many stimuli such as the sensations from your clothes, can be ignored, to allow us to concentrate on more urgent stimuli (like the information in this book).

> **REMEMBER THIS**
> Perception involves reception and interpretation of stimulus information.

2 FEATURES OF SENSORY SYSTEMS

A sensory system must allow the following processes to take place:

- **Transduction**. This is the process by which a stimulus initiates a nerve impulse. Stimuli vary widely in their nature; they can be physical, light, heat or chemical, and so on. All, however, result in the generation of a nerve impulse (see Chapter 15). For example, when physical pressure is applied to the skin, sensory cells inside specialised receptors in the skin depolarise, and the physical stimulus is transduced, or changed, into the electrochemical event known as the action potential.

- **Transmission**. A nerve impulse is transmitted along a sensory neurone into the central nervous system (CNS) (see Chapter 16). The brain recognises the strength of a stimulus not by the size of the impulse but by its frequency. The stronger the stimulus, the higher the frequency of impulses.

- **Information processing**. The CNS processes information that it receives from several parts of the body and makes a decision about the best course of action to take. We have superb memories compared to other animals and so we are able to combine new sensory information with past experience before deciding on an appropriate course of action.

RECEPTORS

A receptor is part of the nervous system that has become adapted to receive stimuli. Some receptors are modified neurones, while others are separate cells connected to neurones. Some receptors, such as those in the skin, occur alone or in small groups. Others are concentrated within a specialised sense organ such as the eye or the ear.

Fig 19.1 Different species have different sense organs. Anyone who has taken a dog for a walk will know that their pet lives in a different world. They can drag you from lamp-post to lamp-post, obsessed with smells that we can't even begin to detect, even if we wanted to. Their hearing is much more acute, too. Dogs can hear much higher frequencies than we can. On the other hand, their colour vision is not as good

? QUESTION 1

1 Look at the cells in Fig 19.2. What type of receptor cells are they? Classify them according to stimulus, location and complexity, as in Table 19.2.

Sensory receptors are vital. They sense the external environment and they also provide information to the CNS about the body's internal environment. Receptors that detect changes in the outside world are positioned at or near the body surface. These include the specialised sense organs, the eye and the ear, the smell receptors of the nose, and the taste receptors of the tongue. The human head, like that of most animals, is basically a set of sense organs in front of a mass of nerves (the brain) that is capable of processing all the incoming information. In contrast, receptors that respond to internal stimulation, such as stretch receptors in blood vessels or muscles, are found deep inside the body.

Types of receptor

Receptors can be classified in three ways: by the type of stimulus they respond to, by their location within the body, and by their level of complexity (Table 19.2).

? QUESTION 2

2 Look at Table 19.2. What type of receptors are the photoreceptors of the eye: are they exteroceptors, proprioceptors or interoceptors?

Table 19.2 Classification of receptors

Receptor name	Action	Example
Classified by stimulus		
photoreceptor	responds to light	eye
chemoreceptor	responds to chemicals	taste bud
thermoreceptor	responds to changes in temperature	temperature receptors in skin
mechanoreceptor	responds to mechanical (touch) stimuli	nerve fibres around hairs
baroreceptor	responds to changes in pressure	baroreceptors in carotid artery
Classified by location		
interoceptor	responds to stimuli within the body	baroreceptors in carotid artery
exteroceptor	responds to stimuli outside the body	eye and ear
proprioceptor	responds to mechanical stimuli conveying information about body position	muscle spindle
Classified by complexity		
general senses	single cells or small groups of cells	temperature receptors in skin
special senses	complex sense organs	eye and ear

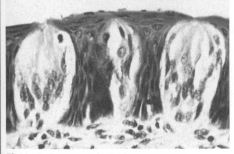

Fig 19.2 Taste buds in the human tongue

There are also two basic types of receptor, and both can be any of the classes of receptor described in Table 19.2. Simple or primary receptors consist of a single neurone whose axon carries a nerve impulse into the CNS. The temperature and pressure receptors in the skin are both examples of **primary receptors**.

Secondary receptors receive information from a cell that is not part of the nervous system and then pass the information on to a nerve cell for transmission to the CNS. Taste buds in the tongue are secondary receptors.

THE PACINIAN CORPUSCLE

When people shake hands, there is a subtle exchange of information. A firm, confident handshake communicates friendliness but 'I'll take no nonsense', while a feeble, cold, clammy, loose grip says 'I don't really want to be here'. This is only possible because of the simplest receptor in the human body – the **Pacinian corpuscle**. This is a **mechanoreceptor** that responds to pressure or any kind of mechanical stimulus.

The corpuscle itself is a nerve ending surrounded by several concentric capsules of connective tissue, separated by a viscous (stiff) gel. In its resting

REMEMBER THIS

The Pacinian corpuscle is named after Filippo Pacini, a nineteenth century Italian anatomist.

state, the Pacinian corpuscle in cross-section resembles a dartboard with the nerve ending at the bull's eye (Fig 19.3). Touching the hands or feet makes this epidermal receptor deform into an oval cross-section including the nerve ending. The viscous gel then moves and allows the nerve ending to resume its normal shape. If the pressure is released, the corpuscle as a whole returns to its original shape and the nerve ending is again deformed, signalling the end

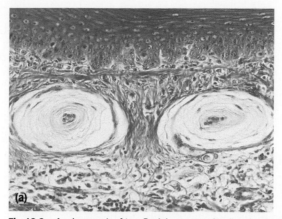

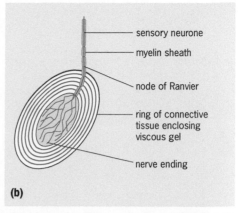

of the hand or foot pressure. The viscous gel flows back and everything returns back to the normal resting stage. The nerve sends the signal to the brain, alerting it to the exact location of the pressure. Pacinian corpuscles can detect vibrations and touch in terms of frequency, duration and intensity.

Fig 19.3a A micrograph of two Pacinian corpuscles (touch receptors) in human skin. They are abundant at fingertips. In this touch receptor, pressure is detected by the end of the neurone at the centre of the connective tissue capsule which is layered like an onion. When the neurone is stimulated, action potentials are propagated along the neurone. The brain interprets these impulses as pressure on that part of the body. **19.3b** A simple diagram of a cross-section through a Pacinian corpuscle

HOW DO RECEPTORS WORK?

How do cells detect stimuli? How, for instance, can we distinguish between a light and a heavy touch? The answer lies in the fact that primary receptors have proteins in their cell surface membranes that are sensitive to the type of stimulus. Whatever the stimulus, transduction follows the same basic pattern, which is summarised in Fig 19.4. A receptor protein is activated by a specific stimulus. In turn, the protein opens or closes specific ion channels in the cell surface membrane. This change in ion movement causes depolarisation of the membrane; this is the **generator potential**.

If the depolarisation goes beyond the threshold level for that cell, it triggers an action potential (see Chapter 15). All receptors therefore act as **biological transducers**: they create action potentials in neurones in response to physical and chemical stimuli from the environment.

The greater the stimulation of the sense cell, the greater the depolarisation. Once the generator potential exceeds the threshold level, an action potential is produced and is transmitted along the axon. It is important to remember that all action potentials are of the same magnitude – you can't have powerful and weak nerve impulses. The body is able to tell the difference between strong and weak stimuli by the frequency of the action potential created.

Receptors are very sensitive – they have to be in order to detect small changes in the environment. The human nose can detect ethanoic acid (acetic acid, the acid in vinegar) at a concentration of 5×10^{11} molecules per litre of air; a dog's nose, though, can detect the same smell at a concentration about 2 million times less, 2×10^5 molecules per litre.

Body senses also **amplify** the incoming signal. A stimulus is often quite weak – just a few photons of light when viewing a distant star for example, or a few molecules of a chemical. But the action potential travelling from the eye to the brain, for example, has about 100 000 times as much energy as the few photons of light that triggered it.

The endocrine system also shows amplification. One molecule of the hormone adrenaline, for example, can release millions of glucose molecules in the cascade effect (see Chapter 18).

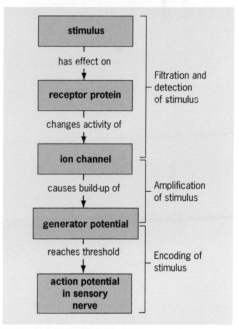

Fig 19.4 Flow chart summarising the function of receptor cells

✓ **REMEMBER THIS**

A stimulus alters the electrical properties of the membrane of a receptor cell. The potential difference across the membrane is reversed: it goes from –70 mV to +40 mV. We say that it has depolarised. When a membrane depolarises there is the formation of a nerve impulse, or moving action potential, that is transmitted along a nerve fibre. Find out more about action potentials in Chapter 15.

3 VISION

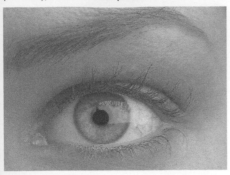

Fig 19.5 Human eyes are located in a bony socket (the orbit), and have other protective features

Fig 19.6 The human eye. The eyeball itself is a sphere about 2.5 cm in diameter. It has three basic layers, the inner retina containing the light-sensitive cells, the pigmented choroid layer and the tough outer coat, the sclera

The eyes are literally our window on the world (Fig 19.5). This most sensitive and intricate of sense organs has fascinated both scientists and poets for centuries. In this section we look at the basic structure of the eye, before going on to study focusing and the correction of defects. Finally, we look at vision at the molecular level.

STRUCTURE OF THE HUMAN EYE

The eye is an important and delicate structure. The eyes of humans and other mammals are contained in the orbit, a bony socket of the skull. They are protected by the bone and also by a pad of fat at the rear of each eyeball that helps to cushion it from shocks. Other features of the eye are also protective:

- The eyebrows prevent sweat running from the forehead into the eye.
- The eyelashes prevent the entry of airborne particles such as flying insects, often adding to the efficiency of the eye blink reflex.
- Tears constantly bathe the surface of the eye, removing the surface film of dust and dirt and killing bacteria.
- The conjunctiva, a thin, transparent membrane which covers the cornea, also protects the eye from airborne debris.

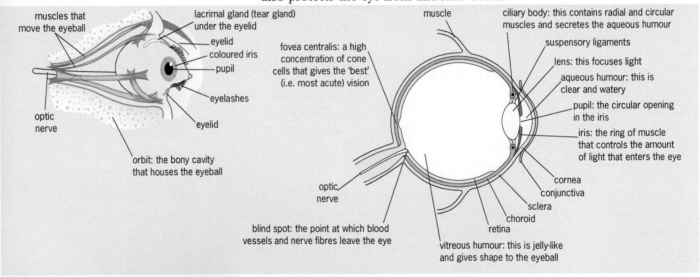

The main body of the eye is divided into two parts by a **biconvex lens**. This lens contains transparent **lens fibres** (long thin cells that have lost their nuclei) together with an elastic **lens capsule** made of glycoprotein. The lens is flatter at the front than at the back and is soft and slightly yellow. With age it becomes flatter, yellower, harder and less elastic; it can also become cloudy (Fig 19.7). There is no blood in the lens – substances diffuse in and out of it from the surrounding fluid.

The eyeball has three cavities: the region between the cornea and the iris, the region between the iris and the lens, and the cavity that fills all the space behind the lens. The first two cavities are filled with a fluid called the **aqueous humour**, which is thin and watery. The third cavity, the largest (it takes up about 80 per cent of the volume of the eyeball) is filled with a thicker fluid called the **vitreous humour**. This transparent gel has the consistency of egg white and contains 99 per cent water and 1 per cent collagen fibres and hyaluronic acid. The vitreous humour preserves the spherical shape of the eyeball and helps to support the **retina**.

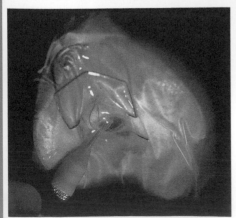

Fig 19.7 As people get older, the lens in their eyes can become cloudy, interfering with normal vision. In this routine operation, an ultrasound probe is being used to break up a cloudy lens before a plastic lens is fitted

The retina

The retina is the light-sensitive layer at the back of the eye. It is a complex structure with a deep layer of light-sensitive cells called **rods** and **cones**, together with a middle layer of **bipolar neurones** which connect the rods and cones to a surface layer of **ganglion cells** (Fig 19.8). Fibres from the ganglion cells join up to form the **optic nerve**. Only the back of the retina is **photosensitive**. The front surface extends over the **choroid** and forms the inner lining of the **iris** and **ciliary body**.

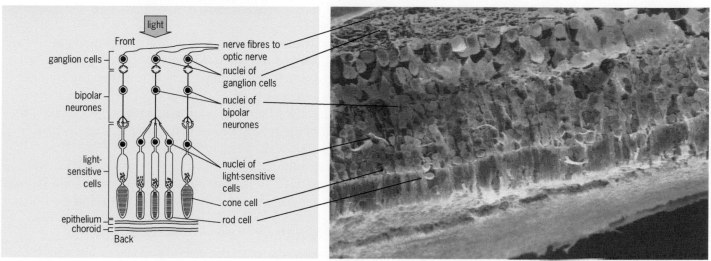

Fig 19.8 Detailed structure of the retina showing photoreceptors and their connections

The sensitive part of the retina has the area of a 10 pence coin and acts as a projection screen onto which images are directed. Strangely, the nerves that come from the retina pass in front of the light-sensitive cells, but these do not interfere with vision.

The choroid

The **choroid** contains many blood vessels. Blood supplies the cells of the eye with nutrients and oxygen and removes waste. The choroid is dark-coloured because of a high concentration of the pigment melanin in its cells. These pigmented layers prevent internal reflection of light rays and so prevent us seeing a confused and blurred image.

At the front of the eye, the choroid expands around the edge of the lens to form the ciliary body. The smooth muscles in the ciliary body alter the shape of the lens (Fig 19.9). The lining of the ciliary body secretes aqueous humour, the fluid that fills the front of the eye.

The iris is also an extension of the choroid, partially covering the lens and leaving a round opening in the centre, the pupil. Its function is to control the amount of light entering the eye. The iris contains **radial** and **circular muscles** that can alter pupil size. Iris muscles are controlled by the autonomic nervous system (see Chapter 16).

The sclera

The **sclera** forms the 'white' of the eye. It is a tough external coat mainly of collagen fibres. Six exterior muscles attached to the sclera enable the eye to look up, down and side-to-side. The central one-sixth of the sclera is colourless and transparent, forming a clear window, the cornea, through which light gets into the eye.

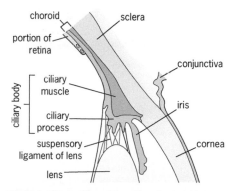

Fig 19.9 The position of the ciliary body and the suspensory ligaments in the eye. When we focus on objects that are distant from us, the ciliary muscles relax, increasing the tension on the suspensory ligaments. This pulls the lens into a thinner and flatter shape. When we look at objects closer to us, the ciliary muscles contract, the tension on the ciliary ligaments decreases and the lens becomes fatter

HOW THE EYE WORKS

The process of seeing is complex and involves five different stages:

- Light enters the eye.
- An image is focused on the retina.
- Energy in the light that makes up the image is transduced into an electrical signal.
- Nerve impulses carry information about the image into the brain.
- The brain decodes the information and perceives the image.

Let's look at each of the stages in more detail.

Light enters the eye

Light enters through the cornea and then passes through the aqueous humour, the pupil, the lens and the vitreous humour, before it reaches the retina. To operate well in conditions of different light intensity, the eye must control how much light reaches the retina. It does this by changing the diameter of the pupil. In bright light, the pupil **constricts** (gets smaller) to prevent overstimulation of the retina and the perception of 'dazzle'. In dim light, the pupil **dilates** (gets wider), letting in more light.

The size of the pupil is controlled by the muscles in the iris (Fig 19.10). These are themselves controlled by the autonomic nervous system and so pupil adjustment is a reflex response, not under the conscious control of the brain.

Focusing light on the retina

If you do not focus a camera correctly, the picture you take is blurred. Similarly, the eye must focus light to produce a high-resolution image on the retina. The eye focuses an image by refracting or bending light using the cornea and the lens, forming an upside down or inverted image on the retina.

Because of the composition and curvature of the cornea, most of the refraction of light occurs in this structure. The lens is important in the fine focusing of light onto the retina. The cornea is fixed but the lens is adjustable, allowing **accommodation**, a reflex that makes the eye focus on objects that are at different distances from the eye.

Bright light
Relaxation of radial muscles and contraction of circular muscles cause pupil to constrict

Dim light
Contraction of radial muscles and relaxation of circular muscles cause pupil to dilate

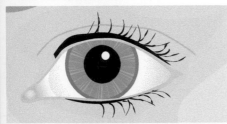

Fig 19.10 The eye is prevented from being dazzled by a speedy reflex that makes the pupil smaller. To achieve this **constriction** of the pupil, radial muscles in the iris relax and circular muscles contract. In dim light, the opposite happens: radial muscles contract and circular muscles relax, causing the pupil to enlarge

? QUESTION 3

3   The eye is often compared to a camera. Which part of the camera corresponds to the retina of the eye?

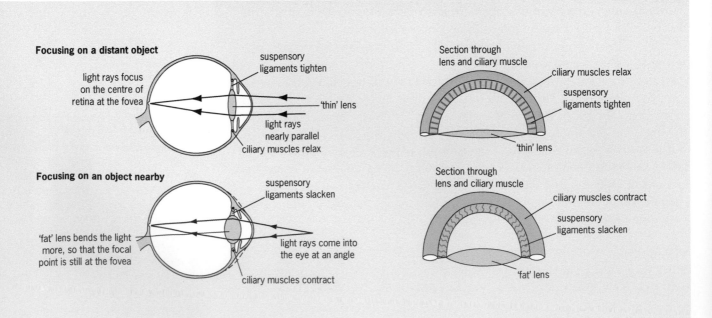

Focusing on a distant object
light rays focus on the centre of retina at the fovea
suspensory ligaments tighten
'thin' lens
light rays nearly parallel
ciliary muscles relax

Section through lens and ciliary muscle
ciliary muscles relax
suspensory ligaments tighten
'thin' lens

Focusing on an object nearby
'fat' lens bends the light more, so that the focal point is still at the fovea
suspensory ligaments slacken
light rays come into the eye at an angle
ciliary muscles contract

Section through lens and ciliary muscle
ciliary muscles contract
suspensory ligaments slacken
'fat' lens

Fig 19.11 The eye can focus on near and distant objects by altering the shape of the lens. This process is called accommodation

Accommodation (focusing)

Generally, when we open our eyes in the morning, they are not focused on nearby objects. At rest, the ciliary muscles have relaxed, pulling the lens flat. In this state we focus on distant objects.

When we focus on an object that is only a metre or so away, the ciliary muscles of the eye contract involuntarily (Fig 19.11). This reduces the tension on the ligaments that hold the lens in place, and the lens becomes 'fatter'. The focal length changes, and the image is focused. When we then look at an object much further away, the ciliary muscles relax, again automatically. The tension on the ligaments supporting the lens increases, and the lens becomes 'thinner', allowing our vision to **accommodate**.

PROBLEMS WITH THE LENS SYSTEM

In some individuals the lens and cornea do not focus correctly (Fig 19.12). If you are short-sighted, a condition known as **myopia**, you focus light from an object in front of the retina. This produces a blurred image. Short-sightedness can result from an elongated eyeball or a thickened lens. It is corrected by using a **diverging** or **concave** lens.

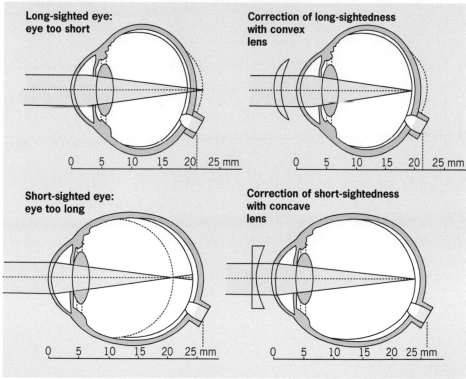

Fig 19.12 In some people, an image forms in front of the retina (this is myopia) or behind the retina (this is hypermetropia). In these cases the eyeball is either too long or too short. Astigmatism can result from an irregular cornea or an irregular lens

If you are long-sighted you are said to have **hypermetropia**, and you focus light behind the retina. Again, objects appear blurred. Long-sightedness results from a shortened eyeball or a lens that is too thin, and is corrected using a **converging** or **convex** lens.

A more complicated visual defect is **astigmatism**. In astigmatism, the surface of the cornea is irregular, the object appears blurred because some of the light rays are focused and others are not (this is similar to the distortion produced by a wavy pane of glass). Astigmatism is corrected using a **cylindrical** lens that bends light rays in one plane only.

> **? QUESTION 4**
>
> 4 Imagine you are standing at a bus stop reading a book. What happens **a)** to the ciliary muscles, **b)** to the lens and **c)** to the pupil of your eye as you look up to see the bus approaching in the distance?

Fig 19.13 Primates have both binocular and colour vision. Binocular vision allows animals to judge distance. This is vital when jumping from branch to branch. It is thought that colour vision allows food to be chosen with greater accuracy; many fruits use colour to signal when they are ripe. Humans, being primates, owe our vision to our ancestry

BINOCULAR VISION

Humans and some other mammals, notably primates and predators such as cats, have **binocular vision** (Fig 19.13). We have eyes at the front of our head, so the field of vision seen by each eye is similar but not identical. Normally, the brain is able to compute the incoming signals so that we see only one image.

Binocular vision gives us a big advantage. When two sets of signals are decoded by the brain, we get the impression of distance and depth and of objects being three-dimensional. This **stereoscopic vision** allows us to judge distance and depth accurately. It is easy to see why primates need binocular vision; when living in the trees, any animal that cannot judge distance is likely to fall sooner rather than later.

VISION AT THE MOLECULAR LEVEL

When light reaches the back of the retina it is focused on to the **photoreceptors**, cells that are specialised to detect light. These cells contain pigment molecules that are bleached by light. Bleaching of the pigment results in a generator potential and, when the threshold is reached, action potentials are created and nerve impulses begin to travel towards the brain.

The two photoreceptors in the human eye are the rods and cones (Fig 19.14 and Table 19.3).

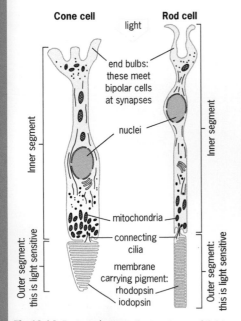

Fig 19.14 Rods and cones: the two types of light-sensitive cell in the human eye

Table 19.3 Comparing rods and cones

Feature	Rod cells	Cone cells
shape	rod-shaped outer segment	cone-shaped outer segment
connections	many rods connect with one bipolar neurone	only a single cone cell per bipolar cell
visual acuity	low	high
visual pigment(s)	contain rhodopsin (no colour vision)	contain three types of iodopsin, responding to red, blue and green light
frequency	120 million per retina – 20 times more common than cones	7 million per retina – 20 times less common than rods
distribution	found evenly all over retina	found all over retina but much more concentrated in the centre, particularly the yellow spot or fovea centralis
sensitivity	sensitive to low light intensities (used in dim light)	sensitive to high light intensities (used in bright light)
overall function	vision in poor light	colour vision and detailed vision in bright light

? QUESTION 5

5 It has long been known that a deficiency of vitamin A causes the condition of night blindness. Give an explanation for this.

Photoreception in rod cells

Rods contain a reddish-purple compound called **rhodopsin**, or visual purple. This consists of two joined compounds – **opsin**, a protein, and **retinal**, a light-absorbing compound derived from vitamin A. It is important to know what happens to rhodopsin in the dark and when it is exposed to light.

In the dark, sodium ions flow from the surrounding fluid into the outer segment of the rod cell via **non-specific cation channels**. The sodium ions then diffuse along the rod cell and into the inner segment, where they are actively pumped back out of the cell. The overall movement of sodium ions produces a slight depolarisation of the whole rod cell membrane. The potential difference across the membrane is about –40mV, compared with the normal –70 mV resting potential. This slight depolarisation brings about continual release of a neurotransmitter, thought to be **glutamate**, which crosses the synapse and prevents bipolar cells from depolarising. Bipolar cells are the 'on' or 'off' cells that connect rod and cone cells to nerve cells in the optic nerve. The overall result is that, in the dark, we don't see anything because no signals are coming from the rod or cone cells into the brain.

When light falls on the rhodopsin molecule, it breaks down into retinal and opsin. The opsin activates a series of reactions, which close the cation channels. This stops the movement of sodium into the outer segment of the rod cell but the inner segment continues to pump sodium out. The inside of the cell becomes more negative and the release of glutamate stops. No glutamate means there is nothing to prevent depolarisation of the bipolar cell and an action potential is generated and transmitted down the optic nerve. These nerve impulses pass to the brain for decoding and we see an image. In the absence of further light stimulation, the retinal molecule goes back to its original shape and then recombines with opsin to form rhodopsin. It is worth remembering that this resynthesis process takes time.

Rhodopsin is very sensitive to light and so rods can function effectively in dim light. In bright light our rhodopsin is broken down quicker than it can be re-formed, so that most of it is bleached for most of the time. In this state we are said to be light adapted. If we enter a dark room from a brightly lit one we cannot see anything because much of our rhodopsin is bleached. However, as we resynthesise the pigment and become dark adapted, things become clearer and we are able to see in dim light, even if it is only in shades of grey.

Photoreception in cone cells

The function of the cones is to see colour and detail, although they can only do this in bright light. In dim light they do not function at all. Cone cells contain the pigment **iodopsin**, a photosensitive pigment made up of **photopsin**, a different protein from that found in rods, and retinal. The events that occur in cone cells stimulated by light are basically the same as those that occur in rod cells. There are, however, three different types of cone. Each one contains a slightly different pigment and responds to a different wavelength of light. One responds to red light, one to green light and one to blue light, so allowing us to see in colour (Fig 19.15).

The colour that we 'see', or more accurately **perceive** (interpret in our brain), depends on which cones are stimulated. When all the cones are fully stimulated we see 'white'. When very few are stimulated we see black. Stimulation of separate types produces red, green or blue, and all colours in between are produced by combinations of different levels of stimulation of the three types together. This model of how three different types of cone can produce the range of colour vision that we experience is known as the **trichromatic theory of colour vision**.

> **? QUESTION 6**
>
> 6 Which cones are stimulated by
> **a)** blue light at 490 nm,
> **b)** violet light below 440 nm? (Look at Fig 19.15.)

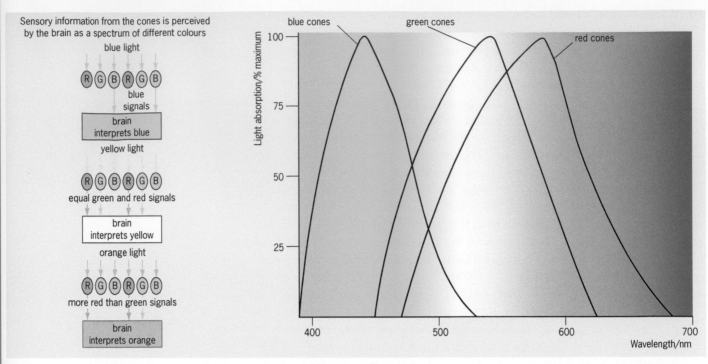

Sensory information from the cones is perceived by the brain as a spectrum of different colours

Fig 19.15(a) There are three types of cone responding (optimally) to three wavelengths of light. The colour we perceive depends largely on the proportion of the stimulation of the three types. If a single type of cone is missing, the individual will not be able to distinguish between certain colours – red–green is the commonest – and is said to be colour blind

Fig 19.15(b) Colour blindness can be diagnosed by means of Ishihara tests. A person with normal vision will see the pink and red loop, while a red–green colour-blind person will not.

Visual acuity: the ability to see detail

From what distance can you read a newspaper? Can you read the bottom line of the standard optician's eye-test board? Tests such as these are a measure of **visual acuity**, which is defined as the ability to resolve detail. For instance, two lines such as these ═══════════ will obviously appear as two lines, but if you prop the book up and walk backwards, a point will be reached when they look like one grey line.

An interesting feature of visual acuity is that we only have a very narrow field of accurate vision. Focus on any particular word on this page and you see it in detail, but those on either side are unclear. We see only a very small proportion of our field of vision in detail – and that is the part that is focused on the fovea, and therefore that stimulates the cone cells.

Overall, rods are more sensitive than cones. Rods can function in poor light, but they cannot perceive detail; you can't read small print in dim light. The cones have a higher visual acuity than the rods, but they need bright light. Why is there such a difference? The answer lies largely in a phenomenon known as **retinal convergence**.

Figs 19.16 and 19.17 illustrate retinal convergence. There are over 120 million rods and about 7 million cones, but only about 1 million neurones in the optic nerve. Clearly, each rod and each cone cannot all have their own neurone. In fact, many rods converge into one neurone, while only a few cones are associated with each neurone. In the centre of the fovea it is thought that each cone actually has its own neurone.

Fig 19.16 (right) The concept of retinal convergence. Generally, the further out you go from the centre of the fovea to the edge of the retina, the greater the degree of convergence. Two separate stimuli (such as two dots or lines) may stimulate two separate cones, but not the one in between. Therefore, the brain will see these as two separate images. The same two stimuli falling on the rods will not stimulate separate neurones, because a group of rods all feed into the same neurone. In this case, the brain will perceive only one stimulus, and see only one line or dot

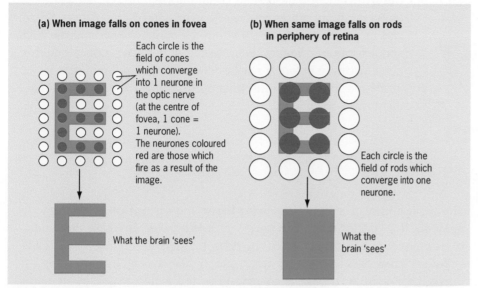

Fig 19.17 Why can we see in detail only when we look straight at something? Taking the letter E as an example, it can be seen that if the image falls on the cones it will stimulate some neurones but not others. However, if it falls on the rods it will simply stimulate lots of adjacent neurones, and all we will see is a splodge, not a clear letter E

TRANSMISSION OF NERVE IMPULSES TO THE BRAIN

If you look back at Fig 19.8, you can see that the photoreceptor cells are linked to a set of nerve cells in the retina called bipolar cells. These link with a second type of nerve cell called ganglion cells. It is the axons of ganglion cells that take information from the eyes to the brain. Ganglion cell fibres are bundled together to form the optic nerve.

We cannot see an object whose image falls on the retina at the point where axons of ganglion cells leave to form the optic nerve. Since it contains no receptor cells, any light striking this small area is not sensed; hence its name, the blind spot (Fig 19.18).

Fig 19.18 To demonstrate the blind spot, hold your book at arm's length with the two symbols straight in front of your eyes. Close your left eye and concentrate on the cross with your right eye. Bring the book slowly towards you. Keep looking at the cross on the left: eventually, as the dot falls on the blind spot of your right retina, the image of the dot on the right will disappear

HOW THE BRAIN PERCEIVES AN IMAGE

Action potentials do not differ very much from one another and so there are only two possible ways in which stimuli going into the brain can be coded. One is the rate at which action potentials arrive and the other is their destination in the brain.

The rate at which sensory neurones fire tells the brain the strength of the stimulus: specific neurones tell the brain which part of the retina has been

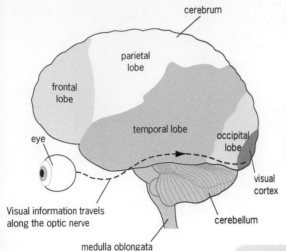

cerebrum

parietal lobe

frontal lobe

temporal lobe

occipital lobe

eye

visual cortex

Visual information travels along the optic nerve

cerebellum

medulla oblongata

Fig 19.19 The location of the visual cortex in the human brain

stimulated. The brain analyses this information to assess the nature of the original stimulus. This process, called **visual processing**, is immensely complex.

Specific areas of the cerebral cortex are associated with different sensory functions (see Chapter 16). The **visual cortex** in the **occipital lobes** deals with visual information (Fig 19.19). However, body senses must not be considered in isolation. Sensory and motor systems interact within the control regions of the CNS, and sensory information is processed at various points along the nerve pathway before it reaches the brain. In the brain it is processed further with information from other senses and stored memories.

4 HEARING AND BALANCE

The ear is a miniature **receiver**, **amplifier** and **signal-processing system**. It is divided into three parts – an **outer ear**, **middle ear** and **inner ear** – and is connected to the brain by the auditory nerve (Fig 19.20).

Sound waves pass through the outer and middle ear and into the inner ear through the **oval window**. Pressure on the oval window squashes the fluid in the inner ear, and compression waves travel through the canals of the

Fig 19.20 The structure and function of the human ear

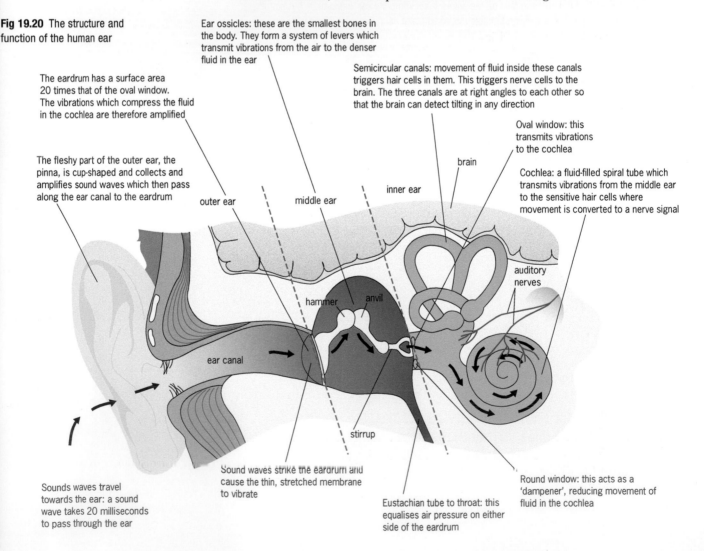

Ear ossicles: these are the smallest bones in the body. They form a system of levers which transmit vibrations from the air to the denser fluid in the ear

The eardrum has a surface area 20 times that of the oval window. The vibrations which compress the fluid in the cochlea are therefore amplified

The fleshy part of the outer ear, the pinna, is cup-shaped and collects and amplifies sound waves which then pass along the ear canal to the eardrum

Semicircular canals: movement of fluid inside these canals triggers hair cells in them. This triggers nerve cells to the brain. The three canals are at right angles to each other so that the brain can detect tilting in any direction

Oval window: this transmits vibrations to the cochlea

Cochlea: a fluid-filled spiral tube which transmits vibrations from the middle ear to the sensitive hair cells where movement is converted to a nerve signal

brain

inner ear

outer ear

middle ear

auditory nerves

hammer

anvil

ear canal

stirrup

Sounds waves travel towards the ear: a sound wave takes 20 milliseconds to pass through the ear

Sound waves strike the eardrum and cause the thin, stretched membrane to vibrate

Eustachian tube to throat: this equalises air pressure on either side of the eardrum

Round window: this acts as a 'dampener', reducing movement of fluid in the cochlea

Fig 19.20 (Cont.)

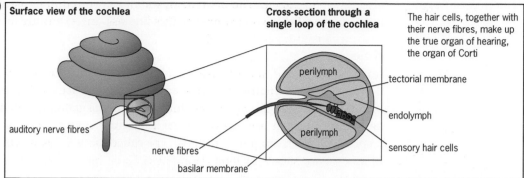

| Surface view of the cochlea | Cross-section through a single loop of the cochlea | The hair cells, together with their nerve fibres, make up the true organ of hearing, the organ of Corti |

auditory nerve fibres

nerve fibres

basilar membrane

perilymph

tectorial membrane

endolymph

perilymph

sensory hair cells

cochlea. The round window eventually receives these movements of fluid and dampens them down (this prevents them from being reflected back into the cochlea) by bulging into the middle ear. Sensitive hair cells in the cochlea fire off signals, sending action potentials along the auditory nerve to the brain.

5 BEHAVIOUR

SIMPLE BEHAVIOURS

One of the simplest types of behaviour is **orientation**. This enables organisms to move in order to find favourable environments and to avoid unfavourable ones. Animals such as insects show two types of simple orientation:

- **Taxes**: movements in a specific direction that are directed by a stimulus such as light or food.
- **Kineses**: more random movements that are not directed by a stimulus.

Taxes and kineses are usually regarded as **reflexes**. Reflexes also occur in higher animals – including mammals and humans; these are covered in more detail in Chapter 16.

Taxes

In many orientation behaviours, the animal moves directly towards or directly away from a stimulus. Maggots (fly larvae) have very simple eyes and can respond to the direction of light (Fig 19.21). Honey bees move towards a scent which is associated with food. These types of directional response are called taxes (singular: taxis).

A positive taxis involves movement towards a stimulus, a negative taxis involves movement away from the stimulus. As Table 19.4 shows, taxes are

Table 19.4 Summary of some common taxes. These are normally classified either by the type of stimulus that triggers them or by the type of response they elicit

Name of taxis	Type of stimulus	Type of response
telotaxis		fixing on a stimulus without scanning from side to side
klinotaxis		moving head repeatedly from left to right to judge stimulus intensity
geotaxis	gravity	
thermotaxis	heat	
phototaxis	light	
anemotaxis	air currents	
rheotaxis	water currents	
chemotaxis	chemicals, e.g. salts	

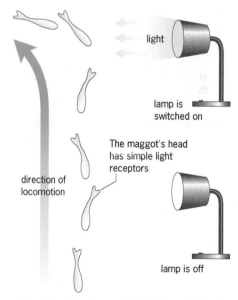

Maggot shows avoidance reaction when light is turned on

light

lamp is switched on

direction of locomotion

The maggot's head has simple light receptors

lamp is off

The maggot has to move its head from side to side to compare light levels on either side of its body

Fig 19.21 The maggot has simple eyes that can distinguish changes in light intensity but it cannot tell from which direction the light is coming. It moves its head from side to side to judge on which side the light is brightest

also classified according to the type of stimulus involved (light, water, gravity) or to the type of behaviour shown (straight-line responses, turning movements).

Kineses

Imagine a ladybird, inactive during a cold night. It begins to stir at dawn. As the light level and temperature increase, the animal becomes more active and begins to move around and locate food. Conversely, woodlice are active in the dark and damp but become less active in brighter, drier conditions. These responses, which are not directional, are called kineses. Some common types of kineses are shown in Table 19.5.

Table 19.5 Some common kineses. Like taxes, these are normally classified either by the type of stimulus that triggers them or by the type of response they elicit

Name of kinesis	Type of stimulus	Type of response
photokinesis	light intensity	
chemokinesis	chemical gradients	
hygrokinesis	humidity levels	
klinokinesis		turns faster when stimulus is more intense
orthokinesis		moves with greater speed when stimulus is more intense

MORE COMPLEX BEHAVIOURS

We can divide all animal behaviour into two categories: **instinctive behaviour,** or **innate behaviour**, which is inherited, and **learned behaviour**, which is modified by experience.

Human behaviour

Consider two examples of human behaviour:

- If children see pictures of small children's faces and adult faces they prefer the adult face. A change in preference occurs at 12 to 14 years of age in girls when they express a preference for baby pictures. Boys display the same trend two years later.
- If we sit too close to a stranger in a library then we tend to use physical 'barriers' such as books or bags to keep the stranger at a distance. If this person gets too close we often move seats.

These are typical of the types of behaviour studied in **human ethology** – the biology of human behaviour. It asks the question 'why do we behave the way we do?'

Ethologists view human behaviour in the same way as that of other animals: they think it is adaptive and has evolved to suit its purpose. In the two examples above, there is strong evidence to link children's behaviour with child rearing and adult behaviour with protection.

One goal of human ethologists is to look for patterns of behaviour shared by all peoples of the world. These are seen in examples of behaviour that involve body movements, rather than speech. Gestures such as smiling and raising the eyebrows, and behaviours such as grooming and hugging seem to be common to all human societies.

We all think that we have a free choice in our behaviour, but do we really? We may be under the control of our genes or be shaped by our surroundings. There is evidence for both points of view.

SUMMARY

After completing this chapter you should know that:

- A **sensation** is a general state of awareness of a **stimulus**; **perception** involves the **interpretation** of **sensory information**.
- A **receptor** is a **biological transducer**: it converts energy from one type of system (e.g. chemical) to another system (e.g. electrical). A receptor is classified according to the type of stimulus it receives (e.g. photoreceptor), its location (e.g. interoceptor) or its level of complexity.
- The **Pacinian corpuscle** is a simple receptor in the skin that detects pressure and vibration.
- The eyeball contains three outer layers, the **sclera**, **choroid** and **retina**.
- Incoming light rays are **refracted** (bent) by the cornea and by the **lens** of the eye.
- **Accommodation** is the ability to focus the eye automatically, to see near and distant objects.
- **Rod** and **cone** cells are stimulated by the effects of light on two pigments, **rhodopsin** and **iodopsin**. Breakdown of these compounds alters the electrical properties of the cell membranes.
- There are three types of cone cell pigment. Each responds to a different range of wavelengths of light. Humans have **trichromatic** vision: our cone cells detect three colours, blue, green, red.

- **Stereoscopic vision** enables us to perceive depth and distance.
- The human ear has two main functions, to detect sound and to maintain balance.
- Ethology is the study of animal behaviour under natural conditions.
- Patterns of behaviour are controlled by the nervous system; simple behaviours include **taxes** and **kineses**.
- A **taxis** is an orientation response related to the direction of the stimulus. A positive taxis describes movement *towards* a stimulus; a negative taxis, movement *away* from a stimulus.
- A **kinesis** is an orientation behaviour where the response relates not to the direction of the stimulus, but to its intensity.
- **Instinct** or **innate behaviour** is unlearned. It has survival value. There are also various types of learned behaviour.

 Practice questions and a How Science Works assignment for this chapter are available at www.collinseducation.co.uk/CAS

20

Human reproduction

20 HUMAN REPRODUCTION

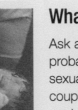

Sometimes sex in humans does result in a new baby, but more often than not, it doesn't. Sex in humans is more than just making babies

What is sex for?

Ask anybody what the purpose of sex is and, after a few funny looks, you'll probably get the answer 'to make babies'. Yet, although millions of acts of sexual intercourse take place across the globe every day, the vast majority of couples have no intention of making a baby. In practice, many of them use contraception to prevent pregnancy.

In the same way, if you ask what breasts are for, you might get the answer 'for breast-feeding'. However, when in close proximity to a topless woman, the average man probably does not spend much time thinking about how many children she could successfully breast-feed. All mammals produce milk, but humans are the only ones with permanently swollen hemispheres of fatty tissue.

Clearly, sex in humans is about more than just making babies. In fact, we seem rather obsessed with sex. Humans are one of the few species that have sex when there is no egg to be fertilised, and statistics tell us that the average human has sex several thousand times in their lifetime.

As you read through the account of human reproduction that follows, have a think about the function of sex in humans. If it is for more than basic reproduction, what is its purpose? Is it simply for pleasure? Or it is a form of 'social cement', strengthening the bond between people, encouraging them to fall in love and stay together? You decide.

1 HUMAN LIFE SPAN

As far as sex is concerned, humans are unique. We develop bodies capable of reproducing at around 12 or 13 years old – a good few years before most of us become independent from our parents. For most female mammals, mating occurs only when there is an egg to be fertilised and most ovulations result in pregnancy. The female spends her life being pregnant or lactating, or both.

Fig 20.1 Different human cultures have different concepts of what makes a family. Here, four generations of one family have been photographed together; you can see the family resemblances and the effects of ageing quite easily

The ceaseless cycle of reproduction is brought to an end by death. Humans are quite different. The average woman in the UK ovulates about 400 times, but has only two children during her life. We take a detailed look at human reproduction in the rest of this chapter.

Old age is another part of human life that is alien to most animals (Fig 20.1). Predators who get a bit slower catch fewer prey, get weaker and eventually starve. Ageing prey become easy targets for a predator having a lazy day. *Homo sapiens* is in a unique situation. When our physical capabilities start to fail, we find less demanding things to do. Top sports people, for example, might stop performing at the highest level in their mid-30s but then they can progress into training, management or broadcasting. Eventually though, age catches up with us.

As you read the next two chapters, you might like to reflect on the human animal; our attitudes to sex, a 20-year childhood, a retirement often longer than the working life and an ever-increasing wish to prolong our own life span. What a strange creature indeed.

Human life span from birth onwards is discussed in Chapter 21.

2 THE NEED TO REPRODUCE

None of us is immortal and so, in common with all other types of organism, we must reproduce in order for our species to survive.

An interesting way of looking at the whole business of reproduction is to think of organisms as disposable containers for the genes they contain. None of us will live forever, but our genes, mixed with those from other people, might survive long after we are forgotten.

SEXUAL REPRODUCTION IN ANIMALS

Sexual reproduction involves the fusion of **gametes**, or sex cells. Female sex cells, called **egg cells** or **ova**, fuse with the male gametes, called **spermatozoa** or **sperm**, in a process known as **fertilisation**. The resulting cell, the **zygote**, develops into a new individual.

In sexual reproduction, the genetic material of two different individuals is mixed and combined to produce an individual that is genetically different from either parent. This produces variation within a population. All individuals, unless they have an identical twin, are **genetically unique**: different from all other individuals in the group.

Reproduction in mammals

Mammals show a great range of reproductive strategies that reflect their circumstances, but the basic life cycle of most mammals follows a general pattern. After mating, the fertilised egg cell(s) develop inside the uterus, nourished by the placenta. The **gestation period** of a mammal, the time between fertilisation and birth, often reflects the metabolic rate of the organism. It is shorter in small, short-lived mammals such as mice, and longer in large, long-lived mammals such as elephants.

Other factors complicate the length of gestation. For example, it is also limited by the size of the fetal skull (which must fit through the female's pelvis at birth) and the mobility of the mother. In species where the mother needs speed and agility to survive, the gestation period tends to be shorter. Birth is followed by a period when the mother gives close protection and suckles her young. The length of upbringing varies greatly: it depends on factors such as degree of maturity at birth and the amount of learning required to survive.

REMEMBER THIS

The need to reproduce, so passing genes on to the next generation to ensure the survival of the population, is one of the basic features of living organisms. In many organisms, the urge to reproduce can override the urge to live.

Variation is discussed in Chapter 22.

REMEMBER THIS

Diploid cells contain two sets of chromosomes, haploid cells contain only one set. In animals, **somatic** (body) cells are diploid while gametes (sex cells) are haploid.

The onset of sexual maturity marks the change from **juvenile** to **adult**. Most female mammals conceive as soon as they become sexually mature and enter their first **oestrus** (season). From then on, throughout their reproductive life, they are either pregnant, feeding young or both. A female's reproductive life is usually brought to an end only by death. This has been the harsh fact for humans, too, until fairly recently. Even now, in societies where contraception is either unavailable or not used for cultural reasons, many women reproduce continuously and only a minority of them experience old age.

The role of hormones in mammalian reproduction

For many animals, the timing of reproduction is vital. If there is a severe variation in the seasons, animals must ensure that their offspring are born when food is plentiful and conditions are mild. **Hormones** that control and synchronise reproductive events are produced by three different organs: the **hypothalamus**, the **pituitary gland** and the **gonads** (sex organs).

The hypothalamus is a major point of contact between the nervous and **endocrine** (hormone) systems. Internal and external stimuli reach the hypothalamus, which responds by releasing hormones. They, in turn, control the rest of the endocrine system (Fig 20.2).

Day length is a classic example of an external stimulus. When the hours of daylight reach a threshold level, perhaps indicating the arrival of spring, the hypothalamus responds by releasing a hormone called a **gonadotrophin releasing factor** (**GnRF**). This controls the activity of the nearby pituitary gland, stimulating it to release hormones called **gonadotrophins**.

Gonadotrophins have a direct effect on the gonads (the ovaries and testes). These respond by releasing **steroid sex hormones** such as **oestrogen**, **progesterone** and **testosterone**.

> **Control of the timing of reproduction in sheep is discussed in Chapter 6.**

> **The hypothalamus is discussed in Chapter 16.**
> **Steroid hormones are discussed in Chapter 3.**

? QUESTION 1

1 The hypothalamus controls the level of sex hormones via a negative feedback mechanism. Suggest how the hypothalamus would respond to a fall in oestrogen levels.

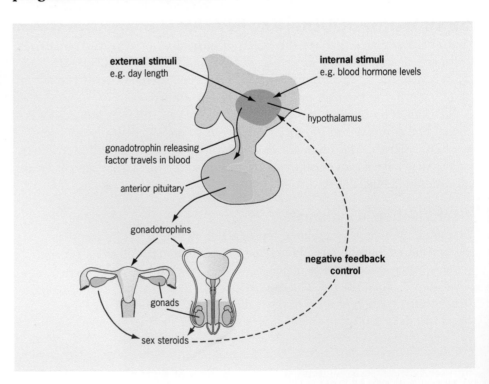

Fig 20.2 The hormonal control of reproduction. The hypothalamus can respond to external and internal nervous stimuli and is also sensitive to hormone levels in the blood. The hypothalamus controls the levels of several different circulating hormones using a negative feedback mechanism

3 HUMAN SEXUAL REPRODUCTION

Sexual activity in humans has evolved to achieve more than just fertilisation. Although humans become sexually mature in their early to mid-teens, many societies have cultural and legal controls aimed at restricting sexual activity before the late teens. Although beyond the scope of this book, it is interesting to consider whether human sexual behaviour may have evolved to strengthen the emotional bond between a couple. This 'social cement' may make it more likely that couples will stay together and so provide a stable environment for raising children.

PUBERTY

Puberty is the time between childhood and adulthood: it marks the process of sexual maturing. We still don't know exactly what triggers the onset of puberty, but we know that, early on, GnRF is secreted from the hypothalamus (Fig 20.2). GnRF travels the short distance to the anterior pituitary, where it stimulates the release of gonadotrophins, **follicle stimulating hormone (FSH)** in females and **interstitial cell stimulating hormone (ICSH)** in males (Fig 20.3). These two hormones are chemically identical, but have different names because they have different effects in the two sexes.

In girls, FSH targets the ovaries, which it stimulates to produce **oestrogen**. This steroid hormone is responsible for many of the female sex characteristics. In addition, oestrogen stimulates the ovaries to start releasing egg cells (ovulation) and this leads to the first monthly period, or **menstruation**, an event known as the **menarche**.

To start with, periods tend to be irregular and unpredictable, and may sometimes occur without ovulation. Within about a year, hormone levels have increased to the point where they stimulate the regular development of follicles. This makes periods more regular and, unfortunately for many girls, painful. Period pain is mainly due to the hormone progesterone, which causes uterine cramps.

In boys, ICSH targets the **interstitial cells** of the testes. These are embedded in the connective tissue between the seminiferous tubules – (Fig 20.4b, overleaf). ICSH stimulates these testis cells to secrete **testosterone**, the steroid hormone that stimulates development of male sex characteristics.

The age of onset of puberty

The average age for the onset of puberty is 12 to 13 in girls, 13 to 15 in boys. Interestingly, this is much earlier than in previous centuries. Two hundred years ago, the average age of the menarche in girls was 16, around four years later than it is now. The reason for this is almost certainly the improvement in diet. A better diet enables us to grow faster and so reach the same stage of maturity at an earlier age. In girls, the proportion of body fat appears to be important: a girl who is a dedicated athlete (a gymnast for example) and has a high muscle:fat ratio may find that the menarche is delayed. Also, girls who have started their periods but who then crash diet and lose a lot of weight may find their periods stop for a while.

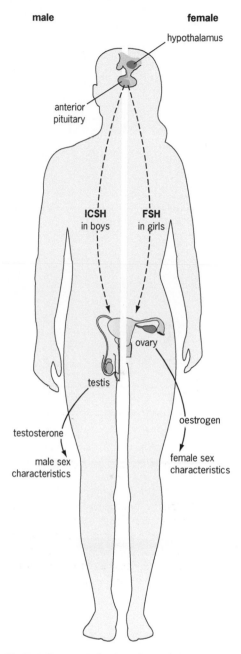

Fig 20.3 The onset of puberty is controlled by hormones. Testosterone causes the penis, scrotum and testes to enlarge and mature. It also causes enlargement of the **larynx** (voice-box) which deepens the voice. Body hair is more extensive, with coarse hair appearing on the chest and face. Ironically, in later life, the same hormone leads to male-pattern baldness. A man's shoulders and chest tend to be large and there are often changes in facial structure, with the nose and chin becoming more prominent. The muscles tend to be better developed and generally the male has greater physical strength than the female.

Oestrogen causes the ovaries, oviducts, uterus and vagina to mature and brings about the development of secondary sex characteristics. Breasts grow, pubic and underarm hair grows, the shape of the pelvis changes and the body fat is redistributed so that a woman tends to have wider hips and a more curved shape than a man

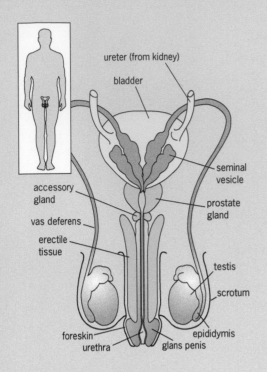

ureter (from kidney)
bladder
accessory gland
vas deferens
erectile tissue
seminal vesicle
prostate gland
testis
scrotum
foreskin
urethra
glans penis
epididymis

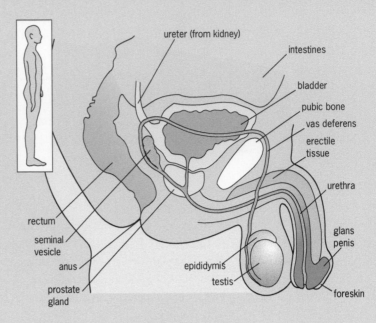

ureter (from kidney)
intestines
bladder
pubic bone
vas deferens
erectile tissue
urethra
glans penis
foreskin
testis
epididymis
rectum
seminal vesicle
anus
prostate gland

Fig 20.4a The main parts of the male reproductive system, seen from the front and the side

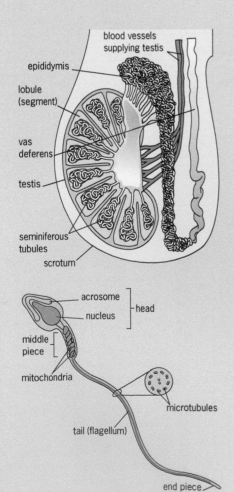

blood vessels supplying testis
epididymis
lobule (segment)
vas deferens
testis
seminiferous tubules
scrotum

acrosome
nucleus
head
middle piece
mitochondria
microtubules
tail (flagellum)
end piece

spermatogonia
spermatocytes
spermatids

lumen of tubule
Sertoli (nurse) cells
tails of sperm

Fig 20.4b The microscopic structure of the testis showing a section through a seminiferous tubule. The testis contains many seminiferous tubules. Inside each of these at any time, many thousands of sperm are maturing

THE MALE REPRODUCTIVE SYSTEM

The overall structure of the male reproductive system is shown in Fig 20.4. The male system secretes testosterone, makes and stores sperm and delivers them into the female's body.

Spermatozoa are made in the testes, a pair of organs that are held in a pouch of skin called the **scrotum**. It may seem odd that such delicate and vital organs are relatively unprotected outside the body, but there is a reason for this. The process of sperm production, **spermatogenesis**, is most efficient at around 35 °C, two degrees cooler than the core of the body. Men whose testes do not descend into the scrotum cannot produce healthy sperm.

The penis

The penis introduces sperm into the female's body so that internal fertilisation can occur. Some animals do not have a penis – most birds and reptiles, for example – and their attempts at fertilisation are a little more haphazard. The male produces semen from his genital opening and simply rubs it on to the female's genital opening.

The testes and spermatogenesis

The production of sperm in human males is a continuous process, beginning at puberty and continuing well into old age: men in their 90s have fathered children. Spermatogenesis centres around the process of meiosis, and occurs at a remarkable rate: over a thousand human sperm are made every second. Look again at Fig 20.4b and you will see that each testis is composed of a series of **lobules**. They contain **seminiferous tubules**, the structures in which sperm production takes place. This process, shown in Fig 20.5, has three main phases:

- **Multiplication**. As large numbers of sperm are needed, cells of the **germinal epithelium** divide by mitosis to produce many **spermatogonia** (sometimes called sperm mother cells).
- **Growth**. The spermatogonia grow into **primary spermatocytes**. At this stage the cells are still diploid ($2n$).
- **Maturation**. The diploid primary spermatocytes undergo meiosis. After the first division they become **secondary spermatocytes** and when meiosis is complete they have become haploid **spermatids**. In the final part of the maturation process, spermatids differentiate into the familiar **spermatozoa** (sperm).

Throughout their development, sperm cells are closely associated with **Sertoli** or **nurse cells**, from which they obtain nutrients. In the lumen of the seminiferous tubule (Fig 20.4b) – the tails of the spermatozoa are clearly visible: their heads are attached to Sertoli cells.

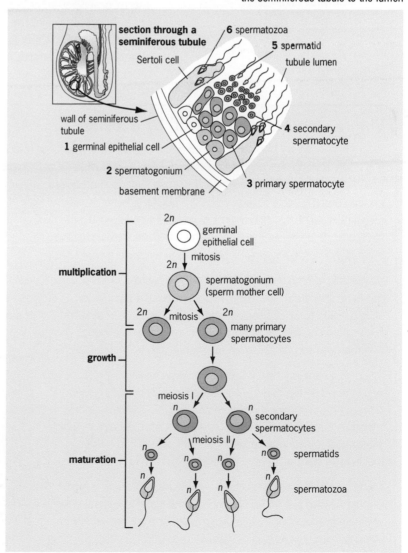

Fig 20.5 The process of spermatogenesis. Sperm cells develop as they pass from the outer wall of the seminiferous tubule to the lumen

> **? QUESTION 2**
>
> 2 What do you think are the advantages of producing small numbers of non-motile female gametes and large numbers of motile male gametes?

The parasympathetic nervous system is discussed in Chapter 16.

SEXUAL AROUSAL IN MALES

Men become sexually aroused by thinking about sex, or as a result of physical stimulation, or a combination of both. Nerve impulses from the brain pass down parasympathetic nerves and cause arterioles leading to the penis to dilate. The penis receives more blood than can drain away, spongy **erectile tissue** in the shaft becomes filled with blood, and an **erection** results.

Flaccid (non-erect) penises vary greatly in size, largely depending on how much blood is retained in the spongy tissue. When erect, about 90 per cent are between 14 and 16 cm long. The end of the penis, the **glans**, is particularly sensitive, and continued stimulation from rhythmic thrusting eventually leads to a series of reflexes known as **ejaculation**. Stored spermatozoa are propelled along the **vas deferens** by powerful peristaltic waves. As they pass various accessory glands, different secretions are added to the sperm and the final ejaculate is a milky fluid called **semen** (Table 20.1).

Human males ejaculate, on average, about 5 cm^3 of semen, which contains between 50 and 200 million sperm. Most sperm never get anywhere near an egg cell, even after unprotected sex. For most of the monthly cycle, the cervix is blocked by a plug of mucus that sperm cannot penetrate. Only at around ovulation time does the mucus consistency change, allowing sperm to pass through easily. An infection can also cause a blockage of the Fallopian tubes which will prevent fertilisation.

Table 20.1 The constituents of semen

Gland	Secretion	Purpose
seminal vesicles	fructose	energy for spermatozoa
	mucus	lubrication
	protein	forms clots, which alter consistency of semen
	prostaglandins	stimulate peristalsis in the female system
prostate	alkaline chemicals	neutralise acid in vagina
	clotting agent	clots the protein from the prostate, forming a gelatinous mass
Cowper's gland	clear fluid	cleans urethra prior to ejaculation

HOW SCIENCE WORKS

Viagra – a drug discovered by accident

Viagra is now a very popular drug that treats impotence in men (Fig 20.6). It stimulates an erection in men when they want to have sex. Men fail to have erections because they don't produce enough of a molecule called cyclic GMP (cGMP). Normally, sexual stimulation causes large amounts of cGMP to be released into the blood. This relaxes the smooth muscle in the arterioles at the base of the penis, allowing blood to flow in to cause an erection.

Viagra mimics the shape of the cGMP molecule, which is the natural substrate of the enzyme phosphodiesterase. When Viagra binds to phosphodiesterase, the enzyme is unable to break down cGMP, allowing it to accumulate in the blood. This has the desired effect and an erection is achieved.

Viagra is a relatively new drug and it has an interesting background. In the mid 1980s, researchers were looking for a drug to treat high blood pressure and angina. They tried to find a molecule that would inhibit the enzyme phosphodiesterase, which would lead to an increase in the amount of available cGMP, resulting in vasodilation and so relieving the symptoms of angina.

Many potential small molecules were screened and the most promising was put into a Phase I clinical trial. No serious side-effects were found in healthy volunteers and a Phase II trial in patients with severe angina was started. Unfortunately, the drug failed to show much effect. However, at the same time further Phase I trials were carried out using higher doses of Viagra. This time there were side-effects – the male volunteers in the trial began having frequent erections.

The drug was then trialled for its potential to treat male impotence. In the first trial, in 1994, 10 out of the 12 patients reported an improvement in erectile function. Further trials were carried out on a larger sample, and about 90 per cent of the patients reported an improvement. The drug was finally licensed in 1998.

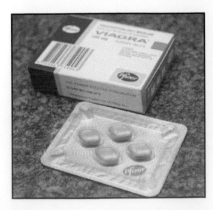

Fig 20.6 Viagra, sildenafil citrate, is now one of the fastest selling drugs in the world

THE FEMALE REPRODUCTIVE SYSTEM

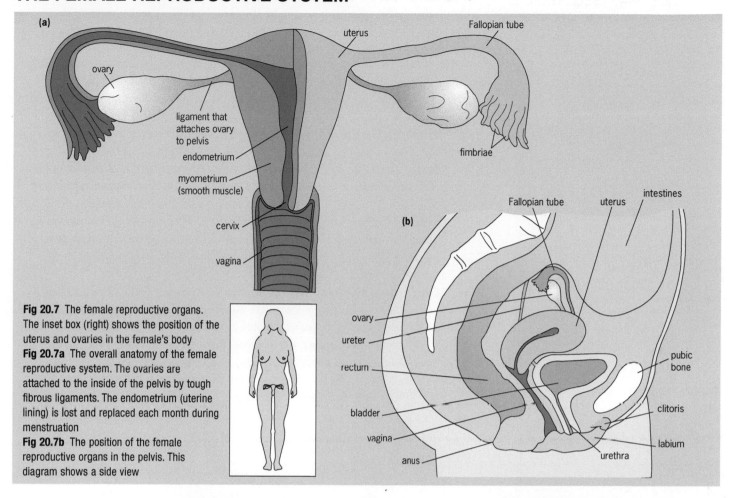

Fig 20.7 The female reproductive organs. The inset box (right) shows the position of the uterus and ovaries in the female's body

Fig 20.7a The overall anatomy of the female reproductive system. The ovaries are attached to the inside of the pelvis by tough fibrous ligaments. The endometrium (uterine lining) is lost and replaced each month during menstruation

Fig 20.7b The position of the female reproductive organs in the pelvis. This diagram shows a side view

Fig 20.7 shows the structure of the female reproductive system. This includes a pair of **ovaries**, which are the primary female sex organs. These produce egg cells, or **ova**, and also secrete the hormones **oestrogen** and **progesterone**. Each ovary is about 3 to 4 cm across and is attached to the inside of the pelvis by a **ligament** (a tough band of connective tissue). The oviducts, or **Fallopian tubes**, connect the ovaries to the **uterus** (womb). Each tube ends in finger-like **fimbriae**, which move close to the ovary at the time of ovulation.

The uterus is a compact muscular organ that nourishes, protects and ultimately expels the fetus. The human uterus is able to expand from about the size of a small orange with a capacity of about 10 cm³ to accommodate a full-term baby. This is a 500-fold increase in the capacity of the uterus. The bulk of the uterine wall is made from smooth muscle and is known as the **myometrium**. The lining of the uterus, the **endometrium**, consists of two layers. The underlying layer is a permanent **basement membrane** which produces the surface layer, the **decidua**. This layer is built up every month and shed during **menstruation**.

The **cervix**, or neck of the uterus, is a narrow muscular channel that is usually blocked by a plug of mucus. During sexual arousal the muscles of the **vagina** (a muscular tube that leads to the outside of the body) relax and glands in the vagina secrete lubricating mucus. This allows the male's penis to enter without discomfort. During childbirth, the cervix dilates to around 10 cm in diameter to allow the baby to pass through.

? QUESTION 3

3 An egg cell, the female gamete is much larger than its male counterpart, the sperm. How do the amounts of DNA in the nucleus of an ovum and sperm compare?

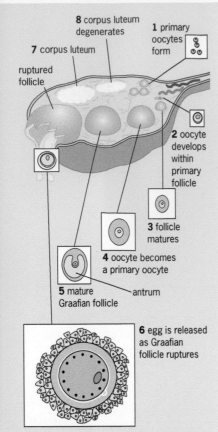

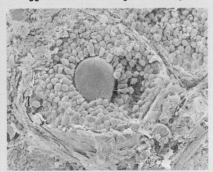

Fig 20.8a Section through an ovary showing the egg cell at different stages of development

Fig 20.8b Scanning electron micrograph of a primary oocyte showing an egg surrounded by follicle cells

Oogenesis

Throughout this section we refer to the female gamete as an **egg cell**. The egg cell is surrounded by several layers of cells and the complete unit is called a **follicle**.

The production of egg cells, **oogenesis**, takes place within the ovaries of the developing female fetus. At birth, a girl already has about 2 million **primary oocytes**. Most of these degenerate during childhood and by puberty there are only about 200 000 left. Of these, only about 450 ever mature fully – one per month throughout the female's reproductive life. As in spermatogenesis, the process of oogenesis (Figs 20.8 and 20.9) is divided into three phases:

- **Multiplication.** As the female embryo grows, **primordial germ cells** in the **epithelium** (outer layer) of the ovary go through a series of mitotic divisions to produce a population of larger cells called **oogonia**.
- **Growth.** Oogonia move towards the middle of the ovary where they grow and go through further mitotic divisions to become **primary oocytes**. Each oocyte is surrounded by a layer of follicle cells. Together they form a **primary follicle**.
- **Maturation.** From puberty onwards, a few primary follicles mature each month. Usually, only one completes its development, the rest degenerate. The remaining primary follicle grows larger, becoming an **ovarian follicle**. Its cells secrete **follicular fluid**, producing droplets which join together to form a fluid that fills the space known as the **antrum**. The mature follicle, the **Graafian follicle**, is almost 1 cm in diameter. It protrudes from the wall of the ovary just before ovulation.

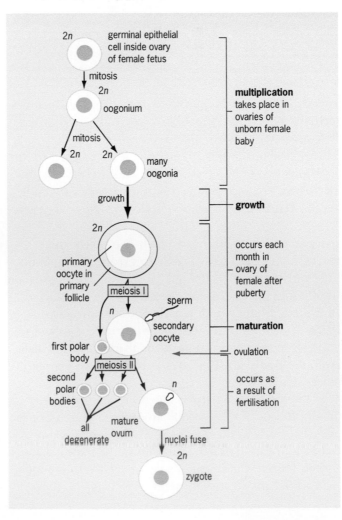

Fig 20.9 The process of oogenesis begins in the ovaries of a developing fetus but is not completed unless the egg cell is fertilised

Inside the developing follicle, the oocyte begins its first meiotic division. There is no need for more than one egg cell, so the second set of chromosomes formed at meiosis I is discarded, passing into a small cell (with very little cytoplasm) known as the **first polar body**. This appears to have no function, but it often completes the meiotic division, producing two similar cells; both later break down. After meiosis I, the egg cell is known as a **secondary oocyte**. It then begins the second meiotic division but gets no further than metaphase. The division is completed only if the egg cell is penetrated by a sperm.

When penetration of the egg cell occurs, meiosis II is completed and the egg cell becomes the mature **ovum**. This produces another 'spare' set of chromosomes, the **second polar body**, a cell that also degenerates.

> ### REMEMBER THIS
> Meiosis produces four daughter cells. In oogenesis, one ovum is formed; the other three cells degenerate.

SCIENCE IN CONTEXT

Contraception

The twentieth century has seen a revolution in 'family planning'. People can regulate the number and spacing of their children by using one or more of the wide range of effective methods of contraception now available (Table 20.2).

Table 20.2 The efficiency of the commoner methods of contraception

Method	Mode of action	Failure rate (pregnancies per 100 women/year)
the pill	prevents ovulation	0–3
IUD (coil)	prevents implantation	0.5–6
morning after pill	prevents egg release; alters uterine lining so a fertilised egg cannot implant	10
condom	prevents sperm from reaching cervix	3–20
diaphragm + spermicidal gel	prevents sperm from reaching cervix, gel kills sperm	3–25
spermicidal gel alone	kills sperm	3–30
coitus interruptus	penis withdrawn before ejaculation	10–40
the rhythm method	abstain from sex near to time of ovulation	15–35
vaginal douche	washes sperm out of vagina	80
male sterilisation (vasectomy)	vas deferens cut so that semen contains no spermatozoa	0–0.15
female sterilisation (tubal ligation)	prevents egg from passing into uterus	0–0.05
nothing	no contraception at all – a control for the other methods	85

'The pill'

The pill – or, to use its full name, the combined oral contraceptive pill – has been used widely since the 1960s. It contains a mixture of artificially produced oestrogen and progesterone. The pill is effective because it raises the blood hormone levels to those encountered during the secretion phase of the menstrual cycle, or those of early pregnancy. The high level of progesterone inhibits the secretion of FSH, preventing the development of new follicles and therefore egg cells. The high progesterone levels also inhibit menstruation, so the pill is usually taken for 21 days and then not taken for 7 days. This allows menstruation to take place but does not leave enough time for any follicles to develop. Strictly speaking, this is not normal menstruation but a 'withdrawal bleed' caused by the drop in progesterone.

The 'mini-pill'

This variation of the pill contains low doses of hormones, not enough to prevent ovulation, but enough to change the consistency of the mucus at the cervix. Instead of becoming permeable to sperm at ovulation, the mini-pill causes continued secretion of the thick, non-ovulatory mucus, which prevents sperm getting through.

The morning after pill

An emergency contraceptive that can be taken within 72 hours of unprotected sex to minimise the risk of pregnancy. It works best when taken within 24 hours. The pills contain the female hormone called levonorgestrel – which is an ingredient of ordinary contraceptive pills. This prevents the ovaries from releasing an egg and it also alters the lining of the womb, so a fertilised egg cannot implant. This is a drastic method of contraception and it causes bad nausea and vomiting – it is a last resort rather than a regular method.

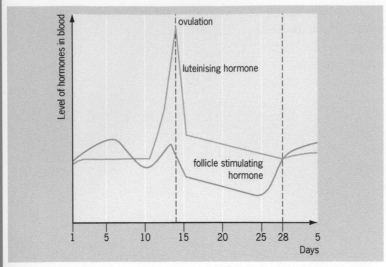

Fig 20.10a How levels in luteinising hormone and follicle simulating hormone change during the menstrual cycle

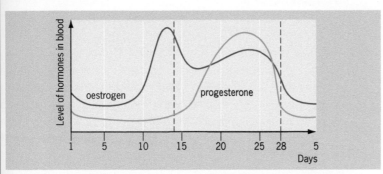

Fig 20.10b How oestrogen and progesterone levels change during menstruation

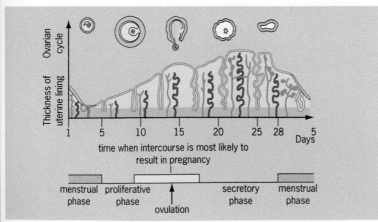

Fig 20.10c Stages of follicle development and changes in the thickness of the endometrium. Note that pregnancy is more likely to occur after unprotected sex between days 9 and 17, in a cycle where ovulation occurs on day 14

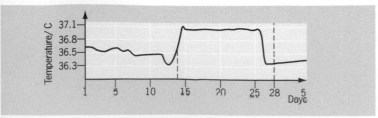

Fig 20.10d A woman's body temperature rises just after ovulation. In couples finding it difficult to conceive, taking daily temperature measurements is a good way to pinpoint ovulation

Ovulation

On around day 14 of the human female's menstrual cycle, an egg cell (a secondary oocyte) is released from the ovary in a process called **ovulation**. Pressure in the antrum builds up and ruptures the Graafian follicle, forcing the egg cell out. The egg cell is released into the body cavity, but very few get lost since the fimbriae of the Fallopian tubes hover close to the ovaries, and the ciliated lining of the tubes creates a current that gently sucks in the egg cells.

Remarkably, a woman can become pregnant even if she has lost an ovary on one side and a Fallopian tube on the other. This suggests that the functioning fimbriae actively seek out the productive ovary by moving right across to the other side of the woman's body.

THE HUMAN FEMALE MENSTRUAL CYCLE

The average length of the human menstrual cycle is 28 days, but variations from 24 to 35 days are normal, and greater variations are not uncommon. The cycle is divided into four phases, Fig 20.10c:

- the **proliferative phase** in which the endometrium regenerates;
- the **ovulation phase** in which the ovum (egg cell) is released;
- the **secretory phase** in which the endometrium secretes nutrients in preparation for implantation;
- the **menstrual phase** in which the endometrium is shed from the body.

By convention, the first day of the period (the most obvious event) is called day 1 of the menstrual cycle. Fig 20.10 shows the timing of the phases listed above. It is important to relate the events of the menstrual cycle to the hormonal changes shown. You will need to refer to these diagrams as you read the following text.

The proliferative phase

On day 2 of the cycle, the pituitary gland releases **follicle stimulating hormone (FSH)**. This stimulates the development of several ovarian follicles. At around day 6, one of the follicles dominates and begins to secrete oestrogen. The others degenerate. The remaining follicle develops into a **Graafian follicle** and continues to secrete oestrogen until day 14 of the cycle. Consequently, blood oestrogen levels rise. Oestrogen causes **proliferation** (growth) of the endometrium to replace the layer lost during the previous menstruation. After 14 days, the repair is complete.

The ovulation phase

On day 12 to 13, the blood oestrogen level reaches a threshold level, which triggers the release of **luteinising hormone (LH)**, from the anterior pituitary gland. The rapid increase in LH levels triggers ovulation around day 14. The ovum lives for only 24 to 36 hours. During this time, it moves only a few centimetres from the ovary.

The secretory phase

Luteinising hormone has a second effect: it causes the Graafian follicle to develop into a **corpus luteum** ('yellow body'). The name comes from the yellow appearance of the secretory cells that develop inside the 'remains' of the Graafian follicle. The corpus luteum secretes oestrogen and progesterone. Progesterone causes spiral-shaped blood vessels to grow into the endometrium. This thickened lining begins to secrete nutrients and mucus to prepare for an embryo to be implanted. During this phase, the high levels of progesterone inhibit the production of FSH. As long as progesterone levels are high, the endometrium is maintained and no new follicles are stimulated. The 'contraceptive pill' takes advantage of this inhibition (see Science in Context box on page 307).

If the egg cell is not fertilised, the corpus luteum lasts for about 10 to 12 days and then degenerates, ceasing to secrete progesterone. This is a key event because the inhibition of FSH is lifted. The endometrium is no longer protected and the cycle can start again.

The menstrual phase

The drop in progesterone and oestrogen levels causes the uterine capillaries to rupture, and the endometrium is lost from the body through the cervix, together with some blood. Renewed secretion of FSH begins around day 2, and the cycle begins again.

> **? QUESTION 4**
>
> 4 Many women do not have a 28-day cycle. Although the second half of the cycle (ovulation to menstruation) is usually 14 days, the first half can vary considerably. If a woman's cycle is 35 days, on which day is she likely to ovulate?

SCIENCE IN CONTEXT

The menopause and hormone replacement therapy

The menopause is a natural event that occurs when the ovaries stop working. For most women, this happens between the ages of 45 and 54. This time is often difficult and traumatic. The lowered levels of oestrogen and progesterone are directly responsible for the unpleasant symptoms of the menopause:

- circulatory problems such as hot flushes and night sweats;
- psychological problems, such as depression, anxiety and insomnia;
- skeletal problems. Oestrogen inhibits reabsorption of bone, and after the menopause, bone loss can be as great as 7 per cent per year. This condition, osteoporosis, affects the spongy bone particularly and the sufferer is more likely to break a bone;
- oestrogen is also thought to give women some protection against some types of heart disease. Women below menopausal age are less likely than men to have heart disease, but afterwards they catch up.

Hormone replacement therapy can help to reverse many of these symptoms and effects. The basic idea behind HRT is simple: to restore hormone levels to those of the early follicular phase of the menstrual cycle using tablets, implants or transdermal patches (Fig 20.11).

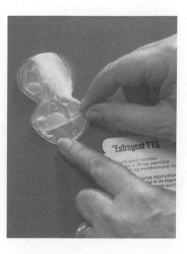

Fig 20.11 One method of administering the hormones in HRT: a transdermal patch (a patch attached to the skin) that releases controlled amounts into the bloodstream. The hormones used in HRT are extracted from the urine of pregnant mares or made artificially

THE EVENTS OF PREGNANCY

If the egg cell is fertilised, the menstrual cycle is interrupted and the female's body changes in response to the events of pregnancy.

Fertilisation

Fertilisation is a complex sequence of events that begins when the sperm reaches the egg cell (oocyte) as shown in Fig 20.12.

It may also help you to look back at Fig 20.9. The secondary oocyte is surrounded by several layers of follicle cells, the **corona radiata**, and a layer of glycoprotein, the **zona pellucida**. Before a sperm can penetrate the layers surrounding the oocyte, it undergoes a process called **capacitation**. The actual mechanism is poorly understood, but it seems to set off the **acrosome reaction**, which allows the sperm head to enter the oocyte. The **acrosome** is a bag of digestive enzymes on the tip of the sperm head (Fig 20.12). During the acrosome reaction, the bag splits, releasing the enzymes, which digest a pathway through any remaining follicle cells and the zona pellucida.

As soon as the outer membrane of the first sperm penetrates the cell surface membrane of the oocyte, a rapid reaction occurs. Many **cortical granules** fuse with the zona pellucida, forming a **fertilisation membrane**. This reaction starts at the point of entry of sperm head and spreads rapidly over the surface of the oocyte, preventing entry of other sperm. So, only one sperm enters the diploid secondary oocyte, even though many reach it at the same time. Entry of the sperm nucleus triggers the completion of meiosis II in the female nucleus, leading to the formation of the second polar body and a haploid mature ovum.

Almost immediately afterwards, a spindle forms and the paternal and maternal chromosomes come together, forming a diploid zygote. Within 12 hours, the first mitotic division takes place. Cell division is now rapid, forming a bundle of cells called a **morula** (Latin for the blackberry it resembles). As divisions continue, this becomes a **blastocyst** and moves slowly along the Fallopian tube through the action of cilia, which create a steady current of fluid towards the uterus (Fig 20.13).

It takes around 6 or 7 days to complete the journey down the Fallopian tube. When the blastocyst arrives at the uterus on day 21 of the cycle, the lining must be in just the right condition to accept it. For a successful pregnancy there must be exact timing between the preparation of the endometrium and the development of the embryo.

Implantation

Pregnancy begins not with fertilisation but when the embryo **implants** in the wall of the uterus. This happens about a week after fertilisation and is not always successfully completed. Many women trying to conceive may have a 'near miss', when an egg is fertilised but fails to implant.

Implantation begins when the blastocyst makes contact with the endometrium (Fig 20.13), usually on the back wall. The outer layer of the blastocyst, the **trophoblast**, causes an **inflammatory-type response** (normally a response to damage) and this causes an outgrowth of the endometrium at this point. The placenta develops where the trophoblast and the endometrium interact.

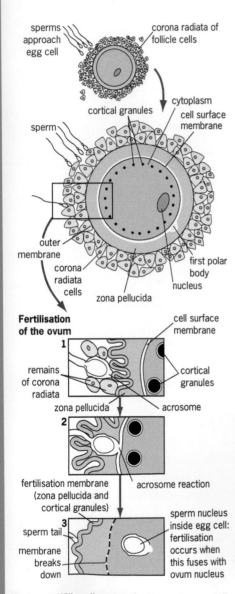

Fig 20.12 With a diameter of 100 μm, the egg cell is the largest cell in the human body. Most of the volume is taken up by inert food material that will fuel the embryo during its first few cell divisions. A spermatozoon must pass through any remaining follicle cells (the corona radiata) and the zona pellucida, before it can fertilise the egg cell. The head of the sperm enters the body of the egg cell, but the tail remains outside

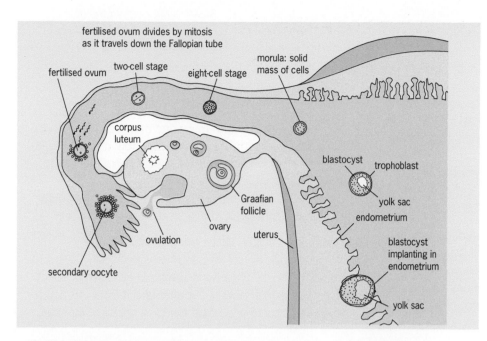

Fig 20.13 Following fertilisation, the embryo undergoes several mitotic divisions as it passes slowly down the Fallopian tube. When the embryo reaches the uterus, it implants in the endometrium

If implantation is successful, the embryo begins to secrete **human chorionic gonadotrophin** (**hCG**). This hormone forces the corpus luteum in the ovary to continue to secrete progesterone, thereby maintaining the endometrium and inhibiting FSH production.

The **chorion**, one of the membranes that later grows and surrounds the embryo, develops **villi** (projections) that burrow into the endometrium. These are thought to break down the mother's blood vessels, causing the chorionic villi to become bathed in maternal blood.

Human chorionic gonadotrophin is discussed in Chapter 6.

THE PLACENTA

The placenta is a temporary organ that allows the blood systems of the fetus and the mother to come into close contact, without actually mixing (Fig 20.14). The placenta allows nutrients and oxygen to pass to the fetus from the mother, and allows metabolic waste back into the mother's blood (see Table 20.3 on page 314).

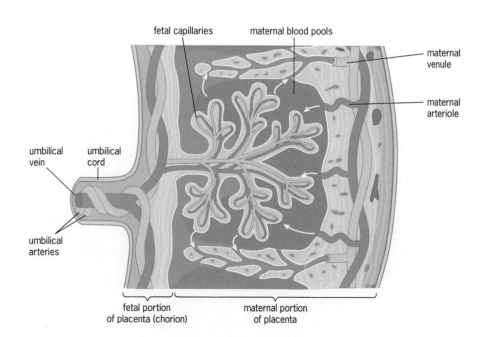

Fig 20.14 The fine structure of the placenta. The capillaries within the chorionic villi are bathed in pools (lacunae) of maternal blood. The placenta is a large disc-shaped organ that is usually attached to the back wall of the uterus. At birth, contraction of the uterine muscle causes the chorionic villi to split away from the endometrium and the placenta is delivered after the baby: hence its common name of afterbirth. Most mammals eat the placenta after the birth of their young. It is an important source of nourishment at a time of great need

Infertility and its treatment

Nine out of ten couples that use no contraception and actively try to conceive are successful within a year. But one in ten couples may face the distressing problem of infertility. Many are 'sub-fertile' rather than infertile, and can be helped by the various methods described here. However, for a few, nothing works. Remaining childless when you desperately want a family can be very traumatic and couples may need help and counselling to come to terms with their inability to reproduce.

There are several possible causes of infertility and there are now many treatments, as a result of a great deal of research over the last 20 years. This has not been without controversy but it has helped many couples who would otherwise have been unable to conceive.

Causes of infertility

Doctors accept that a couple may need infertility treatment if they have been trying to conceive for at least a year without any success. The first objective is to establish whether either partner has an obvious problem that is preventing conception. Couples are also recommended to try to establish when exactly the woman ovulates during the month using temperature charts and looking for changes in the consistency of vaginal mucus or by using one of the over-the-counter ovulation prediction kits that are now available. Having sex at the most fertile time could be all that is needed. However, some couples still fail to conceive.

Female infertility may be caused by:

- blocked Fallopian tubes;
- altered hormone levels leading to a failure to ovulate or implant. Failure to ovulate, known as anovulatory infertility, is usually due to a failure to secrete the right balance of hormones;
- cervical mucus that halts, repels or kills sperm.

Male infertility may be caused by:

- a low sperm count. Samples that are found to have fewer than 20 million sperm per cm^3 are said to be abnormally low;
- production of large numbers of abnormal sperm (more than 4 per cent);
- production of antibodies that make the sperm stick together.

Ovulation induction

A woman sometimes does not ovulate because the balance of hormones in her body is abnormal. To restore the hormone levels and initiate follicle development and ovulation, she can have treatment with artificial gonadotrophins or drugs that stimulate the natural secretion of gonadotrophins. One such drug, clomiphene, works by increasing FSH secretion.

After such treatment, the response of the ovaries can be followed by ultrasound, which shows how many follicles are developing in each ovary. A follicle is considered to be ready for ovulation when it reaches 17 mm in diameter. The endometrium is also checked – it should be at least 8 mm thick at this time.

When a ripe follicle is detected, ovulation can be stimulated artificially by injecting human chorionic gonadotrophin (hCG), a hormone that has a similar effect to LH. Ovulation should occur after about 36 to 48 hours, and so intercourse should be timed to coincide with this.

Intra-uterine insemination (IUI)

In around 20 per cent of infertility cases, there seems to be no problem with either partner. Surprisingly, in some of these cases, intra-uterine insemination proves successful. The basic idea behind IUI is to introduce the partner's semen into the uterus.

The steps in IUI are as follows:

- follicle development is stimulated and monitored;
- treatment to induce ovulation is given;
- a fresh sperm sample is introduced into the uterus.

In vitro fertilisation (IVF)

A woman undergoing IVF treatment must endure weeks of hormone treatments, followed by the uncomfortable process of egg cell collection (Fig 20.15). Up to about 20 egg cells may be recovered. Her egg cells are fertilised by her partner's sperm '*in vitro*' and, when the tiny embryos have developed, two of these are put back into her uterus. IVF has, at best, about a 30 per cent success rate for pregnancy achieved by implantation.

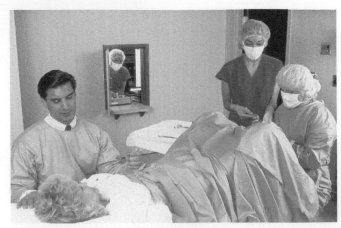

Fig 20.15 Egg collection treatment in progress

Infertility and its treatment (cont.)

In vitro (= 'in glass') fertilisation is what most people think of as 'test-tube baby' treatment. In IVF, the egg cells and sperm are taken from the couple and fertilised in a dish. The process of fertilisation normally takes 12 to 15 hours and after this the new embryos start to develop. Cell division is taken as a sign that fertilisation has been successful. Then, the tiny embryos (no more than balls of eight to 16 cells at this stage – see Fig 20.16) are placed into the uterus.

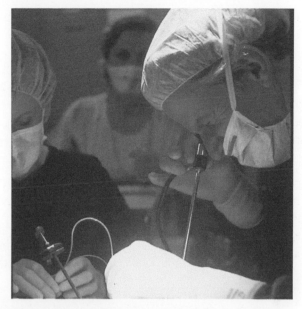

Fig 20.17 A mixture of egg cells and sperm is being inserted into the woman's Fallopian tube using a keyhole method

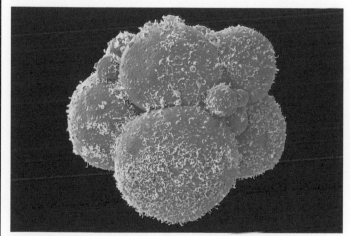

Fig 20.16 A human embryo at the eight-cell stage. The eight large cells (red) are covered with microvilli (yellow). The smaller cells will degenerate

Gamete intra-Fallopian transfer (GIFT)

GIFT involves stimulating the ovaries and collecting the egg cells in much the same way as in IVF treatment. The important difference is that in GIFT the egg cells are mixed with sperm and immediately introduced into the Fallopian tubes (Fig 20.17), without waiting to see if fertilisation occurs. The advantage of this procedure – which would seem to be less controlled than IVF – is a 5 per cent better success rate. This is possibly because the Fallopian tube is the natural site for fertilisation: it may secrete chemicals that stimulate the process.

Ethical and moral considerations

Several aspects of IVF are very controversial. The procedure generates an excess of embryos that are never implanted, but they can be frozen and stored. Very early stage human embryos are an excellent source of stem cells for stem cell research and therapy. However, many people object to human embryos being used in this way. They argue that these balls of cells are living, potential human beings, and should never be used experimentally.

Another moral dilemma is whether IVF and other infertility treatments should be available on the NHS. The procedures are expensive and some people argue that money would be better spent on, for example, cancer treatment. They also point out that the human population is already spiralling out of control, and ask if using large amounts of medical resources to add to it is justified.

The placenta is an organ adapted to maximise the exchange of materials so, as you would expect, it has a large surface area provided by the **chorionic villi**. A close look at the cells of the villi shows that the membranes are folded into microvilli and also contain many mitochondria: these two features maximise the processes of diffusion and active transport. There are many small vesicles in the cells of the villi, suggesting that substances are being absorbed by pinocytosis.

The placenta is an important **endocrine organ**. It secretes the hormones that maintain pregnancy, taking over from the corpus luteum at about 12 weeks. The placenta secretes progesterone (which maintains the endometrium), oestrogen (which inhibits the ovulatory cycle), human chorionic gonadotrophin and **human placental lactogen** (which stimulates the development of the mammary glands).

Diffusion, active transport and pinocytosis are discussed in Chapter 4.

Table 20.3 Exchange of materials across the placenta

Mother to fetus	Fetus to mother
oxygen	carbon dioxide
glucose	urea
amino acids	other waste products
lipids, fatty acids and glycerol	
vitamins	
ions: Na, Cl, K, Ca, Fe	
alcohol, nicotine, many drugs	
viruses	
antibodies	

Pregnancy, labour and birth

In humans, pregnancy lasts for an average of 40 weeks. Some of the main stages of growth of a baby are shown in Fig 20.18.

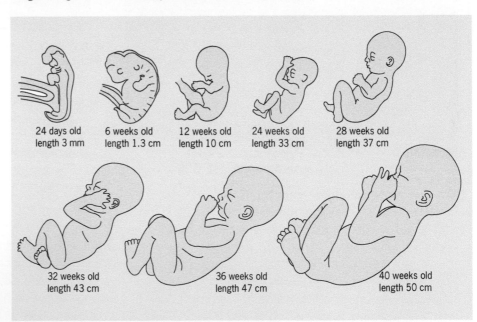

24 days old
length 3 mm

6 weeks old
length 1.3 cm

12 weeks old
length 10 cm

24 weeks old
length 33 cm

28 weeks old
length 37 cm

32 weeks old
length 43 cm

36 weeks old
length 47 cm

40 weeks old
length 50 cm

Fig 20.18 Stages in the growth of a baby in the uterus

From around the 12th week of pregnancy, progesterone secreted by the placenta inhibits uterine contractions. The level of progesterone rises steadily until just before birth, usually around 38 weeks after conception, when it starts to fall dramatically. This lifts the inhibition of uterine contractions. The mother's anterior pituitary begins to secrete **oxytocin** and the placenta secretes prostaglandins, two hormones that actively promote contractions. Oxytocin stimulates uterine contractions at about 40 weeks. The resulting tension in the muscle and pressure on the cervix are stimuli that bring about further secretion of oxytocin, causing more powerful contractions.

There are three stages of labour:

● Stage 1. The cervix dilates (opens) to a diameter of 10 cm.
● Stage 2. The fetus is pushed out of the uterus (Fig 20.19).
● Stage 3. The placenta and umbilical cord are expelled (Fig 20.20).

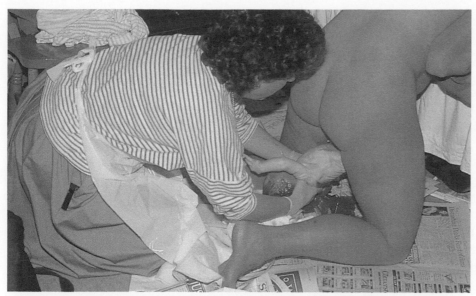

Fig 20.20 The placenta is a vital organ to a developing baby but as soon as the baby is born, the placenta becomes instantly unnecessary. Some mammals (and apparently humans) eat the placenta so that the nourishment does not go to waste

Fig 20.19 Childbirth is one of the most intense human experiences. Childbirth used to be a time of great danger to both mother and child, but advances in medicine have greatly reduced the risk. Today, in the developed world, 99 out of every 100 babies born survive beyond their first birthday (200 years ago, only 54 out of every 100 lived)

Lactation

All mammals produce milk from specialised **mammary glands**. Humans are unique in having permanent breasts but these are normally composed only of fatty tissue. Throughout pregnancy, however, oestrogen and progesterone stimulate the development of milk-producing tissue.

After birth, the first fluid released from the breasts is called **colostrum**. This contains no fat and very little sugar, but has important antibodies that 'lend' the baby immunity until it has time to develop its own. After about three to four days, normal milk is produced.

Milk is produced constantly and, although a certain amount of leaking from time to time is usual, milk flow only happens when the baby suckles (Fig 20.21). The stimulus of sucking at the nipple causes the posterior pituitary gland to secrete oxytocin. This hormone travels in the blood and causes contraction of the muscular **myo-epithelial cells** surrounding the milk glands or alveoli. This squeezes the milk out of the alveoli, through the milk ducts and into the infant's mouth. This mechanism, called the **let-down reflex**, takes several seconds to take effect but is quite powerful. If the unsuspecting infant lets go of the nipple, it can get a jet of milk in the eye.

Throughout the period of lactation, the pituitary continues to secrete **prolactin**, a hormone that maintains the milk ducts and, to some extent, inhibits ovulation. This inhibition is called **lactational anoestrus** and reduces the risk of conception so soon after birth. However, it is not always effective and breast-feeding cannot be relied on as a contraceptive method.

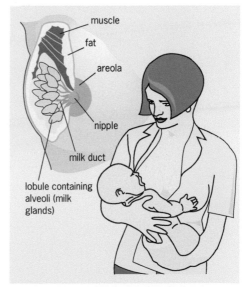

Fig 20.21 During pregnancy the milk-producing glands grow, substantially increasing the size of the breasts. When the new baby suckles, the physical stimulation leads to a series of hormonal events that increase the amount of milk that is made

SUMMARY

After reading this chapter, you should know and understand the following:

- **Sexual reproduction** involves two sexes which produce **haploid gametes**. Males produce **spermatozoa** by **spermatogenesis**, females make egg cells by **oogenesis**. Both processes involve a special type of cell division called **meiosis**.

- At fertilisation, a new **diploid zygote** is formed. The zygote, which is genetically unique, grows and develops by a series of mitotic divisions.

- The timing of reproductive events in humans is controlled by hormones. The **hypothalamus** controls the **pituitary gland**, which releases **gonadotrophins**. These stimulate the gonads to secrete steroids: **oestrogen**, **progesterone** and **testosterone**.

- The human menstrual cycle lasts, on average, for 28 days. The cycle begins with **menstruation**, which occurs as the next egg cell is prepared inside the ovary. By day 14 a new endometrium has grown. Ovulation takes place on day 14 and the egg cell is then available to be fertilised. If the egg cell is not fertilised, the endometrium breaks down and the cycle repeats.

- In humans, the egg cell is fertilised in the Fallopian tube. The **zygote** begins mitotic divisions, becoming a **morula** and then a **blastocyst** as it moves along the tube towards the uterus. When the blastocyst has implanted in the endometrium, it is known as an **embryo**.

- The developing embryo (which is called a **fetus** after about 8 weeks) is nourished via the placenta, a temporary organ that does the job of the fetal lungs, intestines and excretory system.

- Birth is brought about by a series of hormonal changes that begin uterine contractions. In mammals, the infant is nourished by milk produced by the mother.

Practice questions and a How Science Works assignment for this chapter are available at www.collinseducation.co.uk/CAS

21

Growth and ageing

21 GROWTH AND AGEING

Life is a sexually transmitted fatal condition

In the time it takes to read this sentence, 50 million of your cells will have died. This is an old biological 'fact', but serves as a stark reminder of our mortality. The brain, like all the other organs in the body, degenerates as we get older. In most people, this results in a gradual loss of memory. Many older people struggle with their short-term memory – they forget everyday things – but have very clear and lucid memories of what happened in their childhood.

Alzheimer's disease accelerates memory loss and causes a decline in many other mental abilities. Eventually, people with Alzheimer's might not remember who they are or who the people around them are, and this can lead them to behave aggressively or even violently. Few things in life can be as distressing as watching a close relative decline in this way.

The causes of Alzheimer's disease are not yet properly understood. Several years ago it was suggested that accumulation of aluminium in the brain tissue could be a contributory factor, but this has now been more or less discounted. Today, researchers are devoting more time and effort to investigating the genes that make it more likely for people to develop Alzheimer's disease. Three key genes have been identified that increase the risk of early-onset Alzheimer's disease (which starts well before the age of 60) and one particular form of the apolipoprotein E (APOE) gene increases the risk of late-onset Alzheimer's (which starts after the age of 60).

There is still no cure for Alzheimer's but drugs that inhibit the enzyme that destroys neurotransmitters in the brain help the symptoms. The Food and Drug Administration in the USA has approved a drug called donepezil, which inhibits cholinesterase, to treat Alzheimer's disease, whether mild, moderate or severe.

As the proportion of elderly people increases in many populations in the western world, there are serious implications for our societies. Older people suffer more from degenerative diseases such as Alzheimer's and will need to be cared for, stretching the resources of healthcare services

1 GETTING BIGGER AND GETTING OLDER

In this chapter, we look at how the body grows from birth until the late teens, and then we study some of the events that occur as the body starts to decline in old age.

WHAT IS GROWTH?

We all know what growth is, but we need a biological definition:

> **Growth is an increase in size brought about by the addition of more body tissue.**

Growth therefore excludes temporary changes. Women who suffer from pre-menstrual tension can retain water in their tissues and appear to 'grow' by gaining a couple of kilograms in the few days before their period, but this is not growth as such.

Human growth is **diffuse**: it occurs all over the body, not just at one growing point. A 10-year-old is not only taller than a 3-year-old but his or her organs – heart, liver, pancreas, etc. – are all larger. Not all of these organs grow at the same rate though, and in the first part of this chapter we look at **growth patterns**. It is important to understand how internal factors such as hormones and external influences such as nutrition work together to control growth.

Fig 21.1 In the wild, mammals like the Grant's gazelle rarely reach old age. Their usual fate is to be killed and eaten by carnivores or to die of disease long before this. Today, thanks to a combination of better living standards, diet, modern medicine and other factors, humans grow old and 'wear out'. Some of the same changes that we experience as we age are seen in the animals we protect, such as our pets

AGEING IN HUMANS

Many animals die long before they reach their full lifespan (Fig 21.1). Most are lucky to live to reproduce. A longer life, however, has its disadvantages. Organs degenerate and systems become less efficient. In the second part of this chapter we look at some of the unwelcome changes that affect the body as it gets older.

2 MEASURING GROWTH

Growth is three dimensional; people get taller, wider and the distance between their back and front increases. To give an accurate picture of overall growth, three basic criteria are commonly used:

- an increase in **body mass**;
- an increase in **height**;
- an increase in **supine length** (the distance between the top of your head and the bottom of your feet when lying flat on the floor).

BODY MASS

Body mass is quick and easy to measure. Body mass changes slightly, even from day to day and hour to hour – we weigh more when we have just eaten a large meal or when we have a full bladder, for example – but these fluctuations are minor. Larger gains in mass occur during childhood, and we use standardised charts so that children's sizes can be plotted to make sure that any potential problems in growth are spotted early (Fig 21.2).

Women gain weight as they go through pregnancy and, as people get older, they can gain weight by eating too much food and doing too little exercise. Homeostatic mechanisms try to keep our weight stable, but we can override this control – all too easily, judging by the epidemic of obesity currently sweeping the developed countries (see the Science in Context box on page 322).

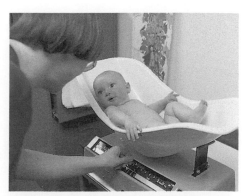

Fig 21.2 We monitor the way babies grow by weighing them regularly

Homeostasis is discussed in Chapter 11.

HEIGHT

Like body mass, height is easy to measure. The person stands straight with heels flat on the ground and facing forwards. A horizontal bar is moved down until it touches the top of the head and the height is read off from a suitably positioned scale (Fig 21.3). Height also gives a good indication of growth over time (Fig 21.4).

Using height as the only measure of body size has its disadvantages. It cannot, for example, give information about the changes in the width of the shoulders or hips, changes that are just as important as increases in height. And it's no good for assessing the growth of babies who are too young to stand.

SUPINE LENGTH

You can only measure a person's height if they are old enough to stand. In very young children, length is measured instead. The infant is placed as flat on its back as possible. The ankles are gently pulled to stretch the child and straighten its legs. The feet are turned up vertically. The supine length can now be measured. Adults are usually between 1 and 3 centimetres taller lying down than standing up because gravity causes a slight compression of the skeleton in the upright position.

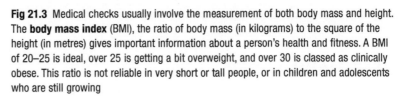

Fig 21.3 Medical checks usually involve the measurement of both body mass and height. The **body mass index** (BMI), the ratio of body mass (in kilograms) to the square of the height (in metres) gives important information about a person's health and fitness. A BMI of 20–25 is ideal, over 25 is getting a bit overweight, and over 30 is classed as clinically obese. This ratio is not reliable in very short or tall people, or in children and adolescents who are still growing

Fig 21.4 Typical height curves for girls from the ages of 2 to 18 years. The spread of values is the result of large studies that measured the growth patterns in thousands of children. The centre line is the average curve; the shaded areas show the normal range of height at each age

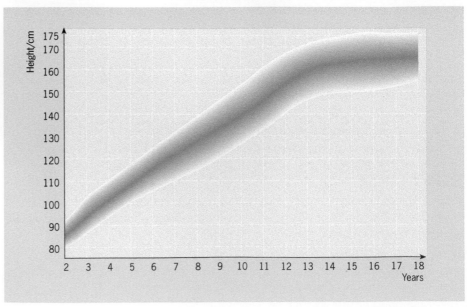

TECHNIQUES TO ASSESS AVERAGE BODY SIZE AND GROWTH

In 1759, a Frenchman, Count Philibert de Montbeillard, decided to measure the height of his newborn son. He took measurements every six months from birth until the boy was 18 years old. Collecting data like this – measuring a person or group of people many times – is called a **longitudinal study**. Studying your own child over time is easy, because they live with you, but researchers who study hundreds or thousands of people to get an idea of average growth patterns find maintaining contact with all their subjects quite difficult. People move away or forget to turn up for appointments.

An alternative approach is to use a **cross-sectional study**. This approach could be used to investigate the growth of human males between 10 and 18 years of age. The study would choose different age groups – a group of 10-year-olds, a group of 11-year-olds, and so on. Each group would be large – perhaps a hundred boys – and each of them would be measured once only. By working out the average height of each age group and then plotting a graph or age against average height, this would reveal the overall pattern of growth.

The two methods can give different information and it is important to take this into account when interpreting data. Look at the graph in Fig 21.5.

? **QUESTION 1**

1 Look at Fig 21.5. How many individuals were used in the longitudinal study?

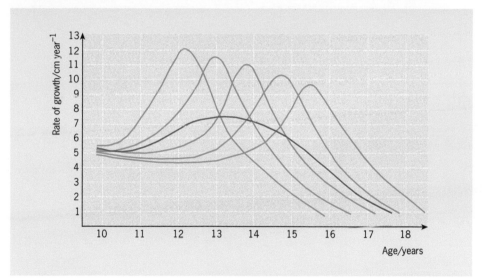

Fig 21.5 The blue curves show the rates of growth of different boys obtained from a longitudinal study. These curves differ in shape from the red curve. This curve was obtained from a cross-sectional study

The blue curves show the rate of growth in centimetres per year for five boys. This was a longitudinal study in which the height of each boy was measured at regular intervals over the full period of time. You can easily see the peak of growth that occurs at puberty. The red curve shows what happens if the average height of all the individuals in each age group is calculated and plotted – a cross-sectional study. Not really the same are they? Note how the red curve doesn't have the same 'puberty peak'.

REMEMBER THIS

The line in a cumulative growth curve can go up or flatten off; it never dips. Children get taller or they stay the same height, but they don't shrink. Growth rate, on the other hand, can decrease as well as increase. Fig 21.6b shows that a child's growth rate is higher during the first three years of life than later on, and shows the peak in growth rate that occurred in this child at puberty (see also Fig 21.5).

Representing growth

There are two main ways to represent growth: a **cumulative growth curve** and a curve showing **growth rate**. Fig 21.6 shows an example of each type of curve. Fig 21.6a, a cumulative growth curve, shows the height of a child in centimetres from birth to the age of 18. Fig 21.6b shows growth rate. This shows the number of centimetres that the child grew each year; this is calculated easily by subtracting the child's height in the previous year from the height in the current year.

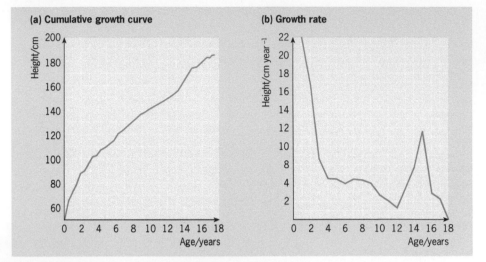

Fig 21.6 Representing the growth of a single male child: **a** is the cumulative growth curve showing changes in his total height; **b** is a graph showing his growth rate

When growth becomes a health hazard

The 2008 *Guinness Book of Records* lists Manuel Uribe as the world's fattest man (Fig 21.7). The Mexican weighs 560 kilograms (1234 pounds). Amazingly, before his weight was officially recorded, he had had stomach surgery and had lost 200 kg. He hopes to lose more to escape his bedridden existence.

You might think that to get as obese as this must take an enormous effort and a continuous eating binge. But, in daily terms, the excess dietary intake required to put on weight isn't really that much. If this man had started out as a 70 kg 16-year-old, he would have needed to put on only 37 grams per day to reach his final weight at the age of 45. That means eating just one chocolate bar or packet of crisps that he didn't burn off. Doesn't sound that difficult, does it?

Losing weight after you have gained it is, on the other hand, much more of a problem. The developed world, where obesity is prevalent, now has a booming industry based around people's need to diet. The medical profession now recognises the health dangers of obesity – diabetes, cardiovascular disease, gall stones are just a few of the disorders that are much more common in overweight people. Some dieters are more successful than others, and the press reports, from time to time, the 'success story' of someone who has lost half of their body weight.

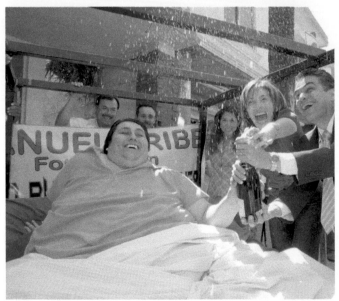

Fig 21.7 Manuel Uribe from Monterrey, Mexico. To celebrate his entry into the *Guinness Book of Records* for 2008, Uribe was loaded onto a trailer and taken for a ride through the streets of his hometown. It was one of the few times he has left his bed for 20 years

3 GROWTH PATTERNS

Different organisms show differences in their patterns of growth. A crocodile is about 15 centimetres long when it hatches from its egg and can grow to a length of over 5 metres. It increases in length because all parts of it grow at more or less the same rate. As a result, an adult crocodile has very similar body proportions to a young crocodile. Its pattern of growth is described as **isometric**. Growth in a mammal such as a human follows a different pattern. The various parts of the body grow at different rates. Mammalian growth is described as **allometric**.

THE PATTERN OF HUMAN GROWTH

Humans are mammals, so it is no surprise that our growth patterns are more like those of an elephant than a crocodile. Human growth is allometric and limited. Different parts of the body grow at different rates and we grow to a maximum height that is determined by both our genetic make-up and our environment. If we have two tall parents, we are more likely to be tall, and a child that has plenty of food grows taller than a child who is malnourished or suffers starvation.

Fig 21.8 shows the growth in height of a human female and a human male. Although the diagrams suggest that growth stops during adolescence, there is actually still a bit of growth later – but no more than 2 or 3 millimetres. Until the age of about 30, the length of the spine increases as small amounts of bone are added to the vertebrae. However, we usually say that in Europe and North America, growth stops at around the age of 15.5 years for the average girl and 17.5 years for the average boy.

? **QUESTION 2**

2 Would you expect the ratio of tail length to total body length to change as a crocodile grows? Give a reason for your answer.

✔ **REMEMBER THIS**

Different organisms also show different growing periods. Some, like the elephant, reach maturity, then stop growing. Their pattern of growth is limited. Others – the crocodile, for example – continue to grow, often slowly, throughout their lives. These organisms show unlimited growth.

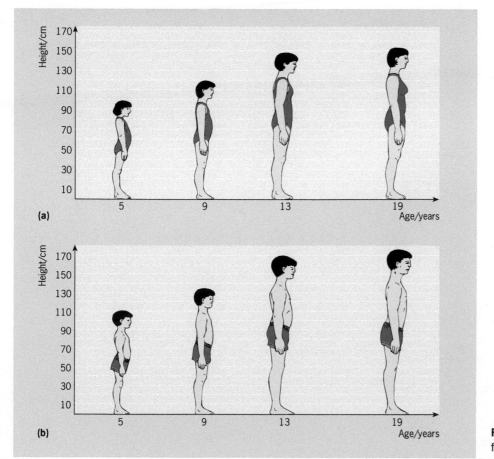

Fig 21.8 Growth in height of **a** a typical human female and **b** a typical human male

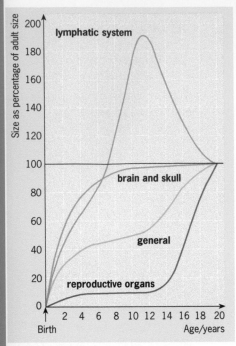

Fig 21.9 Growth curves of different parts of the human body compared with the overall growth curve

Allometric growth in humans

Allometric growth is clearly visible in Fig 21.8; the diagram shows a change in body proportions. Although organs such as the kidney and liver have a growth curve similar to that of the body as a whole, there are notable exceptions. These are shown in Fig 21.9.

The brain and skull grow very rapidly after birth, reaching 90 per cent of adult size soon after the age of 6. The lymphatic system also develops early. It has a very different growth curve from the rest of the body, reaching full size well before puberty and then decreasing. Finally, the reproductive organs grow very slowly in childhood. Maximum growth of these organs occurs during puberty.

THE CONTROL OF HUMAN GROWTH

Human growth is complex. It involves, directly or indirectly, all the systems of the body, and takes place over a long period of time, so growth obviously has to be *co-ordinated*. This is achieved by **hormones**.

Before you read this section, it would be helpful to look back at Chapter 18 and remind yourself of the ways in which hormones can affect the activities of their target cells. There are at least 12 different hormones that have a direct effect on growth at different times during a person's life. The functions of some of these are summarised in Fig 21.10.

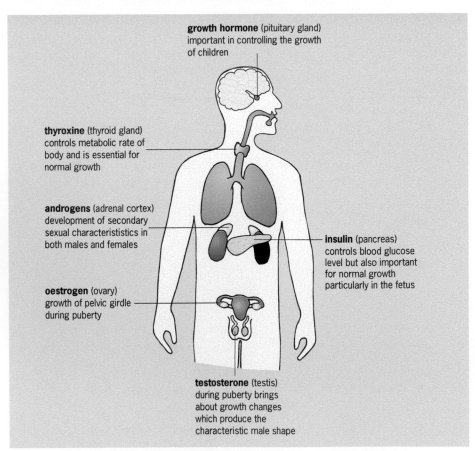

Fig 21.10 The main endocrine glands associated with human growth

Growth hormone

Growth hormone is produced by the anterior lobe of the **pituitary gland**. Human growth hormone has a rather unusual property for a mammalian hormone: it works in humans but in no other species. We say it is **species specific**.

Growth hormone is not necessary for the growth of a baby in its mother's uterus but, at birth, it takes over the control of growth. It continues to influence growth throughout childhood but then its production falls during early adulthood. Recent studies have suggested that elderly people who take growth hormone supplements are fitter and stronger, so in adults, the small amount of growth hormone that is still produced may play a role in tissue repair.

Like other hormones produced by the pituitary gland, growth hormone is secreted in pulses. For most of each day, the concentration of growth hormone in the blood is so low that it can hardly be detected, but several times each day, levels rise dramatically. Growth hormone does not itself cause growth to occur – it stimulates the release of **somatomedin**, another hormone from the liver. It is this second hormone that stimulates growth: it stimulates protein synthesis in cartilage cells at the ends of bones and in muscle cells.

Thyroxine

Thyroxine, a hormone produced by the thyroid gland, is also essential for normal growth. As well as controlling **basal metabolic rate**, and stimulating the process of growth in general, it also has a specific effect on protein synthesis and growth of the skeleton.

PUBERTY

Puberty is the period of life when the **sex organs** grow and start to produce **gametes**. **Secondary sexual characteristics** develop, such as the distribution of body hair and the differences in body shape that distinguish males from females (Fig 21.11).

The role of hormones in puberty

The events of puberty are controlled by hormones, although we still do not understand the full story. To find out about the exact function of a particular hormone, we would have to monitor changes in hormone concentration in the blood and then link these changes to the various aspects of puberty; but it is very difficult to measure hormone secretion accurately.

As we saw in the last section, hormones can be released into the blood in pulses rather than in a steady stream. So we would have to collect small blood samples from a large enough sample of people over a long period before we could know what was happening. Another problem is that many of the hormones involved in controlling puberty are broken down very rapidly.

REMEMBER THIS

Tumours in the pituitary can cause over-production of growth hormone in adults, causing a condition called acromegaly. See the Science in Context box on page 276.

? QUESTION 5

5   Before growth hormone was produced by genetic engineering, what would have been the only source to use for treating human growth disorders?

Basal metabolic rate is discussed in Chapter 7.

? QUESTION 6

6   A blood sample is collected and found to have a very low concentration of a particular hormone. Suggest three different ways of explaining this.

The hormonal changes that control the onset of puberty are described in detail in Chapter 20.

Fig 21.11a Being a choirboy is not a job for life: at puberty, the male voice breaks (suddenly gets deeper) as the size of the larynx increases

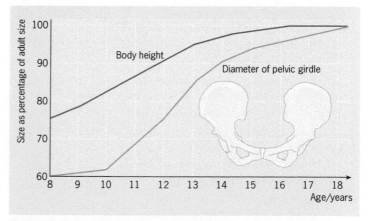

Fig 21.11b During birth, the baby passes through the mother's pelvic girdle. This graph shows that the increase in the diameter of the pelvic girdle of females occurs relatively late in puberty

Fig 21.12 The proportion of old people in the populations of many less developed countries is likely to rise as we approach 2020

4 AGEING

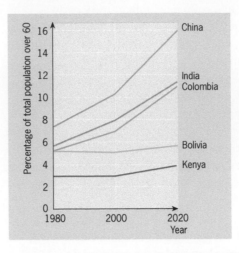

As we pointed out at the start of this chapter, few wild animals reach old age. There are many reasons for this, but predation and disease are mainly responsible. In humans, the situation is very different. Because of improvements in the standards of living in many parts of the world over the past 200 years, more people live to a ripe old age. In developed countries, we are already experiencing the effects: a 'top heavy' population. In the next few years, the same population changes will occur in the less well-developed nations. Fig 21.12 shows how, in China, India, Colombia, Bolivia and Kenya, the percentage of people over 60 is likely to increase in the years to 2020.

Old age affects individuals through changes in cell structure and organ function. An ageing population also affects society. An increase in the number of people living to old age places a huge strain on budgets for health and social care – one of the great challenges of the twenty-first century.

CELL STRUCTURE, ORGAN FUNCTION AND AGEING

Biologists who study the effects of ageing in humans would like to be able to define what is 'normal'. Illness and incapacity are so common among old people that complete freedom from disease is extremely rare. In addition, there is a lot of individual variation.

Let's look at the breathing system, for example: we could say that there is a 50 per cent reduction in lung capacity by the time a person reaches the age of 80. But, with so much variation, some individuals will be almost unaffected, while others will have such a severe reduction in lung capacity that they are barely able to move.

Fig 21.13 shows some of the ways old age affects organs and systems.

? QUESTION 7

7   It is expected that the total number of old people in Kenya will rise considerably, even though their percentage in the population will change little. Explain why.

✔ REMEMBER THIS

Many diseases become more common as we age – cancer and Alzheimer's disease are good examples. In the UK there are fewer than 100 cases of bowel cancer diagnosed each year in men aged 40. This increases to 500 cases per year in men aged 60 and over 3000 men aged 80 are diagnosed with colon cancer each year. The figures show a similar increase in women, but their risk of developing colon cancer is lower than men at each age.

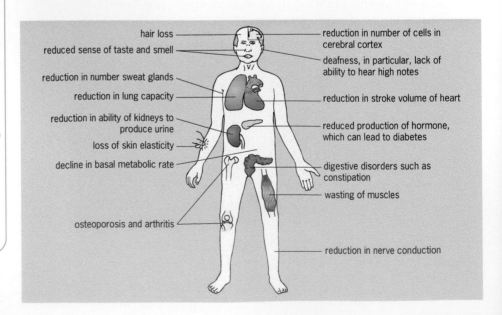

Fig 21.13 Some of the many effects of ageing. It is important to remember that different individuals can be affected to different extents

Alzheimer's disease, dementia and senility

We often hear people say, "Oh, I must be going senile in my old age…" when they forget something or do something stupid. Senility and dementia are not inevitable in old age but they affect far more older people than young people. Alzheimer's disease is the most common form of dementia and it affects just under half a million people in the UK and 14 million people worldwide.

The disease that we know as Alzheimer's disease was first described by the German neurologist Alois Alzheimer. It is a physical disease of the brain in which brain structure breaks down and a chemical imbalance of neurotransmitters develops. Tangles of protein form plaques in the brain that build up – at first these cause no symptoms, but as they increase, people with the disease show symptoms. Sufferers become forgetful, they can't find the right word, they get confused and they become frustrated and depressed at their inability to remember recent events. In its advanced stages, Alzheimer's disease is completely disabling and people with it need 24-hour care.

Alzheimer's is a disease that is age related. It affects one in 14 people over 65 and one in six people over 80. Presenile dementia occurs in people less than 65 and is usually due to other factors such as genetics and environmental factors such as diet.

Genetic inheritance of Alzheimer's disease

There are a few families where there is a very clear inheritance of the disease from one generation to the next. This is often in families where the disease appears relatively early in life. In the vast majority of cases, however, the effect of inheritance seems to be small. If a parent or other relative has Alzheimer's disease, your own chances of developing the disease are only a little higher than if there were no cases of Alzheimer's in the immediate family.

Environmental factors

There has been a lot of speculation that taking in too much aluminium in the diet can lead to Alzheimer's disease, but there is little evidence to back up this claim. It has now been generally discounted and it is not really known which environmental factors increase the risk of developing the disease. People who have had severe head or whiplash injuries appear to be at increased risk of developing dementia. Boxers who receive continual blows to the head are also at risk. Research has also shown that people who smoke and those who have high blood pressure or high cholesterol levels increase their risk of developing Alzheimer's.

Treatments for Alzheimer's disease

There is no cure for Alzheimer's and the disease is progressive – it always gets worse, never better. People may have periods of time with no deterioration, but the changes in the brain are irreversible. Some drugs have been licensed, as there is some evidence that they can help treat some symptoms of the disease. Acetyl cholinesterase (AChE) inhibitors are prescribed because Alzheimer's disease causes cholinergic neurones to be destroyed. AChE-inhibitors theoretically reduce the rate at which the transmitter acetylcholine (ACh) is broken down and so increase the concentration of ACh in the brain, slowing the rate at which people forget things. Some people find they work well but the value of the drugs is still controversial.

Fig 21.14 In 1901, Alois Alzheimer, a German psychiatrist, identified the first case of Alzheimer's disease, in a 50-year-old patient Auguste D and followed her to her death in 1906, when he first reported the case to the world

AGEING BONES AND JOINTS

As we get older, the skeleton suffers from a lifetime of wear and tear. The physical stresses on the body throughout life can lead to two important medical conditions: **osteoarthritis** and **osteoporosis**.

Osteoarthritis

Osteoarthritis is one of the commonest conditions associated with old age (Fig 21.15). X-ray evidence suggests that nearly 80 per cent of those over 65 years old have some arthritic damage to their joints. In most of them, this is not severe enough to cause painful symptoms.

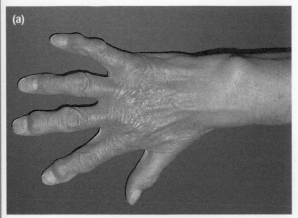

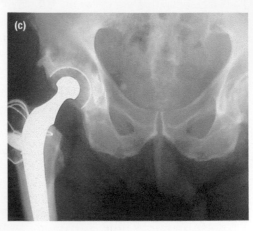

Fig 21.15 Osteoarthritis is not a single disease. It describes the various conditions that result from joint failure. Joints commonly affected are **a** those in the hands, and **b** in the knees as shown by this X-ray image, and in the hips. **c** If the hips are severely affected, surgery might be required to replace the hip joint with an artificial one

Many different factors cause osteoarthritis to develop and the way they interact is not fully understood. Experts now mostly agree that the old 'wear and tear' hypothesis is not backed up by evidence. For example, people who exercise well into old age have a much lower risk of developing osteoarthritis than their 'couch-potato' counterparts. Some researchers are now concentrating on cells in the bone and **cartilage** at joints, which send signals to each other using a huge variety of molecules, since abnormal signals may start the process of destruction in cartilage and/or bone. Whatever the actual cause, the result is loss of cartilage at the **articular surfaces** of bones that form the joints (Fig 21.16). In addition, there is an increase in bone-forming cells called **osteoblasts**. New bone may be laid down, deforming the shape of the joint.

Fig 21.16 In a healthy hip joint **a**, cartilage covers the articulating surfaces of the bones which form the joint. In the hip joint of a person suffering from osteoarthritis **b**, this cartilage has been lost

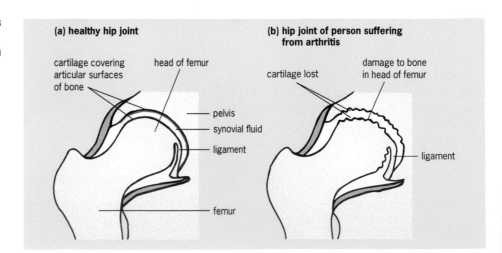

Although we cannot reverse the changes that accompany the development of osteoarthritis, we can do a lot to ease the symptoms. Keeping mobile is now recommended, and anti-inflammatory drugs and painkillers are given in the early stages. Later, a joint replacement may be necessary.

Osteoporosis

The density of bone varies with age, as Fig 21.17 shows. Young people up to the age of about 20 show a steady increase in bone density. After 20, there is a gradual decrease in density. In men, this loss of bone is not usually a matter for concern. However, women's bone density is usually less than men's, and they suffer further reduction during the menopause. Once bone density becomes very low, the affected person is said to suffer from osteoporosis. The bones lack strength, and even very little stress can fracture them.

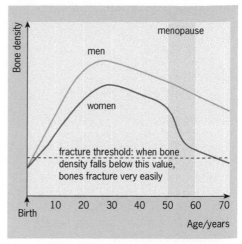

Fig 21.17 The effect of age on bone density

Using X-rays to see further

We normally associate X-rays of bones with fractures, but a hospital X-ray department investigates many conditions that affect the skeleton. Fig 21.18 shows some computer records from the radiography department of a large hospital.

The pie charts in Fig 21.19 show the age on admission of males and females with different conditions.

Fig 21.18 A hospital record of patients receiving X-rays

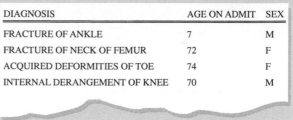

DIAGNOSIS	AGE ON ADMIT	SEX
FRACTURE OF ANKLE	7	M
FRACTURE OF NECK OF FEMUR	72	F
ACQUIRED DEFORMITIES OF TOE	74	F
INTERNAL DERANGEMENT OF KNEE	70	M

- Some conditions are the result of gender-specific behaviour. Young boys tend to be rougher in their play than girls of the same age and so they break their arms and legs more often. It is also likely that older women referred to the hospital with misshapen toes wore tight, fashionable shoes when they were younger.
- Congenital conditions are present at birth. Children with these conditions are X-rayed more often.
- Osteoarthritis is a disease of the elderly and is a major reason for X-ray investigations in the 70–79 age group.
- Older women tend to break bones often because their sense of balance has declined, but osteoporosis is also a major factor. Fractures of the neck of the femur are particularly common and many elderly women receive hip replacements.

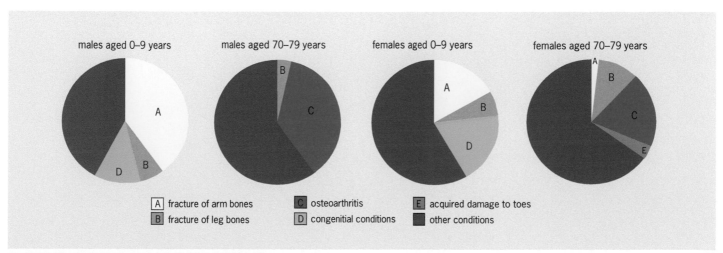

Fig 21.19 Pie charts drawn from data like that in Fig 21.18

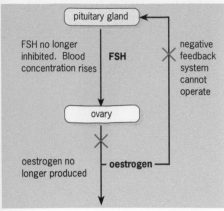

Fig 21.20 Some of the hormonal changes that accompany the menopause

text in diagram:
pituitary gland

FSH no longer inhibited. Blood concentration rises

FSH

negative feedback system cannot operate

ovary

oestrogen no longer produced

— oestrogen —

AGEING AND THE REPRODUCTIVE SYSTEM

Fertility in men gradually declines with age. A man's general state of health usually determines how old he is when he loses the ability to father children. Some healthy men father children when in their 70s.

In women, however, the inability to produce children is a more marked event, usually occurring around the age of 50. This is known as the **menopause**. In the four or five years before the menopause, as the ovaries gradually fail, there is a progressive decline in the number of menstrual cycles in which an ovum is produced.

A reduction in the amount of oestrogen that the ovaries produce is the cause of many of the unpleasant symptoms of the menopause. **Hormone replacement therapy** (HRT) involves supplying extra oestrogen and can be used to control these symptoms. Fig 21.20 summarises the hormonal events associated with the menopause.

> **Reproduction is discussed in detail in Chapter 20.**

SUMMARY

After reading this chapter, you should know and understand the following:

- **Body mass**, **height** and **supine length** can be used to measure body size and body growth. There are advantages and disadvantages with all three criteria.
- Growth curves may be plotted using data collected from either **longitudinal** or **cross-sectional** studies. They can show **cumulative growth** or **growth rate**.
- Human growth is **allometric** and **limited**.
- Growth is controlled by hormones. These include **growth hormone** and **thyroxine**.
- **Puberty** is the period of life when the **sex organs** grow and start to produce **gametes**. The events

of puberty are initiated and controlled by changes in hormone concentrations.
- Ageing brings about changes in cell structure and organ function.
- Fertility in men gradually declines with age. In women, however, the inability to produce children is associated with the menopause.
- **Osteoarthritis** and **osteoporosis** are conditions that commonly affect the skeletal system of older people.

 Practice questions and a How Science Works assignment for this chapter are available at www.collinseducation.co.uk/CAS

22

Genetics: the basics

GENETICS: THE BASICS

The purple sea urchin – can you see a family resemblance?

Do you share genes with a purple sea urchin?

In the first decade of the twenty-first century, the complete genomes of humans and many other animals have been completely sequenced. This has led to some interesting comparisons. It will probably come as no surprise that 98.7 per cent of your basic genome is shared with the chimpanzee, as apes are the closest living relatives of humans. It may come as a bit more of a shock that the purple sea urchin is also a close relative. About 70 per cent of sea urchin genes correspond to human genes and sequencing its genome has turned up a few other surprises.

This simple organism turns out to have a very advanced immune system – it is born being able to recognise bacteria and viruses and fight them off, while the human immune system has to learn this and often needs help from vaccines to manage it effectively. The sea urchin, which has no ears or eyes, was also thought to be completely blind – but its nice collection of visual perception genes says otherwise. Sea urchins also have genes associated with muscular dystrophy and Huntington's disease – both serious human genetic diseases. Research is now going on to use the sea urchin as a genetic toolbox to find out more about these diseases and maybe to develop new drugs to treat them.

? **QUESTION 1**

1 Some simple unicellular organisms such as bacteria and amoeba can reproduce by binary fission; they simply split in half. Is it correct to say that such organisms never die?

1 WHAT ARE GENES?

Genes are the 'instructions' that control everything that happens inside a cell. Physically, genes are different sized stretches of DNA that code for the manufacture of proteins such as enzymes. Enzymes organise the chemistry that goes on in our bodies and are behind all the life processes, including growth, repair and reproduction. Throughout our lives, the genes remain in the nuclei, directing the activities of each of our cells.

Children grow and develop according to the genes they have inherited from their parents. It's a harsh fact, but nothing lives forever: the inevitable fate of all organisms, often sooner rather than later, is death. However, although we die, our genes can live on in our offspring.

 REMEMBER THIS
In organisms that reproduce sexually, a set of genes from the male combines with one from the female to form a unique individual (see Chapter 20).

Fig 22.1 Residents of 'The Island', a sci-fi film, who discover they are clones, bred for the benefit of their creators

2 STUDYING GENETICS WITH THIS BOOK

Studying genetics means learning about DNA and what it does. In this section, we divide our study of this remarkable molecule into five areas:

- The **cell cycle** is the study of how cells divide. Discovering just how a single fertilised egg can develop into an organism as complex as a human is one of the central challenges of biology. When the control of cell division goes wrong, it can lead to cancer.
- **Molecular biology** is the branch of science that deals with all aspects of the DNA molecule: its structure and function. How are genes used to make products? Why are some genes active when others aren't? How can DNA copy itself? How do mutations occur?
- **Genetics** is the study of inheritance: the way in which genetic information is passed on from one generation to the next. Why are only some genes passed on? Why are some characteristics always shown while others are hidden?
- **Evolution** looks at how species change and develop with time. What is the mechanism of this change? How long does the process take? Can we observe evolution in action? What do we know about human evolution?
- **Genetic engineering** is the popular name for the technology that manipulates the DNA molecule. How can we change DNA? How can we transfer DNA between organisms? What are the benefits of genetic engineering? Can we cure genetic disease? Should we pursue this technology at all?

3 THE POTENTIAL OF GENETICS

Few scientific subjects make the headlines as often as advances in genetics and DNA technology. We are witnessing an accelerating revolution that is likely to have far-reaching effects. We know what the set of human chromosomes looks like (Fig 22.2). We have also mapped the complete human genome and are now embarking of the daunting task of finding out where the genes are, what they do and how they interact with each other. We may have the 'book of humans', but it's huge and written in a foreign language.

The implications of this progress are enormous:

- Two out of every three people die for reasons connected to their genes. Although only a small percentage inherit a lethal genetic disease, many more inherit a *tendency* to develop a condition such as heart disease, premature senility or cancer. We call this a **genetic predisposition**. An in-depth knowledge of the human genome will make it possible to screen babies before or soon after birth for genetic abnormalities and tendencies to develop such diseases. But this raises ethical and moral questions, not just scientific ones: would you want to know just how likely you are to develop cancer in middle age? Would you want anyone else to know – life insurance companies, potential employers, friends or partners? Who would have access to your DNA files?

Cell division is discussed in Chapter 23.
Molecular biology is discussed in Chapter 24.
Genetics is discussed in Chapter 25 and
Evolution is discussed in Chapter 26.
Genetic engineering is discussed in Chapter 27.

✔ **REMEMBER THIS**

Chromosomes condense and become visible during cell division (Chapter 23).

Normal cell division – mitosis – duplicates the DNA and then divides it accurately into two so that the two new cells formed are genetically identical.

Meiosis is the type of cell division that produces haploid cells such as eggs and sperm. This process halves the amount of genetic material that goes into each new cell and shuffles it at the same time so that no two sex cells are the same.

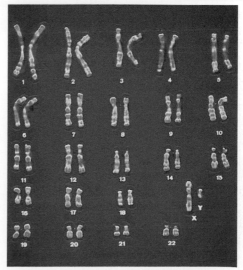

Fig 22.2 A complete set of human chromosomes. Each body cell contains two sets of 23 chromosomes and each chromosome contains up to 4000 genes. The total amount of DNA in a cell is called the genome. The human genome consists of about 60 000 genes

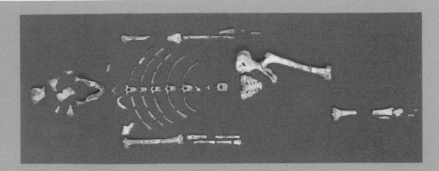

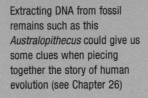

Extracting DNA from fossil remains such as this *Australopithecus* could give us some clues when piecing together the story of human evolution (see Chapter 26)

This model shows the double helix structure of DNA

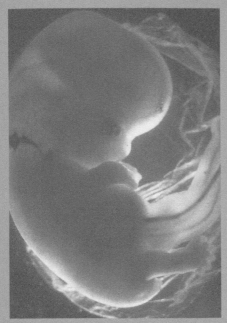

A human embryo 7 weeks after conception. We have developed the technology to clone animals as advanced as humans, but this area of genetics is an ethical minefield

Wheat has been bred selectively to produce a greater yield. Many new types are easier to harvest and are also resistant to disease

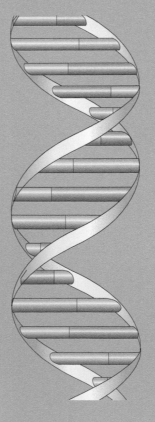

This fermenter houses genetically engineered bacteria which contain the human growth hormone gene. Inside the fermenter, the bacteria multiply and produce large amounts of human growth hormone

Fig 22.3 The applications of our knowledge of DNA

Male and female sperm can be separated according to the amount of DNA they contain. The sperm are stored in liquid nitrogen until needed. Farm animals can be artificially inseminated to produce offspring of the required sex

- We can isolate faulty genes, such as those which cause cystic fibrosis, and replace them with multiple copies of healthy genes. Although still at the experimental stage, this **gene therapy** holds much promise for the future.

- We can isolate genes for useful products, such as insulin and growth hormone, and place those genes into another organism (Fig 22.3). This is **recombinant DNA technology**. Bacteria already produce human insulin and other products on a large industrial scale, and it is also possible to transfer such genes into higher animals such as sheep so that they produce milk containing the required protein.

- We may be able to halt some of the genetic causes of ageing, and allow more people to live even longer.

Only the bravest scientists would dare to speculate about where our genetic knowledge might eventually take us. It could be to a greatly improved world where suffering is reduced while quality of life and life expectancy are improved. Genetic engineering and agricultural techniques might give us food-producing species that can live in the harshest of conditions, so providing food in areas of most need.

But can we be this optimistic? There are still many things we don't know about living organisms and it is impossible to predict exactly the effects of interfering with an organism's DNA. Already, companies are trying to patent genetic material and its products, leading to hot debate about whether or not such natural products can be owned, like a brand name or an invention. Will genetic advances be used solely to make money, and therefore be available only to those who can pay?

The ethical implications of some very recent developments also need to be considered. The first mammals have now been cloned but the process is unreliable. This has led to a heated debate about whether or not we can, or should, clone humans.

> **Recombinant DNA technology is discussed in Chapter 6.**

4 WHAT GOES ON IN THE NUCLEUS?

The nucleus of every human cell contains DNA (Fig 22.4). Each nucleus contains a set number of long, elaborately coiled DNA molecules that condense into chromosomes. Genes, the individual instructions, are dotted along the chromosomes.

In an adult organism, most cells are not dividing: the chromosomes are uncoiled, and active genes are producing proteins to control the activities of the cells. Not all genes are active at the same time: controlling which genes are active and which are not enables different cells to carry out different functions. For instance, all cells in the human body contain two copies of the insulin gene. Only in certain specialised cells in the pancreas, however, are the genes switched on and used to make insulin.

GLOSSARY OF GENETICS TERMS

Like many branches of science, genetics comes with its own vocabulary, seemingly designed to make some very straightforward ideas inaccessible to the average person. You will need to know the following terms:

Diploid: cells or organisms containing two sets of genes. Human body cells, for instance, are all diploid.

Haploid: cells or organisms containing one set of genes. Human sex cells (egg cells and sperm) are haploid.

Gene: a length of DNA that codes for a particular polypeptide. Some proteins consist of more than one polypeptide; these are coded for by more than one gene.

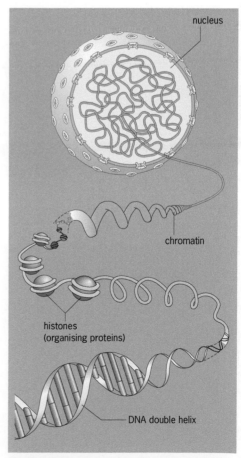

Fig 22.4 Most of the DNA in the cell is locked away in the nucleus where, in effect, it forms a reference library. The genes, which act as instructions for all cell functions, must be passed on when the organism reproduces

Allele: one form of a gene. There may be two or more alleles of any particular gene. Diploid organisms contain two alleles, one on each pair of chromosomes. For example, in pea plants, the gene for flower colour may have a red allele and a white allele.

Genotype: the genetic make-up of an individual.

Phenotype: the observable characteristics an organism possesses. For example, in peas, the genotype **Aa** may result in a phenotype of red flowers. Many alleles are expressed, or used, and these contribute to the characteristics of the individual. Other genes – recessive alleles in the presence of a dominant version – are not expressed and so remain hidden.

Dominant: the allele that, if present, is expressed. If the red allele in the pea plant is dominant, the pea has a red flower, even if it has a white allele as well.

Recessive: the allele that is expressed only in the absence of the dominant. In the pea, if the white allele is recessive, the plant has white flowers only if it has two white alleles.

Homozygous: when both alleles of a particular gene are the same. In the pea example, an individual would be homozygous if it had two red alleles or two white alleles. We usually give letters to denote alleles, for example **A** for the dominant allele and **a** for the recessive allele. Individuals with genotype **AA** or **aa** are homozygous.

Heterozygous: the two alleles of a gene are different. In a pea plant, a heterozygous individual could have a red allele and a white allele, so their genotype would be shown as **Aa**.

5 VARIATION: WHY IS EVERYBODY DIFFERENT?

The group of people in Fig 22.5 shows what we all know: that everyone is different. Height, weight, intelligence, personality, eye colour, size of feet – the list is endless. No two people have the same combination of features. Even identical twins show subtle differences. So what makes us the way we are?

Each human baby is born with a unique set of genes – we call this its **genotype**. However, only some of these genes affect the observable characteristics of the organism – its **phenotype**. Recessive genes, for

Fig 22.5 Like most species, humans show great variation between different individuals

example, remain hidden, masked by dominant versions. The other factor that contributes to the phenotype is the environment – the circumstances in which the organism grows affect the way in which genes are expressed.

INTERSPECIFIC AND INTRASPECIFIC VARIATION

As you saw at the start of the chapter, humans and chimps have 98.7% of their genome in common. This difference, just in the actual structure and sequence of the genes, is called **interspecific variation**. This variation in the genome between species is often very low when species are closely related in evolutionary terms. However, as you know, chimps and humans are not identical, and they don't behave in the same way. So what accounts for this apparent anomaly?

The answer is gene expression. There are millions of different ways in which the same thousand genes can be expressed, turned off and interact with each other. Genes in humans are expressed very differently from the corresponding genes in the chimp. Different members of the same species can also show big differences in the way they express the same genes. This is why there are such big differences between individuals (Fig 22.5). Biologists use the term **intraspecific variation** to describe the differences in gene expression that occur between members of the same species. Intraspecific variation between one person and another is very high. The same is true of chimpanzees: there is very high intraspecific variation between individual chimps.

WHY IS VARIATION NECESSARY?

The simple answer is: survival. Variation in a population has great survival value because it greatly increases the chance that at least some individuals will adapt to changing conditions. To illustrate this, consider a **clone**, a population of genetically identical individuals. All is well when conditions are favourable, but when a serious disease comes along it is very likely that all of the organisms in the population will die. In a more varied population that has developed as a result of sexual reproduction and variation, there is a much greater chance that some of the individuals could be resistant to the disease. These individuals survive to breed, ensuring that the population, and perhaps the whole species, continues to exist.

HOW VARIATION OCCURS

Variation arises because individual genes **mutate**. **Mutations** are changes in the genetic material of an organism, often caused by faults in the copying of the DNA. Once there is some variation to work on, the differences are maximised by the process of sexual reproduction. Using humans as an example we can see how this happens.

First, gametes are produced by meiosis. This is a special type of cell division which not only produces haploid cells, but shuffles the genes so that all eggs or sperm made by a particular individual are different. In meiosis, there are two key events that produce variation:

- **Crossover during prophase I.** Blocks of genes are swapped between chromosome pairs, bringing some new genes together and separating others.
- **Independent assortment during anaphase I.** Either chromosome from each pair can pass into the gamete. As a human cell has 23 pairs of chromosomes, 2^{23} different combinations are possible.

Secondly, following meiosis, **fertilisation** joins gametes from two separate individuals to form a genetically unique zygote from which the new diploid organism develops.

REMEMBER THIS

The phenotype of an organism = its expressed genes + effects due to the environment that it is exposed to.

REMEMBER THIS

A population is a group of interbreeding individuals of a particular species in a particular place (see Chapter 26). A species usually consists of many different populations. Some endangered species become particularly vulnerable because they consist of just one isolated population, restricted to a particular area.

REMEMBER THIS

'Survival of the fittest' is more correctly called **natural selection**. In the long term, natural selection, acting on the variation in a population, is the main driving force behind the process of **evolution**.

Evolution is discussed in more detail in Chapter 26.

Copying DNA is discussed in more detail in Chapter 25.
Cell division is discussed in more detail in Chapter 23.

REMEMBER THIS

Often, continuous variation is **quantitative** (you can put a numerical value on it) and it is **polygenic** (controlled by several genes). Discontinuous variation is more often qualitative (you can describe it, but not measure it) and is controlled by one or only a few genes. In Chapter 25 we look at the principles of inheritance, using examples of discontinuous variation.

Continuous and discontinuous variation

There are two main types of variation in a population: **continuous variation** and **discontinuous variation** (Table 22.1):

- Continuous variation is so-called because the factors, or variables, can be any value inside a given range. Examples of continuously varying characteristics include height and weight in humans (Fig 22.6a), and the number of leaves in plants.
- Discontinuously varying features do not display a range: they are either one thing or the other. The ABO human blood group system is a classic example of discontinuous variation: people are either A, B, AB or O. They cannot be anything in between. Height in pea plants is another example: pea plants can be either tall or dwarf (Fig 22.6b).

Table 22.1 Examples of continuous and discontinuous variation

Continuous	Discontinuous
most dimensions in animals and plants such as height, weight, length of bones, number of leaves	ability to roll tongue in humans
tolerance to adverse conditions, such as cold, heat, dehydration	ability to taste phenylthiourea
human fingerprints	flower colour in peas
	ABO blood group in humans
	coat colour in cats

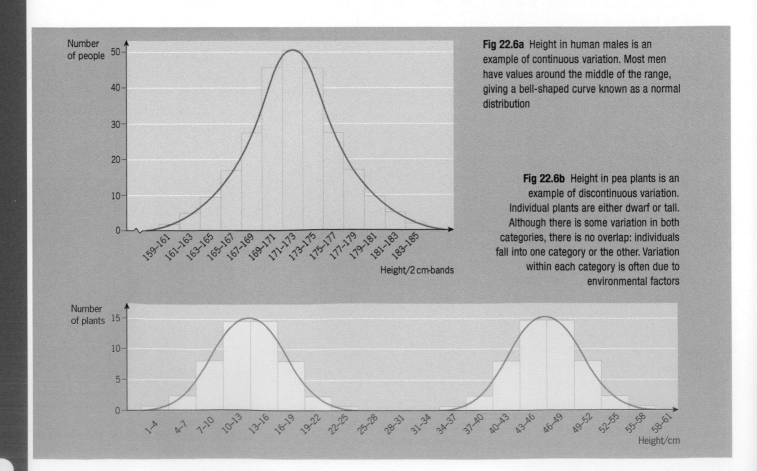

Fig 22.6a Height in human males is an example of continuous variation. Most men have values around the middle of the range, giving a bell-shaped curve known as a normal distribution

Fig 22.6b Height in pea plants is an example of discontinuous variation. Individual plants are either dwarf or tall. Although there is some variation in both categories, there is no overlap: individuals fall into one category or the other. Variation within each category is often due to environmental factors

6 A BRIEF HISTORY OF GENETICS

The history of genetics gives an insight into the changing nature of scientific progress. Early scientific discoveries and breakthroughs were few and far between, often made by individuals who observed and studied nature and put forward radical theories to explain what they had seen. Mendel and Darwin below, are classic examples.

More recent discoveries have been made by teams of dedicated people working together, using the scientific method and latest technology to build on knowledge accumulated by others. Table 22.2 outlines some of the landmarks in this most dynamic of scientific areas.

Table 22.2 Some landmark discoveries in genetics. In many of these, the scientists mentioned by name led a much larger team of scientists who worked towards the discovery described.

Year	Scientist(s)/institution	Discovery
1858	Charles Darwin, Alfred Russel Wallace	Jointly announced their theory of evolution by natural selection or 'survival of the fittest'.
1859	Charles Darwin	Published his book *The Origin of Species.*
1866	Gregor Mendel	Published his laws of genetics following studies on pea plants. His findings were ignored.
1900	Hugo de Vries, Carl Correns, Eric von Tshermak	Discovered meiosis, providing an explanation for Mendel's laws. Mendel's work is rediscovered by all three and gains recognition.
1905	Nettie Stephens, Edmund Wilson	Independently discovered the principles of sex determination: XX = female, XY = male.
1910	Thomas Hunt Morgan	Proposed the theories of sex linkage, mutation, linkage and chromosome maps following work on *Drosophila* (fruit fly).
1928	Fred Griffiths	Proposed that some 'transforming principle' had changed a harmless strain of bacteria into a lethal one. The hunt for DNA began.
1941	G. W. Beadle, E. L. Tatum	Irradiated the bread mould *Neurospora*, producing mutations which suggested that genes code for enzymes.
1944	Oswald Avery	Purified the 'transforming principle' in Griffiths' experiment, showing it to be nucleic acid (DNA).
1950	Erwin Chargaff	Discovered that the base pairing ratios in DNA were always the same, whatever the organism (Chargaff's principles: A – T, C – G).
1951	Rosalind Franklin	Obtained high-quality X-ray diffraction studies of DNA, showing that it has a helical structure.
1952	A. D. Hershey, M. Chase	Used bacteriophages to show that DNA, not protein, is the material of heredity.
1953	Francis Crick, Jim Watson	Built on the work of Chargaff and Franklin to work out the 3D structure of DNA (Fig 22.7).
1958	M. Meselson, F. W. Stahl	Used radioisotopes of nitrogen to prove the semi-conservative mechanism of DNA replication.
1958	Arthur Kornberg	Purified DNA polymerase from *E. coli* and used it to make DNA from nucleotides in a test tube.
1961	F. Jacob, J. Monod	Propose a mechanism for switching genes 'on' or 'off': the operon hypothesis of metabolic control.
1966	Marshall Nirenberg, Gobind Kharana	Cracked the genetic code: particular triplets of bases on DNA code – via mRNA – for the 20 amino acids.

? QUESTION 2

2 Look at Fig 22.6b on the previous page. A pea plant is 34 cm in height. Is it a tall or a dwarf plant?

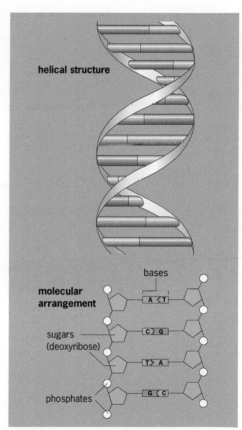

Fig 22.7 The discovery of the structure and function of DNA is one of the greatest scientific achievements of the twentieth century. The DNA molecule stores genetic information and, with the help of enzymes, makes exact copies of itself, time after time

Table 22.2 Some landmark discoveries in genetics, cont'd.

Year	Scientist(s)	Discovery
1970	H. O. Smith, K. W. Wilcox	Isolated the first restriction enzyme, *Hind*II, which cuts DNA.
1972	Paul Berg, Herb Boyer	Produced the first recombinant DNA molecules.
1973	Annie Chang, Stanley Cohen	Showed that DNA could be inserted and cloned inside a bacterium.
1977	Fred Sanger	Developed a method for sequencing DNA.
1977	Genentech	The first genetic engineering company is founded, using recombinant DNA to make pharmaceuticals.
1983	James Gusella	Located the gene responsible for Huntington's disease on chromosome 4 (see Chapter 25).
1984	Cary Mullis	Developed the polymerase chain reaction (PCR) in which minute samples of DNA can be copied, prior to analysis.
1984	Alec Jeffreys	Developed DNA profiling (or 'fingerprinting').
1989	Francis Collins	Identified the gene (*CFTR*) responsible for cystic fibrosis on chromosome 7.
1990	French Anderson	First attempts at gene therapy: T cells of a 4-year-old girl were exposed to viruses containing working copies of her defective gene.
1994	Calgene, California, USA	Genetically engineered 'Flavr savr' tomatoes went on sale.
1995		Transgenic sheep made to express human genes in their mammary glands, so that they produce milk containing valuable pharmaceuticals.
1997	Roslin Institute, Edinburgh	First mammal cloned. Dolly the sheep fuels controversy about genetic research.
2001		First stem cells extracted from cloned human embryo. Gaur, an almost extinct bovine species, cloned using a cow as a surrogate mother. Pigs cloned for first time, bringing the possibility of pig organs for transplant into human patients a step closer. Researchers use cow stem cells to construct a complete cow kidney in the laboratory.
2001	Texas A&M University	First cat ('cc') cloned.
2002	House of Lords select committee	Gave scientists in the UK the go-ahead to create human embryo clones under strictly controlled conditions.
2003 (April)	The Human Genome Project	Complete human genome sequence published.
2003 (December)	Eric Lander and Richard Wilson	Chimpanzee genome sequenced.
2004	Multiple centres	Honey bee, chicken, rat, dog and cow genomes sequenced.
2005	International HapMap Consortium	Publish comprehensive catalogue of human genetic variation, for researching the genes involved in common diseases such as asthma, diabetes, cancer and heart disease.
2007 (January)	Richard Mayeux, Lindsay Farrer and Peter St.George-Hyslop	Identified *SORL1* gene as risk factor for Alzheimer's disease.
2007 (June)	Multiple centres	Scientists publish a series of papers that challenge the traditional view of our genetic blueprint.

23

Cell division

23 CELL DIVISION

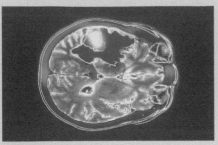

A magnetic resonance imaging (MRI) scan of a section through the brain of a 42-year-old woman. A tumour, coloured yellow, can be seen in the left hemisphere. This is a **metastatic** tumour; it has spread from another tumour elsewhere in the body

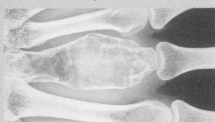

Coloured X-ray of hand showing a tumour growing on the bone on one finger

Cell division out of control = cancer, or does it?

In a healthy human adult, cells divide only when they should. Some cells, such as those that line the gut, are replaced at a remarkable rate. Other cells, such as muscle cells, live longer and need to be replaced far less often. Occasionally, the systems that control cell division break down, and a cell that should be stable divides uncontrollably. Soon, a mass of tissue called a tumour forms. Tumours can be either benign or malignant.

A benign tumour is not cancerous. Cells divide within a small, confined area and the growth is often surrounded by a membrane. New cells form in the centre of the mass. This type of tumour can be dangerous if it presses on important organs, or if it grows very large. It is usually treated by surgery. Because the tumour cells are confined and do not invade other tissues, recovery and survival rates are good.

A malignant tumour is commonly known as cancer. The tumour grows at the edges, spreading out and invading the surrounding tissues (supposedly like a crab – hence the name cancer). This type of tumour is much more dangerous. Vital tissues and organs can be destroyed quickly and this can lead to death. Even if the tumour is removed from the body, actively dividing cells may have already broken off and set up secondary growths elsewhere in the body.

1 SOME BASIC PRINCIPLES

Cancer and other lifestyle diseases are discussed in Chapter 31.

An understanding of cell division is of great importance in the study of human biology. It is the process by which we grow and develop. As adults, our body uses the main type of cell division for maintenance and repair. When control of cell division is lost, the result can be cancer. Another type of cell division halves the chromosome number in eggs and sperm (the sex cells, or **gametes**) so that when fertilisation occurs, the resulting zygote has the normal double number.

Humans, like all animals, are eukaryotic organisms: in eukaryotic cells, DNA is organised into chromosomes and is enclosed within a double nuclear membrane. Before going on to study the mechanism of cell division in eukaryotic cells, familiarise yourself with some of the basic terms and concepts by looking at Table 23.1.

AN INTRODUCTION TO CELL DIVISION

Our bodies are made from two different types of cell: body (or somatic) cells and sex cells (or **gametes**). The nucleus of each body cell contains two sets of chromosomes and so has two sets of genes. Body cells are described as diploid (from Greek, meaning 'double number') (Fig 23.1).

Mitosis is the process of normal cell division. In mitosis, the chromosomes are copied and then divided equally between the two new daughter cells. So each mitotic division produces two cells, both diploid, and each with exactly the same genes as the parent cell.

? QUESTION 1

1 'DNA makes RNA makes proteins.' Explain this statement.

Table 23.1 Some basic terms important in cell division

Term or feature	What it is or what it means
gene	A length of DNA that codes for the production of a particular polypeptide or protein.
genome	The name given to the full set of genes in a cell. The human genome consists of about 30 000 genes.
diploid	Diploid cells contain two versions of every gene.
haploid	Haploid cells contain one version of every gene.
chromosome	A long single molecule of DNA, organised around proteins called histones. The largest human chromosomes contain about 4000 genes. Chromosomes exist in cells all the time but they can be seen only during cell division, when they condense and separate.
homologous chromosomes	Chromosomes exist in pairs – humans have 23 pairs. Homologous chromosomes have the same genes at the same positions, but not necessarily the same versions of each gene.
chromatid	During cell division, the DNA of a cell is replicated (copied). When the chromosomes condense, they therefore appear as double structures: each unseparated chromosome within such a pair is called a chromatid.
sister chromatid	When two chromatids are genetically identical (as in mitosis), they are called sister chromatids.
bivalent	A bivalent is a pair of homologous chromosomes that line up together, as they do during meiosis.
locus	The position of a gene on a chromosome.
transcription	The process in which a molecule of mRNA is assembled on an active gene; mRNA thus becomes a mobile copy of the gene.
translation	The process of converting the code on the mRNA into a protein. This is achieved by protein synthesis: amino acids are joined in a particular order to make a protein such as an enzyme.

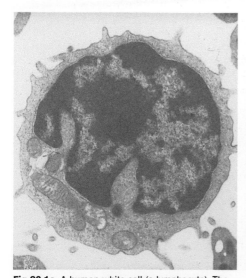

Fig 23.1a A human white cell (a lymphocyte). The nucleus of this cell contains all the genes necessary to make a whole new organism, a clone of the person from whom this cell was taken. Although the technology exists to clone humans, the area is fraught with huge moral, ethical and legal problems

Fig 23.1b Dolly the sheep, born in 1997, was the first-ever mammal to be cloned from a diploid adult body cell

Fig 23.1c A micropipette puts the nucleus from an adult cell into a non-fertilised egg cell that has had its own nucleus removed

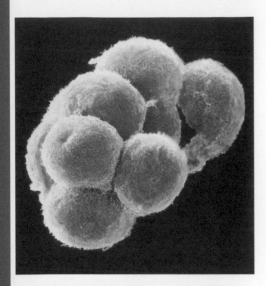

Some cells in the human body are not diploid. Gametes, the sperm and eggs, contain only one copy of each gene as they have only one set of chromosomes. These cells are haploid and are produced by a special type of cell division called **meiosis**.

A male and female gamete join together at fertilisation, to form a new diploid cell called a **zygote**. This one cell divides by mitosis to produce a complete new organism (Fig 23.2).

Fig 23.2 This human embryo is 4 days old, but the fertilised egg has already divided mitotically four times, producing a cluster of 16 cells. The cells will soon specialise to form the tissues and organs of the body

Reproduction and growth are discussed in Chapters 20 and 21.

SCIENCE IN CONTEXT

Cell turnover

As children, we grow because the cells in our bodies divide mitotically, and cell production outnumbers cell death. When we reach our adult size, our cell population stays relatively constant. The rate at which cells die and are replaced is known as **cell turnover**.

Research has shown that different tissues and organs have very different turnover rates. Brain cells, for instance, are not usually replaced, and after our 20s there is a slow but steady decline in number. This is not normally something to worry about – you won't run out. Other cells have a low turnover. Liver cells, for example, might divide only once every one to two years although, unlike the brain, they can regenerate if some cells are damaged or destroyed. Most of the high turnover cells are epithelial cells (Figs 23.3 and 23.4). These are the cells of the body that most frequently give rise to tumours. Cells of the bone marrow (see Chapter 30) and cells in the testes (Fig 23.5) also divide very frequently.

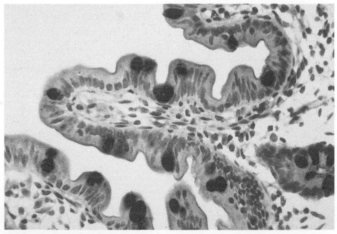

Fig 23.4 Intestinal epithelial cells have a very rapid turnover: the entire gut lining is replaced every couple of days. Much of the content of these cells is digested and reabsorbed in an efficient recycling system, but even so, gut cells form a significant proportion of faecal matter

Fig 23.3 Scanning electron micrograph of human epidermis. Dead skin cells are constantly being lost and replaced by mitosis. A large proportion of household dust consists of human skin cells

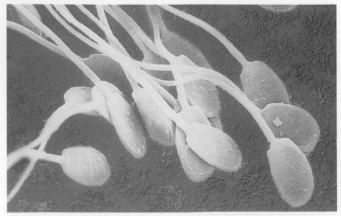

Fig 23.5 The production of sperm is a remarkably intense process. The average healthy human male produces about 1000 new sperm every second. The process of sperm production, spermatogenesis, which involves meiosis, is described in Chapter 20

2 THE CELL NUCLEUS

The DNA of a cell is contained in the nucleus. In a cell that is not actively dividing, DNA exists as **chromatin** (see Fig 3.41). This has a granular, or grainy, appearance (Fig 23.6). Separate chromosomes are present, but the DNA is so spread out that we cannot tell one from another.

When a cell is about to divide, the chromosomes condense and separate. They become visible under the light microscope as dark, rod-like structures. Each pair of chromosomes is known as a **homologous pair**. This means that they both contain genes at the same positions, or **loci**. You received one chromosome of each pair from your mother and one from your father. So, although homologous pairs contain the same genes, they do not necessarily carry the same versions of each gene. This is the key to understanding why organisms vary. Alternative forms of genes are called **alleles**.

The appearance, number and arrangement of chromosomes in the nucleus are referred to as the **karyotype**. The human karyotype consists of 23 pairs of chromosomes (Fig 23.7). All diploid human cells contain 23 pairs of chromosomes and this is often given the notation $2n = 46$. This means that a diploid cell has 46 chromosomes (a haploid cell has 23). Other organisms have different numbers of chromosomes (Table 23.2).

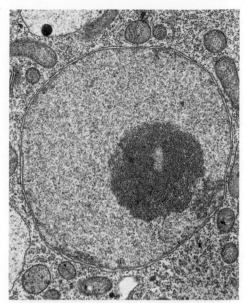

Fig 23.6 In a non-dividing cell, the chromosomes are spread out as chromatin and appear as pale granules. In this form, the DNA is used as a template to make mRNA in the process of transcription. The mRNA passes to the cytoplasm and is itself used as a template for protein production in the process of translation. In this cell, the nucleolus is also clearly visible: it is the dark sphere inside the nucleus

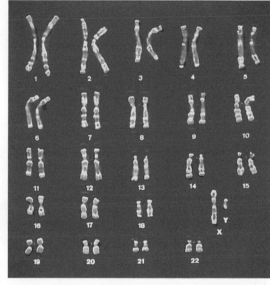

Fig 23.7 The 23 pairs of chromosomes in the human karyotype. Karyotypes are used to detect chromosomal abnormalities and to perform sex tests on athletes. The presence of XX confirms the athlete is female, the presence of XY shows the athlete is male

> Alleles are discussed in more detail in Chapters 22 and 25.

> **? QUESTION 2**
>
> 2 Which notation would you use to describe the number of chromosomes in human sperm?

Table 23.2 The number of chromosomes in the cells of some plants and animals, compared with humans. Note that there is no connection between chromosome number and the complexity of an organism. Some ferns have hundreds of chromosomes

Species	Number of chromosomes
penicillin mould (*Penicillium notatum*)	5
broad bean (*Vicia faba*)	12
lettuce (*Lactuca sativa*)	18
yeast (*Saccharomyces cerevisiae*)	34
cat (*Felis cattus*)	38
human (*Homo sapiens*)	46
potato (*Solanum tuberosum*)	48
chimpanzee (*Pan troglodytes*)	48
horse (*Equus caballus*)	64
chicken (*Gallus gallus*)	78
dog (*Canis familiaris*)	78

> **✔ REMEMBER THIS**
>
> How can a potato have more chromosomes in its cells than humans do? It's quality, not quantity, that counts. All organisms have the same DNA but it's what the DNA codes for that makes the difference. When you translate potato DNA, you get a potato plant. When you translate human DNA, different proteins are made, and eventually a human being results.

3 THE STAGES OF THE CELL CYCLE

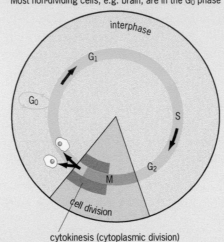

Key

G_1 – growth of cytoplasm and organelles
S – synthesis of DNA
G_2 – second growth phase
M – mitotic phase (nucleus and cytoplasm
divide to produce two diploid cells)

Most non-dividing cells, e.g. brain, are in the G_0 phase

interphase

G_1

G_0

S

G_2

M

cell division

cytokinesis (cytoplasmic division)

Fig 23.8 The four stages of the cell cycle: the three stages of interphase and the stage of active division in which mitosis occurs

The cell cycle is the complete sequence of events in the life of an individual diploid cell. The cycle starts and ends with cell division and consists of the stages of mitosis plus **interphase**, the interval between divisions in which the cell carries out its normal functions. Fig 23.8 shows the main stages of the life cycle of a cell.

Cells that do not divide, such as those in muscle and nerve tissue, are always in interphase: this is the normal state for a functioning cell. In interphase, the DNA in the nucleus is unwound, and active genes are read to produce proteins such as enzymes.

A new cell has three options:

- It can remain stable, in interphase for many months or years. Brain and other nerve cells rarely divide, if ever.
- It can undergo mitosis within a short period of time. Skin cells and cells that line the gut all have a very rapid turnover (see the Science in Context box on cell turnover on page 344).
- It can undergo meiosis. Specialised germ cells in the ovary and testes include meiosis in their cell cycle and produce egg cells or sperm.

INTERPHASE

We start looking at the cell cycle at the point when a cell has just divided. The cell is now in interphase. Interphase has three distinct stages: G_1, S and G_2.

G_1 – the first growth phase

Just after it has been produced by division of its parent, a new cell is in the early part of the first **g**rowth phase, G_1, sometimes called G_0. Cells that do not divide remain at this point in the cell cycle because they do not need to go further: they never replicate their DNA. 'New' cells are relatively small, with a full-sized nucleus but relatively little cytoplasm.

During G_1, protein synthesis starts and the volume of cytoplasm and the number of organelles increase rapidly. The process is actually quite complicated, not least because some organelles such as mitochondria (which have their own DNA) divide independently of the cell nucleus. In later G_1, the cell takes on more 'normal' proportions: the nucleus begins to look smaller as the surrounding cytoplasm increases in volume.

The S phase

The **S** phase – DNA **s**ynthesis phase – follows G_1. In this phase, the cell's DNA **replicates** (copies itself). Predictably, a cell only enters the S phase if it is going to divide. The point at which DNA replication starts is called the **restriction point**. After this, the cell becomes **restricted**, or locked into an automatic sequence that inevitably moves on to cell division.

? **QUESTIONS 3–4**

3 Name the three main stages of interphase.

4 Without DNA replication, what would happen to the genetic material each time cell division took place?

G$_2$ – the second growth phase

Before the actual mitotic cell division (see below), the cell enters a second, shorter **g**rowth phase, **G$_2$**, in which the proteins necessary for cell division are synthesised.

MITOSIS

Mitosis is a continuous sequence of events but, for clarity, they are divided into four distinct stages: **prophase**, **metaphase**, **anaphase** and **telophase**. The stages of mitosis are illustrated in Fig 23.9. You will need to refer to this constantly as you read the text.

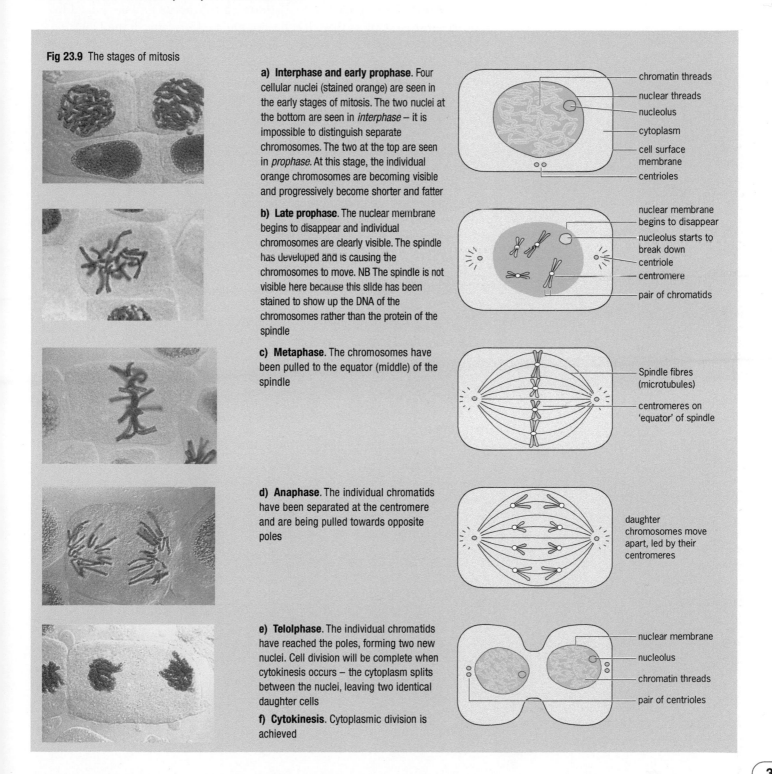

Fig 23.9 The stages of mitosis

a) Interphase and early prophase. Four cellular nuclei (stained orange) are seen in the early stages of mitosis. The two nuclei at the bottom are seen in *interphase* – it is impossible to distinguish separate chromosomes. The two at the top are seen in *prophase*. At this stage, the individual orange chromosomes are becoming visible and progressively become shorter and fatter

- chromatin threads
- nuclear threads
- nucleolus
- cytoplasm
- cell surface membrane
- centrioles

b) Late prophase. The nuclear membrane begins to disappear and individual chromosomes are clearly visible. The spindle has developed and is causing the chromosomes to move. NB The spindle is not visible here because this slide has been stained to show up the DNA of the chromosomes rather than the protein of the spindle

- nuclear membrane begins to disappear
- nucleolus starts to break down
- centriole
- centromere
- pair of chromatids

c) Metaphase. The chromosomes have been pulled to the equator (middle) of the spindle

- Spindle fibres (microtubules)
- centromeres on 'equator' of spindle

d) Anaphase. The individual chromatids have been separated at the centromere and are being pulled towards opposite poles

- daughter chromosomes move apart, led by their centromeres

e) Telolphase. The individual chromatids have reached the poles, forming two new nuclei. Cell division will be complete when cytokinesis occurs – the cytoplasm splits between the nuclei, leaving two identical daughter cells

f) Cytokinesis. Cytoplasmic division is achieved

- nuclear membrane
- nucleolus
- chromatin threads
- pair of centrioles

REMEMBER THIS

Exam hint – remember the mnemonic: **IPMAT** – **I**nterphase, **P**rophase, **M**etaphase, **A**naphase, **T**elophase.

The movement of chromosomes during cell division is controlled by microtubules, which form the **spindle**. In a non-dividing cell the microtubules are found as two bundles of fibres, the **centrioles**, in an area of cytoplasm known as the **centrosome**, or **microtubule organising centre** (**MTOC**). During cell division the centrioles move to opposite sides of the nucleus, from where they form the spindle.

Prophase

Chromatin begins to condense into chromosomes: Fig 23.9a. At this stage, each chromosome has replicated itself and now consists of two identical **chromatids**. These are known as **sister chromatids** and are joined by a **centromere**. Unless there has been a mutation (a fault in the DNA replication), these sister chromatids are genetically identical. The subsequent stages of mitosis organise and split the pairs of chromatids, so that one chromatid of each pair goes into each daughter cell.

As the chromosomes condense, other changes occur in the cell:

- The nucleolus begins to break down.
- The centrioles move to opposite sides of the nucleus.
- The centrioles begin to assemble the spindle.
- The nucleolus disappears.
- The nuclear membrane begins to break up.

From prophase onwards, most 'normal' cell activity, such as protein synthesis and secretion, is halted until division is over.

Metaphase

The beginning of metaphase (meta = middle) is marked by the complete disappearance of the nuclear membrane, which breaks down into separate vesicles, moves into the surrounding cytoplasm and joins with the endoplasmic reticulum: Fig 23.9c.

The spindle (Fig 23.10) becomes fully developed and fills the space that was occupied by the nucleus. Then the most obvious event of metaphase happens: the chromatid pairs attach themselves to individual spindle fibres and align themselves on the equator of the spindle.

Anaphase

At the start of anaphase (ana = apart), as Figs 23.9d and 23.10 show, the chromatids are pulled apart by movements of the spindle fibres: sister chromatids are pulled to opposite poles. The newly separated chromatids are now called chromosomes, and are single structures. If you watch a film of mitosis (highly recommended for learning purposes), you will see that anaphase is the most obvious event.

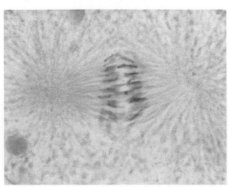

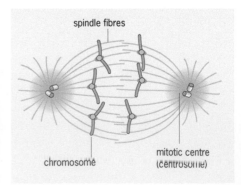

Fig 23.10 The spindle is a cradle of microtubule fibres which organise the chromosomes during cell division

Telophase

In telophase (telo = final) the chromosomes reach the poles of the spindle: Fig 23.9e. They then unravel and become indistinct, forming the familiar chromatin of interphase. The nuclear membrane re-forms and the nucleolus reappears. Soon afterwards, transcription resumes and the cell restarts protein synthesis, endocytosis and other normal cytoplasmic functions. This marks the end of mitosis.

Cytokinesis – division of the cytoplasm

In the events of mitosis just described, we have looked only at the splitting of the nucleus of the cell. Splitting of the cell itself is called **cytokinesis** and usually begins during anaphase. In most animal cells, microtubules form a furrow in a ring around the cell. These gradually constrict until the cells separate, as shown in Fig 23.11.

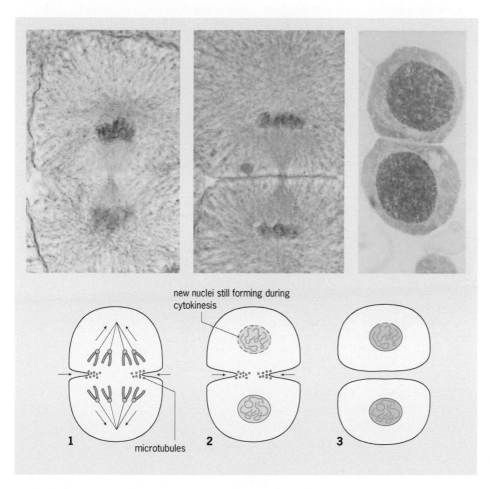

new nuclei still forming during cytokinesis

microtubules

Fig 23.11 Cytokinesis in animal cells. As the events of mitosis come to an end, movement of microtubules causes a constriction around the centre of the cell. Eventually the cytoplasm divides, leaving two new daughter cells

MEIOSIS: MAKING THE GAMETES

Meiosis is the type of cell division that halves the genetic material in cells. For this reason, it is known as a **reduction division**. In humans, as in all animals, meiosis makes haploid gametes (sex cells).

Meiosis is a remarkable process that produces haploid gametes and shuffles the genes, so that each gamete produced is genetically different. This is why children born to the same parents are usually not identical: they are produced by the fusion of a genetically unique sperm and egg. Identical twins are an exception because they originate from the same fertilised egg.

> **? QUESTION 5**
>
> 5 **a)** Where in the body does meiosis occur in i) women and ii) men?
> **b)** Why is meiosis sometimes referred to as reduction division?

THE STAGES OF MEIOSIS

Meiosis has two separate divisions. Both divisions have stages that are given the same names as in mitosis, but they have a number to denote whether they refer to the first or second division, for example anaphase I, prophase II.

It is important to remember here that a normal diploid cell contains two copies of each chromosome, one from the 'mother', and one from the 'father'. Just before meiosis, the cell's DNA becomes organised into chromosomes and replicates to form pairs of chromatids in preparation for division, just as it does before mitosis. So at the first stage of meiosis, the cell contains 92 (2 × 46) chromatids. The 23 chromosomes originally from the mother have been duplicated, making 23 identical pairs of chromatids, as have the 23 chromosomes originally from the father.

An overview of meiosis is shown in Fig 23.12. There is a basic difference between meiosis and mitosis: in meiosis, the homologous chromosomes pair up; in mitosis this does not happen. During the first part of meiosis, each homologous pair can swap sections of DNA, so that each pair becomes 'genetically mixed'.

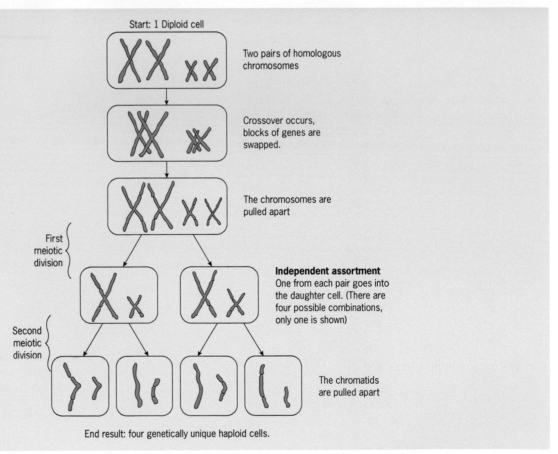

Start: 1 Diploid cell

Two pairs of homologous chromosomes

Crossover occurs, blocks of genes are swapped.

The chromosomes are pulled apart

First meiotic division

Independent assortment
One from each pair goes into the daughter cell. (There are four possible combinations, only one is shown)

Second meiotic division

The chromatids are pulled apart

End result: four genetically unique haploid cells.

Fig 23.12 An overview of meiosis in an imaginary animal cell with two pairs of homologous chromosomes

The individual stages of meiosis are shown in Table 23.3. In the first meiotic division (prophase I to telophase I), one of each homologous pair of genetically mixed chromosomes passes into each new cell. In the second meiotic division, the individual chromatids are pulled apart so that one chromatid (now a chromosome) goes into each daughter cell. The end result of meiosis is four haploid cells, each containing a single set of chromosomes. Each cell is genetically unique.

Table 23.3 The stages of meiosis in an imaginary animal cell showing just one of the pairs of chromosomes

Stage of meiosis First division	What is happening	What it looks like
interphase	Just before meiosis, DNA replicates, so cells which contained two copies of each chromosome now have four. Chromosomes not yet visible.	nucleolus
early prophase I	Chromosomes become visible. Centrioles move to opposite sides of cell.	centrioles — early prophase I
mid prophase I	Each homologous pair of chromosomes comes together to form a bivalent.	mid-prophase I
late prophase I	Each chromosome in a bivalent forms two chromatids. Genetic mixing occurs: chiasmata, the points of crossover, are visible.	
metaphase I	The bivalents arrange themselves on the equator of the spindle.	spindle — metaphase I
anaphase I	The chromatid pairs from each homologous chromosome split apart and move to opposite poles of the cell.	anaphase I
telophase I	Cytokinesis begins, two new cells form, each has two copies of each chromosome. These chromosomes are genetically different from those in the original cell.	telophase I
interphase	A resting time (length varies between cell types).	
Second division		
prophase II	A new spindle forms, at right angles to the first.	
metaphase II	Chromosomes, each of which is a pair of chromatids, align themselves on the equator of the spindle.	metaphase II (1 cell only)
anaphase II	Chromatids are pulled apart to form two chromosomes that then move to opposite poles of the cell.	anaphase II (1 cell only)
telophase II	Cytokinesis begins. Four haploid cells, each with only a single chromosome, have been formed. Each chromosome is genetically different.	telophase II

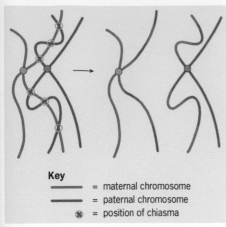

Key

— = maternal chromosome

— = paternal chromosome

✸ = position of chiasma

Fig 23.13 Chiasmata form between homologous chromosomes during late prophase I of meiosis. At these points, parts of the maternal chromosome separate and join with the paternal chromosome, and vice versa. This produces new chromosomes that are genetically different from each other, and from both the maternal and paternal chromosomes from which they are derived

Independent assortment is discussed in Chapter 25.

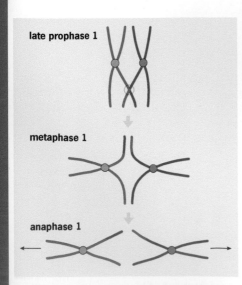

Fig 23.14 Chiasmata form during late prophase I and crossover takes place. The newly formed chromatid pairs separate during metaphase I and then the two pairs move to opposite poles of the cell during anaphase I

Prophase I

Prophase I is more complex than the corresponding phase in mitosis, and can be subdivided into early, middle and late.

Early prophase I starts when the chromosomes condense and the nucleolus disappears.

In **mid-prophase I**, the homologous chromosomes, one from each parent, pair up. Each pair forms a bivalent. This does not happen in mitosis. When a chromosome pair is exactly aligned, it is said to be at **synapsis**. Next, each chromosome in a pair divides into two chromatids, giving four chromatids per bivalent.

In **late prophase I**, **recombination** – or crossover – takes place (Fig 23.13). One (or both) of the chromatids of the two homologous chromosomes breaks off at certain points and fuses with a chromatid of the other chromosome in the bivalent, forming joints called **chiasmata** (the singular, *chiasma*, means 'crosspiece'). This process ensures that blocks of genes are swapped between maternal and paternal chromosomes. The position of chiasma formation varies, even within the same species, and this produces a large variety of new allele combinations.

Prophase ends with the chromatids of each bivalent (pair of homologous chromosomes) entwined and joined by chiasmata.

Metaphase I

In metaphase I, the nuclear membrane disappears and the spindle is fully developed. The bivalents move to the equator of the spindle in the same way as individual chromosomes do during the matching phase in mitosis.

Anaphase I

The chromatid pairs of a bivalent are pulled apart (Fig 23.14) because of the action of spindle fibres, a process that separates the entwined chromatids. At the end of anaphase I, the chromatid pair from one of the original homologous chromosomes is positioned at one pole of the cell and the chromatid pair from the other homologous chromosome is at the other pole. Either the maternal or the paternal chromosome can pass into either cell. This is the process that allows **independent assortment of alleles**. It is another reason why meiosis increases variation.

Occasionally, this phase is not completed successfully and a pair of chromosomes fails to separate. The result is that both homologous chromosomes pass into one daughter cell, the other receiving neither. This situation can lead to conditions such as Down's syndrome.

Telophase I

The spindle disappears and the nuclear envelope re-forms around the two sets of chromosomes. At the same time cytokinesis separates the cytoplasm, forming two daughter cells, ready for the second meiotic division.

Interphase

The length of the resting interphase between the two meiotic divisions varies widely. It is sometimes short or even non-existent. If there is no interphase, the chromosomes remain condensed and the cell passes straight from telophase I into prophase II. In human females, however, ova may remain in interphase for decades. Basic egg cells are made in a girl before birth, but they do not complete meiosis until just before ovulation, anything up to 50 years later.

At the end of meiosis I, there are two cells, each containing two copies of each chromosome on a chromatid pair. The second meiotic division separates the chromatids, so that each daughter cell formed is haploid (has one set of single chromosomes).

The second meiotic division

Prophase II

For each chromosome, the chromatid pair attaches itself to the new spindle, which forms at right angles to the first.

Metaphase II

Each chromatid pair lines up on the equator of the spindle.

Anaphase II

The chromatids are pulled apart to form two chromosomes that move to opposite poles.

Telophase II

The spindle disappears, the nuclear membrane re-forms, chromosomes unravel and cytokinesis produces two separate cells.

The end-product of meiosis

Meiosis produces four genetically different haploid cells, known as a **tetrad**. Genetic variation has been produced in three ways:

- The homologous chromosome pairs originate in different organisms, one maternal and one paternal, and so are genetically different.
- Blocks of genes are swapped between the chromatids of homologous chromosomes as the chiasmata form during prophase I.
- Each daughter cell can receive a copy of either chromosome from a pair, and each copy may have undergone crossover and have different genes from the other three (independent assortment).

? QUESTION 6

6 **a)** How many bivalents form in a human cell undergoing meiosis?

✔ REMEMBER THIS
The importance of genetic variation should become apparent as you study evolution. Find out more in Chapter 26.

SCIENCE IN CONTEXT

Down's syndrome

In the United Kingdom, two children in every 1000 are born with Down's syndrome (Fig 23.15a), a genetic condition that arises from a fault in meiosis. Individuals with this condition have an extra chromosome 21 (Fig 23.15b) and usually have physical and mental disabilities. They often have a small mouth with a normal-sized tongue (making eating and speech difficult), reduced resistance to disease and heart abnormalities.

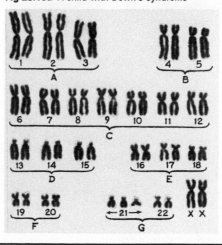

Fig 23.15a A child with Down's syndrome

The reason for the extra chromosome is usually a failure of the chromosome 21 pair to separate during anaphase I of meiosis (see Table 23.3). Both chromosomes 21 pass into one daughter cell, and the other gets no copy at all. If a gamete with no chromosome 21 forms a zygote, the embryo fails to develop. But, if the gamete containing both chromosomes is fertilised, the resulting baby later develops Down's syndrome.

One reason for the non-separation may be the length of time the egg spends in the ovary. In females, egg cells do not complete their meiotic division until they are 'selected' for ovulation (see Chapter 20). The longer the eggs remain in the ovary, the 'stickier' the chromosomes become and the greater the probability that they will not separate during anaphase I. This explains why the chance of having a child with Down's syndrome increases as women get older. Women below the age of 25 have less than a 1 in 2000 chance, but this rises to 1 in 50 for women over 45.

The age of the father is also important: recent studies have shown that in a significant minority of cases the extra chromosome comes from the father – that is, a normal ovum is fertilised by a sperm containing two chromosomes 21.

Fig 23.15b The complement of chromosomes (the karyotype) of a person with Down's syndrome, showing the extra chromosome 21

LEARNING MEIOSIS

Meiosis is a complex process that many people find difficult to learn. Here are some hints that you may find useful. First, divide your study of meiosis into three sections by looking at three questions:

- **What is the point of meiosis?**
  It makes gametes by shuffling the genes to produce haploid cells that are all genetically different from either of the parents. Each fertilisation (combination of an egg and a sperm) produces an individual that is genetically unique.

- **What do each of the meiotic divisions achieve?**
  Start by learning an overview of each division: Table 23.3 is a good place to start. In the first meiotic division, homologous chromosomes pair up, divide into chromatids, swap blocks of genes and then separate. In the second division the individual chromatids separate. The end result is four genetically different haploid cells from one diploid original.

- **What are the stages of the two divisions?**
  Once you are confident that you can answer the first two questions, you can put some flesh on the bones. Describe one stage at a time and then draw what you have just described. Try using different colours for paternal and maternal chromosomes. Alternatively, you could make some chromosomes out of modelling clay and work through the meiotic sequence yourself. Many students find this a valuable exercise.

SUMMARY

After reading this chapter, you should know and understand the following:

- Almost all the cells in the human body are **diploid**: they contain a double set of chromosomes. The only exceptions are the gametes; sex cells – egg cells and sperm – are **haploid** and contain only one set of chromosomes.

- The cell cycle consists of four stages: G_1, in which the cell grows; **S**, in which the DNA is doubled; G_2, a second growth phase; and finally **mitosis** or **meiosis**. G_1, S and G_2 are collectively known as **interphase**.

- Mitosis is normal cell division, the process by which we grow and develop. One diploid cell divides to produce two genetically identical diploid cells.

- You can remember the stages of mitosis using the nmemonic IPMAT: Interphase, Prophase, Metaphase, Anaphase, Telophase.

- Non-dividing cells are said to be in interphase. There are no chromosomes visible and the active genes in the DNA are being used to direct the activities of the cell. If the cell is going to divide, the cell prepares itself during interphase and the DNA is replicated.

- In prophase, the chromosomes condense and become visible. The nuclear membrane disappears and the **spindle** forms. The chromosomes are double: two identical chromatids are joined at the **centromere**.

- In metaphase, the chromosomes arrange themselves in the middle of the spindle (remember, meta = middle).

- In anaphase, the double chromosomes are pulled apart (remember, ana = apart).

- In telophase, the nuclear membrane re-forms around the two new sets of chromosomes and the cytoplasm divides forming two new, genetically identical cells.

- Meiosis is a special type of cell division which produces sex cells or gametes. It is also known as a 'reduction division'. One diploid cell will divide meiotically to produce four genetically different haploid cells. Vitally, meiosis creates new allele combinations.

Practice questions and a How Science Works assignment for this chapter are available at www.collinseducation.co.uk/CAS

24

DNA, genes and chromosomes

24 DNA, GENES AND CHROMOSOMES

Resveratrol anyone?

Live long and prosper

You are probably inundated with information about how harmful drinking alcohol can be, but one group of scientists are convinced that taking large quantities of one of the compounds found in red wine could be the answer to a long and healthy life. Red wine and the red grapes that are used to make it contain a molecule called resveratrol. Studies have shown that mice given resveratrol and a high-fat diet didn't gain weight and lived longer than those just fed the stodge. This might explain why the French enjoy a high-fat diet yet suffer less heart disease than Americans.

Resveratrol is a gene inducer that turns on expression of a gene called *SIRT-1* in mice, and possibly also in humans. This gene is highly expressed in animals and humans fed a low-calorie diet and it does seem to be linked with living longer. One scientist, David Sinclair of Harvard University in the USA, is convinced that taking high amounts of resveratrol, equivalent to drinking about 15 glasses of red wine a day, could prevent human ageing and extend lifespan. He has founded a pharmaceutical company to investigate this further. However, approval cannot be given for a drug that just prevents ageing as the US Food and Drug Administration, the body that approves drugs, does not recognise ageing as a 'disease'. Resveratrol may have other beneficial effects though and the company has begun clinical trials to see if an improved version of resveratrol can help control glucose levels in people with diabetes.

1 WHAT IS A GENE?

The nucleus of a eukaryotic cell has a set number of chromosomes – every human body cell has 46. Each chromosome is a single, elaborately coiled DNA molecule that has individual genes dotted along its length (Fig. 24.1). A gene is a length of DNA that codes for the synthesis of one polypeptide.

We used to think that a gene contained all the information needed to build a complete protein, such as an enzyme. However, many proteins consist of more than one polypeptide chain, so these molecules are coded for by more than one gene. For instance, haemoglobin molecules consist of four polypeptide chains, two alpha and two beta, so it takes two genes to make a complete haemoglobin molecule: one is copied twice for the alpha chains, the other twice for the beta chains. The position of a gene on a chromosome is known as its **locus** (plural: loci).

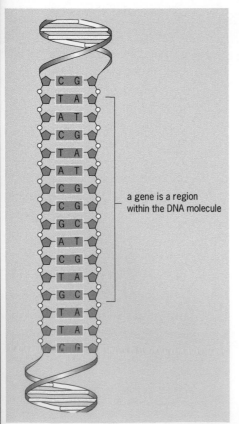

a gene is a region within the DNA molecule

Fig. 24.1 Genes are sequences in the DNA molecule. The sequence of bases shown here represents one very short gene. In the genetic code, a group of three bases codes for one amino acid, so a protein consisting of 500 amino acids requires a gene of at least 1500 base pairs long, probably more. When unravelled, chromosomes are incredibly long – each one consists of up to 4000 genes and these make up only about 10 per cent of its total length (the rest is non-coding DNA)

GENES IN ACTION:
THE CENTRAL CONCEPT OF MOLECULAR BIOLOGY

Can we sum up the way genes work in one sentence? It is a tall order, but the underlying theme, or **central dogma** (Fig 24.2), of genetics is:

The DNA of a gene codes for the production of messenger RNA which, in turn, codes for the production of a polypeptide.

In this chapter, we look at how genes work, and at how they are copied to allow cell division, reproduction and growth.

After Crick and Watson worked out the structure of DNA in 1953, two vital questions remained:

- How does information get from the DNA in the nucleus to the site of protein synthesis in the cytoplasm?
- How does the base sequence of the DNA translate into the amino acid sequence in a polypeptide?

To answer the first question, Crick, Brenner and Monod developed the **messenger hypothesis**. This states that a specific molecule, named **messenger RNA** (mRNA), is copied directly from the DNA sequence of the gene. mRNA is therefore a mobile copy of a gene. It can move from the nucleus to the cytoplasm. Here, ribosomes use mRNA as a template (pattern) to assemble amino acids in the correct order to make the required polypeptide.

Crick also answered the second question: he proposed that an 'adaptor' molecule existed which had two specific binding sites. At one side, the molecule fitted a specific DNA base sequence and at the other, a specific amino acid. This molecule was later discovered and named **transfer RNA** (tRNA) (look ahead to Fig 24.6 on page 361).

2 THE GENETIC CODE

Along the DNA molecule the base sequence is continuous, but it is read in blocks of three. Each block of three, or triplet, is called a **codon**. Each codon codes for a particular amino acid. A particular sequence of bases therefore codes for a specific amino acid sequence. When the amino acids are assembled into a polypeptide, they interact to twist and bend the chain into its final shape. When the polypeptide chain is complete, it forms all or part of a protein which performs a vital role in the organism.

You will have noticed that there are two strands of bases in the DNA molecule. One strand, called the **template strand** or **sense strand**, contains the all-important genetic code for any particular gene. The corresponding part of the other strand is simply there to stabilise the molecule. However, in the same DNA molecule, different genes are present on different sides and the enzymes that transcribe DNA can use either side as the sense strand.

REMEMBER THIS

The genome is defined as 'all the DNA sequences in an organism'. It is a complete set of genes, together with the non-coding DNA in between. The human genome is thought to consist of over 3 billion base pairs. Remarkably, the entire genome is present in every one of our body cells.

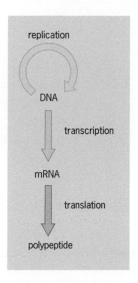

Fig 24.2 The central concept of molecular biology. The base sequence on a particular gene is copied, or transcribed, to produce a messenger RNA template. This mRNA template passes out of the nucleus and moves to the ribosomes where it is read, or translated, into a polypeptide

REMEMBER THIS

Retroviruses such as the human immunodeficiency virus (HIV) are an exception to the central dogma. Retroviruses contain RNA from which they make a DNA copy using the enzyme **reverse transcriptase**.

REMEMBER THIS

The base pairings are always the same. In DNA, A bonds with T and G with C. In RNA, T is replaced by U, so A bonds with U, and G with C (see Chapter 3).

Nucleic acids – molecules with jobs

A favourite topic for biology examiners is to ask the candidate to relate structure to function. On a molecular level, the nucleic acids are classic examples. This feature builds on your knowledge of the basic structure of nucleic acids covered in Chapter 3. The differences between DNA and RNA are shown in Table 24.1 and Fig 24.3.

Table 24.1 The differences between DNA and RNA

	DNA	mRNA	tRNA
Sugar	deoxyribose	ribose	ribose
Bases	C, G, A, T	C, G, A, U	C, G, A, U
Strands	two	one	one
Size of molecule	enormous	small compared with DNA, varies with size of gene	small, constant size
Life span	long term	short term	short term
Site of action	nucleus	nucleus and cytoplasm	cytoplasm

DNA

How is the structure of DNA related to its function?

- It contains a sequence of bases that codes for a sequence of amino acids. Cells use this genetic code to make the proteins that build organisms. The base sequence is read in blocks of three, e.g. GAU – this is a codon.

- It is long, so can store a lot of information.

- It has complementary base pairing that allows replication and transcription.

- It has a stable sugar–phosphate backbone to give strength.

- It has relatively weak hydrogen bonds between the bases. This allows the two strands to come apart – again allowing replication and transcription.

Messenger RNA

mRNA is simply a mobile copy of a gene, carrying a base sequence that is complementary to – not the same as – the gene on which it was made. mRNA molecules are long, ribbon-like molecules consisting of a single strand of nucleotides. Molecules of mRNA vary in length – there are as many different ones as there are genes.

Transfer RNA

This is the adaptor molecule, connecting the genetic code to the protein. tRNA molecules have an **anticodon** on one end and an amino acid binding site at the other. The role of tRNA is to pick up a particular amino acid, bring it to the ribosome and hold it in place so it can be added to the growing polypeptide. tRNA binds to the mRNA by means of the anticodon. It could be argued that because there are 20 different amino acids, there are 20 different types of tRNA molecule. Another way of looking at it is that there are 61 different codons (64 minus 3 **stop codons**) so there are 61 different types of tRNA.

Fig 24.3 Comparison of the three types of molecule

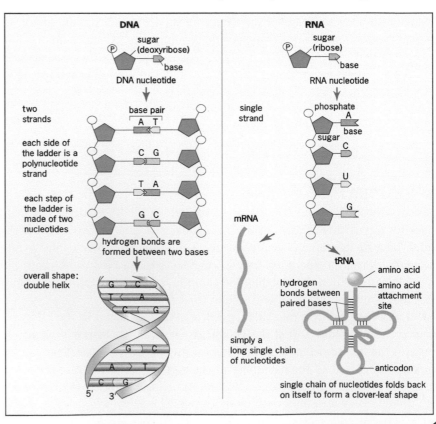

TRANSCRIPTION

DNA can be thought of as a permanent reference library: it does not leave the nucleus and so cannot be used directly for protein synthesis. Instead, the genetic code in a gene is transferred, or **transcribed**, on to a smaller RNA molecule, **messenger RNA**, which can then pass to the ribosomes to act as a template for protein synthesis. In effect, mRNA is a mobile copy of a gene.

Before transcription can begin, the DNA helix must unwind and the two halves of the molecule must come apart, so exposing the base sequence. This process begins when the DNA is 'unzipped' by specific enzymes and the enzyme **RNA polymerase** attaches to the DNA molecule at the **initiation site**. This is a base sequence at one end of the gene which effectively says: start here. Once in place, the enzyme moves along the gene (Fig 24.4), assembling the messenger RNA molecule by adding the matching nucleotides, one at a time.

Once formed, the mRNA copy begins to peel away from the DNA. When RNA polymerase has passed a particular region of DNA, the double helix rewinds. When the whole gene has been transcribed, the complete mRNA molecule leaves the nucleus. As on a production line, RNA polymerase molecules can follow each other along the gene, assembling several copies of mRNA together.

QUESTION 1

1 The RNA polymerase enzyme moves along a DNA sequence reading ATACGCTAT.
a) What is the corresponding sequence on the RNA molecule?
b) How many amino acids are encoded in this sequence?

The contents of a cell and the chemicals necessary for life are discussed in Chapters 1 and 3.

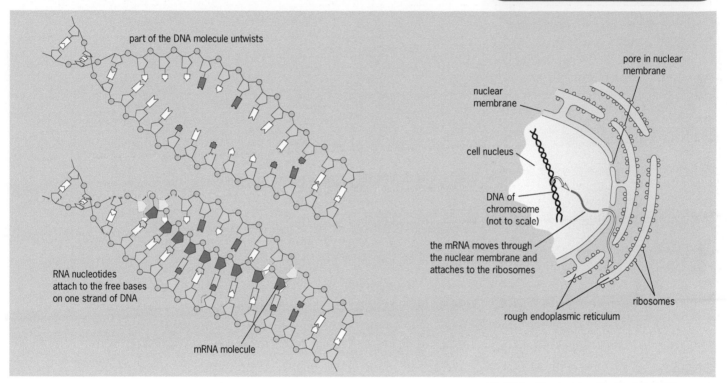

part of the DNA molecule untwists

RNA nucleotides attach to the free bases on one strand of DNA

mRNA molecule

pore in nuclear membrane

nuclear membrane

cell nucleus

DNA of chromosome (not to scale)

the mRNA moves through the nuclear membrane and attaches to the ribosomes

ribosomes

rough endoplasmic reticulum

Fig 24.4a The process of transcription. The process is controlled by enzymes that are not shown here. Starting at the initiation site, helicase enzymes separate the two sides of the DNA molecule, and then the enzyme RNA polymerase calalyses the assembly of mRNA, one nucleotide at a time, until the gene has been transcribed

Fig 24.4b The sites of transcription and translation in the cell. Transcription happens in the nucleus and involves the assembly of a molecule of messenger RNA. Translation happens on the ribosomes and involves the assembly of a protein according to the code on the mRNA

TRANSLATION

Translation occurs when the base sequence on the mRNA is used to synthesise a polypeptide. The process is called translation because the information contained in the gene – and delivered by the mRNA – is translated into a polypeptide (Table 24.2 on the next page).

Protein synthesis is one of the major activities of cells, and it is highly likely that you are already familiar with many of the organelles and chemicals involved. The roles of individual organelles in the process are listed in Table 24.3.

Translation occurs on **ribosomes**, structures that hold all the components together as an amino acid chain is made. The mRNA strand passes out of the nucleus and attaches to the ribosome. At the point of attachment, two tRNA molecules (Fig 24.5) deliver the required amino acids and hold them in position so that they can be added to the growing polypeptide.

The mechanism of translation is shown in detail in Fig 24.6. The process continues until an mRNA **stop codon** – which has no tRNA molecule – moves into the ribosome's first binding site. At this point, all the components separate, releasing the completed polypeptide.

QUESTION 2

2 How do mammals ensure that a regular supply of amino acids reaches all of their cells?

REMEMBER THIS

Genes code for the manufacture of polypeptides. Fig 3.25 on page 46 shows the importance of protein in the human body.

REMEMBER THIS

Many people find protein synthesis a difficult topic to learn. A valuable exercise is to make the different components – either from modelling clay or just paper – and work through the events shown in Fig 24.6b.

You will need to make a ribosome, an mRNA molecule with several codons written on it, and enough tRNA molecules and amino acids to match to all the codons on the mRNA.

Table 24.2 The genetic code: the base sequence in each mRNA triplet code translates to a particular amino acid. This same code is used by all organisms

First base	Second base				Third base
	G	**A**	**C**	**U**	
G	GGG glycine	GAG glutamic acid	GCG alanine	GUG valine	G
	GGA glycine	GAA glutamic acid	GCA alanine	GUA valine	A
	GGC glycine	GAC aspartic acid	GCC alanine	GUC valine	C
	GGU glycine	GAU aspartic acid	GCU alanine	GUU valine	U
A	AGG arginine	AAG lysine	ACG threonine	AUG methionine	G
	AGA arginine	AAA lysine	ACA threonine	AUA isoleucine	A
	AGC serine	AAC asparagine	ACC threonine	AUC isoleucine	C
	AGU serine	AAU asparagine	ACU threonine	AUU isoleucine	U
C	CGG arginine	CAG glutamine	CCG proline	CUG leucine	G
	CGA arginine	CAA glutamine	CCA proline	CUA leucine	A
	CGC arginine	CAC histidine	CCC proline	CUC leucine	C
	CGU arginine	CAU histidine	CCU proline	CUU leucine	U
U	UGG tryptophan	UAG **stop**	UCG serine	UUG leucine	G
	UGA **stop**	UAA **stop**	UCA serine	UUA leucine	A
	UGC cysteine	UAC tyrosine	UCC serine	UUC phenylalanine	C
	UGU cysteine	UAU tyrosine	UCU serine	UUU phenylalanine	U

Table 24.3 The main organelles and chemicals involved in protein synthesis

Organelle/molecule	Role in protein synthesis
nucleus	houses the DNA
nucleolus	manufactures the ribosomes
ribosome	site of translation – where polypeptide assembly occurs
endoplasmic reticulum	isolates, stores and transports polypeptides
Golgi body	modifies and packages polypeptides
DNA	stores the genetic information
messenger RNA	is a mobile copy of a gene on DNA
transfer RNA	each brings a specific amino acid to the ribosomes

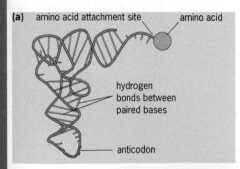

(a)

amino acid attachment site
amino acid
hydrogen bonds between paired bases
anticodon

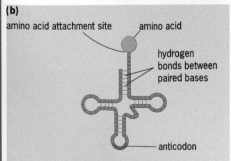

(b)

amino acid attachment site
amino acid
hydrogen bonds between paired bases
anticodon

Fig 24.5 Two ways of representing the tRNA molecule. **a** The 3D shape of the molecule, stabilised by hydrogen bonds as the single nucleotide chain folds back on itself
b A schematic diagram: each type of transfer RNA transports a specific amino acid to the ribosomes, the site of protein synthesis. At one end of the molecule is the anticodon which attaches to the corresponding codon on the messenger RNA. At the other end of the molecule is the amino acid attachment site

After translation, polypeptides are processed according to their final destination:

- Those that are to be exported from the cell, such as digestive enzymes, are threaded through pores in the endoplasmic reticulum to accumulate on the inside. Here they are processed and packaged before being secreted (see pages 12 and 14).

- Polypeptides that will form membrane proteins follow the same route as those for export, but they remain on the cell surface membrane rather than being released.

- Polypeptides that will be used inside the cell, such as those that form haemoglobin, remain free in the cytoplasm.

The role of transfer RNA

Transfer RNA molecules are smaller than mRNA molecules, containing only about 75 to 80 nucleotides. Each consists of a single strand of nucleic acid folded back on itself to form a 'clover-leaf' shape (Fig 24.5). Transfer RNA molecules bring specific amino acids from the cytoplasm to the ribosome so that they attach to the growing polypeptide (Fig 24.6).

Fig 24.6a The ingredients needed for protein synthesis

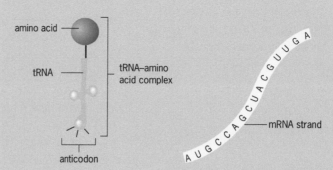

P site: site of attachment of the growing polypeptide

A site: each new amino acid is added here

large subunit of the ribosome

small subunit of the ribosome

amino acid

tRNA

tRNA–amino acid complex

anticodon

mRNA strand

Fig 24.6b The process of protein synthesis

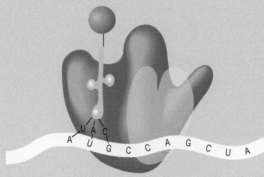

1 The two subunits of the ribosome come together. The mRNA strand binds to the ribosome. Then an amino acid–tRNA complex with an anticodon complementary to the first codon on the mRNA binds to the P site

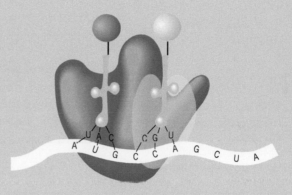

2 In the next step, an amino acid–tRNA complex with an anticodon complementary to the next codon on the mRNA strand binds to the A site

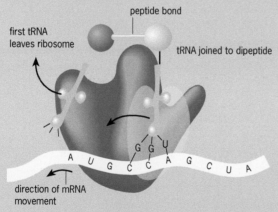

peptide bond

first tRNA leaves ribosome

tRNA joined to dipeptide

direction of mRNA movement

3 A peptide bond forms between the two amino acids which have been brought together on the ribosome. The bond between the first amino acid and the first tRNA is broken and the first tRNA molecule leaves the ribosome. The ribosome moves along the mRNA strand and the second tRNA molecule, now joined to a dipeptide, shifts across from the A site to the P site

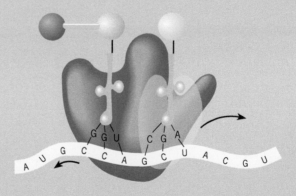

4 A third amino acid–tRNA complex with an anticodon that matches the third codon on the mRNA strands comes into the A site

The sequence of events from **3** to **4** is repeated until the entire length of the mRNA strand has been translated. The polypeptide is then complete

The process by which an amino acid binds to a tRNA molecule is controlled by an enzyme. The process also involves the splitting of ATP. This is important because ATP gives the tRNA–amino acid complex enough energy to form a peptide bond when the amino acid is added to the growing polypeptide.

At one end of the tRNA molecule is the **anticodon**, a three-letter base sequence that matches the codon on the mRNA molecule. At the other end is

? QUESTION 3

3   Is protein synthesis an anabolic or a catabolic reaction?
(Hint: it's a vital part of body building.)

REMEMBER THIS

The DNA code is said to be **degenerate**, which means that there can be more than one codon for each individual amino acid.

? **QUESTION 4**

4 Why must several different tRNA molecules carry the same type of amino acid?

Fig 24.7 A polysome is a cluster of ribosomes which can simultaneously translate a different section of the mRNA molecule. The situation is similar to having a message on a long ribbon, with several people reading different bits at once

a particular amino acid. So, for instance, if the codon on the mRNA reads AUG, which codes for the amino acid methionine, it needs a tRNA molecule with an anticodon of UAC which arrives at the ribosome carrying a methionine molecule.

There are 64 different codons, but three of them code for 'stop translating'. When such a codon arrives at the ribosome, no tRNA is needed. This means that tRNA molecules must match with 61 different codons. There are only 20 different amino acids, so some amino acids are translated from more than one codon. For example, the codons GGG, GGA, GGC and GGU all code for the amino acid glycine.

Like messenger RNA, transfer RNA is also made by transcription. In eukaryote DNA there are many genes whose sole function is to make tRNA. In any cell, most genes occur only once, but there are hundreds of genes that code for tRNA synthesis, to provide the vast numbers of these workhorse molecules that are needed. That tRNA genes do not code for proteins is another exception to the central dogma.

Polysomes and the rate of translation

We have seen that the mRNA molecule is a long single strand of nucleotides, and that the code is translated into a polypeptide at the point of contact with a ribosome. To achieve protein synthesis at a reasonable speed, ribosomes occur in clusters, called **polysomes** or **polyribosomes**, which all translate a different bit of the mRNA at the same time (Fig 24.7).

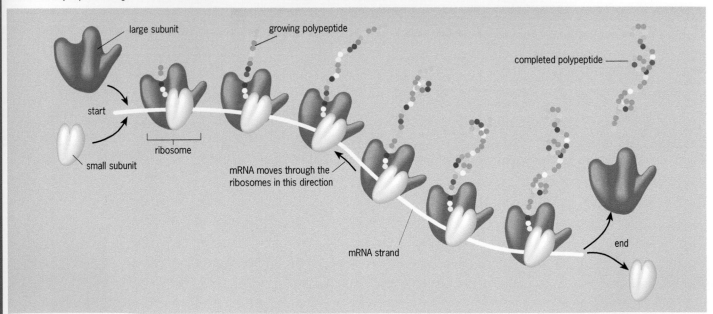

large subunit — growing polypeptide — completed polypeptide — start — ribosome — small subunit — mRNA moves through the ribosomes in this direction — mRNA strand — end

3 A JOURNEY ALONG A CHROMOSOME

If one single chromosome were unravelled, like pulling the wool out of a piece of knitting, we would be left with one long DNA strand. Although this is one single molecule, it is several *centimetres* long and contains up to 4000 genes.

How do you go about investigating such a molecule? One way is to work backwards: if you can isolate the mRNA molecules that are in the cytoplasm, you can trace their origin in the DNA.

A clever technique called **nucleic acid hybridisation** is used to pinpoint the origin of the mRNA, giving the location of the gene itself. In this technique, DNA strands are denatured: when heated to 87 °C, hydrogen bonds that hold the two strands together are broken, and the individual strands separate. If messenger RNA strands from the cytoplasm are added to the mixture, they stick to the region of DNA on which they were originally made.

Studies using this technique have unexpectedly revealed that there is some non-coding DNA in genes (Fig 24.8). Most eukaryote genes contain regions, called **introns**, which do not find their way into the mRNA molecule at all. The parts of the gene that are expressed in the mRNA are called **exons** because they are the expressed regions of the gene.

Generally, a gene consists of a DNA sequence which is present only once. The non-coding sequences, however, tend to consist of repeated or 'stuttered' sequences. Humans all have the same repeated sequence, but in different individuals it is repeated a different number of times and in different places. This is the basis of DNA profiling – see the Assignment for Chapter 3 on the CAS website.

The function of non-coding DNA is not known. It is probably not useless: much of the repeated DNA in a chromosome occurs around the region of its centromere and may have an important role in maintaining the physical stability of the chromosome as a whole.

The DNA code can be compared to a page of writing in a book. If the letters were strung together continuously, we would find it difficult to read: the gaps between words help us to make sense of the sentences.

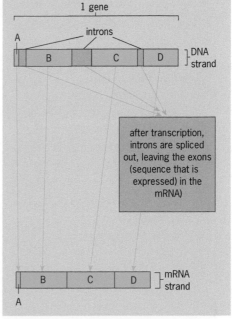

Fig 24.8 Even in a gene, there are non-coding lengths of DNA called **introns**. Only the **exons** are encoded into the mRNA molecule, and therefore expressed

4 THE CONTROL OF GENE EXPRESSION

Most multicellular organisms start life as a zygote: a single fertilised cell. This cell has a complete set of genes. As the cell divides by mitosis, the DNA is copied faithfully, so each new cell also contains the same complete set of genes. How, then, do cells differentiate? How, for example, do some animal cells specialise into muscle, nerve or skin?

The answer lies in **gene expression** – how different genes are 'switched on', or **expressed**, in different cells (Fig 24.9).

Fig 24.9 A spectacular example of the power of gene expression is seen in the life cycle of some insects such as the peacock butterfly. The genes present in the fertilised egg are still present in the adult, but a different set is activated to give rise to the caterpillar and to control its metamorphosis into a pupa and then an adult – which looks like a completely different organism

Insulin and the pancreas are discussed in Chapter 11.

REMEMBER THIS

If we could control gene expression, we would be able to make new tissues and organs for repair and transplants. See the section on stem cells in Chapter 27 for more about the potential of controlling gene expression.

At any one time, the average cell is probably using only 1 per cent of its available genes. Some of the expressed genes are used for 'housekeeping' – they code for the proteins needed by all cells, such as the enzymes involved in respiration. In contrast, other genes are expressed only in specialised cells. The genes for making insulin, for example, are expressed only in the β cells of the islets of Langerhans in the pancreas.

CONTROLLING TRANSCRIPTION

The control of transcription is the key to the selective activation of genes. A gene is switched on, or **induced**, when transcription begins.

Transcription starts when the enzyme **RNA polymerase** and several proteins known as **transcription factors** bind to an area 'upstream' of the gene called the **promoter region**. The proteins assemble into a **transcription initiation complex** (**TIC**). Once this complex is in place, transcription can begin (Fig 24.10).

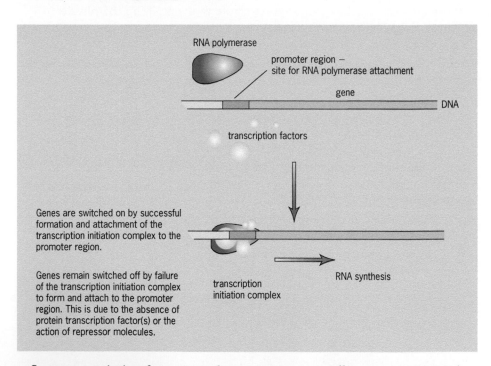

Genes are switched on by successful formation and attachment of the transcription initiation complex to the promoter region.

Genes remain switched off by failure of the transcription initiation complex to form and attach to the promoter region. This is due to the absence of protein transcription factor(s) or the action of repressor molecules.

RNA polymerase

promoter region – site for RNA polymerase attachment

gene

DNA

transcription factors

RNA synthesis

transcription initiation complex

Fig 24.10 Controlling transcription

Some transcription factors are always present in a cell, some are present but in an inactive form, and some are not made until a key stage in a cell's life. Genes remain switched off until the correct transcription factors are present in their active form. Transcription factors can be activated by **signal proteins**, which may be **hormones**, **growth factors** or other regulatory molecules, or by environmental signals.

In addition, transcription can be prevented by the presence of certain repressor molecules, which attach to the promotor region and prevent the formation of the TIC.

Pioneering work on gene induction was done by Jacob and Monod in the early 1960s. They worked on the bacterium *Escherichia coli. E. coli* can digest lactose (milk sugar) by making the enzyme β galactosidase. When there is no lactose, the enzyme is not made because the β galactosidase gene is not active. However, the presence of lactose causes the gene to be induced. Jacob and Monod found that in the absence of lactose there was a repressor molecule that bound to the promotor region and prevented the formation of the TIC. When lactose was present, the repressor molecule was removed and the gene was activated.

Gene induction in humans happens all the time and it is extraordinarily complex. One set of genes that have been studied in detail are the stress response genes. There are lots of these in the genome and normally they are inactive. If a cell becomes stressed – because it is exposed to a carcinogen, or to a high salt concentration, or because it gets short of oxygen, these genes are then induced to try to rescue the cell. Many of them code for enzymes that repair cell structures. Researchers in Cambridge in the UK have shown that when human cells in culture are stressed they start to express about 20 of these stress genes very quickly.

During development of the embryo, different genes are induced at different times and working out what genes are active and how they are activated is one of the great challenges left in biology. Cells in a developing embryo 'know' where they are because of chemical signals they receive from surrounding cells. These signals also control whether each cell divides or differentiates.

siRNA

A new class of RNA molecules has recently been found that has a variety of functions in the cell, including the control of transcription. They can act as repressor molecules because they prevent gene expression. Called **small interfering RNA** (**siRNA**), they are a class of small molecules, 20–25 nucleotides long and double stranded. Most notably, siRNA is involved in the **RNA interference** (**RNAi**) pathway where the siRNA interferes with the *expression* of a specific gene.

siRNAs have great potential in medicine and genetic engineering, because they can be made artificially and introduced into cells to bring about the specific **knockdown** of a particular gene, e.g. one that causes a particular type of cancer. Essentially, any gene of which the sequence is known can be de-activated by a tailor-made siRNA molecule.

5 DNA REPLICATION

Whenever a cell divides, the DNA must be copied. Otherwise, there would be no reproduction and no growth. When cells divide, each new daughter cell must have a complete set of genes. **Replication**, the mechanism of DNA copying, is therefore of fundamental importance.

Just three years after Crick and Watson published their model of DNA structure in the journal *Nature*, their theory about DNA replication was confirmed by Arthur Kornberg. He put DNA into a mixture containing all four **nucleotides** (each of the four bases attached to a phosphate and a sugar), together with the enzyme DNA polymerase, and showed that the DNA could replicate without any other factors present.

In the mixture, each DNA strand unwinds and acts as a template for the construction of a new strand. The exposed strand acts as a template on which the free nucleotides arrange themselves in exactly the same sequence as the intact strand they replace. This model is called **semi-conservative replication** because each of the resulting strands of DNA contains one strand from the original DNA and one newly synthesised strand: half has been conserved and half is new.

Strong support for the semi-conservative model of DNA replication came from the work of Meselson and Stahl (Fig 24.11). By growing bacteria with bases containing the radioisotope ^{15}N (sometimes called heavy nitrogen) and then following its progress, they showed that each new DNA strand contains half the original strand.

> **? QUESTION 5**
>
> 5 Predict the outcome of the third generation of DNA molecules in Meselson and Stahl's experiment. Of the eight DNA strands produced, how many would be hybrids and how many would be all ^{14}N strands?

365

Fig 24.11 Meselson and Stahl's experiment to support the semi-conservative replication model of DNA replication

1 To start with, many generations of *E. coli* bacteria are grown in a medium containing nucleotides labelled with the heavy nitrogen radioisotope, ^{15}N. Eventually, almost all of the bacterial DNA contains ^{15}N, and so is denser than normal DNA. A control culture of the same bacterium is grown with 'normal' ^{14}N nucleotides for comparison

2 The bacteria grown in ^{15}N are transferred to ^{14}N medium. After 0, 20 and 40 minutes, samples of the DNA are extracted, placed in a solution of caesium chloride and centrifuged. This separates out the DNA according to its density and allowed Meselson and Stahl to distinguish between DNA containing ^{15}N and ^{14}N

3 The results confirm that, initially, all the DNA contained ^{15}N. After 20 minutes the DNA had replicated, producing two strands of ^{15}N/^{14}N hybrid DNA. After 40 minutes, the second generation strands consisted of half ^{15}N/^{14}N hybrids and half all-new ^{14}N strands

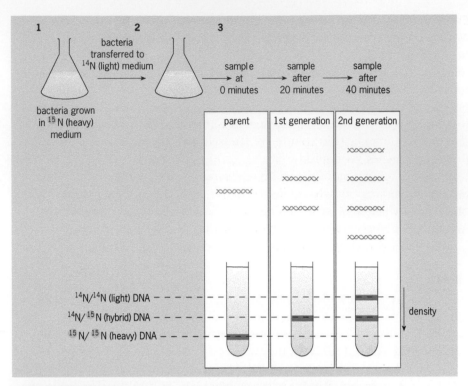

Cell division is discussed in Chapter 23.

THE MECHANISM OF SEMI-CONSERVATIVE REPLICATION

The basic mechanism of DNA replication is shown in Fig 24.12a. DNA is copied before the cell divides and the basic idea is simple: the helix unwinds and each strand acts as a template for the manufacture of a new, identical strand.

In practice, DNA replication is rather more complex than it appears, and is a good illustration of the importance of enzymes (Fig 24.12b). Two **helicases** use energy from ATP to separate the DNA strands. Next, **DNA binding proteins** attach to keep the strands separate. **DNA polymerase** enzymes then move along the exposed strands, synthesising the new strands, often in short segments. Finally, **DNA ligase** enzymes join the segments together, completing the new DNA strands.

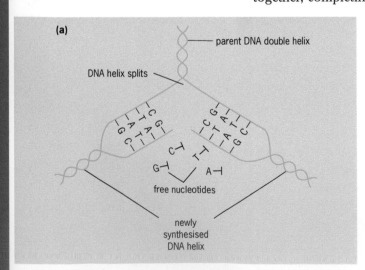

Fig 24.12a When DNA replicates, the double helix unwinds and each exposed strand acts as a template for the synthesis of a new strand. This is a simplified diagram showing how free nucleotides are added to produce two new DNA helices

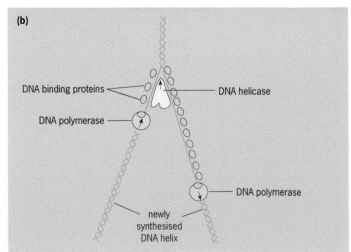

Fig 24.12b DNA replication is actually more complicated and several enzymes other than DNA polymerase are involved. DNA helicase unwinds the parent helix and then this is 'held open' by DNA binding proteins so that the DNA polymerase can gain access to the parent strand

DNA proofreading

Mistakes in DNA replication occur regularly. The chances of the wrong base being added are between 1 in 10^8 and 1 in 10^{12}. This may seem low, but so many nucleotides are added that this rate means approximately one mistake occurs for every 10 genes that are copied. Clearly, such a rate would cause the death of most organisms. This disaster is prevented by the polymerase enzymes themselves. They can detect when the wrong base has been inserted, and pause to correct the mistake. In this way 99.9 per cent of mistakes are corrected. Uncorrected mistakes give rise to **mutations**.

There are many different DNA repair mechanisms in cells to put right the genetic damage that inevitably accumulates during the life of an organism. The condition xeroderma pigmentosum illustrates the importance of these mechanisms: those affected are liable to get skin cancer because their cells cannot repair genetic damage caused by exposure to ultraviolet light (Fig 24.13).

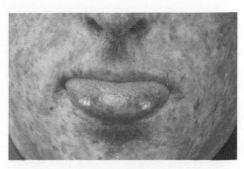

Fig 24.13 Xeroderma pigmentosum is a rare genetic disease characterised by dry, freckled skin which is extremely sensitive to sunlight. One theory suggests that individuals with this condition, who have to stay indoors, could have given rise to the vampire legends of central Europe

HOW SCIENCE WORKS

How do you compare the DNA of different organisms?

The technique of DNA hybridisation (Fig 24.14) can be used to compare the similarity of the DNA of different species, and is proving to be a useful tool in the search to piece together evolutionary trees.

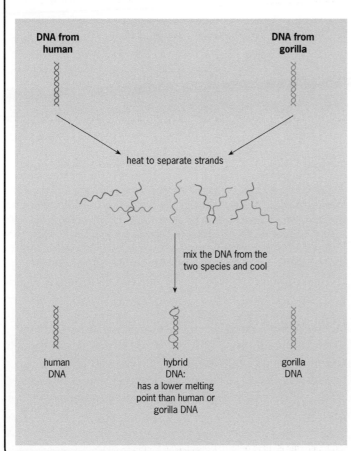

DNA from human

DNA from gorilla

heat to separate strands

mix the DNA from the two species and cool

human DNA

hybrid DNA: has a lower melting point than human or gorilla DNA

gorilla DNA

Fig 24.14 The principle of DNA hybridisation

DNA is much more stable than protein because it is held together by regular hydrogen bonds along the whole of its length. However, when DNA is heated to about 87 °C, known as the melting point of DNA, the hydrogen bonds break and the two halves of the molecule separate. When the temperature drops below 87 °C, the two strands join up: they **re-anneal**.

This is how the basic process of DNA hybridisation works:

- DNA from two different species is cut into small fragments and heated to 87 °C so that the two strands separate. The separated strands from each species are then mixed together. As the temperature drops, the DNA strands re-anneal. Some of the molecules formed will be hybrids: they contain one strand from each species.

- The temperature at which the hybrid strands re-anneal tells us about the degree of similarity between the species. When two strands from the same species re-anneal, they do so at 87 °C. Closely related DNA re-anneals at temperatures slightly lower but close to 87 °C. This indicates that they have many base sequences in common and so form many hydrogen bonds along the length of the molecule. In contrast, DNA from more distantly related species must cool to much lower temperatures before they can anneal as there are fewer hydrogen bonds to hold the strands together.

For example, human DNA can be hybridised with gorilla DNA or rabbit DNA. The melting point for non-hybridised DNA from all three single species is 87 °C. The re-annealing temperature for the human/rabbit DNA is lower than for the human/gorilla DNA. This shows that humans and gorillas have more DNA sequences in common than humans and rabbits.

POLYMORPHIC GENES

Genes that have two or more alleles are called polymorphic. Many genes have no alleles – they code for essential proteins and any mutation in them is selected against. For instance, the genes that code for the enzymes in respiration must be copied exactly time after time, or the organism dies. Some genes, however, can cope with a bit of variation. The genes that code for eye colour or hair colour are obvious examples. In humans, it has been estimated that 31.7 per cent of our genes are polymorphic.

MUTATIONS

A mutation is a spontaneous change in the genetic material of an organism (Fig 24.15b). It usually occurs when DNA is copied and cells divide. Generally, there are two types of mutation:

- a gene mutation,
- a chromosome mutation.

Both of these may occur in **somatic cells** (normal body cells) or **germ cells** (sex cells).

Fig 24.15a To many people, this is the image conjured up by the word mutant. But in biology, a mutation is a fault in the copying of genetic material and is usually lethal or makes no difference to the organism. Only in rare cases will mutation give rise to new variation, and it is sure to be less spectacular than science fiction writers would have us believe

Fig 24.15b An example of a real-life mutation. A chance mutation in a single gene of the speckled moth *Biston betularia* produced a fortunate change from the normal speckled coloration, seen in the moth on the left, to the melanic form seen on the right. In sooty areas, this new form was better camouflaged and so had a better chance of avoiding predators. More of the black moths passed on their genes to the next generation, making the melanic form dominant in industrialised areas

Cancer and other lifestyle diseases are discussed in Chapter 31.

Somatic cell mutations are not passed on to the next generation. Generally, organisms accumulate somatic mutations as they get older, and this is thought to be one of the causes of cancer. In contrast, when a mutation arises in an egg or sperm cell which then goes on to become a zygote, the mutation is passed on to all the cells of the new individual. Germ cell mutations alter the genome of an organism.

Gene mutations

A gene mutation is a change in the base sequence of DNA. We have already seen that when DNA replicates, two identical strands are formed. Occasionally, however, a fault can occur. In a typical gene of, say, 1400 base pairs, a change in just one of the base pairs has the potential to make the whole gene useless, and can prove lethal to the organism. A change in a single base is known as a **point mutation**.

How can such a tiny change be so disastrous? It is because a change to any base will alter a codon so that it will probably code for a different amino acid. In turn, this 'wrong' amino acid can affect the way the polypeptide chain folds, so changing the shape of the whole protein molecule. It might then be unable to function. (Remember: many genes code for enzymes or parts of enzymes whose proper functioning relies on precise shape.) Table 24.4 shows the different types of gene mutation that can occur, using words instead of codons. In all cases, the original meaning of the message is lost.

Table 24.4 Different types of gene mutation

Type of gene mutation	Effect on code (using words rather than codons)
normal code	THE FAT OLD CAT SAW THE DOG
addition (frame shift)	THE EFA TOL DCA TSA WTH EDO G
deletion (frame shift)	THE ATO LDC ATS AWT HED OG
substitution	THE FAT OLD BAT SAW THE DOG
duplication	THE FAT TOL DCA TSA WTH EDO G
inversion	THE TAF OLD CAT SAW THE DOG

From Table 24.4 we can see that the two least damaging types of mutation are substitutions and inversions, because these alter only one of the seven amino acids. In some cases, if the new amino acid occurs in a non-vital part of the chain, or has similar properties to the correct one, then the mutation might not actually matter at all.

In contrast, deletions and additions are potentially much more damaging because they cause **frame shifts**: only one base may be added or lost, but this causes the whole sequence to shift along, altering all the codes. In this case, the protein made is completely different from the original and is unlikely to function as it should. The Assignment on the CAS website for this chapter deals with sickle cell anaemia, a disease caused by a fault in just one base.

How common are gene mutations?

The rate at which genes mutate varies between species and between genes at different loci. At a rough estimate, there is an average mutation rate of about 10^{-5} per locus per generation, meaning 1 in 100 000. So, in a population of one million individuals, you could expect 10 to have inherited a mutation at any particular gene. The rate of gene mutation (and chromosome mutation) can be greatly increased by environmental factors, known as **mutagens**, such as ultraviolet, X-rays, alpha and beta radiation, and chemicals such as mustard gas and cigarette smoke.

Chromosome mutations

During mitosis (normal cell division) chromosomes are copied, condensed and pulled apart. The end result should be that each daughter cell receives a complete set of perfect chromosomes. But faults can occur. A chromosome mutation occurs when the structure of a whole chromosome – or set of chromosomes – is altered in some way.

? QUESTION 6

6 From the information given in Table 24.4, which type of mutation is likely to cause the fewest changes to a protein?

✔ REMEMBER THIS

In questions about the significance of mutation, the examiners are usually looking for the following specific points:
Mutation → difference base sequence → different amino acid sequence → protein shape difference → protein won't function in organism

369

Sickle cell anaemia

A small minority of people, particularly those of African origin, inherit sickle cell anaemia (SCA), a condition in which their haemoglobin acts abnormally. SCA is a genetic disease: it results from a gene mutation that is inherited. Carriers of one sickle cell allele are usually perfectly healthy, and in some circumstances even have an advantage over people who possess two normal alleles. However, individuals with two sickle cell alleles have the disease.

The faulty allele makes abnormal haemoglobin S instead of the normal haemoglobin. Problems arise when oxygen tensions are low, as they are in actively respiring tissues. At low oxygen tensions, the abnormal haemoglobin tends to come out of solution and crystallise, distorting the red blood cells into crescent shapes (Fig 24.16). This has two main consequences. First, sickle cells tend to form clots, which can block blood vessels. Secondly, the spleen destroys abnormal sickle cells at a much greater rate than normal, causing anaemia.

Haemoglobin molecules consist of four polypeptide chains: two α-globins and two β-globins. DNA analysis shows that sickle cell anaemia results from a fault in just one base in the β-globin gene. The base sequence of the first seven codons of this gene reads:

Normal base sequence: GUA–CAU–UUA–ACU–CCU–GAA–GAG

Sickle cell sequence: GUA–CAU–UUA–ACU–CCU–GUA–GAG

You can see that there is a point mutation (or substitution) in the sixth codon. This tiny change leads to a different amino acid being built into the protein, one that changes its shape and properties.

The inheritance of sickle cell anaemia

Generally, people have two copies of each gene, and this is the case with sickle cell anaemia. The two alleles involved are:

Hb^A = coding for normal haemoglobin

Hb^S = coding for sickle cell haemoglobin

So the three possible genotypes are:

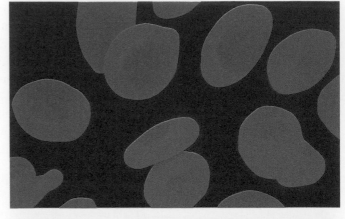

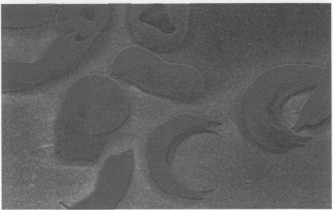

Fig 24.16 Top: Normal red blood cells. **Bottom:** Abnormal red blood cells from someone with sickle cell anaemia. Haemoglobin S tends to polymerise, forming crystal-like structures that distort the red blood cells

Hb^{A}Hb^A	normal
Hb^{A}Hb^S	carriers of the disease but with few, if any, symptoms
Hb^{S}Hb^S	have the disease

The sickle cell gene is far commoner than you would expect for one which is selected against. The reason for this is that the carriers usually show a higher resistance to malaria than normal. So while **Hb^{A}Hb^A** individuals are susceptible to malaria, and **Hb^{S}Hb^S** individuals tend to die from sickle cell anaemia, **Hb^{A}Hb^S** individuals have neither disadvantage and are more likely to survive and reproduce. Hence the **Hb^S** allele does not disappear from the gene pool.

Fig 24.17 shows the four basic types of chromosome mutation: deletion, duplication, inversion and reciprocal translation.

While gene mutations normally affect just one gene, chromosome mutations involve the disruption of whole blocks of genes. For this reason, many chromosomal mutations give rise to **syndromes**, which are complex sets of symptoms with a single underlying cause.

In addition to the structural mutations outlined above, another common failure is **non-disjunction**. This occurs when pairs of chromosome fail to separate during anaphase I of meiosis. Some gametes get both chromosomes, while others receive neither. Down's syndrome, for example, occurs when a gamete containing two copies of chromosome 21 joins with a normal gamete containing one, giving the affected individual three copies of chromosome 21.

Down's syndroms is discussed in Chapter 23.

REMEMBER THIS
Meiosis is also known as reduction division. See Chapter 23 for more detailed information about this important process.

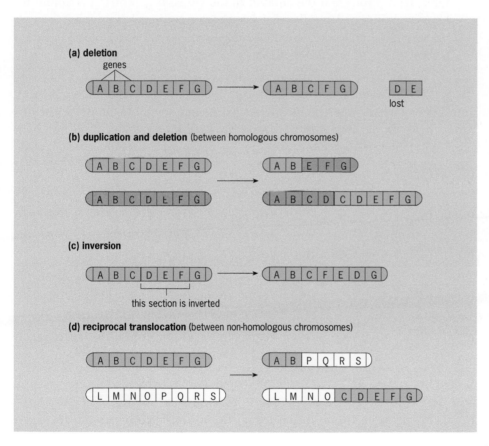

Fig 24.17 The main types of chromosome mutations
a Deletion occurs when a chromosome splits and a fragment is lost as the parts rejoin. In this case, a whole set of genes is lost, and so this is often fatal
b A duplication occurs when a section of chromosome is copied twice. This can occur when pairs of homologous chromosomes break at different places and then reconnect to the wrong partners. Thus, one of the partners would have two copies of a particular sequence, but the other would have neither. This can have serious consequences
c An inversion results when a broken segment is reinserted in the wrong order. Any genes contained in this region will be transcribed and translated backwards, and will therefore not make the correct proteins
d A reciprocal translocation occurs when two non-homologous chromosomes exchange segments. This makes the chromosome pairs of unequal size, and this can cause problems in meiosis, with daughter cells receiving the wrong chromosomes

Polyploidy

Occasionally, a mutation causes a whole set of chromosomes to be changed. For instance, a fault in meiosis might result in some gametes having two sets of chromosomes while others receive none. When such a diploid gamete joins with a normal one, the result is a **triploid** zygote – one with three sets of chromosomes. In another case, a fault in mitosis after fertilisation could result in the chromosomes doubling but not separating, leaving four sets in a single **tetraploid** cell. Neither of these embryos could develop.

The possession of multiple sets of chromosomes is known as **polyploidy**. This phenomenon is very common in flowering plants and has played a vital role in their evolution. Botanists estimate that over half of the world's flowering plant species are polyploid, including some of the world's most economically important crops: sugar beet, tomatoes and tobacco.

? QUESTION 7

7 Polyploid cells can be induced experimentally by adding the chemical colchicine, a substance extracted from the crocus. Colchicine inhibits the formation of the spindle during cell division. Suggest why colchicine results in polyploid cells.

Natural selection and evolution are discussed in Chapter 26.

Why mutations are important

A mutation can have one of three consequences:

- It can be lethal: an organism with a lethal mutation cannot survive and might not even develop beyond the zygote. Genes code for vital proteins such as enzymes, so many gene mutations mean that the protein is the wrong shape, and so cannot function metabolically. Chromosome mutations which involve the loss of whole blocks of genes are usually lethal.

- It can have no effect. It might occur in a non-coding part of DNA, or in a gene that is not expressed. Alternatively, it can result in a different amino acid being incorporated into the protein, but one that does not alter its ability to function.

- It can be beneficial. Occasionally, a mutation produces an improvement in phenotype. Statistically, this is a very rare event, but given the number of DNA replications and the number of individual organisms, it is bound to happen sooner or later. Beneficial mutations are hugely important because, ultimately, they are the source of all variation. Natural selection favours these 'improved' mutants.

Fig 24.15b on page 368 shows an example of a beneficial mutation. Following the Industrial Revolution, a mutation in the peppered moth produced a black (melanic) form that was better camouflaged against sooty backgrounds. The black moth had an advantage because it was hard for predators to see and so survived to pass on its genes to the next generation. Consequently, the mutated gene spread and the genome of the species was altered.

SUMMARY

After reading this chapter, you should know and understand the following:

- A chromosome is a large, densely staining body consisting of a single, supercoiled DNA strand. The genes along a chromosome are found at particular points called **loci**.

- A gene is a length of DNA that codes for the synthesis of a particular polypeptide or protein. Only about 10 per cent of a chromosome consists of genes; the rest is non-coding DNA.

- The genetic code is the sequence of bases contained in the DNA molecule. The code is read in groups of three bases, called **codons**. Each codon codes for a particular amino acid.

- The central dogma (principle) of molecular biology is that genes are read by **messenger RNA** which in turn moves out into the cytoplasm where it codes for the assembly of a protein.

- At the ribosome, messenger RNA attracts **transfer RNA** molecules with the right amino acids attached. The amino acids are held together so that peptide bonds form; in this way the protein is assembled.

- DNA replication occurs when the two strands separate and the enzyme **DNA polymerase** assembles two new strands on the originals. This is called semi-conservative replication as each old strand is conserved as half of the new DNA.

- A fault in DNA replication leads to a **gene mutation**. The base sequence is altered so that the gene no longer codes for the correct protein. Chromosome mutations can result in whole blocks of genes being lost, or translated backwards for example.

- Most mutations are harmful, some have no effect, but a few can be beneficial. Beneficial mutations might can give the organism an advantage; in this way, mutations can give rise to variation and hence drive evolution.

 Practice questions and a How Science Works assignment for this chapter are available at www.collinseducation.co.uk/CAS

25

How genes are inherited

The dilemmas of Huntington's disease

About eight people in every 100 000 are born with the genetic disorder Huntington's disease. This condition, which involves premature degeneration of the brain, is caused by a single faulty gene on chromosome 4. The slow decline into dementia that develops is fatal.

Diagnosis is difficult because early symptoms are vague: sufferers can have bursts of temper, become more clumsy than usual or have difficulty remembering things. Also, the symptoms do not usually become apparent until the sufferer is in their 30s or 40s, usually after they have passed the gene on to the next generation.

Until recently, people with a family history of the disease knew that they were at risk, but they could never be sure whether they were actually carrying the gene. This led to a dreadful dilemma: should they go ahead and have children and risk passing the gene on, or should they decide to remain childless, only later to experience the anguish of finding out that they were free of Huntington's disease when it was too late to try for a baby?

Today, tests are available that can tell people whether they are carriers of the Huntington's gene. A sample of DNA can even show if an unborn baby carries the disease. This does not make the decisions involved any easier, but it does give affected families accurate information on which to base those decisions.

Woody Guthrie, the famous American folk singer, suffered from Huntington's disease. His son, Arlo, was born before his father knew he had a disease that could be passed on to his children. Luckily, Arlo is unaffected: he has not inherited his father's faulty gene.
The degeneration caused by Huntington's disease can be revealed by a brain scan. A protein called huntingtin, produced by a faulty gene that was discovered in 1993, damages nerve cells in some areas of the brain, including the cerebral cortex. The gene persists because affected people tend to reproduce before symptoms appear

1 MENDELIAN INHERITANCE

Some important terms and concepts relating to inheritance are covered in Chapter 22. If you are new to genetics, read that chapter before you start to study this one.

Why don't we look exactly like our parents? Why are brothers and sisters all different (unless identical twins) when born to the same parents? As Fig 25.1 shows, there is often a marked family resemblance between parents and their children and between siblings. However, all new-born babies, except identical twins, are genetically unique. Every new-born baby has around 30 000 genes and because humans – like most animals – are diploid organisms, a baby has two versions of each gene. It inherits one version from its mother and one version from its father.

 REMEMBER THIS

A gene is a length of DNA that acts as a template for the production of messenger RNA. mRNA forms a mobile copy of a gene, travelling from the cell's nucleus to the cytoplasm. Ribosomes use this template to synthesise a specific amino acid chain. Complete proteins are made from one or more of these chains.

Fig 25.1 The tendency for certain features to be inherited is shown clearly in this family. However, no two people ever inherit exactly the same set of genes unless they are identical twins

It is difficult to imagine what happens when 30 000 different genes are mixed. Not surprisingly, although a vast amount of genetics research is under way, we are still a long way from completely understanding what happens when those mixed genes are passed on to the next generation. Some genes are activated while others remain hidden, some master genes switch on whole blocks of other genes.

How can we begin to make sense of it all? Amazingly, the man who first worked out the underlying rules did so without any knowledge of genes, chromosomes, DNA or cell division. Known as the father of genetics, **Gregor Mendel** (Fig 25.2) worked with pea plants. These are much simpler organisms than humans and provide a good starting point to understand some of the basic rules of **Mendelian inheritance**.

THE WORK OF GREGOR MENDEL

Gregor Mendel, a Czech monk, was the first to work out the basic laws that govern the inheritance of genes.

Fig 25.2 Gregor Mendel (1822–1884). Mendel's discoveries, like those of many scientists, were due to a combination of hard work, good scientific method, a touch of genius and a significant amount of luck

Before Mendel's work became known, it was widely thought that inheritance was due to 'blending' of different features. So, for example, a tall person married to a short person would produce children who would eventually grow to an intermediate height. Mendel showed it was wrong to assume that characteristics blended and demonstrated how individual characteristics are passed down to the next generation.

For years, Mendel bred and studied the edible pea plant (Fig 25.3). He chose these plants because they had easily observable features, they were easy to cultivate, they had rapid life cycles and their pollination could be controlled. His breeding experiments centred on the inheritance of a few features such as plant height, flower colour and pea shape/appearance.

Mendel was lucky because he chose features that were controlled by single genes. In addition, pea plants tend to fertilise themselves and produce pure breeding homozygous individuals. Had Mendel worked with more complex features or a different species, he might not have been able to work out the underlying laws of inheritance.

Mendel conducted his experiments using a good scientific method. He controlled his experimental conditions carefully, ensuring that plants were pollinated only by the pollen he transferred. Importantly, he repeated experiments as many times as was practicable, to make sure the results were reliable. Mendel was trained in mathematics and he applied statistical methods to his results in a way that was very advanced for the time.

His genius really showed in the way he interpreted his results. Mendel proposed the **particulate theory**, in which he stated that there are individual units of heredity that cannot be diluted, only hidden. Mendel came up with the idea that each pea plant has two factors for each character, one inherited from each parent. Only one unit can be present in the gamete and so be passed on to the next generation. Mendel's 'particles' are, of course, genes.

In 1866, Mendel published a detailed account of his findings, but their significance was not appreciated by the scientific community. In 1884, Mendel died without ever receiving true recognition for his work. In 1900, his work re-emerged thanks to the independent work of three other European scientists (see Table 22.2 in Chapter 22). In the time between 1866 and 1900, the process of meiosis had been discovered and this provided a mechanism that would explain Mendel's observations. All three scientists cited Mendel's original work, which had suggested the basic mechanism of inheritance.

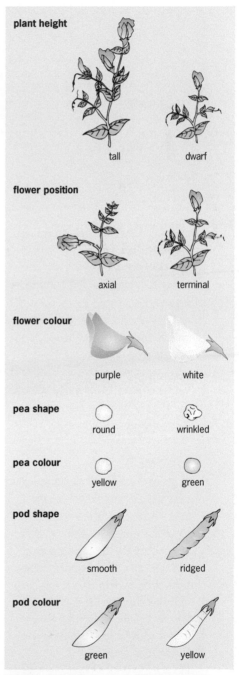

Fig 25.3 The features of pea plants studied by Mendel. Pea plants tend to self-fertilise, producing true-breeding strains. For example, white-flowered plants contain only white-flowered alleles: they are not 'hiding' any purple alleles and so always produce white flowers

Fig 25.4 An overview of meiosis. In a diploid cell, the DNA is first duplicated and then blocks of genes are swapped between homologous chromosomes. Then, in the first meiotic division, the pairs of double chromosomes are separated. Finally, in the second meiotic division, the chromatids are pulled apart and each cell receives a single chromosome

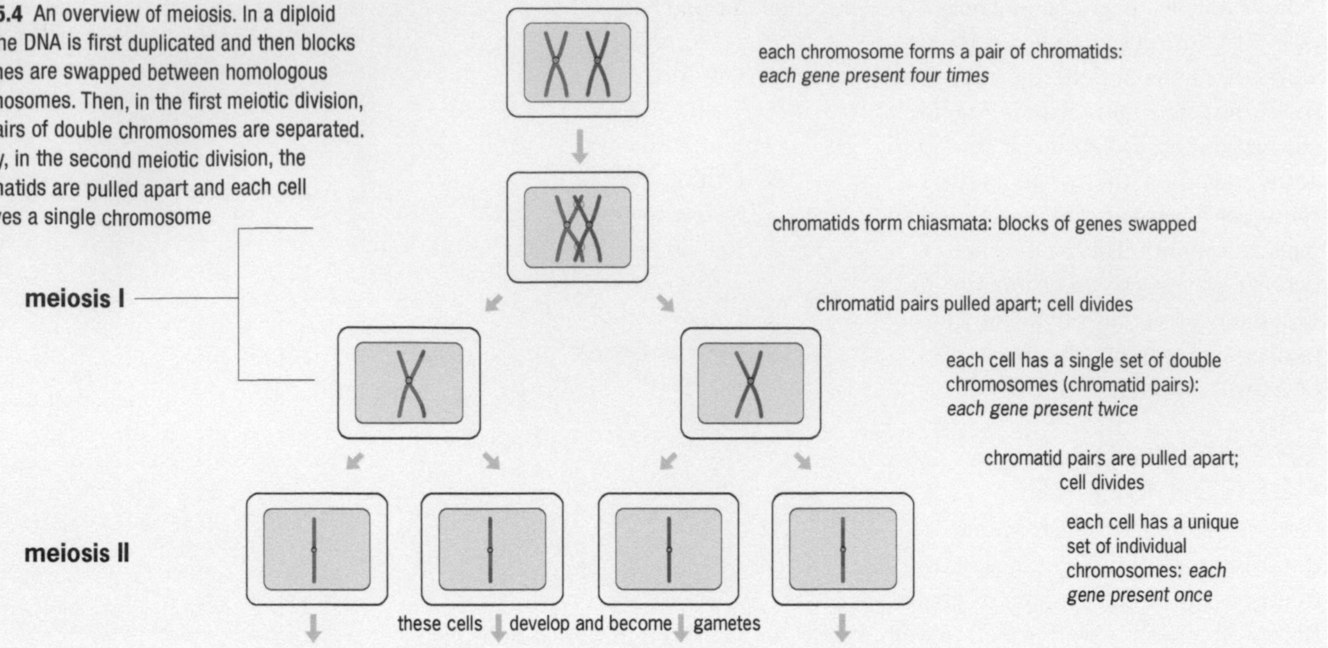

each chromosome forms a pair of chromatids: *each gene present four times*

chromatids form chiasmata: blocks of genes swapped

chromatid pairs pulled apart; cell divides

meiosis I

each cell has a single set of double chromosomes (chromatid pairs): *each gene present twice*

chromatid pairs are pulled apart; cell divides

meiosis II

each cell has a unique set of individual chromosomes: *each gene present once*

these cells ↓ develop and become ↓ gametes

Meiosis and Mendel's findings

Mendel's findings can be summarised into two laws, both of which are explained by the process of meiosis (Fig 25.4).

● Mendel's **law of segregation** states:

> **An organism's characteristics are determined by internal factors that occur in pairs, only one of which can be present in a single gamete.**

In other words, an organism has two versions of each gene, but only one of these versions is passed on to the offspring. This is because meiosis separates homologous pairs of chromosomes, so that only one of each pair passes into the gamete.

● Mendel's **law of independent assortment** states:

> **Each of a pair of contrasted characteristics can be combined with each of another pair.**

Again, this is explained by meiosis; any one of a pair of chromosomes can pass into the gamete with any member of another pair. The significance of this is seen in dihybrid (two-gene) inheritance.

MONOHYBRID INHERITANCE: HOW SINGLE GENES ARE PASSED ON

Monohybrid (single gene) inheritance concerns the inheritance of different alleles (usually two) of a single gene. Like Mendel, we start with the pea plant. Peas have several easily observable features that are controlled by single genes (see Fig 25.3).

For example, pea plants have one gene for height. The height gene has two alleles: tall (**T**) and dwarf (**t**). Pea plants are diploid and so have two alleles. There are therefore three possible genotypes:

TT – homozygous for **T** (homo = same)
Tt – heterozygous (hetero = different)
tt – homozygous for **t**

Consider what happens when a homozygous tall plant is crossed with a homozygous dwarf plant (Fig 25.5). All the gametes from the tall plant contain a T allele and all those from the dwarf plant have a t allele. These combine at fertilisation to give offspring all with the genotype Tt. The **first generation**, known as the F_1, are all tall, because tallness is dominant. However, although they look identical to the tall parent plant, they are different in one very important respect: they are heterozygous and not homozygous.

If two of these heterozygous plants are crossed, half of the gametes from each parent are **T** and half are **t**, giving us four possible genotypes in the **second generation**, the F_2:

25% are **TT**

25% are **tT**

25% are **Tt**

25% are **tt**

The first three genotypes give tall plants (the **T** allele is present) but a quarter of the plants are dwarf (they are homozygous for the **t** allele).

It is important to realise that chance plays a very important part in genetics. If we took four F_2 pea plants we might expect to get three tall plants and one dwarf, but it is highly likely that we would get something else, such as four and none or two and two. The greater the number of offspring, the higher the chance that the numbers reflect the expected ratio. This is why Mendel had to carry out hundreds of crosses before he became convinced of the underlying ratio – see the Assignment for this chapter on the CAS website.

The test cross

It is often important to know whether an individual displaying a dominant trait is homozygous or heterozygous for a particular allele. For example, how can we tell if a tall pea plant is pure breeding, or if it is carrying a hidden dwarf allele? To find out we do a **test cross**. This involves crossing the tall plant with a homozygous dwarf plant (**tt**). If the tall plant is homozygous (**TT**), all the offspring produced are tall. However, if it is a heterozygote (**Tt**), about half the offspring are tall and half are dwarf.

MONOHYBRID INHERITANCE IN HUMANS

Clear-cut examples of monohybrid inheritance in humans are relatively rare, and often involve genetic disease where people inherit one or more faulty alleles. Genetic diseases are often recessive because faulty alleles that fail to make an important protein can be masked by normal ones that function properly. Table 25.1 shows examples of monohybrid inheritance in humans.

In contrast, some genetic disorders such as Huntington's disease (see the Science in Context box at the start of this chapter) are caused by dominant alleles. The alleles concerned code for a product that actively causes damage; the symptoms are not due to an allele not doing its job. Such alleles are dominant because the presence of a normal allele cannot mask the symptoms.

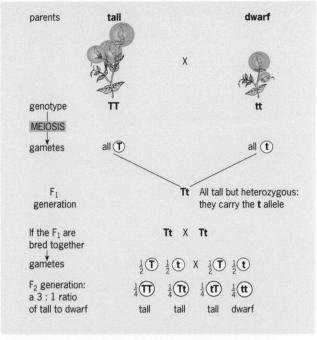

Fig 25.5 Monohybrid inheritance in pea plants. Pure breeding tall and dwarf pea plants are homozygous (**TT** and **tt**). When crossed, the plants of the first (F_1) generation are all tall but they are heterozygous (**Tt**), carrying the allele for dwarfness. If these heterozygous individuals are selfed (bred together) the dwarf form reappears in the next (F_2) generation in a ratio of 1:3 (i.e. 1 in 4). We say that the heterozygotes do not breed true

> **?** **QUESTIONS 1–2**
>
> **1** In pea plants, yellow seeds are dominant to green. If a heterozygous yellow-seeded plant were crossed with a green-seeded plant, what ratio would you expect in the F_1 generation?
>
> **2** Why is a homozygous dwarf plant used in a test cross? Why not use a homozygous tall plant?

REMEMBER THIS

The ability to roll the tongue is often cited as a simple example of inheritance in humans. However, although people seem to fall neatly into one of two categories – either you can roll your tongue or you can't – there is considerable debate about whether or not this skill can be learned, and whether it is controlled by one gene or more.

Table 25.1 Examples of monohybrid inheritance in humans

Traits	Features
Dominant traits	
Huntington's disease	see page 374
freckles	
dimple in chin	
Recessive traits	
sickle cell anaemia	Haemoglobin polymerises, distorting red blood cells into a sickle shape. This leads to circulatory blockages and anaemia. See the Assignment for Chapter 24 on the CAS website and the How Science Works box on page 370.
attached ear lobes	
phenylketonuria (PKU)	Inability to process the amino acid phenylalanine. Protein intake must be monitored to prevent build-up of the amino acid to harmful concentrations. Can be fatal if not treated.
haemophilia (also sex linked – see below)	Inability to produce a critical blood-clotting factor, carried on the X chromosome. Haemophilia is now treated by supplying the affected individuals with the clotting factor.
albinism	Inability to make the pigment melanin.
rhesus blood group	Presence (Rh^+) or absence (Rh^-) of rhesus protein on red blood cells.
cystic fibrosis	Excessive mucus production, especially in lungs and pancreas. Breathing and digestion are affected and sufferers are very susceptible to lung infections.
galactosaemia	Inability to convert galactose to glucose in the liver. All lactose must be avoided to prevent fatal levels of galactose from accumulating.
Tay–Sachs disease	Inability to produce critical lipases, which results in fatty accumulations in the brain.
lactose intolerance	Inability to breakdown the disaccharide lactose to glucose and galactose. Leads to vomiting, diarrhoea, flatulence. Avoiding lactose prevents symptoms.

SCIENCE IN CONTEXT

Cystic fibrosis: an example of monohybrid inheritance in humans

The disease cystic fibrosis, covered in detail in Chapters 27 and 28, is caused by the mutation of a single allele (Fig 25.6). A normal allele (**C**) makes a membrane protein essential for the proper functioning of certain epithelial cells. The mutated allele (**c**) does not code for a functional protein.

Most people have two healthy alleles: their genotype is **CC**. In the UK population, however, around 1 person in 25 carries a faulty allele. These carriers (**Cc**) are perfectly healthy because they also possess a normal allele, which makes the working protein.

However, in 1 couple in 625 (1 in 25 x 1 in 25) both partners are carriers. There is a one in four chance that any child they have could inherit both faulty alleles and therefore be a cystic fibrosis sufferer. As 1 cystic fibrosis child in 4 is born to 1 couple in 625, approximately 1 child in every 2500 has this condition.

Fig 25.6 The inheritance of cystic fibrosis. If both parents are carriers, there is a 25 per cent chance that any child will be a cystic fibrosis sufferer. There is also an equal chance that a child will inherit no faulty alleles and that his/her descendants will be completely free from the disease. There is a 50 per cent chance of the child being a carrier of cystic fibrosis

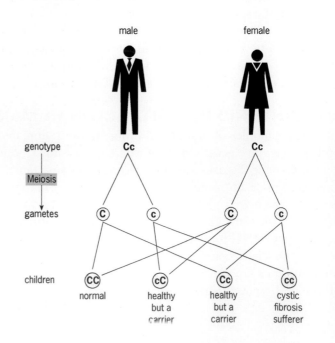

Inbreeding and outbreeding

Everyone agrees that brothers and sisters or other closely related people should not have children together. In most human societies such **inbreeding** is illegal for social, moral and biological reasons.

Biologically, inbreeding is undesirable because it reduces genetic variation in the offspring. Inbreeding promotes homozygosity: inbred individuals are homozygous for many more alleles than individuals produced by outbreeding. So, faulty recessive alleles are more likely to be paired up, resulting in a much higher proportion of genetic defects and disease.

In practice, inbreeding tends to happen in small, isolated human and animal populations, or as a consequence of artificial selection. Pedigree dog and cat breeding is a good example of this. To maintain the most desirable features, champion animals are often bred with near relatives. Analysis of the pedigree certificates of the champions in a particular breed will often show that the same individuals crop up again and again. The consequence of this artificial selection is a relatively high proportion of genetic defects.

Outbreeding (the breeding of unrelated individuals) is more desirable as it produces more new genetic combinations and greatly reduces the chance that faulty genes will be expressed. Many organisms have developed mechanisms which prevent inbreeding (Fig 25.7). If pedigree animals are cross-bred, rather than inbred (cross a Siamese cat with a Persian, for example), you often get a fitter, healthier animal. We say that such offspring show **hybrid vigour**, and cross-breeding is a favourite tool of plant breeders who find that hybrids often grow faster and show better disease resistance.

Fig 25.7 When woodlice lay a batch of eggs, the offspring are either all male or all female. This means that brothers and sisters cannot mate, thus reducing the chances of inbreeding

DIHYBRID INHERITANCE: HOW TWO GENES ARE PASSED ON

What happens when a tall, purple-flowered pea plant is crossed with a dwarf, white-flowered individual? Do you always get tall plants with purple flowers, or can the features re-combine, producing tall, white-flowered plants, and dwarf, purple-flowered ones? The simple answer is yes, you can get these recombinants, thanks to the process of meiosis.

The inheritance of two separate genes is called **dihybrid inheritance**. If, as is usual, the two genes are on separate chromosomes, Mendel's second law (page 376) applies: each of a pair of contrasted characteristics can be combined with each of another pair. So, an organism with a genotype of **AaBb** can produce gametes of **AB**, **Ab**, **aB** or **ab**. In other words, one allele from each pair passes into the gamete, and all four combinations are possible. This is due to **independent assortment** during meiosis (Fig 25.8).

? QUESTIONS 3–4

3 There are approximately 700 000 live births every year in the UK. How many of these babies would you expect to have cystic fibrosis?

4 Suggest why there is less inbreeding in most human societies today than in previous centuries.

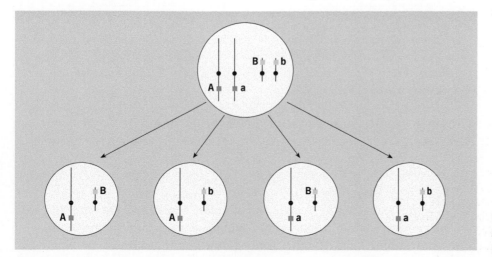

Fig 25.8 Consider a diploid cell with the genotype **AaBb**. Meiosis produces four different types of gametes by independent assortment

For an example, let's return to pea plants. In addition to tall and dwarf, we can consider a gene for seed texture. The allele **R** for round seeds is dominant to **r**, the allele for wrinkled seeds.

Fig 25.9 shows a cross between a pure breeding, tall, round-seeded plant (**TTRR**), and a pure breeding dwarf, wrinkled-seeded plant (**ttrr**). One allele from each pair is passed into the gametes. All of the gametes from the tall, round-seeded plant are **TR** and all those from the dwarf, wrinkled-seeded plant are **tr**. All of the F$_1$ generation have the genotype **TtRr** and are tall with round seeds.

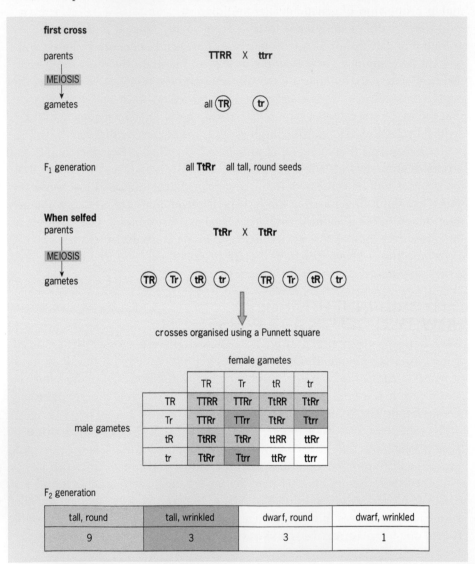

female gametes				
	TR	Tr	tR	tr
TR	TTRR	TTRr	TtRR	TtRr
Tr	TTRr	TTrr	TtRr	Ttrr
tR	TtRR	TtRr	ttRR	ttRr
tr	TtRr	Ttrr	ttRr	ttrr

F$_2$ generation

tall, round	tall, wrinkled	dwarf, round	dwarf, wrinkled
9	3	3	1

Fig 25.9 Dihybrid inheritance in peas. The grid used to work out the phenotype ratios in the F$_2$ generation is called a Punnett square

When we cross two **TtRr** individuals, things get more complex! Each plant produces four different gametes. Fertilisation is random, producing 16 different combinations in the F$_2$ generation. To organise our thoughts we can use a **Punnett square** (see Fig 25.9).

Dihybrid inheritance questions in examinations

Many examination questions require the candidate to work out the ratios of offspring resulting from a particular cross. If you have not practised these questions they can be very time consuming, leading to stress and silly mistakes.

The example given above (**TtRr** × **TtRr**) is the most complex dihybrid situation you can get. This is because both male and female are heterozygous for both alleles, and so produce 4 different gametes each, resulting in 16 different combinations. All other genotypes will produce fewer combinations.

If you practise these combinations you can learn to work out the gametes and resulting ratios in your head without resorting to Punnett squares and counting the results.

Consider the cross **TtRR × Ttrr**. The first individual produces the gametes **TR** and **tR** in equal amounts. The second one produces **Tr** and **tr** in equal amounts. So, there are four possible combinations: the same number as in a monohybrid cross.

You should see that the offspring will be one quarter **TTRr**, two quarters **TtRr** and one quarter **ttRr**, giving three tall plants with round seeds to one dwarf plant with round seeds. You would not get any dwarf plants with wrinkled seeds.

Some exam questions give the ratios of offspring phenotypes and ask you to work backwards to the original genotypes. If you do not get the expected ratios, consider **linkage** as an explanation (see below).

2 WHEN MENDEL'S LAWS DON'T APPLY

LINKAGE AND CROSSING OVER

So far, we have considered the inheritance of genes on different chromosomes that could be separated at meiosis. We say that genes that are on the same chromosome are **linked**. Such genes cannot obey Mendel's laws because they do not undergo independent assortment. They are inherited together unless separated by crossing over during prophase I of meiosis (Fig 25.10).

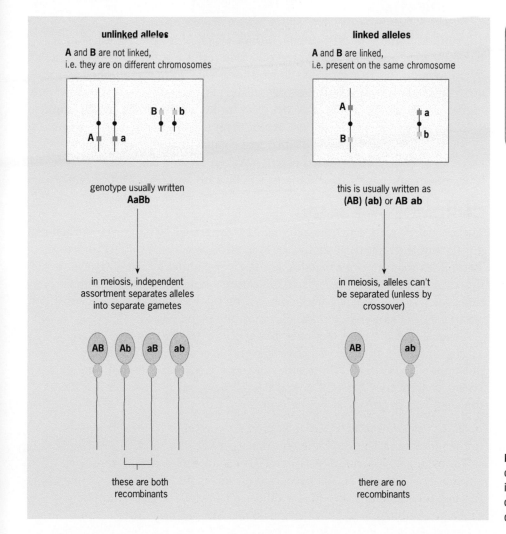

unlinked alleles

A and **B** are not linked, i.e. they are on different chromosomes

genotype usually written
AaBb

in meiosis, independent assortment separates alleles into separate gametes

AB Ab aB ab

these are both recombinants

linked alleles

A and **B** are linked, i.e. present on the same chromosome

this is usually written as
(AB) (ab) or **AB ab**

in meiosis, alleles can't be separated (unless by crossover)

AB ab

there are no recombinants

Fig 25.10 When two genes are close together on the same chromosome they are almost always inherited together: they are linked. The chance of being separated by crossover depends on the distance between them

> ✔ **REMEMBER THIS**
> Humans have over 30 000 genes present on just 23 pairs of chromosomes. So, any single gene is linked with a few thousand others because they are all on the same chromosome.

To illustrate linkage, let's consider the sweet pea plant (*Lathyrus odoratus* – a different species from the edible pea we have been looking at). In this species, the allele for purple flowers is dominant to red flowers, and the allele for elongated pollen grains is dominant to round pollen. If pure breeding plants with purple flowers and elongated pollen grains are bred with plants with red flowers and round pollen grains, the F$_1$ generation all have purple flowers and elongated pollen.

When the F$_1$ are **selfed** (bred together) you would expect a 9:3:3:1 ratio, but this does not happen. The F$_2$ generation are mainly like the parents (purple and elongated or red and round) but there are a few recombinants (red and elongated or purple and round). This is because the genes for flower colour and pollen shape are linked (present on the same chromosome) and so are inherited together. Any recombinants we get must therefore be due to crossing over.

Consider the set of results shown in Table 25.2.

Table 25.2 An example of a phenotypic ratio that indicates linkage

Phenotype	Approx. expected numbers if no linkage (9:3:3:1)	Observed numbers in F$_2$ generation
purple, elongated	405	336
purple, rounded	135	31
red, elongated	135	28
red, rounded	45	325
total	720	720

The total number of offspring is 720, of which 59 (31 + 28) are recombinants. Eight per cent of the offspring result from crossing over and so, by convention, we can say that the **loci** for flower colour and pollen shape are 8 units apart on the chromosome. If the percentage of recombinants were higher, we would know that the loci were further apart on the chromosome.

When genes are not linked, we expect them to be separated 50 per cent of the time because there is a 50 per cent chance that alleles on separate chromosomes will be inherited together. When looking for linkage, geneticists look for a frequency of recombinants significantly less than 50 per cent.

CHROMOSOME MAPS

The frequency with which linked genes are separated by crossover can tell us a lot about the relative positions, or loci, of these genes on the chromosome. Two alleles that are close together are rarely separated by crossing over, while those located on opposite ends of the chromosome are separated almost every time. Analysis of crossover frequency can be used to work out the relative positions of alleles and so make chromosome maps (Fig 25.11).

Genes Y and Z have a crossover frequency of 10 per cent and so are 10 units apart. A third cross between Y and X determines the relative positions of X, Y and Z. If the order of genes is XYZ (as shown), the crossover frequency between X and Y is 13 per cent. If the order is XZY, the crossover frequency is 33 per cent.

REMEMBER THIS

The term **locus** is used to describe the position of a gene on a chromosome.

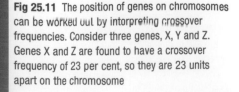

Fig 25.11 The position of genes on chromosomes can be worked out by interpreting crossover frequencies. Consider three genes, X, Y and Z. Genes X and Z are found to have a crossover frequency of 23 per cent, so they are 23 units apart on the chromosome

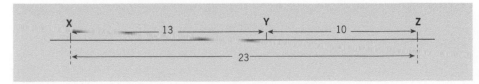

EXAMPLE

An A-level genetics question involving linkage

This example is a modified question from an OCR Modular Biology examination in Applications of Genetics.

The mosquito *Aedes aegypti* has a spotted and a spotless form on a grey or yellow body. The dominant allele is spotless, **S**, and the recessive allele is spotted, **s**. The allele for grey body **G**, is dominant to the allele for yellow body, **g**. A cross was made between homozygous spotless, grey-bodied mosquitoes and spotted, yellow-bodied mosquitoes. The F_1 individuals were then crossed with the double recessive strain and the numbers of the resulting phenotypes were counted. The results are shown below.

Phenotype	Number
Spotless, grey bodied	442
Spotted, yellow bodied	458
Spotless, yellow bodied	46
Spotted, grey bodied	54

Q State the genotype and the phenotype of the F_1 using the above symbols.

A The question is asking for the genotype of the first generation. You know that the parents are homozygous spotless, grey (genotype: **SSGG**) and spotted, yellow (genotype: **ssgg**). The genotype of the F_1 must be **SsGg**, giving a phenotype of spotless, grey bodied.

Q Using a genetic diagram, explain fully the results shown in the table.

A The table below shows the result of a cross between an F1 individual, **SsGg**, and a double recessive which must be **ssgg**.

So we have:

Parents with genotype	Spotless, grey SsGg		×	Spotted, yellow ssgg	
Gametes	SG Sg sG sg		×	all sg	
Expected ratio	25% SsGg	25% Ssgg		25% ssGg	25% ssgg
Phenotypes	Spotless, grey	Spotless, yellow		Spotted, grey	Spotted, yellow
	1 :	1 :		1 :	1

You should see that the expected results are not the same as those given in the table of results on the left, so there must be an extra complication. The key feature from the table is that there are more of the parental phenotypes than expected, and fewer recombinants. This should scream *LINKAGE* at you!

To produce a really thorough answer you should also work out the percentage of recombinants.

The total number of crosses in the table is 1000, of which 100 (46 + 54) are recombinants. This leaves you with the simple calculation to show that 10 per cent of the mosquitoes are recombinants, and so the two genes are 10 units apart on the chromosome.

3 SEX DETERMINATION

What decides whether a baby will be a boy or a girl?

In humans, sex determination depends on the inheritance of special **sex chromosomes**. Every human cell contains 23 pairs of chromosomes: 22 pairs of **autosomes** and one pair of sex chromosomes (Fig 25.12). Females have two large X chromosomes and so we say they are the **homogametic** sex. Males have one X chromosome and one smaller Y chromosome, and are therefore the **heterogametic** sex.

In the female, all eggs receive an X chromosome during meiosis (Fig 25.13). However, meiosis in the male produces sperm, half of which receive a Y chromosome and half an X. So as the sperm swim for the female's egg it's a real race: if a Y sperm fertilises the female's egg, the offspring is male. If an X sperm wins, the baby will be a female.

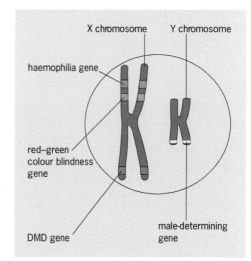

Fig 25.12 This diagram shows the sex chromosomes from a male: females have two identical X chromosomes. Note that the Y is much smaller than the X, and carries fewer genes. The Y chromosome does, however, carry the male-determining gene. (The human karyotype of 22 pairs of autosomes and 1 pair of sex chromosomes is shown in Fig 22.2.)

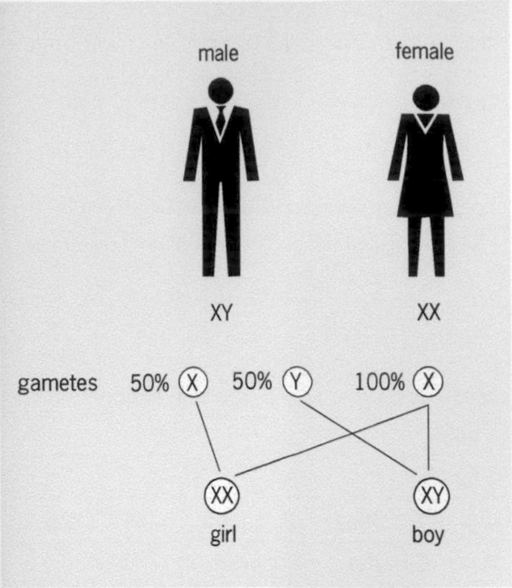

In humans, about 114 boys are conceived for every 100 girls. We do not yet fully understand the reason for this. Interestingly, however, more male embryos die and so at birth the proportions are down to 106 to 100. By puberty the numbers are equal and in old age the females outnumber males by two to one.

In the 1990s, a group of genetic researchers demonstrated that the Y chromosome carries just 60 functional genes, including a male-determining gene called **SRY**. All embryos are female unless the active SRY imposes maleness upon it. When a male embryo and a female embryo share the same blood supply, it is possible that the SRY can produce hormones that can force an XX embryo to develop as a male. This is an extremely rare situation in humans but it is common in cattle.

Androgen insensitivity syndrome (AIS) is a condition that can cause a genetically male child to develop outwardly as a female. The receptor proteins for the male hormones (androgens) don't work properly, so the embryo fails to read the signals that should turn it into a male. This syndrome may go undetected until the teens, when puberty fails to happen. Genetic analysis will reveal that the 'girl' is in fact a male. Although she will look and act like a girl, she will not menstruate and will be unable to have children.

Fig 25.13 Sex determination in humans. Half of the sperm carry an X chromosome, half carry a Y

Blood clotting is discussed in Chapter 30.

4 SEX-LINKED INHERITANCE

Genes carried on the sex chromosomes are said to be **sex-linked**. Human females have two X chromosomes, which, like the autosomes, carry two alleles of every gene. Females therefore have two sets of sex-linked alleles. In males, however, the Y chromosome is smaller and cannot mirror all of the genes on the X chromosome, so males have only one set of most sex-linked alleles. This is why males suffer from the effects of X-linked genetic diseases more often than females. There are very few Y-linked diseases or conditions, probably because the Y chromosome carries so few genes.

A classic example of a sex-linked trait transmitted by the X chromosome is **haemophilia**. People suffering from this disease do not make factor VIII, an essential component in the complex chain reaction of blood clotting. In addition to the problems caused if they injure themselves, haemophiliacs suffer from internal bleeding as a result of normal activity. Bleeding at the joints during even light exercise is a particular problem. However, haemophiliacs can usually live a full and active life by having regular injections of factor VIII.

Haemophilia is caused by a sex-linked recessive allele. Females have a pair of alleles but males possess only one. So, if a male inherits the haemophilia allele, he has the disease since he cannot possess another healthy allele to mask its effect.

Using the following notation:

X^h = the haemophilia allele on the X chromosome
X^H = the healthy allele on the X chromosome
Y = the Y chromosome which does not carry either allele

we can see that the possible genotypes are:

$X^H X^H$ = healthy female
$X^H X^h$ = healthy, carrier female
$X^H Y$ = healthy male
$X^h Y$ = haemophiliac male

? QUESTION 5

5 What would the notation be for a female haemophiliac? Suggest why they are much rarer than male haemophiliacs.

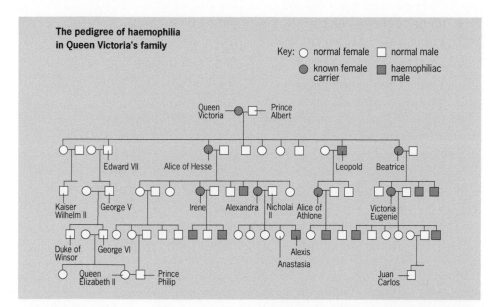

The pedigree of haemophilia in Queen Victoria's family

Key:
○ normal female
□ normal male
⬤ known female carrier
◼ haemophiliac male

Queen Victoria — Prince Albert

Edward VII   Alice of Hesse   Leopold   Beatrice

Kaiser Wilhelm II   George V   Irene   Alexandra   Nicholai II   Alice of Athlone   Victoria Eugenie

Duke of Winsor   George VI   Alexis Anastasia   Juan Carlos

Queen Elizabeth II   Prince Philip

Fig 25.14 Queen Victoria (1819–1901) had nine children and passed the haemophilia allele on to two daughters and one son. Her eldest son, Edward VII, did not inherit the faulty allele and so the disease is not carried by his descendants, the present-day British Royal Family

Fig 25.15 Duchenne muscular dystrophy affects mainly boys. The body muscles are gradually replaced by fibrous tissue and become weak, usually confining sufferers to a wheelchair by the age of 10. Few live beyond 20. One male child in 4000 is born with this disease, sometimes caused by a spontaneous mutation, so the mother may not even be a carrier. Gene therapy holds new hope for the future

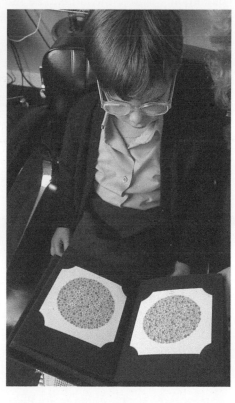

? **QUESTION 6**

6 If a female carrier of the haemophilia allele marries a normal male (as was the case of Queen Victoria and Prince Albert) and they have a son, what is the probability that he will have haemophilia?

Fig 25.16 A person with red–green colour blindness cannot distinguish between red, green, orange and yellow. On these Ishihara test charts, a person with normal vision would see numbers but a person with red-green colour blindness would see only a random pattern of dots. This condition is caused by a defective allele that codes for the one of the three groups of light-sensitive cone cells in the retina. Eight per cent of males but only 0.4 per cent of females suffer from colour blindness

We can study the inheritance of haemophilia using a **pedigree diagram**. The word pedigree means 'of known ancestry' and applies as much to humans as it does to domestic animals. The pedigree of the Royal Family can be traced back to 1066 and shows that Queen Victoria was a carrier of the haemophilia allele and passed it on to four of her nine children (Fig 25.14). Other sex-linked traits in humans include Duchenne muscular dystrophy (DMD, Fig 25.15) and red–green colour blindness (Fig 25.16).

5 MULTIPLE ALLELES

Sometimes there are more than two alleles for a particular gene. Such genes are said to be **polymorphic.** It has been estimated that 31.7 per cent of human genes are polymorphic.

A classic example of a polymorphic gene is the ABO blood group system in humans. The entire human population can be classified into A, B, AB or O depending on the proteins they carry on the membrane of their red blood cells. The ABO system is controlled by one gene (**I**) with three alleles, I^A, I^B and I^O. The I^A allele codes for A proteins, I^B for B proteins and I^O for no relevant proteins. I^A and I^B are codominant over I^O. Each individual inherits two alleles which combine to produce the blood group as shown in Table 25.3.

Blood group testing can sometimes be used to disprove (but not prove) parentage. Table 25.4 shows, for instance, that a man of blood group AB must pass on either an I^A or an I^B allele, and so cannot possibly be the father of a child with blood group O. DNA fingerprinting, of course, provides a much more accurate (if more expensive) test which can say with a great deal of certainty who *is* the father. The Assignment for Chapter 3 on the CAS website looks at the process of DNA fingerprinting.

Table 25.3 The genetics of the ABO blood group system

Genotype	Phenotype (blood group)
$I^A I^A$ or $I^A I^O$	A
$I^B I^B$ or $I^B I^O$	B
$I^A I^B$	AB
$I^O I^O$	O

? QUESTION 7

7 If a child has the blood group B and his mother is blood group O, can his father be a man with blood group A? Say why.

Table 25.4 Possible blood groups of children born to parents of particular blood groups

		Paternal blood group			
		A	**B**	**AB**	**O**
	Genotypes	$I^A I^A$ or $I^A I^O$	$I^B I^B$ or $I^B I^O$	$I^A I^B$	$I^O I^O$
Maternal blood group	**A** $\quad I^A I^A$ or $I^A I^O$	A, O	A, B, AB, O	A, B, AB	A, O
	B $\quad I^B I^B$ or $I^B I^O$	A, B, AB, O	B, O	A, B, AB	B, O
	AB $\quad I^A I^B$	A, B, AB	A, B, AB	A, B, AB	A, B
	O $\quad I^O I^O$	A, O	B, O	A, B	O

6 POLYGENIC INHERITANCE

So far in this chapter, we have been concerned with **discontinuous inheritance**: all-or-nothing features controlled by single genes. Many features, however, are **continuous** and are controlled by several genes that act together. We describe the inheritance of such characteristics as **polygenic inheritance**.

Skin colour in humans is an example of polygenic inheritance. The depth of skin colour depends on the amount of melanin present. This is determined by at least three genes. As an example of this type of inheritance at work, imagine that the alleles **A**, **B** and **C** contribute cumulatively to the overall amount of melanin but the alleles **a**, **b** and **c** do not. So, someone with the genotype **AABBCC** would have the darkest skin, while the genotype **aabbcc** would confer a very pale skin colour.

As Fig 25.17 shows, a cross between two purely homozygous individuals would result in an intermediate (heterozygous) F_1 generation, but the F_2 generation would show a wide variation in skin colour depth. The graph shows a **normal distribution**: most individuals have skin colours in the midrange. This range is one of the characteristic features of continuous variation.

✔ REMEMBER THIS

The inheritance of height in humans is polygenic.

Continuous variation is discussed in Chapter 22.

? QUESTION 8

8 Non-identical twins are born to a couple who both have the genotype **AaBbCc**. Is it possible for one baby to be born with white skin and the other to have black skin? Explain your answer.

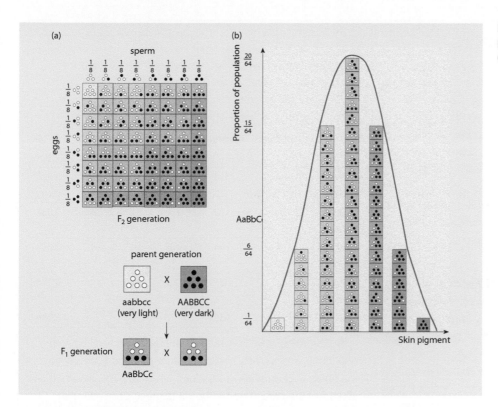

7 EPISTASIS (WHEN GENES INTERACT)

Up to now, we have discussed inheritance and the effect of genes that act on their own. In practice, however, many genes interact with each other: the presence of one allele often affects the expression of others. We call this interaction **epistasis**. A classic example of epistatic gene interaction is coat colour in mice. The natural coloration of wild mice is called **agouti** and is produced from banded hairs (Fig 25.18).

Two genes are involved, each with a dominant and a recessive allele. We will represent them by the notation **Aa** and **Bb**, where **A** and **B** are dominant over **a** and **b**. The allele **A** codes for the ability to produce hair pigment: **AA** and **Aa** mice have pigmented hairs but all **aa** individuals are albinos. The **B** allele codes for the ability to make hair with graduated coloration: **BB** and **Bb** mice have graduated hair, **bb** mice have hair that is all one colour.

Only mice that have both dominant alleles **A** and **B** show the agouti coloration. These can have the genotype **AABB**, **AaBB**, **AABb** or **AaBb**. Mice that are **aa** cannot make pigment, so, whether they are **BB**, **Bb** or **bb**, no bands are visible. Mice that can make pigments but not banded hair (genotypes **AAbb** or **Aabb**) are plain black.

> **? QUESTION 9**
>
> 9 Two agouti mice, genotypes AaBb, are bred together. What phenotypic ratio would you expect in the next generation?

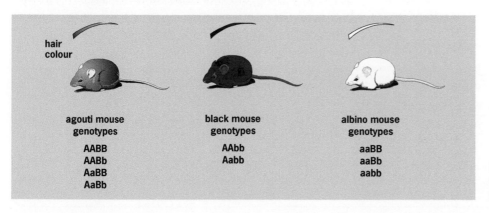

agouti mouse genotypes	black mouse genotypes	albino mouse genotypes
AABB	AAbb	aaBB
AABb	Aabb	aaBb
AaBB		aabb
AaBb		

Fig 25.18 The base of the hair of a mouse with agouti coloration is dark brown/black fading to lighter brown towards the tip

The chi-squared test in genetics

The chi-squared test is a simple statistical method that scientists use to tell if their results are significant or due to chance. For instance, if you roll a die, you have a one in six chance of getting any particular number. If you rolled the die six times, you would probably not get each number once; you may get three twos or some other combination. However, the more times you roll the die, the greater the probability that all six numbers will be represented evenly.

If you are playing a game with someone who gets a higher proportion of sixes than normal, you might suspect them of playing with a loaded die. You could use the chi-squared (χ^2) test to find out if they are cheating.

The nature of science and the null hypothesis

Most people accept that there is a link between smoking and lung cancer but, in science, it is virtually impossible to prove anything absolutely. What you can do is **gather support** for your hypothesis.

You could divide lung cancer victims into smokers and non-smokers. You could then use the χ^2 test to decide if there was a significant difference in the number of smokers who developed lung cancer compared with the non-smokers.

First, you must develop a null hypothesis. This could be: 'There is no difference in the incidence of lung cancer between smokers and non-smokers'. You could then perform the χ^2 test and come up with the conclusion that 'there is a very low probability that these results are due to chance'. In this case you have rejected the null hypothesis and supported the idea that there is a link between smoking and lung cancer.

The test itself

The χ^2 test compares the observed results with those you would expect. That is, do the actual results fit the results that we had expected?

The formula for the χ^2 test is:

$$\chi^2 = \sum \frac{(O - E)^2}{E}$$

In this formula

O = the observed result
E = the expected result
Σ = the sum of

Use a table to progressively work out all the steps needed by the formula. In this worked example, a geneticist performed one of Mendel's classic pea experiments and got the following results.

Yellow, round   318
Yellow, wrinkled   103
Green, round   106
Green, wrinkled   33
Total   560

The geneticist wanted to know whether these results did actually reflect the expected 9:3:3:1 ratio, or whether they deviated so significantly that the 9:3:3:1 hypothesis was in doubt. She did the following calculations:

Category	Hypothesis	Observed	Expected	$O - E$	$(O - E)^2$	$\dfrac{(O - E)^2}{E}$
yellow, round	9	318	315	3	9	0.028
yellow, wrinkled	3	103	105	−2	4	0.038
green, round	3	106	105	1	1	0.009
green, wrinkled	1	33	35	−2	4	0.114
total		560				= 0.189

So the value of χ^2 is 0.189. What next? The hard work is now done and we only need one more piece of information; the number of **degrees of freedom,** which is a measure of the **spread of the data**. The number of degrees of freedom is always one less than the number of categories of information. So in this case, it is four categories minus one, which equals three degrees of freedom.

We now look up the value of 0.189 with three degrees of freedom on a table of χ^2 values. Table 25.5 is an extract from the χ^2 tables. We can see that our value of 0.189 lies between the 0.98 and 0.95 column. From this we can say that the probability that our results are significant is between 95 and 98 per cent. Put another way, it tells us that the probability that these results are due to chance is between 2 and 5 per cent. By convention, scientists take the cut-off point as 5 per cent. If the test shows that there is a less than 5 per cent probability that your results are due to chance, you can reject your null hypothesis.

Chi-squared in examinations

If you have worked your way through the above examples, you may have concluded that chi-squared involves a lot of arithmetic and is time consuming. The examiners have come to the same conclusion and therefore it is unlikely that you will be set a full investigation to work through from scratch: they want to test your biology, not your maths. However, this test is a very useful tool and examiners will want you to appreciate its uses, and possibly to do some simple calculations.

Table 25.5 Extract from χ^2 tables

	Probability (e.g. 0.5 = 50%)									
	0.99	0.98	0.95	0.9	0.7	0.5	0.3	0.2	0.1	0.02
1 D of F	0.00016	0.0063	0.0039	0.0158	0.148	0.455	1.07	1.64	2.71	5.41
2 D of F	0.0201	0.0404	0.103	0.211	0.713	1.39	2.41	3.22	4.60	7.82
3 D of F	0.115	0.185	0.352	0.584	1.42	2.37	3.66	4.64	6.25	9.84

SUMMARY

After reading this chapter, you should know and understand the following:

- Humans, like most organisms, are **diploid**; each cell carries two versions, or **alleles**, of each **gene**. If both alleles are the same, the organism is **homozygous** for that allele. If the pair are different, the organism is **heterozygous**.

- During **meiosis**, only one of the alleles passes into the **gamete**. This is Mendel's first law.

- **Monohybrid inheritance** is the study of one gene. A cross of **AA** × **aa** gives all **Aa** in the **first generation (F_1)**. A cross of **Aa** × **Aa** gives a 3:1 phenotypic ratio in the **second generation (F_2)**.

- **Dihybrid inheritance** involves two separate genes, usually on separate chromosomes. Homozygous individuals **AABB** × **aabb** produce offspring that are all **AaBb** in the F_1 generation. However, owing to **independent assortment** during meiosis, **AaBb** individuals produce gametes of **AB**, **Ab**, **aB** and **ab**. If bred together, **AaBb** individuals produce 16 different genotypes, and 4 different phenotypes with a ratio of 9:3:3:1.

- Genes present on the same chromosome are said to be **linked**. Such genes are inherited together unless separated by crossover in prophase I of meiosis. The greater the distance between the genes on the chromosome, the more often they are inherited separately.

- Humans have 23 pairs of chromosomes: 22 pairs of **autosomes** and one pair of **sex chromosomes**. The X chromosome is larger than the Y and carries more genes.

- The sex of a baby is determined at the moment of conception. Because gametes are produced by meiosis, half of the sperm carry a Y chromosome and half carry an X. If a Y sperm fertilises the egg, a male results; X gives a female.

- Genes carried on the sex chromosomes are said to be **sex-linked**. Females have two complete sets of sex-linked genes, but males have only one copy of all but the few genes carried on the Y chromosome.

- Some diseases such as haemophilia are sex-linked; caused by defective genes on the X chromosome. Females have two alleles and so can be carriers. Males have only one allele and so cannot be carriers: they are either healthy or they have the disease.

- Some genes have more than two alleles. The human ABO blood group is an example of such a **multiple allele**. In this case, one gene has three alleles, I^A, I^B and I^O

- Many characteristics are controlled by many different genes. We describe such characteristics as **polygenic**; they often produce a range of values. Human height is a good example of a polygenic characteristic.

- The presence of one allele can affect the expression of an allele of a different gene. This type of gene interaction is called **epistasis**. For example, one allele may give an organism the ability to make pigment, while another allele determines the colour and/or the distribution of that pigment.

Practice questions and a How Science Works assignment for this chapter are available at www.collinseducation.co.uk/CAS

26

Evolution

26 EVOLUTION

Adam and Eve: were they really the start of the human race?

Creation or evolution?

Humans are a very strange species. We stand on two legs, naked apart from hair on the top of our heads and at a few strategic places. We are covered in sweat glands, we have a flat face and a long neck. Males have very large genitals compared to body size, while females have permanently swollen breasts. An alien zoologist looking at humans for the first time might be surprised that we dominate the Earth.

In the absence of any survival equipment, such as sharp claws or teeth, great speed or protective armour, why have we been so successful? The answer lies, of course, in our brains. Given our body size, we have the largest brain of any animal. Coupled with dextrous hands, our brains allow us to use language, co-operate and make tools to a far greater extent than any other species.

But where did humans come from? Until fairly recently, the answer, for Christians, was clear. 'And the Lord God formed man of the dust of the ground, and breathed into his nostrils the breath of life; and man became a living soul' (Genesis 2:7). Women followed soon after: 'And the rib, which the Lord God had taken from man, made he a woman, and brought her unto the man' (Genesis 2:22).

Until relatively recently most people believed either this or another explanation based on religious beliefs. Some still do, but Charles Darwin's theories on evolution changed fundamental scientific thinking. It is clear that humans evolved from an ape-like ancestor.

1 WHERE DO LIVING THINGS COME FROM?

To most human eyes, the living world seems fairly static. Like produces like: elephants give birth to baby elephants, frogs make frogs. It seems that no new species appear. There are literally millions of different species on Earth. Where did they come from?

THEORIES ON THE ORIGIN OF SPECIES

Ideas about the origins of species – including ourselves – have been central themes in philosophical and religious thought for centuries. Until the nineteenth century, virtually everybody in the Christian world looked to the Bible. The book of Genesis provided the answers: all animals and plants were created at the same time, by the great Creator. Even the eminent taxonomist Linnaeus, who in 1742 published his system for classifying all living plant species known at the time, said 'there are as many species as God created in the beginning'. Other religions had different versions of the same creationist idea.

The idea that species were not fixed, and that the infinite variety of living things had developed through a slow process of evolution, seems to have arisen first in early Greek philosophy and reappeared from time to time throughout the following centuries. It failed to gain lasting acceptance because no explanation could be found to show how the process might have come about. In addition, in some cultures, any 'non-religious' thought was strongly discouraged, even on penalty of death. A catalyst in the development of the theory of evolution was the discovery of fossils (Fig 26.1). These were clearly the remains of extinct creatures. Why were they no longer alive?

Fig 26.1 The discovery in the nineteenth century of fossils such as this *Triceratops* fuelled the idea that the living world was not static. Different layers of rock held different sets of organisms and this led to the acceptance that the living world was not fixed: plants and animals could change with time

One Christian explanation said that extinct species were the victims of a great global catastrophe, and that a new set of organisms had been created to replace them. It was even suggested that only the passengers aboard Noah's Ark survived a great global flood. When it was pointed out that different rocks contained different types of plants and animals, it seemed that there would have had to be many such floods, and so the **catastrophe theory** gradually lost credibility.

An early attempt to explain the mechanism of evolution came from the French naturalist Lamarck, who had made extensive studies of different life forms. He was convinced of the process of evolution, but the mechanism he proposed in 1809 to explain it was wrong. Lamarck supposed that organisms adapt to their environment by developing new structures and losing old ones. Such differences, acquired during the animal's life, were then passed on to the next generation. As an example he cited the webbed feet of water birds. Lamarck suggested that, in an effort to swim, the birds extended their toes and so stretched the skin between them. The stretched condition was then inherited and this process was repeated until it produced a fully webbed foot. With no knowledge of genes, chromosomes or DNA, this idea of the **inheritance of acquired characteristics** was perfectly reasonable for the time.

> **? QUESTION 1**
>
> 1 What was wrong with Lamarck's theory of evolution?

HOW SCIENCE WORKS

How old is the Earth?

Accurately estimating the age of the Earth is central to the theory of evolution. In 1650, James Usher (then the Archbishop of Armagh) announced that his study of Hebrew literature had led him to the conclusion that the date of creation was the year 4004 BC. Dr Lightfoot, vicar of the University church at Cambridge, took it a step further and proclaimed that the date of creation was 9 a.m. on 23 October, 4004 BC. Baron George Cuvier later stated that the Earth must be much older, more like 70 000 years.

Faced with the large array of different fossils, nineteenth-century geologists decided that the Earth was much older still. In 1830, Charles Lyell suggested that the true age of the Earth would turn out to be millions, rather than thousands of years. He was the first to propose that old rocks could be uncovered or brought to the Earth's surface (by volcanic activity, for example), so revealing the remains of long-dead species.

Ultra-sensitive dating techniques that measure the decay of radioactive isotopes have now put the age of the Earth at about 4 600 000 000 (4.6 billion) years. This is an important breakthrough because, although our brains cannot comprehend such a time scale, it fits in well with the idea of progressive evolution. The Assignment for this chapter on the CAS website should help you to put this geological time scale into perspective.

> **REMEMBER THIS**
> A working definition of evolution is 'a change in the genetic composition of a population over time'. It has also been defined as 'a gradual process in which new species develop from pre-existing ones'. But as we shall see in this chapter, populations can evolve without changing into new species.

> **REMEMBER THIS**
> The cornerstone of Darwin's theory is that the mechanism of evolution is natural selection – this is known as **Darwinism**. The application of twentieth-century knowledge about chromosomes, genes and DNA to Darwin's theories is called **neo-Darwinism**.

2 CHARLES DARWIN AND THE THEORY OF EVOLUTION

In a book with the shortened title of *The Origin of Species*, Charles Darwin included these four ideas in his theory of evolution:

- **The living world is changing**, not static.
- **The evolutionary process is usually gradual**, not a series of jumps such as catastrophes and re-creations.
- **The common ancestor theory**. This suggested that closely related species evolved from one basic ancestor. Darwin said that man and apes evolved from a common ancestor, but the popular press took this to mean that humans had recently evolved from apes, leading to cartoons such as the one in Fig 26.2.
- **The theory of natural selection** as a mechanism to explain evolution.

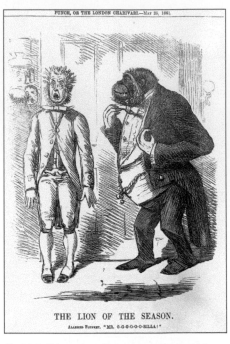

Fig 26.2 This cartoon first appeared in *Punch* in 1861. It shows how poorly people of the time understood Darwin's theories

HOW SCIENCE WORKS

Charles Darwin

Charles Darwin (Fig 26.3) was the son of a doctor. After studying medicine at Edinburgh he changed his mind and prepared for a career in the Church. However, despite going to Cambridge to study theology (religion), Darwin retained his interest in biology and geology.

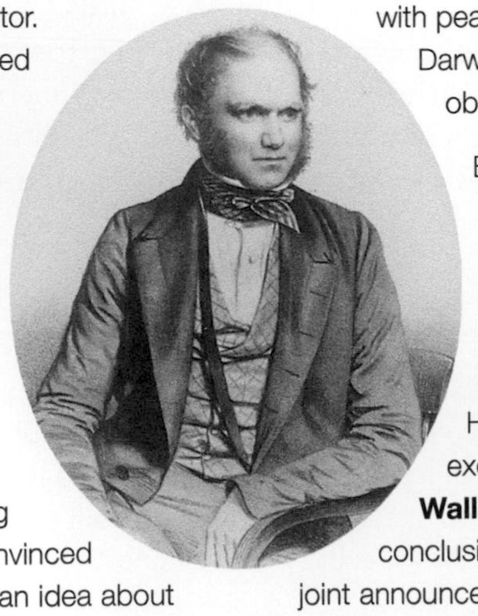

Fig 26.3 Charles Darwin, 1809 to 1892

Darwin was offered the post of naturalist on the HMS *Beagle*, embarking on a scientific survey of South American waters. It was observations he made during this trip that first stimulated the young Darwin into contemplating the origins of species. As well as becoming convinced of the process of evolution, he also developed an idea about how it might happen. This trip was not only the turning point in Darwin's life; it was also the start of the greatest ever revolution in our perception of the living world, and of our place within it.

On returning home, Darwin buried himself in his studies, determined to prove or disprove his ideas about the evolution of species. Like Lamarck, Darwin developed his theories with no knowledge of genes, chromosomes or DNA. Mendel's work with peas (see Chapter 25) was going on during Darwin's lifetime but it remained hidden in an obscure journal.

Even though he managed to build up an overwhelming case for evolution, Darwin was reluctant to publish his theories, knowing the conflict they would cause with the Church. For more than 20 years Darwin carried on gathering evidence to support his theories. He was finally pushed into publication after exchanging letters with **Alfred Russel Wallace**: he too had come to the same conclusion about evolution. They agreed to make a joint announcement, and the Darwin–Wallace paper was presented to the Linnaean Society of London in July 1858.

In November 1859, Darwin published his revolutionary work, *The Origin of Species by Means of Natural Selection*. Few books before or since have caused quite such a controversy. Although Church leaders largely remained quiet, many people bitterly attacked it because it questioned the theory of creation. Darwin was called 'the most dangerous man in England'.

NATURAL SELECTION: 'SURVIVAL OF THE FITTEST'

Darwin's observations of the living world led to three basic conclusions:

- Generally, organisms produce far more offspring than can possibly survive.
- Living things are locked in a struggle for survival. They compete for food, space, mates, etc. In short, they are trying to eat and not be eaten, so that they can reproduce.
- Individuals of the same species are rarely identical: they show **variation**.

There was nothing particularly original about these observations, or in the idea that organisms evolved. What Darwin did, however, was to put all these factors together and suggest **natural selection** as the mechanism for evolutionary change.

Natural selection is commonly simplified to 'survival of the fittest'. Biologists define fitness as the ability of an organism to survive and reproduce. The fittest organisms survive to produce more offspring than less fit ones. In this way, the characteristic of a population can gradually change from generation to generation.

To illustrate the principle, imagine a population of blackbirds in a woodland. Like most organisms, they reproduce sexually and there is variation in the population. Some have better reflexes than others, forage for food more successfully, others have a better immune system. Normally, there is competition for resources such as food, mates and nesting sites. In this situation, the fittest birds survive and produce more offspring than others. The fitter birds may gather more food and so rear larger families than less fit individuals (Fig 26.4). Or they may rear two clutches of eggs in the time it takes others to produce one. The vital point here is that the *fittest individuals pass their genes on to more of the next generation*.

The severity of an organism's circumstances is called the **selection pressure**. The greater the selection pressure, the faster the evolutionary process. When organisms have a short life cycle, the process can be very swift indeed. In the 1950s, farmers in the UK and Australia took the drastic step of introducing a deadly disease, myxomatosis, into the ever-expanding rabbit population (Fig 26.5). This virus killed over 99 per cent of the rabbits: this is selection pressure at a massive level. Only the rabbits whose genes gave them resistance to myxomatosis were able to survive and pass on their genes to the next generation.

REMEMBER THIS

Fitness in biology is a measure of reproductive success. So, a fit flowering plant produces and disperses more seeds than its competitors. Fit fungi are more successful because they produce more spores.

Fig 26.4 Blackbirds that can gather the most food tend to raise the greatest number of offspring, and so pass on their genes to more of the next generation

Fig 26.5 When the European rabbit is introduced into countries where it is not endemic, it tends to out-compete the native herbivores. With few predators around, the rabbit population can reach plague proportions. When this has happened in the UK and Australia, farmers faced ruin. The deadly myxoma virus was introduced into the rabbit population, killing 99 per cent. The remaining 1 per cent were naturally resistant to the virus and are the ancestors of today's, largely resistant, rabbit population

? QUESTION 2

2 Although we have waged war against species such as locusts, weevils and beetles, we have never succeeded in making any pest species extinct. On the other hand, many larger species such as rhinos, whales and giant pandas are facing extinction, despite our efforts to save them. Give two reasons for this difference.

NATURAL SELECTION IN ACTION

Many people tend to think that evolution is something that happened in the past and that the world of today is the finished product. But this is far from the truth: evolutionary forces are still at work and natural selection is continually changing the characteristics of populations, often with great speed. Consider the following examples:

- The widespread use of antibiotics such as penicillin rapidly led to the development of resistant strains of bacteria. Penicillin works by inhibiting one of the enzymes that the bacteria need in order to manufacture cell walls. However, many resistant strains can make an enzyme, penicillinase, that breaks down the antibiotic.

- The use of warfarin as a rat poison has led to the development of warfarin-resistant rats. Warfarin is an anti-coagulant that kills rats by inducing **haemorrhage** (internal bleeding). Resistant rats have a modified enzyme in their blood-clotting system that allows their blood to clot in the presence of warfarin.

- Copper tolerance in the grass *Agrostis tenuis*. Copper is a metabolic poison that prevents many plants from growing on copper-polluted land (Fig 26.6). However, resistant plants have evolved. These can transport copper out of their cells, so that it accumulates in the cell wall and does not interfere with the plant's metabolism. (In non-polluted areas, the normal type of grass out-competes the copper-resistant strain.)

The organisms in these examples have a short life cycle and high reproductive capacity, and so can respond quickly to the selection pressure placed on them. The effects of selection pressure on organisms that live longer and have a lower reproductive capacity take much longer to become apparent.

Fig 26.6 This disused copper mine at Parys Mountain in Anglesey (above) is devoid of life. Copper is a metabolic poison that kills virtually all plants. The grass *Agrostis tenuis* (right) is one of the few that has evolved copper resistance

DIFFERENT TYPES OF NATURAL SELECTION

From what you have learned about natural selection, it would be tempting to assume that it is just a mechanism for change. However, this is not always the case: it can be a means of keeping things static. Many organisms, crocodiles and sharks for example, have not changed significantly for millions of years.

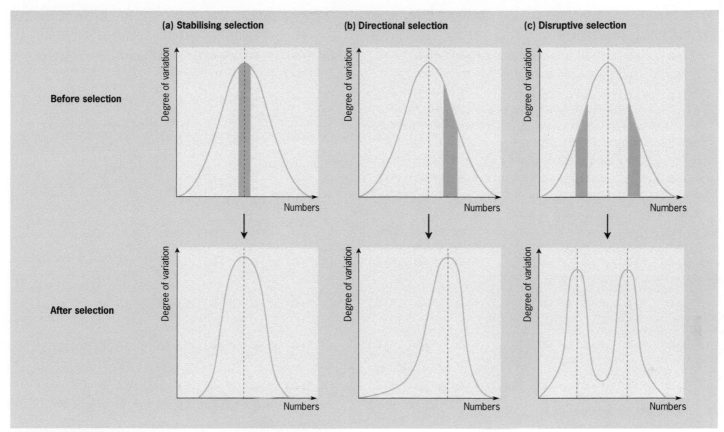

Fig 26.7 The different types of natural selection. Individuals in the shaded areas in the top row of graphs have a selective advantage

There are basically three different types of natural selection:

- **Stabilising selection** (Fig 26.7a). In a favourable and unchanging environment, stabilising selection can lead to a 'standardising' of organisms by selecting against the extremes. An example of this is human birth weight. Studies show that particularly large or small babies have a higher mortality than those in the mid-range.

- **Directional selection** (Fig 26.7b). This mechanism tends to occur following some kind of environmental change that causes selection pressure. The organisms with a particular extreme of phenotype may have an advantage. At some stage, directional selection must have operated to produce the long neck of the giraffe. Probably, in times of food shortage, only the tallest individuals could reach enough food to survive and pass on their genes to the next generation.

- **Disruptive selection** (Fig 26.7c). This type of selection acts at both ends of the distribution, favouring individuals with extreme rather than average characteristics. Rarer than the other two types, disruptive selection is important in the formation of new species, where natural selection acts differently on different populations. An example of this is seen in Darwin's finches (see page 399) where disruptive selection has acted on beak size. In some situations, those with longer, thinner beaks would have succeeded in exploiting insects as a food source, while those with shorter, stronger beaks would have succeeded in cracking seeds. Those with average beaks were adapted for nothing in particular and would have reproduced less successfully.

? QUESTION 3

3 From your knowledge of the living world, give three examples of directional selection.

Artificial selection

Humans have practised artificial selection, or selective breeding, for thousands of years. We haven't needed a scientific understanding of genetics to do this: animals or plants with the most desirable features are simply crossed or mated (Figs 26.8 and 26.9). For example, if cows with the highest milk yield are impregnated by the healthiest and fastest-growing bulls, many of the next generation are big, healthy cows that produce large quantities of milk. Those that do not have the desired features are not used for breeding. In this way, artificial selection rather than natural selection is used to bring about great changes in the phenotype in a short time: animals and plants can be 'domesticated' in just a few generations.

Selective breeding is still important in many different areas of food production:

- Increasing the growth rate and the meat, milk or egg production of livestock.
- Increasing the disease resistance of crop plants (increasing the disease resistance of animals by selective breeding is not as reliable).
- Changing the muscle to fat ratio of livestock, so that more lean meat is produced.
- Increasing the yield and nutritive value of crop plants such as wheat.
- Increasing the tolerance of plants and animals to drought, heat or pollutants.

Fig 26.9 Like all pedigree animals, this champion Abyssinian cat is the product of many generations of artificial selection. Breeders consider the wedge-shaped face a desirable characteristic and so have concentrated on this feature. Breeding animals in this way can lead to problems, and some people strongly disagree with breeding animals for superficial 'perfection'

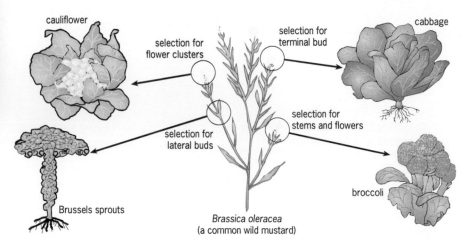

Fig 26.8 Many familiar vegetable types have arisen from the artificial selection of the wild mustard, *Brassica oleracea*. In the original wild population, plants show variation in the size and shape of leaves, stems, terminal buds, lateral buds and flower clusters. Selection for individual features rapidly produces the familiar crops you see here

3 THE EVIDENCE FOR EVOLUTION

Today, there are few scientists who seriously doubt the theory of evolution, or that it happens by natural selection. Here we consider some lines of evidence.

FOSSILS

Palaeontology, the study of fossils, supports the theory of evolution. Most fossils are found in sedimentary rocks, formed by layers of silt that preserve the bodies of organisms before they have a chance to decay in the normal way. In many places (see Fig 26.10) it is clear that different layers, or **strata**, represent different time periods. By studying the fossils in these strata, we find obvious evidence for the gradual evolution of some species.

GEOGRAPHICAL DISTRIBUTION

A study of the distribution of species presents strong evidence that organisms evolved 'into' their ecosystem rather than being placed there during a single act of creation.

During his time in South America, Darwin encountered a classic piece of evidence for evolution. On the South American mainland, finches have short, straight beaks to crush seeds. Other species have taken advantage of the other food sources. On the Galapagos Islands, however, things are different. One way or another, the finches managed to get to the islands, while many other species of birds did not. In the absence of competition, the finches have evolved to take advantage of the range of food sources on offer (Fig 26.11). Woodpecker finches, for instance, have long beaks for extracting insects from holes in trees.

Fig 26.10 The Grand Canyon in Arizona. The horizontal rock strata can be clearly seen here and span some 500 million years of geological history

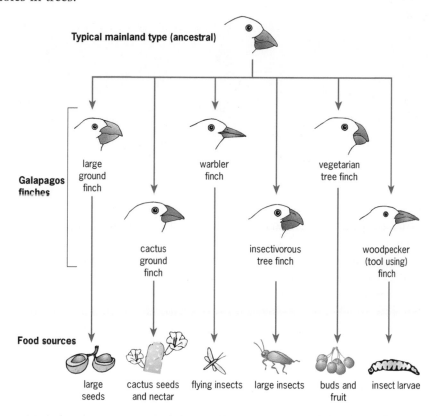

Fig 26.11 Darwin's finches on the Galapagos Islands. In a process called adaptive radiation, one common ancestor has evolved into many different species of finch, each exploiting a particular food source, such as seeds, insects and buds. Interestingly, Darwin did not immediately recognise the significance of all of these different species; he simply collected a large number of specimens in a bag and presented them to a bird specialist on his return home

HOW SCIENCE WORKS

Evolution on islands

As Darwin found out on his voyage, islands are particularly interesting places to study evolution. Young, volcanic islands such as the Galapagos (Fig 26.12) are often colonised by plants whose seeds have been blown there by the wind, and by flying animals which have been driven off course by storms. In later years, of course, colonisation by non-flying animals (such as rats and cats) is likely to be a result of human activity.

The evolution of birds on islands can be very spectacular. Free from predators, birds often lose the power of flight and sometimes grow very large. Before the Maori settlers arrived from Polynesia, New Zealand was a bird-dominated ecosystem. Huge flightless birds, **moas**, some as tall as 3.5 metres, evolved to fill the niches normally occupied by mammals. They were easy prey for the Maoris, however, who hunted them to extinction within a few centuries. Ironically, the only survivor of the moa family, the kiwi, has become the country's national emblem.

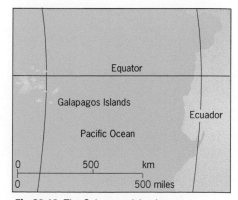

Fig 26.12 The Galapagos Islands are an archipelago situated near the Equator in the Pacific Ocean, about 600 miles west of Ecuador

COMPARATIVE ANATOMY

The anatomy (body plan) of plants and animals can reveal a lot about evolutionary relationships. Fig 26.13 shows one of the classic examples of comparative anatomy, the **pentadactyl limb** (five-fingered limb). This structure appears in various forms throughout the vertebrates: the feet of reptiles and amphibians, the human hand, the bat's wing, the seal's flipper, for example.

Fig 26.13 The vertebrate pentadactyl limb provides strong evidence for evolution. In all these examples, the basic bone layout is the same (in the diagram, the same type of bone is given the same colour in each animal) but the limbs have become modified in different ways according to the way the animal moves

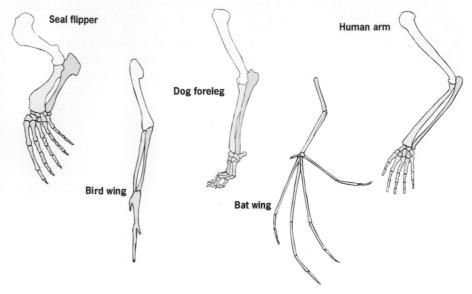

This basic five-fingered plan has been modified in a variety of ways to suit the animal's mode of locomotion. Structures with a common origin but with a different function are said to be **homologous**. So, the wing of a bat is homologous to a human hand. The possession of the pentadactyl limb in so many different organisms is persuasive evidence that these animals have evolved from a common ancestor.

In contrast, totally unrelated structures can become adapted for the same function. The wings of insects and the wings of birds, for example, have a totally different anatomy and origin in the embryo, but are similar in shape because they perform the same job. Such structures are said to be **analogous**.

? **QUESTION 4**

4 Are the following pairs of structures analogous or homologous?
   a) Trout's fin and dolphin's flipper.
   b) Horse's leg and bat's wing.

Fig 26.14 Convergent evolution occurs when unrelated species solve the same evolutionary problem in a similar way
(a) Cacti, which are native only to the Americas, have evolved to cope with desert living. The leaves have become spines which lose no water and protect the plant. The fleshy stem expands and stores water and has taken over the job of photosynthesis

(b) You might guess that this euphorbia from Africa was a cactus, but you would be wrong. The two plants have evolved in different continents and are unrelated; they look very similar because they have developed the same features to survive desert conditions

The process by which unrelated species evolve to resemble each other is called **convergent evolution** (Fig 26.14). A classic example of this is seen in seals (mammals), penguins (birds) and fish. Because of their need to swim efficiently, all have evolved the same streamlined shape. The forelimbs of the seal and the wings of the penguin have both evolved to resemble fins.

Fig 26.14(c) (left) The flipper of the whale and the fin of the shark (right) are another classic example of convergent evolution. The two animals are completely unrelated and the internal structure of forelimbs is totally different. However, they are both the same shape because of the job they have to perform

EMBRYOLOGY

Studying the development of an embryo reveals some interesting evolutionary secrets. The scientist Ernst Haeckl said 'ontogeny recapitulates phylogeny' which means that the evolutionary development of a species (its **phylogeny**) is mirrored by the changes of each individual from birth to maturity (its **ontogeny**).

A number of vertebrate embryos follow a very similar path from fertilisation through several stages of early development – so similar that it is often difficult to tell them apart (Fig 26.15). For example, all vertebrate embryos develop branchial arches; these are relics of the gills possessed by aquatic ancestors. However, although the study of the developing embryo provides further evidence for evolution, it does not give us much information about the actual evolutionary process.

Fig 26.15 Four developmental stages in embryos of fish, turtles, chicken and humans. You can see that the early stages are very similar in all four. For example, in the drawings in the second column, all four embryos have branchial arches, the 'folded' areas just under the head at the front. Only in the later stages do differences become obvious

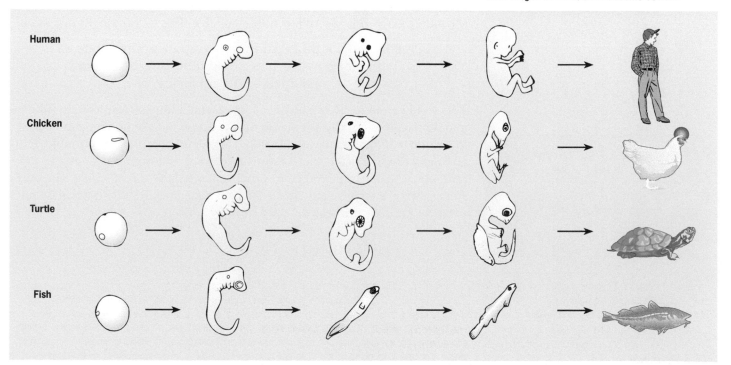

COMPARATIVE BIOCHEMISTRY AND BEYOND

Many animal genomes have been completely sequenced in the last few years and more are published all the time. Scientists are now able to compare organisms at the gene sequence level and gain new information about evolution. The science of **comparative genomics** looks at the similarities and differences between the sequences of corresponding genes to show how closely related two species are.

There are also several online databases that enable researchers to add their data about the protein structure of individual proteins from different animal species. This is called **comparative proteomics** and it aims to look for the minor differences between highly conserved protein molecules to, again, help map out the evolutionary relationships between different species.

Like humans, individual animals also have a unique genetic fingerprint and this can be used to compare animals from the same species. Analysis of genetic fingerprints looks for subtle differences in genes, which are similar but not identical between different individuals. For example, one research project is tracking populations of snow leopards in Central Asia. Researchers set 'hair snares' to collect hair samples from the leopards that pass by and then analyse the DNA from the hairs. By comparing the profiles they get, they can tell how many different leopards are in the same area. They can even plot family trees, showing how related leopards move from one area to another. This is helping researchers maximise their conservation efforts.

4 HOW A NEW SPECIES EVOLVES

We have looked at natural selection and seen that it can lead to stability or change. In this section, we examine how natural selection can lead to the formation of a new species.

WHAT IS A SPECIES?

We instinctively think that we know what the word 'species' means, but an accurate definition is difficult, if not impossible. This is a common definition:

> **A species is a population or group of populations of similar organisms that are able to interbreed and produce fertile offspring.**

This would seem to be reasonable; a horse and a donkey can mate to produce a mule (Fig 26.16), which is sterile, so the horse and the donkey can be said to belong to different species. However, this definition runs into trouble. There are many closely related species, wolves, coyotes and domestic dogs, for example, which can interbreed to produce perfectly fertile offspring. The only reason why the wolves and coyotes remain distinct is that the hybrids do not compete as successfully as the pure strains.

Fig 26.16 'And nothing to look forward to either.' The mule is the product of a mating between a donkey and a horse, two species that have different numbers of chromosomes. It cannot reproduce because it is sterile. Because of its stamina, the mule has been used as a beast of burden for thousands of years

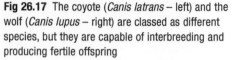

Fig 26.17 The coyote (*Canis latrans* – left) and the wolf (*Canis lupus* – right) are classed as different species, but they are capable of interbreeding and producing fertile offspring

The difficulty we have in defining the term species stems from the very nature of the process of evolution. New species evolve from pre-existing ones, and this is usually a slow transition, which may take many generations and thousands of years. There are many living species that are still in the process of separating from each other. The wolf and the coyote almost certainly evolved from a common ancestor and for all practical purposes are separate species (Fig 26.17). If left to evolve naturally over the next few thousand years, the two species may well accumulate enough genetic differences to prevent interbreeding.

So, we can modify the definition to:

A species is a collection of recognisable organisms which shares a unique evolutionary history and is held together by cohesive forces of reproduction, development and ecology.

Although more of a mouthful, it is a more satisfactory description because it recognises that all members of a species are usually (but not always) recognisable, and that they share the same 'biology', occupying the same niche in the ecosystem and having particular social structures, courtship, mating habits, etc.

THE DEVELOPMENT OF NEW SPECIES

Speciation, the development of new species, happens when different populations of the same species evolve along different lines. It is thought that speciation usually happens in the following way:

- Part of a population becomes isolated in some way that prevents them from breeding with the rest of the population. Very often, the breakaway population will, by chance, have a different genetic constitution from the original (see genetic drift – page 407).
- The two populations experience different environmental conditions, so that natural selection acts in different ways, causing the genetic make-up of the two populations to differ.
- Eventually, genetic differences accumulate to a point where individuals from separate populations that come together again can no longer interbreed.

It is worth mentioning that new species can also arise from a hybridisation between two different species. This particularly applies to plants, and plant breeders take advantage of this ability to produce new crop varieties.

There are two basic methods of speciation:

- **Allopatric speciation** (*allo* = different, *patris* = country), where the two populations are physically separated.
- **Sympatric speciation**, where the two populations are reproductively isolated in the same environment.

 REMEMBER THIS

Populations are groups of interbreeding individuals of the same species occupying a particular geographical region. Examples of populations may be the water fleas in a pond, all the oaks in a forest or the elephants in a game reserve.

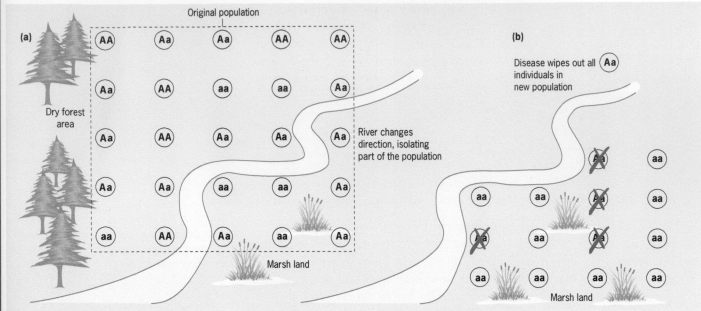

Fig 26.18 The principle of geographical isolation
(a) The original population of mice has a 50:50 distribution of the alleles **A** and **a**, i.e. **A** is as common as **a**. A river changes its course and cuts off a few individuals. In the new, isolated population the allele frequency, by pure chance, is 4:10. This change in allele frequency is called the founder effect. The fewer the individuals in the breakaway founder population, the more dramatic the change in allele frequency is likely to be
(b) The founder population starts to grow, but they are in different conditions from the original population. The alleles **A** and **a** were neutral up to now, but then a water-borne disease spreads through the new population, which are confined to marsh land. The **aa** genotype confers resistance on those who carry it. By natural selection, the **A** allele is selected against, and so the **a** allele becomes even more frequent. In this way, the genetic make-up of the two populations diverges

How populations become isolated

Usually, species evolve when populations become isolated by physical barriers such as stretches of water or mountain ranges. This is known as **geographical isolation** (Fig 26.18).

Evidence suggests that geographical isolation is the most common way in which populations become reproductively separated. However, speciation can happen even when two populations overlap. For instance, two populations of frogs may have a different mating call, and the females may respond only to calls of males in 'their' population. Table 26.1 summarises the different isolating mechanisms.

Table 26.1 A summary of isolating mechanisms

Mechanism	How it works
Prezygotic mechanisms, i.e. gametes don't come into contact	
geographic	two populations do not come into contact due to some physical barrier, e.g. mountain range, water
temporal	different timing, e.g. flowering times or mating seasons do not coincide
behavioural	individuals of other populations are not acceptable as mates, e.g. wrong courtship, mating call or pheromones
mechanical	structural differences prevent mating, e.g. the shapes of the genitals in insects
Postzygotic mechanisms, i.e. even if mating takes place	
hybrid non-viability	hybrid zygotes fail to develop to sexual maturity
hybrid sterility	hybrids do not produce functional gametes
hybrid weakness	hybrids have reduced survival or reproductive rates

5 EVOLUTION IN ACTION

The process of evolution is generally slow, and it is no surprise that for centuries people thought that the living world was static. However, there are situations where you can see the process in action. Organisms with short life spans and high reproductive capacities, such as bacteria or insects, can evolve in a very short time indeed.

Also of great interest to evolutionary biologists is a **polymorphism**, a situation where two or more different forms, **morphs**, of a species exist in a population. Each morph confers a different survival advantage in a different situation. Polymorphisms such as that of wing colour in the peppered moth allow us to see how natural selection leads to changes in allele frequency, and so we can observe evolution in action.

THE PEPPERED MOTH

Fig 26.19 shows both forms of the peppered moth *Biston betularia*. This moth was common in the UK at the time of the Industrial Revolution. In areas of heavy industry, many buildings and trees became covered in soot. The speckled moth lost its camouflage in these areas and so was eaten more often by its predators.

Luckily for the moth, a mutation occurred around this time that produced a black moth known as the **melanic form**. In industrialised areas, the melanic moth had a selective advantage. However, in 'cleaner' areas the reverse was true: the speckled moth thrived and the melanic form was rarely seen. Both morphs belong to the same species, but their survival depends largely on the environment.

Fig 26.19 The two forms of the peppered moth, normal (speckled) and melanic (black). Against the background of the bark of a tree shown in this photograph, the speckled moth on the right is almost invisible while the melanic form on the left is far more likely to be seen and caught by a predator. However, it becomes invisible on a black surface such as a sooty wall, and so is more likely to survive in an industrial environment

HOW CAN WE STUDY EVOLUTION?

If evolution is so slow, how can we observe it happening? An important principle here is that evolution involves a *change in allele frequency*. For instance, when myxomatosis was introduced into the rabbit population, the alleles for resistance were selected for, and so they increased in frequency. In any particular situation, if we can show that allele frequencies are changing, we know that evolutionary forces are at work.

Evolutionary studies are usually based on interbreeding populations, which are known as **demes** or **Mendelian populations**. The sum total of all the genes circulating within such a population is called the **gene pool**.

POPULATIONS AND ALLELE FREQUENCY

When the gene pool remains more or less the same, a population is not evolving. Natural selection may be acting, but it is often a force for stability rather than change. However, if the gene pool of a population is changing, natural selection is acting as a force for change and this is evolution in action.

To show how we measure allele frequency, imagine a population of 500 rats. In the population are two alleles for coat colour: black (**B**) is dominant over brown (**b**). The rats are diploid – each individual has two alleles – so the population contains 1000 coat colour alleles. If 600 of the alleles are **B**, then we can say that the frequency of **B** is 60 per cent, or 0.6 (statisticians prefer decimals). As there are only two alleles, it follows that the frequency of **b** must be 40 per cent or 0.4. (Remember that the sum of the allele frequencies must be 100 per cent, or 1.)

If one gene is dominant to the other – as in this case – the symbol p is used for the frequency of the dominant allele **B**, and q is used for the frequency of **b**. It is clear that $p + q$ must equal 1.

The values of p and q change when a population evolves. To find out if this is happening, we use the **Hardy–Weinberg principle**.

> **REMEMBER THIS**
> $p + q = 1$
> Frequency of dominant allele + frequency of recessive allele = total number of alleles of that gene.

Fig 26.20 This person is an albino. The white hair, pink eyes and pale skin are the result of an inability to make the pigment melanin

The Hardy–Weinberg principle

The principle, named after the British mathematician G.H. Hardy and the German biologist W. Weinberg, states that there will be no change in the frequency of alleles in a population so long as the following conditions are met:

- The population is very large, diploid and reproduces sexually.
- Mating is totally random – there is no tendency for certain genotypes to mate together.
- All genotypes are equally fertile, so there is no natural selection.
- There is no emigration or immigration.
- No mutations occur.

We can use the Hardy–Weinberg rule to estimate allele frequencies using just the information about the frequency of a particular genotype. Consider the example of albinism in humans. One in 10 000 humans are born albinos – they have two copies of a recessive gene which means they cannot make the pigment which gives colour to hair, skin and eyes (Fig 26.20). The frequency of the allele for albinism in the population is difficult to estimate because carriers of the allele are not albinos. So, how can we find the frequency of the albinism allele?

If we call the normal allele **A** and the albinism allele **a**, most of the population are **AA**, some are carriers: **Aa** or **aA**, and a small number are albinos: aa. Using the Hardy–Weinberg formula we can estimate the percentage of the population who carry the **a** allele. Remember that p and q represent the frequency of the **A** and **a** alleles respectively. The Hardy–Weinberg equation is:

$$p^2 + 2pq + q^2 = 1$$

Put into words, it says:

The frequency of AA individuals, added to the frequencies of Aa, aA and aa, must equal 1.

We also know that $p + q = 1$, and we can use these two equations to work out the proportion of heterozygotes in the population, and any other frequencies involved:

1 in 10 000 is a frequency of 0.01%, or 0.0001 expressed as a decimal.
1 in 10 000 has the genotype **aa**, so the frequency of **aa**; qq or q^2, has the value of 0.0001.

So $q = \sqrt{0.0001} = 0.01$.

If $q = 0.01$, then p must be $1 - q$, which is 0.99.

So, in other words, 99 per cent or 0.99 of the alleles in the human population are **A**.

From this we can calculate that the number of carriers of the albino allele $2pq$ is:

$$2pq = 2 \times 0.99 \times 0.01 = 0.0198 \approx 0.02$$

So, almost 2 per cent of the population carry the albinism allele.

When allele frequencies change

The Hardy–Weinberg principle states that allele frequencies remain constant as long as certain conditions apply. It follows that when these conditions are not met, the allele frequencies can change and the population can evolve. Consider what happens when the conditions listed above are not met.

- The **population is small**. If the population is small, chance plays a large part in determining which alleles pass to the next generation. The smaller the population, the greater the probability that the genes of one generation will not be accurately represented in the next. The name given to a change in allele frequency caused by chance is **genetic drift**. These are two important causes of genetic drift: **bottlenecks**, when a population declines and then recovers; and the **founder effect**, when a population starts from just a few individuals (see below). As Figs 26.18 and 26.21 show, this can result in a drastic change in the allele frequency.

- **Mutations occur**. Mutations are ultimately the origin of all genetic variation. They are rare, and most are harmful or neutral. Occasionally, however, a mutation gives an organism a survival advantage, or the environment changes so that previously harmful/neutral mutations become advantageous. For example, a mutation in a bacterial gene that leads to the production of a modified enzyme could make the bacterium resistant to an antibiotic. That individual would survive, even in the presence of the antibiotic, and pass its resistance gene on to the next generation.

- **Individuals move in or out of populations**, that is, immigration or emigration. If the new individuals breed, this movement may lead to gene flow. When a few individuals move into a new environment and start a population, we see the founder effect. Like the beads in Fig 26.21, the individuals in the founder population are probably not representative of the original population. A group of six human survivors on a desert island, for example, might have two red-haired people and two blondes. A population founded by these people would have a much higher percentage of red-heads and blondes than the population in their country of origin. In the most extreme case, a new population can be founded by one individual, such as a self-fertilising plant or a pregnant female animal.

- **Mating is non-random**. In many cases, individuals choose mates with a particular genotype. In this case, mating is not random, it is **assortive**. Such assortive mating can change allele frequency.

- **Natural selection occurs**. When there is competition for resources – and this is usually the case in an ecosystem – natural selection acts and only the most successful individuals pass on their genes.

> Genetic variation is discussed in Chapter 22. Mutation is discussed in Chapter 24.

REMEMBER THIS

Scientific names are important. If you call an organism by its scientific name, any biologist in the world, whatever language they speak, will know what you are talking about. The bird known as the sparrow in America is a completely different species from the one we are familiar with in Britain, but confusion can be avoided as the two birds have completely different scientific names. Most scientific names have Latin or Greek origins.

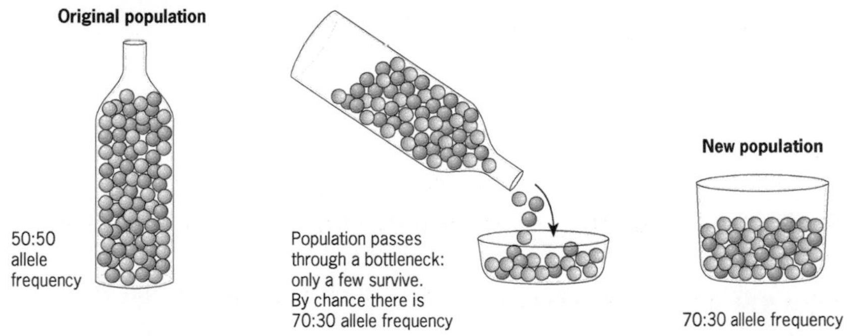

Original population

50:50 allele frequency

Population passes through a bottleneck: only a few survive. By chance there is 70:30 allele frequency

New population

70:30 allele frequency

Fig 26.21 An illustration of genetic drift. In the original 'population' there are 1000 beads; half red, half blue. If 10 beads are extracted at random, there is a high probability that there won't be 5 of each. If these 10 'organisms' then become the founders of a new population, the proportion of colour (the allele frequency) will be significantly changed, in this example to 7:3

REMEMBER THIS

Organisms do not evolve according to a set pattern, just to make the taxomist's life easier. A detailed study of a particular organism might show that it falls neatly into Kingdom, Phylum, Class, Order, Family, Genus and Species. Or it might not. This is why you sometimes find extra groups such as sub-phyla, super-families or sub-species.

6 EVOLUTION: OUR PLACE IN THE GREAT SCHEME OF THINGS

There are millions of different species on Earth, and we have given all the ones we know about a scientific name. The gorilla is simply *Gorilla gorilla*, the chimpanzee is *Pan troglodytes* while we have given ourselves the name *Homo sapiens*, which means 'wise man'. A scientific name tells us both the **genus** and the **species**. The generic name has a capital latter while the specific name does not. When written, scientific names should always be in italics or underlined.

QUESTION 5

5   Think up a mnemonic (memory aid)
    to help you learn the sequence
    Kingdom, Phylum, Class, Order,
    Family, Genus, Species: KPCOFGS.

✔ **REMEMBER THIS**

Taxonomic groups do not overlap.
For instance, there are no animals
that are both reptiles and
amphibians. An animal must
belong to one class or the other;
for example, a toad is an
amphibian and an iguana is a
reptile. Neither animal can be part
reptile and part amphibian.

We classify organisms according to the similarities between them and their evolutionary relationships. Understanding relationships between species allows us to piece together the story of life on Earth. Classification is therefore a bit like a huge family tree over a massive time scale. Just like an address has a house number, a road, a town, etc, we classify organisms into progressively narrower groups. Table 26.2 shows this 'address', which we more correctly call a **taxonomic hierarchy**. Taxonomy is the science of naming and classifying organisms and the term hierarchy reflects the layered structure of the family tree of living organisms that consists of groups within groups.

HUMANS AND THEIR CLOSE RELATIVES

Humans vary in size, shape, colour, blood groups and in thousands of other ways, but we are all one species, *Homo sapiens*. *Homo* is a genus, a set of species that are grouped together because of their close similarity. There are no living species considered to be close enough to humans to be put in our genus, so we are the only species described as *Homo*. However, there have been human ancestors that belonged to the genus *Homo*. Humans are also the only species in the family Hominidae. However, the basic body plan shared by humans and the great apes is so similar that biologists have put us all in the super-family of Hominoidae. The word **hominoid** describes any ape-like or human-like creature..

Fig 26.22 Chimpanzees show the classic ape characteristics; a sloping face with protruding jaw and heavy brow ridges.

Table 26.2 The taxonomic hierarchy: how we classify human beings

Taxonomic group	Name	Explanation
Kingdom	Animalia	There are five kingdoms, animals (Animalia), plants (Plantae), fungi, protoctista and bacteria (Prokaryotae). We are animals.
Phylum	Vertebrata	We are animals with backbones
Class	Mammalia	There are five classes, fish, amphibians, reptiles, birds and mammals. The defining features of mammals are fur/hair and milk production
Order	Primates	Primates are adapted for life in the trees. Key primate features are covered in the main text
Family	Hominidae	Human-like creatures – includes several extinct species
Genus	*Homo*	No other living species is sufficiently like humans to be included in this genus
Species	*sapiens*	We are the thinking species, modern man

HUMAN EVOLUTION

So how did humans evolve? It is widely accepted that humans evolved in Africa, from the same ancestors that gave rise to the great apes: chimps (Fig 26.22), gorillas and orang-utans. We did not, however, evolve from these species, as is often thought. Piecing together a complete evolutionary picture is difficult. If you took all the remains that have ever been found of ancestral humans, they would all fit into an average-sized living room. However, we do have some vital fossil evidence that has allowed us to piece together a likely **lineage**, or line of development (see the How Science Works box opposite).

The first primates

The first primates to evolve appeared about 70 million years ago. We owe a lot of our human features to our primate ancestry.

Ramapithecus

One thing that all fossil remains of early humans have in common is that they are incomplete. Fragments of jawbone or skull may be all that is found, and scientists are left trying to guess what the rest of the skeleton looked like, and where it fits in the great scheme of things. People can make great claims about a new specimen being another important 'missing link', only to find that they were mistaken when more complete evidence comes to light.

One classic case was that of *Ramapithecus*. The first incomplete specimens of *Ramapithecus* were found in Nepal in 1932. The finder claimed that the jaw was more like a human's than any other fossil ape then known. In the 1960s this claim was revived. At that time, it was believed that the ancestors of humans had diverged from other apes 15 million years ago. Biochemical studies upset this view, suggesting that there was an early split between orangutan ancestors and the common ancestors of chimps, gorillas and humans. Humans had separated from African apes about five million years ago, not 15 million or 25 million.

Meanwhile, more complete specimens of *Ramapithecus* were found in the 1970s, which showed that it was less human-like than had been thought. It began to look more and more like *Sivapithecus* – an ancestor of the orangutan. It could be that *Ramapithecus* was just the female form of *Sivapithecus*. They were definitely members of the same genus. It is also likely that they were already separate from the common ancestor of chimps, gorillas and humans, though fossils of this presumed ancestor have not yet been found. *Ramapithecus* is no longer regarded as a likely ancestor of humans.

Primates have the following key features:
- Flexible limbs with gripping hands. Hands and feet have **opposable thumbs** – thumbs that can oppose all the other digits, giving an excellent grip.
- Fingernails rather than claws. This allows some protection, but doesn't interfere with the sensitive fingertips. Primate hands are highly dextrous. This makes it easier to obtain food and it allows more effective grooming. It has also helped us develop the ability to make tools.
- Flat faces with forward-facing eyes. This allows stereoscopic vision, giving judgement of distance, an essential skill for a life in the trees. Primates also have colour vision, which is believed to help them find ripe fruit.
- Large brains. Primate brains have large areas that process information from the eyes, ears, mouth and skin.
- Socially, primates tend to live in family groups, giving birth to single offspring that are dependent on their parents for a long time.

Primate fossils

The oldest primate fossils are about 70 million years old. These look more like modern tree-shrews, but have certain primate features. Over the next 30 million years various primate species evolved and became extinct, until the first anthropoids appeared about 40 million yeas ago. These were like modern monkeys, living in trees and with a well-developed social structure.

In 1924, the fossilised skull of a child was found in a quarry at Taung in South Africa. After much speculation, this specimen was named *Austrolopithecus africanus* which means 'Southern ape from Africa' (Fig 26.23). Since then, several hundred austrolopithecines have been discovered, mainly in Africa. It seems highly likely that humans evolved from this type of 'subhuman'. The outline of human evolution that is shown in Table 26.3 and in Fig 26.24 is highly speculative, and can change with the discovery of one new fossil. A comparison of the skeletons and dentition of a gorilla, and early hominid (Lucy) and a modern human, is shown in Fig 26.25.

Making sense of fossils

This photograph shows the fossil remains of Lucy, a young girl who lived approximately 3.5 million years ago. Looking at it can illustrate quite well how fossils are able to tell us a lot more than they appear to at first glance.

How can we tell this fossil is a 12-year-old girl?

You can sex a human, or humanoid, skeleton from the shape of the pelvis. Females have a wider pelvis, with a larger gap to allow a baby's head to pass through. Size and bone development indicate age.

How do we know the age of these remains?

Fossils are formed in sedimentary rocks. When this girl died, her body was covered with layers of silt and and sand in conditions that protected it from the normal processes of decay. The soft tissues such as the skin, muscles and internal organs do decay, but the bones undergo a chemical reaction that turns them into rock. In time, the sediments bury the fossils deeper and deeper. We can estimate the age of the skeleton by comparing the layer in which it was found with other strata (stratigraphy), but for a more accurate measurement we can use **potassium/argon dating** – see the How Science Works box on page 412.

What did Lucy look like when she was alive?

From the skull fragments we can see some obvious features: the sloping face, protruding jaw and prominent eyebrow ridges; she was obviously not a modern human. We can reconstruct the soft tissues in the same way that forensic artists do with the skull of a murder victim. If you know the muscle structure and the average depth of the soft tissues at strategic points on the skull, you can use clay, and a lot of skill, to reconstruct the face.

How large was her brain?

This is easy – you just look at the space left inside the skull. You can estimate the volume by filling the cranium with water or sand and then emptying into a measuring cylinder. You can even make a latex cast of the brain cavity and measure its volume more accurately by displacement.

What did she eat?

There are several clues about diet. Generally, the more vegetation in the diet, the stronger the muscles which pull the jaw from side to side, and the flatter the molars. This girl has teeth very similar to ours, so she probably had a varied, omnivorous diet. Sometimes microscopic examination of the scratches on the enamel can tell us a lot about the diet of a fossilised human.

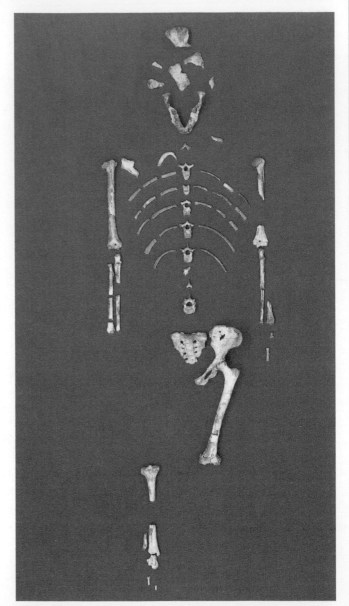

Fig 26.23 This is Lucy, the most complete specimen of *Australopithecus afarensis* ever found. She was discovered in 1974 in what is now Ethiopia, and named after the Beatles song 'Lucy in the sky with diamonds' that the archaeologists Don Johanson and Tom Gray were playing at the time of the discovery.

Overall, the great skill in piecing together fossil evidence comes from knowing where to look, and in recognising hominid remains when you see them. There are incredibly skilled people who can look at a pile of rubble and spot a tooth, jaw fragment or vertebrae of a primitive human, without mistaking it for modern ape or human remains. Who knows what remains are yet to be discovered?

How do we know that she walked upright?

First, the hole where the spinal cord goes into the skull (the foramen magnum: Latin for 'big hole') is under the skull so that the head pivots on the spinal column. Also, the leg bones are thicker than those of the arm, and the pelvic girdle is stronger than the shoulder girdle. This implies that the weight was placed mainly on the legs rather than evenly distributed.

Table 26.3 The history of the hominids

Time period	What was going on?
8–4.4 million years ago	This period is sometimes called the 'fossil void', which is frustrating because, at some time in this period, the hominids, the family that gave rise to man, appeared
4.4 million years ago	*Australopithecus ramidus* appeared. Discovered in Ethiopia in 1992, this is the earliest known hominid. Evidence suggests that *A. ramidus* was bipedal (walked on two legs) and had a small brain but we know little about his lifestyle
4–3 million years ago	The time of *Australopithecus afarensis*. The first, almost complete skeleton, Lucy (Fig 26.23) was discovered in 1974. Lucy was like a modern human from the waist down, suggesting bipedalism, but her upper half was distinctly ape-like. Lucy and her relatives evolved in a time of climatic change, when the great forests gave way to extensive grasslands. The largely forest-dwelling ape-men and women were forced to survive out in the open. In this situation, natural selection would favour individuals who were more intelligent, showed greater social co-operation and were better at using tools. Some experts think that this is what drove evolution to produce modern humans
3–2.2 million years ago	*Australopithecus africanus* existed. A specimen found in 1924 in South Africa shows a distinctly human jaw and dentition, but the cranial capacity was still small: more ape-like than human
2.3–1.7 million years ago	*Homo habilis*, 'handy man', appeared. This is the earliest creature to be assigned to the genus *Homo*. *Homo* is distinguished from *Australopithecus* by progressive changes: a larger brain, a higher skull, a flatter face and a more rounded jawbone. It is possible that *H. habilis* made primitive tools, such as sharp-edged stones that could cut up animal carcasses. *H. habilis* was replaced by, or evolved into, *Homo erectus*
1.8–0.5 million years ago	*Homo erectus* was a larger species with a larger brain. This species was almost as big as modern humans, although the skull still had the primitive, heavy-browed ape-like form. This species existed for over a million years.
0.4–million years ago	*Homo sapiens* appeared in Europe and Africa. Two sub-species are *Homo sapiens neanderthalensis* (Neanderthal man) and *Homo sapiens sapiens* (ourselves). Neanderthals had slightly larger brains than *H. sapiens sapiens*; they were large, muscular, barrel-chested people with long, globe-like skulls and long noses. They made tools and there is evidence to suggest that they buried their dead. The most recent Neanderthal fossils are only 32 000 years old. Studies of DNA suggest that *Homo sapiens sapiens* first appeared in Africa about 150 000 years ago: just a moment ago in the geological time scale (see the How Science Works box on page 412). They began to migrate from Africa about 100 000 years ago. Why Neanderthals died out remains a mystery.

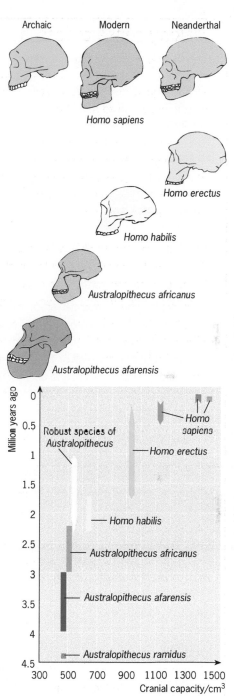

Fig 26.24 The major evolutionary stages of the hominids.

Top: The development of skull shapes from *Australopithecus afarensis* to *Homo sapiens sapiens*

Bottom: The place in time and cranial capacity of the hominids

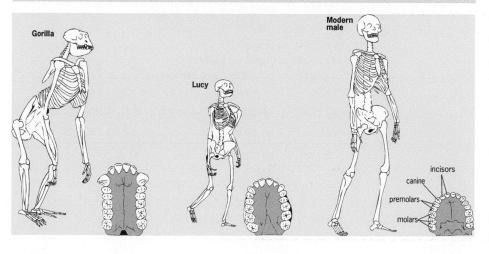

Fig 26.25 A comparison of the skeletons and jaws of a gorilla, Lucy and a modern human

The history of the world

Rocks, and the fossils they contain, can now be dated by radiometric dating techniques, which measure levels of certain isotopes. We can use this technique to construct a geological time scale – a biological history of the world. This box outlines the process of dating using radioactive isotopes and puts the history of life on Earth into perspective.

Dating rocks and fossils

The principle behind radiometric dating is simple: some elements exist as radioactive isotopes that decay at a predictable rate. Measurement of this decay allows us to estimate the age of the rock.

For example, when volcanoes erupt, the lava contains radioactive potassium, ^{40}K. This decays steadily into argon (^{40}Ar). The half-life of ^{40}K is 1.3 billion (1.3×10^9) years. So, in 1.3 billion years, half of the potassium decays. Half of the remaining potassium will decay in the next 1.3 billion years, and so on. By measuring the ratio of radioactive potassium to argon, scientists can estimate the age of the rock strata to the nearest 50 000 years.

In many places, layers of volcanic lava alternate with sedimentary rocks that contain fossils. If we can date the lava, we can also estimate the age of the fossils. Some fossils of primitive bacterium-like organisms are thought to be around 3 500 000 000 years old.

Carbon dating

This process can be used to date samples that still contain some organic material. Cosmic radiation is constantly bombarding the atmosphere, turning $^{12}CO_2$ into $^{14}CO_2$. This becomes fixed into organic molecules by the process of photosynthesis and then passes up the food chain. When an organism dies, the ^{14}C decays into ^{14}N, with a half-life of 5730 years. Carbon dating can be used to measure the age of any carbon-containing material up to 50 000 years old.

The Earth is currently estimated to be about 4 600 000 000 years old. This time scale is impossible to comprehend, but if we condense this huge span into one year we can begin to put it into perspective.

The Earth calendar

Date	Event
1 Jan	Earth forms
26 Feb	oldest rocks – nothing was permanent before this time
23 March	first life – bacteria-like organisms obtain energy from organic molecules (food) by the process of glycolysis
April–May	organic food runs out; photosynthesis evolves, makes food and releases oxygen as a by-product
30 June	oxygen atmosphere fully developed; aerobic respiration evolves some time after this
17 Sept	first eukaryotes
19 October	first co-ordinated multicellular colonies (like sponges)
5 Nov	first invertebrates (worms and jellyfish)
15 Nov	rapid increase in invertebrate diversity
20 Nov	first fish
21 Nov	first coral reefs
29 Nov	first colonisation of land – vascular plants appear
4 Dec	first amphibians
6 Dec	first trees
7 Dec	first insects
8 Dec	first reptiles
17 Dec	first mammals
19 Dec	first birds and dinosaurs
21 Dec	first flowering plants
25 Dec	last of the dinosaurs
26 Dec	first primates
31 Dec	
14.00	first upright bipedal hominids
21.00	*Homo erectus*
22.45	recent Ice Age starts
23.00	evidence that primitive humans can cook food
23.48	Neanderthal man appears
23.54	*Homo sapiens* appears
23.55	Australian aborigines settle
23.58.49	agriculture invented
23.59.28	beginning of recorded history
23.59.29	pyramids built
23.59.41	Buddha born
23.59.45	Christ born
23.59.53	Battle of Hastings
23.59.57	start of the scientific method
23.59.58	Australia 'discovered' by Cook; the Industrial Revolution
23.59.59	most of the knowledge contained in this book is discovered

SUMMARY

When you have read this chapter, you should know and understand the following:

- **Evolution** is defined as a change in the genetic composition of a population over time. New species develop from pre-existing species in this way.
- The basic mechanism of evolution is **natural selection**. This acts on the genetic variation that occurs in a population. Individuals whose genotype gives them an advantage over others are more likely to pass their genes on to the next generation.
- A **population** is a group of interbreeding individuals of the same species in the same geographical area.
- New species are formed when populations become isolated in some way so that they cannot interbreed. Natural selection then acts on the different populations, changing the frequency of **alleles**. With time, the differences between the two populations can become so great that an individual from one cannot interbreed with an individual from the other.
- Humans can speed up the evolutionary process by **artificial selection**. Breeding organisms with desirable features, such as cattle with a high milk yield, changes the allele frequency rapidly. Within only a few generations, organisms are produced which are very different from the original wild stock.

- The evolutionary process is fastest in species which have a large population size, a large reproductive capacity (fecundity) and a rapid life cycle. Many disease-causing and pest organisms fall into this category.
- Many endangered species have a low reproductive capacity and a long life cycle.
- Evolution happens when there is a change in the allele frequency. The **Hardy–Weinberg principle** states that, in a large population, the allele frequency remains constant unless some outside force acts to change it. These forces are chance (the smaller the population, the greater the effects of chance), migration of organisms, selective mating and natural selection.
- When the conditions for the Hardy–Weinberg equilibrium are not met, the allele frequency changes and the population evolves.
- Humans evolved in Africa from ape-like ancestors. Key stages include bipedalism (walking on two legs), an increase in brain size, the use of tools, the development of language, social co-operation and the evolution of agriculture.

 Practice questions and a How Science Works assignment for this chapter are available at www.collinseducation.co.uk/CAS

27

DNA technology

27 DNA TECHNOLOGY

Carbon copy cat

This is 'Rainbow', cc's genetic donor. Note the differences in coat colour between cc and Rainbow

In February 2002, the first cloned cat, a kitten called cc (the typists' abbreviation for 'carbon copy') made her world press debut. Cute though she was, cc highlighted some of the technological and ethical issues surrounding cloning technology.

cc was cloned by transplanting DNA from Rainbow, a female three-coloured (tortoiseshell) cat into an egg cell whose nucleus had been removed, and then implanting this embryo into Allie, the surrogate mother. cc is genetically identical to her genetic donor, Rainbow, but she isn't an exact physical copy, underlining the delicate balance between genetics and environment. Coat colour in cats is determined partly genetically, and partly by environmental factors, such as conditions in the womb.

cc also demonstrates just how difficult and 'wasteful' cloning is. In their first attempt, researchers obtained the cells used to make the clone from the skin cells of a donor cat. Eggs from other cats were used for the next step. Their chromosomes were removed and replaced with the DNA from the skin cells, creating cloned embryos that were then transplanted into surrogate mothers.

In total, 188 such nuclear-transfer procedures were performed, which resulted in 82 cloned embryos that were transferred into seven recipient females. Only one cat became pregnant, with a single embryo. But this pregnancy miscarried and the embryo was surgically removed after 44 days. In the next attempt the scientists tried using cells from ovarian tissue to receive the DNA from the cat to be cloned.

cc, the first-ever cloned cat shown here at seven weeks old, was born on 22 December 2001, at the College of Veterinary Medicine, Texas A&M University. The announcement of the successful cat cloning was delayed until the animal had completed its shot series and its immune system was fully developed

Five cloned embryos made in this way were implanted into a single surrogate mother. Pregnancy was confirmed by ultrasound after 22 days and a kitten was born on 22 December 2001, 66 days after the embryo was transferred. So, out of 87 implanted cloned embryos, cc is the only one to survive. This is comparable to the success rate in sheep, mice, cows, goats and pigs, but would be unacceptably low for people.

REMEMBER THIS

✓ Genetic engineers manipulate DNA to allow individual genes to be transferred from one organism to another. Often the two organisms are from completely different species.

1 THE NEW GENETICS

Genetics has come a long way since the days of Mendel and his pea experiments. A vast amount of scientific progress has been made in the last 25 years but these advances have been fraught with moral and ethical dilemmas. Learning about the 'new genetics' cannot be done without considering these and doing some deep thinking (Fig 27.1). So, what is involved in the new genetics?

TECHNIQUES FOR DISCOVERING THE GENETIC BASIS OF DISEASE

If you want to understand the causes of a particular genetic disease, you must first identify and then locate the gene or genes involved. Some of the techniques used to locate genes are described in Section 2.

All lifestyle diseases have some sort of genetic component. One person may smoke cigarettes for 60 years and not develop lung cancer, but another person does. Why the difference? It must be in the genes/alleles an individual possesses. For example, several different genes are associated with increased breast cancer risk: if a woman carries a variant allele of one of these genes, her risk of developing breast cancer at a young age is greatly increased. There are many ethical dilemmas here:

- How much should people be told?
- How much do they want to know?
- Who owns this information? The individual? The doctors?
- How much should financial institutions such as mortgage and life insurance companies be told?

Environmental factors influence how genes are expressed; just because someone has a particular gene sequence, it cannot be used to definitely say what may happen to them in the future.

Gene-hunting techniques are also used to identify genes in order to extract them from one organism for transfer to another – the process by which transgenic organisms are produced.

CLONING GENES AND MAKING TRANSGENIC ORGANISMS

It has been possible for some time to take a single gene from one organism and put it into the DNA into another, so that the gene is 'adopted' and expressed by the recipient organism. To date, this has been most successful in plants and lower animals such as invertebrates, although some mammalian transgenic organisms have been developed. Some of the techniques used in gene cloning are described in Section 3.

GENE THERAPY

If someone has a genetic disease, they may have a variant allele in just one gene, but this will be present in every one of their body cells. If that allele makes a non-functional protein, it is theoretically possible to treat their disease by gene therapy. The basic principle is the same as that used to produce a transgenic organism. A copy of the healthy allele of the gene, which can code for the functional protein, could be transferred into body cells to restore normal gene expression. Human gene therapy has not met with great success, although many researchers hope that the setbacks and problems may be overcome in the future.

GENOME AND GENE SEQUENCING

Finding a particular gene, or cloning it, is easier if you have a map to look at to start with. It wasn't until the 1990s that DNA sequencing technology developed and became sufficiently automated for whole chromosomes and whole genomes to be sequenced. A major advance was the publication of the complete sequence of the human genome in April 2003, the culmination of 13 years' intensive effort. Several other mammalian genomes have now also been sequenced (see Table 22.2).

Fig 27.1 'You were so keen to see what you could do, you didn't stop to think about whether or not you should.' This is a scene and a quote from the film *Jurassic Park*. It is quite unlikely that we will ever be able to recreate dinosaurs, but the quote reflects an important point: all genetic research has the potential for abuse and we should think very carefully about its consequences

The use of transgenic organisms in farming is covered in detail in Chapter 6. Cancer and heredity are discussed in Chapter 31.

Vast computer databases are now available online to scientists all over the world to give them information about human genes and the variations that can exist in those genes. In 2006, the International HapMap Consortium published a comprehensive catalogue of human genetic variation, for researching the genes involved in common diseases, such as asthma, diabetes, cancer and heart disease.

WHOLE ORGANISM CLONING

Dolly the sheep was the first mammal to be cloned in 1997 and cc the cat was cloned in 2002. 2005 saw the first dog clone – Snuppy, an Afghan pup born in South Korea. The following animals have also been successfully cloned (year of first cloning in brackets): goats (2000), cows (2001), mice (1998), pigs (2000), cats (2002), rabbits (2003), mules (2003), rats (2003), horses (2003), deer (2003), ferrets (2006) and a gaur (2001). The guar was the first endangered animal to be cloned but it died shortly after birth.

Fewer 'ordinary' species are now cloned – the point of the early clones was to perfect the technique. As we describe in Section 4, in the future, animal cloning will have two main purposes:

- **Conservation of endangered species.** If animals threatened with extinction can be cloned, this could be used to rescue or save a species from total disappearance.
- **Production of pharmaceuticals or organs for transplantation.** This is controversial, but it is possible to clone sheep and cows and then modify the embryos at an early stage so that the cloned animal produces drugs or useful proteins in its milk. One suggestion is that cows could be modified to produce milk very similar to human breast milk to feed to babies.

Note than no primates have been cloned so far. It took over 300 attempts to get Dolly – the other attempts resulted in deformed, abnormal fetuses. Not surprisingly, the ethical dilemmas of human cloning are enormous. Although it is technically possible to clone a human embryo, and this is done to produce stem cells, it is outlawed in most countries to implant a cloned human embryo into a female's uterus for it to develop into a baby (Figs 27.2, 27.3 and 27.4). There have been wild claims that this has already been done, but none has been proven. Just the practice of using embryos to harvest stem cells has created huge public outcry and controversy.

STEM CELL CLONING

Stem cells are cells in the body that are capable of differentiating into any type of body cell, given the right signals. Adults have very few stem cells, but they are plentiful in early stage embryos. This is why very early stage embryos left over from IVF treatment are used to harvest stem cells for research, a controversial process that is highly regulated. Section 7, which concentrates on stem cell cloning, concludes this chapter.

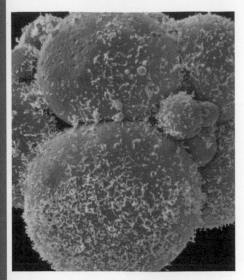

Fig 27.2 An 8-celled embryo – known as a morula at this stage – just 3 days after fertilisation. At this stage all the cells are stem cells. Is it right to take these cells to help in research and medical treatment, or is the embryo an individual human being with rights?

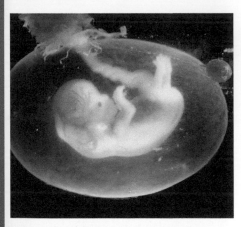

Fig 27.3 The embryo is now 7–8 weeks old. It is not referred to as a fetus until week 12. Are the rights of this embryo any different now that it is 7 weeks older than the morula in Fig 27.2?

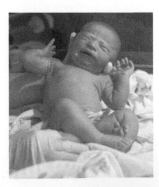

Fig 27.4 A newborn baby girl, 39 weeks after conception. What does the future hold for this baby? Will it be a long and healthy life in which many diseases are controlled or cured? Or will it be one cut short by diseases for which treatment was available but too expensive?

Genetic screening and diagnosis

The Science in context box on page 374 describes how the singer Woody Guthrie, along with other sufferers, only found that he had Huntington's disease when the symptoms appeared. By that time, he was about 40 and had already had a son to whom he could have passed the disease. Now it is possible to screen newborn babies for an increasing list of genetic abnormalities.

All babies in the UK are tested routinely for phenylketonuria. This condition and the heel prick test that detects it (Fig 27.5) are described in Chapter 28. Cystic fibrosis, sickle cell anaemia and Tay–Sachs disease can also be detected by screening techniques.

A simple blood test can tell Tay–Sachs carriers from non-carriers. Blood samples can be analysed by either enzyme assay or DNA studies. The enzyme assay is a biochemical test that measures the level of Hex-A (a vital enzyme called hexosaminidase A) in a person's blood. Carriers have less Hex-A in their body fluid and cells than non-carriers.

Screening can now also be done for genes that increase the risk of developing some types of cancer. Knowing that you have a gene that makes it more likely that you will get cancer, or that you might be able to pass on a genetic disease to your children, can be difficult to cope with and genetic screening tests are usually done in conjunction with lengthy counselling.

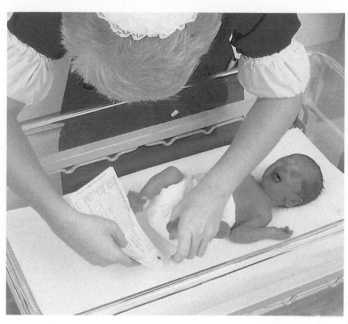

Fig 27.5 Soon after birth, babies have the Guthrie (heel prick) test which screens for phenylketonuria. This genetic disease can cause brain damage if untreated, but a special diet, if followed carefully, can prevent any problems

2 LOCATING GENES

The first problem that researchers tackle is finding the gene they are interested in. Every human body cell contains 46 chromosomes, each having several thousand genes. So how, for example, do you go about finding the insulin gene that you suspect is involved in asthma? Let's look at three different approaches.

- **Isolate the gene from the rest of the genome.** This has been compared with looking for a needle in a haystack, when the needle is also made from straw. However, there are several sophisticated techniques that can be used to map chromosomes. For instance, if you know part of the base sequence in the required gene you can make a genetic probe. This is a single-stranded piece of matching DNA labelled with a radioactive or fluorescent marker. When mixed with the genomic DNA (DNA from the genome) – in the right conditions – the probe seeks out the complementary base sequence and binds to it, showing you exactly where the gene is (Fig 27.6).

    These techniques can now be combined with the vast database produced by the Human Genome Project (see Section 5). This database of all the genes in the human genome is freely available to all scientists, and they can look up likely genes and gain predictions of function, using computer programs that construct the most likely

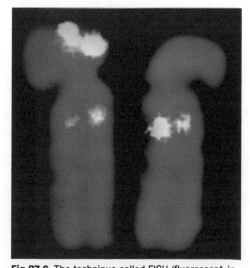

Fig 27.6 The technique called FISH (fluorescent *in situ* hybridisation) was used here – a genetic probe containing a fluorescent marker seeks out the gene. This micrograph shows a deletion on chromosome 17. On the normal chromosome on the left the normal gene shows up as two green dots. These are absent on the chromosome on the right

DNA, genes and chromosomes are discussed in more detail in Chapter 24.

? QUESTION 1

1 Suggest what the enzyme reverse transcriptase does.

✓ REMEMBER THIS

DNA is common to all organisms, and it always has the same basic structure. This allows us to combine genes from organisms as diverse as humans and bacteria.

protein from the DNA sequence. The possibilities are extending so fast, it is impossible to give more than a brief overview here.

- **Use messenger RNA.** If you can isolate the mRNA molecule that is a copy of the gene that you are trying to find, you can use it as a genetic probe as described above. You can also use it to make a DNA copy using the enzyme reverse transcriptase. This means essentially that you make an 'artificial' gene. For example, you could find the mRNA that codes for insulin in the cells where the insulin gene is expressed – the cells in the islets of Langerhans. DNA made from mRNA is called complementary DNA, or cDNA. Artificial genes are shorter than those in the genome because they contain no introns.

- **Work backwards from the protein.** If you work out the amino acid sequence of the desired protein – a relatively easy process – you can make a piece of DNA that codes for it. It would obviously be a tedious process to make a large gene, but now there is equipment that can synthesise artificial DNA quickly and easily. The end-product of this process is also a short, artificial gene made of cDNA.

3 CLONING GENES AND MAKING TRANSGENIC ORGANISMS

The first step in cloning a gene, once you have located it, is to remove it from the host genome. Enzymes are the tools of the trade here.

MANIPULATING DNA USING ENZYMES

Many different enzymes are used routinely in gene cloning, including:

- Exonucleases are used to remove a terminal base from a DNA sequence – this is useful in the analysis of base sequences.
- DNases make random cuts in DNA, chopping it into various sized fragments (this is also useful for DNA fingerprinting).
- Kinases add groups such as phosphates to DNA: this helps with labelling and analysis, particularly when radioactively labelled phosphates are used.
- Polymerases synthesise nucleic acid molecules from nucleotides. This is essential for the polymerase chain reaction (PCR) that is used to amplify DNA sequences for analysis. Reverse transcriptase is a polymerase that makes DNA from RNA. It occurs naturally in retroviruses such as the human immunodeficiency virus, HIV.

Restriction enzymes and DNA ligases are particularly useful.

Restriction enzymes

Restriction enzymes can be thought of as 'molecular scissors' – they cut DNA strands at specific points (Table 27.1). More properly called **restriction endonucleases**, they are made by bacteria in response to attack by viruses called bacteriophages. Their name reflects their function: they restrict damage by chopping bacteriophage DNA into smaller, non-infectious fragments, and they make their cuts inside the nucleic acid molecules.

There are many different endonuclease enzymes, produced originally by different species of bacteria. Each enzyme cuts DNA at a different base sequence, a point known as the recognition site. The enzymes make staggered cuts in the DNA (Fig 27.7), commonly called sticky ends. These can combine with complementary sticky ends, and this allows lengths of DNA that have been removed from one organism to be spliced – inserted – into the DNA of another. EcoRI, for example, is a common endonuclease isolated from the

Table 27.1 The target sites of some restriction enzymes

Enzyme	Bacterial origin	Recognition site
EcoRI	E. coli	G|AATTC CTTAA|G
HindIII	H. influenzae	A|AGCTT TTCGA|A
BamHI	B. amyloliquefaciens	G|GATCC CCTAG|G

bacterium *Escherichia coli*. This is a very popular enzyme with genetic engineers because it is readily available. It is also reliable: it cuts DNA at its recognition site with precise accuracy, showing very little **star activity** (cutting DNA at sites other than the recognition site).

You might be asking, 'If restriction enzymes cut DNA, why is the bacterium's own DNA not damaged?' The answer lies in a process called **methylation**: the bacteria add a methyl ($-CH_3$) group to the sequences in their own DNA that could act as recognition sites. This prevents the restriction enzymes doing any damage, presumably because the methylated region no longer fits into the active site of the enzyme.

REMEMBER THIS

The prefix *endo-* means inside; the prefix *exo-* means outside. An endonuclease cuts up a DNA strand by making cuts along the whole length of the molecule. In contrast, an exonuclease cuts DNA by chopping off the nucleotides at the ends of the molecule.

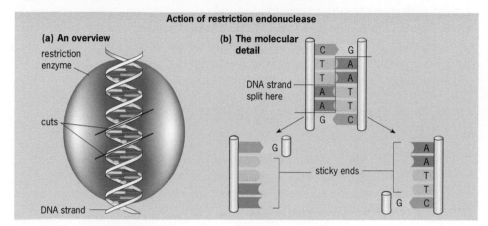

Fig 27.7 Endonuclease enzymes are wonderful tools for the genetic engineer because not only do they cut DNA, but they make staggered cuts, leaving 'sticky ends' which make re-joining DNA strands much easier. Endonucleases cut at the sugar–phosphate bonds in the DNA molecule, leaving only the hydrogen bonds intact. At room temperature and above, these bonds are not enough to hold the molecule together, and so it splits, producing a 'sticky end'

SCIENCE
IN CONTEXT

The polymerase chain reaction

The first recombinant DNA molecules were produced in the early 1970s when bacteria were made to copy foreign DNA along with their own. However, in 1983 an American called Cary Mullis cloned DNA without bacteria using the relevant enzymes. This process, the polymerase chain reaction (PCR), is gene cloning in a test tube. It allows DNA to be amplified (copied many times).

The idea behind PCR is very simple (Fig 27.8). You mix together your original piece of DNA with the enzyme DNA polymerase in a solution of nucleotides. Next, add some primers, short pieces of DNA which act as signals to the enzymes, effectively saying 'start copying here'. You then heat the DNA so that it denatures into two strands. The thermostable DNA polymerase gets to work and produces two identical strands of DNA (see Chapter 24 for the detailed mechanism of DNA replication).

In a second cycle of reactions, the two strands become four, and so on. Typically, the cycle is repeated about 20 times and within an hour you have millions of copies of the original piece of DNA.

This technique has many applications. Forensic scientists use it to amplify DNA samples from spots of blood or hair roots, to obtain enough material for forensic analysis by DNA profiling. PCR also allows archaeologists to study small samples of DNA from historical material such as the preserved bodies found in peat bogs or ice graves.

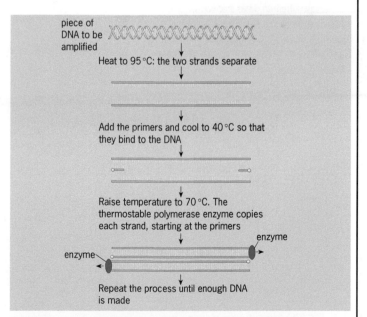

Fig 27.8 The polymerase chain reaction. The process commonly uses TAQ polymerase from the thermophilic bacterium *Thermophilus aquaticus* (hence TAQ). The enzyme has an optimum working temperature of about 80 °C. The bacterium was first found in hot springs in Yellowstone National Park, and has since yielded several useful enzymes

? QUESTION 2

2 Different restriction enzymes have differently sized recognition sites. Those in Table 27.1 are described as 'six cutters' because their recognition sites are six bases long, but four or five cutters are also common. Why do you think four cutters produce a larger number of fragments than five cutters?

DNA replication is discussed in Chapter 24. Plasmids and bacteria are discussed in Chapter 1.

Fig 27.9 The principles of recombinant DNA technology. A human gene that codes for a useful product, such as insulin, is transferred into a bacterium which then multiplies to form a large population. The bacteria all express the human gene, and large amounts of the gene product can be recovered

DNA ligases

The sticky ends of the DNA strands produced by the action of restriction enzymes are joined together by enzymes called DNA ligases (ligate means to join/bind). The normal function of a DNA ligase enzyme is to join strands of DNA during replication. They are also involved in DNA repair. Genetic engineers use them as 'molecular glue' and they are essential partners for restriction enzymes (Fig 27.9). An example is T4 DNA ligase, which is made by the bacterium *E. coli*.

AMPLIFYING THE DNA OF A CHOSEN GENE

Armed with restriction enzymes, ligases and other enzymes, genetic engineers can isolate a single gene that they wish to clone. They can use the polymerase chain reaction (PCR) to make multiple copies of the gene so that they have enough material to work with (see Science in Context box on previous page).

USING VECTORS TO TRANSFER GENES INTO BACTERIA

Although DNA fragments can be multiplied by PCR, it is often much easier to put the DNA into organisms such as bacteria which will then adopt the new DNA as their own, and copy it for you. This is called gene cloning, or recombinant DNA technology (Fig 27.9).

All the bacteria express the gene and large amounts of the gene product can be made by culturing the bacteria in large industrial fermenters (Fig 27.10).

Putting a gene from a mammal or a human into a bacterium is not an easy task. The first step is to attach the gene to a vector, or carrier. One such vector is a plasmid (Fig 27.11), a tiny circular piece of DNA that occurs in bacteria. Like restriction enzymes and ligases, plasmids can be bought commercially.

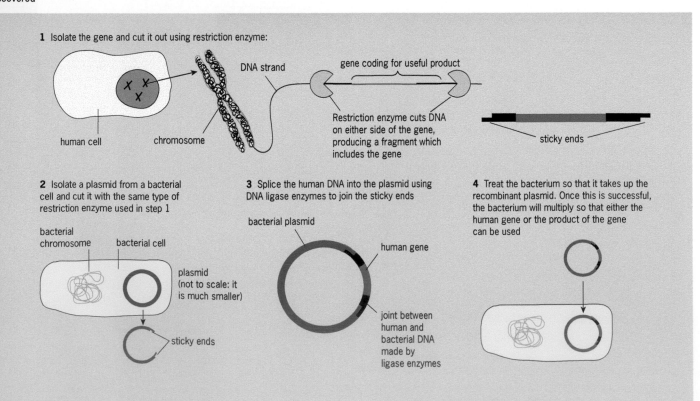

1 Isolate the gene and cut it out using restriction enzyme:

DNA strand · gene coding for useful product · human cell · chromosome · Restriction enzyme cuts DNA on either side of the gene, producing a fragment which includes the gene · sticky ends

2 Isolate a plasmid from a bacterial cell and cut it with the same type of restriction enzyme used in step 1

bacterial chromosome · bacterial cell · plasmid (not to scale: it is much smaller) · sticky ends

3 Splice the human DNA into the plasmid using DNA ligase enzymes to join the sticky ends

bacterial plasmid · human gene · joint between human and bacterial DNA made by ligase enzymes

4 Treat the bacterium so that it takes up the recombinant plasmid. Once this is successful, the bacterium will multiply so that either the human gene or the product of the gene can be used

The plasmid is cut using the same endonuclease enzyme used to cut the human DNA to obtain the gene. Using the same endonuclease is important – the DNA must be cut at the same base sequence to produce complementary sticky ends that will join up. The gene and the plasmid are mixed together and then a DNA ligase is added to join up the sticky ends. The DNA molecule produced is circular, like the original plasmid.

In the second step, the genetic engineer must induce the bacterium to accept the plasmid. One technique involves soaking the bacteria in ice-cold calcium chloride and then incubating them at 42 °C for 2 minutes. Nobody seems to know exactly why this works, but it does. Bacteria which have accepted the plasmid now contain recombinant DNA, and are, by definition, transgenic organisms.

The conversion process is not very reliable: for every bacterium that takes up the recombinant plasmid, about 40 000 do not. So how do you tell which ones are recombinant? A clever trick here is to insert two genes into the plasmid: the one you want to use and one that makes the recombinant bacteria easy to detect. For instance, you can add a second gene that confers antibiotic resistance and then culture the bacteria on agar plates containing the antibiotic. Only the bacteria with the modified plasmid will survive and grow.

An alternative is to add a second gene that codes for an enzyme that metabolises a coloured substrate. When bacteria are grown on agar plates made with the substrate, colonies that have taken up the plasmid are a different colour from colonies of non-recombinant bacteria.

Once you have identified colonies of recombinant bacteria, you can start pure cultures. Transgenic bacteria are often described as 'sick' because they multiply more slowly than normal bacteria: the population doubles every 30 minutes instead of every 20 minutes. Within 10 hours, one transgenic bacterium can produce over a million copies of itself, each one containing a working clone of the original gene.

Modified bacteriophages also make good vectors. They reproduce by inserting their own DNA into the DNA of a host bacterium. Bacteriophages can transfer larger amounts of DNA than plasmids but, at present, their use is limited.

TRANSFERRING GENES INTO EUKARYOTIC CELLS

Usually, a gene is transferred into a different organism so that it can be expressed, making greater volumes of product. This happens only when the gene is able to use the mRNA, ribosomes and Golgi body, etc. for protein synthesis. Some eukaryotic genes are not expressed effectively by prokaryotic cells, and so must be transferred into another eukaryotic cell. Yeast, a single-celled fungus, is often used.

Producing recombinant eukaryotic cells is more difficult than producing recombinant bacteria. The cell walls of fungal cells are a major barrier. But they can be digested away with suitable enzymes to form a protoplast, a cell without a cell wall. In this 'naked' state, the yeast cell will accept plasmids.

Other methods of transferring DNA into eukaryotic cells include the following techniques.

- **Microprojectiles**. It is possible to shoot DNA into host cells. Tiny pellets of metal are coated with DNA and fired at high speed at the target cells. Remarkably, some of the cells recover and accept the foreign DNA.
- **Electroporation**. This involves exposing the host cells to rapid, brief bursts of electricity to create temporary gaps in the cell surface membrane, through which foreign DNA can enter.

Fig 27.10 Transgenic bacteria or yeasts are cultured in large cylindrical sterile fermenters, although the process is not strictly speaking fermentation. These containers were originally used for making wine and the name has stuck

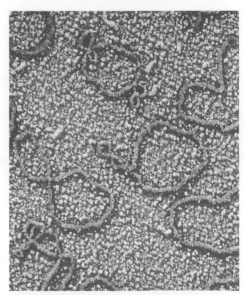

Fig 27.11 False-colour transmission electron micrograph of plasmids of bacterial DNA from the bacterium *Escherichia coli* (*E. coli*). This plasmid, called pBR322, is the one most commonly used in genetic engineering work

 REMEMBER THIS
The word *protocol* is often used in genetic engineering. It means 'standard procedure' and so a cloning protocol would be a standard method for cloning a gene.

Endocytosis is discussed in Chapter 4.

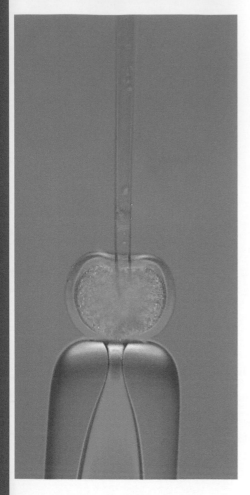

- **Liposomes**. In this more subtle method, DNA is inserted into liposomes – spheres formed from a lipid bilayer. The liposomes fuse with the surface membrane of the host cell and introduce the DNA into the cytoplasm.
- **Calcium phosphate precipitation**. Plasmids are mixed with calcium phosphate. As this precipitates, the grains that form contain DNA. These grains can enter cells by endocytosis.
- **DNA injection**. DNA can be inserted directly into a cell using a very fine pipette called a micromanipulator that avoids the inevitable hand tremor (see Fig 27.12). Tracey, a transgenic sheep, was created when the gene for alpha-1-antitrypsin was manually injected into the cells of a sheep embryo. This method leaves much to chance: many embryos have to be treated before one takes up the foreign DNA.

USING RECOMBINANT PROTEINS IN MEDICINE

The protein products of cloned human or mammalian genes expressed in bacteria or eukaryotic cells like yeasts are called recombinant proteins. We saw in Chapter 6 how bovine growth hormone can be produced as a recombinant protein. Proteins important for human medicine can also be manufactured in large quantities in this way – such as human insulin.

As the Science in Context box below shows, sheep have been used sucessfully to produce proteins that can be used in human medicine.

Fig 27.12 Injecting DNA directly into a sheep cell. The cell is held steady by gentle suction from the pipette on the left, while the even finer pipette on the right penetrates the cell surface membrane. This technique is used to produce transgenic and cloned animals

SCIENCE IN CONTEXT

Farming or pharming?

A new and highly experimental use of genetic engineering is 'pharming'. This involves farming, but its objective is not to produce food. The idea is to genetically engineer crop plants and animals usually used for food so that they can produce medicines (Fig 27.13). The technique is difficult and expensive because it requires a transgenic organism that is bred to carry a human gene that makes a particular protein – it is possible, for example, to genetically engineer a cow to produce human haemoglobin in its milk. Once transgenic embryos have been produced in the laboratory, these are then implanted into surrogate mothers and carried to term, resulting in transgenic cows that secrete human haemoglobin in their milk.

Milk from a cow is relatively easy to obtain in large amounts; the secreted haemoglobin is then extracted and purified. This process has been used to produce other blood components, human growth hormone and large quantities of human proteins needed for research.

Fig 27.13 Transgenic sheep. These sheep have a human gene which makes them produce α-1-antitrypsin (AAT) in their milk. People who lack AAT suffer from emphysema

4 GENE THERAPY

The new technology of gene therapy promises to revolutionise medicine in the twenty-first century (Fig 27.14). The idea that it can be used to treat genetic diseases such as cystic fibrosis and Tay–Sachs disease has been around for several years but it is taking longer than expected to develop gene therapy into an acceptable and safe treatment.

Recently, scientists have also begun to look beyond just the diseases caused by a defect in a single gene and are considering gene therapy as a potential treatment for all sorts of problems, from cancer and heart disease to AIDS.

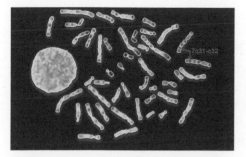

Fig 27.14 Causes and treatment of cystic fibrosis Chromosome pair 7 is highlighted in pink, with the location of the cystic fibrosis gene mutation arrowed

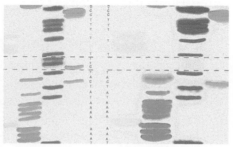

Electrophoresis of the gene sequence shows that base sequence TTC is present in an unaffected person (left) and absent in someone with cystic fibrosis

The healthy gene is isolated before being inserted into a virus. Cystic fibrosis patients use an inhaler with the virus that delivers the healthy gene to lung cells

GENE THERAPY FOR GENETIC DISEASES

The first person to be given gene therapy was Ashanti DeSilva. She became the first patient with severe combined immunodeficiency (SCID) (Fig 27.15) to be treated in September 1990, when she was 4 years old. A team in the USA led by French Anderson gave Ashanti four infusions of cells containing the working gene that she lacked. Over the four months of the treatment, her condition improved.

With the help of follow-up treatments, she has now been transformed from a small girl who was constantly ill and could not leave the house, to a normal, healthy and lively teenager. She continues to do well and, apart from needing regular injections of one of the enzymes she cannot make naturally, she lives a normal life.

> ✔ **REMEMBER THIS**
> A simple definition of **gene therapy** is the treatment of a genetic disease by giving individuals with the disease copies of healthy genes.

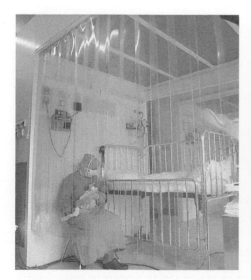

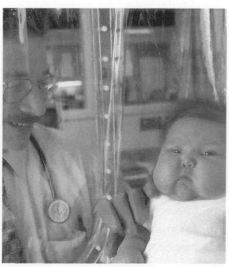

Fig 27.15a A SCID baby lives in a sterile compartment and is handled by medical staff wearing full body gowns to avoid passing on any infections

Fig 27.15b This Apache baby is having gene therapy: some of his bone marrow cells are removed, the gene for the enzyme he lacks is inserted into them and they are returned to his body. Then he can produce normal immune cells

Early in 2000, 10 years after the first successful gene therapy trial to treat SCID, researchers in Paris managed to treat two babies, aged 8 and 11 months with gene therapy. The babies both had SCID, but a slightly different form from Ashanti DeSilva. Their treatment involved taking out some bone marrow and selecting out a set of blood stem cells. Stem cells are the cells that differentiate into all sorts of blood cell types, including the white cells that make up the immune system. By introducing the gene that the babies were missing into these stem cells, the researchers managed to get the correct gene into all of their immune cells. Both babies developed a fully functional immune system with a few weeks and can now live normal lives. A similar process was used in the summer of 2001 on 18-month-old Rhys Evans at Great Ormond Street Hospital in London.

THE PROCESS OF GENE THERAPY

Researchers use one of several approaches in gene therapy to correct the function of a disease-causing allele:

- A normal allele can be inserted somewhere into the genome. When expressed, the protein it produces is functional and makes up for the allele already there that produces a non-functional protein. This is a common method.
- A normal allele can be swapped with a faulty allele by homologous recombination.
- The faulty allele can be repaired returning the allele to its normal function.
- The degree to which an allele is turned on or off can be changed.

How are genes introduced?

The most common method of gene therapy is to introduce a healthy copy of an allele into the genome to replace the faulty allele – but how is this done?

A carrier molecule called a vector must be used to deliver the therapeutic allele; it cannot get inside the host genome by itself. The most usual vector is a virus (Fig 27.16). Viruses have evolved to insert their DNA (or RNA) into genomes. If an allele for gene therapy is inserted into a viral genome and then the virus is used to infect a patient, the virus inserts the allele into the patient's genome along with its own DNA.

The viruses used are not pathogenic – they don't cause disease in humans, but the procedure is not without risk. An 18-year-old man in a gene therapy trial in 1999 died because of a massive allergic reaction to the adenovirus vector he was given (Fig 27.17).

Common virus vectors used in gene therapy:

- **Retroviruses**. Viruses that can create double-stranded DNA copies of their RNA genomes, e.g. HIV.
- **Adenoviruses**. Viruses with double-stranded DNA genomes, e.g. the common cold virus.
- **Adeno-associated viruses**. Small, single-stranded DNA viruses that can insert their genetic material at a specific site on chromosome 19.
- **Herpes simplex viruses**. Double-stranded DNA viruses that infect neurons, e.g. herpes simplex virus type 1, which causes cold sores.

DNA can also be introduced without virus vectors. The simplest method is the direct introduction of therapeutic DNA into target cells. This is difficult to do and its use is very limited. An alternative is to use liposomes, artificial lipid spheres which carry the DNA inside them. The sphere can blend with the cell membrane and take the DNA inside a cell.

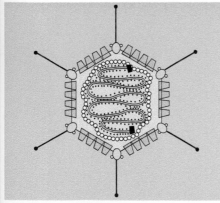
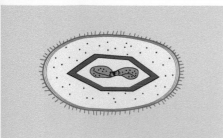

Fig 27.16 Adenovirus (top) and herpes simplex virus (below). Adenoviruses can cause severe chest infections, while herpes simplex causes cold sores

DOES GENE THERAPY EVER WORK?

The first gene therapy clinical trials that began in 1990 met with some success when SCID patients were treated. However, despite many other gene therapy trials since, there have been few other successes. As of 2007, no gene therapy treatment has been approved for use in human patients. Trials are still going on, but researchers are now more cautious. The field suffered a serious setback with the death of a patient during a trial in 1999 and confidence in the techniques was again shattered in 2003. A gene therapy trial treating more SCID patients in the USA had to be stopped after a child in a similar French trial was found to have developed leukaemia as a result of the gene therapy.

Gene therapy has not achieved success for several reasons:

- It is difficult to get an introduced allele or gene to become stable in the host genome – as cells divide, they tend to lose it and patients have to undergo multiple rounds of gene therapy.
- Patients start mounting an immune response to the viral vectors, so repeated rounds of therapy become impossible.
- Viral vectors can be unpredictable when introduced into living cells – although treated to remove their ability to cause disease, they can revert.
- Most diseases, apart from the few genetic diseases caused by problems in one allele, involve several genes. The chances of being able to fix one-gene problems are remote. Fixing problems involving several genes looks impossible at the moment.

US halts gene tests after youth dies

Effects of experimental treatment 'not known'

Duncan Campbell
in Los Angeles

The controversial use of an experimental gene therapy treatment has been suspended throughout the United States after the death of a teenager.

Some medical experts are suggesting that experimental forms of treatment are being put into operation too quickly, before all possible tests have been carried out.

Jesse Gelsinger, 18, from Arizona, died on September 17 in the University of Pennsylvania hospital in Philadelphia. He had been suffering from a serious, potentially fatal, metabolic disease.

Since his death, all such treatment has been halted by federal officials until the precise cause of death is established.

Thousands of patients in the US have been treated with variations of gene therapy.

The doctors treating Gelsinger injected a genetically engineered virus into his liver at the highest dosage allowed by the federal regulations governing such treatment.

Now federal officials plan to ask 100 other researchers throughout the United States who have been involved in such treatment to inform them if there have been any serious side-effects of their experiments.

"This was a tragic and unexpected event," James Wilson, director of the university's institute of gene therapy, told the Washington Post after Gelsinger's death.

"I hope in a month we'll have looked at every angle so we can share what we have learned form this."

Another 17 patients who have been undergoing the same treatment as Gelsinger – although not necessarily at such high dosages – have reportedly suffered no side effects.

The experiment involves

using a genetically modified adenovirus, one of the viruses which cause the common cold, to correct a defective gene in the patient. It has so far only been used on the very ill who have not appeared to respond to any form of more conventional treatment.

The father of the dead youth, Paul Gelsinger, said he hoped that his son's death would aid the fight to discover a cure for the condition.

"I lost a hero," he said of his son, who had fought a long and debilitating battle against the disease.

The death coincides with a serious debate among doctors in the US about the speed with which genes are being used in different forms of treatment when little is still known about their effects.

There have been suggestions that some methods are being employed before the full results of the potential of gene treatment is known from laboratory experiments.

There is also a continuing argument, driven by the religious right, about the ethics of using live genes.

> 'This was a tragic and unexpected event. We'll share what we have learned from this'

Fig 27.17 Jesse Gelsinger was only 18 when he died in 1999 after reacting badly to an adenovirus vector in a gene therapy trial. This was a serious setback for gene therapy

What is cystic fibrosis?

Cystic fibrosis (CF) is a genetic disease that is due to one defective allele. About one person in 25 carries the defective allele, but they also carry the normal allele and do not suffer from the disease.

The normal allele codes for a protein called cystic fibrosis transmembrane regulator (CFTR). This essential membrane protein in epithelial cells transports chloride ions out of the cells and into mucus. Normally, when chloride ions are secreted, sodium ions follow, and this decreases the water potential of the epithelial mucus. Water follows outwards by osmosis, making normal, watery mucus, which can be moved by the cilia (tiny hairs) lining the airways.

Cystic fibrosis sufferers make a protein that differs in just one of its 1480 amino acids. Although slight, this fault prevents CFTR from functioning normally. Chloride ions and sodium ions cannot be secreted, and so the mucus becomes much thicker than normal. Dead epithelial cells containing long DNA molecules also accumulate in the mucus, adding to the general congestion of the airways.

Sticky mucus is also a real problem in the pancreas. The mucus blocks the pancreatic duct, preventing the secretion of pancreatic juice.

Treatment of CF includes regular physiotherapy in which the chest is massaged to dislodge the mucus. Even so, infections are common and most CF sufferers have to take a variety of antibiotics according to the infection they have at the time.

> Cystic fibrosis inheritance is discussed in Chapter 25.

Gene therapy for cystic fibrosis?

As cystic fibrosis is due to a defect in a single gene, it is a good candidate for gene therapy. In 1989, the cystic fibrosis allele was located on chromosome 7. The base sequence of the healthy gene was then compared to the defective allele, and the nature of the fault was narrowed down at the molecular level. This opened up the exciting possibility that if healthy genes could somehow be introduced into the epithelial cells, they might be expressed and so make the correct membrane protein, solving the problem for a time.

The basic steps are as follows:

1   The CFTR gene is isolated, and cut out.
2   The gene is cloned many times by PCR.
3   The genes are encapsulated, either by putting them into liposomes (spheres made from lipid) or viruses.
4   The gene particles are inhaled, so that they can pass into the epithelial cells of the lung to be incorporated into the DNA of the cells.

Once in place, if all goes well, the healthy genes are transcribed, making the correct protein.

Germline gene therapy and cystic fibrosis

Replacing the gene in body cells would still mean that the eggs or sperm of the sufferer would carry the defective alleles, so this could be passed on to future generations. To prevent this, one future possibility is germline gene therapy. This is even more fraught with ethical problems as the original alleles in the zygote would be changed. The individual would grow and develop with healthy alleles in all cells and all of the individual's children would inherit the healthy allele.

Germline gene therapy sounds attractive but there are problems, so much so that this area of research is banned in the UK, and in many other countries. Ethically, there is the familiar 'where do you stop?' problem. A new treatment for cystic fibrosis or haemophilia would be desirable, but having the technology might allow for all sorts of temptations. Our knowledge of the human genome could reveal, for example, genes that code for intelligence, height or skin colour. Tampering with these characteristics would lead to obvious problems.

Biologically, germline gene therapy is potentially dangerous because we know very little about how genes function in the embryo. Tampering with genes in the zygote could have effects that might become apparent only in later life.

5 GENOME AND GENE SEQUENCING

Sequencing individual genes or whole genomes involves discovering the exact order of the bases cytosine, guanine, adenine and thymine in the DNA strand or strands that make them up. The structure of DNA was worked out by Francis Crick and James Watson in the early 1950s but it took about 30 years before researchers were able to develop a reliable sequencing method.

DIDEOXY METHOD OF GENE SEQUENCING

The most popular method for gene sequencing is called the dideoxy method, or Sanger method, which was invented by Frederick Sanger. He received his second Nobel Prize for this in 1980. The basic method is shown in Fig 27.18.

The DNA to be sequenced is prepared as a single strand. The single strand of DNA then acts as a template to build up a piece of double-stranded DNA as new bases are added by the enzyme DNA polymerase I.

But, instead of doing this just with ordinary bases, the DNA is supplied with DNA polymerase I and two different mixtures of bases:

1 A mixture of all four normal (deoxy) nucleotides in sample quantities:
 - dATP
 - dGTP
 - dCTP
 - dTTP

2 A mixture of all four **dideoxy** nucleotides, each present in limiting quantities and each labelled with a tag that fluoresces a different colour (Fig 27.18):
 - **ddATP**
 - **ddGTP**
 - **ddCTP**
 - **ddTTP**

Because all four normal nucleotides are present, the second strand of DNA is built until, by chance, the DNA polymerase inserts a dideoxy nucleotide (one of the bases with the coloured tag). The polymerase enzyme is then blocked, as it can't handle a base with this shape. It no longer adds any more normal bases – the process of making double-stranded DNA stops, leaving the strand unfinished. If the ratio of normal nucleotide to the dideoxy version is high enough, some DNA strands will become quite long before insertion of the dideoxy version stops the process.

Finally, the end result is a mixture of strands of double - stranded DNA, all of different lengths. All possible lengths are represented and analysis using gel electrophoresis separates out each length. By looking at which coloured tag is at the end of each length, the sequencer can work out the sequence of bases in the DNA.

This is only one method of sequencing and, today, the whole process is automated and very fast, which is why so many gene sequences and whole genome sequences have been worked out in the last few years.

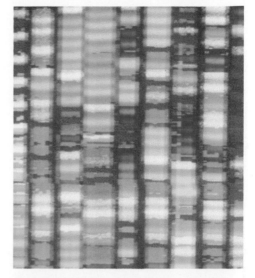

Fig 27.19 Computer analysis of DNA sequencing autoradiograms, used to decode the base sequence of lengths of DNA

Fig 27.20 Automated electrophoresis equipment used to study the structure of DNA

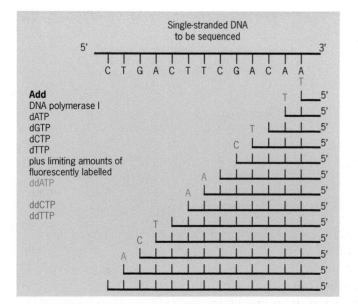

Fig 27.18 Sanger's dideoxy method of gene sequencing

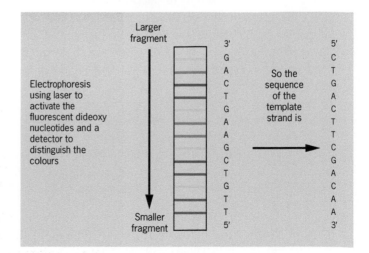

THE HUMAN GENOME PROJECT

A project to completely sequence the entire human genome began in 1990s and is still yielding important results. Work on the project took 13 years and did not proceed without setbacks and controversy (see the How Science Works box below).

A draft sequence was released in 2000, and this was improved and released as the full sequence in 2003. This revealed 20 000 to 25 000 human genes, as well as the regions controlling them. The resulting DNA sequence maps are now being used as tools explore human biology.

The DNA sequences announced in 2003 were only rough drafts for each human chromosome. This was like a sketch that is later built up into a detailed painting by adding detail. Researchers are now concentrating on doing just that. The final human chromosome to be sequenced in much more detail, chromosome 1, was finished in 2006. A project started in the same year to produce a map of human allele variants, which will be really useful for research into human disease.

HOW SCIENCE WORKS

Human conflict and the human genome map

The image of a scientist as an unfeeling and slightly weird boffin is a common stereotype, but scientists are very human, as the background to the enormously successful project to sequence the human genome demonstrates.

The draft human genome, the sequence of most of the human chromosomes, was announced jointly in June 2000 by the Human Genome Project (HGP), a publicly funded international programme, and Celera Genomics, a US private company co-founded by Applera Corp and Dr Craig Venter (Fig 27.21a). Simultaneous publication of the research behind the announcement followed in February 2001. This public show of togetherness masked a bitter dispute that rumbled on for years.

The two groups used slightly different techniques to read the 3.2 billion base pairs of DNA that make up the human genome. The HGP method broke the genome up into manageable segments that were mapped before sequencing took place. Celera took a more sledgehammer approach: it broke the genome into tiny pieces and then used massive computing power to reassemble the read fragments. The HGP camp were put on the defensive because Celera managed to produce a draft sequence in a fraction of the 10 years it took the HGP, not to mention all the public money that they spent. Celera were criticised for the quality of their work, and for 'cribbing' essential information from the publicly available HGP work, which was updated as work progressed. Many scientists hoped that the joint announcement of the results would be the end of the rows, but a subsequent paper in a leading medical journal contained an article from HGP scientists, again criticising the quality of the Celera sequence.

The Sanger Centre, headed by Dr John Sulston (Fig 27.21b) represented the UK contribution to the HGP.

Fig 27.21a The US geneticist Dr Craig Venter, founder of The Institute of Genomic Research (TIGR) in Maryland, USA. He is standing in the DNA sequencing laboratory at TIGR

Fig 27.21b The British molecular biologist Dr John Sulston, director of the Sanger Centre in Cambridge, UK. He is standing in the main sequencing room

LOOKING TO THE FUTURE

As well as adding more detail, the story of the human genome is far from over. The genome sequence is a bit like a map – it is getting more detailed, but the really interesting part will be the journeys that we will be able to take using that map.

Other genomes from different mammals, lower animals and prokaryotes have also been sequenced and vast projects are underway to see how entire genomes are related. Doing this will enable us to compare sets of expressed RNAs or proteins, gene families from a large number of species, variation among individuals, and the classes of genes that regulate the expression of other genes.

In the next decade we may be able to find:

- The exact locations and sequences and expression of individual genes.
- The mechanisms by which genes are regulated.
- Exactly how the DNA sequence is organised into genes – where are the non-coding sequences and what do they do?
- The detailed structure and organisation of chromosomes and the function of proteins associated with DNA in chromosomes.
- How gene expression, protein synthesis and post-translational events are controlled.
- More about why some genes have been conserved in different organisms throughout millions of years of evolution.
- How SNPs (single-base DNA variations among individuals) correlate with health and disease.
- If we can tell from the variations seen between gene sequences in different individuals whether or not they will develop particular diseases.
- Which genes are involved in complex traits and multigene diseases.
- More about how genes interact in the development of a human being from a fertilised egg.

6 WHOLE ORGANISM CLONING

Cloning is a difficult topic for anyone; it is not only complex biology, but it is also a tricky subject ethically. What you feel about cloning may not just depend on your knowledge of science – it may also depend on your religious, moral or other views. We don't have enough space in this book to give more than a brief introduction.

CLONING WHOLE ORGANISMS

In 1997, when scientists succeeded in cloning a sheep, there was a lot of fuss in the media. Was it ethically and morally right to clone animals as advanced as mammals? Would this lead to scientists attempting to clone humans?

The definition of a clone is 'a genetically identical copy' and the term can apply to individual strands of DNA, individual cells or whole organisms. Cloning is, in fact, a common natural process: all organisms made by asexual reproduction are clones of their parent, unless there is a mutation. Even humans make clones – they are more commonly called identical twins.

There is much commercial interest in the cloning process because it can produce exact copies of organisms with desirable characteristics. If, for example, you have a 'perfect' apple tree that produces large amounts of delicious, healthy, blemish-free apples, you can clone it rather than risk losing its valuable genotype in the lottery of sexual reproduction.

Cloning plants is easy, and can be done with relatively little equipment in schools and colleges. Cloning mammals, however, is a lot more difficult but technically feasible. There are two approaches to cloning whole organisms, embryo splitting and nuclear transfer.

REMEMBER THIS

Eugenics is the application of genetics to improve human characteristics. It is generally frowned upon, since making genetic alterations to select for cosmetic features, such as blonde hair and blue eyes, is medically unnecessary. However, deciding where eugenics ends and germline gene therapy and embryo selection for medical reasons begin is a difficult ethical problem.

Embryo splitting

Eggs and sperm or fertilised embryos are collected from prize specimens of cattle, for example. Cells can be pulled apart when the embryo is at the 8-cell or 16-cell stage and each fragment can then develop into a complete embryo. Each embryo is a clone of the original. The process can be repeated many times and then the cloned embryos can be introduced into the uteruses of normal cows. After the normal gestation period, several very average animals give birth to a herd of prize specimens.

Nuclear transfer

Cloning is based on nuclear transfer, the same technique scientists have used for some years to copy animals from embryonic cells. Nuclear transfer involves the use of two cells. The recipient cell is normally an unfertilised egg taken from an animal soon after ovulation. Such eggs are poised to begin developing once they are appropriately stimulated. The donor cell is the one to be copied. A researcher using a high-power microscope holds the recipient egg cell by suction on the end of a fine pipette and uses an extremely fine micropipette to suck out the chromosomes, sausage-shaped bodies that incorporate the cell's DNA (look back to Fig 27.12). (At this stage, chromosomes are not enclosed in a distinct nucleus.) Then, typically, the donor cell, complete with its nucleus, is fused with the recipient egg. Some fused cells start to develop like a normal embryo and produce offspring if implanted into the uterus of a surrogate mother.

TECHNICAL AND ETHICAL PROBLEMS WITH CLONING ORGANISMS

Dolly the sheep and cc the cat were created by the process of nuclear transfer, a process that many people feel will give rise to cloned humans at some stage in the future. As we saw at the start of this chapter, the technique sounds simple, but we can't do it reliably. It took over 200 attempts to get Dolly and almost 100 to get cc, the cloned cat. Many laboratories report that only 1 to 2 per cent of cloned embryos survive to become live offspring. Even some clones that survive through birth die shortly afterwards. In 2001, a cloned gaur (an endangered species of wild ox from Asia) lived only a few days before dying from a common dysentery.

This unreliability has two main implications for human cloning that intends to produce living babies. First, each attempt requires an individual egg. Sheep don't tend to complain when you stimulate their ovaries and harvest the eggs, but humans are a different matter. Very few women will give up eggs for any reason other than their own pregnancy, so finding 200 ova for each attempt is always going to be a problem. Some women in the USA may have been persuaded to do this by a cult aiming to clone a human, but whether they will actually do so remains to be seen. Secondly, in humans, failure brings great distress. Miscarriages and malformed offspring always cause great distress to all concerned in the process.

In 2001, the UK government passed a law that allowed scientists to produce human embryo clones solely for the purpose of making replacement tissues that could be used in transplant procedures (Fig 27.22). This was challenged by the pro-life lobby, but in February 2002, a House of Lords Select Committee upheld this decision, and the Human Fertilisation and Embryology Authority, which regulates such research is now likely to give researchers licences to experiment on early stage human embryos.

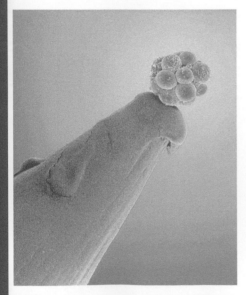

Fig 27.22 A coloured scanning electron micrograph (SEM) of a human embryo at the 10-cell stage on the tip of a pin. The ball of cells (orange) of the embryo is known as a *morula,* a cluster of almost identical, rounded stem cells, each containing a central nucleus. This 10-cell embryo is about three days old. It is at the early stage of transformation from a single cell to a human composed of millions of cells

In Australia, the Prohibition of Human Cloning for Reproduction and the Regulation of Human Embryo Research Amendment Act 2006 came into operation on 12 June 2007. The Assisted Human Reproduction Act, which prohibits human cloning, for any purpose, whether for reproductive, therapeutic, or research purposes was passed in Canada in 2004. Penalties for contravening the prohibition are severe: a maximum fine of $500,000 or 10 years in prison, or both. Legislation in the USA has banned all human cloning in the past, but embryo stem cell research is still allowed. Most countries ban human cloning that intends to produce full-term pregnancy but some unscrupulous scientists can find loopholes and there are repeated stories in the media of cloned babies being born. None has been confirmed.

7 STEM CELL CLONING

If the objective is not to make a human baby, then why clone? The answer comes down to stem cells. These cells are capable of differentiating into any type of adult body cell. Theoretically, you could say that every cell in the body has the capacity to do this, as they all contain the same genome. However, once a cell has gone through a pathway of differentiation, and has become, say, a brain cell, there is little you can do to shift it back again.

WHAT ARE STEM CELLS?

Stem cells are undifferentiated cells that can turn into any type of body cell, given the right conditions. They have two important properties.

- They can **self-renew**. Stem cells can go through unlimited rounds of cell division without differentiating.
- They have **unlimited potency** – they can differentiate into any body cell type.

In reality, not all stem cells have unlimited potency. They have different levels of potency, which describe their exact potential for differentiation. This reflects their source.

Totipotent stem cells are the top of the tree – they can differentiate into any type of cell – including the cells that make up an embryo and the cells that develop into the placenta and umbilical cord. Totipotent stem cells are the cells produced just after the egg and sperm cell fuse – only the first few divisions produce totipotent stem cells.

Pluripotent stem cells are the descendants of totipotent cells. They can differentiate into cells from any of the three germ layers of the embyro, but they cannot develop into placenta or umbilical cord tissue.

Multipotent stem cells are partly differentiated but they can differentiate further to produce cells within a closely related family. For example, haematopoietic stem cells can differentiate into the different cell types found in the blood but cannot differentiate into, for example, neurones.

Unipotent stem cells are well differentiated but they can self-renew. They are able to divide to reproduce themselves many times – which makes them different from non-stem cells, which have limited capacity for cell division.

? QUESTION 3

3 There is a common misconception that people can have their cells frozen so that they can be cloned after their death to 'live again'. How would you explain to someone that this is utter rubbish?

WHAT KINDS OF STEM CELL ARE THERE?

There are three types of stem cell that occur in humans and other mammals:

- **Embryonic stem cells**. These are all the cells in the early stage embryo that come directly from blastocysts.
- **Adult stem cells**. A select number of cells in the adult body still act as stem cells – some cells found in the nasal epithelia in the nose still act as stem cells.
- **Cord blood stem cells**. Stem cells found in the blood and tissues of the umbilical cord.

WHAT CAN STEM CELLS DO FOR US?

Stem cells hold the key to giving the body what it needs to regenerate healthy tissue. With an ageing population in the developed world, the rate of degenerative disease is increasing massively. Stem cells offer the possibility of replacing neurones in the brain of a stroke victim, or repairing a break in the spinal cord, or repairing the cells of the pancreas so that they can make insulin again.

Studying totipotent human embryonic stem cells is also expected to tell us a lot about how humans develop. How exactly do undifferentiated cells become differentiated? Which genes are turned on and off, when does this happen and what controls it? Understanding this enormously complex process might also help us understand what happens when it goes wrong and cancerous cells divide out of control, losing their differentiated features.

Stem cells might be used to test new drugs – clones of stem cells can be produced that are identical – an important requirement for reducing the variables when comparing one drug with another.

Stem cells could be used to generate new human tissues and organs for transplantation. At the moment, donated tissues and organs are used but they create the problems of rejection – and demand is well in excess of supply. Stem cells are already being used to grow skin cells in culture to treat burn victims but it may be possible to renew some of the brain cells that go wrong in Alzheimer's disease. Maybe one day, stem cell technology and the cloning technology that produces transgenic animals will result in animals with organs that are suitable for human transplantation.

WHY STEM CELLS ARE AN ETHICAL NIGHTMARE

There is a long way to go with ironing out technical problems before many of the potentials of stem cells can be realised. But that is not the only problem. The whole practice of obtaining and using human stem cells is fraught with ethical and moral dilemmas.

The main problem is that most research uses embryonic stem cells, as these are totipotent. The process of removing stem cells from an embryo destroys it – and some people think this is destroying a living human being. Pro-life organisations are opposed to obtaining stem cells from embryos.

Another difficulty is the source of the embryos. Human embryos for research and harvest of stem cells are obtained from three main sources:

- Embryos left over from IVF treatments are used to harvest stem cells. These embryos cannot survive without being transplanted and many more embryos are fertilised than are needed. Ethical problems include obtaining consent of the parents, and checking whether all the embryos are healthy and free from genetic disease.

- Embryos created by *in vitro* techniques in which donated human eggs are fertilised to create early stage embryos just to harvest stem cells. As you might expect, the people who think that destroying an embryo is bad think this is even worse. They say scientists are playing God – creating life and then destroying it, for their own gain.

- Even more controversial is the potential practice of human cloning to create embryos. Scientists are able to introduce a body cell into the shell of a fertilised egg and get it to divide into a blastocyst. This is technically a human clone and is a rich source of human stem cells. But some people argue that this is the slippery slope to cloning humans for reproduction and are completely opposed to it.

FINALLY ...

The problem with writing a chapter like this is that it will be out-of-date before the ink is dry at the printers.

To keep up, and to find information to help you make your own mind up about cloning, refer to good journals, such as *Scientific American* and *New Scientist* (both are available online), and look at the BBC online Science and Technology news site.

SUMMARY

After reading this chapter, you should know and understand the following:

- Genetic engineers isolate, cut out and transfer genes between organisms. Genetic engineering technology has many applications, including making human products such as insulin on a large scale.

- Strands of DNA are cut using **restriction endonuclease** enzymes, which always cut DNA at a particular base sequence. Strands of DNA can be joined by using DNA **ligase enzymes**.

- In order to transfer DNA into a living cell such as a bacterium, a **vector** (carrier) is needed. Common vectors are **plasmids** (circles of DNA found in bacteria) and viruses.

- Once the foreign DNA is inside the host cell, it is adopted by the host. Genes in the foreign DNA are expressed using the host cell's synthetic machinery (ribosomes, mRNA, etc.). In addition, the foreign DNA is **cloned** (copied) every time the host cell divides.

- The Human Genome Project has now produced the complete sequence of all 30 000 genes in the human genome. Knowing this sequence will provide a big boost for research in many areas of biology and medicine.

- The **polymerase chain reaction** (PCR) allows samples of DNA to be cloned in a test tube using the necessary enzymes, instead of inside a host cell.

- **Gene therapy** is a technique that replaces faulty genes with healthy ones. It provides great hope for sufferers of genetic diseases such as cystic fibrosis.

- Many recent developments in this field have made cloning whole animals and organs possible. However, although the technical problems have been overcome, the ethical difficulties still pose a great problem for this area of research.

 Practice questions and a How Science Works assignment for this chapter are available at www.collinseducation.co.uk/CAS

28

Health and disease: an introduction

28 HEALTH AND DISEASE: AN INTRODUCTION

Living to a grand old age is now a more attainable goal for more people than it was a hundred years ago

Getting old is now normal

Currently, a man living in the UK can expect to live until he is 84; a woman until she is 87. A hundred years or so ago, things were very different: in 1900, average life expectancy was between 45 and 55 in this country. So what accounts for this massive change?

If you ever take a look around cemeteries or churchyards, you will notice that some people from the 1800s did live to their 70s or 80s or beyond. The maximum age to which a human being can live does not seem to have extended very much in the past 100 years. What has changed is the likelihood of dying when young. Infant mortality, the number of children under 5 that did not survive, was very high in the 1900s. Then, about 150 children in every 1000 died at birth and perhaps another 200 did not make it to their 5th birthday. Today, fewer than 20 children in every 1000 die before they are 5. The average life expectancy has increased because of factors such as better living conditions, better medicine and better diet.

Today we know that these sorts of environmental factors combine with our genes to determine how long we live. Researchers have been studying very long-lived people to see if they can identify which genes are important. Recently, a cluster of genes on chromosome 4 was found to be more common in people who reached the age of 90, but finding out exactly what these genes do is likely to take time.

1 DEFINING HEALTH AND DISEASE

Until this century, the commonest causes of death were the 'catchable' diseases – those carried by bacteria, viruses and other organisms too small to detect. Then the microscope showed that diseases such as dysentery and tuberculosis were caused by **infectious agents** whose spread could be worked out and then controlled. Later came antibiotics, vaccines and other sophisticated medicines.

Today, the commonest causes of death in the western world are all **non-communicable**, caused in part by our lifestyle and in part by the fact that modern medicine prevents most of us from dying from diseases that are **communicable**. In this section of the book, we look first at infectious diseases, and then at the ways in which the body defends itself against these potentially lethal invaders. We progress to the biggest killers, the lifestyle diseases, before finishing with genetic diseases and some thoughts for the future.

WHAT DO WE MEAN BY HEALTH?

Health is difficult to define accurately; we can say that it isn't being ill or having a disease, but laying down a set of characteristics that describes being healthy is more of a problem. Health can be thought of as a dynamic state in which a person is able to cope well with their internal and external environments in order to maintain a state of wellness. The internal environment includes our genes and our psychological approach to life, as well as our exposure to microorganisms that cause infectious diseases. The external environment includes our physical living conditions, the relationships we have with other people and whether we have enough money to eat a healthy diet and have a healthy lifestyle.

The World Health Organization defines health like this, recognising that mental as well as physical health is important:

Health is a state of complete physical, mental and social well being.

However, few of us attain this sort of health for long periods of time. What if you lose an arm in an accident? When you have had time to heal the injury, you cannot grow another arm and be complete in a physical sense, but you are not really ill. What about someone with a genetic disorder such as Down's syndrome? A definition of health must be flexible enough to include such special cases.

WHAT DO WE MEAN BY DISEASE?

Defining disease is more straightforward. Diseases arise when the body's functions are disturbed or when its structures are altered. Diseases can be defined as communicable if they are caused by infectious agents and can be passed from one person to another. Non-communicable diseases are caused by a broad range of environmental and genetic factors. It is important to remember that the development of a particular disease and the way it responds to treatment vary from person to person.

2 THE MAIN CAUSES OF DISEASE

INFECTIOUS DISEASES

Communicable or infectious diseases are those that can pass from one person to another. These diseases are caused by microorganisms such as bacteria and viruses (Figs 28.1, 28.2 and 28.3, overleaf). Although common in many developing countries, infectious disease is much rarer in modern Europe because of effective treatments such as vaccines and antibiotics.

LIFESTYLE DISEASES

Non-communicable diseases seem to be connected with our lifestyle. Cancer, for example, is a disease that can affect many different parts of the body. This disease clearly does not have a single cause but our risk of developing it depends on how we live. Smoking tobacco is the major cause of lung cancer, and too much sunbathing can lead to skin cancer. Eating a diet rich in fruit and vegetables can reduce our risk of many cancers.

GENETIC DISEASES

Cystic fibrosis is an **inherited disease**. It is due to a fault in a single gene. The normal allele of the gene makes a protein responsible for transporting chloride ions across cell membranes. If a person inherits one faulty allele, they are fine because they have a normal allele that still makes the protein. People develop the disease only if they inherit two faulty alleles; this makes cystic fibrosis a **recessive characteristic**.

People with no normal allele cannot make any functional transport protein and, as a result, produce extremely thick mucus. This affects organs such as the lungs and the pancreas. So, cystic fibrosis has a genetic cause.

Cystic fibrosis is discussed in more detail in Chapter 25.

? QUESTION 1

1 Suggest why communicable diseases are not as common in Europe as they were 200 years ago.

REMEMBER THIS

An **endemic** disease is one that is always present in the population.

An **epidemic** is an outbreak of disease that spreads rapidly through a population, affecting a large number of people.

A **pandemic** is an epidemic that spreads through a large geographical area – often the whole world.

Find out more about *Plasmodium*, malaria and the *Anopheles* mosquito in Chapter 29.

DEGENERATIVE DISEASES

Degenerative diseases are problems that develop as a result of the natural ageing process. Arthritis, Alzheimer's disease and some types of diabetes and heart disease can result from age-related degeneration of body systems.

HOW DISEASE-CAUSING FACTORS INFLUENCE EACH OTHER

A few diseases can be put definitely into only one of the categories just described. Other diseases are a result of interplay between different factors. People who inherit some specific genes can be at an increased risk from some cancers, particularly those that affect the colon or the breast. Viruses seem to be involved in the development of cervical cancer. Degenerative diseases can occur earlier in life, and their appearance can be related to lifestyle. Late-onset diabetes, for example, is more likely in people who are overweight. People can compensate for a genetic background that makes it more likely that they develop heart disease by changing their lifestyle – by eating a healthy diet and taking regular exercise, they might avoid heart disease.

SCIENCE
IN CONTEXT

Communicable diseases: infections

Many communicable diseases are caused by microorganisms (Figs 28.1, 28.2, 28.3).

Fig 28.1 Smallpox has now been totally eradicated. In 1967, when the World Health Organization began its campaign to eliminate this dreadful viral disease, there were 10 million cases spread through 30 different countries. The last case was recorded in 1977 in Somalia

FATHER THAMES INTRODUCING HIS OFFSPRING TO THE FAIR CITY OF LONDON.
(A Design for a Fresco in the New Houses of Parliament.)

Fig 28.2 This *Punch* cartoon shows three diseases caused by bacteria: diphtheria, scrofula (a form of tuberculosis) and cholera. Between them, they affected large numbers of people in London and other parts of the United Kingdom in the mid-nineteenth century

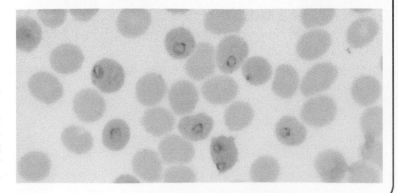

Fig 28.3 *Plasmodium*, the malarial parasite, is a small one-celled organism belonging to the Kingdom Protoctista. It is carried by the *Anopheles* mosquito and invades human red blood cells, seen here as ring-like organisms. Worldwide, malaria kills about 2.7 million people every year

Non-communicable disease

Lifestyle diseases

Lifestyle diseases are those that result from the conditions in which we live (Fig 28.4).

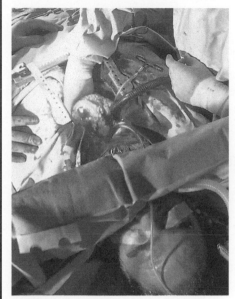

Fig 28.4 Heart disease is the greatest single cause of death in England and Wales. It accounts for nearly one in three of all deaths. Many environmental factors are linked to a greater risk of heart disease. These include stress, smoking, a high-fat diet and lack of exercise

Genetic diseases

It has been estimated that approximately 2 per cent of the UK population have some form of genetic disease (Fig 28.5).

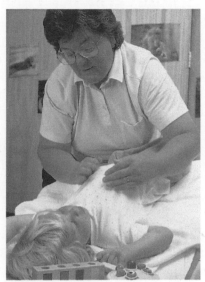

Fig 28.5 Cystic fibrosis is caused by a single gene. It is a recessive characteristic. Individuals who inherit two defective alleles cannot make a protein responsible for transporting chloride ions across cell membranes. As a result, they produce extremely thick mucus, which affects the functioning of organs such as the lungs. Physiotherapy helps to disperse this mucus

Degenerative diseases

Age brings about gradual decline in the efficiency of many of the systems in the human body (Fig 28.6).

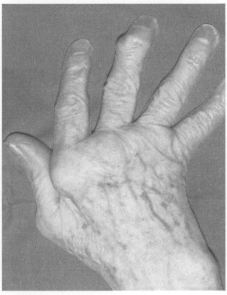

Fig 28.6 Osteoarthritis is a painful condition that affects the joints of many older people. The cartilage at joints is lost and this may result in damage to the underlying bone, so that the sufferer has difficulty in moving the joints

Malaria is a severe disease that affects many people living in the tropics. It is caused by *Plasmodium*, a single-celled parasite that is transmitted from the blood of one person to that of another by *Anopheles* mosquitoes. People who can afford preventive medicines, mosquito nets over beds and insect repellents are less likely to be bitten by mosquitoes than the poorer members of the community who cannot pay for such protection.

In addition, people who live in areas where malaria is common often have genes that cause them to make slightly different sorts of haemoglobin. This changes the red blood cells so that the malaria parasite cannot infect them as easily and these people have some resistance to developing the symptoms of malaria. However, the change in the red blood cells causes problems; the red cells lose their biconcave shape and start to look like crescents, or sickle blades. Such people have the disease sickle cell anaemia, which can itself be life-threatening.

So, with malaria, there is a link between lifestyle, infection and genetics. Whether or not a particular person catches malaria and then dies from it depends on all three factors.

Agriculture has influenced disease in three ways:

- The cultivation of crops led to changes to the environment. For example, simple irrigation systems that made more water available for crops, also provided the perfect breeding grounds for the larval forms of the parasites that cause malaria and schistosomiasis.

Sickle cell anaemia is discussed in more detail is Chapter 24.

Malaria and schistosomiasis are discussed in more detail in Chapter 29.

● Closer contacts between humans and their domestic animals allowed new diseases to emerge, such as smallpox, plague and influenza.
● Trading links allowed diseases to spread rapidly.

3 IDENTIFYING A DISEASE

LOOKING FOR THE OBVIOUS

Before a doctor can treat a disease effectively, he or she must identify the condition that is making the patient ill. Someone usually goes to the doctor with a collection of symptoms. For example, a woman finds that she gets tired very easily. Her husband thinks she looks rather pale and she notices that she gets out of breath when she exerts herself. The doctor observes the patient and listens to her describing these symptoms, and suspects that she is suffering from anaemia.

Anaemia is not a disease itself, but it has a cause. It could be due to:
● a shortage of iron in the diet – iron is an important part of haemoglobin molecules;
● loss of blood – this might result from heavy periods or perhaps an ulcer that bleeds constantly;
● destruction of red blood cells caused by sickle-cell anaemia, an inherited condition. An affected person has an abnormal form of haemoglobin that transports oxygen with reduced efficiency. The red blood cells of someone with this condition have a shorter life span than normal red blood cells.

The doctor looks for clinical signs that might show which one of these is most likely. Treatment can begin only when an accurate diagnosis has been made.

IDENTIFYING PANCREATITIS

The pancreas is an important organ located just below the stomach (Fig 28.7). It has two main functions. It produces the hormones insulin and glucagon that regulate blood glucose concentration, and it secretes digestive enzymes. The enzymes include trypsin, which breaks down proteins; lipase, which digests fats; and amylase which hydrolyses starch to maltose.

Pancreatitis is a severe inflammation of the pancreas, caused when the enzymes that normally break down food in the intestine start to break down and destroy the tissue of the pancreas itself. It can be acute (when it comes on suddenly) or chronic (long term).

Pancreatitis can be difficult to identify. The pancreas is in the first loop of the small intestine and is close to the liver and stomach, so pain in this area could be caused by any of pancreatitis, liver disease, stomach ulcers or gallstones. A correct diagnosis is vital, and this is where biochemical tests come in handy.

> ✓ **REMEMBER THIS**
> **Symptoms** are indications of a disease that are noticed by the patient.
> **Signs** are indications of a disease that are picked up by a doctor but not necessarily by the patient.

Fig 28.7 The pancreas is situated in the first loop of the small intestine and near to a number of other important organs. Diseases that affect any of these organs could give similar symptoms

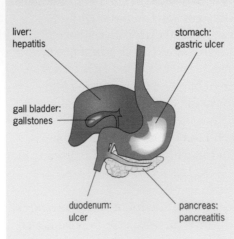

liver: hepatitis
stomach: gastric ulcer
gall bladder: gallstones
duodenum: ulcer
pancreas: pancreatitis

> **? QUESTION 2**
>
> 2 Trypsin is an enzyme secreted by the pancreas. Explain why:
> **a)** you would expect to find trypsin in the faeces of a healthy person;
> **b)** the amount of trypsin in the faeces will decrease in a person with chronic pancreatitis.

If the pancreas tissues are damaged rapidly, the enzymes in the cells escape into the blood. Tests involve checking whether lipase and amylase are in a blood sample. If they are, it is a good indication that the patient has acute pancreatitis.

If the condition is long term, however, the pancreas tissue is more slowly and progressively destroyed, and the ability to secrete enzymes declines gradually. This is chronic pancreatitis and is identified by measuring the amounts of enzymes in the patient's faeces. Recently, researchers have found that chronic pancreatitis is common in people who have a mutation in the gene that is faulty in cystic fibrosis. Although their mutation causes a less severe set of symptoms, it might soon be possible to develop a genetic screening technique to identify people at risk from developing pancreatitis later in life.

PHENYLKETONURIA: AN INHERITED ENZYME DEFICIENCY

Phenylketonuria is an inherited condition that affects approximately 1 in every 10 000 people. All new-born babies in the UK are tested for phenylketonuria. In the heel prick or **Guthrie** test, the midwife takes a blood sample from the heel of the baby and tests the sample for the presence of phenylalanine, an amino acid.

In a healthy baby, phenylalanine is converted into tyrosine by the enzyme phenylalanine hydroxylase (Fig 28.8). This enzyme is missing in a baby with phenylketonuria, and phenylalanine and other metabolic substances made from it build up in the baby's blood. These substances interfere with brain development, so it is very important that the condition is identified as soon as possible after birth and then treated with a low phenylalanine diet.

Suppose the enzyme phenylalanine hydroxylase is missing. Let's look at the possible effects on the biochemical pathway, and how we could use these effects to show that the patient has phenylketonuria.

- First, there would be no enzyme. So we could do a direct test for the enzyme and show that it was absent from the patient's blood.
- If phenylalanine hydroxylase were missing, its substrate, phenylalanine, would not be converted into tyrosine, so the amount of phenylalanine should be higher than normal, and the amount of tyrosine should be lower or absent. We could test for raised phenylalanine levels and reduced tyrosine levels.
- Finally, the biochemical pathway from phenylalanine to phenylpyruvate might be affected. There is more phenylalanine present and, since it cannot be converted to tyrosine, it might be converted to phenylpyruvate instead. We could test for an increase in phenylpyruvate.

DIABETES AND TESTING FOR GLUCOSE

People with diabetes are unable to control the concentration of glucose in their blood. The blood glucose concentration rises and glucose appears in the urine. A hundred years ago, the condition was always fatal. Now, as long as it is correctly diagnosed, people can be treated successfully and lead normal lives.

Monitoring the amount of glucose in the blood and urine is an essential part of this treatment. Early methods relied on chemical tests such as Benedict's test in which glucose is used as a reducing agent. More modern methods make use

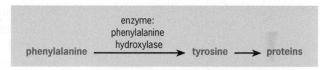

Fig 28.8 Part of the biochemical pathway by which phenylalanine is converted to other substances in the body. Each step in this biochemical pathway is controlled by a different enzyme

? QUESTION 3

3 Suggest how phenylketonuria might be treated by altering the amounts of phenylalanine and tyrosine in the diet.

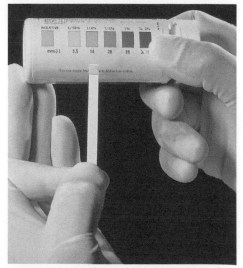

Fig 28.9 The Clinistix™ test indicates glucose level in urine by comparison of the test stick with a colour chart

Diabetes is discussed in more detail in Chapter 11.

Clinistix™ and the Glucowatch®
are discussed in more detail in
Chapter 6.

? QUESTION 4

4 Suggest two advantages
of people with diabetes
being able to test a urine sample
for glucose with Clinistix™ rather
than with Benedict's solution.

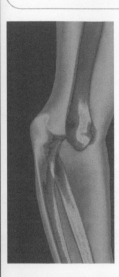

Fig 28.10 An X-ray of a
dislocated elbow joint

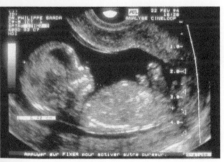

Fig 28.11 An ultrasound image of a human fetus

of enzymes. Enzymes are highly specific and sensitive. This makes them ideal
for analysing small samples. Today, people with diabetes use Clinistix™ to
test for the presence of glucose in urine (Fig 28.9). Enzymes are also involved in
the Glucowatch® Biographer, which measures the amount of glucose in the
blood.

LOOKING INSIDE THE BODY

Sometimes, the only way to make an accurate diagnosis is to look inside
the body. Cutting it open is one way, but there are always risks with surgery.
The risks are rare, but it is safer to use **non-invasive** methods where
possible. Several techniques now produce images of what is going on inside
the body. The images sometimes look very strange to the untrained eye,
but a skilled technician or doctor can interpret them.

X-rays

X-rays have a much shorter wavelength than visible light. They penetrate
different tissues in the body to different extents. Although they are most often
used to look at bones and teeth (Fig 28.10), X-rays can show other organs
if a suitable contrast medium is used. For example, if someone drinks a
barium meal (a solution containing salts of the heavy metal barium), the
barium absorbs the X-rays and details of the stomach and small intestine
show up on the final image of X-rays taken shortly afterwards.

The disadvantage of X-rays is that they damage DNA. For this reason,
X-rays are not used to examine the reproductive organs or to scan the
developing baby inside a pregnant woman. The people who use X-ray
equipment have to take safety precautions to protect themselves.

Ultrasound

Ultrasound is high-frequency sound, so high that the human ear cannot detect
it. An ultrasound beam passes into the body and bounces off dense tissue.
The reflected waves produce an image that shows internal organs in great
detail. Unlike X-rays, ultrasound does not damage DNA and so can be used
to check the development of a baby in the uterus (Fig 28.11).

Endoscopy

An endoscope is a long tube with a light at one end and a means of
displaying an image at the other. It can be used to examine the inside of
hollow body parts such as the oesophagus and stomach (Figs 8.2–8.6,
pages 132–134) or the trachea and bronchioles. Many modern endoscopes
rely on fibre optics. This means that they are small in diameter and flexible.
Endoscopes are used for a variety of purposes such as inspecting the external
ear and eardrum for signs of damage or infection, or for looking at the inside
of the gut.

? QUESTION 5

5 Explain which of the methods for looking inside the body would be most suitable
for:

a) examining the ovaries of a woman undergoing fertility treatment in order to see
the developing follicles;

b) examining the inside of a knee joint before performing a cartilage operation.

SUMMARY

After reading this chapter, you should know and understand the following:

- **Health** is difficult to define, but it is more than just an absence of disease.
- **Communicable** or infectious diseases may be spread from person to person and are caused by **microorganisms**.
- **Non-communicable** diseases have many causes. They can result from our **lifestyle** or from our **genes**, or they can occur as we age as a consequence of the process of **degeneration**.
- Infectious diseases became increasingly important as agriculture developed and humans began to live in large settled communities.

- Diseases are characterised by distinct **signs** and **symptoms**.
- Diagnosis may be confirmed by a range of **biochemical tests**. Many of these rely on **enzymes**.
- **X-rays, ultrasound** and **endoscopy** allow us to investigate the internal structure and functioning of the body.

Practice questions and a How Science Works assignment for this chapter are available at www.collinseducation.co.uk/CAS

29

Infectious disease

Ebola: an emerging disease

We are used to hearing about new treatments, new vaccines and new medical advances on an almost daily basis. But it comes as something of a shock when a new disease appears. Everyone knows about AIDS, but other new diseases have appeared within the past 30 years. Ebola virus, for example, first emerged in Zaire (now Democratic Republic of Congo) and Sudan in 1976. Over 500 cases were reported and infection proved very dangerous: almost 88 per cent of victims died in Zaire and 53 per cent perished in Sudan. Ebola haemorrhagic fever occurred again in 1978, in virtually the same place in Sudan, but only three other cases have been reported since.

The 'shock, horror, gasp' factor of ebola fever drove journalists into a frenzy. Reports, articles and even films predicted the mayhem that would follow if ebola virus hit the USA, even though the number of people who have ever died from ebola is tiny. But important questions need to be answered: where has this virus been hiding? Why did it suddenly emerge to cause disease? Could there be a large epidemic?

The animal reservoir for ebola virus seems to be a species of monkey that lives in areas of dense forest that are rarely visited by people. Yes, a few places like this still exist, and as people start to go into them, they come into contact with such monkeys. The virus is transmitted and what was a harmless organism in the monkey is deadly in its new host.

In Zaire and Sudan, ebola virus spread quickly between people because of close contact. The centre of the epidemic in Zaire involved a missionary hospital where needles and syringes were re-used without sterilisation. Most of the staff of that hospital died and there were a few cases involving people taking care of the sick or preparing bodies for burial, but the epidemic was self-limiting. Because of its relatively short incubation period and the severe symptoms it causes, it is unlikely ever to cause a worldwide epidemic.

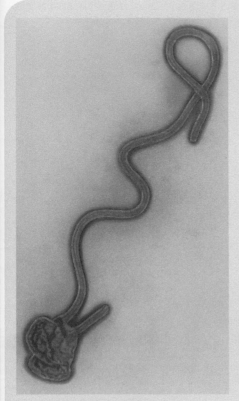

An electron micrograph showing an ebola virus particle. Ebola virus particularly targets liver cells and macrophages: massive destruction of the liver is a hallmark feature of infection

1 MICROORGANISMS AND DISEASE

Disease-causing microorganisms have been responsible for enormous numbers of human deaths. They have frequently altered the course of history. The species that probably had the greatest effect on human populations was *Yersinia pestis*, the bacterium responsible for the great outbreaks of plague such as the Black Death and the Great Plague. The Black Death swept through Europe in the middle of the fourteenth century, killing almost a quarter of the entire population. The Great Plague of 1665 had a devastating effect on the population of London and other British cities and towns (Fig 29.1).

Plague is not a disease confined to the rat-infested slums of the Middle Ages. It still exists today. Every year there are about 15 cases in the western USA, and the disease is still able to terrify. An outbreak of plague in India in 1994 brought headline fears in many newspapers of a new epidemic. But it didn't happen: only a handful of people died. A disease that once killed millions of people can now be readily controlled using **antibiotics** and **antiseptics** and by preventing the conditions that lead to its **transmission**.

In this chapter, we look at the different organisms that cause infectious disease.

WHAT MAKES MICROORGANISMS DANGEROUS?

Microorganisms are very common in and on the human body. There are more bacteria living on your skin than there are people in the world. Most bacteria cause you no harm; the few that do are called **pathogens**. It is important to find out which microorganisms are responsible for which diseases because, until the infection has been identified, it is difficult to give effective treatment.

The nineteenth-century biologist Robert Koch put forward a set of basic principles for identifying a pathogen: these principles are known as **Koch's postulates**.

- The microorganism must always be present when the disease is present. It should not be present if the disease is absent.
- It should be possible to isolate the microorganism from an infected host and grow it in culture.
- When cultured microorganisms are introduced into a healthy host, the disease should develop.
- It should be possible to isolate the microorganism from the new host.

By following these principles, researchers have identified the microorganisms responsible for many diseases. But there are problems in applying Koch's ideas, especially when humans are involved. For example, it would be unethical to introduce a suspected pathogen to humans to see whether or not they developed a disease. In addition, some people fail to show disease symptoms after being infected. They might be resistant to that disease or they might be **carriers** who can pass on the disease to others, even when they appear disease-free themselves (Fig 29.2).

Fig 29.1 Part of the stained glass window in the church at Eyam in Derbyshire. It tells the story of how plague was brought to the village in a bundle of cloth from London. This bundle was opened by the tailor George Viccars who became the first of the 267 out of the total of 350 people in the village to die from plague

How the immune system responds to infection is discussed in Chapter 30.

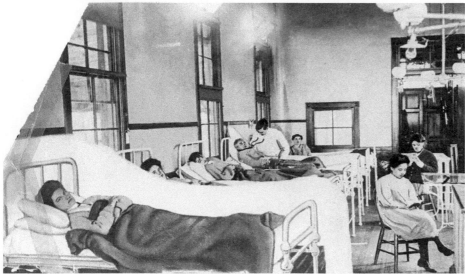

Fig 29.2 Typhoid fever is a severe disease of the digestive system caused by the bacterium *Salmonella typhimurium*. Carriers of typhoid fever are people who are infected but show no signs of the disease, the bacteria living in the gut pass out of the body in the faeces. Typhoid Mary, who lived in New York in the twentieth century, was a carrier of *Salmonella typhimurium*. She gained a bad name because, unwittingly, she shared her infection with many of the people she came into contact with. While she remained the picture of health, they sickened and died

? QUESTION 1

1 Suggest why it might have been difficult to use Robert Koch's ideas to prove that typhoid fever was caused by *Salmonella typhimurium*.

The body's strategies to resist infection are discussed in Chapter 30.

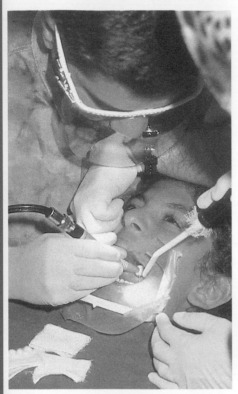

Fig 29.3 During dental treatment, the patient is at risk from infection (through the wounds in the gums), but so is the dentist. The patient can pass on airborne infections and there is a slight risk of blood-to-blood transmission

? QUESTION 2

2 Study the photograph of the dentist at work in Fig 29.3. What precautions are being taken to avoid transmission of any infection:
a) from the dentist to the patient?
b) from the patient to the dentist?

? QUESTION 3

3 Everyone gets colds, but measles, mumps and diphtheria are now very rare in the UK. Can you guess why this is? (There is a clue on page 478 in Chapter 30.)

HOW DO MICROORGANISMS GET INTO THEIR HOST?

Your body provides an ideal environment for pathogens to multiply rapidly. It's warm and wet and there are plenty of nutrients. Fortunately for you, your skin prevents most of them taking advantage of the ideal conditions inside. The outer layer of skin consists of dead, dry cells made of an indigestible protein called **keratin**. Microorganisms find it difficult to penetrate this layer to enter the body. They are more likely to use the slightly easier entry route offered by the **mucous membranes** of the breathing system (Fig 29.3), the gut, the urinary system or the reproductive system.

In order to cause disease, pathogens must be transmitted from one person to another. Many of them cannot survive for very long outside the body. The main **transmission routes** are:

- through the air in droplets of **mucus** or **saliva**;
- in contaminated food or water;
- by sexual contact with another person;
- through breaks in the skin.

Airborne infections

Many microorganisms are transmitted through the air. These include influenza, colds and measles, which are all caused by viruses; and the bacterial diseases tuberculosis, diphtheria and meningitis. When an infected person coughs, sneezes or even just breathes out, tiny droplets of mucus and saliva are expelled (Fig 29.4). These droplets carry microorganisms with them. The larger ones fall to the ground rapidly, but the smaller ones can remain suspended in the air for long periods of time. In poorly ventilated, crowded conditions, diseases such as influenza can spread very rapidly.

Fig 29.4 The power of a sneeze. When someone sneezes, the droplets that explode from the mouth and nose start their journey at over 150 kilometres per hour

Contaminated food and water

Every year, over 5 million people die from infections that cause diarrhoea. Many victims are small children. In Bangladesh, three out of five children under 5 die from infant diarrhoea. The microorganisms responsible spread in water or food contaminated by human faeces.

Direct contact

Direct contact between the skin of one person and that of another is unlikely to produce infection. Sexual contact is another matter, however. Microorganisms that spread like this include the human immunodeficiency virus (HIV), the virus that causes genital herpes, the bacterium responsible for gonorrhoea, the spirochaete that causes syphilis and *Candida*, the fungus that causes thrush (Fig 29.5).

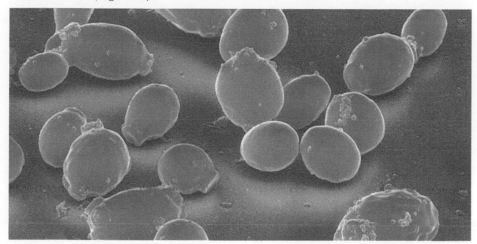

Fig 29.5 Coloured scanning electron micrograph of the fungus that causes thrush, *Candida albicans*. This yeast-like fungus is common in the mouth, vagina and gut

Through the skin

The outer layer of the skin consists of tough cells that protect the body very effectively against infection. But this layer can be broken, and cuts and grazes form an entry point for bacteria. For example, tetanus is a serious disease that affects the nervous system. The bacteria that cause tetanus are common in soil and can enter the body through puncture wounds and animal bites.

Mosquitoes, ticks and lice all feed by puncturing the skin. In doing this, they transmit microorganisms directly into the blood. One of the most important insect-borne diseases is malaria, which is transmitted by *Anopheles* mosquitoes (Fig 29.6).

THE DIFFERENCE BETWEEN INFECTION AND DISEASE

Infection is not the same as having a disease. Before the disease develops, there is an **incubation period** in which the pathogen multiplies in the cells and tissues of the host. Think about what happens when you have been in a room with someone who is suffering from a cold. The infected person might sneeze and expel tiny droplets of mucus and saliva carrying the cold viruses on them. You inhale these particles and the viruses get into your throat. There they attach to the epithelial cells in the mucous membrane. Like all cells in the body, these cells have cell surface membranes that contain many different protein molecules. The type of proteins present depends on the cell type, and on the genetics of the individual. The virus attaches only if the right protein is present in the cell surface membrane.

Fig 29.6 An *Anopheles* mosquito feeding on human blood. Over 1.5 million people per year die from malaria

Fig 29.7 Coloured transmission electron micrograph of human immunodeficiency virus (HIV) budding from an infected T lymphocyte human blood cell. Four viruses are shown at different stages of budding

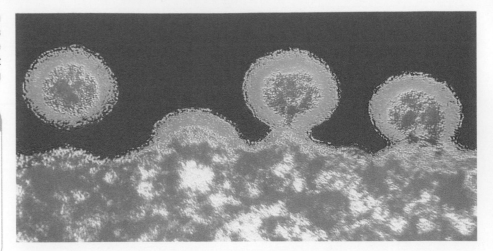

The immune system is discussed in Chapter 30.

? QUESTION 4

4 The proteins of the cell surface membrane are determined by genes. Use this information to explain why some people are susceptible to particular diseases while others are not.

? QUESTION 5

5 Treatment for diphtheria involves the use of an antibiotic such as penicillin and an antitoxin. Explain why it is necessary to use both an antibiotic and an antitoxin.

The virus then enters the cell and begins a cycle of reproduction and colonisation of new cells (Fig 29.7). But infection is not entirely one-sided. The body has a complex immune system. This might prevent the symptoms of the disease from ever developing. Even if some illness does follow, the immune system works to allow us to recover quickly.

THE EFFECTS OF DISEASE

Infectious disease has two main effects. First, the pathogen can damage the cells of the body directly. The human immunodeficiency virus, HIV, gives rise to the symptoms of AIDS mainly as a result of destroying large numbers of a particular type of white cell. As a result, the immune system stops working effectively and this allows other infections or certain types of tumour to develop. Similarly, some of the symptoms of malaria result from a loss of red blood cells or the blocking of capillaries that supply particular organs in the body.

Alternatively, the pathogen can cause the release of substances that injure the body. Many bacteria, for example, produce **toxins**. **Exotoxins** are released by bacteria as they grow. **Endotoxins**, which form part of the bacterial cell wall, escape into the body when the bacterium dies and the cell wall breaks down.

Diphtheria is a good example of a bacterium that causes disease because it produces a toxin. It was a dangerous bacterial infection that used to kill many children until about 50 years ago. Since then, antibiotics and an effective vaccine have virtually eliminated it.

Diphtheria has an incubation period of 2 to 6 days, and then a sore throat and fever develop. A membrane forms across the throat and makes it difficult to breathe (Fig 29.8). As the bacteria multiply, they release a toxin into the blood. This toxin inhibits protein synthesis and results in damage to many parts of the body, particularly to the heart and the nerves. Interestingly, this toxin is produced only in bacteria of the species *Corynebacterium diphtheriae* that are themselves infected by a **bacteriophage**. The toxin is actually produced by one of the bacteriophage genes.

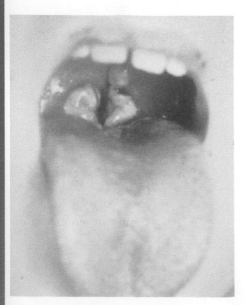

Fig 29.8 The infected throat of a child with diptheria

2 INFECTIONS CAUSED BY BACTERIA

Prokaryotes are discussed in Chapter 1.

Bacteria are **prokaryotes**. Each bacterial cell is a mass of cytoplasm enclosed by a cell surface membrane and a cell wall. Bacterial cells have no nucleus and their DNA is in the form of a loop in the cytoplasm. There are no membrane-bounded organelles such as mitochondria, endoplasmic reticulum and lysosomes that are found in eukaryotic cells. Fig 29.9 shows a typical bacterium, and summarises how different bacteria can cause disease.

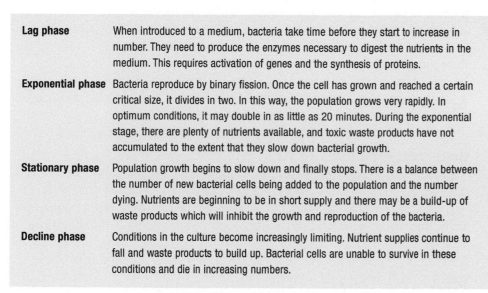

Bacterial DNA This is arranged in a loop

Non-living capsule found in some species of bacteria. Microorganisms with capsules are not readily destroyed by phagocytes: they are described as virulent

Plasmid A small circle of DNA which codes for the production of endotoxins and antibiotic resistance

Bacterial cell wall This is a complex structure (not made of cellulose like the cell walls of plant cells). The cell wall of the bacterium causing TB contains a lot of lipid and is resistant to drying out, enabling the bacterium to survive outside the body

Cytoplasm containing large numbers of ribosomes. These are involved in synthesising proteins such as enzymes and structural proteins

Cell surface membrane

Fig 29.9 Some of the strategies that bacteria use to invade the human body and cause disease

Bacteria don't need much to grow and reproduce. Give a single bacterial cell carbohydrate or some other source of carbon, a supply of nitrogen and a few mineral salts, and it can divide to produce a population of millions. It is easy to mix all these substances together in the laboratory to produce an artificial medium on which to culture bacteria. It's even easier to provide them with some leftover food in the home. In both cases, the bacteria start to increase rapidly.

A typical population curve is shown in Fig 29.10. The actual rate of growth depends on environmental conditions. Low temperatures or a shortage of nutrients, for example, limit the rate of growth.

? QUESTION 6

6 Explain how storing food in a refrigerator will affect the population growth curve of bacteria on a piece of chicken.

Fig 29.10 A population growth curve for the bacterium *Escherichia coli* growing on an artificial medium. Bacterial food poisoning results from eating foods that have been contaminated with certain types of bacteria or with the toxins that they have produced. Food that gives rise to this type of food poisoning has often been stored under conditions in which bacteria can multiply rapidly

Lag phase	When introduced to a medium, bacteria take time before they start to increase in number. They need to produce the enzymes necessary to digest the nutrients in the medium. This requires activation of genes and the synthesis of proteins.
Exponential phase	Bacteria reproduce by binary fission. Once the cell has grown and reached a certain critical size, it divides in two. In this way, the population grows very rapidly. In optimum conditions, it may double in as little as 20 minutes. During the exponential stage, there are plenty of nutrients available, and toxic waste products have not accumulated to the extent that they slow down bacterial growth.
Stationary phase	Population growth begins to slow down and finally stops. There is a balance between the number of new bacterial cells being added to the population and the number dying. Nutrients are beginning to be in short supply and there may be a build-up of waste products which will inhibit the growth and reproduction of the bacteria.
Decline phase	Conditions in the culture become increasingly limiting. Nutrient supplies continue to fall and waste products to build up. Bacterial cells are unable to survive in these conditions and die in increasing numbers.

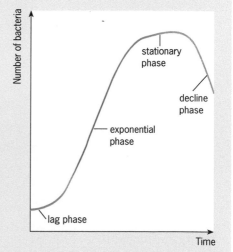

TUBERCULOSIS

Tuberculosis (TB) is a disease caused by an infection with the bacterium *Mycobacterium tuberculosis* (Fig 29.11). Globally, each year, there are 3.8 million new cases of TB, and 2.6 million of them prove fatal.

In the UK and other developed countries, the incidence of tuberculosis fell dramatically during the twentieth century. Many factors contributed to this decline. There was a general improvement in living standards during this time, and advances in medicine such as the discovery of effective antibiotics and vaccines have made tuberculosis treatable and preventable. However, there are now signs that the number of cases is rising again. What has gone wrong?

There are more TB cases in the UK because alcoholics, people with HIV, recent immigrants and healthcare workers who care for the first three groups are at increased risk. The disease is common in people in hostels for the homeless, prisons, and centres for immigrants arriving from areas with high rates of HIV infection.

In the UK, BCG vaccination (with live but weakened tubercle bacteria) is no longer routinely given to all children of secondary school age. Doctors have decided it makes more sense to target BCG vaccination for people who are at greatest risk of the disease. The vaccine is now recommended for:

- children under one year of age living in areas where the incidence of TB is 40 cases per 100 000 people or higher;
- children under one year of age whose parents or grandparents were born in a country with an incidence of TB of 40 cases per 100 000 people or higher;
- children with risk factors for TB who have not previously been vaccinated;
- new immigrants from countries with a high incidence of TB who have not already been vaccinated;
- contacts of people diagnosed with TB affecting the lungs;
- healthcare workers, veterinary staff, staff working in prisons, residential homes, shelters for the homeless or hostels for refugees;
- people intending to live, travel or work in countries with a high incidence of TB for more than a month.

The global impact of TB

The fact that TB is increasing again in the UK may worry you but what is more worrying is the impact that TB has in developing countries. Worldwide, the poorest people who suffer problems such as HIV infection are also the most common victims of this treatable and preventable disease.

- TB kills 5000 people a day and between 2 and 3 million people each year, 98 per cent of whom live in the developing world.
- One-third of the world's population is infected with TB.
- 750 000 women die each year of TB.
- Hundreds of thousands of children will become TB orphans this year.
- One out of every three HIV/AIDS patients has TB.

Signs and symptoms of TB

Tuberculosis can affect many of the organs in the body. In almost 80 per cent of cases it affects the lungs. Someone with primary tuberculosis has a fever, tends to lose weight and often has a persistent cough. They might cough up blood because the bacteria destroy the lung tissue: a chest X-ray shows this damage clearly (Fig 29.12). The condition can be confirmed by identifying the bacteria in a sputum sample (Fig 29.11) or in a small piece of lung tissue removed with the help of an endoscope.

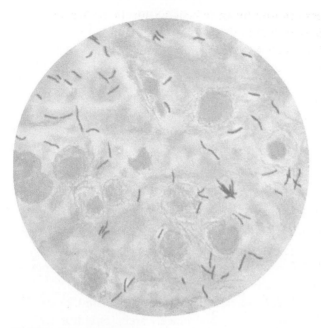

Fig 29.11 TB bacilli in sputum

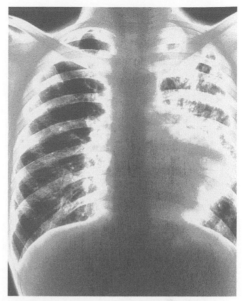

Fig 29.12 In this chest X-ray, the fluffy yellow areas are where the bacteria that cause tuberculosis have damaged the lung tissue

Treatment and control of TB

There is no completely effective vaccine against TB. **BCG vaccine** (bacille Calmette–Guérin) was developed in 1921. It is very useful in preventing certain types of TB and is given to people at high risk of catching the disease. The vaccine is a live but **attenuated**, or altered, strain of *Mycobacterium tuberculosis*. The vaccine provides cross-protection against leprosy. Although it is used widely, the BCG vaccine cannot be given to people with HIV and an alternative is needed urgently. Following successful safety tests, a Phase II trial of the new MVA85A vaccine against TB began in the summer of 2007.

Once the disease has been caught, it can remain hidden, or latent, with the infected person showing no symptoms. When it is diagnosed, it can be treated with antibiotics. Combinations of drugs are used to discourage the development of strains that are resistant to particular antibiotics. Since tuberculosis is infectious, health workers trace all the people who have been in close contact with patients. These contacts can then be screened for the disease and treated if necessary, and even people with latent infections are treated so they cannot pass the disease on.

Despite careful use of antibiotics, drug-resistant TB strains are now common and there are even multi-drug-resistant strains that don't respond to any antibiotics currently available. 134 000 people will die in 2008 from multi-drug-resistant TB.

SALMONELLA FOOD POISONING

There are many different species of bacteria that belong to the genus *Salmonella*. Only about 10 of them cause food poisoning, known as **salmonellosis**. When food contaminated with the bacteria is eaten, the bacteria pass into the intestine. Other bacteria that cause food poisoning do not multiply once they are inside the body – they cause illness as they release toxins. *Salmonella* species actually enter the cells lining the small intestine and multiply. As the population increases, some bacteria die and release

? QUESTION 7

7   The symptoms of salmonellosis usually take between 12 and 72 hours to develop. The symptoms of other forms of food poisoning develop much more quickly. Explain why.

✔ REMEMBER THIS

If you want to avoid salmonellosis or other food poisoning, avoid buying and eating food after the recommended date. Store perishable food in a cool place or, preferably, in a refrigerator. Do not re-freeze food that has already been frozen and thawed once. Avoid the risk of contamination by maintaining good personal hygiene and by keeping raw and cooked foods separate from each other. Make sure that cooking and re-heating is thorough at sufficiently high temperatures.

endotoxin. The endotoxin released as the bacteria die irritates the lining of the intestine, and causes the unpleasant and typical symptoms of food poisoning or **gastroenteritis**: nausea, vomiting, diarrhoea and abdominal pain. Bad personal hygiene and poor food preparation skills can spread *Salmonella* food poisoning.

Typhoid fever

Typhoid fever is caused by another species of *Salmonella*. It is a much more severe disease than salmonellosis and spreads in conditions of poor sanitation (Fig 29.13). The patient often has a long-lasting fever, becomes delirious, and the spleen becomes inflamed. The wall of the intestine can be so badly damaged that it bleeds.

Treatment and control of salmonellosis and typhoid fever

Most cases of food poisoning are left to run their course. Sometimes, loss of body fluid may be severe, particularly in infants. Then, applying **oral rehydration therapy**, ORT, or using an **intravenous drip**, will replace body fluids. Antibiotics may be used to treat typhoid fever.

The best way of dealing with *Salmonella* or any other organism that causes food poisoning is to avoid getting it in the first place: careful attention to basic food hygiene removes much of the risk.

Oral rehydration therapy (used for severe dehydration in cases of salmonellosis and typhoid fever) is discussed in Chapter 13.

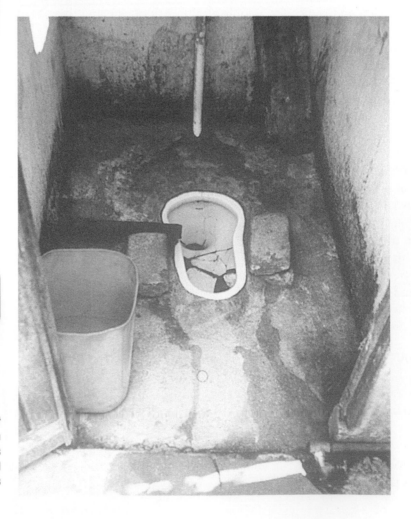

Fig 29.13 Typhoid is found mainly in the poorest of developing countries but insanitary conditions can take people by surprise. This latrine is like this because of a hurricane and local flooding, which led to an outbreak of typhoid in Mauritius

3 INFECTIONS CAUSED BY VIRUSES

It is difficult to decide whether viruses are living organisms at all. They do not show several of the characteristics that define a living organism; for example, they don't respire or feed. But they are more than just a collection of very complex chemicals. Their origins are also rather uncertain. Some biologists think they are stray bits of DNA that have 'escaped' from the genomes of higher plants and animals.

Viruses vary considerably in size but they are all very small (Fig 29.14). The virus responsible for polio, for example, is one of the smallest and is only 20 to 30 nm in diameter. The herpes virus is much larger; its diameter is about 250 nm. In addition to their small size, viruses have a very simple structure. They consist of a piece of genetic material, DNA or RNA, surrounded by a protein coat and sometimes also by a membrane.

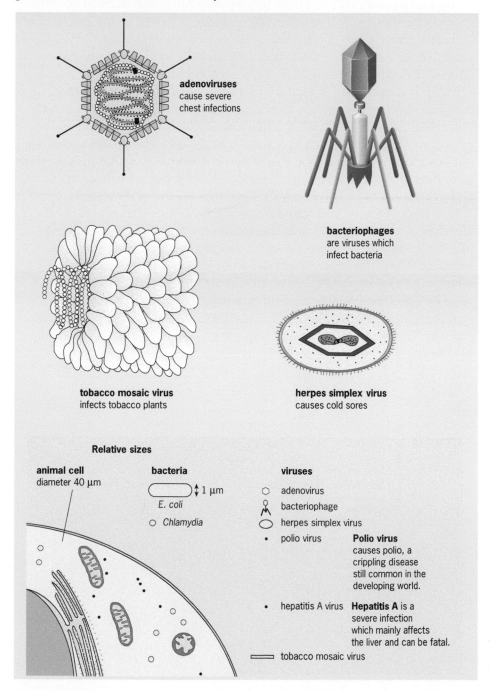

adenoviruses
cause severe
chest infections

bacteriophages
are viruses which
infect bacteria

tobacco mosaic virus
infects tobacco plants

herpes simplex virus
causes cold sores

Relative sizes

animal cell
diameter 40 μm

bacteria
⬭ ↕ 1 μm
E. coli
○ *Chlamydia*

viruses
○ adenovirus
⬦ bacteriophage
◯ herpes simplex virus
• polio virus **Polio virus** causes polio, a crippling disease still common in the developing world.
• hepatitis A virus **Hepatitis A** is a severe infection which mainly affects the liver and can be fatal.
▭ tobacco mosaic virus

Fig 29.14 In the blue box, the bacterial cell *E. coli* and an animal cell have been drawn to the same scale. Viruses are generally much smaller than bacteria, but there is some overlap. Herpes simplex, one of the larger viruses, is similar in size to *Chlamydia*, one of the smaller bacteria. The magnified views of four of the viruses show the appearance of individual virus particles (not to scale)

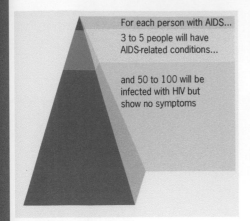

Fig 29.15 The HIV pyramid

For each person with AIDS...

3 to 5 people will have AIDS-related conditions...

and 50 to 100 will be infected with HIV but show no symptoms

In order to multiply, a virus must take over the host cell. It uses the cell's organelles and biochemical processes to make more virus particles. Viruses therefore have a great capacity to harm the cells and tissues of their host, and they cause a huge range of diseases, including influenza and AIDS.

AIDS

AIDS, short for **acquired immunodeficiency syndrome**, is characterised by the destruction of vital cells in the immune system. It is caused by the human immunodeficiency virus (HIV). The disease was first recognised in the early 1980s and the virus itself was identified in 1986. It has spread dramatically (Fig 29.15).

A total of 39.5 million (34.1 million–47.1 million) people were living with HIV in 2006 – 2.6 million more than in 2004. This figure includes the estimated 4.3 million (3.6 million–6.6 million) adults and children who were newly infected with HIV in 2006. Sub-Saharan Africa continues to bear the brunt of the global HIV epidemic. Two-thirds (63%) of all adults and children with HIV globally live in sub-Saharan Africa, mainly in southern Africa. One-third (32 per cent) of all people with HIV globally live in southern Africa and 34% of all deaths due to AIDS in 2006 occurred there.

HIV is a **retrovirus**: it carries the enzyme **reverse transcriptase**. Once the virus infects a cell, often a T lymphocyte, this enzyme allows the cell to make virus DNA from virus RNA. The virus DNA is then inserted into the cell's own DNA. Here it acts as a gene. It might do nothing for a long time but eventually it will cause the production of more virus RNA. This will result in thousands of new viruses that burst out through the cell surface membrane and infect other cells. This life cycle is summarised in Fig 29.16.

HIV is often transmitted from one person to another during sex, but the virus can also spread when drug users share dirty needles, when infected blood is transfused, or when contaminated blood is used to make blood products. When HIV enters the body, it can be many years – maybe 10 or 15 – before AIDS develops. A very small number of people are known to have been HIV-positive since the mid-1980s and have not so far developed the syndrome. Perhaps they never will, and AIDS researchers are trying to find out why. This might provide really important clues about how AIDS can be held in check in other people.

Fig 29.16 HIV infects vital cells in the immune system, notably the T helper lymphocytes. This diagram shows how a single virus can infect a host cell and multiply to produce many thousands of new viruses

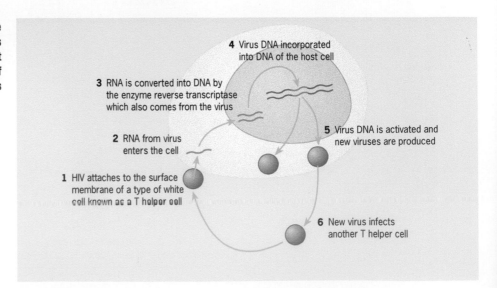

4 Virus DNA incorporated into DNA of the host cell

3 RNA is converted into DNA by the enzyme reverse transcriptase which also comes from the virus

2 RNA from virus enters the cell

5 Virus DNA is activated and new viruses are produced

1 HIV attaches to the surface membrane of a type of white cell known as a T helper cell

6 New virus infects another T helper cell

Table 29.1 The stages in the development of AIDS

Stage	Number of T helper cells per mm³ of blood	Main features of stage
A	over 500	No symptoms but patient has antibodies to HIV in the blood.
B	200–499	Patient usually has fever or diarrhoea which lasts for a month or more. He or she might have infections such as thrush, caused by *Candida*, a type of fungus.
C	less than 200	Likely to have one or more of the following diseases known to be related to AIDS: • tuberculosis • Karposi's sarcoma, a form of cancer in which large black or brown tumours appear on the skin (Fig 29.17) • pneumonia

Signs and symptoms of AIDS

Table 29.1 shows the main stages in the development of AIDS from first infection with HIV.

Control and treatment of HIV and AIDS

HIV is capable of infecting another person only when it is passed inside lymphocytes or macrophages. Apart from blood itself, semen and vaginal secretions contain large numbers of these cell types, making body fluids high-risk sources of infection.

Practising safe sex – using condoms during every sexual encounter – is very important to prevent spread of HIV. Using a barrier method of contraception not only prevents pregnancy, it cuts down the chance that HIV in one person's semen or vaginal secretions will get into his or her partner's blood. Drug users who share dirty needles are also at risk. Programmes that provide free, sterile needles on demand are designed to cut down HIV transmission by this route.

Drug treatments for HIV offer many people the chance to control the virus and stay healthy for much longer. Treatment options have had a huge impact on the lives of people with HIV and those who care for them, through reductions in AIDS-related illnesses, admissions to hospital and death rates. Treatment has also enabled some people with HIV to go back to work and plan for the future. However, the treatments do not work equally well for everyone. They can have side-effects and drug resistance can be a problem.

Highly active antiretroviral therapy (HAART) is the most common form of drug therapy for HIV in the developed world. There are five main groups of drugs involved:

- nucleoside reverse transcriptase inhibitors (NRTIs);
- non-nucleoside reverse transcriptase inhibitors (NNRTIs);
- nucleotide reverse transcriptase inhibitors;
- protease inhibitors;
- fusion inhibitors.

Usually, three different drugs from at least two of these groups are taken together, two to four times a day. The drugs control the virus by stopping it from making copies of itself inside the cells of the body. People taking drug treatment for HIV will probably need to take it for the rest of their lives. Stopping drug treatment, even for short periods of time, can cause the virus to become resistant to drugs.

Treatment does not stop someone with HIV from being able to pass on the virus through unprotected sex or sharing needles or injecting equipment.

REMEMBER THIS

Worldwide, approximately one in every 100 adults aged 15 to 49 is HIV-infected. In sub-Saharan Africa, about 8.4 per cent of all adults in this age group are HIV-infected. In 16 African countries, the prevalence of HIV infection among adults aged 15 to 49 exceeds 10 per cent.

An estimated 5 million new HIV infections occurred worldwide during 2001; that's about 14 000 infections each day. More than 95 per cent of these new infections occurred in developing countries. In 2001 alone, HIV/AIDS-associated illnesses caused the deaths of approximately 3 million people worldwide, including an estimated 580 000 children younger than 15 years.

Worldwide, more than 80 per cent of all adult HIV infections have resulted from heterosexual intercourse.

(Data from the National Institutes of Health, USA.)

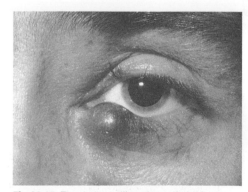

Fig 29.17 The dark swelling shows that this person has Karposi's sarcoma, a form of cancer, which affects blood vessels in the skin. Karposi's sarcoma is much more common in people with AIDS than it is in the rest of the population

REMEMBER THIS

An effective vaccine is thought to be a long way in the future because HIV can change its surface proteins and evade host immune responses.

Measles and the controversy over vaccination

If you are fit and well-nourished, measles is an unpleasant childhood infection that makes you ill for two weeks, and then you get better (Fig 29.18). However, it affects some people much more seriously, particularly if they are weakened by malnutrition or other diseases. In many developing countries, measles is a big killer. Fortunately, a measles vaccine has been available for some time and children in the UK and other countries in the developed world don't usually catch this disease.

In 1998, a study of children who had received the combined MMR vaccine – for measles, mumps and rubella (commonly known as German measles) – suggested that some children developed bowel disease after the vaccination, and then went on to develop autism. This is a condition in which children, and adults, have difficulty relating to other people, and show unusual behaviour. But was the vaccine responsible?

A great controversy has since raged, with some parents believing passionately that the vaccine is to blame for their children's problems. Another study that was already going on at the time has now come up with the closest we will probably ever get to the 'truth' about how safe the MMR vaccine is. In January 2001, researchers announced the results of following nearly 2 million children in Finland over a period of 14 years. They found no link between the vaccine and the incidence of either bowel disease or autism. This number of subjects in a study is enormous, and leaves very little to chance, so most experts now conclude that the vaccine is not a cause of autism.

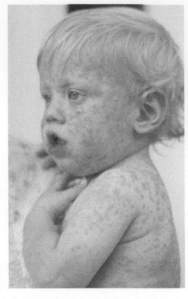

Fig 29.18 A child with measles

? QUESTION 8

8 Suggest how a knowledge of the way in which the influenza virus is spread could be used to limit the spread of a flu epidemic.

Fig 29.19 The electron micrograph shows influenza viruses leaving an infected cell. The drawing shows the main features of an influenza virus

INFLUENZA, AN EXAMPLE OF A COMMUNICABLE DISEASE

Outbreaks of influenza or flu occur every year, with huge epidemics, or pandemics spreading throughout the world every 30 years or so. The disease is caused by a virus that affects the epithelial cells that line the nose and throat (Fig 29.19). After a short period of incubation, flu causes headache, fever and aching muscles and joints. In most years, a bit of rest and drinking plenty of fluids allows most people to get better within about a week. However, the old and the very young can be very ill and thousands of people with die from 'flu every winter.

Flu passes between people rapidly because it is airborne. When we breathe out, sneeze or cough, tiny droplets of mucus and saliva fly into the air. If you

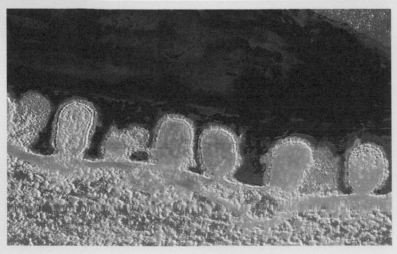

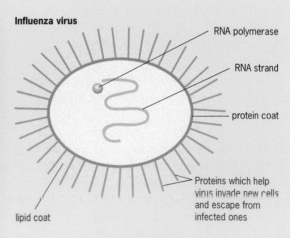

Influenza virus

RNA polymerase

RNA strand

protein coat

Proteins which help virus invade new cells and escape from infected ones

lipid coat

have flu, these droplets contain flu viruses and this is how they are passed on to other people.

A global disease

The influenza virus has different forms. Two of these, **Type A** and **Type B**, are associated with major epidemics.

Influenza epidemics are most often caused by Type A viruses. Type B is not as common but outbreaks of Type B influenza occur on a global scale. Such large-scale epidemics are known as **pandemics**.

One pandemic broke out in Europe in 1556 and lasted until 1560. Historians estimate that during this period almost 20 per cent of the population of the United Kingdom died of influenza. The worst flu pandemic that is recorded in recent history occurred in 1918–1919, just at the end of the First World War. Experts estimate that this very virulent form of flu killed between 40 and 60 million people worldwide. It affected just as many people who were in their 20s as the elderly and the young, spreading rapidly in the troubled conditions of the Western Front (Fig 29.20).

Protection against influenza

Why is it that you can have a single attack of measles and then be immune for life, but you can get influenza many times? A person who has had influenza becomes immune to the strain of virus that caused it. However, the influenza virus has a very high rate of mutation. This means that different strains arise, each with slight differences in its outer protein coat. The body might be immune to the strain responsible for an outbreak of influenza one year, but this immunity does not protect against other strains that cause later outbreaks of flu.

Influenza vaccines are available (Fig 29.21). They are usually given to those at risk such as elderly people and those with asthma and heart conditions. The actual vaccine produced is effective only against the two or three strains of the virus common that year. It is necessary to have another injection with a different vaccine the following year since new strains will probably have evolved.

4 PARASITES AND DISEASE

Bacteria and viruses are **microorganisms**; they are too small to be seen with the naked eye. However, some larger organisms also cause disease. These are the **protoctists** and the **parasitic worms**. In Europe, they are relatively rare, but on a global scale they have an enormous impact. Over half of the world's population is infected by at least one of the four parasitic worms shown in Fig 29.22a–d (overleaf). The roundworm *Ascaris lumbricoides* is the subject of the Assignment for this chapter on the CAS website.

It is all too easy to concentrate on the horrors of parasitic infections. As biologists we should not forget that parasites as a group provide one of the best examples of **adaptation**.

? **QUESTION 9**

9   Explain why the 1918–1919 influenza pandemic spread much more rapidly than the pandemic that occurred in 1556.

Fig 29.20 A major flu pandemic killed about 50 million people in the period immediately after the First World War. There was little doubt that the speed at which the disease spread from country to country and continent to continent was partly due to troops returning home

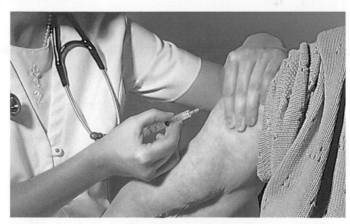

Fig 29.21 Vaccination against influenza is offered to the most vulnerable members of the community

Tapeworms are discussed in Chapter 35.

? **QUESTION 10**

10   What particular problems are likely to be encountered by a parasitic blood fluke living in a vein?

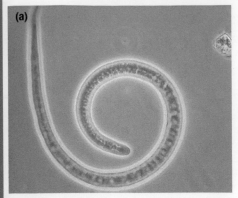

Fig 29.22a Roundworm

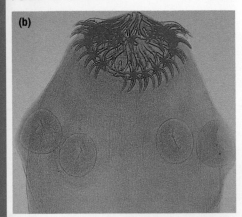

Fig 29.22b Tapeworm: head end

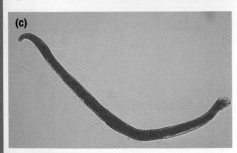

Fig 29.22c Hookworm

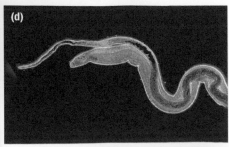

Fig 29.22d Blood fluke *Schistosoma*: male (fat) and female (thin) flukes

Plasmodium life cycle and malaria are discussed in more detail in Chapter 35.

A useful definition of a parasite is as follows:

A parasite is an organism that lives in or on a host organism. It gains a nutritional advantage from this relationship while the host suffers a disadvantage.

It might seem that this is a very easy way of making a living, but think about the conditions a parasitic worm faces, living in your intestine. There are obvious advantages in lying in a soup of pre-digested food. There is no need to produce digestive enzymes, a complex nervous system is unnecessary because food does not need to be searched for, and a locomotory system is almost a disadvantage since the worm might move to less favourable areas of the gut.

On the other hand, there are disadvantages. Parasites are made of the same complex macromolecules as the food that you eat, so they must avoid being digested and must try to stay put despite the constant contractions of the gut muscles. It's also difficult to move from the intestine of one person into the intestine of another. Because of this, life cycles can be very complex and many involve **intermediate hosts**. Since the chances of completing the life cycle and reaching a new host are small, many parasites produce vast numbers of offspring.

MALARIA

Malaria is caused by four species of the protoctist *Plasmodium* – *P. vivax*, *P. falciparum*, *P. ovale* and *P. malariae* (Fig 29.22e). They are transmitted by the bite of the *Anopheles* mosquito.

During the 1960s, malaria was successfully eradicated from the warmer areas of southern Europe and the USA, and its incidence was reduced significantly in many countries in the tropics. But the disease is now on the increase again. The World Health Organization (WHO) estimates that each year 270 million new malaria infections occur worldwide, along with 110 million cases of illness and 1.5 million deaths. Malaria is endemic in at least 100 countries of the tropical and subtropical regions of the world; there are an astounding 2 billion people at risk of malaria infection, and malaria is the cause of a quarter of all childhood deaths in Africa.

There are several reasons why the disease is so widespread: the cost of maintaining control programmes is high, insecticide resistance has become more common in the mosquito that transmits malaria, and the malarial parasite itself has developed drug resistance.

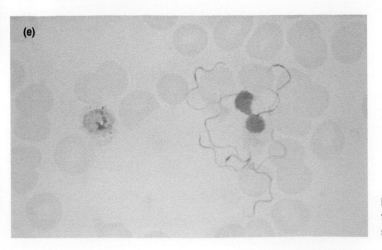

Fig 29.22e Malaria: thread-like sporozoites in blood

Signs and symptoms of malaria

One of the most characteristic features of malaria is the pattern of fever (Fig 29.23), with temperature peaks at regular intervals. These correspond to a new generation of parasites bursting out of the red blood cells, destroying them and causing anaemia. The parasites can block capillaries, limiting the blood supply to organs in the body: the brain, digestive system and kidneys can all be affected.

Control and treatment of malaria

One way to combat malaria is to break the cycle of infection. This can be done by killing the mosquitoes with insecticides, clearing the pools of stagnant water in which the larvae live and preventing the insects from biting humans. Where possible, the risk can be reduced by sleeping under mosquito nets (Fig 29.24), wearing long-sleeved shirts and trousers after dark when mosquitoes are most active, and using mosquito repellents. But all these measures are too expensive for many local people. Also, they are not certain to work.

Anti-malaria drugs

The anti-fever properties of the bitter bark of the tree *Cinchona ledgeriana* were known in Peru before the fifteenth century. **Quinine**, the active ingredient in the bark, was first isolated in 1820. **Chloroquine** is now the most commonly used anti-malaria drug in Africa. It lowers fever and reduces the number of parasites in the body. Resistance to chloroquine, however, is widespread and has been rising steadily since the early 1980s (Fig 29.25). For this reason, it is used in combination with other drugs. Research continues into resistance-reversing drugs, but so far no useful drugs have emerged.

Drugs based on **artemisinin** are important treatments for severe malaria. These are derived from the Chinese plant *Artemesia annua* (or quinghaosu). They are effective, fast-acting, but relatively expensive drugs.

Artemether and quinine are the anti-malarial drugs most suitable for the treatment of severe malaria in children.

Fig 29.23 A temperature chart for a patient with malaria. Note the peaks that occur at regular intervals. This cyclic pattern of fever is characteristic of malaria

Fig 29.25 The map shows that malaria is confined to tropical and subtropical regions. Chloroquine has been used very successfully in the control of malaria but resistance to it is now a major problem

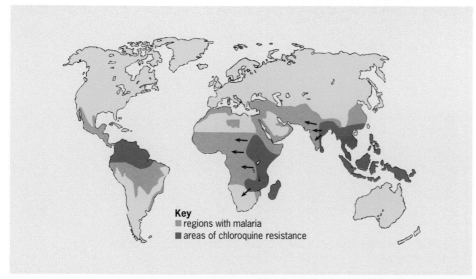

Key
- regions with malaria
- areas of chloroquine resistance

Fig 29.24 The use of mosquito nets soaked in insecticide can be very effective in controlling malaria. Unfortunately, the cost may be beyond the resources of people living in malarial areas

✔ REMEMBER THIS

An infusion of quinghaosu (*Artemesia annua*) has been used for at least the past 2000 years in China. Its active ingredient artemisinin has only recently been scientifically identified.

Will there ever be a vaccine for malaria?

Despite decades of research effort, no one has yet managed to develop a vaccine that is effective against a protoctist such as *Plasmodium* so patients continue to suffer (Fig 29.26). Several potential malarial vaccines have been developed, but many have run into problems. The vaccine developed by Manuel Patarroyo in Colombia has been tested in people and seemed to be the answer everyone was waiting for. However, further trials showed that the vaccine was not protective, and arguments about it have raged ever since.

Currently, molecular biologists are using two main tactics to find a suitable vaccine. Some are trying to identify antigens that appear on the parasite's surface at various stages of its life cycle. The idea is to put several *Plasmodium* genes into the harmless **vaccinia virus**, and then introduce it as a recombinant vaccine into the body. Once inside, it divides a few times and sets up a good immune response to *Plasmodium*. Some of these recombinant vaccines look promising and may be tested in humans very soon.

The other approach is to develop a DNA vaccine for malaria. The aim is to isolate the *Plasmodium* genes that provoke an immune response during infection, and to put them into a plasmid. This is a ring of DNA obtained from bacteria. The DNA itself is introduced into the body where it uses the human cell machinery to make the proteins coded for in the genes. The person's immune system detects these proteins, and makes an immune response. It is this response that could combat the real infection if the person is later exposed to *Plasmodium*.

Another new approach, announced late in 2001, is to express proteins that can stimulate immunity to *Plasmodium* parasites in goat's milk. Researchers showed that it was possible to produce transgenic mice that could express proteins from *Plasmodium* in their milk and experiments were then started to try to produce goats to do the same thing. The advantage of this strategy is that a herd of goats could probably produce enough vaccine for the whole of sub-Saharan Africa.

Trials of various types of malarial vaccine are in progress and new approaches are being investigated. Even so, it could be at least five years before we have a real hope of a useable vaccine for humans.

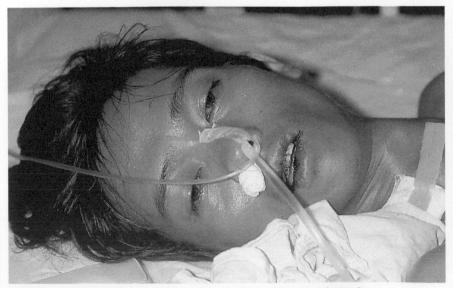

Fig 29.26 An effective malarial vaccine would cut the misery of malaria infection for millions of people

SCHISTOSOMIASIS

Schistosomiasis, also known as **bilharzia**, results from infection by one of three species of a blood fluke called ***Schistosoma***. The three species are: *Schistosoma mansoni* (Africa and South America), *Schistosoma haematobium* (Africa and the Middle East) and *Schistosoma japonicum* (Asia). Like many other parasitic worms, these flukes have a very complex life cycle (Fig 29.27).

Schistosomiasis is generally believed to rank second after malaria as the most important parasitic disease in the world. A total of 500–600 million people are at risk of infection and more than 200 million people in 74 countries succumb to the disease. Of these, 800 000 die each year and 120 million others have severe symptoms.

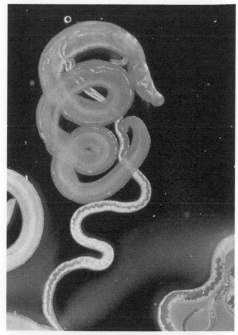

Fig 29.27 Schistosomiasis and its cycle of infection (left). Part of the life cycle takes place in humans and part in water snails. Unfortunately the snails are very difficult to kill. In the photo (above) the male schistosome is blue and the female is white

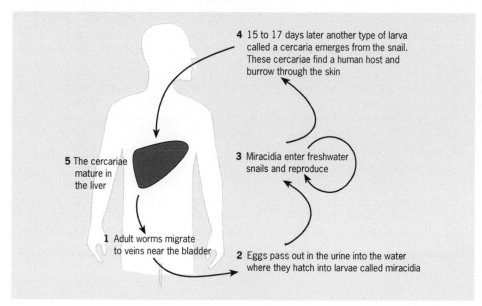

4 15 to 17 days later another type of larva called a cercaria emerges from the snail. These cercariae find a human host and burrow through the skin

5 The cercariae mature in the liver

3 Miracidia enter freshwater snails and reproduce

1 Adult worms migrate to veins near the bladder

2 Eggs pass out in the urine into the water where they hatch into larvae called miracidia

Signs and symptoms of schistosomiasis

The body responds to an infection with schistosomes by trying to 'wall off' the eggs that are produced by the worms within the veins. This scar tissue can cause the liver to enlarge or can lead to blockages in the bladder and kidney. The severity of the damage is directly related to the numbers of parasites in the body. Since the parasite does not reproduce inside the human host, clinical signs do not appear until the number of worms in the body reaches a critical level.

Treatment and control of schistosomiasis

The main drug used to treat schistosomiasis is highly effective and has few side-effects. However, it does not prevent reinfection – only continued, repeated treatments will keep the disease under control – and it is expensive. This creates an impossible situation because the people most affected by the disease are the ones least able to afford treatment. Increased use of drugs also increases the likelihood that drug-resistant strains will evolve.

Tackling the snail that carries the schistosome is also difficult. The chemicals used to kill the snails are thought to be harmful to the environment. Furthermore, the increased use of irrigation worldwide has aided the spread of schistosomiasis by expanding the habitat of the freshwater snail (Fig 29.28).

Fig 29.28 The best way to control the disease is for people to avoid the freshwater areas that form the snail's breeding ground. This is good in theory, but if all supplies of freshwater are affected, where can the local people bathe and wash clothes? Until the poverty and poor living conditions of many of the areas where schistosomiasis is rife are tackled, people have no choice but to expose themselves to repeated infection

? QUESTION 11

11 The cercariae emerge from the water snails during the daytime. How is this behaviour an advantage to the parasite?

HOW PARASITES ARE ADAPTED FOR THEIR LIFESTYLE

The parasites that cause malaria and schistosomiasis illustrate how specialised parasites become as they adapt to living in the blood system of their human host. The two parasites and their adaptations are compared in Table 29.2.

Table 29.2 Adaptations to a parasitic lifestyle: *Plasmodium* and *Schistosoma* compared

Adaptation	*Plasmodium*	*Schistosoma*
Attaching to the host	The malarial parasite spends much of its adult life inside the cells of its host. Attachment is not a problem.	Adult worms live in the veins around the bladder. Because they are small they could easily be washed away by the blood. Suckers help them attach to the walls of the veins.
Resisting host defence mechanisms	The blood is a dangerous place for any parasite because of the white cells of the immune system (see Chapter 30). Within an hour of entering the body, malarial parasites are inside liver cells. Here they hide from the immune system. The antigens on the surface of a malarial parasite change frequently and the immune system cannot keep up.	Adult worms have a remarkable way of avoiding the host's immune system: they disguise themselves by covering up with molecules from the host's red blood cells.
Reproduction and reproductive organs	There are many reproductive stages in the life cycle of the malarial parasite. The parasites reproduce in the mosquito, but this insect is small compared to a human. Once in their main host, malaria parasites also reproduce in the liver and in the blood cells, causing a rapid increase in parasitic load.	The life cycle of this worm depends very heavily on chance. A person must urinate into freshwater; the miracidia larvae that hatch from the eggs must find a snail host and the cercariae must encounter a human shortly after emerging from the snail. The production of large numbers of offspring is therefore the best way of ensuring survival. Each egg laid by an adult worm can produce up to 200 000 cercariae.
The life cycle	Usually, blood from one person does not mix with that of someone else. So malaria parasites cannot depend on person-to-person contact for transmission. A second host is used. In malaria, the female *Anopheles* mosquito is the organism. When she bites someone with malaria, she takes in infected red blood cells. These parasites complete their life cycle in her body, finally migrating to her salivary glands. When she bites someone else, she injects a small amount of infected saliva, and malaria is passed on.	Part of the life cycle of *Schistosoma* is spent inside its human host; part is spent inside a freshwater snail. In many parts of the tropics where schistosomiasis is endemic, humans spend much of their time around areas of freshwater. They use it for washing clothes, irrigating crops, watering animals and bathing. This increases the chances of the parasite being passed from one person to another.
Reduction of body systems	The malaria parasite spends most of its life inside one of its host's red blood cells. It does not have to move to find food, and the medium in which it lives has the same water potential as its own cytoplasm. It has therefore lost the capacity to locomote and to regulate cell water content – features of many free-living single-celled organisms.	Schistosomes do not have a complex nervous system because conditions they find in the blood are remarkably constant. They cannot move either: a locomotory system is also unnecessary since the parasites' survival depends on it being anchored in the host.

? QUESTION 12

12 What is the advantage to the malarial parasite in living in a medium that has the same water potential as its own cytoplasm?

SUMMARY

After reading this chapter, you should know and understand the following:

- Robert Koch produced a series of principles that enabled biologists to identify the microorganism that causes a specific disease. These principles are known as **Koch's postulates**.

- In order to cause disease, a pathogen must gain access to the body. This involves penetrating the body's defences.

- **Transmission** of pathogens can take place through the air, in contaminated food or water, by sexual contact or through the skin.

- For infection to cause disease, the pathogen must multiply inside the host. Disease can result from **direct damage** to the cells and tissues of the host or from indirect damage caused by **toxins** produced by the infecting microorganism.

- Some species of bacteria and viruses are pathogenic. So are some species of larger organisms such as protoctists and worms. These are **parasites** and they show many **adaptations** to their way of life.

 Practice questions and a How Science Works assignment for this chapter are available at www.collinseducation.co.uk/CAS

30

Defence against disease

30 DEFENCE AGAINST DISEASE

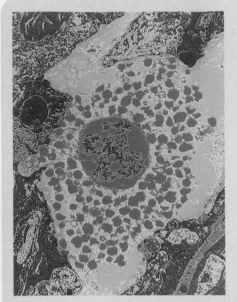

A false-colour electron micrograph of a mast cell, a type of white cell involved in allergic reactions. The small red vesicles contain histamine, the chemical that produces many of the symptoms of allergy. Many people take antihistamines to reduce their allergy to such things as pollen, house dust or animals

Death by allergy

A few years ago, a major medical journal reported the tragic case of a teenage girl who died within 20 minutes of taking a health supplement containing 'royal jelly', the substance made by queen bees. Sudden death in young people usually makes headlines, particularly when it seems to be related to a trivial event such as taking a normally harmless food supplement. But similar deaths have also occurred as a result of a bee sting.

The victims all have in common a very rare and severe allergy. Just as some of us suffer from hay fever in the summer because we react to pollen in the air, some people over-react to the 'foreign' proteins produced by bees.

Normally, when someone is stung by a bee, the result is a painful red swelling that disappears again in a couple of days. In someone who is allergic to bee protein, the effects can be catastrophic. As soon as the poison gets into their bloodstream, many of their white cells release large amounts of histamine and other chemicals that cause severe inflammation. Fluid builds up in the tissues and the smooth muscles all contract. The whole body is affected and both blood volume and blood pressure quickly plummet. Sometimes, the airways in the lungs narrow so much that the affected person can no longer breathe. These symptoms, collectively known as anaphylactic shock, can happen very quickly and can be fatal unless immediate treatment with adrenaline and antihistamines is available.

1 INTRODUCING IMMUNITY

We are all surrounded by bacteria, viruses, fungi and other organisms that are capable of invading our bodies and causing disease (Fig 30.1). We are able to overcome infections by these **pathogens** (disease-causing organisms)

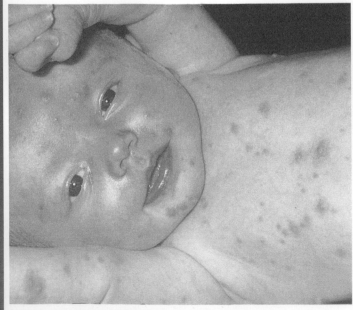

Fig 30.1a This baby boy has chicken pox, caused by the varicella-zoster virus

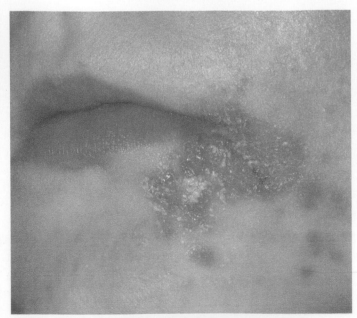

Fig 30.1b This girl has impetigo, a highly infectious bacterial infection

because we have an **immune system**. This is a complex system involving many different cells and tissues that allows us to develop **immunity** – resistance to infections.

Common pathogens include bacteria, fungi, viruses and protoctists. This last group includes microscopic parasites such as *Plasmodium*, which causes malaria, and larger parasitic animals such as tapeworms. A pathogenic organism is able to:

- break through the physical barriers of the body and enter tissues or cells;
- resist the efforts of the immune system to destroy it, long enough to multiply inside the host's body;
- get out of one host and into another;
- damage the host's tissues – either directly, or indirectly by means of **toxins** (poisons) that it releases. Some bacterial **exotoxins**, such as the one produced by *E. coli* O157, which has caused epidemics of gastroenteritis in the past few years (page 456) are very powerful and can be fatal.

The easiest way to start understanding the immune system is to look first at its overall functions, rather than concentrating on the individual parts. Fig 30.2 summarises the main lines of defence that an organism comes up against when it tries to infect a healthy person. We look at each of these in more detail in the sections that follow.

Our study of the immune system covers:

- The ability of the immune system to distinguish between invading organisms and the tissues of its own body. This concept of **self** and **non-self** is important in determining whether the immune system keeps the body healthy, or whether it is overcome by infection.
- Problems of the immune system. In the Science in Context box on page 475, we see how the body can react against its own tissues, causing **autoimmune disease**.
- How medicine today can manipulate the immune system to provide life-saving treatments such as vaccines, blood transfusions, organ and tissue transplants.

QUESTION 1

1 What conditions inside our bodies make it ideal for the growth of microorganisms?

REMEMBER THIS
Pathogens are disease-causing organisms. Such diseases are communicable or infectious: they can pass from one person to another. Other diseases, such as diabetes and cancer are non-communicable.

Fig 30.2 An overview of the body's defences: non-specific responses are general responses to damage. They include inflammation and phagocytosis of debris. Specific responses are targeted against individual types of microorganism

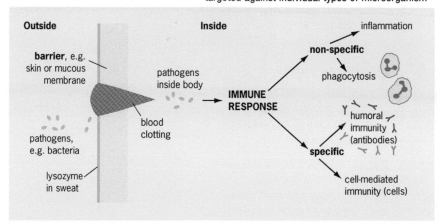

2 THE BODY'S BARRIERS TO INFECTION

One of the most obvious ways to avoid infection is to stop potential pathogens getting into the body in the first place. The four main strategies that the body uses are summarised below:

- **Mechanical** defence. **Nasal hairs** filter the air that is drawn into the upper airways. **Cilia**, which line the airways, sweep bacteria and other particles away from the lungs.
- **Physical** defence. The skin, made from **stratified squamous epithelium**, forms a tough, impermeable barrier that normally keeps out bacteria and viruses. The **mucous membranes** that line the entry points to the body such as the nose, eyes, mouth, airways, genital openings and anus produce fluids and/or sticky mucus. These fluids trap microorganisms and stop them attacking the cells underneath.

Types of tissue are discussed in Chapter 2.

QUESTION 2

2 Smoking leads to paralysis of the cilia that line the airways. Suggest why this causes problems.

- **Chemical** defence. Fluids such as sweat, saliva and tears contain chemicals that create harsh environments for microorganisms. Sweat contains **lactic acid** and the enzyme **lysozyme**, both of which slow down bacterial growth. Stomach acid kills many microorganisms that manage to get that far. When we are injured, blood clots at the injury site, sealing the breach to prevent entry of bacteria (see the next section).
- **Biological** defence. Normally, a vast number of non-pathogenic bacteria live on the skin and mucous membranes. These do not harm the body but they out-compete pathogenic bacteria, preventing them from gaining a foothold from which to launch a full-scale infection.

3 BLOOD CELLS AND DEFENCE AGAINST DISEASE

BLOOD CLOTTING

Whenever blood vessels break, blood leaks out and **clots** (Figs 30.3 and 30.4). We can see this happening when we cut ourselves, but blood can also clot deep inside the body. Clotting enables the body to avoid blood loss and, at the surface, to prevent infection. It is important that blood clots only when it should, because when a blood clot, a **thrombus**, blocks a vital blood vessel, it can cause a fatal heart attack or stroke.

Fig 30.3 Blood clotting is a cascade reaction: a complex set of chemical reactions that occur one after the other

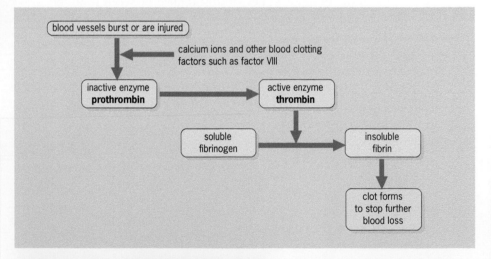

✔ **REMEMBER THIS**

The process of thrombosis causes blood to clot inside vessels. The clot itself – a thrombus – can be fatal if it prevents blood from reaching vital tissues such as heart muscle. A coronary thrombosis is a common cause of heart attack while a thrombosis in the brain can cause a stroke.

? **QUESTION 3**

3 The saliva of several species of leech contains an anticoagulant. What are anticoagulants and why does the leech need them? Could 'leech spit' have any medical uses?

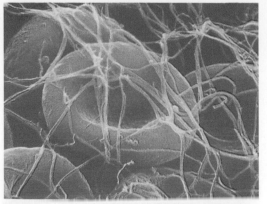

Fig 30.4 A scanning electron micrograph of a blood clot. You can see red blood cells trapped in a fine mesh of fibrin. Blood clotting is a complex process that is set off when blood vessel walls are damaged and finishes when soluble fibrinogen is converted into an insoluble fibrin mesh

Fig 30.3 outlines the major steps involved in the control of blood clotting. When blood vessels are injured or burst, a **cascade reaction** is initiated. The activation of one molecule leads to the activation of many more. In the final steps of the process, an inactive enzyme, **prothrombin**, is converted to active **thrombin**. This form of the enzyme converts soluble **fibrinogen** into insoluble **fibrin**. The fibrin fibres form a mesh that traps red blood cells and forms a clot (Fig 30.4).

If any of the factors in the cascade are missing, the blood cannot clot. This occurs in the condition **haemophilia**, a sex-linked genetic disease in which the sufferer (usually a male) cannot make factor VIII.

The genetics of haemophilia are discussed in Chapter 25.

CLASSIFICATION OF WHITE CELLS

Fig 30.5 is a blood smear showing some white cells – **leucocytes** – along with many red blood cells. The red cells outnumber the white by about 700 to 1. White cells are made in the bone marrow and are found throughout the body. They can move and are able to squeeze between cells, passing freely in and out of the circulation. Individual types of white cell are classified according to their appearance, origin or function (Table 30.1).

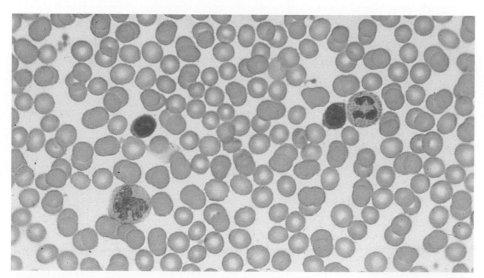

Fig 30.5 In this blood smear, the white cells stand out because their nuclei are stained. The numerous red cells, which have no nuclei, do not take up the purple stain

Table 30.1 Different types of white cell. Neutrophils, eosinophils and basophils have a granular cytoplasm and are therefore called granulocytes. The other types are called agranulocytes

Cell type	Diagram	How to recognise them	Relative abundance %	Function
neutrophils		lobed nucleus	57	phagocytosis
lymphocytes		large round nucleus; little cytoplasm	33	specific immunity; B cells make antibodies. T cells involved in cell-mediated immunity
monocytes		large, kidney-shaped nucleus	6	phagocytosis; monocytes develop into macrophages (general 'rubbish collecting' cells)
eosinophils		stain red with eosin	3.5	associated with allergy
basophils		stain with basic dye	0.5	release chemicals such as histamine that are responsible for inflammation

? QUESTION 4

4 Use Table 30.1 to identify the white cells shown in Fig 30.5.

✓ REMEMBER THIS

White cells spend only about 10 per cent of their time in the blood, and so cannot really be called white blood cells. Leucocyte is the general name for all white cells.

4 THE NON-SPECIFIC IMMUNE RESPONSE

When the body is damaged by cuts, scratches or burns, or is attacked by a pathogenic organism that manages to breach its defences, it produces a **non-specific immune response**. It is called a non-specific response because it occurs in response to tissue damage itself, not to the cause of the damage.

INFLAMMATION

Inflammation is a rapid reaction to tissue damage. Whether it is in response to a cut, insect bite or a heavy blow such as a sport injury, the classic signs of inflammation are always the same:

- **redness**: blood vessels dilate, increasing blood flow to the area;
- **heat**: also caused by the extra blood flow;
- **swelling**: the extra blood forces more tissue fluid into damaged tissues;
- **pain**: swollen tissues press on receptors and nerves. Also, chemicals produced by cells in the area stimulate the nerves.

Inflammation (Fig 30.6) is triggered by damaged cells. Ruptured cells and some white cells (**mast cells** and **basophils**) release 'alarm' chemicals such as **histamine**. These substances dilate blood vessels and the increased blood flow leads to the classic signs of inflammation. The 'alarm' chemicals also attract white cells that remove bacteria and debris by **phagocytosis**.

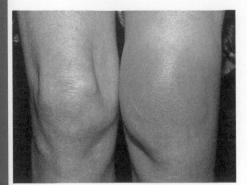

Fig 30.6 The familiar effects of inflammation: reddened skin, swollen tissues and tenderness

Why inflammation is useful

Inflammation prevents the spread of infection and speeds up the healing process. It also provides a way of telling the rest of the immune system what is going on. When microorganisms are phagocytosed, fragments of their cells (particularly molecules that were originally on their surface) are processed by the phagocytes. Some of these surface molecules, which we call **antigens**, allow the specific immune system to recognise and remember the type of microorganism that has tried to invade the body.

Antigens stimulate the specific immune system to produce cells and chemicals that bind specifically to that antigen, and to no others. Find out more about specific immunity in the next section of this chapter.

> Phagocytosis is discussed in Chapter 4.

Phagocytosis

Neutrophils are the commonest type of white cell. Together with monocytes, they are known as **phagocytes** because of their ability to 'eat' pathogens by phagocytosis (Fig 30.7). In this process, the white cell engulfs the pathogen, takes it into a vacuole inside its cytoplasm and then digests it with **lytic** enzymes.

Phagocytosis involves the following steps:
- The membrane of the neutrophil extends and surrounds the bacterium.
- The bacterium ends up inside the cell cytoplasm, contained within a vacuole.
- Lysosomes that contain digestive enzymes and free radicals fuse with the vacuole that has the bacterium inside.
- The bacterium is digested into fragments within the vacuole.
- Part of the vacuole buds off, taking some of the bacterial fragments with it.
- The fragments then attach to proteins of the major histocompatibility complex (MHC).
- When the vacuole fuses with the cell membrane, the bacterial fragments, bound to MHC proteins become part of the cell membrane and stick out from the surface of the neutrophil.

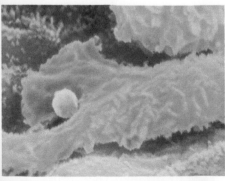

Fig 30.7 This phagocyte (a neutrophil) is engulfing a speck of dust. These white cells occur in the bloodstream and throughout the tissues. Each phagocyte (which is spherical when not active) can ingest between 5 and 25 bacteria before the toxic products of breakdown kill the cell. At sites of infection there may be many dead, liquefied white cells that, together with dead bacteria and other debris, form pus

Autoimmune disease: when the body attacks its own tissues

In people who have an autoimmune disease, the mechanism that enables the immune system to tell what is self and what is non-self breaks down. T and B cells (types of lymphocyte) begin to attack the body's own cells and tissues. This is the underlying cause of **multiple sclerosis**, **insulin-dependent diabetes**, **myasthenia gravis** and **rheumatoid arthritis**.

In multiple sclerosis, T cells attack the myelin sheath around nerves. This severely limits nerve function, resulting in loss of movement and sometimes in blindness.

In insulin-dependent diabetes, the body makes antibodies that destroy the β cells in the islets of Langerhans in the pancreas. This means that the pancreas becomes unable to produce insulin and the affected person can no longer control their blood glucose.

In myasthenia gravis, the body makes antibodies that attack the motor end plates, the specialised synapses that connect motor nerves to muscles. If the motor end plates become damaged, the muscles cannot contract. The first symptom of this disease, which affects one in

30 000 people (mainly female), is rapid tiring during exertion. The muscles become progressively more unresponsive and the affected person can have difficulty breathing.

People with rheumatoid arthritis suffer from swollen and deformed joints (Fig 30.8).

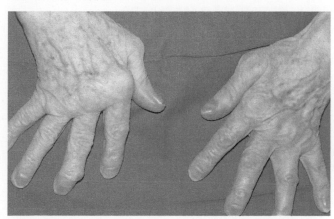

Fig 30.8 This person suffers from rheumatoid arthritis. The cartilage at the joints has been attacked by the immune system, causing swollen and painful joints

---

- The antigens of the bacterium are 'presented' on the outside of the neutrophil. They can now be 'seen' better by other cells of the immune system, enhancing the immune response against the bacteria.

The MHC comprises a group of genes in the human genome. It is important in the development of immunity. The proteins coded for by the genes are expressed on the outer surface of all body cells and are commonly called 'self antigens'. These allow our immune system to tell what is self and what is not. MHC proteins are attached to antigens from invading bacteria, viruses and parasites to stimulate the cell-mediated immune response (see below).

5 SPECIFIC IMMUNITY

The specific immune system protects the body from 'invasion' by microorganisms and parasites and also makes sure that the body's defences do not turn on its own tissues. The specific immune response is made up from two different systems that co-operate closely.

- **Humoral immunity**, also called **antibody-mediated immunity**, involves only chemicals: no cells are directly involved. The chemicals, called **antibodies**, attack bacteria and viruses before they get inside body cells. They also react with toxins and other soluble 'foreign' proteins. Antibodies are produced by white cells called **B lymphocytes**, or **B cells**.
- **Cell-mediated immunity**, as the name suggests, involves cells that attack 'foreign' organisms directly. Activated **T lymphocytes**, or **T cells**, kill some microorganisms, but they mostly attack infected body cells. The body uses cell-mediated immunity to deal with multicellular parasites, fungi, cancer cells and, rather unhelpfully, tissue transplants.

 REMEMBER THIS

An antigen is a molecule or part of a molecule, for example a protein, that is detected by the specific immune system as 'foreign': not part of the host's body. The immune system responds to an antigen by producing a very specific protein, an antibody, which reacts with that antigen, and that antigen only.

HUMORAL AND CELL-MEDIATED IMMUNITY

Fig 30.9 summarises the differences between humoral and cell-mediated immunity and Fig 30.10 shows how antigens from pathogens stimulate antibody production by B cells.

Fig 30.9 Lymphocytes are made in the bone marrow by stem cells. Two distinct types of lymphocyte develop, T lymphocytes and B lymphocytes (T cells and B cells). Both T and B cells migrate from the bone marrow to the lymph glands, but the T cells go via the thymus while the B cells go straight from the bone marrow

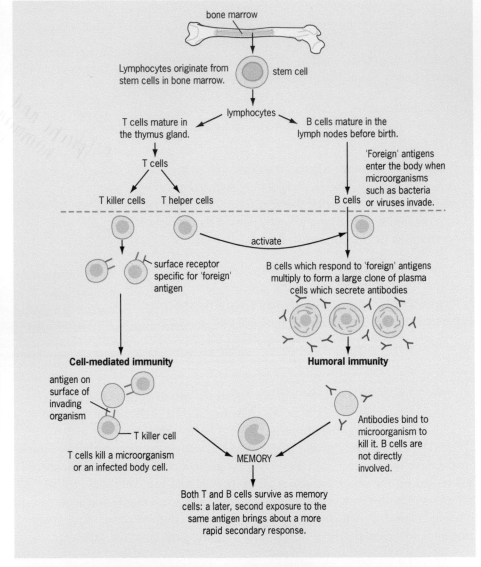

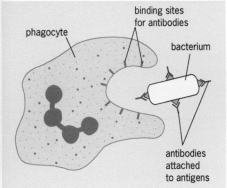

Fig 30.10 Pathogens such as bacteria are covered in antigens, which stimulate B cells to produce antibodies. Many antibodies coat the pathogen, labelling it as foreign to stimulate attack by phagocytes

B CELLS AND HUMORAL IMMUNITY

B cells are produced in the **bone marrow** and are distributed throughout the body in the **lymph nodes**. B cells respond to the 'foreign' antigens of a pathogen by producing specific antibodies. Antibodies are complex proteins that are released into the blood and carried to the site of infection. B cells do not fight pathogens directly.

An antibody, or **immunoglobulin**, is a Y-shaped protein molecule that is made by a B lymphocyte in response to a particular antigen. Fig 30.11 shows the overall structure of an antibody molecule. Antibodies interact with the antigen and render it harmless.

When a pathogen tries to invade the body for the first time, each of its antigens activates one B cell, which divides rapidly to produce a large population of cells. All the new cells are identical (we say they are **clones**) and they all secrete antibodies specific for the invading pathogen. When the infection is over, most of the newly made B cells die: their job is done. We describe this sequence of events as a **primary immune response**.

REMEMBER THIS

A clone is a set of genetically identical individuals. In immunology, a clone refers to a population of B cells, all of which can produce the same antibody because they are genetically identical.

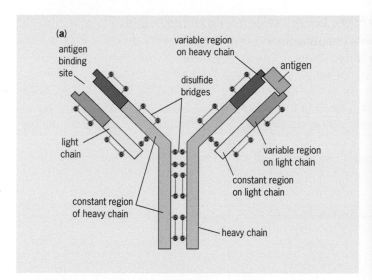

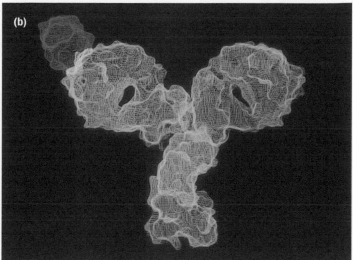

So that the body can respond more quickly next time, some of the activated B cells persist in the body for several years. These **memory cells** 'remember' what the pathogen is like and, if it tries to invade again, they divide rapidly to produce an even greater number of active B cells, all capable of secreting the specific antibody. This response is called a **secondary immune response** and is very much quicker and more effective than the primary response (Fig 30.12).

This ability of the immune system is central to **vaccination**. A vaccine stimulates the body to produce a primary immune response to a particular pathogen, without becoming infected by it. A subsequent booster produces a secondary response. Later, if the pathogen tries to invade, the body can mount a very fast response and the person does not become ill.

Fig 30.11a Antibodies are Y-shaped molecules consisting of four polypeptide chains, two heavy and two light. Much of the molecule is constant but the tips of the Y are variable and match precisely part of a particular antigen molecule

Fig 30.11b A computer-generated 3-D image of an antibody molecule (green) bound to an antigen (red)

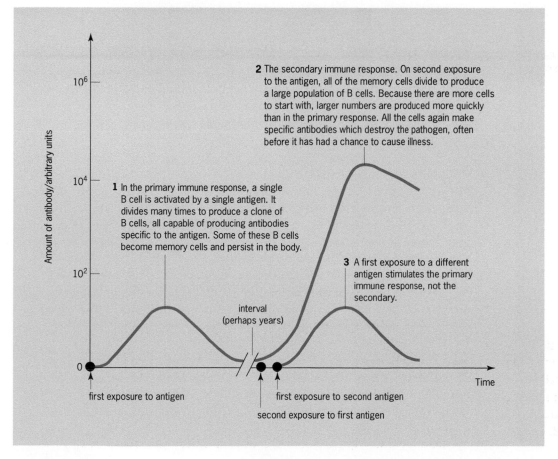

Fig 30.12 The immune system can remember antigens

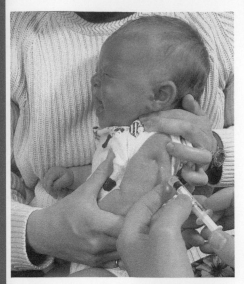

Fig 30.13 Vaccinations are an effective way of stimulating the body's own defences so that we need not suffer the infectious diseases, such as measles, mumps and whooping cough, that used to be a common feature of childhood

Vaccinations

Several infectious diseases overwhelm the normal primary immune response and so can be fatal on first exposure. Thankfully we are able to speed up the specific immune response by giving vaccines against the pathogens that cause them (Fig 30.13). The basic idea behind a vaccine is that it contains some form of the pathogen, so that it stimulates memory cells to develop, ready to destroy the real pathogen should it be encountered. Obviously, the vaccine can't simply be the pathogen itself, or the toxins it makes. Somehow, the vaccine must be made less **virulent** – less able to produce disease. Examples of this are shown in Table 30.2.

Table 30.2 Recommended vaccination schedule in the UK

Vaccination	Type of vaccine	Age due (UK vaccination schedule)
diphtheria	killed organism	2, 3 and 4 months, 3–5 years
tetanus	modified toxin	2, 3 and 4 months, 3–5 years
whooping cough	killed organism	2, 3 and 4 months, 3–5 years
polio	live, non-virulent	2, 3 and 4 months, 3–5 years
haemophilus influenzae type B (Hib) for meningitis	purified bacterial capsule	2, 3 and 4 months, 3–5 years
measles	live, non-virulent	12–18 months, 3–5 years
mumps	live, non-virulent	12–18 months, 3–5 years
rubella	live non-virulent	12–18 months, 3–5 years
CG (bacille Calmette–Guérin) for tuberculosis	live, non-virulent	10–14 years
hepatitis B	genetically engineered antigens	for people at risk, e.g. health professionals
meningococcal group C conjugate vaccine	purified bacterial components of *Neisseria meningitidis* group C	2, 3 and 4 months, but also for all young adults (under 25)

HOW MOTHERS GIVE IMMUNITY TO THEIR BABIES

When babies are born, they emerge from the protective environment of their mother's uterus. They are exposed to many potential pathogens. Healthy babies have a fully functional immune system and can mount primary immune responses to many different antigens straight away. However, babies also get a bit of extra help from their mothers.

Antibodies pass across the placenta and, after birth, the supply continues through breast milk. Babies have very porous intestines that can absorb these large proteins directly into the bloodstream without digesting them. These large pores close by the age of one year. We call this kind of immunity – passed from one person to another – **passive immunity**. It does not last long, because the antibodies are broken down within a few days, but it can help a baby to fight off common pathogens.

Passive immunity is also used to treat some types of poisoning, such as snakebites. **Antiserum**, blood that contains antibodies specific to a particular snake venom, is produced in horses, purified and then given to people who have been bitten by a snake.

ALLERGIES

An **allergy** or **hypersensitivity** is a reaction to the presence of a normally harmless substance called an **allergen**. Some common allergens are pollen, house dust mites, animal fur and feathers, fungal spores, insect bites and penicillin.

? QUESTION 5

5   What do you think it means when products such as make-up are labelled hypo-allergenic?

The commonest symptoms of an allergy are sore eyes, runny nose, sneezing and asthma. Many of these symptoms result from an inflammation of the mucous membranes, caused by **mast cells**, which release chemicals such as histamine. Many anti-allergy treatments suppress mast cells or neutralise histamine – chemists sell many **antihistamines** in the pollen season. For more information about asthma, see the Assignment for Chapter 9 on the CAS website.

T CELLS AND CELL-MEDIATED IMMUNITY

Like B cells, T cells respond to specific antigens. When a pathogen first infects the body, each individual antigen stimulates a single T cell. This divides to form a **clone**, in the same way that B cells do. Some of the **activated T cells** become **memory cells** and persist in the body, ready to mount a secondary response if the pathogen attacks again. The others, however, do not produce antibodies. They develop further to become one of three types of T cell:

- **T helper cells** are so called because they help with, or rather control, the rest of the specific immune response. They cause B cells to divide and then to produce antibodies, they activate the two other sorts of T cell (see below), and they activate macrophages, so the macrophages ready to phagocytose pathogens and debris.
- **T killer cells** attack infected body cells and the cells of some larger pathogens (e.g. parasites) directly. The two cells face each other, membrane-to-membrane, and the killer T cell punches holes in its opponent. The infected cell or parasite loses cytoplasm and dies.
- **T suppressor cells** are a sort of safety cut-out mechanism. When the immune response becomes excessive, or when the infection has been dealt with successfully, these T cells damp down the immune response. Obviously, this is a good idea: if the body continued to make antibody and stimulate more and more T and B cells to divide, even when there was no need, this could damage the body and would be, at best, a waste of resources.

STRETCH AND CHALLENGE

Presenting antigens

As we saw on pages 474–475, bacteria that are engulfed by phagocytic white cells are fragmented and their antigens are expressed on the surface of the infected cell in combination with proteins from the MHC. This is a vital step in setting up a cell-mediated immune response. T helper cells are activated when they come into contact with an antigen-presenting cell and they are then specific for the antigen that has been presented to them.

Activated and specific T helper cells then produce chemicals called cytokines that circulate in the blood, activating B cells and also T killer cells. The cytokines also stimulate the specific B cells and T killer cells to divide, forming large clones of active immune cells. The T killer cells are specific for the presented antigens and so recognise and kill only body cells that are expressing these antigens in combination with MHC proteins. This ensures that only infected cells are destroyed. The T killer cells attach themselves to cells expressing MHC-linked antigens and then release chemicals that cause the infected cell to lyse (split open).

---

6 BLOOD TRANSFUSIONS AND ORGAN TRANSPLANTS

In the past 100 years, advances in modern medicine have led to the technology and knowledge that allow doctors to give one person's cells, tissues and organs to another person. One of the main objectives in developing these life-saving treatments has been to overcome the natural reaction of the immune system to destroy transplanted cells and tissue, which it 'sees' as non-self.

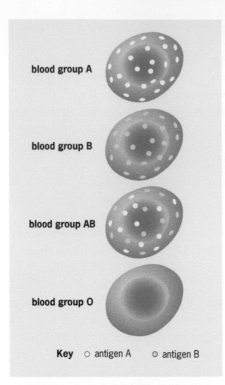

Fig 30.14 People can be placed in one of four groups according to which red blood cell antigens they have

blood group A

blood group B

blood group AB

blood group O

Key   ○ antigen A   ○ antigen B

BLOOD GROUPING AND TRANSFUSIONS

Towards the end of the nineteenth century, medical scientists realised that accidents were often fatal simply due to blood loss. Losing large volumes of blood sent victims into shock and they died, even though their injuries were otherwise not too severe. Many women, in particular, often died when they lost blood when giving birth. Different doctors tried transfusing blood from a healthy person into the injured person, to try to restore their blood volume.

Sometimes this worked and sometimes it didn't. When it failed, the results were disastrous. We now know that if the **blood groups** (Fig 30.14) of the **donor** (the person giving the blood) and the **recipient** (the person receiving it) are different, the recipient's immune system may react against the donated blood, producing a massive and deadly immune response.

In the early 1900s, an Austrian scientist, Karl Landsteiner, discovered that red blood cells from different people had different sets of antigens on their surface. The entire human population can be placed into one of four main blood groups: **A**, **B**, **AB** and **O**, according to the antigens on their red blood cells.

From Fig 30.14, you can see that people with blood group O have no antigens on the surface of their red blood cells. Blood from these people cannot cause an adverse immune response if it is transfused into people from the other three blood groups.

We say that people with blood group O are **universal donors**. By the same reasoning, people with blood group AB cannot have any antibodies to either of the blood group antigens and so cannot mount an immune response if they are given blood by people in any of the other three groups. We say that people with AB blood are **universal recipients**.

Blood transfusions are dangerous when the recipient has antibodies in their blood that react with antigens present on the surface of red cells in the donor blood (Fig 30.15). So, for example, a person of blood group A cannot donate blood to someone with group B. The recipient in this case has anti-A antibodies which react with the A antigens on the red blood cells of the donor blood. Table 30.3 gives a full list of possible transfusions.

REMEMBER THIS

For examination purposes, a quick way to remember safe blood transfusions is as follows: O is the universal donor, so can be given to anyone. AB is the universal recipient, so can receive from anyone. Apart from that, only like-to-like transfusions (e.g. A to A) are safe. All others cause problems.

? QUESTION 6

6 Why do you think some of the early transfusions worked, even though they were done before we knew about the importance of blood groups?

Fig 30.15 We can find out what blood group someone has by mixing a sample of their blood with anti-A and anti-B antibodies. Blood from people with group A antigens agglutinates (forms clumps) with anti-A, blood from people with group B antigens agglutinates with anti-B. AB blood agglutinates with both, O with neither. In the diagram, the agglutination reaction produces a granular appearance as red cells clump together. In the smooth, dark samples, agglutination has not taken place

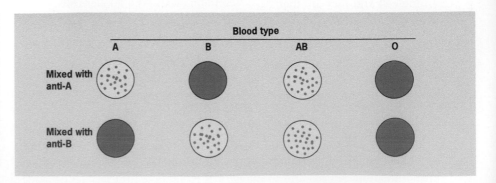

Table 30.3 Which transfusions are possible?

Blood group	Antigens present on red blood cells	Antibodies present in blood	Can donate blood to	Can receive blood from
A	A	anti-B	A, AB	A, O
B	B	anti-A	B, AB	B, O
AB	AB	none	AB	A, B, AB, O
O	none	anti-A, anti-B	A, B, AB, O	O

? QUESTION 7

7 Look at Table 30.3. Why can someone with blood group AB donate blood only to someone with the same blood group?

TRANSPLANTS AND GRAFTS

Transplantation involves taking cells, tissues or organs from one individual and placing them into another individual. Since the first organ transplant operations in the 1950s, patients have received hearts, lungs, skin, corneas, kidneys and various other organs. The success rate of these operations has improved steadily.

The biggest problem facing most transplant recipients is that of **rejection**. The cells of transplanted tissue are covered in antigens that stimulate the specific immune response of the recipient, notably the cell-mediated response brought about by T cells. Symptoms of rejection include the degeneration of blood vessels in the transplanted organ and the destruction of whole cells followed by their replacement with 'scar' tissue. A patient whose transplant is rejected becomes seriously ill: they lose the function of the organ concerned and the massive immune response puts an incredible amount of stress on their already weakened body.

Rejection can be minimised by tissue typing or by using drugs to suppress the immune system.

? QUESTION 8

8 What is the difference between agglutination and blood clotting?

Tissue typing

Like red blood cells, other body cells also have surface antigens that are different in different people. Body cells, such as cells in the kidney or liver, have many antigens that differ from person to person. This is why matching the tissue type of the donor and recipient is much more complex than matching blood groups. In practice, the cells from two people are never identical, so the aim is to match people who have as few differences as possible.

Immunosuppression

If the recipient of a transplant has a fully functional immune system, even a closely matched organ will be rejected to some extent. To counter this, transplant patients are given immunosuppressive drugs. The most effective is currently **cyclosporin**. This chemical, isolated from a fungus, inhibits the action of T cells and so has a profound effect on cell-mediated immunity. Since the introduction of cyclosporin in the early 1980s, the success rates of some transplants have risen to over 90 per cent.

The problem with immunosuppressive drugs is, of course, that they reduce the body's ability to fight off infection. Transplant patients are more prone to infection from normally harmless microorganisms, particularly viruses. Although the dose of immunosuppressive drugs can be reduced after a few months if there is little sign of rejection, people who have had transplants must continue to take them for life.

? QUESTION 9

9 Suggest why, although penicillin is effective against a wide range of bacterial diseases, it is no use for treating diseases caused by fungi.

7 ANTIBIOTICS AND THE TREATMENT OF INFECTION

Throughout human history, millions of people have died from bacterial infections. Today, we still suffer from such infections but we now have antibiotics, safe and effective drugs that can be used against the bacteria that cause disease.

The first antibiotics developed in the 1930s and 1940s were chemicals that one microorganism produced to kill another. For example, penicillin, the first antibiotic to be produced in large enough quantities to be use in combating bacterial infections, is produced by a type of fungus. Since then, antibiotics have been extracted from a range of unusual sources including snowdrop bulbs and toad skin; yet others can be synthesised.

HOW SCIENCE WORKS

Producing penicillin on a large scale

Penicillin is produced commercially using a strain of fungus called *Penicillium chrysogenum*. This is an aerobic organism that needs a high level of oxygen to grow in culture. For this reason, it is grown commercially in fermenters with a relatively small volume – 40–200 dm^3. These are easier to aerate effectively than much larger fermentation tanks. (Commercial fermentation is discussed in Chapter 6.)

The commercial production of penicillin is shown in Fig 30.16. The fungus produces most penicillin at temperatures between 25 and 27 °C. During the first 40 hours or so, the fungus is growing in the culture and increasing in biomass. Once it has reached a certain concentration in the culture, it starts to produce large amounts of penicillin, but this continues for only a short time before the fungus starts to run out of nutrients and begins to die. To get the most out of each fermentation, new nutrients are added after 40 hours to stimulate a further increase in biomass. About 20–40 per cent of the culture is removed, to be processed so the penicillin can be purified, and new culture medium replaces it. This process, called **batch fill and draw**, can be carried out up to ten times before the fermentation tank needs to be cleaned out completely.

Fig 30.16 Commercial production of penicillin. The process involves several stages; scaling up of the original fungal culture, fermentation, batch fill and draw, filtration, and then three steps of downstream processing. The final product is 99.5 per cent pure

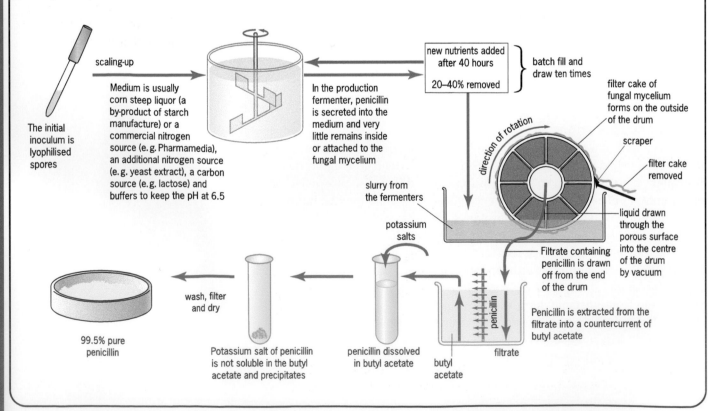

scaling-up

The initial inoculum is lyophilised spores

Medium is usually corn steep liquor (a by-product of starch manufacture) or a commercial nitrogen source (e.g. Pharmamedia), an additional nitrogen source (e.g. yeast extract), a carbon source (e.g. lactose) and buffers to keep the pH at 6.5

In the production fermenter, penicillin is secreted into the medium and very little remains inside or attached to the fungal mycelium

new nutrients added after 40 hours

20–40% removed

batch fill and draw ten times

direction of rotation

filter cake of fungal mycelium forms on the outside of the drum

scraper

filter cake removed

slurry from the fermenters

liquid drawn through the porous surface into the centre of the drum by vacuum

potassium salts

Filtrate containing penicillin is drawn off from the end of the drum

99.5% pure penicillin

wash, filter and dry

Potassium salt of penicillin is not soluble in the butyl acetate and precipitates

penicillin dissolved in butyl acetate

butyl acetate

penicillin

filtrate

Penicillin is extracted from the filtrate into a countercurrent of butyl acetate

ANTIBIOTICS AND ANTIBIOTIC RESISTANCE

Before the 1940s, having a serious bacterial infection often meant death. Antibiotics such as penicillin seemed to be wonder drugs at first, bringing people back from the brink of death. Today we have thousands of different antibiotics that work in hundreds of different ways. The mechanism, of action of four common ones is summarised in Fig 30.17.

Antibiotics are, however, no longer seen as wonder drugs. Since the 1940s, resistant strains of bacteria have arisen that can cause infections that are very difficult to treat, as they no longer respond to an antibiotic. One of the best-known antibiotic-resistant bacteria is MRSA – methicillin-resistant *Staphylococcus aureus*.

But how did this happen? Bacteria, like all living organisms, vary. In any population of bacteria, the majority will be killed by an antibiotic. But it only takes one bacterium to have a gene mutation that enables it to survive the effect of an antibiotic to create billions of descendant bacteria – all resistant to the antibiotic. Bacteria also swap antibiotic resistance genes by passing plasmids containing the genes from one bacterium to another – so the resistance spreads even more quickly.

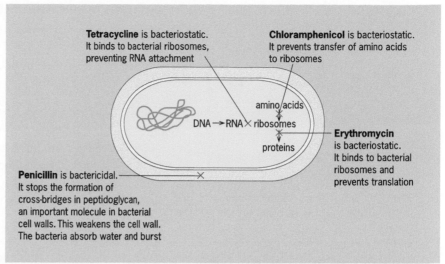

Tetracycline is bacteriostatic. It binds to bacterial ribosomes, preventing RNA attachment

Chloramphenicol is bacteriostatic. It prevents transfer of amino acids to ribosomes

amino acids

DNA → RNA × ribosomes

proteins

Erythromycin is bacteriostatic. It binds to bacterial ribosomes and prevents translation

Penicillin is bactericidal. It stops the formation of cross-bridges in peptidoglycan, an important molecule in bacterial cell walls. This weakens the cell wall. The bacteria absorb water and burst

Fig 30.17 Some antibiotics kill the microorganisms they target. These are said to be bactericidal. Bacteriostatic antibiotics do not kill microorganisms; they prevent them multiplying and give the immune system a better chance to overcome the infection

What are the types of antibiotic resistance?

A bacterium becomes resistant to an antibiotic when one of its alleles becomes altered. There are four main mechanisms by which this can lead to resistance:

- The altered allele occurs in the gene that codes for the protein that is a target for the antibiotic. If the altered protein is a different shape, it can no longer bind to the antibiotic. If there is no binding, the antibiotic can have no effect on any biochemical pathways inside the bacterium.
- Some antibiotic-resistant bacteria stop the antibiotic reaching its target molecule inside the cell either by preventing the antibiotic from entering the cell or by pumping it out faster than it can get in. This can happen if the mutation occurs in an allele that codes for a membrane receptor or transport protein.
- If the target of an antibiotic is an enzyme, bacteria that produce an alternative version of the enzyme can bypass the antibiotic. This enzyme still carries out the same function as the original in the cell, but the antibiotic does not affect it. Having both forms of the same enzyme gives the bacteria a great advantage – they can survive equally well whether or not the antibiotic is present.
- A mutation might enable the bacterium to produce an altered enzyme that is then able to react with the antibiotic and disable it.

These types of mutation are happening all the time in bacteria – normally, they just arise and are lost. However, when antibiotics are present in large amounts for a long time, this exerts a selection pressure that results in lots of antibiotic resistant bacteria.

How much of a problem is antibiotic resistance?

Globally, antibiotic resistance is now a major public health problem. Over-prescribing of antibiotics by doctors is thought to be a major cause. Many still

? QUESTION 10

10 A man gets a sore throat so he goes to the doctor who prescribes antibiotics. The man takes half the course and feels better, so he stores the rest for next time, thinking he can save himself a trip. What is the problem with this behaviour?

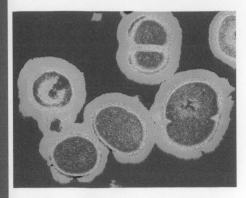

Fig 30.18 Methicillin-resistant *Staphylococcus aureus* (MRSA) isolated from a knee abscess

give antibiotics, particularly to children, for sore throats and earache, which are mostly caused by viruses. Another factor is that people don't finish courses of antibiotics – they stop taking them as soon as they feel better. This allows bacteria even with partial resistance to re-establish an infection. Antibiotics used in farming that get into the human food chain also don't help.

Hospitals now face serious problems as MRSA and other antibiotic-resistant strains of bacteria can cause terrible infections in elderly patients and in those in intensive care, who are already weak. The UK has one of the highest rates of hospital-acquired infection in Europe: MRSA costs the NHS £1 billion per year and, together, MRSA and antibiotic resistant *Clostridium* infections killed 16 500 people between 2004 and 2007.

SUMMARY

After studying this chapter, you should know and understand the following:

- We are surrounded by potential **pathogens** (disease-causing organisms) such as bacteria and viruses which can cause disease if allowed to enter and multiply inside the body.
- The body's defence system consists of barriers to keep microorganisms out, and mechanisms to detect and destroy those that do enter. Together, these mechanisms are the **immune response**.
- The immune system is capable of non-specific responses and specific responses. Non-specific mechanisms (**inflammation** and **phagocytosis**) occur in response to any invading microorganism. Specific mechanisms allow the body to recognise and fight individual types of microorganism. There are two specific responses: **cell-mediated immunity** and **humoral immunity**.
- **Lymphocytes** (types of white cells) are responsible for the specific immune response. **T cells** mature in the thymus gland and **B cells** come directly from bone marrow. Both are able to recognise foreign antigens that come from pathogens.
- B cells are responsible for humoral immunity: they secrete specific proteins called **antibodies**.
- T cells are responsible for cell-mediated immunity and help to control the overall specific response. **T killer cells** attack pathogens or infected cells.

T helper cells activate B cells, telling them to secrete antibodies. **T suppressor cells** damp down the immune response when the infection is over.

- **Autoimmune diseases** occur when the body's immune system attacks its own tissues. Examples include rheumatoid arthritis, multiple sclerosis and myasthenia gravis. **Allergies** occur when the body 'over-reacts' to a normally harmless substance such as pollen.
- When cells or tissues are taken from one individual and given to another, they are often recognised as 'foreign' and destroyed. This process is often called **graft rejection**. To minimise the chance of rejection in blood transfusions and organ transplants, it is important to match blood and tissue types.
- **Antibiotics** are important drugs for treating infections caused by bacteria. In the time since antibiotics were first developed and used in the 1940s, bacteria have evolved to become resistant to them. Antibiotic-resistant strains of many bacteria are becoming common and represent a problem for doctors and researchers.

 Practice questions and a How Science Works assignment for this chapter are available at www.collinseducation.co.uk/CAS

31

Lifestyle diseases

31 LIFESTYLE DISEASES

Mad cow disease – was it all hype?

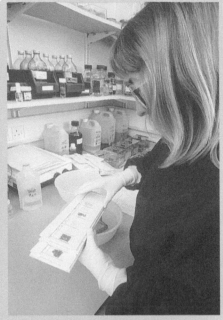

In 1996, a handful of deaths were linked with a type of dementia called Creutzfeldt–Jacob disease (CJD). CJD had been known for a long time, but, with an average age of 27, these victims were much younger than the people who usually developed CJD. All the signs pointed to a new variant form, now called variant CJD (vCJD).

When the UK government announced there was a chance that vCJD had been passed to people in the beef they ate, there was an incredible furore. Was everyone who ate beef regularly at risk of vCJD? The resulting debate demonstrates how difficult it is to pinpoint a lifestyle risk factor for a disease that takes several years to develop.

Bovine spongiform encephalopathy (BSE) was first identified in cattle in 1986, and scientists concluded that it had developed from a similar disease in sheep, called scrapie. Why the scrapie agent jumped species isn't really understood. But the practice of recycling slaughtered sheep and, later, infected cows, to make protein supplements for cows, helped to establish a massive infection of BSE in British cattle. Many were culled to stem the epidemic and to lessen the possibility of infected meat entering the human food chain.

By the end of 1996, scientists had analysed the chemistry of the agent that causes vCJD and showed that it is closely related to BSE in cattle. Research since then has demonstrated that some of the affected people definitely contracted vCJD from eating beef. New regulations have been introduced to prevent such transmission happening again, but the long incubation times of CJD make it impossible to be sure that more people are not infected.

Fortunately, data collected in the 10 years after 1996 showed that the vCJD outbreak in the UK declined after 2000. The peak number of deaths was 28 in this year, followed by 20 in 2001, 17 in 2002, 18 in 2003, 9 in 2004, 5 in 2005, and none in 2006. The mass epidemic that was feared never actually happened.

A technician preparing microscope slides of human brain slices to be checked for vCJD. The disease is caused by a prion – this is a self-replicating molecule that is a protein rather than a nucleic acid

REMEMBER THIS

Lifestyle diseases are diseases that are influenced by our lifestyle.

Degenerative diseases are diseases that affect us as we age: sometimes there is nothing we can do to prevent them. Other degenerative diseases are caused, or at least made worse by our lifestyle.

Genetic predisposition – some lifestyle diseases and some degenerative diseases are more likely in people with specific alleles. Against this genetic background, lifestyle still plays a part.

1 WHAT ARE LIFESTYLE DISEASES?

In this chapter, we look at a group of diseases influenced by the sort of life we lead. Such **non-transmissible diseases** or **lifestyle diseases** are the major cause of death in the more developed countries (Fig 31.1).

Cardiovascular disease, which affects the heart and circulatory system, many types of **cancer** and some diseases of the respiratory system can all affect people as a result of things they eat, drink or smoke, and the way they live. Getting enough exercise at all times of life, for example, can help to reduce the risk of developing cardiovascular disease. However, the interaction between all these factors is complex, which is why it is difficult to predict who will develop lifestyle diseases.

In this chapter, our study of lifestyle diseases includes:

- **Lifestyle aspects** that are linked with disease: smoking, alcohol, poor diet and lack of exercise. These factors affect a person's chance of developing a particular condition, so they are often called **risk factors**.
- **Cardiovascular diseases**. In their most severe form, heart attacks and strokes can be fatal. Less severe forms of cardiovascular disease don't kill but they disable people to such an extent that their **quality of life** is very poor.
- **Cancers**. A cancer is a mass of cells that multiply in an uncontrolled way, invading surrounding healthy tissue. Often cancers form tumours, or swellings.

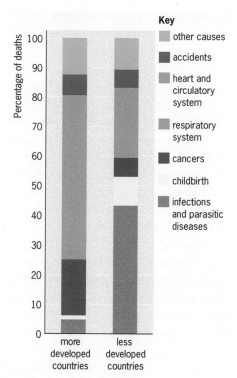

Key
- other causes
- accidents
- heart and circulatory system
- respiratory system
- cancers
- childbirth
- infections and parasitic diseases

Fig 31.1 This bar chart shows the percentage of deaths from a number of different causes in more developed and less developed countries

QUESTION 1

1 Suggest why a higher percentage of people die from cancers in more developed countries.

RISK FACTORS FOR LIFESTYLE DISEASES

Some bacteria are responsible for specific forms of heart disease and some viruses are associated with particular types of cancer, but the main causes of non-transmissible disease are features of our lifestyle. A risk factor is an aspect of our lifestyle that increases our chances of developing a disease. Table 31.1 shows some of the most important ones.

Table 31.1 Some examples of risk factors

Factor	Level of risk
smoking	There is good evidence that cigarette smoking is directly responsible for about 90 per cent of all cases of lung cancer. Smoking also makes us more likely to succumb to diseases such as bronchitis and emphysema, and it increases the risk of developing cancers in other parts of the body.
drinking alcohol	Regular drinking of small amounts of alcohol could have a beneficial effect on the circulatory system. However, there is plenty of evidence that shows that regular heavy drinking is linked to an increased risk of developing cancer of the mouth, throat, oesophagus and liver. Between 2 and 4 per cent of all cancers are thought to be due directly to alcohol consumption.
exposure to ultraviolet light	Sunbathing is dangerous! Over the past 25 years, the incidence of skin cancers has doubled. Most of the increase is thought to be due to greater exposure to ultraviolet rays in sunlight (Fig 31.2). Pale-skinned people can burn easily and a handful of bad episodes of sunburn in childhood can double your chances of getting skin cancer later in life. Also, experts now believe that the protective ozone layer in the upper atmosphere, which acts as a barrier to the dangerous UV rays in sunlight, has been seriously eroded by pollution.
what we eat	The links between diet and heart disease and between diet and cancer are more complex. Some foods are associated with greater risk of developing some diseases. A high-fat diet, for example, seems to increase risk of developing cardiovascular disease, but other environmental, lifestyle and genetic factors are also important, and teasing out which factors have the greatest effect is very difficult.

Fig 31.2 The 'pale and interesting' look went out of favour in the early part of the twentieth century and tanned skin was considered healthy. However, in the last 25 years of the twentieth century, cases of skin cancer rose dramatically and the tanned body became less fashionable. Beach holidays are still popular as we move through the first decade of the twenty-first century, but there are signs that this sort of over-tanned look is losing favour. Perhaps 'pale and interesting' is on its way back?

QUESTION 2

2 Suggest two reasons that could explain why skin cancer is more common today than it was in 1950.

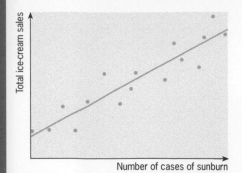

Fig 31.3 Ice-cream sales plotted against number of cases of sunburn shows a correlation. But does it prove that ice-cream causes sunburn?

? QUESTION 3

3 It has been well proven that cigarette smoking causes lung cancer. People who smoke know this. What factors contribute to carrying on this dangerous habit?

The vascular system is discussed in Chapter 10.

Investigating risk factors

Proving that a risk factor causes a specific disease is not easy. Sometimes two things seem to be linked or **correlated**, but this does not mean that one causes the other. Fig 31.3 shows what happens if we use data from a British seaside town in summer and plot the number of cases of sunburn against ice-cream sales.

The number of cases of sunburn seems to increase with the number of ice-creams sold: there is a clear positive correlation between the two sets of data. However, we all know that eating ice-cream does not cause sunburn. A third factor, the hot weather, is involved. When it's hot, people are more likely to lie in the sun and get sunburnt, and they are also more likely to buy ice-cream.

Medical researchers try to identify risk factors so that people have reliable information to know what to avoid (whether they choose to or not is another matter). Pinning down risk factors to particular diseases is very difficult because we are continually altering our lifestyles. We might eat a good diet and take regular exercise in our teens and then fall into bad habits in our 20s. Our exposure to environmental factors changes, often beyond our control. New preservatives could be added to food we eat, or there can be changes in the levels of pollution that contaminate the air we breathe. Any of these changes can alter our chances of developing a particular disease.

2 ATHEROSCLEROSIS AND CARDIOVASCULAR DISEASE

Cardiovascular disease is a major lifestyle disease in most developed countries. It claims about 175 000 lives per year in the UK and almost a million lives per year in the USA. As its name suggests it involves the **cardiovascular system** – the blood vessels that transport blood throughout the body.

Cardiovascular disease is more common in people over 50, but it can happen in younger people. It shows itself in many ways but it has a common underlying physiological cause: the narrowing and clogging up of arteries as a result of **atherosclerosis**. We look at this process in detail later in the chapter, but first we look at the different forms of cardiovascular disease.

TYPES OF CARDIOVASCULAR DISEASE

Cardiovascular disease is a general term. The symptoms and effects of different forms depend on which blood vessel is affected. The severity of the condition depends on the extent to which the vessel is blocked. Generally, partial blockages in small arteries or veins are less life-threatening than complete blockage of a major vessel. A summary of different forms of cardiovascular disease is given below and illustrated in Fig 31.8 on page 491.

Heart attack

A heart attack, or **myocardial infarct**, happens when the blood supply to the heart muscle is interrupted. The cause of the interruption is usually a blockage of one of the coronary arteries. This can be due to a blood clot that has travelled to the heart from elsewhere in the body, or it can be due to atherosclerosis. When blood cannot get oxygen to the heart muscle, the affected part of the heart dies. A heart attack causes severe pain in the chest and left arm. Mild heart attacks are dangerous but do not kill; careful treatment using drugs to 'thin' the blood to prevent another blockage, followed by a change in lifestyle can help to prevent another attack.

If the heart attack is severe, and a large part of the heart muscle is affected, the heart can experience **ventricular fibrillation**: instead of co-ordinated beating, the individual muscle fibres contract in an unsynchronised way. The affected part, usually the left ventricle, sometimes becomes unable to pump blood and the heart stops. People who have had a severe heart attack can be **resuscitated** using electric shocks delivered to the chest (Fig 31.4); this restarts the heart and enables it to re-establish a normal beating rhythm.

Fig 31.4 Paramedics resuscitating a heart-attack victim. When someone has a major heart attack, prompt emergency treatment is necessary, first to save their life and, second, to give them a good chance of recovery.

Stroke

A stroke, or cerebral haemorrhage, results if the supply of blood to the brain is interrupted. The symptoms vary according to the size and location of the area affected. There might be no more than a tingling sensation or weakness down one side of the body, but at its most severe, a stroke can lead to permanent paralysis (Fig 31.5) or death.

Other conditions

A **thrombosis** is the blockage of an artery caused by a blood clot. It can occur anywhere in the body.

An **embolism** occurs when a mobile blood clot (called an embolus, hence embolism) moves through the circulatory system and then lodges in a vein and blocks it. If a pulmonary vein (from lungs to heart) becomes blocked, this causes a pulmonary embolism that can be fatal.

Angina pectoris is a severe pain in the chest that occurs when the coronary arteries are unable to supply sufficient oxygen to the heart muscles.

Thrombophlebitis occurs when a vein is damaged and develops a blood clot that blocks it. Blood clots in the leg veins can occur in patients who are forced to remain in bed for long periods.

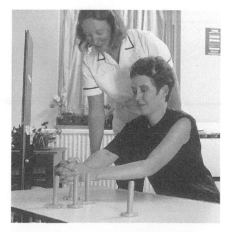

Fig 31.5 One-sided paralysis can result from a stroke. Here a 29-year-old woman, who had a stroke in the right side of her brain, is being encouraged to pick up and move each peg by gripping it between both hands. This is to help her regain some use of her left arm.

THE LINK BETWEEN ATHEROSCLEROSIS AND CARDIOVASCULAR DISEASE

To understand how and why cardiovascular disease develops, we need to look at three things:

- How does an atheroma form?
- How does atherosclerosis lead to cardiovascular disease?
- What are the risk factors associated with atherosclerosis and how do they increase the chance of atheroma formation?

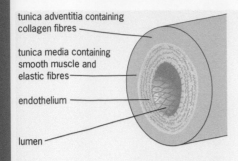

tunica adventitia containing collagen fibres

tunica media containing smooth muscle and elastic fibres

endothelium

lumen

Fig 31.6 Cross-section through a large artery. The space through which the blood flows is called the **lumen**. Surrounding this is a layer of smooth flat cells called the **endothelium**. Outside this lining is the **middle layer** or **tunica media**, which contains a high proportion of smooth muscle and elastic fibres. Finally, there is an **outer layer** or **tunica adventitia** containing fibres of the structural protein collagen

THE FORMATION OF AN ATHEROMA

Atheromas, areas or **plaques** of fatty material, develop on the inside of arterial walls. Remind yourself of the structure of an artery wall by looking at Fig 31.6. As plaques build up, they can affect the circulatory system by:

- blocking the flow of blood through the artery;
- initiating the formation of blood clots. Blood clots that form inside coronary arteries are particularly dangerous. They can break free and lodge in the smaller coronary arteries, causing some of the heart muscle to die. This is what initiates a heart attack. If blood clots interfere with the blood supply to the brain, the result is a stroke.

Heart failure

The term heart failure suggests a condition in which the heart suddenly comes to a complete stop. This doesn't happen – heart failure usually develops slowly, often over years, as the heart gradually loses its pumping ability and works less efficiently. The underlying cause is usually coronary artery disease.

How serious the condition is depends on how much pumping capacity the heart has lost. Nearly everyone loses some pumping capacity as they get older, but the loss is significantly more in heart failure.

There are two main types of heart failure:

- Systolic heart failure occurs when the heart's ability to contract decreases. The heart cannot pump with the force required to push enough blood out through the aorta. Blood coming into the heart from the lungs can back up and cause fluid to leak into the lungs, a condition known as **pulmonary congestion**.

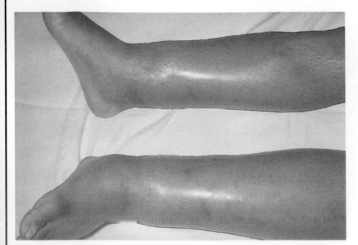

Fig 31.7 This person has a bad case of oedema

- Diastolic heart failure occurs when the heart has a problem relaxing. The heart can't fill with blood because the muscles of the atria and ventricles have become stiff. People with this form of heart failure accumulate fluid, especially in the legs, ankles and feet. Their condition is called **oedema** (Fig 31.7).

Tiredness, breathlessness and excess fluid retention are common symptoms but because heart failure usually develops slowly, these obvious signs often don't appear until the damage to the heart is serious. This is partly because the heart adjusts to cope with the effects of heart failure by getting bigger, developing thicker muscle and by contracting more often.

Heart failure is treated by a combination of lifestyle changes and drug treatment. Patients are advised to stop smoking, lose weight if necessary, avoid drinking alcohol, and change their diet to reduce the amount of salt and fat they eat. Regular, moderate exercise is also helpful, but this needs to be done only with a doctor's approval.

However, heart failure is always a serious condition because it poses an increased risk of sudden death, through cardiac arrest. Drugs are usually used to control the symptoms and to try to prevent them getting worse:

- Diuretics help reduce the amount of fluid in the body and are useful for patients with fluid retention and high blood pressure.
- The drug digitalis increases the force of the heart's contractions, helping to improve circulation.
- Recent studies have shown that drugs known as angiotensin-converting enzyme (ACE) inhibitors can be useful. ACE inhibitors improve survival among heart failure patients and can slow, or perhaps even prevent, the loss of heart-pumping activity.

Arterial blood supplies all the body's organs with nutrients, including triglycerides and cholesterol. These are lipids and have several important functions in the body: they form cell membranes and they act as energy stores. Lipids combine with proteins to form **lipoproteins** before they can be transported in the blood.

Lipoproteins that contain cholesterol are engulfed by white cells called **phagocytes** that occur in the artery wall, just under the endothelium. If the blood contains large amounts of cholesterol-rich lipoproteins, large numbers of stuffed phagocytes accumulate. Their rich 'filling' gives them a foamy appearance and they are, in fact, often referred to as **foam cells**. When patches or streaks of foam cells start to appear on the inside of artery walls, this is recognised as the first sign of atherosclerosis. At this stage, there is little effect on the flow of blood through the artery and no discernible symptoms.

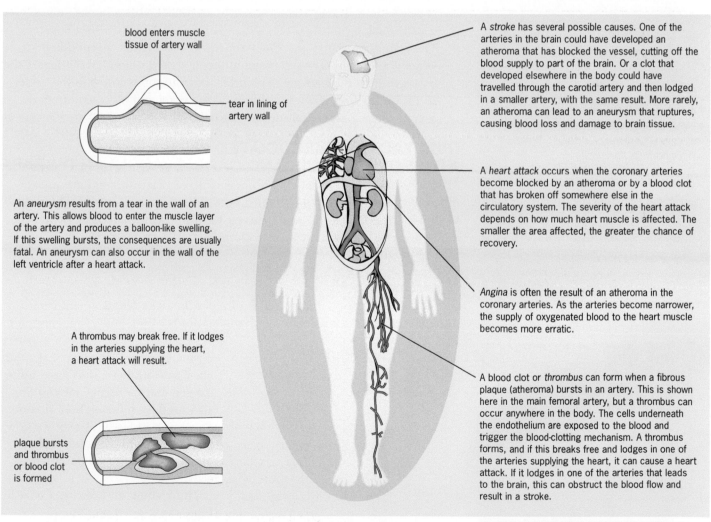

blood enters muscle tissue of artery wall

tear in lining of artery wall

An *aneurysm* results from a tear in the wall of an artery. This allows blood to enter the muscle layer of the artery and produces a balloon-like swelling. If this swelling bursts, the consequences are usually fatal. An aneurysm can also occur in the wall of the left ventricle after a heart attack.

A thrombus may break free. If it lodges in the arteries supplying the heart, a heart attack will result.

plaque bursts and thrombus or blood clot is formed

A *stroke* has several possible causes. One of the arteries in the brain could have developed an atheroma that has blocked the vessel, cutting off the blood supply to part of the brain. Or a clot that developed elsewhere in the body could have travelled through the carotid artery and then lodged in a smaller artery, with the same result. More rarely, an atheroma can lead to an aneurysm that ruptures, causing blood loss and damage to brain tissue.

A *heart attack* occurs when the coronary arteries become blocked by an atheroma or by a blood clot that has broken off somewhere else in the circulatory system. The severity of the heart attack depends on how much heart muscle is affected. The smaller the area affected, the greater the chance of recovery.

Angina is often the result of an atheroma in the coronary arteries. As the arteries become narrower, the supply of oxygenated blood to the heart muscle becomes more erratic.

A blood clot or *thrombus* can form when a fibrous plaque (atheroma) bursts in an artery. This is shown here in the main femoral artery, but a thrombus can occur anywhere in the body. The cells underneath the endothelium are exposed to the blood and trigger the blood-clotting mechanism. A thrombus forms, and if this breaks free and lodges in one of the arteries supplying the heart, it can cause a heart attack. If it lodges in one of the arteries that leads to the brain, this can obstruct the blood flow and result in a stroke.

Fig 31.8 The biological bases for the different forms of cardiovascular disease

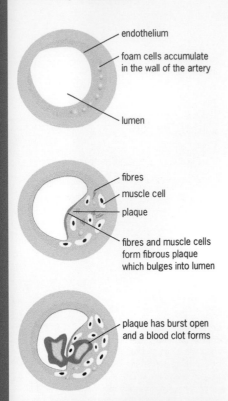

endothelium

foam cells accumulate in the wall of the artery

lumen

fibres

muscle cell

plaque

fibres and muscle cells form fibrous plaque which bulges into lumen

plaque has burst open and a blood clot forms

Fig 31.9 Areas of fatty material called atheromas build up in the walls of arteries. This is one of the first signs of cardiovascular disease. Eventually it can almost block the artery

HOW DOES ATHEROMA LEAD TO VASCULAR DISEASE?

Gradually, usually over many years, muscle cells and fibres begin to grow over the affected patches, giving rise to a **fibrous plaque**. This bulges into the lumen of the artery and starts to obstruct the flow of blood (Fig 31.9). This can have one of several effects:

- The artery can become very narrow. Depending on the position of the narrowed artery, the oxygen supply to vital parts of the body can be reduced. If, for example, the arteries feeding the heart muscle are affected, the result can be **angina**.
- The blood vessel can become completely blocked. Blockage of an artery that takes oxygenated blood to a vital organ can lead to many different problems. If the coronary arteries become blocked, the result can be a heart attack. If arteries in the brain are affected, this can lead to a stroke.
- The blood vessel can burst. A burst artery, or **aneurysm**, inside the body is very serious. If one of the main arteries near the heart bursts, this is usually fatal.
- The plaque can burst into the blood vessel, damaging the endothelium. When the cells underneath come into contact with blood, this triggers the blood-clotting mechanism. A blood clot or thrombus forms in the vessel. If this breaks free, it can travel around the body and block an artery or a vein, causing an embolism or a thrombus.

RISK FACTORS ASSOCIATED WITH ATHEROSCLEROSIS AND CARDIOVASCULAR DISEASE

As we have seen, the underlying process of atherosclerosis can have many different effects, depending on which part of the body is involved. The risk factors for **cardiovascular diseases** are also complex because many different factors affect the way this whole body system functions. We shall look at the three main ones that have been identified.

Hypertension

Blood vessels are more than a series of pipes that take blood to and from different organs. They react to the changing needs of the body, supplying, for example, more blood to the muscles during exercise. When we make our muscles work hard, they need more oxygen and more glucose and they also produce more carbon dioxide. The graph in Fig 31.10 shows how the supply of blood to different organs changes as we do different activities.

Think about an exercising muscle. How can it get more blood? Although blood vessels are flexible, they don't just expand and let more blood through. If this were to happen, the blood would just flow more slowly and no extra oxygen would actually reach the muscle. An increase in blood pressure is needed. This speeds up the blood

QUESTION 4

4 Explain why an exercising muscle needs more glucose.

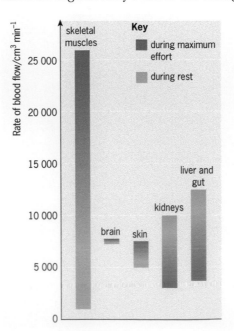

Fig 31.10 The blood flow to different organs in the body changes as we exercise. The flow to muscles and the skin increases. The flow to the brain stays the same. The blood supply to organs like the kidneys and the gut decreases

flow so that more blood reaches the muscle in a given time, supplying more oxygen and glucose to muscle cells and removing waste products more rapidly. Contraction of the smooth muscle in the walls of the body's arteries helps to bring about this increase in pressure.

High blood pressure is perfectly normal as long as it results from short-term fluctuations in blood flow. Problems start when the blood pressure stays higher than normal while someone is at rest. This person is suffering from high blood pressure, or **hypertension**.

Hypertension sets up a sequence of events that becomes a vicious circle. This is summarised in Fig 31.11. Smooth muscle in the artery walls responds to an increase in blood pressure in the same way that skeletal muscle responds to extra exercise – it increases in size. This results in a thickening of the artery wall and a narrowing of the lumen. The volume of blood passing through the arteries remains the same, causing the blood pressure to rise further. At higher pressure, the blood also exerts a greater force on the endothelial cells that line the artery and these cells are more likely to become damaged. Small flaws in the lining of the artery increase the risk that an atheroma will form. This, in turn, makes it more likely that a thrombosis will develop.

Two things increase the risk of hypertension. A high salt intake is one. If you increase the salt concentration of your blood, its water potential falls. A lower water potential means that blood volume increases as water is drawn from the cells of the body into the plasma. The higher the volume of the blood, the greater its pressure.

The other factor is smoking.

? **QUESTION 5**

5 Give the advantage to an exercising person of:
a) an increased blood flow to the muscles,
b) a decreased blood flow to the kidneys.

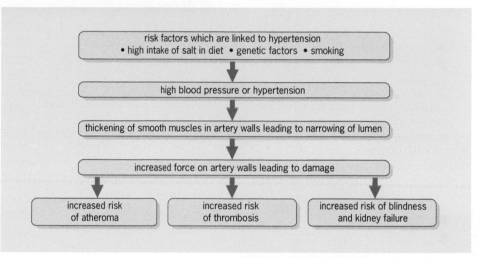

Fig 31.11 This flow chart shows some of the ways in which hypertension can affect the body

The link between smoking and atherosclerosis

If you smoke, you are much more likely to develop atherosclerosis. However, there is no straightforward explanation of how smoking actually causes the formation of atheroma. Several mechanisms could be involved:

- Cigarette smoke contains nicotine. Nicotine is a **vasoconstrictor**: it narrows the blood vessels and leads to an increase in blood pressure. High blood pressure can damage the artery walls and so increase the likelihood of atheroma formation.
- Smoking increases the level of cholesterol-rich lipoproteins in the blood. This accelerates the formation of the fatty streaks that are the first sign of atherosclerosis.
- Smoking increases the amount of fibrinogen circulating in the blood. This increases the chance of a blood clot forming if an atheroma breaks up in the artery.

The link between diet and atherosclerosis

Although newspapers and magazines can give the impression that all cholesterol is bad, it is an essential component of cell surface membranes. Cholesterol is used by the body to synthesise hormones such as oestrogen and testosterone. It becomes a danger only when it is present in high concentrations in the plasma.

Cholesterol is a lipid and, like other lipids, it is insoluble in water. In order to be transported in the blood it is combined with a protein, forming a complex molecule called a **lipoprotein**. There are different sorts of lipoprotein, which have varying amounts of cholesterol and other lipids. **Low-density lipoprotein (LDL)** is commonly called 'bad' cholesterol. This type of lipoprotein contains high amounts of cholesterol and can contribute to the formation of atheromas. **High-density lipoprotein (HDL)** is commonly called 'good' cholesterol because not only does it not contribute to atherosclerosis, it actually helps to scavenge some of the cholesterol from LDL particles, making them less dangerous.

It is best to have a total blood cholesterol level of 200 milligrams per cubic centimetre of blood. Of this total, the amount of LDL should be low – between 100 and 130 mg cm^{-3}. HDL should make up the rest. If someone's total blood cholesterol is higher than 200 mg cm^{-3} or if LDL is over 130 mg cm^{-3}, or if HDL is lower than 35 mg cm^{-3}, they are at increased risk of developing atheromas.

> **? QUESTION 6**
>
> 6 LDL receptors are proteins. Use this information to suggest why some people may be genetically more susceptible to heart attacks than others.

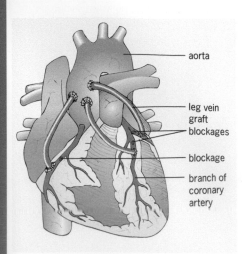

Receptor on cell surface membrane

LDL particle in blood

Liver cells with receptors on the cell surface membrane into which LDLs fit

Surplus LDLs pass through endothelium where they are engulfed by phagocytes to form foam cells

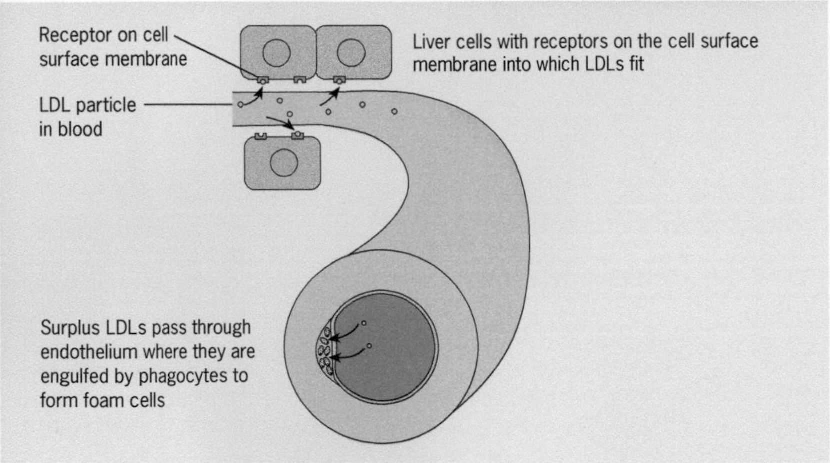

Fig 31.12 How LDL causes atheromas

The sequence of events by which cholesterol contributes to the formation of an atheroma is summarised in Fig 31.12. LDL particles that are excess to the requirements of the liver and other organs leave the blood and penetrate the walls of blood vessels, particularly in places where the wall has been damaged. Phagocytes then engulf them but are unable to break down the cholesterol that they contain. This accumulates in the cytoplasm of the phagocytes which then become foam cells (see page 491), The appearance of foam cells is the first recognisable sign of atherosclerosis.

TREATING CARDIOVASCULAR DISEASE

Although emergency resuscitation is the first treatment needed to save the life of someone who has just suffered a major heart attack or stroke, there are many different treatments for underlying atherosclerosis.

Heart bypass surgery

A **coronary artery bypass graft operation**, sometimes called CABG or 'cabbage', is done to reroute blood around clogged coronary arteries to improve the supply of blood and oxygen to the heart. The surgeon sometimes takes pieces of vein from the leg of the patient, sews one end into the aorta and attaches the other end to a coronary artery below the blocked area (Fig 31.13). In this way, the blockage is bypassed and normal blood supply is restored. It is also possible to detach one of the arteries that lead into the chest wall and then to graft it into the coronary artery. The number of grafts made is indicated by the way the surgeon describes the operation – a double heart bypass involves two grafts, triple and quadruple bypass operations use three and four grafts.

> **? QUESTION 7**
>
> 7 Look at Table 31.2. Name two risk factors for heart disease.

aorta

leg vein graft blockages

blockage

branch of coronary artery

Fig 31.13 The principle of a triple heart bypass

Angioplasty

Angioplasty is used to treat angina or to reduce the risk of a heart attack. In **balloon angioplasty**, a thin tube with a deflated balloon at the end is introduced through the skin and fed into the coronary artery that has become narrowed by atherosclerosis (Fig 31.14). The surgeon checks its progress using a real-time high-speed X-ray image and, once the end of the tube is in the right place, the balloon is inflated gently. This pushes through the obstruction, making the lumen of the artery bigger.

Sometimes in balloon angioplasty, a short length of rigid tubing called a stent is left in the artery. This keeps the vessel open and prevents further narrowing at that point. Although angioplasty is most often done in coronary arteries, it can be carried out in any blood vessel in the body.

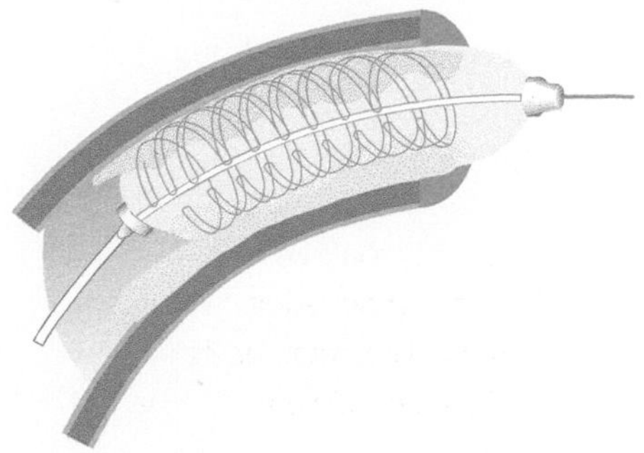

Fig 31.14 Balloon angioplasty and stent

Common drug therapies

Hundreds of drugs are used in the treatment of heart and vascular disease. Table 31.2 provides an overview of some of the main classes.

Table 31.2 Some of the main classes of drugs used in the treatment of heart and vascular disease

Drug type	Drug	Used to treat	Effect on the body
coronary vasodilator	isosorbide dinitrate and other nitroglycerins	angina	Relaxes veins in the body. The amount of blood that returns to the heart is reduced and so the heart has to work less hard.
beta-blockers	Inderal	angina; to prevent heart attacks in people with coronary artery disease; to reduce death rate in people who have suffered a heart attack	Blocks receptors on the heart that respond to adrenaline and related hormones, preventing the heart from working harder when these hormones are released.
diuretics	bumetanide	high blood pressure; congestive heart failure	Causes kidneys to retain less water, reducing excess fluid in the body.
calcium channel blockers	Cardizem	angina; coronary artery disease; high blood pressure	Relaxes the muscular walls of the arteries. In coronary arteries, this enables better blood supply to heart muscle. In arteries in the rest of the body, it reduces blood pressure.
cholesterol-lowering drugs	fluvastatin	high blood cholesterol and high LDL	Reduces total blood cholesterol, targeting LDL specifically by inhibiting an enzyme that is involved in LDL production. Used in combination with a low-fat, low-cholesterol diet.
anti-arrhythmics	digoxin	heart rhythm problems; angina; high blood pressure	Different problems have different effects. Digoxin increases the strength of contraction of the heart muscle as well as helping to prevent abnormal heart rhythms.
anticoagulants	warfarin, aspirin	post-heart attack or stroke treatment to prevent further attacks.	Make blood less likely to clot, reducing the danger of thrombosis and thrombophlebitis.

Send in the clot-busters

When someone has had a heart attack or a stroke and is recovering, part of the treatment is designed to prevent further attacks (Fig 31.15). Giving up smoking, eating less fat and doing more exercise are priorities, but they take time to have an effect. While they are most at risk, heart attack or stroke victims, particularly those with quite severe underlying circulatory problems, are often prescribed clot-busting drugs such as warfarin or aspirin. These drugs 'thin' the blood so that it is less likely that a clot will form and break away from the blood vessel wall to cause an arterial blockage in the heart or brain. Several large clinical trials have shown that a small amount of warfarin or half an aspirin tablet taken every day decreases the chance of having a second heart attack or stroke.

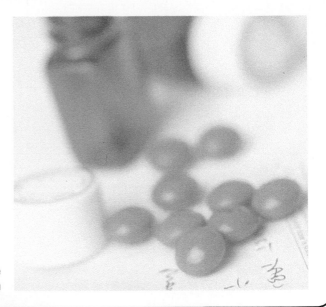

Fig 31.15 Pills of the anticoagulant drug warfarin

PREVENTING CARDIOVASCULAR DISEASE

Treatment for cardiovascular disease is improving constantly, but that old saying 'prevention is better than a cure' holds true. Once the arteries have become clogged with atheroma, the process is irreversible, and surgery and drug therapy can only treat the symptoms it causes.

Changing our lifestyle to prevent build-up of atheromas in the long term is the answer, and education programmes and public health advice stress three main points:

- Never smoke. If you do, give up, or at least cut down.
- Eat a balanced diet that contains less than 25 per cent of its calories as fat. Saturated fat should be no more than 10 per cent. Eat at least five portions of fresh fruit and vegetables each day.
- Take regular exercise. Exercise that is aerobic – that leaves you out of breath – should be done at least three times a week for at least 20 minutes.

3 CANCER

Cancer has been a major cause of death and disease throughout recorded history, but it is now the second biggest cause of death in the western world. There are two probable reasons for this apparent increase. First, many people are surviving longer, free from other diseases, and so have an increased chance of developing cancer. Second, we are now more likely to be exposed to carcinogens (cancer-causing agents) such as cigarette smoke and sunlight.

Cancer is not a single disease: over 200 different types of cancer have been identified. This has led to some bewildering jargon, with tumours named according to the tissue they are in, their growth pattern, their effect on the patient, their response to treatment or after the person who discovered them.

However, all cancers are basically similar: they all result from uncontrolled cell growth. The most common sites for cancer in the human body are shown in Fig 31.16. The number of people in the UK who die from each type each year is shown in Table 31.3.

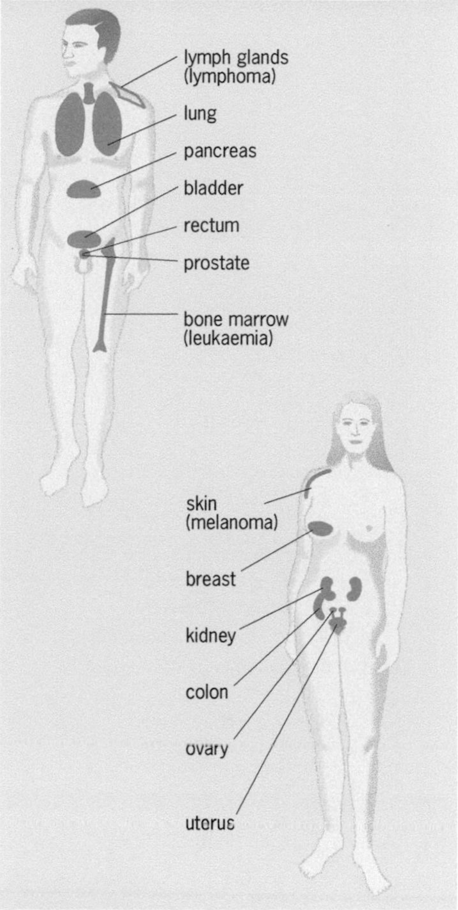

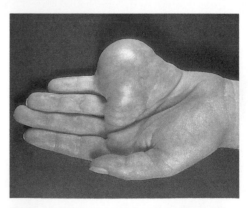

Fig 31.16 Common cancer sites in males and females. Note that non-Hodgkin's lymphoma (Table 31.3) is a cancer of cells in the blood. The blood system and lymphatic system are both affected

Table 31.3 Annual UK death rates in men and women for the ten most common types of cancer			
Men		**Women**	
lung	28 312	breast	34 604
skin (non-melanoma)	19 934	skin (non-melanoma)	18 265
prostate	15 654	bowel	15 651
bowel	15 641	lung	13 701
bladder	9 232	ovary	5 951
stomach	6 920	stomach	4 493
non-Hodgkin's lymphoma	3 899	uterus	4 191
oesophagus	3 565	cervix	4 173
pancreas	3 373	bladder	3 662
leukaemia	3 238	pancreas	3 564

WHAT IS CANCER?

Cancer occurs when there is a breakdown in the cellular control mechanism that puts the brakes on cell division. So cells that should be stable begin to divide, forming a tumour. A **tumour** is a swelling that can occur almost anywhere in the body. It is made up of a mass of abnormal cells that divide continuously.

Some tumours are **benign**. Although they can grow to the size of a grapefruit, they do not actually destroy the surrounding tissue or spread to other organs (Fig 31.17). Other tumours are malignant. They destroy the surrounding tissue and their cells often break away and spread through the blood or lymph system into other sites, where they form **secondary tumours**. A malignant tumour is what we usually describe as cancer.

The rate of cell division varies greatly. Some tumours develop quickly, others can take ten years to reach a noticeable size. Cancer cells usually fail to differentiate: they cannot specialise for the particular function of the tissue they grow in.

Fig 31.18 (overleaf) shows how cancerous cells spread and look different from normal cells. In Fig 31.18 and Fig 31.19, notice that the cancerous cells are smaller, with clear, enlarged nuclei – the classic signs of active cell division.

Fig 31.17 Benign tumours are not usually as big as this, but even those that are can be treated successfully by surgery

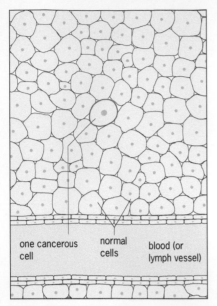

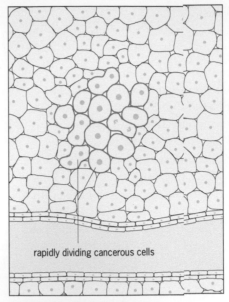

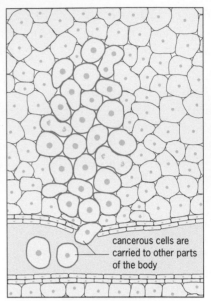

Fig 31.18 How cancers develop
1 One cell starts to divide uncontrollably: it becomes cancerous

2 The cancerous cell divides rapidly, forming a mass of cells – the primary tumour – which squashes out the neighbouring normal cells

3 The process of metastasis. A clump of cells breaks free from the primary tumour and is then carried by the blood or lymphatic system to another part of the body. Fortunately, very few of these clumps of cells, about 1 in 10 000, are able to establish themselves and form a secondary tumour, but that is enough. Nearly 60 per cent of people who are diagnosed with cancer are found to have well-established secondary tumours

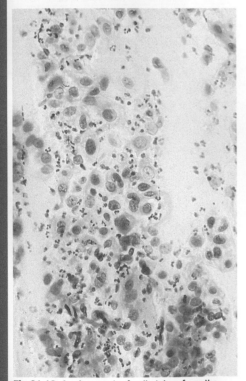

Fig 31.19 A micrograph of cells taken from the cervix of a woman during a routine smear test. The pathologist examines the tissue for signs of abnormal mitosis. Here, the cancerous cells are clearly seen because they stain red and have relatively large nuclei

WHY IS CANCER SO DANGEROUS?

Tumours interfere with the activity of the cells in the tissues of organs that surround them. Benign tumours can compress tissues, preventing normal blood flow or nerve function. Malignant tumours do even more damage, invading surrounding tissues and killing normal cells in the process. Cancer can also spread to other parts of the body by the process of **metastasis** (Fig 31.18).

Many tumours form from epithelial cells, and in order to break away, they must escape from the proteins that bind them to the underlying basement membrane that is characteristic of epithelial tissue (see Chapter 2). They do this by secreting protein-digesting enzymes. They also produce reduced amounts of the adhesion molecules that usually hold cells together.

THE BIOLOGICAL BASIS OF CANCER

Although there are many different types of cancer, we now know that all cancers are the result of DNA damage in cells. If this damage occurs in specific genes, control of cell division in that cell can be lost and the cell can begin to divide continuously (Fig 31.20).

You might want to revise cell division (Chapter 23) before going on to look at how control of the process can break down.

Genes and the control of cell division

In Chapter 23 we saw that different types of cell in the body divide at different rates. Some, such as the neurones in the brain, do not divide at all after we reach the age of about 2 years. Others, such as liver cells, divide only rarely. Others, notably epithelial cells, divide as often as once every couple of days. Even so, all cell division is tightly controlled at the genetic level.

Some genes in the cell code for proteins that activate other genes to produce growth factors, chemicals that start a sequence of events that leads to cell division. Other genes code for proteins that inhibit cell division. Genes are turned on and off according to the cell type and its position in the body. Needless to say, the interplay of all the different genes involved is very complex. However, scientists have identified a few genes that, when they suffer a mutation, can cause the cell to lose control of its division pattern. These cells have been named **proto-oncogenes**, or just **oncogenes**.

The human genome contains many oncogenes. One, the *ras* oncogene, is on chromosome 11. It codes for a chemical called G-protein, which basically acts as an on-switch for cell division. Normal G-protein is usually inactivated very quickly by cellular enzymes. This means that the G-protein on-switch for cell division is only temporary and is switched off most of the time.

If the DNA in the *ras* oncogene is damaged, however, it codes for a G-protein that cannot be inactivated. In a cell with a defective *ras* gene, the switch for cell division always says 'on'. Cells that lose control of cell division can become cancerous, and *ras* mutations are associated with 20 to 30 per cent of all human cancers.

However, mutations in oncogenes are much more common than the cancer rate would suggest. This is because the cell has a second, back-up control. It also has genes called **tumour suppressor genes**. These prevent cells dividing too quickly, until the body's immune system can either kill the rogue cells, or until the cell's enzymes manage to repair the damage in the DNA (Fig 31.21).

The tumour suppressor genes do an excellent job, but their DNA too can be damaged and they can suffer mutations. If these genes fail to work, this back-up system is lost and any cells that start dividing uncontrollably are more likely to go on to develop into cancers. A gene called *p53*, on chromosome 17, has been well studied, and about half of all human cancers have been linked with a mutation in this tumour suppressor gene.

Oncogenes and tumour suppressor genes can both mutate because of DNA damage caused during life. Occasionally, though, faulty forms are passed on to the next generation. People who are born with a gene defect in one of these genes are said to have a genetic predisposition to cancer. They are more likely to develop cancer and often do so at a relatively early age (see page 502 and Table 31.4).

RISK FACTORS AND CANCER

So, two different classes of gene in the cell can fail and can result in cancer. But what does the damage to the DNA of these genes? Many things in our environment are carcinogenic. Some of the risk factors for cancer are connected with our lifestyle, others are present in our environment. In both cases, we can usually modify our cancer risk by changing our behaviour. This chapter looks at just a few of the major lifestyle and environmental factors that affect cancer risk.

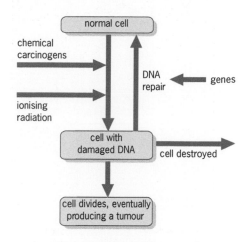

Fig 31.20 Tumours arise when DNA in specific genes is damaged

? **QUESTION 8**

8 How might the ability of some cancers to spread very rapidly to other organs be related to the protein-digesting enzymes that the tumour cells contain?

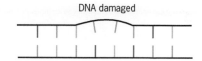

Fig 31.21 Cells have many different ways of repairing DNA. This process is also controlled by genes. Mutations affect genes that repair other, damaged genes, as well as genes that control cell division

Smoking

Smoking is now one of the major causes of lung cancer in the developed world. Each year, lung cancer due to smoking leads to around 40 000 deaths in the UK (a large sports stadium holds about 40 000 people) and to around 175 000 deaths in the USA.

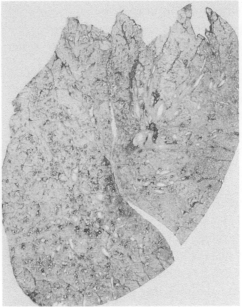

The link between smoking and lung cancer has been recognised for many years but it is only recently that scientists have identified a compound in tobacco smoke that binds to and damages the *p53* tumour suppressor gene. The longer a person smokes and the greater the number of cigarettes they have a day, the greater the chance that the *p53* gene will be damaged.

Fig 31.22 If you smoke, your lungs will eventually become clogged up with tar, and will look like this

HOW SCIENCE WORKS

Eat those greens

There is a great weight of evidence that eating lots of fresh fruit and vegetables decreases the risk of developing cancer – so much evidence, in fact, that governments and health organisations in Europe, the USA and Australia all advise that we each eat five portions of fresh fruit or vegetables every day.

But why are fruit and veg so protective? One theory is that the high levels of vitamins they contain, particularly vitamins A and the B group, act as anti-oxidants. This type of chemical mops up free radicals that form in the body all the time, and that can damage DNA. Unfortunately, several large trials have shown that although taking vitamin supplements does no harm, it doesn't actually lower cancer risk.

Other chemicals in fruit and vegetables are responsible. Three that have already been isolated are these:

- Chemicals called dithiol-thiones that occur in vegetables such as broccoli, cauliflower and cabbage seem to have several effects on the body. They activate liver enzymes that deal with carcinogens and they inhibit tumour growth.
- Sulflorane, another chemical isolated from broccoli, also activates helpful liver enzymes.

- Genistein, a compound found in soya beans, prevents the formation of the blood vessels that tumours need to support their rapid growth.

Many studies are now planned, and scientists expect to develop new drugs based on natural plant compounds that can cut cancer deaths by prevention.

Fig 31.23 These vegetables are known to help reduce the likelihood of developing cancer

Diet

There is strong evidence that some of the things we eat and drink contain substances that are carcinogenic.

- Drinking alcohol is linked to 3 per cent of cancers in the western world, causing higher rates of mouth, throat and oesophageal cancer. Smokers who drink large amounts of alcohol are particularly at risk.
- Heavily salted and smoked foods also increase the risk of developing mouth, throat and oesophageal cancer. About 1 per cent of all cancers are due to eating salted and smoked foods regularly.
- Eating a diet very high in over-cooked meats is thought to increase the risk of stomach cancer, because charring red meat (such as blackening it on a barbecue) produces breakdown products that are carcinogenic.
- Food contaminated with aflatoxins, substances produced by the fungus *Aspergillus flavus*, have been linked to specific cases of liver cancer. In some parts of Africa, groundnuts (peanuts) contaminated with this fungus are a staple food and liver cancer is common. Aflatoxins are chemically modified by liver enzymes. This modified aflatoxin affects the base guanine, which is part of the DNA molecule.

Apart from these specific examples, the cancer risk of eating specific foods is quite difficult to assess. People's diets are extremely varied and many studies involving hundreds of thousands of people are just beginning to uncover some definite patterns. However, the American Cancer Society states that, on the basis of current scientific evidence, one-third of the 564 800 cancer deaths that occurred in the USA in 1996 were related to diet.

At the moment, the generally accepted view is that the diet to avoid is one high in red meat, particularly meat that is charred, high in animal fats, and low in fibre, fresh fruits and vegetables (see the How Science Works box opposite).

Exposure to radiation

We are all exposed to a common source of radiation – the Sun. Exposure to sunlight is linked with **melanoma**, a form of skin cancer that accounts for about 2 per cent of all cancer deaths. It occurs most often in fair-skinned people and is caused by the ultraviolet light in sunlight. Particularly dangerous are the higher frequency UVB rays that can damage DNA: UVB radiation causes 90 per cent of all skin cancers (Fig 31.24).

The cancer risk posed by other forms of radiation seems to be low. Excessive radiation such as the fallout from the atomic bomb blasts at Hiroshima and Nagasaki led to cancers in the survivors, but the low levels emitted by nuclear power plants do not seem to increase cancer risk. There is also little evidence to support the theories that objects such as mobile phones, microwaves and overhead power lines increase the risk of cancer.

Exposure to chemical carcinogens

Substances such as asbestos, benzene, methanal (formaldehyde) and diesel exhaust are carcinogenic. We know this because of instances in the past where people have been exposed to high levels – usually through their work.

Exposure to microorganisms

Specific microorganisms are linked to the development of some cancers:

- The Epstein–Barr virus, which causes glandular fever, has been linked to some cancers of the lymph gland in the western world and with nose and throat cancer in Asia.
- HIV leads indirectly to an increased cancer risk: people with AIDS have a much higher rate of a skin cancer called Kaposi's sarcoma because their immune system cannot weed out rogue cancer cells.

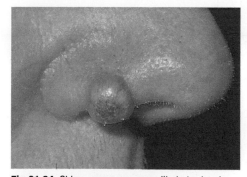

Fig 31.24 Skin cancers are more likely to develop on parts of the body that are exposed to a lot of sunshine

 REMEMBER THIS

Today there is much greater control of work environments, but many work-related cancers are expected in the developing world as countries start industrialisation programmes.

- Infection with genital wart viruses is associated with higher rates of cancer of the cervix. Women who have many sexual partners are more likely to be exposed to wart viruses and so run a higher risk of developing cervical cancer.
- Infection with the bacterium *Helicobacter pylori*, recently found to cause stomach ulcers and gastritis, also causes higher rates of stomach cancer.

Cancer and heredity

About 5 to 10 per cent of cancers arise because of faulty genes that can pass from one generation to the next. In the past 10 years, several such genes have been identified. However, before the genes were recognised, it was clear that some cancers were hereditary, because:

- some families have many members all with the same type of cancer;
- some cancers develop early – in people under 50.

Study of the DNA from members of such families has provided a lot of information about genes and cancer. Table 31.4 shows some of the genes that, when they are faulty, confer a very high cancer risk.

Table 31.4 Genes and cancer risk

	Gene	Tumour type	Gene class
breast cancer	BRCA1	breast, ovary	tumour suppressor
	BRCA2	breast (both sexes)	tumour suppressor
	p53	breast, sarcoma	tumour suppressor
colon cancer	MSH2	colon, endometrium, other	mismatch repair
	MLH1	colon, endometrium, other	mismatch repair
	PMS1,2	colon, other	mismatch repair
	APC	colon	tumour suppressor
melanoma	MTS1 (CDKN2)	skin, pancreas	tumour suppressor
	CDK4	skin	tumour suppressor
neuroendocrine cancer	NF1	brain, other	tumour suppressor
	NF2	brain, other	tumour suppressor
	RET	thyroid, other	oncogene
kidney cancer	WT1	Milms' tumour	tumour suppressor
	VHL	kidney, other	tumour suppressor
retinoblastoma	RB	retinoblastoma, sarcoma, other	tumour suppressor

CANCER – THE PATTERNS OF DISEASE

In the nineteenth century, cancer caused the deaths of fewer people than it does today. We must be careful how we interpret this statement. Infectious diseases were much more important 150 years ago as causes of death than they are now. Because of this, the average age of death was lower than it is today, and fewer people lived to the ages where cancers become increasingly common. In addition, medical treatment was far more basic and methods of diagnosis meant that the disease could be recognised only when it had reached an advanced stage.

If you were to compare data that shows the deaths from different types of cancer in one particular year, with the corresponding figures from 10 years earlier, you would notice differences. Between 1986 and 1996, for example, stomach cancer declined because of the discovery that various gastric problems, including cancer, were caused by infection with *Helicobacter pylori*.

Treating people for this infection cut stomach cancer cases significantly during this time.

You might also notice an increase in the incidence of some cancers. Breast cancer cases, for example, increased between 1986 and 1996 because of advances in methods of screening and diagnosis. Much of the apparent increase was probably due to doctors being able to detect the condition earlier than they would have before 1986.

However, often it is not as easy to account for differences. Most of the time, statistics on cancer death rates and cancer incidence are difficult to interpret.

It might be better to ask some questions: Are the differences actually statistically significant? An apparent increase in cases might be only a reflection of an underlying trend, or the differences could be just the result of chance. Has there been a change in exposure to particular risk factors? Could we explain any changes in the incidence of lung and bladder cancer in terms of changes in smoking habits? Is the picture incomplete, and does this confuse the issue? Why are some cancers more common in one sex than in the other?

Answers to questions such as these are important if we are to understand the factors affecting the trends shown by such data. The Assignment for this chapter on the CAS website provides some more information about how patterns in the incidence of a particular type of cancer can shed light on the factors that cause it.

THE BATTLE AGAINST CANCER

Cancer prevention

Although drugs and food supplements to prevent cancer are a possibility, they are likely to take time to develop. In the meantime, changing your lifestyle could help you reduce the risk of cancer in later life. Four steps seem to be the most important:

- Never smoke. If you smoke, stop.
- Eat a well-balanced diet in which animal fat makes up less than 25 per cent of the calories. Include at least five portions of fresh fruit or vegetables every day, and try to obtain most of your protein from plant sources (e.g. beans, peas, grains). Avoid very salty or smoked foods and charred meats, particularly red meats.
- Take regular exercise and maintain a sensible weight.
- Drink alcohol in moderation.

Screening for cancer

Successful cancer treatment often depends on getting an early diagnosis. The longer a malignant tumour is left, the larger it grows and the more likely it is to spread and affect other organs. Surgery can be used to remove a localised tumour but it is of little use if the condition has spread throughout the body.

Skin cancer is one of the commonest forms of cancer in younger people, but at least it is easily noticed. Surgical removal is very effective and results in 97 per cent of people surviving five years after the operation. Cancer of the colon is a very different matter. There are no visible signs and early symptoms are vague. Most people don't take them seriously and end up visiting their doctor for the first time when the cancer is already quite advanced. Not surprisingly, the five-year survival rate for this form of cancer is much lower, at only 37 per cent.

Regular screening can help to detect cancers early, but the advantages have to be weighed against its high financial cost. The NHS has screening

? QUESTION 9

9 Look back at Section 1 in this chapter, and you will notice that the lifestyle recommendations that help to prevent cancer are much the same as those that help prevent heart disease. Why do you think large numbers of people in the UK and the USA ignore them most of the time?

Fig 31.25a A patient is undergoing routine mammography. This enables tumours to be identified early, and increases the chance of treatment being successful

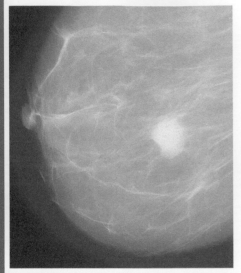

Fig 31.25b This mammogram has been enhanced with the aid of a computer. The denser tissue of the tumour shows up as a white area

? **QUESTION 10**

10 Explain why it is unlikely that there will be a national prostate cancer screening programme.

programmes for two of the most common female cancers, breast cancer and cervical cancer. Breast cancer screening is carried out every three years by mammography (Fig 31.25). It targets women over 50 years old since breast cancer occurs most often in women over this age and because breast tissue in younger women is too dense for mammography to be effective. The breast cancer programme has probably caused a 20 per cent fall in death rate, simply because cancers have been picked up and treated early enough.

The national cervical screening programme was launched in 1988 and has cut deaths from cervical cancer by about 30 per cent. The programme recommends that a smear is taken every three years. Cells are stained and examined under a microscope, and those likely to become cancerous can be identified (look back to Fig 31.19). Early treatment is then possible.

Screening programmes for other cancers might not be so effective. Prostate cancer is common, particularly in older men, and there are methods of identifying it at an early stage. These include using ultrasound or testing the blood for high concentrations of prostate-specific antigen or PSA, an enzyme secreted by cells in the prostate. An enlarged prostate gland produces more PSA than a healthy one.

But is screening cost-effective? Can it be used to prevent cancer deaths? The answer to some extent depends on where you live and what kind of healthcare system you have locally. In the USA, where private medical insurance bears the cost, PSA screening is commonplace, even though there are some doubts about its reliability. Early treatment of prostate cancer is carried out routinely. In the UK, the cost of a general screening programme would be an enormous burden on the NHS and would save relatively few lives: most men with fully developed prostate cancer are very elderly and are usually at far higher risk of dying from some other condition.

Treating cancer

Surgery, chemotherapy and radiotherapy are important cancer treatments (Fig 31.26). They are all constantly being improved by research and have varying success rates, depending on the type of cancer and its site in the body. Drug treatment for leukaemia in children, for example, now cures nine out of ten children. Lung cancer can be much less easy to treat: the lungs cannot be completely removed by surgery, and the tissue is delicate and easy to damage.

New strategies are approaching cancer treatment from several different angles. Table 31.5 describes some of them.

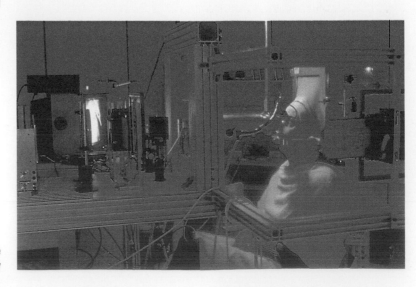

Fig 31.26 This person has a tumour in the left eye and is undergoing proton radiation therapy

Table 31.6 New strategies for treating cancer

Treatment	What it does	Current status
conformal radiation therapy	New technology allows radiation beams to be concentrated only on the tumour, avoiding damage to surrounding healthy tissue. The exact 3D position of the tumour is found using CT scanning or MRI scanning, and the positioning of the treatment beams is determined by computer.	In relatively widespread use: new types of tumours are treated every year.
neutron and proton radiation therapy	X-rays and gamma rays are commonly used in radiotherapies; protons and neutrons can also be used.	Recent studies show that protons can treat small tumours of the spine that lie near vital organs, and that neutrons can treat salivary gland tumours.
immunotherapy	New drugs to treat cancer are being developed all the time, but some of the most promising new drugs recognise substances found on the surface of tumour cells, and attack them while ignoring healthy cells.	Clinical trials are in progress. Some drugs are in use.
differentiating agents	Drugs that do not kill tumour cells but force them to differentiate into cells with a specific function. When this happens, the cells stop dividing and tumour growth stops.	Several new drugs are in early clinical trials.
angiogenesis inhibitors	Chemicals that stop the fast growth of new blood vessels into a tumour. Without a good blood supply, the tumour expands more slowly and can be easier to treat by radiotherapy or more traditional forms of chemotherapy.	Several angiogenesis inhibitors are undergoing clinical trials in patients, and other compounds that show potential are in an earlier stage of research.
oncogene blockers	Scientists have been looking at how mutated oncogenes lead to cancer. They are trying to develop drugs to block the product of the damaged gene, so that the cell does not continue to divide uncontrolled.	Any cancer that arises because of a mutation in an oncogene.
gene therapy	If a tumour suppressor gene becomes damaged, it no longer inhibits cancers from forming. In addition to treating the cancers that result, researchers are looking at the possibility of replacing the damaged gene in tumour cells.	Experimental studies using cells in culture have been quite successful. Clinical trials could soon begin.

4 LUNG DISEASE

In addition to lung cancer and tuberculosis there are three other important diseases that can affect the lungs. These are asthma, emphysema and fibrosis; the last can occur as a result of cystic fibrosis or other lung diseases.

Tuberculosis is discussed in Chapter 29. Cystic fibrosis is discussed in Chapter 25.

ASTHMA

Asthma is a common childhood ailment, affecting at least one in ten children, and a large number of adults. The number of asthma cases has risen dramatically in recent years, fuelling speculation that an increase in air pollution is to blame.

To fully understand why asthma happens, you need to be familiar with the fine structure of the lungs – see Chapter 9.

Put simply, asthma is a difficulty in breathing caused when the smooth muscles of the bronchioles contract, narrowing the airways that lead to the alveoli. Asthma sufferers then find it hard to breathe and have to make far more effort to deliver a normal amount of air to the lungs.

Many factors can cause asthma, including an allergic reaction (reaction of the immune system to a substance that, in non-sufferers, has no effect). Common allergens are house dust mite faeces (Fig 31.27), fur, feathers and pollen.

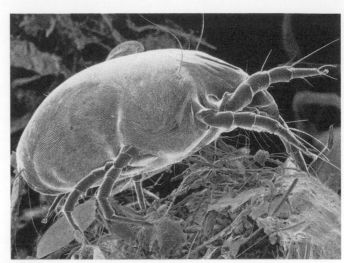

Fig 31.27 House dust mites, *Dermatophagoides farinae*, are normal inhabitants of our mattresses. Feeding on the large deposits of skin which accumulate there, they produce faeces that can cause an allergic reaction that leads to asthma in some people

Fig 31.28 Using an inhaler is easy, once you have mastered the right technique. Many people have had the wrong technique for years, with the result that their inhalers are not as effective as they might be

Steroids are discussed in Chapter 18.

Treating asthma

Several approaches are used together to help control asthma. A doctor's first priority with all but very young children is to explain what is happening inside the lungs. Sufferers will be less frightened and can come to terms with their problem. The worst physical effects of the condition are then controlled by a combination of drugs and prevention.

Generally, two types of drugs are prescribed to asthma sufferers: bronchodilators and steroids. Bronchodilators give instant relief from chest tightness and wheezing because they contain chemicals which relax the bronchiole walls. Many asthma sufferers carry a reliever inhaler around with them. It contains salbutamol or terbutaline, which both act as bronchodilators to relieve the feeling of being 'tight-chested'. This is very useful in the treatment of attacks and is safe to use frequently (Fig 31.28).

Many asthma sufferers also use a preventive inhaler that contains steroids. Steroids such as Becotide™ act by reducing the degree of inflammation of the bronchioles. Steroids are preventative: asthma sufferers take regular doses morning and night to reduce the problem of over-reaction. The amount of steroids taken is minimal but they are effective if taken according to instructions.

EMPHYSEMA

Emphysema is a lung disease that results from gradual but relentless damage to the fragile alveoli in the lungs. When the delicate epithelial cells of the alveoli are damaged, they are replaced with 'scar' tissue that is thick, tough and inelastic. This process is called **fibrosis.** The lungs have a smaller surface area and a thicker wall, so gas exchange is less efficient. The lungs are also less elastic.

Normally, exhaling is largely a passive process, the elastic nature of the lungs forces air out without much help from the muscles of the ribs and diaphragm. Emphysemic patients have to make more of an effort to breathe out.

This takes a lot out of people and emphysema sufferers are weak, lack energy and usually lose a lot of weight. Although it can start simply as mild breathlessness during exercise and a persistent cough, it can leave sufferers housebound and reliant on oxygen supplies, and eventually cause their death.

What causes emphysema?

The main cause is smoking – 80 per cent of cases of emphysema occur in people who smoke. Emphysema usually affects older people as it is due to cumulative damage over a long period, such as a lifetime of smoking. Working in certain industries, such as mining, appears to increase the risk of developing emphysema.

It is common for smokers to get a combination of emphysema and chronic bronchitis, caused by the lungs' inability to clear mucus, which becomes infected. This combination is called chronic obstructive pulmonary disease (COPD). In COPD, surgeons have taken the radical step of giving lung transplants, and another technique called lung volume reduction surgery is used. This cuts away the most diseased parts of the lungs, meaning that the remaining lung will work more efficiently.

Currently, emphysema cannot be cured, although radical surgical methods have been tried with some success. Most patients try to control the symptoms with bronchodilators – the same inhalers asthmatics use – as these help relax the airways and help oxygen get to the lungs. They can also do breathing exercises and try to improve their general fitness.

Hereditary emphysema

Alpha-1-antitrypsin deficiency (A1AD) is a genetic disorder that causes people to develop emphysema. People with the disease cannot make the protein alpha-1-antitrypsin, which is normally found in blood. This protein protects delicate tissues from damage – when it is not there, the alveoli in the lungs are particularly badly affected.

With no protection, the alveoli are more easily damaged by inhaling smoke and pollutants, and emphysema can develop in even quite young people. Fortunately, giving injections of alpha-1-antitrypsin can treat the disease.

> ### ✔ REMEMBER THIS
>
> Emphysema is an example of a disease in which **fibrosis** occurs. Cystic fibrosis is another example. But what is fibrosis?
>
> Fibrosis is the formation of excess fibrous connective tissue occurring in an organ or a tissue after damage. It can occur in other organs as well as the lungs. It is a common feature of cirrhosis of the liver, for example, and of sickle cell anaemia, where fibrosis occurs in the spleen.

SUMMARY

When you have studied this chapter, you should know and understand the following.

- **Cardiovascular disease** (CVD) and **cancer** are the two main lifestyle diseases.
- Lifestyle diseases are **non-transmissible**. They are associated with different risk factors such as smoking and diet.
- The underlying cause of heart disease is the process of **atherosclerosis**. This is commonly called hardening of the arteries.
- The major risk factors for CVD are hypertension (high blood pressure), smoking and diet.

- Although many cases of CVD are preventable and new treatments are being developed all the time, this disease is a major killer in developed countries.
- Cancer occurs when cells start to divide out of control. Many factors – genetic and environmental – cause this to happen.
- The major risk factors for cancer are smoking and diet. Many cases of cancer are preventable.
- Lifestyle lung diseases include lung cancer, asthma and emphysema.

 Practice questions and a How Science Works assignment for this chapter are available at www.collinseducation.co.uk/CAS

32

Ecology: the basics

competition for a mate | rainfall | parasites | number of other species in ecosystem | seasonal changes in climate | plant producers, e.g. grass | competition for food

number of animals in population | how much water is available | hours of sunlight

temperature | altitude of environment

catastrophes – drought, hurricanes, fire, and so on | general climate | interaction with mate

action of decomposers on body after death | concentration of minerals in the soil | time needed to rear young | competition for space and territory

Fig 32.1 Ecology is the study of whole organisms: we learn how an organism interacts with individuals of its own species and other species, and with the physical environment that surrounds it. The jigsaw pieces show just a few of the factors that influence the lives of these zebras in the savannah grasslands of Africa

1 THE SCIENCE OF ECOLOGY

Ecology is a branch of biology. The word ecology was first used by Ernst Haeckel in the 1860s. It comes from two Greek words: *oikos* meaning home and *logos* meaning understanding. His definition of ecology was 'the knowledge of the sum of the relations of organisms to the surrounding outer world, to organic and inorganic conditions of existence'. Put more simply, ecology is the study of organisms in their natural surroundings. It looks at how they are adapted to their environment and how they interact with both the living and non-living world around them.

Each organism and each physical feature of an environment is separate, but they all interact and interlock, forming a complex system that is a bit like a three-dimensional jigsaw puzzle; ecologists study different parts of the puzzle and then try to work out how it is put together (Fig 32.1).

ECOLOGY AND OTHER BRANCHES OF SCIENCE

Ecology is closely related to two other branches of science: **environmental biology** and **environmental science**. The first of these looks more generally at living things in the environment. The second uses the biology, geology and the physical sciences to help us to explore our environment.

It is also impossible to study ecology without bearing in mind the genetics, evolution and behaviour of the organisms at the centre of a study. Ecological genetics and behavioural ecology are beyond the scope of this book but it is

> **REMEMBER THIS**
> The word ecology is loosely used to mean environmental concern or environmental conservation. These topics are important, and they depend heavily on knowledge gained through ecology, but they are not the same thing. We define ecology as: '**the scientific investigation of living organisms in their natural surroundings**'.

important to remember that each organism you study in the context of its environment has developed to fit that environment through the process of natural selection.

Natural selection and evolution are discussed in Chapter 26.

PRACTICAL ECOLOGY

As in other branches of science, progress in ecology depends heavily on scientific investigation. Looking back at our definition of ecology, it should come as no surprise that many of these studies are done outside. Field studies (Fig 32.2) involve a great deal of observation and can include experimental studies, though these are more often carried out in the laboratory where they can be more strictly controlled. Ecologists use these basic approaches together to gain a complete picture of the organism or the environment that is being investigated.

Fig 32.2 Safety and field trips. Ecologists conduct studies outside and must consider safety when working near water or in isolated areas. Here, a student is identifying organisms from a pond, and two students are measuring the rate of water flow in a stream

The exciting thing about ecology is that it is still a relatively young science. In just over 100 years of ecological study, we have barely begun to even scratch the surface. Many environments of the world remain unexplored and there are many organisms that we know of, but whose habits and living conditions are known only sketchily, or are a complete mystery. Unlike biochemistry or physiology, ecology offers an A-level student the chance to do original practical work, and perhaps to make an important scientific discovery (Fig 32.3).

Fig 32.3 Habitats change frequently. Porlock used to have a mile-wide strip of farmland between it and the sea. In 1996, the shingle barrier that formed the sea defences was breached and a salt marsh with saltwater lakes has developed rapidly. This is a new habitat that now supports birds such as egrets. Coastal areas are subject to many changes like this, creating new areas for observing the processes of colonisation and adaptation

Why do human biologists need to study ecology?

You might be wondering what a section on ecology is doing in a human biology textbook. Ecology is sometimes thought of as a study of nature – lots of field trips and wading around in mud looking at insects. But ecology is much more wide-ranging.

The quality of our environment and the way we interact with the outside world are major factors that determine our health. Substances that contaminate our environment in water, food, air and soil contribute to the untimely death of millions of people every year (Fig 32.4).

The problem is worst in the developing world where 4 million babies and young children die from diarrhoea every year because of dirty water and bad food. Over 1 million people die from malaria each year and over 267 million others suffer recurrent symptoms. Treating these illnesses with medicines is of only limited use: to really tackle them effectively, it is crucial to find out what is causing them and take steps to reduce people's risk. That doesn't mean just finding out what bacterium, virus or parasite is causing the disease directly. It means identifying the environmental factors that are allowing the disease to spread.

Many third world countries, for example, have seen huge increases in cases of malaria with the more widespread availability of plastic 'disposable' containers. When these are dumped, they provide the perfect trap for rainwater and then form a wonderful stagnant pool in which the *Anopheles* mosquito can lay its eggs. Recognising this as a problem has helped to prevent many cases of malaria.

Fig 32.4 We have a two-way relationship with our environment, and our actions can have unexpected consequences. These people in the Philippines live by recycling rubbish. The plastic bags around them look harmless enough but they could be lethal

Fig 32.5 The biosphere is the part of the Earth's surface that supports life. It extends from the bottom of the deepest oceans into the part of the Earth's atmosphere that contains breathable air: in total a vertical distance of between 30 and 40 kilometres. It covers most of the hydrosphere, the surface layers of the lithosphere and some of the atmosphere. The hydrosphere is the part of the Earth's surface that is covered by water. The lithosphere is the layer of soil and rock that forms the Earth's crust. The atmosphere is the blanket of gases that envelop the planet, maintaining a mean temperature of about 7 °C. This temperature is crucial to life because it allows most of the water on Earth to exist as a liquid

2 TERMS AND CONCEPTS IN ECOLOGY

The relationships between an organism and the physical features of its environment, and between all the other organisms that live with it, are incredibly complex. Any description of them uses terms that you have probably not come across before. The next section gives an overview of some basic terms and concepts.

THE BIOSPHERE

All the living organisms that we currently know of live on Earth. But not all parts of the Earth support life. The part that does, forms a sort of 'skin', the **biosphere**, which is shown in Fig 32.5.

The biosphere is divided into biomes

The biosphere is made up of lots of different areas that have very different environmental conditions. We call each of these fairly broad areas **biomes**. The major biomes of the world are listed in Table 32.1 and some of these are shown in Fig 32.6. Many of the best-studied biomes are the **terrestrial biomes** (those that exist on land) but there are also **aquatic biomes** (those that exist in water).

? QUESTION 1

1 Look at Table 32.1. Which biomes do you think exist in Britain?

Table 32.1 The major biomes of the world

Biome	Type	Major features
savannah	terrestrial	Tropical grassland with few trees. Example: the Serengeti in Africa.
temperate grassland	terrestrial	Grassland with hot summers and cold winters. Some broad-leaved plants. Example: the prairies in the USA.
desert	terrestrial	Hot and dry. A few highly specialised plants and animals. Example: the Kalahari desert in Africa.
tundra	terrestrial	Cold for most of the year: very short growing season. No trees: some specialised plants. Example: northern Canada.
tropical rainforest	terrestrial	Tropical climate. Lush vegetation. Great diversity of animal and plant life. Example: forests of equatorial Africa.
temperate rainforest	terrestrial	Cool and wet. Tallest trees in the world. Example: redwood forests of North America.
temperate deciduous forest	terrestrial	Hot summers and cold winters. Dominated by broad-leaved deciduous trees. Example: oak woodlands of central and western Europe.
lakes and ponds	aquatic: freshwater	Large bodies of standing freshwater. Example: Great Lakes of North America.
streams and rivers	aquatic: freshwater	Flowing freshwater. Example: the Amazon.
marine rocky shore	aquatic: seawater	The border between the land and the sea. Most rocky shores show **zonation**: there are characteristic bands of different environmental conditions that lead to colonisation by very different plants and animals.
coral reef	aquatic: seawater	The tropical rainforests of the seas. Reefs form in warm, shallow waters and form a biome that contains an incredible diversity of living organisms.

Fig 32.6 The biomes of the world show great extremes of environmental conditions

(a) Arctic tundra

(b) Desert

(d) Tropical rainforest

(c) Coral reef

? QUESTION 2

2 The following components of its ecosystem all affect an aphid living on the underside of a rose leaf. Which are biotic and which are abiotic?

a) the humidity of the air

b) predation by ladybirds

c) wind speed

d) temperature

e) competition for food

f) disease

THE ECOSYSTEM CONCEPT

The **ecosystem** is the basic functional unit of ecology. It is a single working unit that consists of a group of interrelated organisms and their physical environment (Fig 32.7). We can divide any one of the biomes described above into many different ecosystems.

When we study a particular organism in its ecosystem, we look at the physical features that might affect it, such as rainfall, soil type, temperature and so on. These physical or **abiotic** features help to determine what range and type and numbers of living organisms live inside the ecosystem. Of course, we also look at the living or **biotic** part of the ecosystem, to see how our chosen organism relates to the other organisms present. This is particularly important if we are looking at the impact of human activities.

On pages 516 to 518, we look at the abiotic and biotic features that affect the distribution of organisms in an ecosystem in much more detail.

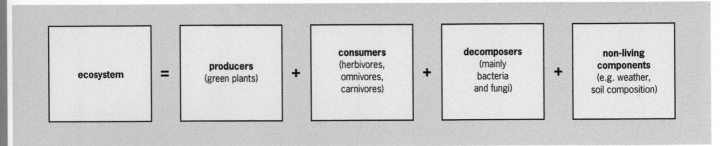

| ecosystem | = | producers (green plants) | + | consumers (herbivores, omnivores, carnivores) | + | decomposers (mainly bacteria and fungi) | + | non-living components (e.g. weather, soil composition) |

Fig 32.7 A schematic diagram showing the essential features of an ecosystem

LEVELS OF ORGANISATION IN AN ECOSYSTEM

Individual living organisms may look randomly arranged in an ecosystem, but they are organised into recognisable units. Fig 33.8 provides an overview of organisation in an ecosystem.

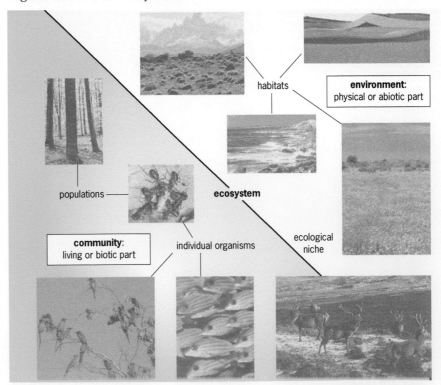

Fig 32.8 An ecosystem has two basic parts: the abiotic part and the biotic part. Within both of these, there are further levels of organisation

From ecosystem to individual

We can study the human body and narrow our investigation from a body system to an important organ, to the level of the individual cell. In a similar way, we can focus in on the components of an ecosystem. Each one contains a **community** of organisms, which consist of many different **populations**. Each population is made up of many **individuals**. It is important to define these basic features of an ecosystem:

- A community is a collection of groups of organisms from different species that live in close association in the same ecosystem. Organisms in a community interrelate.

- A population is a group of individuals of the same species that live in a particular area at any one moment in time. We can look at a population of rabbits or a population of beech trees. Different populations interact in an ecosystem, forming a community.

- An individual is a single organism within a population. Some populations contain organisms that are genetically identical. Most populations of bacteria, for example, are identical because they have developed as a result of asexual reproduction. But many populations contain individuals that are the result of sexual reproduction: these individuals show genetic variation (Fig 32.9).

> The effect of human activity on the environment is discussed in Chapter 36.

> Genetic variation and evolution are discussed in Chapter 26.

Environment, habitat and ecological niche

The divisions listed above relate primarily to the *living* components of an ecosystem. We can also sub-divide an ecosystem with reference to its *non-living* components. The term **environment** describes the overall physical surroundings that occur in a biome or an ecosystem (although it is often used more generally: people speak of the environment of the Earth).

Within the environment of a single ecosystem are **habitats**, individual areas in which particular groups or individual organisms live. The habitat of any organism is its normal home. Some organisms have only one habitat, for example, the giant panda lives only in bamboo forest. Others, such as humans, live in many different ecosystems and biomes and so have many different habitats.

Another important term is **ecological niche**. To describe an organism's niche, we need to show how that organism relates to the physical and biological components of its surroundings. It is not just where it lives, it is also how it lives and what it does.

Fig 32.9 This individual stoat is part of a larger population. Like other members of the group, she is genetically unique. She has evolved to become well adapted to her woodland environment and, if her particular genes allow her to live a long life, she will pass her genes on to several new members of the next generation

 **QUESTION 3**

3   What is the difference between an organism's habitat and its ecological niche?

3 THE BASIC FEATURES OF AN ECOSYSTEM

Let's consider an example of a real ecosystem. Fig 32.10 shows the main features of a familiar aquatic ecosystem, the garden pond. This is simplified (a real pond contains more organisms), but you can see that an ecosystem is made up of physical or abiotic components such as water, mineral nutrients, soil and rock, and living or biotic components such as fish, snails, insects and water plants.

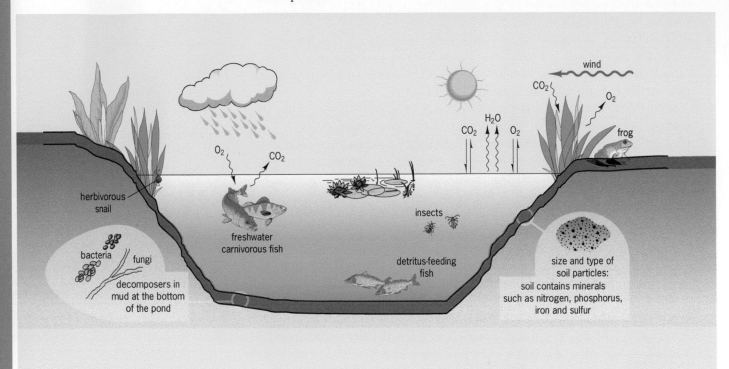

Fig 32.10 A schematic diagram showing a garden pond as an ecosystem. Every living organism in the pond is affected by abiotic factors such as temperature, light levels and mineral availability, and by biotic components – other living organisms

Think about one of the insects in the pond. It can only survive and live there if the physical conditions in the pond suit it. So the pH and the temperature of the water must be within a suitable range. The insect also needs to interact with other organisms: it needs to eat smaller organisms or to feed on the sap of plants; and it needs to be with other insects of the same species so that it can find a mate and reproduce. And there is a 'grey area': some of its physical needs, dissolved oxygen for instance, are only met if the insect lives with organisms that release oxygen as a product of photosynthesis.

THE ABIOTIC COMPONENTS OF AN ECOSYSTEM

The abiotic components of an ecosystem give it a physical form, providing places for organisms to live. Physical conditions help to determine where each organism survives best: plants adapted to dry, almost desert conditions will not thrive if a river changes its course and the environment becomes boggy and wet. The nature of the physical environment in any ecosystem depends on:

- the rock and soil types present;
- the type of landscape: mountain range, flood plain, etc.;
- the position on Earth: altitude, latitude, etc.;
- the climate and weather, including light, temperature, water availability, wind and water movements;
- the potential for catastrophe: fire, flood, etc.

REMEMBER THIS

In a real ecosystem, the abiotic and biotic factors all act together to determine the environmental conditions for each organism. Both interact to constantly change the conditions in the ecosystem. How this happens depends on the ecosystem and the physical conditions. The Science in Context box on the next page gives a thought-provoking example.

Some abiotic factors, such as the position of an ecosystem on Earth, remain unchanged as time passes. Others, such as the landscape and soil type, change very slowly. Although the basic climate of an ecosystem is usually pretty constant, you can get daily and often hourly changes in weather. Catastrophes such as fires and floods are sudden and short-lived, but they can produce severe, long-lasting effects (Fig 32.11).

Fig 32.11 In some ecosystems, catastrophic events are necessary. Some species of plant, such as *Banksia* from Western Australia (left), reproduce only after a serious fire (below). *Banksia* seeds are protected by closed capsules that can be broken open only by the heat of the flames: without exposure to fire, the seeds cannot germinate

SCIENCE IN CONTEXT

A modern-day plague

The abiotic and biotic parts of any ecosystem interact in many different ways, some subtle and some not so subtle. Australia provides one of the most dramatic examples of how disrupting the delicate balance of an ecosystem can devastate both the environment and the populations of living organisms that depend on it.

Two hundred years ago, Australia was hot, dry and covered predominantly by grasslands, with some areas of forest. When European settlers arrived they started to clear the forests and brought in two alien species of mammal: sheep and rabbits. The rabbits in particular multiplied extremely rapidly in conditions that suited them very well, and were soon eating the grass faster than it could grow.

Ten years after just 24 rabbits were introduced, the wild population had grown to millions and started destroying crops and wild vegetation as well as any grass that was left. They out-competed native marsupial species, driving many of them to extinction. This plague continued until the late 1950s, when farmers introduced the myxomatosis

virus (Fig 32.12). This killed over 99 per cent of the rabbits, with a few surviving because they were naturally resistant. Today, Australia is in danger of being overrun by another rabbit plague.

Fig 32.12 A rabbit suffering from the symptoms of myxomatosis, the deadly viral disease that wiped out most of the rabbit population. Most rabbits today are resistant to the virus

THE BIOTIC COMPONENTS OF AN ECOSYSTEM

The biotic environment that surrounds an organism in an ecosystem results from the activities of all the other organisms living there. Complex relationships called intraspecific interactions occur between organisms of the same species. Interspecific interactions happen between organisms from different species (Table 32.2).

Table 32.2 Types of interactions that occur between organisms in an ecosystem

Activity	Type of interaction	What happens
reproduction	intraspecific	Location of, selection of and competition for a mate.
caring for young	intraspecific	Both parents, one parent or, more rarely, older siblings feed, protect and shelter the young.
social behaviour	intraspecific	Animals co-operate to find food and to defend themselves against competitors or predators (Fig 32.13).
competition	intraspecific	Organisms compete for resources such as food, space, light, water and mineral nutrients.
reproduction	interspecific	Animal vectors pollinate some species of plant and help to disperse the fruits and seeds of others (Fig 32.14).
caring for young	interspecific	Rare, but some species rear the young of another species.
mutualism	interspecific	Two organisms both benefit from a long-term association.
parasitism	interspecific	Individuals from one species use another species to provide food and shelter while giving nothing in return. Many hosts actually suffer from the interaction: many parasites cause disease.
predation	interspecific	All animals obtain food by eating another organism.
protection	interspecific	Many species make use of other species to hide from predators. Some copy the markings of other species to mimic warning signals that predators avoid.
competition	interspecific	Organisms in a community compete for resources such as food, space, light, water and mineral nutrients.
defence	interspecific	Organisms have developed various strategies to defend themselves against predators, parasites and competitors.

Fig 32.13 Meercat sentries keep a look out for predators while other members of the group search for food and look after the young

Fig 32.14 Black-chinned humming birds pollinate the penstemon as they drink its nectar

4 AN OVERVIEW OF THIS SECTION OF THE BOOK

In this section of the book, we provide an introduction to ecology at advanced level, with an emphasis on what is important for human biologists to know. In the next chapter there is a thorough overview of the importance of energy and a description of the biochemical processes of respiration and photosynthesis. In Chapter 34 we look at recycling in the natural world, with detail on the carbon cycle and nitrogen cycle. Chapter 35 is a study of populations and, finally, in the last chapter in this section, we look at how humans affect their environment and examine some of the big issues that seem likely to threaten our environment in the future.

33

Energy in living systems

33 ENERGY IN LIVING SYSTEMS

An understanding of energy is of fundamental importance in biology, from the study of individual molecules and reactions to global issues such as the use of fossil fuels, pollution, damage to the ozone layer and the greenhouse effect

The global importance of energy

All living organisms depend on energy. Most of them live by the energy that comes from sunlight. Studying energy and how it flows through organisms and ecosystems is valuable for many topics in human biology.

Part I of this chapter gives an overview of some of the key concepts that you will need to understand in order to study energy flow in living systems. Part II gives a fairly detailed account of photosynthesis. This topic is crucial to understanding how energy flows through ecosystems and it is a core topic on most human biology syllabuses. Part III concludes the chapter with the process of cell respiration. Again we have included details of the biochemical pathways involved in this vital life process, but the information is presented in a step-wise format, so that you can build up your understanding.

With all three topics, remember that you should concentrate on understanding the relevance and overall importance of the process, not the details of individual biochemical reactions and pathways.

PART I: OVERVIEW OF KEY CONCEPTS

1 SOLAR ENERGY, PHOTOSYNTHESIS AND RESPIRATION

This planet, and all the organisms on it, must have energy and almost all of it is supplied by our nearest star, the Sun. A huge amount of sunlight reaches the surface of the Earth but only about 2 per cent of it is actually absorbed by plants (Fig 33.1). The rest of the energy heats up the land, air and water, preventing the planet from freezing.

In **photosynthesis**, plants use radiant energy from the Sun to convert simple inorganic substances (mainly carbon dioxide and water) into larger organic molecules (glucose, starch, lipids and proteins). The plant tissues built from the products of photosynthesis form food for organisms that cannot make their own, such as animals, fungi and most bacteria. Some of these feed on other organisms. Ultimately, therefore, the products of photosynthesis provide the energy which most living things need to carry out the processes of life. This energy is made available to organisms through the process of **cell respiration**.

We look at the biochemistry of photosynthesis and cell respiration later in this chapter, but first we revise some key concepts.

Fig 33.1 Photosynthesis by blue-green bacteria was responsible for the appearance of oxygen in the Earth's atmosphere 2 billion years ago, and photosynthesis by plants maintains the oxygen level at 20 per cent of the atmosphere today

2 WHY DO WE NEED ENERGY?

Humans need a constant supply of energy to maintain their life processes:
- **Growth and repair** of cells and tissues. Energy is required for the biochemical reactions that build large organic molecules from simpler ones. For example, energy is needed to build proteins from amino acids.
- **Active transport**. Energy is required to move some substances in or out of cells. The transport of amino acids from the small intestine into the blood is achieved by active transport. Active transport often takes place against a diffusion gradient and it allows the body to control its internal environment more efficiently.
- **Movement**. All movement requires energy, and human movement occurs on several levels:
  - inside cells, e.g. chromosomes separating
  - whole cells, e.g. sperm swimming (Fig 33.2)
  - tissues, e.g. muscles contracting
  - whole organs, e.g. a heart beating
  - parts or whole organisms, e.g. talking, walking.
- **Temperature control**. Humans are warm-blooded animals, **homoiotherms**, and we use around 70 per cent of the energy from respiration to maintain the body at a constant 37 °C.

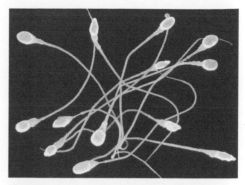

Fig 33.2 Movement requires energy. To swim far enough up into the Fallopian tube to reach an egg, a sperm cell needs enough energy to enable it to swim over 7500 times its own length – the equivalent of a 10 kilometre swim for an adult human

> Active transport and the movement of substances into and out of cells is discussed in Chapter 4. Nitrogen recycling is discussed in Chapter 34 and human nutrition is discussed in Chapter 7.

3 HOW DO LIVING ORGANISMS GET THE ENERGY THEY NEED?

In order to respire and so release energy, organisms need a supply of food. Food can be obtained in two ways:
- Some organisms can make their own food. These organisms are called **autotrophs** (meaning 'self-feeders') and they use energy from the surroundings to make the materials they need. There are two kinds of autotroph, the **photoautotroph** and the **chemoautotroph**. Photoautotrophs, such as green plants, photosynthesise, using sunlight as an energy source. Chemoautotrophs, all of which are bacteria, use energy made available from chemical reactions other than those involved in photosynthesis. Some bacteria involved in nitrogen recycling are chemoautotrophs.
- Some organisms need to obtain ready-made food because they cannot make their own. Such organisms are called **heterotrophs**. Heterotrophs must obtain their food, either directly or indirectly, from organisms that carry out photosynthesis. They either eat their food (ingesting it into a gut, as most animals do) or digest the food first and then absorb the nutrients, as decomposers, mainly fungi and bacteria, do (Fig 33.3).

Fig 33.3 All of these organisms have something in common – they are heterotrophs. They cannot make their own food and must take it, ready made, from an outside source. There are several different ways of doing this. It's obvious that the giraffe is feeding on plants and the lions are devouring a prey animal. Less obvious, is that the fungi are digesting and feeding on the wood of the tree. In the Petri dish on the right, bacteria are feeding on blood agar

4 ENERGY FLOW: WHERE DOES ENERGY COME FROM, WHERE DOES IT GO?

As Figs 33.4 and 33.5 show, as little energy is transferred at each point in a food chain, there is relatively little energy available for the final consumers, usually large carnivores. For this reason, such animals are quite rare.

Think of your own diet. You consume large amounts of food but very little of it is incorporated into your body. Compare your own body mass with an estimate of the mass of all the food you have ever eaten: you can see that a lot of energy must have been lost as heat. This heat is used to keep your body temperature at a constant 37 °C. So much is needed because you are constantly losing heat to the air, your clothes and the objects you touch.

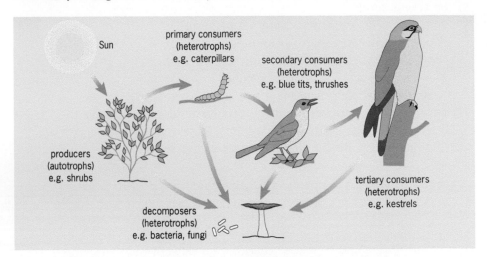

Fig 33.4 Energy flow through a food web

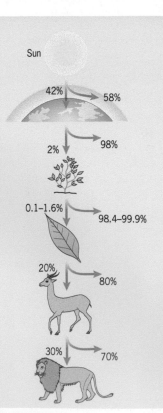

Of the light energy from the Sun that reaches the Earth, only about 42 per cent actually gets through to the surface of the planet. The rest is either reflected by the atmosphere, or is absorbed by it, a process that warms the atmosphere.

Of the light energy reaching the Earth's surface, about 2 per cent is absorbed by plants. The rest goes to produce the heat that warms up the land and the oceans.

Of the energy absorbed by plants, only 0.1–1.6 per cent is incorporated into plant tissue – and even some of this is lost as heat because the plant itself must respire.

Of the plant tissue eaten by herbivores, about 20 per cent (at most) will become incorporated into the tissues of the animal. Most of the rest is lost as heat as the animal respires.

Of the animal tissue eaten by a carnivore, a maximum of 30 per cent becomes incorporated into the body of the carnivore. Most of the rest is lost as heat.

Fig 33.5 The basic idea of energy flow. At the start of any food chain there must be an organism capable of making food – usually a green plant. From there, the energy is passed along the chain within food molecules. Some of this energy is eventually used to make the cells, tissues and organs of heterotrophic organisms

5 ENERGY TRANSFER IN ECOSYSTEMS

In photosynthesis, green plants and some algae and bacteria use the energy from sunlight to power chemical reactions that combine simple inorganic molecules to form complex organic compounds. We say that the autotrophs in an ecosystem are **producers**: the products that they make form the food that ultimately feeds all other organisms in the ecosystem.

The heterotrophs in an ecosystem meet their need for energy by feeding on other organisms. We say that they are the **consumers**. Generally, **primary consumers** are **herbivores**: they obtain their energy by eating producers. **Secondary consumers** are **carnivores** or **scavengers**: they eat primary consumers. **Tertiary consumers** are **carnivores** that prey on other meat-eaters. This chain of dependence is usually called a **food chain** (Fig 33.6). The distinct levels of each chain are called **trophic levels**.

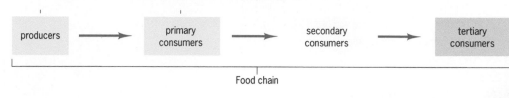

Fig 33.6 A simple flow chart summarising the overall structure of a food chain that has four trophic levels

As one organism eats another, there is a transfer of energy from the bodies of the producers (or consumers) into the bodies of consumers through the trophic levels of a food chain. Energy transfer occurs in one direction only: it is not recycled, and much of the energy escapes from the ecosystem as heat (look ahead to Fig 33.13).

Another major group of heterotrophs in an ecosystem are the **decomposers**. These are mainly bacteria and fungi. They break down the bodies of dead animals and plants, usually by extracellular digestion (Fig 33.7) and absorb the soluble products. Transfer of energy from producers and consumers to decomposers is also one-way, with a high proportion eventually lost as heat. The process of decomposition allows mineral nutrients to be cycled in the ecosystem.

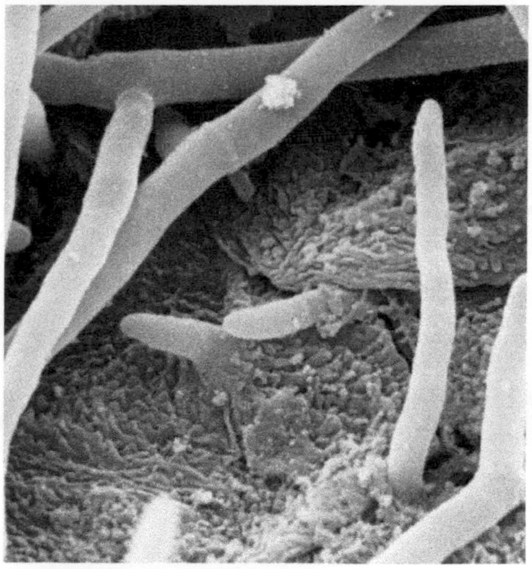

FOOD CHAINS AND FOOD WEBS

An example of a food chain is shown in Fig 33.8. Most animals have a varied diet. A food chain is therefore only part of a much larger feeding picture: the **food web**. Take the top freshwater carnivore, the pike. It does not feed on stickleback alone but also on roach and insects such as dragonfly nymphs and pond skaters.

Most rivers and lakes include a wide variety of producers, ranging from the microscopic plankton to the larger pondweeds, rushes and flowering plants.

> **REMEMBER THIS**
> Humans are **omnivores**: we eat a mixture of plant and animal food. An omnivore can be a primary, secondary and even a tertiary consumer all at the same time.

> **? QUESTION 3**
> 3 Look at Fig 33.6. What do the arrows in a food chain represent?

Fig 33.7 Fungi grow into their food (in this case, human skin), secreting enzymes which digest the food externally. The soluble products of digestion are then absorbed through the walls of the hyphae

> **Mineral recycling is discussed in Chapter 34.**

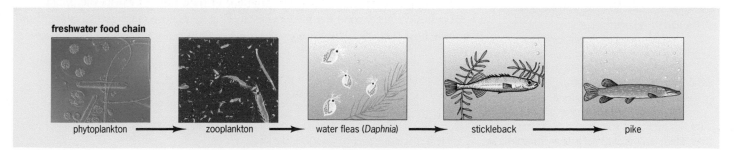

freshwater food chain

phytoplankton → zooplankton → water fleas (*Daphnia*) → stickleback → pike

Fig 33.8 An example of a freshwater food chain with five trophic levels (most food chains have fewer trophic levels than this)

Food webs and pesticides

In developing a new pesticide, scientists look for a chemical that kills only the pest that we want to get rid of, harms no other species, breaks down into something harmless after a short time, does not lead to the development of resistance, and is cheap. This is a pretty tall order: no pesticide yet developed is perfect, but modern pesticides are much safer than some of those developed and used in the 1950s and 1960s.

DDT (dichlorodiphenyltrifluoroethane) was used as a pesticide in many countries after World War II. It breaks down slowly, remaining in the environment for between two and five years.

During the 1950s and 1960s, it became obvious that DDT was affecting whole food webs. Fig 33.9 shows what happened when DDT was transferred up the trophic levels of a food chain in an ecosystem in the USA. DDT is not excreted and concentrates in the fatty tissues of organisms. So, although the concentration of DDT in zooplankton, the primary consumers in the chain, was only 0.04 parts per million, its concentration in the tissues of a top carnivore such as the osprey or bald eagle was 25 parts per million. The bodies of the large birds of prey converted the DDT into a substance that made their eggshells very fragile. As a result, very few birds managed to breed and their numbers fell quickly.

In 1972, DDT was banned and since then, the numbers of the large predatory birds have recovered. However, there are still problems because of the illegal use of DDT.

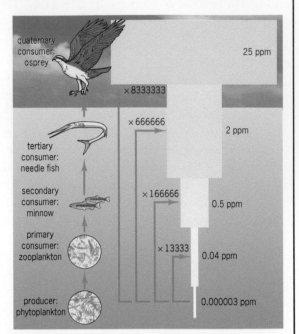

Fig 33.9 Bioamplification of DDT through an aquatic food chain. DDT accumulates in the fatty tissues and cannot be excreted. The fat of the osprey contains over 8 million times more DDT than the water at the bottom of the food chain in which the producers live

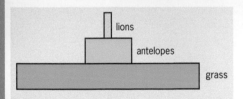

Fig 33.10 A simple pyramid of numbers. The width of each box represents the relative number of organisms present in each trophic level at any one particular time. In this example, many grass plants are needed to sustain all the antelopes that one individual lion needs to eat to survive

Fig 33.11 Two examples of inverted pyramids of numbers. Neither is completely inverted; they simply do not conform to the expected pyramidal shape because one of the levels is distorted
(a) One sycamore tree can feed many thousands of aphids, making the first trophic level of this pyramid of numbers much smaller than the second
(b) This pyramid is complicated by the top trophic level, which involves a parasite. Many parasites can infect one host. *Wuchereria bancrofti* is a nematode worm that causes the disease elephantiasis in humans (see Fig 10.14 on page 166)

PYRAMIDS OF NUMBERS

Food chains and food webs tell us a great deal about the feeding relationships that occur between organisms in an ecosystem, but they are **qualitative** rather than **quantitative**. That means that we know which organisms are part of the chain or web, but we have no idea of the exact numbers of organisms involved at each trophic level.

Clearly, numbers are very important. There are many more individual producers than primary consumers, and, as you carry on along a food chain, the numbers of organisms at each trophic level continue to decrease. At the same time, the body size of individual organisms usually increases. It is difficult, if not impossible, to find out exactly how many producers are eaten by a primary consumer and so on, but we use a pyramid of numbers to show the general idea (Fig 33.10).

Two odd things can happen when you use **pyramids of numbers**. Sometimes the base of the pyramid is very narrow, indicating that there are fewer producers than primary consumers. This seems not to make sense, but these examples arise when a few large plants such as trees produce food for thousands of tiny plant-feeders, such as aphids (Fig 33.11a). In pyramids of numbers that represent food chains that have a population of parasites at the top, the upper level can be much larger than the level below it. This is because many parasites can feed on one host (Fig 33.11b). We say that these pyramids are **inverted pyramids**.

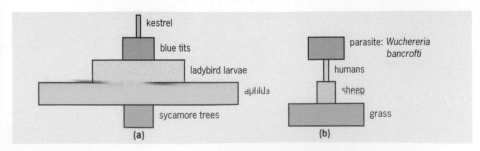

PYRAMIDS OF BIOMASS

Ecologists often use **pyramids of biomass** (Fig 33.12a), to avoid the problem of inverted pyramids of numbers. Pyramids of biomass show the mass of all the organisms at each trophic level – dry mass is the most useful measure. This is obviously extremely difficult: obtaining the dry mass of organisms means killing them, drying them and measuring the remains. Not surprisingly, very few pyramids of biomass have ever been determined in this way.

However, using samples of organisms and their wet mass allows ecologists to make estimates and devise models. But, even this type of pyramid can be inverted, as Fig 33.12b shows.

To obtain an even better model of an ecosystem, one that avoids inverted pyramids altogether, we must look at the way energy is transferred between the different trophic levels.

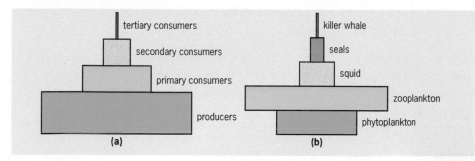

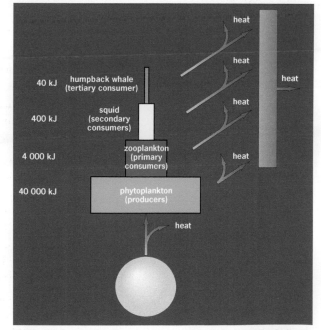

Fig 33.12a A generalised pyramid of biomass for a terrestrial ecosystem

Fig 33.12b In the waters of the Antarctic, the mass of zooplankton is five times that of the phytoplankton. If we draw up a pyramid of biomass for one of the food chains, we get an inverted pyramid. The zooplankton eat the phytoplankton so quickly that the latter never get the chance to attain a large biomass. Instead, phytoplankton have a high turn-over rate, reproducing very quickly. So the biomass of organisms produced each year is large, although the number alive at any one time is often less than the number of consumers

PYRAMIDS OF ENERGY

Fig 33.13 shows a typical pyramid of energy. It shows the amount of energy that is transferred along the food chain and how much is lost as heat at each level. The more levels in the chain, the less energy there is available at the top. Consequently, very few food chains have more than four or five levels. It also explains why the top carnivores such as the big cats and large whales are the first to become endangered when their ecosystems come under pressure from humans.

Efficiency of energy transfer between different trophic levels

As we saw earlier in this chapter, the energy transfer between organisms at different trophic levels of a food chain is never anything like 100 per cent efficient. In fact, less than 30 per cent of the energy available reaches the organisms in the next level. There are three main reasons for this:

- Not all the organisms in one trophic level are eaten by organisms from the next level. Many die and provide food for decomposers.
- Not all food that is eaten is digested and not all digested food is absorbed – some is lost as faeces.
- Most of the energy gained from the food that is absorbed is lost through the processes of respiration and excretion (production of metabolic waste).

Fig 33.13 A pyramid of energy for an Antarctic food chain similar to the one shown as a pyramid of biomass in Fig 33.12b. As you go up the pyramid, there is a 90 per cent loss of energy at each trophic level. This energy, of course, does not disappear: it is 'lost' as heat. For a more detailed investigation of energy transfer, see the Assignment for this chapter on the CAS website

? QUESTION 4

4 List the foods you ate at your last meal. Sketch out the food chains that produced them and, for each food chain, say which trophic level you, the consumer, represent.

PART II: PHOTOSYNTHESIS

6 THE IMPORTANCE OF PHOTOSYNTHESIS

It is difficult to overstate the importance of photosynthesis – the process that feeds the world and replenishes the atmosphere with oxygen. Organisms that photosynthesise are able to capture the energy of the Sun and use it to make carbohydrate molecules such as sugars and starches, and, indirectly, other essential compounds such lipids, proteins and nucleic acids. It is no exaggeration to say that photosynthesis turns thin air into food.

Photosynthesis is a complex process, but it can be summarised as:

$$\text{carbon dioxide} \quad \text{and} \quad \text{water} \xrightarrow[\text{chlorophyll}]{\text{light}} \text{glucose} \quad \text{and} \quad \text{oxygen}$$

$$6CO_2 \quad + \quad 6H_2O \quad \rightarrow \quad C_6H_{12}O_6 \quad + \quad 6O_2$$

You shouldn't think of this as a chemical equation, even though it looks a lot like one. In reality, the carbon dioxide molecules never react directly with water molecules, so this is more a summary of photosynthesis, showing what goes in and what comes out.

SCIENCE IN CONTEXT

Artificial photosynthesis

Plants are able to convert sunlight into a high-energy fuel in a complex yet very efficient process that scientists have so far been unable to duplicate. In recent years, however, progress has been made towards creating a type of artificial photosynthesis. The benefits in the long term could be huge: cheap, clean energy sources.

A team at the Massachusetts Institute of Technology in the US has developed a protein that captures energy from a light source and stores it in the form of hydrogen gas. Creating a molecule to replace a leaf – 'photosynthesis in a beaker' – could help revive interest in the Sun as a source of energy.

One possible spin-off of this technology is that in the future Armed Forces operations in the field could rely on biological methods to produce electricity through photosynthesis, rather than relying on fossil fuels and bulky, heavy batteries.

Over millions of years of natural selection, plants have evolved to make optimum use of the different wavelengths of light, and plants convert 98 per cent of the sunlight they receive into energy. Conversely, current artificial solar energy systems are only 10–15 per cent efficient. Using plant proteins to produce electricity – an area of research known as biological photovoltaics – could make the process much more efficient; perhaps by 40–50 per cent.

It is possible that in the future soldiers could have photosynthetic coatings on their Kevlar helmets that could produce enough energy for their electronic equipment. Other equipment and vehicles could also be covered with these solar converters. An added benefit is that the protein coatings would make whatever they coat more difficult to detect by electronic means since they would mimic the natural environment.

Fig 33.14 The light infantry? In future, helmets may by able to photosynthesise and produce enough energy to power a soldier's equipment

7 PHOTOSYNTHESIS FACTS

- Organisms that can photosynthesise include higher plants (Fig. 33.15), algae (including the seaweeds) and photosynthetic bacteria.
- Organisms that can photosynthesise are called **producers**, or **autotrophs** ('self-feeders').
- The process involves absorbing the light energy of the sun and converting it into chemical energy. The compounds made, such as sugars, starch, lipids and protein, contain this energy in their chemical bonds.
- Globally, photosynthesis makes billions of tonnes of organic compounds every year; one estimate puts global production at 70×10^{12} kg per year.
- Up to 50 per cent of all photosynthesis takes place in the sea, in the tiny algae known as phytoplankton (Fig 33.16).
- Virtually all ecosystems are supported by photosynthesis – it is the most important way energy gets into an ecosystem.
- Photosynthesis needs light, chlorophyll, carbon dioxide, water and a suitable temperature. The rate of photosynthesis is limited by the factor in shortest supply.
- Photosynthesis is a complex step-by-step process – each step is catalysed by a particular enzyme.

Fig 33.15 Where did this tree come from? We know that the material to make our bodies comes from the food we eat, but what about plants? This tree may weigh upwards of 3000 tonnes, but the vast majority of the organic compounds in the wood didn't come from the soil – they were made by photosynthesis using carbon dioxide from the atmosphere. A huge amount of water was also needed from the roots, but most of the dry mass comes, quite literally, from thin air

Fig 33.16 When you think of photosynthesis on a global scale you probably think of rainforests or fields of crops, but the tiny algae in the oceans produce just as much food as land plants. These are diatoms, one type of single-celled algae that are responsible for producing huge amounts of food in the world's oceans. As one of the many species that makes up phytoplankton (plant plankton), they support vast marine and freshwater ecosystems

8 PHOTOSYNTHESIS AND RESPIRATION

If you look at the overall summary for respiration, where:

glucose and oxygen → carbon dioxide and water + energy

you can see that photosynthesis and respiration are effectively the reverse of each other. Photosynthesis traps energy in organic molecules, and respiration releases it. Overall, photosynthesis is the reduction of carbon dioxide to form organic molecules, while respiration is the oxidation of organic molecules to form carbon dioxide. It is important to remember that plants also respire day and night, but in order for them to accumulate organic molecules and to grow, their overall rate of photosynthesis must exceed their rate of respiration.

9 WHAT HAPPENS IN A LEAF?

In most plants, the leaf is an organ of photosynthesis. As Fig 33.17 shows, the leaf is perfectly adapted to maximise the process. The most actively photosynthesising cells in the leaf are the **palisade cells**, and the whole structure of the leaf is geared towards providing these cells with what they need and taking away their waste products.

Fig 33.17 Journey to the centre of a leaf. The chlorophyll pigments are housed on discs called thylakoids that are piled into grana inside chloroplasts. Most chloroplasts are found in the palisade and spongy mesophyll cells in the centre of leaves

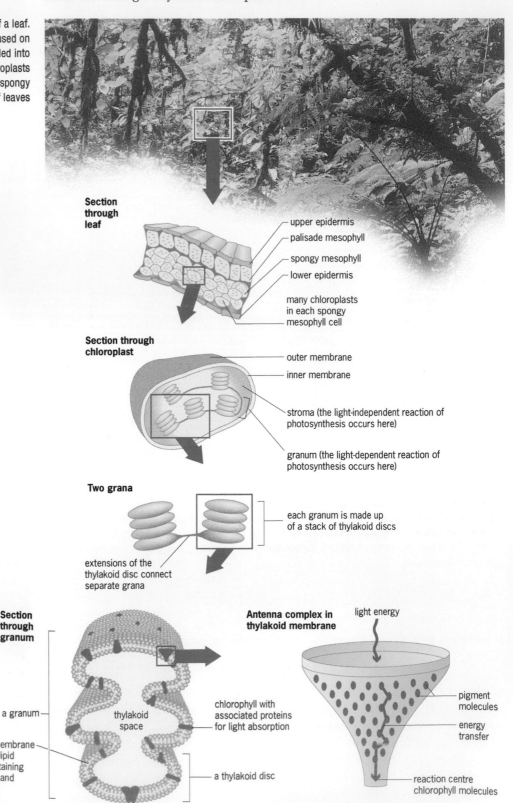

Section through leaf

- upper epidermis
- palisade mesophyll
- spongy mesophyll
- lower epidermis
- many chloroplasts in each spongy mesophyll cell

Section through chloroplast

- outer membrane
- inner membrane
- stroma (the light-independent reaction of photosynthesis occurs here)
- granum (the light-dependent reaction of photosynthesis occurs here)

Two grana

- each granum is made up of a stack of thylakoid discs
- extensions of the thylakoid disc connect separate grana

Section through granum

- a granum
- thylakoid membrane (a phospholipid bilayer containing chlorophyll and proteins)
- thylakoid space
- chlorophyll with associated proteins for light absorption
- a thylakoid disc

Antenna complex in thylakoid membrane

- light energy
- pigment molecules
- energy transfer
- reaction centre chlorophyll molecules

Adaptations of the leaf (shown in Fig 33.17) include:

- It is thin, so that gases can diffuse in and out quickly.
- It has a large surface area to maximise light absorption.
- It has a thin layer of wax – the **cuticle** – to reduce evaporation from the upper surface.
- It has a rich network of **xylem** fibres to deliver water and minerals, and **phloem** fibres to remove the product of photosynthesis (mainly as sucrose). This vascular tissue also gives the leaf more rigidity.
- The cells of the upper epidermis have no chloroplasts, so that more light gets through to the palisade cells.
- The palisade cells are deep and tightly packed. The chloroplasts within them can move so that all have a chance of obtaining all the light they need.
- Guard cells surround pores called **stomata** (singular = stoma; 'hole') opening in the day to allow carbon dioxide in, then closing at night to reduce water loss.
- Loosely packed **spongy mesophyll** cells create air spaces, making gas exchange more efficient between the atmosphere and the palisade cells. The air spaces also trap a little carbon dioxide, allowing some photosynthesis even when the stomata are closed.

CHLOROPLASTS

Chloroplasts are the organelles of photosynthesis. In the leaves of higher plants there can be as many as 100 or more chloroplasts in each cell, but more commonly 20–30. Their detailed structure is shown in Fig 33.18.

? QUESTION 5

5 In order to make photosynthesis in the palisade cell as efficient as possible, what must be supplied and what must be taken away? Use the basic photosynthesis equations to help you.

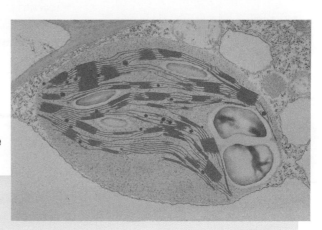

Fig 33.18 False colour transmission electron micrograph of chloroplast. The grana are connected by thin membranes called intergrana lamellae. Note the starch grains and lipid globules. The features are explained in the diagram below

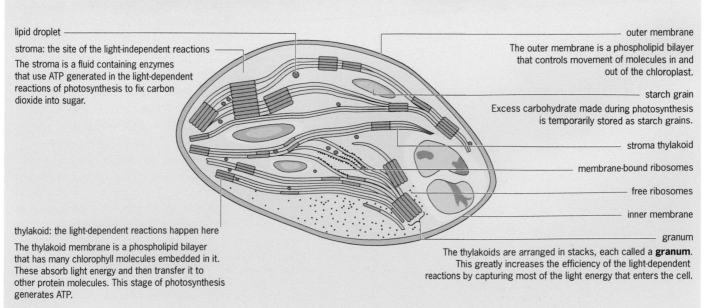

lipid droplet

stroma: the site of the light-independent reactions
The stroma is a fluid containing enzymes that use ATP generated in the light-dependent reactions of photosynthesis to fix carbon dioxide into sugar.

thylakoid: the light-dependent reactions happen here
The thylakoid membrane is a phospholipid bilayer that has many chlorophyll molecules embedded in it. These absorb light energy and then transfer it to other protein molecules. This stage of photosynthesis generates ATP.

outer membrane
The outer membrane is a phospholipid bilayer that controls movement of molecules in and out of the chloroplast.

starch grain
Excess carbohydrate made during photosynthesis is temporarily stored as starch grains.

stroma thylakoid

membrane-bound ribosomes

free ribosomes

inner membrane

granum
The thylakoids are arranged in stacks, each called a **granum**. This greatly increases the efficiency of the light-dependent reactions by capturing most of the light energy that enters the cell.

Chloroplast facts

- Chloroplasts are usually shaped like biconvex discs, about 5 μm across and 2 μm deep. The flattened shape increases the rate of the exchange of materials.
- They have a double outer membrane.
- There is an elaborate system of membranes inside the chloroplast that provides a large surface area for housing the chlorophyll pigments. The disc-shaped membranes that house the chlorophyll are called **thylakoids**, which are stacked into **grana** (singular = granum). Usually, the thylakoids are at right angles to the incoming light.
- The fluid surrounding the grana is called the **stroma**.
- Chloroplasts have their own DNA and ribosomes so they can synthesise their own proteins.
- It is thought that chloroplasts may have originated as free-living algae that became incorporated into a larger cell – see the endosymbiont theory on page 22.

CHLOROPHYLL AND LIGHT

In order to understand the role of chlorophyll in photosynthesis it helps to know something about the nature of light. Fig 33.19a shows the electromagnetic spectrum – the range of different types of radiation. Most of the radiation that reaches the Earth from the Sun is in the region of 400–700 nm wavelength, so not surprisingly this is what our eyes have evolved to see; it is the **visible spectrum**. When stimulated by the whole 400–700 nm range, we perceive white light but, as Fig 33.19b shows, this can be divided up into the familiar seven colours. Chlorophyll has also evolved to make use of this available light.

What we think of as chlorophyll is actually a mixture of compounds that fall into two basic types; the **chlorophylls** and the **carotenoids** (Fig 33.20). The two main chlorophylls, **a** and **b**, are almost identical molecules that absorb light mainly in the red and blue areas of the spectrum. The carotenoids, such as **β-carotene**, harvest light from different wavelengths, thus allowing plants to make much better use of the available light. Energy absorbed by the carotenoids is eventually passed on to the chlorophyll molecules before being used in photosynthesis. The carotenoids also act as a safety mechanism, protecting the leaf from the damaging effects of too much light.

REMEMBER THIS

The word *pigment* is a general name for a coloured compound. Haemoglobin and the different types of chlorophyll are all pigments.

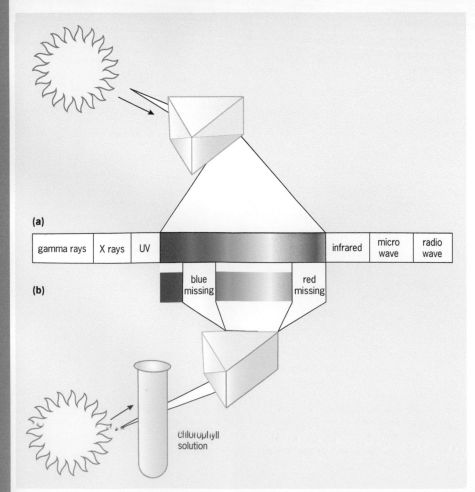

(a)

| gamma rays | X rays | UV | | | infrared | micro wave | radio wave |

(b)

blue missing

red missing

chlorophyll solution

Fig 33.19a The electromagnetic spectrum ranges from the long wavelength (and lowest energy) radio waves to the shortest wavelength (high-energy) gamma rays. Visible light forms just a small part of the spectrum, between ultraviolet and infared

Fig 33.19b White light splits into its component colours when passed through a prism. When white light is first passed through chlorophyll, the red and blue light is absorbed, which is why plants appear green

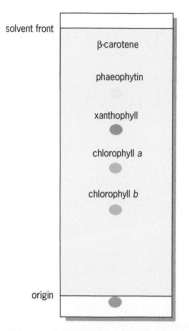

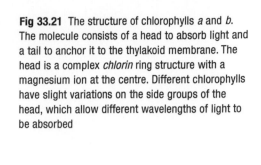

Fig 33.21 The structure of chlorophylls *a* and *b*. The molecule consists of a head to absorb light and a tail to anchor it to the thylakoid membrane. The head is a complex *chlorin* ring structure with a magnesium ion at the centre. Different chlorophylls have slight variations on the side groups of the head, which allow different wavelengths of light to be absorbed

Fig 33.20 If you grind up a leaf with a little sand (to break up the cell walls) and a suitable solvent such as propanone, you can run a paper or thin layer chromatogram (see page 38 in Chapter 3) to separate out the different pigments that make up chlorophyll

Key

→ co-ordinate linkage

X is CH_3 in chlorophyll *a*

X is CHO in chlorophyll *b*

The structure of **chorophylls *a*** and ***b*** is shown in Fig 33.21. The ring structure surrounding the magnesium ion contains five smaller rings consisting of alternating single and double bonds. In reality each bond is partly double; a **conjugated system**. These bonds absorb light, so the more of them there are, the greater the light absorbance. Chlorophyll *a* absorbs strongly in the 660–700 nm range: red light. It also absorbs to a lesser extent in the blue range.

The wavelengths of light that stimulate photosynthesis are shown by an **action spectrum**, obtained by varying the wavelength of light a plant receives and then measuring the rate of photosynthesis. When the chlorophyll pigments are extracted, and their absorption at the different wavelengths is measured, we get an **absorption spectrum**. Fig 33.22 shows that the two graphs are very similar, showing that the chlorophylls are the pigments responsible for photosynthesis.

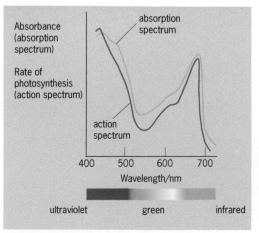

Fig 33.22 The action spectrum shows the wavelengths of light that stimulate photosynthesis, while the absorption spectrum shows the wavelengths of light absorbed by isolated chlorophyll pigments

10 AN OVERVIEW: THE FIRST STEP TO UNDERSTANDING PHOTOSYNTHESIS

Photosynthesis is often seen as a difficult topic, but approaching it in the right way will help you make sense of it. First, you need an overview or framework that you can fit the details on to later. Photosynthesis can be divided into two easy steps as summarised below and in Fig 33.23:

● **Step 1** The light-dependent reaction. Light hits chlorophyll, which emits two high-energy electrons. These electrons pass through a series of electron transfer reactions that make ATP and reduced NADP. The electrons lost by the chlorophyll are replaced when water is split, a process that also produces oxygen as a by-product.

● **Step 2** The light-independent reaction. ATP and reduced NADP are used to power a series of reactions known as the **Calvin cycle**. Overall, these reactions reduce carbon dioxide into the all-important carbohydrate.

For some A-level specifications, all the detail you need (and more) has been covered in the chapter so far and the summary at the end. Other specifications require the more in-depth account that follows.

Fig 33.23 An overview of photosynthesis

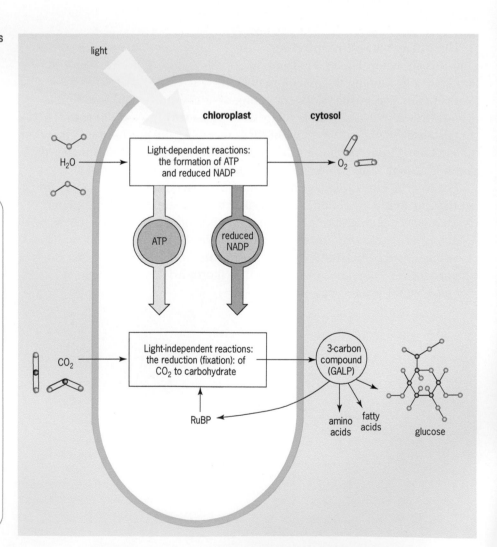

REMEMBER THIS
What is reduced NADP?

Reduced NADP is a coenzyme that carries electrons. It is a powerful reducing agent, and is vital in the reduction of carbon dioxide into carbohydrate. Reduced NADP is similar to NADH in respiration, which performs basically the same function. To avoid confusion, remember 'P for photosynthesis': NADH in respiration, reduced NADP in photosynthesis (NB in reality the P stands for phosphate). While NAD$^+$ is reduced by an electron, NADP is reduced by an electron and a proton (i.e. a hydrogen atom).

11 THE LIGHT-DEPENDENT REACTION

In order to understand the light-dependent reaction, you need to be familiar with the fine structure of the chloroplast, and in particular the arrangement of the chlorophyll molecules on the thylakoid membranes. As Fig 33.17 shows, chlorophyll molecules are arranged in funnel shapes, known as **antennae complexes**. The mouth of the funnel is occupied by several different types of pigment, allowing light to be harvested across more of the visible spectrum.

When a photon of light hits any of the molecules of chlorophyll at the mouth of the complex, the electrons they contain absorb the energy and are raised from a low-energy **ground state** to an **excited state**. As the energy moves from molecule to molecule down the antenna complex, electrons are first excited, then fall back to the ground state as they pass their energy on. The energy passes on down the antenna complex until it eventually reaches two chlorophyll *a* molecules at the bottom, known as the **special pair**.

The stages in the light-dependent reaction (Fig 33.24) are as follows:

- **Stage 1** A photon of light hits photosystem II and is channelled down to the reaction centre, which emits high-energy electrons.
- **Stage 2** The electrons are picked up by electron acceptors. In doing so, the electron acceptors become reduced while the chlorophyll is oxidised. The electrons are then transferred along a series of electron carriers in the thylakoid membrane. The energy in the electrons is used to make ATP – this is **photophosphorylation**.
- **Stage 3** The electrons, having fallen to a lower energy level, are absorbed by the chlorophyll molecules in photosystem I.
- **Stage 4** A photon of light hits the reaction centre in photosystem I, which also emits two high-energy electrons. Again, the electrons are channelled down a series of electron carriers but this time the electrons combine with H^+ ions to reduce NADP into reduced NADP.

✔ **REMEMBER THIS**

The light-dependent reaction takes place on the thylakoid membranes, while the light-independent reaction takes place in the stroma of the chloroplasts.

? **QUESTIONS 6–7**

6 What is the difference between oxidation and reduction?

7 Which one involves a molecule gaining energy?

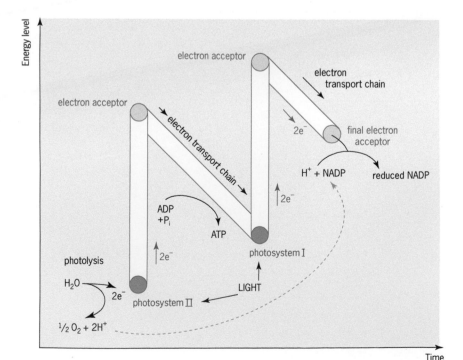

Fig 33.24 Schematic diagram of the light-dependent reaction

REMEMBER THIS

The electrons that leave the special pair of chlorophyll molecules need to be replaced in order for the reaction to continue. Where do the electrons come from?

● For photosystem I the electrons come from the electron transport chain.

● For photosystem II they come from the splitting of water, a process called **photolysis**. Water is split into protons and electrons, leaving oxygen as a by-product. It is this process that has led, over billions of years, to the oxygen-rich atmosphere we have today.

$$2H_2O \rightarrow 4H^+ + 4e^- + O_2$$

REMEMBER THIS

Of the two stages of photosynthesis, the light-independent reaction is much more reliant on enzymes and therefore much more temperature dependent.

CYCLIC AND NON-CYCLIC PHOTOPHOSPHORYLATION

Photophosphorylation refers to the production of ATP using the energy from light. There are two methods of doing this: cyclic and non-cyclic. ATP production as described above is called **non-cyclic phosphorylation**, because the electrons are not recycled. Electrons that start off in water molecules are passed through the process until they end up in reduced NADP, which is then used to make carbohydrate.

However, ATP can also be made by **cyclic phosphorylation**. In this case some of the electrons from photosystem I are channelled back on to the first electron transport chain, and their energy is used to make ATP before returning to photosystem II. This process therefore makes ATP but not reduced NADP. No water is needed to provide electrons, and so no oxygen is produced.

12 THE LIGHT-INDEPENDENT REACTION

In fact, the light-independent reaction happens in the light because it needs the ATP and NADPH that the light-dependent reaction produces.

The light-independent reaction takes place in the stroma (fluid part) of the chloroplast. The process is also known as the **Calvin cycle** after Melvin Calvin, the American scientist who first worked out the details in the 1950s.

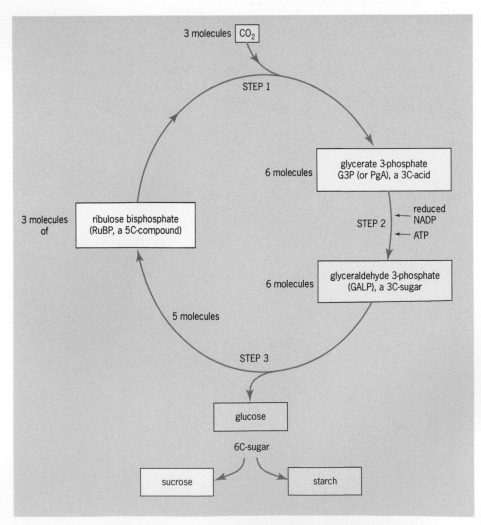

Fig 33.25 The light-independent stage of photosynthesis. See text for an explanation of each step

The steps in the Calvin cycle are shown in Fig 33.25.

- **Step 1** Carbon dioxide diffuses into the stroma of the chloroplast and combines with a 5-carbon sugar called **ribulose bisphosphate** (RuBP). This forms an unstable 6-carbon compound that immediately splits into two 3-carbon compounds called **glycerate 3-phosphate** (G3P). The enzyme that catalyses this reaction, **ribulose bisphosphate carboxylase** (shortened to **rubisco**), is the most abundant enzyme on Earth, accounting for up to 60 per cent of the protein in many leaves.
- **Step 2** The energy in ATP and the reducing power of reduced NADP are used to convert G3P into glyceraldehyde phosphate (GALP).
- **Step 3** GALP is used in two ways: five out of every six molecules are used – via a series of reactions – to make more RuBP to continue the cycle. One molecule in every six is used to make glucose.

Note that the carbon dioxide puts in one single carbon atom for every turn of the cycle. Thus the cycle must turn six times to make one glucose molecule. If the glucose were allowed to accumulate, it would lower the water potential of the cell, so it is converted into starch almost as soon as it is made, and this accounts for the large starch grains that can be seen in actively photosynthesising chloroplasts.

Plants move carbohydrate around in the form of sucrose. The plant's main storage compound, starch, is a large molecule that cannot pass across the cell membrane, so it must be converted into a smaller compound before it can pass into the phloem tissues of the plant.

PHOTORESPIRATION

It is thought that photosynthesis evolved in an atmosphere that contained much less oxygen and more carbon dioxide than does today's atmosphere. The enzyme that catalyses the reaction between RuBP and carbon dioxide, rubisco, can also catalyse a reaction between RuBP and oxygen – so oxygen and carbon dioxide compete for the same active site. When oxygen levels are high and carbon dioxide levels low, the oxygen reaction is favoured:

$$O_2 + RuBP \xrightarrow{\text{rubisco}} \text{glycolate (a 2-carbon compound)} + G3P$$

Compare this to the usual reaction of the Calvin cycle:

$$CO_2 + RuBP \xrightarrow{\text{rubisco}} 2 \times G3P \text{ (2} \times \text{3-carbon compound)}$$

You can see that one molecule of G3P is made instead of the usual two in the Calvin cycle. The glycolate cannot be fixed into carbohydrate, and instead passes into the mitochondria where it is respired.

Owing to its dependence on light, and the fact that it takes in oxygen and releases carbon dioxide, this process is called **photorespiration**. This process is wasteful to a plant because it means that some of the carbon dioxide that could be fixed into carbohydrate is lost. It has been estimated that photorespiration makes photosynthesis 30–40 per cent less efficient. However, some plants have overcome this problem – see the section on **C4 plants** below.

13 VARIATIONS ON THE THEME OF PHOTOSYNTHESIS

The mechanism of photosynthesis described above is known as the **C3 pathway** because the first stable product of the Calvin cycle is a 3-carbon compound, glycerate 3-phosphate. Most familiar plant species are C3 plants but there are two important variations on the theme: **CAM plants** and **C4 plants**.

Global warming makes plants move

Climate change in the UK is causing many species of plant to change their distribution. Some may even have become extinct because of the impact of global warming.

A survey by the Botanical Society of the British Isles showed that as the mean central England temperature rose from 9.05 °C in 1987 to 10.51 °C in 2004, orchids and ferns moved to new ground in the north, while other species moved higher up mountains to escape the heat.

Researchers have been surprised at the rapid and dramatic changes that have been seen in plants in the UK. Although some plants have been winners, growing well in new areas, and some have just held their own by moving habitats, others have lost out. The worst affected are plants with heavy seeds – they can't disperse themselves as quickly when climate change makes its effects felt. Goldenrod, a member of the dandelion family, has done very badly: there are 15 per cent fewer plants now than 10 years ago.

The fear is that as the global temperature continues to increase, there will be more change in plant flora in the UK – and extinction of some species will be inevitable.

CAM PLANTS DO IT IN THE DARK

C3 plants, the plants we have discussed so far, would not survive long in the desert. Their open stomata would prove to be a liability, they would rapidly lose water, dry out and die. In drier habitats, some plants have overcome this problem by opening their stomata to take in carbon dioxide only at night. This drastically reduces water loss as the plants then keep their stomata tightly closed during the hours of heat and daylight. But how do they store their carbon dioxide to make it available for the light-dependent phase of photosynthesis? The key is that they use carbon dioxide to make an organic acid, which they store during the night and then break down in the day to release the essential carbon dioxide.

Plants that photosynthesise in this way are called **CAM (crassulacean acid metabolism)** plants. The carbon dioxide that enters the leaves at night is fixed into the organic acid **malate** by the enzyme phosphoenolpyruvate (PEP) carboxylase, and is then stored in the vacuoles until the daytime.

When the sun rises, the stomata of CAM plants close, preventing the loss of precious water. The malate is broken down to provide carbon dioxide according to the simple reaction:

$$\text{malate} \rightarrow \text{pyruvate and } CO_2$$
$$\text{(C4)} \rightarrow \text{(C3)} \qquad \text{(C1)}$$

CAM plants are well suited to climates where there are high day and low night temperatures, especially where there is occasional drought. They tend to be native (Fig 33.26).

Overall, CAM plants overcome the problem of water loss by:

- opening their stomata at night and closing them during the day;
- having fewer stomata than temperate plants;
- having a low surface area-to-volume ratio, which reduces evaporation.

These measures result in CAM plants losing as little as 50–100 g of water for every gram of dry mass produced, compared with 400–900 g for C3 plants.

C4 PLANTS LIKE IT HOT

In the tropics many species, including some commercially important plants such as sugar cane, maize (Fig 33.27) and millet, exhibit remarkably rapid growth compared with C3 plants. They have achieved this mainly by overcoming the problem of photorespiration (see above). How have they managed to do this?

In 1966 two Australian scientists, Hatch and Slack, discovered that C4 plants were much more efficient at taking up carbon dioxide.

The key to C4 photosynthesis comes from the modified structure of the leaf. Surrounding the vascular bundles (the xylem and phloem tissue) is a layer of **bundle sheath cells** that have different chloroplasts.

The stages in C4 photosynthesis as shown in Fig 33.28 are as follows:

- **Stage 1** Carbon dioxide is fixed in the mesophyll cells. Instead of having RuBP and the enzyme rubisco, C4 plants use PEP (phosphoenolpyruvate) and the enzyme PEP carboxylase. This enzyme has a higher affinity for carbon dioxide than does rubisco, and it does not react with oxygen. The big advantage is that all the available carbon dioxide is fixed; none is lost on photorespiration. The 4-carbon compound oxaloacetate gives C4 plants their name:

$$\text{PEP} + CO_2 \xrightarrow[\text{PEP carboxylase}]{} \text{oxaloacetate (4C)}$$

- **Stage 2** Oxaloacetate is converted into malate (or a similar compound), which is shunted across into the chloroplasts of bundle sheath cells.
- **Stage 3** The malate is used to make carbon dioxide, pyruvate (a 3-carbon chemical) and hydrogen, which is used to make reduced NADP.

Fig 33.26 The prickly pear, *Opuntia,* is a CAM plant. This species thrives in warm, dry climates. When introduced into Australia in the 1920, this species out-competed the native plants and took over large areas of farmland until cactus moth caterpillars were introduced to control it

Fig 33.27 C4 plants such as maize show some of the fastest growth in the plant kingdom

- **Stage 4** The pyruvate is moved back into the mesophyll cells and used to make more PEP, while the carbon dioxide is now fixed into carbohydrate in the same way as in C3 plants, i.e. using RuBP and the enzyme rubisco. Note that in C4 plants the rubisco is always surrounded by a high concentration of carbon dioxide so that it is not inhibited by oxygen.

Fig 33.28 provides an overall comparison of photosynthesis in C3, C4 and CAM plants.

14 WHAT LIMITS PHOTOSYNTHESIS?

This is a very important question, because if commercial plant growers can increase the rate of photosynthesis, they can make plants grow faster, feed more people and/or make more money. The law of limiting factors states that the *rate of a process is limited by the factor in shortest supply*. For example, if carbon dioxide is in short supply, providing the plant with more carbon dioxide will continue to increase the rate of photosynthesis until some other factor – light intensity for example – becomes limiting. In this section we look at each limiting factor. Table 33.1 gives a brief overview of the way in which farmers can increase the growth of crops.

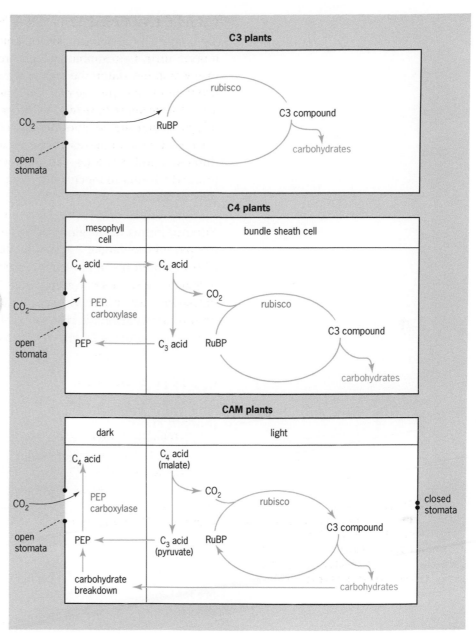

Fig 33.28 Comparisons of C3, C4 and CAM photosynthesis

Table 33.1 How to increase the supply of limiting factors

Factor	How supply can be increased	Is it economically viable… in a glasshouse?	in a field?
temperature	burn a fossil fuel, e.g. paraffin	usually	no
carbon dioxide	burn a fossil fuel, e.g. paraffin	usually	no
mineral ions	add fertiliser	usually	usually
light	artificial light	sometimes, with a valuable crop	rarely
water	irrigate or spray	essential	rarely in UK, essential in some climates

LIGHT INTENSITY

Light intensity is defined as the number of photons falling on a given area in a given time. Geographically, light intensity varies according to latitude – the closer to the equator, the greater the light intensity. Light intensity also varies according to the time of day, the time of year, the amount of cloud cover and whether the plant is shaded by other plants or objects. Generally, as light intensity increases so does the rate of photosynthesis until another factor – often carbon dioxide – becomes limiting. Plants have evolved to cope with different light intensities; for example, plants on the forest floor have to photosynthesise in much lower light intensities than the trees above. How do they manage?

Plants respire all the time, but only photosynthesise when there is light. Photosynthesis will begin after dawn and the rate will increase until it equals the rate of respiration – this is the **compensation point**. After the compensation point, photosynthesis will exceed respiration and so the plant will accumulate organic compounds. Plants that can photosynthesise in low light intensities are shade-tolerant and have a low compensation point. In contrast, shade-intolerant species have a high compensation point and so can only grow in bright environments. They cannot tolerate shading.

WATER SUPPLY

Although water is essential for photosynthesis, this represents only a tiny proportion of the water that flows through the plant in the transpiration stream. A lack of water will normally cause a plant to wilt and affects all other metabolic processes long before the process of photosynthesis is affected directly.

CARBON DIOXIDE

Carbon dioxide is the commonest limiting factor when light intensity is sufficient. The atmosphere contains only about 0.035 per cent carbon dioxide, which is about 35 parts per 10 000. When the carbon dioxide levels are low and oxygen levels are high the enzyme rubisco can be inhibited – see the section on photorespiration and C4 plants above.

TEMPERATURE

Photosynthesis, like all metabolic processes, is temperature dependent. As the temperature increases the kinetic energy of the molecules increases, they 'bounce around' more, so that diffusion of gases in and out of the leaf is faster, and there are more collisions between enzymes and substrates. The optimum temperature for C3 plants is about 25 °C, while for C4 plants it is about 35 °C. Above these temperatures, a metabolic imbalance develops so that photosynthesis becomes less efficient, while extremes of temperature will, of course, denature the enzymes.

INORGANIC IONS

Plants need a variety of mineral ions from the soil in order to remain healthy. Of particular relevance to photosynthesis are nitrogen and magnesium; both are constituents of the chlorophyll molecule. Without these ions, the plant develops yellow leaves, a condition known as chlorosis.

PART III: CELL RESPIRATION

15 CELL RESPIRATION: A VITAL LIFE PROCESS

Understanding cell respiration can be a demanding task. You will probably find it useful to read Part I of this chapter first: it looks at key concepts that you will need to understand before you tackle the details given in this chapter.

All living things obtain the energy they need from **respiration**, a chemical process that breaks down simple food molecules such as glucose. Animals, including humans, digest food to produce these molecules, which are absorbed into the blood and then transported round the body.

Multicellular organisms mainly respire **aerobically** (using oxygen). Oxygen is absorbed by lungs, gills or the body surface and is usually distributed around the body by the **circulatory system**. Only when food molecules and oxygen are together inside cells can the complex process of aerobic cell respiration begin.

SOME IMPORTANT DEFINITIONS

The terms 'respiration' and 'breathing' seem to be used interchangeably by many people, including scientists, and this can lead to confusion. As a basic guide, we use the following definitions.

Respiration is the process that releases the energy in organic molecules such as sugars and lipids. Respiration is a complex, multi-stage process that takes place in *all* cells of *all* organisms, *all* of the time (unless the organism is dormant). You may remember that respiration is one of the seven signs of life. To avoid confusion with breathing, it is often called tissue respiration or cell/cellular respiration. In this book we call it **cell respiration**.

Sources of energy

In humans, as in all organisms, obtaining energy from food is a priority that cannot be ignored. Tragic scenes from parts of the world devastated by famine show the terrible consequences of malnutrition and starvation (Fig 33.29).

We obtain most of our energy by respiring the food that we have eaten recently. Carbohydrate is the primary source of energy, but we also respire small amounts of fat and protein. When the energy provided by recent meals runs out, as it does when we are asleep, for example, our bodies start to respire molecules stored in our tissues. These are replaced when we next eat, usually first thing in the morning when we break our fast.

During prolonged starvation, the body's stored carbohydrate runs out after about a day, and then the body must obtain its energy by respiring fat. An average 70 kg man will have about 10 to 15 kg of stored fat in his body. This is a large energy store, but one that is not readily available (fat can only be respired when some carbohydrate is still available). Increasingly the body is forced to turn to protein, the very fabric of the cells, for the energy it desperately needs.

The body can withstand the loss of about half of its protein only. Complete starvation, where no food at all is available, leads to death within weeks. The exact cause of death varies. It can be an infection such as pneumonia (starvation weakens the immune system) or circulatory problems due to a much reduced blood volume.

Fig 33.29 Famine is still a fact of life for people in many parts of the world

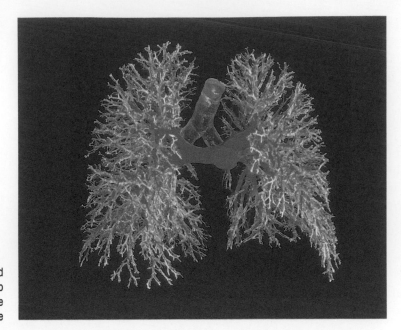

Fig 33.30 Living organisms have developed various organs to increase their ability to exchange gases. In humans, this gas exchange occurs at the lung surface

Gas exchange is discussed in Chapter 9.

Organisms that carry out aerobic cell respiration need to take in oxygen, and they produce carbon dioxide, a waste product that must be removed. Larger organisms have had to develop specialised organs with large surface areas to increase their capacity for gas exchange (Fig 33.30).

Breathing is the mechanical process that supplies oxygen to the body to drive respiration and that removes the carbon dioxide produced. Breathing *ventilates* the gas exchange surfaces.

Aerobic respiration is respiration that requires oxygen.

Most organisms respire aerobically: aerobic respiration releases a relatively large amount of energy and gives the organism a survival advantage over anaerobic organisms.

Anaerobic respiration is respiration without oxygen.

Some organisms, mainly bacteria, can only respire anaerobically. A few more, yeast for example, can turn to anaerobic respiration when there is no oxygen. Some animal tissues (e.g. muscle during strenuous exercise) and plant tissues (e.g. plant roots in waterlogged soil) are able to respire anaerobically if circumstances demand it.

16 AEROBIC CELL RESPIRATION

'Aerobic cell respiration' describes the cell processes that require oxygen to release energy from all types of organic molecules. In many organisms, cell respiration involves the breakdown of a variety of molecules from their food. Humans respire mainly sugars with a small percentage of amino acids and fatty acids but, when the need arises, such as during starvation, the balance can change.

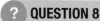 **REMEMBER THIS**

Although our muscles can work anaerobically for a short time, the brain must have a continuous supply of oxygen. A lack of oxygen for just 3 minutes is enough to cause brain damage.

 QUESTION 8

8 Why is it important for organs involved in gas exchange to have a large surface area? What other features would you expect these organs to have?

AEROBIC CELL RESPIRATION OF GLUCOSE

To make a complex topic easier to study, we shall concentrate first on the aerobic respiration of only one type of food molecule: glucose.

For any chemical reaction to take place, energy is required to break existing molecular bonds. The process of forming new bonds may either require or release energy. For there to be a *net release of free energy* (the whole point of cell respiration), the products of cell respiration must be at a lower energy level than the reactants. Chemists say the products are more **thermodynamically stable** than the reactants. The basic equation we use to sum up glucose respiration is:

$$C_6H_{12}O_6 + 6O_2 \rightarrow 6CO_2 + 6H_2O + ENERGY$$

The total energy contained in the reactants (the molecule of glucose and the six molecules of oxygen on the left) is greater than the total energy contained in the products (the six molecules of carbon dioxide and the six molecules of water on the right). The difference is the energy released by cell respiration.

In reality, glucose respiration is a sequence of many different reactions. These can produce up to 36 molecules of a chemical called **ATP** per molecule of glucose. This large amount of energy can do useful work in the cell:

$$C_6H_{12}O_6 + 6O_2 \rightarrow 6CO_2 + 6H_2O + 36ATP$$

The steps involved in the production of ATP depend heavily on redox reactions. These are reactions that involve oxidation of one reactant and reduction of another. Before going on to look at the detailed biochemistry of glucose respiration, take a look at the box on **redox reactions** on page 543, to make sure that you are comfortable with this important concept.

Respiration or combustion?

It is often said that we 'burn up' food in respiration but this is not really true. The energy contained in glucose *could* be released by setting fire to it (Fig 33.31). The glucose would be oxidised to carbon dioxide and water, but the reaction would be rapid and uncontrolled.

During the process of cell respiration, glucose and other organic molecules are broken down in small stages, some of which release energy. This step-by-step breakdown (Fig 33.32) is more gradual than combustion. The whole process of respiration is controlled by enzymes that transfer energy in food molecules to ATP, a substance that is able to power cell processes by supplying on-the-spot, instant and usable energy in controlled amounts.

> The aerobic respiration of other organic molecules and anaerobic respiration are discussed on pages 550 and 555.

? QUESTION 9

9 How is cell respiration similar to the working of a petrol engine?

Fig 33.31 The energy in food can be released by combustion, but almost all of it is lost as heat, raising the temperature of the surroundings to levels that could not be tolerated inside any living cell

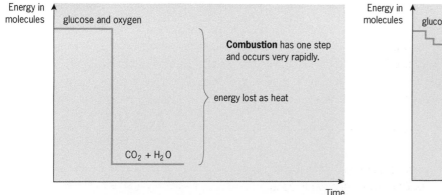

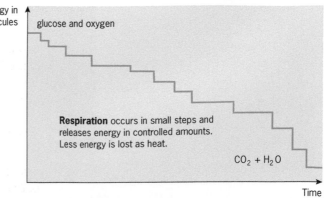

Fig 33.32 The difference between combustion and respiration. Although the difference in energy between the reactants (glucose and oxygen) and the products (CO_2 and water) is the same in both cases, the way in which the energy is released is totally different

All about ATP

ATP stands for **adenosine triphosphate**. ATP is a relatively small, soluble organic molecule that consists of a base (adenine), a sugar (ribose) and three inorganic phosphate (P_i) groups (Fig 33.33). ATP has a high **free energy of hydrolysis**, which means that when ATP is **hydrolysed** (Fig 33.34), a relatively large amount of energy is released. In hydrolysis, ATP reacts with water, producing ADP and inorganic phosphate. Because of its solubility and small size (relative molecular mass 507), ATP can diffuse rapidly around cells and so can supply energy where it is needed. Metabolically active cells, such as those in muscle and in secretory tissue, synthesise and break down many ATP molecules.

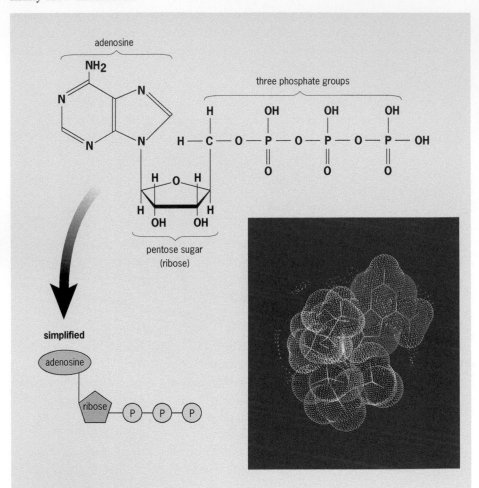

Fig 33.33 The structure of ATP. In the computer graphics image, adenosine is blue, pentose is white and the phosphate groups are red. This molecule is present in all cells and is used to provide energy. The purpose of cell respiration is to re-synthesise ATP at the same rate as it is used up

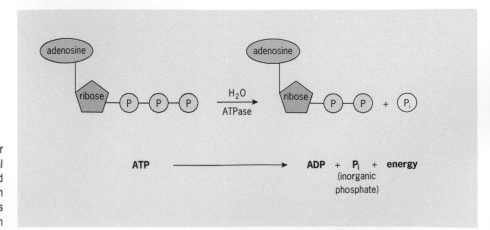

Fig 33.34 The hydrolysis (splitting) of ATP. Under the control of the enzyme ATPase, the terminal phosphate group of ATP is removed and combined with water. This reaction releases free energy which can be used to drive energy-requiring reactions such as muscular contraction

How does ATP release energy?

When ATP loses a phosphate group to become ADP, adenosine *di*phosphate, the reaction is **exergonic** (it *releases* energy). This can be used to do useful work within the cell. The same amount of energy is released when ADP loses another phosphate group to become AMP (adenosine *mono*phosphate). Less energy is released when the last phosphate group is lost.

Many of the reactions within cells are **endergonic**: they *require* energy (Fig 33.35). This energy is supplied by *coupling* them with reactions that involve ATP breakdown.

Endergonic reaction: amino acid A + amino acid B → dipeptide

Exergonic reaction: ATP → ADP + P_i + energy

It is often said that ATP is a 'high-energy molecule' or that is contains 'high-energy phosphate bonds'. Neither statement is really true. Many molecules have more energy than ATP, and many others contain exactly the same phosphate-to-phosphate bonds. ATP is an important molecule in living systems because it can lose its terminal phosphate group *readily*, releasing just enough energy to power biological processes without producing excess heat.

ATP → ADP + P_i + energy (30.6 kJ mol^{-1})

During strenuous exercise, when the muscles quickly use up all their ATP, energy is supplied by a back-up chemical called **creatine phosphate** (CP). Since CP has a higher free energy of hydrolysis than ATP, its breakdown releases enough energy to make more ATP instantly and so allows exercise to continue.

10 To move our muscles, we need energy to be available 'on demand'. Why do we use ATP as an immediate energy source, instead of glucose?

Fig 33.35 For explosive bursts of effort, such as a short sprint, most of the energy comes from ATP already present in the muscles

SCIENCE IN CONTEXT

Redox reactions

Redox is short for *red*uction and *ox*idation, two chemical processes that often occur together in the same reaction.

Oxidation reactions may involve the addition of oxygen or the removal of hydrogen from molecules, but the important underlying rule is that a molecule that is oxidised *loses* electrons.

Conversely, reduction reactions involve a *gain* of electrons. As electrons carry energy, a molecule that has been reduced, and that therefore has gained one or more electrons, will usually carry more energy than the oxidised form of the same molecule.

Reduction and oxidation reactions usually occur together, because if one chemical loses electrons another must gain them (Fig 33.36).

The concept of redox reactions is a vital one in biology. The molecules that make up living things are produced, directly or indirectly, by photosynthesis. This process reduces carbon dioxide to form organic molecules (such as sugars) that contain energy. These can be later oxidised in the process of cell respiration to release energy. Lipids contain more hydrogen than

carbohydrates: they have more C—H bonds than C—O bonds. For this reason, lipids are said to be more highly reduced chemicals than carbohydrates and more energy can be released by the oxidation of their many C—H bonds. Because they contain more energy per gram than other substances, lipids are ideal storage compounds (see Chapter 3).

Cell respiration is the release of energy from food by progressive oxidation. In cell respiration, glucose is oxidised to form carbon dioxide and oxygen is reduced to form water. Most of the ATP created during this process comes from the final stage: a series of redox reactions that occur on the mitochondrial membrane.

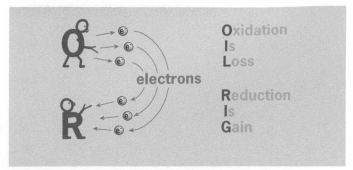

Fig 33.36 Remember – OIL RIG: oxidation is loss, reduction is gain

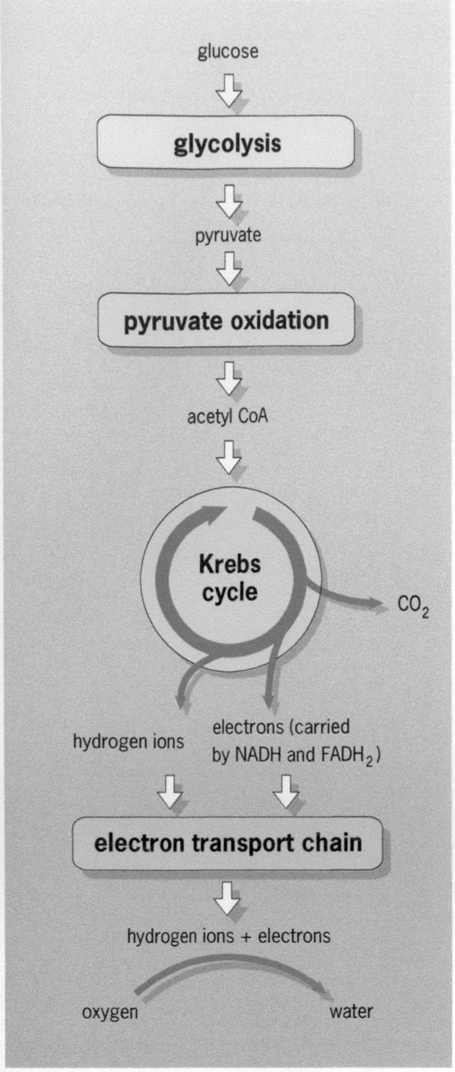

Fig 33.37 The overall process of cell respiration, showing the order of the four main stages

? QUESTION 11

11 What is the difference between substrate-level phosphorylation and oxidative phosphorylation?

THE FOUR MAIN STAGES IN GLUCOSE RESPIRATION

The complete process of aerobic respiration of glucose can be divided into four distinct processes:

- **Glycolysis**; glucose is split into pyruvate.
- **Pyruvate oxidation** (the 'link' reaction); pyruvate is oxidised into acetate.
- The **Krebs cycle**; electrons are stripped off the acetate.
- The **electron transport chain**; the energy in the electrons is used to make ATP.

The way in which the four processes relate to each other is shown in Fig 33.37, and Table 33.2 gives the overall function of each stage. When learning respiration, the two compounds to look out for are ATP and NADH. The production of ATP, as we have seen, is the whole point of the process, but what is NADH?

Overall, respiration involves the removal of electrons from the glucose molecules, and then using the energy in the electrons to make ATP. NADH is an electron carrier.

$$NAD^+ \quad + \quad e^- \quad \rightarrow \quad NADH$$

coenzyme + electron → reduced coenzyme

NAD^+ is a coenzyme that picks up electrons from the processes that release them (glycolysis, link reaction, Krebs cycle) and delivers them to the process that converts them into ATP; the electron transport chain. You might like to think of ATP as cash and NADH as a credit note that can be turned into cash later.

Table 33.2 The overall process of glucose breakdown. Stages 2, 3 and 4 can be thought of as the aerobic (oxygen-requiring) reactions and take place in the mitochondria			
Stage	**Site within cell**	**Overall process**	**Number of ATP molecules produced***
1 glycolysis	cytosol	glucose is split into 2 molecules of pyruvate	2 per glucose
2 pyruvate oxidation	matrix (inner fluid) of mitochondria	pyruvate is converted into acetyl Co-A	none
3 Krebs cycle	matrix (inner fluid) of mitochondria	acetyl Co-A drives a cycle of reactions which produces hydrogen	2 per turn, so 4 per glucose
4 electron transport chain	inner membrane of mitochondria	hydrogen drives a series of redox reactions which release enough energy to make ATP	up to 32 per glucose

*This table indicates that, for each molecule of glucose, 38 molecules of ATP are produced. The actual figure is nearer 36 molecules. See the Example on page 550.

Making ATP by phosphorylation

In the cell, ATP is made by adding a phosphate group to a molecule of ADP. This process is called phosphorylation and it is carried out in two different ways.

Substrate-level phosphorylation is the process in which a phosphate group from a substrate molecule (a molecule other than ATP, ADP or AMP) is transferred to a molecule of ADP, giving a new molecule of ATP. This form of phosphorylation occurs in glycolysis and in the Krebs cycle.

In **oxidative phosphorylation**, ATP is synthesised using free phosphate groups. The energy required is obtained from a series of redox

reactions in the electron transport chain. So whenever you
see 'oxidative phosphorylation', think 'electron transport chain'.

Glycolysis

In this first stage, a glucose molecule is converted, by a series of enzyme-
controlled reactions, into two molecules of pyruvate, a 3-carbon compound.
The process, shown in detail in Fig 33.38, yields relatively little energy (only
two molecules of ATP) but this stage does not require oxygen and takes
relatively little time to complete, so it can provide immediate energy.

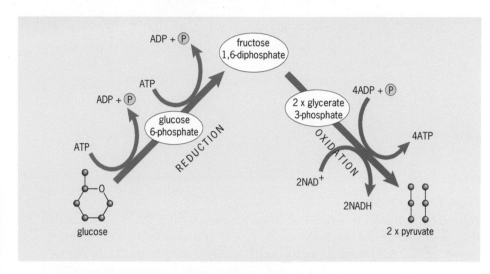

Fig 33.38 Glycolysis is a two-stage process. The 'uphill' part involves raising glucose to a higher energy level by using ATP. In the 'downhill' part, the products are oxidised, yielding two molecules of pyruvate, two molecules of reduced coenzyme and a net gain in ATP

Glycolysis takes place in the **cytosol**, the fluid part of the cytoplasm, so
this basic process takes place all over the cell. The overall process consists of
ten different reactions, each catalysed by a specific enzyme.

The first half of glycolysis actually uses ATP to raise the energy level
of the original glucose molecules by adding phosphate groups to form
fructose 1,6-diphosphate. This must be done if glucose is to have
enough energy to complete the second half of the process.

Fructose 1,6-diphosphate is then split into two molecules of a 3-carbon
compound, **glyceraldehyde 3-phosphate**. These are subsequently
oxidised to pyruvate by a series of five reactions that also produce four
molecules of ATP and two molecules of reduced coenzyme (NADH).

So, overall, per glucose molecule, glycolysis produces:

- two molecules of ATP (four are created but two are used up in glycolysis);
- two molecules of **NADH** (**reduced coenzyme** which later feeds
  electrons into the electron transport chain);
- two molecules of pyruvate (which enters the 'link' reaction if oxygen
  is available).

Pyruvate oxidation: the 'link' reaction

This reaction links glycolysis with the Krebs cycle. It is often thought of as
part of the Krebs cycle but we shall consider it to be a separate reaction.

In the presence of oxygen, pyruvate (the 3C molecule produced by glycolysis)
moves from the cytosol to the mitochondrial matrix where it is oxidised into
acetate (a 2C molecule), producing carbon dioxide as a by-product. This
reaction also produces two molecules of NADH. The acetate is picked up by a
carrier molecule, **coenzyme A**, and **acetyl coenzyme A** is formed.

Overall, the reaction is:

$$\text{2 pyruvate} + 2NAD^+ + 2H_2O \rightarrow \text{2 acetate} + 2NADH + 2H^+ + 2H_2O$$

? **QUESTION 12**

12 What does a reduced coenzyme
have that an oxidised coenzyme
does not?

Pyruvate or pyruvic acid?

The terms pyruvate and pyruvic acid are used interchangeably but do they refer to the same substance? Fig 33.39 shows the difference: pyruvic acid in solution becomes ionised and, like all acids in solution, it loses a hydrogen ion. In the ionised state it is known as pyruvate. The same applies to other organic acids such as lactic acid and acetic acid, which become lactate and acetate respectively.

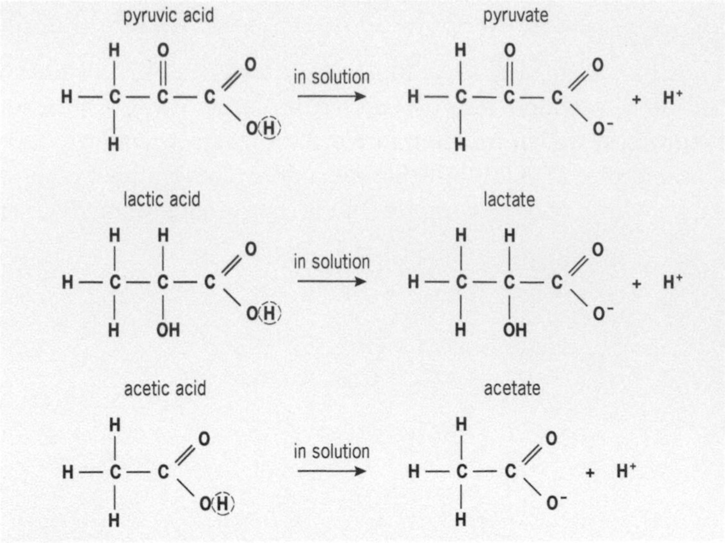

Fig 33.39 Pyruvate, lactate and acetate and their corresponding acids

13 In a human, what happens to the carbon dioxide made by the reactions in the Krebs cycle?

14 Why does the Krebs cycle turn twice for every glucose molecule?

The Krebs cycle

The **Krebs cycle** is a series of reactions named after Sir Hans Krebs, the biochemist who first worked out the details. It is also known as the **citric acid cycle**, the **tricarboxylic acid cycle** or simply the **TCA cycle**. The main purpose of the Krebs cycle is to provide a continuous supply of electrons to feed into the electron transport chain.

The details of the Krebs cycle, which takes place in the matrix of the mitochondria, can be seen in Fig 33.40 on the next page. Overall, per turn, the Krebs cycle produces:

- 3 molecules of NADH;
- 1 molecule of $FADH_2$;
- 1 molecule of ATP (by substrate-level phosphorylation);
- 2 molecules of CO_2;
- 1 molecule of oxaloacetate (to allow the cycle to continue).

The Krebs cycle will 'turn' twice for every glucose molecule that enters it, so, per glucose molecule, six molecules of NADH and two molecules of $FADH_2$ are produced. NADH and $FADH_2$ are particularly important because they carry the electrons which power the next stage of glucose respiration.

NAD and FAD

NAD (**nicotinamide adenine dinucleotide**) and FAD (**flavine adenine dinucleotide**) are **coenzymes**, organic compounds that are catalysts for reactions. As their name implies, coenzymes work closely with enzymes. Unlike enzymes, coenzymes are not proteins. NAD and FAD carry electrons from the **electron donors** in glycolysis, pyruvate oxidation and the Krebs cycle to **electron acceptors** in the electron transport chain in the mitochondrial membrane.

When NAD and FAD accept electrons, they are reduced, becoming NADH and $FADH_2$. In this form they are called **reduced coenzymes**. Reduction of a coenzyme always requires association with a **dehydrogenase enzyme**, an enzyme that removes hydrogen from other molecules. The hydrogen removed by the dehydrogenase enzyme is split into an electron and a hydrogen ion. NAD (or FAD) accepts the electrons produced and the hydrogen ions play an important part in the electron transport chain.

In the human body, NAD is synthesised from vitamin B_3 (nicotinic acid) and FAD is made from vitamin B_2 (riboflavin): this explains why both vitamins are essential components of our diet.

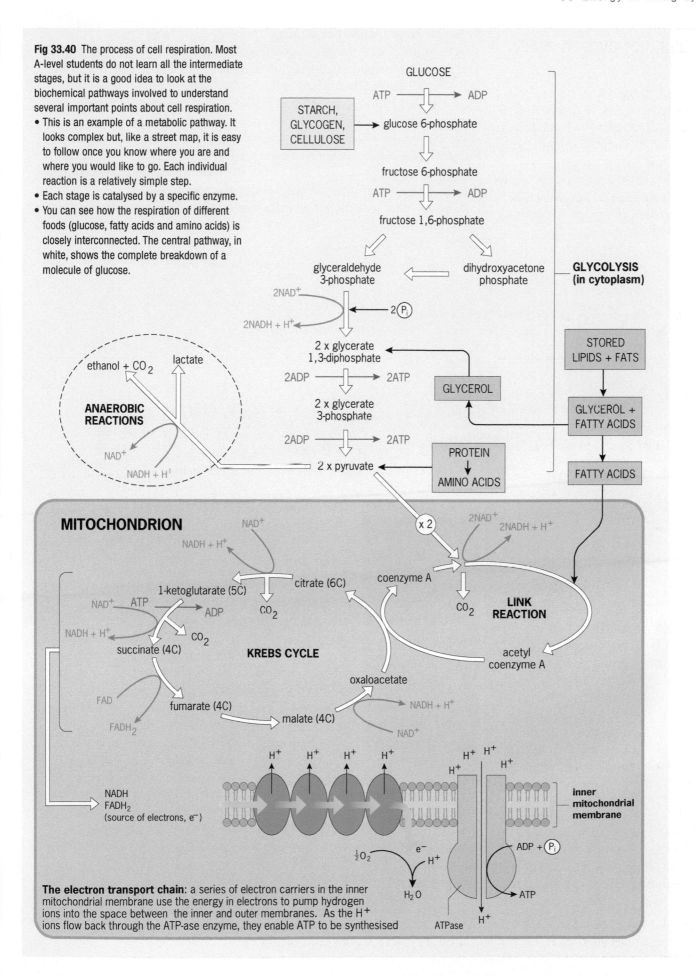

Fig 33.40 The process of cell respiration. Most A-level students do not learn all the intermediate stages, but it is a good idea to look at the biochemical pathways involved to understand several important points about cell respiration.

- This is an example of a metabolic pathway. It looks complex but, like a street map, it is easy to follow once you know where you are and where you would like to go. Each individual reaction is a relatively simple step.
- Each stage is catalysed by a specific enzyme.
- You can see how the respiration of different foods (glucose, fatty acids and amino acids) is closely interconnected. The central pathway, in white, shows the complete breakdown of a molecule of glucose.

GLUCOSE

ATP → ADP

STARCH, GLYCOGEN, CELLULOSE → glucose 6-phosphate

fructose 6-phosphate

ATP → ADP

fructose 1,6-phosphate

glyceraldehyde 3-phosphate

dihydroxyacetone phosphate

GLYCOLYSIS (in cytoplasm)

$2NAD^+$

$2NADH + H^+$

$2 P_i$

2 x glycerate 1,3-diphosphate

2ADP → 2ATP

2 x glycerate 3-phosphate

2ADP → 2ATP

2 x pyruvate

STORED LIPIDS + FATS

GLYCEROL

GLYCEROL + FATTY ACIDS

FATTY ACIDS

PROTEIN → AMINO ACIDS

ANAEROBIC REACTIONS

ethanol + CO_2

lactate

NAD^+

$NADH + H^+$

MITOCHONDRION

NAD^+

$NADH + H^+$

$2NAD^+$

$2NADH + H^+$

x 2

coenzyme A

CO_2

LINK REACTION

citrate (6C)

1-ketoglutarate (5C)

NAD^+ ATP → ADP

$NADH + H^+$ CO_2

succinate (4C)

KREBS CYCLE

acetyl coenzyme A

oxaloacetate

FAD

fumarate (4C)

malate (4C)

$NADH + H^+$

NAD^+

$FADH_2$

H^+ H^+ H^+ H^+ H^+ H^+ H^+ H^+

NADH $FADH_2$ (source of electrons, e^-)

inner mitochondrial membrane

$\frac{1}{2}O_2$

e^-

H^+

H_2O

$ADP + P_i$

ATP

H^+

ATPase

The electron transport chain: a series of electron carriers in the inner mitochondrial membrane use the energy in electrons to pump hydrogen ions into the space between the inner and outer membranes. As the H^+ ions flow back through the ATP-ase enzyme, they enable ATP to be synthesised

The electron transport chain

This final phase of respiration produces the largest number of ATP molecules. The electron transport chain can be thought of as the 'pay day' for all the 'hard work' that has been done in the earlier parts of the process.

As Fig 33.41 shows, the electron transport chain consists of a series of carrier molecules, which will first accept an electron (thereby becoming reduced) and then lose it again (so becoming oxidised). At each of these transfers the electrons lose some energy, which can be used to power the active transport of hydrogen ions across the inner mitochondrial membrane (Fig 33.42 on the next page). This results in a high concentration of hydrogen ions in the outer mitochondrial space and a low concentration in the inner mitochondrial space.

Fig 33.41 The electrons and hydrogen ions made during the first stages of respiration are finally used to synthesise ATP

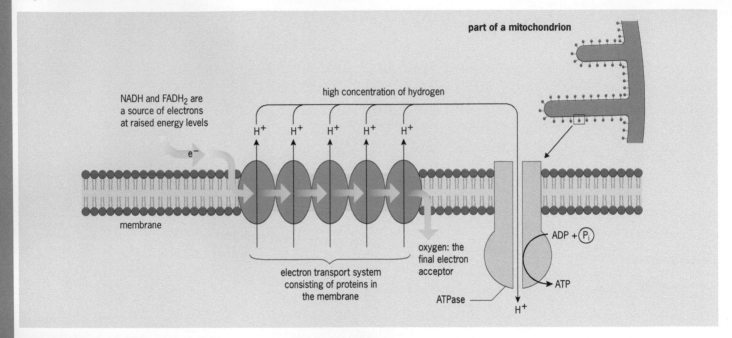

Because of the difference in concentration, hydrogen ions leak back into the inner compartment. The only route that they have is through the middle of the stalked granules – ATPase enzymes. As the hydrogen ions flow down the concentration gradient, enough energy is released to allow free inorganic phosphate molecules to be added to ADP, forming new molecules of ATP. The flow of hydrogen ions is sometimes known as a **proton motive force**.

This model of ATP synthesis by **oxidative phosphorylation** is called the **chemiosmotic hypothesis** and was first proposed by Peter Mitchell in 1961. Mitchell received the Nobel Prize for Chemistry in 1978.

The end-products of the electron transport chain are low-energy electrons and hydrogen ions that combine with oxygen to form water. Although this reaction is at the end of the process, it is a key one: it is why most organisms need oxygen. If there were no oxygen to mop up the electrons and hydrogen ions from the electron transport chain, the pathway could not be completed. This would be a disaster for the cell as the intermediate compounds of glucose respiration would build up and the cell would lose its ability to make most of the ATP it needs.

Looking at the separate processes involved in the breakdown of one molecule of glucose makes it easy to lose track of the overall purpose of glucose respiration: ATP production. The following example gives you the chance to analyse the contribution to ATP production made by each of the different stages.

? QUESTIONS 15–16

15 Muscle cells and cells that manufacture and secrete large amounts of hormones have more mitochondria than, say, a skin cell. Can you say why?

16 The function of a kidney tubule cell is to move substances by active transport. Suggest why these cells are packed with mitochondria.

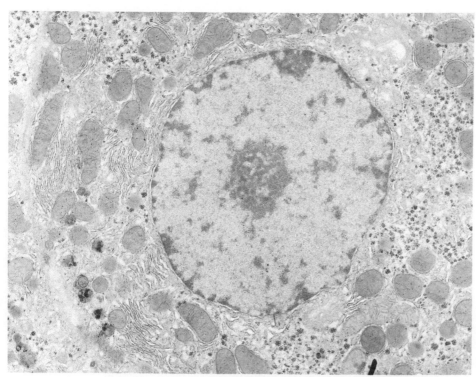

Fig 33.42 Mitochondria are the organelles that house the enzymes, substrates and other chemicals associated with aerobic respiration. Pyruvate oxidation and the Krebs cycle take place in the matrix (fluid) centre of the organelle, while the electron transport chain takes place in the inner membrane itself

Fig 33.42a Micrograph of a pancreatic cell: the large numbers of mitochondria suggest that this is a metabolically active cell. In this case ATP is needed to synthesise the digestive enzymes (proteins)

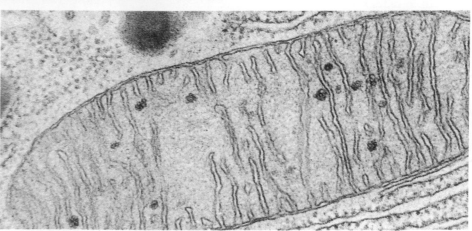

Fig 33.42b Micrograph of a single mitochondrion. Generally, the more metabolically active the cell, the more folds (cristae) there are on the inner membrane

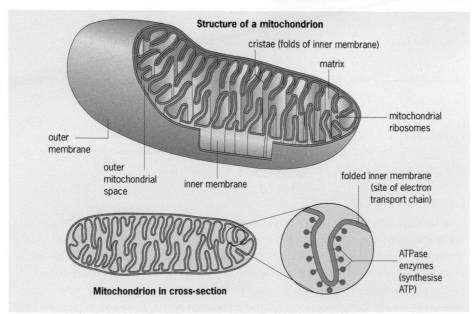

Structure of a mitochondrion

cristae (folds of inner membrane)

matrix

mitochondrial ribosomes

outer membrane

outer mitochondrial space

inner membrane

folded inner membrane (site of electron transport chain)

ATPase enzymes (synthesise ATP)

Mitochondrion in cross-section

Fig 33.42c 3-D illustration of a mitochondrion

EXAMPLE

Q How many ATP molecules are produced, per molecule of glucose, in aerobic cell respiration?

A ATP can be made in one of two ways, by **substrate-level phosphorylation** and by **oxidative phosphorylation** (see page 544). We shall take each process in turn.

Substrate-level phosphorylation

Glycolysis and the Krebs cycle produce ATP in this way. We can complete the following table:

Stage	ATP produced
glycolysis	2 (2 used but 4 made)
pyruvate oxidation	0
Krebs cycle	2 (1 per turn)
total	4

Oxidative phosphorylation

This is the production of ATP by the electron transport chain. It relies on a continuous supply of 'high-energy' electrons from the preceding processes. Before we can calculate the final ATP total, we must calculate the number of electrons being provided by NADH and $FADH_2$.

We can draw up a second table, as follows. Remember, pyruvate oxidation and the Krebs cycle occur twice for every glucose molecule put into the system.

Stage	NADH produced	FADH$_2$ produced
glycolysis	2	0
pyruvate oxidation	2	0
Krebs cycle	6 (3 per turn)	2 (1 per turn)
total per glucose	10	2

We know that:

3 ATP molecules are produced for every molecule of NADH fed into the chain,

and that:

2 ATP molecules are produced for every molecule of $FADH_2$ fed into the chain.

We can now estimate the final ATP total:

10 NAD molecules produce 3 ATPs, giving a total of	30
2 FADH molecules produce 2 ATPs, giving a total of	4
Therefore, the total number of ATP molecules produced by oxidative phosphorylation is	34
The total number of ATP molecules produced by substrate-level phosphorylation is	4

The grand total per glucose molecule is therefore 38 molecules of ATP,

most of which are produced by the electron transport chain.

In practice, the figure is usually 36 molecules of ATP. Because the mitochondrial membrane is impermeable to NADH, the 2 molecules of NADH made during glycolysis cannot carry their electrons directly into the electron transport chain.

To get over this, most cells have a shuttle system in which the electrons are released by the NADH and passed across the mitochondrial membrane where they are picked up by $FADH_2$. As we saw earlier, $FADH_2$ produces only 2 ATPs compared with the 3 made by NADH, thus reducing the final total by 2.

This is the ATP total under *ideal* conditions. Several other factors can reduce ATP production even further, but these are beyond the scope of this book.

AEROBIC CELL RESPIRATION OF OTHER FUELS

So far, we have studied the respiration of glucose, but many organic molecules can be fed in the same central pathway to create ATP.

Lipid breakdown

The lipid stores in the body can be used when carbohydrate is in short supply. Stored triglycerides are broken down into **glycerol** and **fatty acids**. The latter are further broken down to give several 2-carbon acetyl fragments by a process called **beta-oxidation**. These fragments are combined with coenzyme A and so enter the Krebs cycle and, eventually, the electron transport chain. Glycerol is converted into **dihydroxyacetone phosphate** so that it can be used as a fuel in glycolysis.

Protein breakdown

Adult humans need up to 60 g of protein per day. We cannot store protein, so excess **amino acids**, the building blocks of proteins, are degraded in the liver by the process of **deamination**. Enzymes in the liver separate the amine group from the rest of the molecule, leaving an **organic acid**. The amine group is released as free **ammonia**, which enters the **ornithine cycle** to be incorporated into **urea** and excreted from the body in the urine. The organic acids are fed into the Krebs cycle and are respired (Fig 33.40).

> The ornithine cycle is discussed in more detail in Chapter 11.

17 THE RESPIRATORY QUOTIENT

The **respiratory quotient** (**RQ**), is a ratio of gas exchange, worked out by comparing oxygen uptake with carbon dioxide production. The RQ can be used to determine the type of food being respired. The basic equation for the respiration of glucose:

$$C_6H_{12}O_6 + 6O_2 \rightarrow 6CO_2 + 6H_2O + energy$$

shows that, for every glucose molecule respired, six molecules of oxygen are required and six molecules of carbon dioxide are produced. Equal numbers of gas molecules occupy the same volume and so the volume of oxygen taken up by an organism is equal to the volume of carbon dioxide produced.

 REMEMBER THIS
Lipids can fuel aerobic exercise but not glycolysis. Aerobic exercise is therefore a good way to burn excess fat.

STRETCH AND CHALLENGE

The origins of respiration: some speculation

Even the simplest life form needs to release the energy locked in organic molecules and it is thought that glycolysis was one of the first metabolic pathways to evolve.

Since there was very little free oxygen at the time, aerobic respiration was impossible. It seems reasonable to assume that the process of glycolysis, as practised by early, bacteria-like organisms (Fig 33.43), used up much of the available food, giving the organisms a major problem: that of energy supply. Solving this problem proved to be a massive evolutionary step: the development of photosynthesis. Not only did photosynthesis provide a new supply, but it also released free oxygen as a by-product. The oxygen in the atmosphere is due to photosynthesis.

The build-up of oxygen in the atmosphere kick-started another evolutionary leap: the appearance of organisms that could respire aerobically. This probably happened fairly quickly because of the massive survival advantage gained by organisms that respire aerobically to produce the much larger amounts of ATP.

It is at this stage that prokaryotic organisms which could respire aerobically may have developed a close association with large cells, so forming primitive eukaryotic cells. Find out more about the **endosymbiont theory** on page 22.

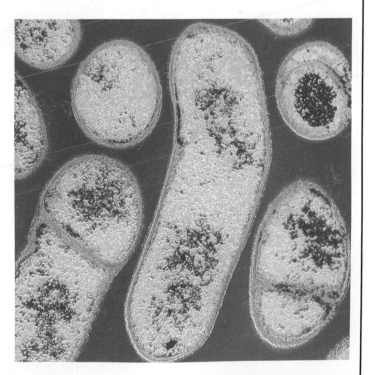

Fig 33.43 *Clostridium botulinum*, the anaerobic bacterium that causes botulism (serious food poisoning), is found in food such as meat and fish that have been badly canned. Anaerobic bacteria such as these are believed to have been the only organisms that could have survived in the oxygen-poor environment 3500 million years ago

WORKING OUT THE RESPIRATORY QUOTIENT

The RQ is worked out using the simple formula:

$$RQ = \frac{\text{volume of } CO_2 \text{ given out}}{\text{volume of } O_2 \text{ taken in}}$$

For the glucose equation, the RQ is: $6 \div 6 = 1$.

USING THE RESPIRATORY QUOTIENT

If we measure the volumes of oxygen and carbon dioxide that are exchanged by an organism and find that the RQ is 1, we can infer that the organism is respiring mainly carbohydrate. Different substrates will produce different RQ values:

- Carbohydrate respiration gives an RQ of 1.
- Protein respiration gives an RQ of 0.8–0.9.
- Lipid respiration gives an RQ of 0.7.
- Fermentation gives an RQ of infinity.

The problem with trying to work out what substrate is being respired by looking at RQ values is that many organisms, including humans, respire more than one substrate. In humans, carbohydrate is the main fuel for respiration, but a small amount of protein is also respired. The RQ figure obtained is therefore variable.

USING A RESPIROMETER

The rate of aerobic respiration can be estimated by measuring how much oxygen is taken up or how much heat is produced. Gas exchange, including oxygen uptake, is measured using a sealed chamber called a **respirometer**.

A simple respirometer

A simple respirometer is shown in Fig 33.44. Organisms placed inside exchange gases and so alter the composition of the air around them.

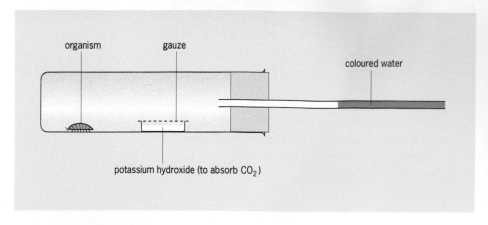

organism   gauze

coloured water

potassium hydroxide (to absorb CO_2)

Fig 33.44 A simple respirometer

To measure gas exchange due to a respiring organism, we could remove the air after a fixed time and analyse it. A simpler method is to place a carbon dioxide absorber such as potassium hydroxide in the respirometer chamber. Over a set time, a volume of oxygen is used up. Carbon dioxide is produced but is absorbed by the potassium hydroxide. So, overall, the volume of gas falls by the volume of oxygen used. The gas pressure inside the chamber falls and coloured liquid is drawn along the tube towards the chamber. The volume of oxygen used up equals the volume of the liquid moved in the fixed time. The following example shows how the respiration rate is calculated from readings taken from the tube.

EXAMPLE

Q a) When measuring the gas exchange of an organism with a simple respirometer, how do you calculate the volume of oxygen used?

b) In what units do you express the rate of respiration?

A a) The position of the liquid along the tube is recorded at the start and end of the set time and the distance the liquid moves is calculated. The volume of liquid (equal to the change in oxygen volume) is a cylinder inside the tube.

Therefore:

$$\text{volume of oxygen used up (in cm}^3) = \pi \times r^2 \times h$$

where r = radius or diameter/2 (in cm) and h = distance moved by the liquid (in cm).

b) The rate of respiration is given as the volume of oxygen used per organism (or per unit mass, e.g. per gram) per unit time; for example, as cubic centimetres of oxygen per gram per hour. So:

$$\text{rate of respiration} = x \text{ cm}^3 \text{ O}_2 \text{ g}^{-1} \text{ h}^{-1}$$

A more complex respirometer

The simple respirometer shows the principles of respirometry, but has these limitations:

- Changes in temperature or atmospheric pressure make gases expand or contract. These changes affect the distance the fluid moves, so the measured distance is not just due to the change in oxygen volume.
- The chemicals used might alter the composition of the gases in the chamber.
- It is difficult to restart the experiment without taking the apparatus apart.
- The volume change calculated using the diameter of the tube may be inaccurate.

To avoid these problems, a more sophisticated respirometer has been developed. This is shown in Fig 34.45.

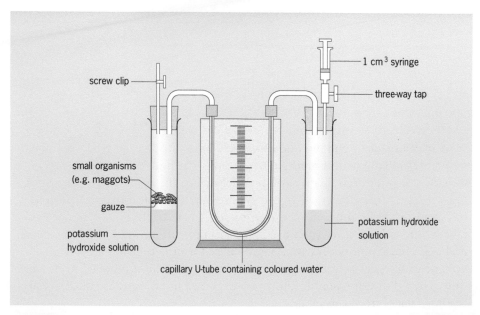

screw clip

1 cm³ syringe

three-way tap

small organisms (e.g. maggots)

gauze

potassium hydroxide solution

potassium hydroxide solution

capillary U-tube containing coloured water

Fig 33.45 A more complex respirometer. The left tube contains the organisms while the right tube acts as a control. In this way, any changes in fluid level that are not due to the organisms, such as changes in room temperature and pressure, are balanced out by the control tube. The experiment can be set up again using the syringe, which will also accurately measure the changes in volume due to the respiring organisms

This apparatus works on the same principle as the simple respirometer, but the syringe allows the volume change to be measured directly. After a set time, the syringe can be pulled up until the coloured water goes back to its starting point. This is a neat way to find the exact volume of oxygen used without having to work out the volume of a cylinder, and it allows the experimenter to move the experiment back to the start at the same time.

EXAMPLE

Using the respirometer

Q The respirometer in Fig 34.45 was set up with 50 g of blowfly (bluebottle) larvae and left for 1 hour. The fluid moved up the left side-arm of the U-tube by 5 cm. It was moved back to its original level by drawing 2 cm³ of air into the syringe.

a) What is the purpose of the syringe and why does it give this respirometer an advantage over the simple respirometer?

b) If there were 10 larvae, what would be the respiratory rate per organism?

A a) The syringe allows movement in the tube to be measured directly as a volume. This removes the need to calculate the volume of the cylinder to find out how much liquid is in the tube.

b) The volume of oxygen used up = 2 cm³. Per larva:
respiratory rate = 2/10 = 0.2 cm³ O_2 h^{-1}

Calculating the respiratory quotient

Q When the experiment was repeated using water instead of potassium hydroxide solution, the fluid moved up 1 cm, not 5 cm.

a) What does this tell us about the volume of carbon dioxide produced?

b) What is the respiratory quotient for the larvae?

A a) Carbon dioxide was not being absorbed, and its volume is one unit less than the oxygen volume. So:

$$CO_2 \text{ out} = 4 \text{ units}$$
$$O_2 \text{ in} = 5 \text{ units}$$

b) $\dfrac{CO_2 \text{ out}}{O_2 \text{ in}} = \dfrac{4}{5} = 0.8$

Blowfly diet

Q a) Assuming the larvae are only respiring one substrate, suggest which one.

b) From what you know about the diet of blowfly larvae, does this seem reasonable?

A a) Proteins

b) Yes: they consume a large amount of protein in rotting meat.

Respirometer design

Q How does the addition of the tube on the right (known as a thermobar) help to overcome the problem of changes in atmospheric pressure and temperature?

A Any pressure changes in the tube on the left are mirrored and cancelled out by those in the right tube.

Cyanide!

Cyanide is a famous poison, often featured in crime fiction and films (Fig 33.46). It is a very effective toxin because it inhibits one of the enzymes of the electron transport chain. This stops the flow of electrons and effectively prevents aerobic respiration, the source of most of the body's ATP. Without ATP, the muscles go into spasm and the body experiences convulsions. Spasm of the muscles that control breathing usually leads to death by asphyxiation (lack of oxygen to the brain).

Fig 33.46 An Agatha Christie novel featuring cyanide poisoning

RESPIRATION, ENZYMES AND TEMPERATURE

Like all metabolic reactions, respiration is controlled by enzymes. The Q_{10} law (page 81) states that the rate of an enzyme-catalysed reaction will approximately double with every 10 °C rise in temperature between 4 and 40 °C. So, with a temperature increase of 10 °C, much more ATP becomes available to power other metabolic processes.

A dramatic example is seen in **ectothermic** animals such as reptiles, which may have wide variations in body temperature during a 24-hour period (Fig 33.47). Their rate of respiration, and therefore their metabolic rate, is roughly twice as fast at 30 °C than it is at 20 °C. Such organisms are only active when their body temperature allows them to react and move quickly.

18 ANAEROBIC RESPIRATION

Anaerobic respiration does not require oxygen, so it can take place in oxygen-poor environments such as stagnant water or deep soil. It can even occur in parts of an aerobic organism that are starved of oxygen, such as muscles during strenuous exercise.

In anaerobic respiration, glucose is broken down into pyruvate but, because no oxygen is available, the pyruvate cannot be broken down any further. Instead of entering the link reaction and the Krebs cycle, the pyruvate is converted to a waste product. The nature of the waste product depends on the organism. Fig 33.48 shows the waste products of anaerobic respiration in different organisms.

Fig 33.47 Reptiles are ectotherms: they cannot control their body temperature in the way in which mammals and birds do. This lizard must absorb sunlight in order to increase its body temperature to the level that allows its enzymes to function at their optimum efficiency

> **? QUESTIONS 18–19**
>
> **18** Why do some plants produce cyanide in their leaves?
>
> **19** If the rate of respiration of a lizard is 400 arbitrary units at 40 °C, what will the rate be at 10 °C?

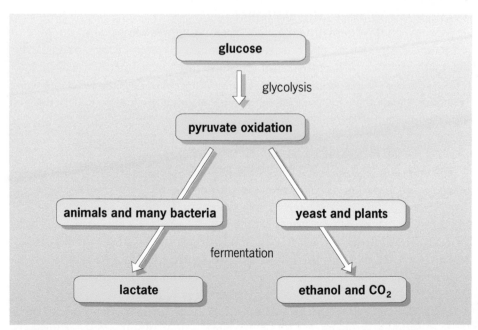

Fig 33.48 The main problem with anaerobic respiration is the build-up of NADH and the shortage of NAD^+. To solve this, H^+ is added to pyruvate to make lactate or ethanol and carbon dioxide. Fermentation is glycolysis plus the reduction of pyruvate to either lactate or ethanol and carbon dioxide

Generally speaking, animals and some bacteria convert pyruvate into **lactate** by a simple reduction reaction. In contrast, plants and fungi such as yeast convert pyruvate to **ethanol** (alcohol) and carbon dioxide.

In animals, vigorous exercise leads to anaerobic respiration in the muscles and the resulting build-up of lactate causes **fatigue**.

Fig 33.49 Fermentation in different microorganisms can be used to make a variety of useful products

Fig 33.50 Yeast is a single-celled fungus that usually respires aerobically. When deprived of oxygen, it switches to anaerobic respiration: sugars are not completely broken down and the by-products are ethanol (alcohol) and carbon dioxide

? QUESTION 20

20 What is the difference between fermentation and glycolysis?

Alcoholic fermentation is discussed in more detail in Chapter 6.

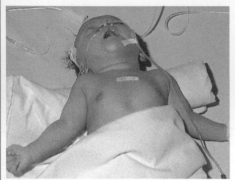

Fig 33.51 A baby with muscle stiffness caused by tetanus. The toxins produced by the anaerobic bacterium, *Clostridium tetani*, are causing the baby's muscles to go into spasm

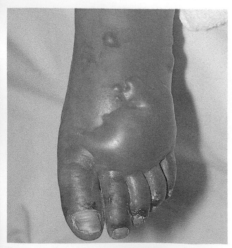

Fig 33.52 This is the dreaded gangrene. Tissue death, caused by a poor blood supply, can lead to infection by the anaerobic bacterium *Clostridium welchii*. This organism causes gas gangrene and the toxins produced can be lethal if not treated quickly

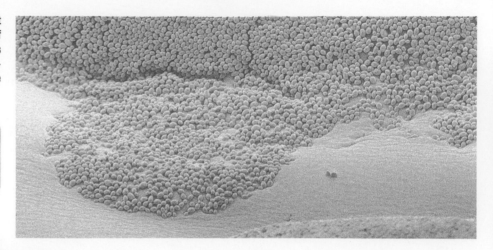

Anaerobic respiration in microorganisms is generally known as **fermentation**. As Fig 33.49 shows, this process can give rise to many useful products. Bacteria that produce lactic acid are used in the manufacture of dairy products such as yoghurt (page 97). The tangy taste is due to the high concentration of this organic acid. In addition to lactic acid and ethanol, other microorganisms can make solvents such as acetone and butanol.

Anaerobic respiration in yeast is also known as **alcoholic fermentation** (see Fig 33.50). This process was discovered by accident and alcoholic drinks such as wine and beer were made for many centuries before the science behind the process was known. For fermentation, yeast simply needs a source of carbohydrate, anaerobic conditions and a suitable temperature. So wine can be made from many organic materials, such as peaches, elderberries or even potato peelings, as well as grapes.

Since alcohol is an intermediate product in the breakdown of sugar, it contains energy that can be released by respiration. Many drinks, particularly beer, are also rich in carbohydrates that have not been turned to alcohol. Heavy drinkers become overweight because they drink large amounts as well as eating meals that, by themselves, provide enough calories to satisfy the body's energy demands.

ANAEROBES AND DISEASE

Many common bacteria can thrive only in anaerobic conditions. Two such bacteria are *Clostridium tetani* and *Clostridium welchii*.

C. tetani usually thrives deep in soil, where there is little oxygen, but is present in small numbers almost everywhere. The bacteria can enter the human body through a deep cut or wound. Once inside, they multiply and produce a toxin that makes muscles go into spasm (Fig 33.51). This is **tetanus**, or lockjaw (named because of its effect on the jaw muscles), which can be fatal if it affects the muscles used for breathing.

C. welchii can cause **gas gangrene**. If it infects areas of dead tissue, such as frost-bitten toes or injured limbs that have a poor blood supply, it turns the tissues black (Fig 33.52). In such 'ideal' conditions, the bacteria multiply rapidly, producing a foul-smelling gas and several lethal toxins. Antibiotics help if the condition is treated early. If not, the affected limb may need to be amputated to prevent the toxins spreading further into the body.

As *C. tetani* and *C. welchii* can *only* respire anaerobically, they are known as **obligate anaerobes**. Organisms such as yeast can respire by either pathway and are known as **facultative anaerobes**. Organisms that thrive only when they are respiring aerobically are called **obligate aerobes**.

SUMMARY

By the end of this chapter, you should know and understand the following:

Part I: Overview of key concepts

- Energy transfer through organisms in an ecosystem is one-way.
- Green plants are autotrophs, organisms that can make sugars from carbon dioxide using energy. Organisms that cannot make their own food are called heterotrophs. Not all autotrophs use light as a source of energy, water as a source of hydrogen or carbon dioxide as a source of carbon.
- In a food chain, energy is passed from producers to primary consumers to secondary and tertiary consumers. Each level of the food chain is called a trophic level. In most ecosystems, many food chains are linked together in a food web.
- There are three main types of ecological pyramid: pyramids of numbers, pyramids of biomass and pyramids of energy. Pyramids of energy show that over 90 per cent of the energy transferred to the next trophic level is lost as heat.

Part II: Photosynthesis

- **Photosynthesis** is a process that captures the light energy of the Sun and uses it to make organic molecules. Initially, glucose is made but plants use this basic carbohydrate to make the other essential organic molecules such as lipids, proteins and nucleic acids. Photosynthesis happens in chloroplasts; specialised organelles that house the chlorophyll pigments and all of the other compounds essential to the process. Photosynthesis is essentially a two-stage process involving the light-dependent reaction and the light-independent reaction.
- The **light-dependent reaction** happens on the **thylakoids,** membranes that contain the chlorophyll molecules.
- When a photon of light hits a molecule of chlorophyll, the pigment becomes excited and emits two high-energy electrons.
- These electrons pass down an electron transport system that synthesises ATP – this is called **photophosphorylation**.

- Energy from the excited electrons is also used to split water; this is **photolysis**.
- Photolysis produces electrons to replace those lost by the chlorophyll and protons (hydrogen ions) and oxygen are made as by-products.
- Electrons are used to make reduced NADP.
- ATP and reduced NADP are essential for the light-independent reaction, and oxygen is released into the atmosphere.
- The **light-independent reaction** happens in the **stroma,** the fluid in the centre of the chloroplast.
- It involves the reduction of carbon dioxide by a series of reactions known as the Calvin cycle.
- ATP and reduced NADP are used to reduce the carbon dioxide. Supplies of these compounds depend on light, so, when it gets dark, the light-independent reaction finishes soon after the light-dependent reaction.
- The essential stages of the Calvin cycle are:
  - Stage 1: carbon dioxide combines with the 5-carbon compound **ribulose bisphosphate (RuBP)**. This reaction is catalysed by the enzyme **rubisco** (ribulose bisphosphate carboxylase).
  - Stage 2: this produces a highly unstable 6-carbon compound that immediately splits into two molecules of **glycerate 3-phosphate (GP)**.
  - Stage 3: ATP and NADPH are used to reduce the GP into **glyceraldehyde-3-phosphate (GALP)**.
  - Stage 4: some of the glyceraldehyde-3-phosphate is used to make carbohydrate, but most of it is used to make more RuBP to continue the cycle. For every glucose molecule produced, five molecules are RuBP are re-synthesised.
- CAM plants save water by opening their stomata at night and storing carbon dioxide in an organic acid before releasing it for photosynthesis in the daytime.
- C4 plants avoid photorespiration, allowing them to make use of all the available carbon dioxide and therefore to grow faster.
- The rate of photosynthesis is determined by the factor in shortest supply; this could be light availability, carbon dioxide level or temperature.

SUMMARY (Cont.)

Part III: Cell respiration

- **Respiration** is the release of energy from food. The energy released is transferred to ATP, a chemical that can provide the cell with energy in small, instant, controllable amounts.

- Respiration consists of many reactions. Each reaction is controlled by a different enzyme. Some of the reactions occur in the cytoplasm and some occur inside the **mitochondrion**.

- Complete **aerobic respiration** of glucose has four stages: glycolysis, pyruvate oxidation, the reactions of the Krebs cycle and the reactions of the electron transport chain.

  - Glycolysis: glucose is broken down into two molecules of pyruvate. There is a net gain of two molecules of ATP. This happens in the cytoplasm of cells.

  - Pyruvate oxidation (the 'link' reaction): in the presence of oxygen, pyruvate enters the mitochondrion and is converted into acetyl coenzyme A.

  - The Krebs cycle: a series of reactions, fuelled by acetyl coenzyme A, which produce two molecules of ATP and many electrons and hydrogen ions.

  - The electron transport chain: a series of redox reactions, fuelled by the electrons and protons from the preceding three processes, and supplied by the coenzymes NAD and FAD. Electron transport releases enough energy to synthesise most of the ATP produced by aerobic respiration.

- Most food chemicals – carbohydrates, lipids and proteins – can be **respired** to provide energy.

- The rate of respiration can be measured as a respiratory quotient, the ratio of carbon dioxide evolved to oxygen absorbed. It can be calculated from measurements made using a respirometer.

- Most organisms respire **aerobically** (in the presence of oxygen). The substrate (e.g. glucose) is broken down completely to produce carbon dioxide, water and a lot of ATP.

- Some organisms and some tissues respire **anaerobically**. In the absence of oxygen, substrates are only partially broken down, leading to waste products such as lactate (in animals and bacteria) or ethanol and carbon dioxide (in plants and yeast).

Practice questions and a How Science Works assignment for this chapter are available at www.collinseducation.co.uk/CAS

34

Recycling in living systems

34 RECYCLING IN LIVING SYSTEMS

Pollution-eating bacteria

Ecosystems are sophisticated recycling 'factories' in which elements and simple molecules are continually extracted from 'waste' chemicals and then used to synthesise new chemicals.

Toxic pollutants that are released into the environment by industrial processes cause problems if they cannot be mopped up or broken down, and recycled by living organisms. Pesticides, some oil components and PCBs (polychlorinated biphenyls) are particularly persistent and can build up to dangerous concentrations as they move through the trophic levels of food chains and webs.

One solution to toxic pollutants is to prevent the use or release of such chemicals. However, this is not practicable as many important industrial processes depend on them. Until recently, scientists had drawn a blank in the search for microorganisms that might have enzyme systems weird enough to break such chemicals down into harmless products. But progress is now being made. In the mid-1990s, American scientists discovered bacteria in the stomach of the bowhead whale that can break down naphthalene and anthracene, two persistent and cancer-causing fractions of oil. A couple of the other species found there prefer to feast on PCBs.

Not only could this finding explain why bowhead whales are particularly resistant to the effects of pollutants; it could lead to the identification and mass culture of bacteria which can help clear up the effects of oil spills and industrial contamination.

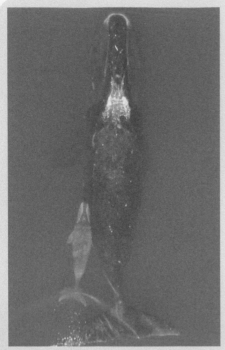

Bacteria that are a normal part of this whale's stomach contents can break down crude oil fractions and PCBs. If these bacteria can be introduced into the stomachs of other species such as dolphins and seals, this could guard them against the worst effects of some pollutants

1 RECYCLING IS NOT NEW

In the last years of the twentieth century, recycling and reusing 'waste' became a normal part of life. However, we humans cannot claim to have thought of it

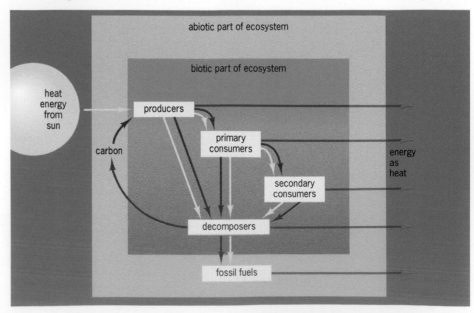

Fig 34.1 The one-way transfer of energy (yellow arrows) and the cyclical movement of mineral nutrients (for example, carbon) in an ecosystem

first. Every ecosystem that has ever existed has practised the same principle. Water molecules are constantly cycling through the environment, evaporating from the surface of the Earth, condensing as clouds, falling again as rain and then cycling through various living organisms. When you turn on the tap, many of the water molecules that come out have been through the kidneys of many people, via water treatment systems. Carbon, nitrogen and other elements also cycle through the environment.

The supply of these vital chemical elements is limited, so they are used again and again. Autotrophs such as plants build them into large, complex molecules, and heterotrophs break the molecules down again, respiring them to gain energy.

The very simple model of an ecosystem in Fig 34.1 illustrates how elements cycle through the compartments of an ecosystem. These processes are all driven by energy from the Sun. As you can see from the diagram, energy is transferred through different organisms in the ecosystem: it is never recycled.

2 MINERAL NUTRIENT CYCLING

All organisms need a supply of energy to carry out the processes of life, but they also need to obtain the mineral nutrients (carbon, nitrogen, phosphorus, etc.) which make up the complex chemicals in their bodies. Unlike energy, mineral nutrients are recycled over and over again, passing between organisms in the same ecosystem, and also other ecosystems of the world. So the carbon, oxygen and nitrogen atoms in our bodies could have been part of the soil a year ago. They may have made up the proteins of some long extinct dinosaur. And one popular anecdote says that, during the course of our lives, each of us breathes out six carbon atoms that were once part of Napoleon Bonaparte.

Producers accumulate mineral nutrients. These then pass along the food chain and are finally released back into the abiotic part of an ecosystem by decomposers. Decomposers replenish supplies of these elements in the soil and in water, and they also help to return them to the atmosphere.

WHY LIVING ORGANISMS NEED MINERAL NUTRIENTS

As we saw in Chapter 3, the chemicals of life, carbohydrates, proteins, fats and nucleic acids, contain mainly carbon, hydrogen and oxygen. Proteins also always contain nitrogen and often sulfur, and nucleic acids contain phosphorus. In addition to these elements, living things need small amounts of calcium, sodium, potassium, iron, copper, chlorine, magnesium and other trace elements.

In this chapter we concentrate on the cycling of carbon and nitrogen.

CYCLING OF INDIVIDUAL MINERAL NUTRIENTS

Mineral nutrients occur in four basic compartments in an ecosystem:
- the organic compartment – living organisms and their debris (faeces, etc.);
- as available nutrients that are held on the surfaces of clay particles in soil or in solution, where they can be taken up by plants and microorganisms;
- as nutrients that are temporarily unavailable because they are bound up in soil and rocks;
- in the atmosphere, as gases that occur in the air.

REMEMBER THIS

Autotrophs make their own food using an external energy source (usually sunlight) and a simple inorganic supply of carbon (usually carbon dioxide).

Photoautotrophs include green plants, algae and bacteria which make their own food by photosynthesis.

Chemoautotrophs are all bacteria: they also make their own food but they may use sources of carbon dioxide and energy that are different from those used by photosynthetic organisms (see Chapter 33).

Heterotrophs cannot make their own food: they must take in food molecules from their surroundings. They use an **organic carbon source** (carbon that was once in living things) so they are totally dependent upon organisms like green plants that are capable of making their own food (see Chapters 7 and 33).

Chapter 7 (Table 7.4 on page 121), gives a list of minerals that the human body needs.

Carbon compounds are discussed in detail in Chapter 3.

The carbon cycle

Life on Earth is based on carbon compounds. Carbon circulates between the abiotic and biotic parts of the environment, as shown in Fig 34.2. Autotrophs capture carbon during photosynthesis, and all living organisms return it to the atmosphere as they respire and also when they die and decompose.

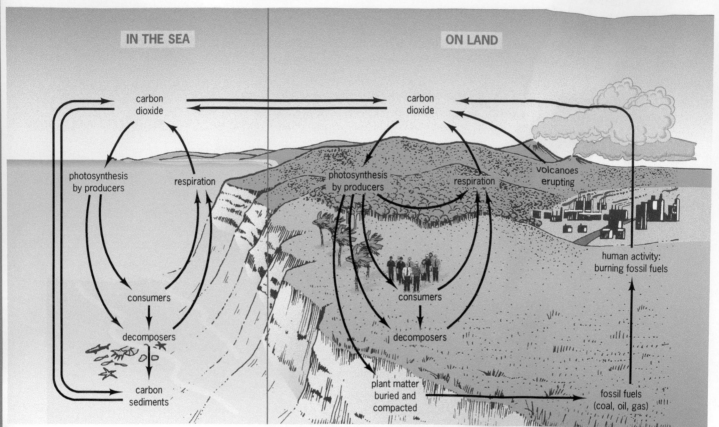

Fig 34.2 The carbon cycle. Photosynthetic organisms fix about 80 billion tonnes of carbon each year. But this represents less than 1 per cent of the total carbon in the biosphere. Carbon can be locked up for long periods of time in limestone rocks, inorganic solutions (containing soluble carbonates) and fossil fuels such as coal: these reservoirs of carbon are often called carbon sinks. When volcanoes erupt, they return carbon dioxide to the carbon cycle

The role of microorganisms in the carbon cycle

By looking at Fig 34.2 or Fig 34.3 you should be able to see that carbon would still cycle through the biosphere for some time if decomposers were removed. However, because of their general role in decomposition, decomposers play a key role in the carbon cycle. Bacteria and fungi feed by secreting digestive enzymes on to organic material to digest it. Some of the 'freed' mineral nutrients are absorbed by these microorganisms. Others escape into the soil.

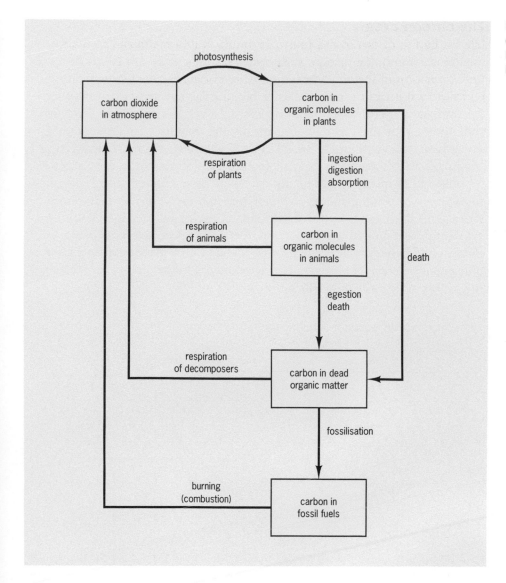

Fig 34.3 An exam/revision version of the carbon cycle. You should find this diagram easy to draw. Learning to draw this, perhaps with your own illustrations and notes, is a good revision exercise

How humans influence the carbon cycle

Humans influence the carbon cycle by two activities not practised by other species. First, humans burn fossil fuels such as coal, oil and gas. This has been going on in a big way since the start of the Industrial Revolution in Europe. Burning such fuels is the only way that their 'locked' carbon is released back into the atmosphere.

The second activity is cutting down forests, particularly rainforests in recent times, to make room for farms, fields and homes. This increases the amount of carbon dioxide in the atmosphere because it reduces the total number of photosynthetic organisms, the 'trappers' of carbon dioxide. The possible effects of these activities are discussed in Chapter 36.

REMEMBER THIS

Decomposers stop organic material accumulating in an ecosystem. They also ensure that essential nutrients, such as nitrates, sulfates and phosphates, are released back into the soil.

The nitrogen cycle

Like carbon, nitrogen is vital to organisms for the formation of proteins, nucleic acids and their products. Although nitrogen is all around us (it makes up 79 per cent of the air we breathe), most living organisms cannot access this supply. Nitrogen gas is unreactive because the two atoms that make up the nitrogen molecule are bound together by a strong triple bond. Nitrogen can, however, be fixed by microorganisms, and then it cycles through ecosystems, as shown in Fig 34.4. Fig 34.5 shows an exam-friendly version.

Nitrogen-rich waste from animals (compounds of ammonia, for example) is converted to nitrates by nitrifying bacteria. These microorganisms also fix nitrogen by converting atmospheric nitrogen into useful products. In terrestrial ecosystems, nitrogen is fixed by free-living nitrifying bacteria in soil and by symbiotic bacteria that live in nodules, the swellings on the roots of clover plants and legumes (plants like peas and beans). In aquatic ecosystems, some blue-green bacteria fix nitrogen. Nitrogen returns to the atmosphere through the action of denitrifying bacteria.

Fig 34.4 The nitrogen cycle. Bacteria play an important role in the movement of nitrogen in the biosphere

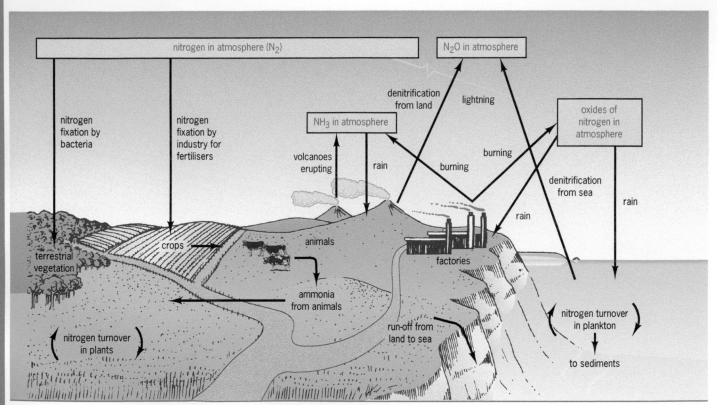

? QUESTION 1

1 Traditionally, farmers used to rotate crops with clover and grass every fourth year. Suggest how this could improve the yield of their crops. Would it have any positive effect on the environment?

These processes, in more detail, are:

- **Nitrogen fixation.** This is carried out mainly by soil bacteria such as *Azotobacter*. Some bacteria such as *Rhizobium* are **symbiotic**: they live in the root nodules of plants such as peas, beans and clover (Fig 34.6).
- **Ammonification.** This is the breakdown of organic nitrogen such as proteins and urea into ammonia. It is also known as **saprotrophic decay** and is usually thought of as 'rotting', or **decomposition**. Bacteria and fungi carry out most of the ammonification in the nitrogen cycle.
- **Nitrification.** Nitrifying bacteria such as *Nitrosomonas* oxidise ammonia to nitrite (NO_2^-). Other bacteria, such as *Nitrobacter* oxidise nitrite to nitrate (NO_3^-). These oxidation reactions release energy that the bacteria use to make food (they are chemoautotrophs).

• **Denitrification**. Bacteria such as *Pseudomonas denitrificans* obtain energy by using the nitrite or nitrate ion as an electron acceptor for the oxidation of organic compounds. They live in conditions with low oxygen levels such as waterlogged soils, and can be important in reducing nitrate levels further.

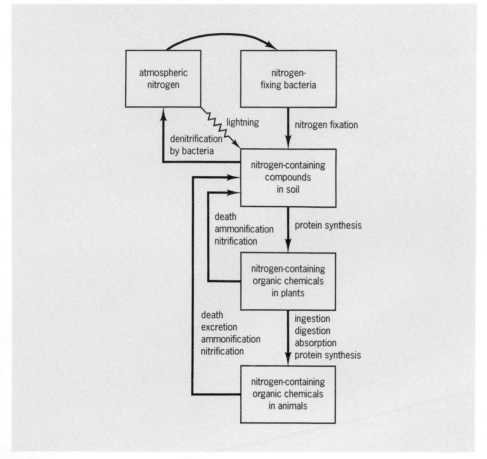

Fig 34.5 An exam/revision version of the nitrogen cycle. You should find this diagram easy to draw. Learn it so that you can reproduce it in exams

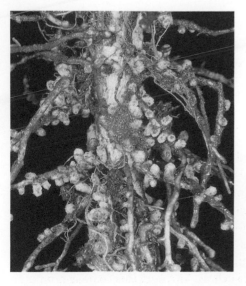

Fig 34.6 Legumes such as peas and beans have nodules on their roots. These house nitrogen-fixing species of the bacterium *Rhizobium*. The bacteria convert nitrogen gas from the atmosphere into ammonia. This passes out of the root nodules, dissolves in soil water and is then taken up by plant roots. The relationship between the bacteria and the plants is therefore symbiotic because both benefit: the plants get a steady supply of nitrogen and the bacteria get a home and a constant supply of sugars from the plant

SUMMARY

After reading this chapter you should know and understand the following:

- Energy transfer through organisms in an ecosystem is one-way. Elements such as carbon and nitrogen cycle between living organisms and the abiotic environment.

- **Decomposers** break down the dead bodies of producers and consumers and play an important role in mineral cycling. Most decomposers are bacteria and fungi.

- The **carbon** and **nitrogen cycles** show how these elements circulate between the biotic and abiotic parts of an ecosystem. You should appreciate the overall cycle and be able to draw a clear summary diagram.

- Microorganisms play an important role in the nitrogen cycle. Different species of bacteria carry out the processes of **nitrogen fixation** (the conversion of atmospheric nitrogen into useful products), **ammonification** (the breakdown of organic nitrogen such as proteins and urea into ammonia), **nitrification** (the oxidation of ammonia to nitrite and the conversion of nitrite to nitrate) and **denitrification** (use of the nitrite or nitrate ion as an electron acceptor for the oxidation of organic compounds).

Practice questions and a How Science Works assignment for this chapter are available at www.collinseducation.co.uk/CAS

35

Populations

35 POPULATIONS

Gatherings of people like this are common today but could not even have been imagined a hundred thousand years ago

The human population explosion

For hundreds of thousands of years, the population of human beings on Earth grew at a steady rate. But in the past 200 years, the world's population has increased from 1 billion to over 6 billion people. World population is currently growing by 80 to 90 million people each year. At 10.25 GMT on 12th September 2007, world population was estimated at 6 744 397 570 people.

This rate of growth is projected to continue, leading to a global population of around 9 billion people by 2028. The rate of increase and the sheer massive numbers of human beings that this population explosion will generate are unprecedented in the history of the planet. Even with our constantly developing technology and ability to adapt, making the Earth's finite resources stretch to accommodate everyone will be a huge challenge.

Trying to predict what will happen to the population further in the future is difficult because it depends on so many factors. If the population carries on increasing at the current rate, a century from now there could be as many as 15 billion people.

But this is the worst case scenario: United Nations projections assume that family planning will become widespread all over the world and that birth rates everywhere will fall. Even taking into account that life expectancy would continue to increase as medicine continues to advance, the UN estimates that population growth will slow down during the early part of this century and will stabilise around the year 2100.

1 HOW POPULATIONS GROW

 REMEMBER THIS

The **biotic potential** is the maximum rate at which a population can increase in size when there are no limits on its growth.

The **environmental resistance** which affects a population is made up of all the factors that act together to limit its size.

When the biotic potential and the environmental resistance reach a balance, the size of the population levels off: it reaches its **carrying capacity**.

Populations are changing constantly. Imagine a new population starting in a particular area with a handful of individuals and ideal environmental conditions. A good example would be a pair of healthy rabbits introduced to a small island that had plenty of edible vegetation and no predators. The population would grow slowly at first: this is phase A in the graph shown in Fig 35.1.

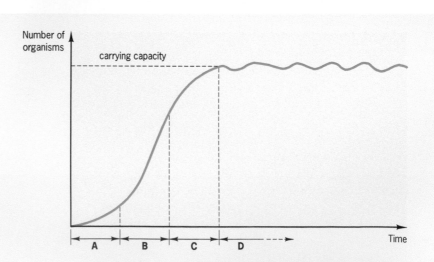

Fig 35.1 This graph shows an idealised curve of population growth. As the curve A to C looks like a flattened S, it is called an S-shaped or sigmoid curve. The text describes phases A to D in detail

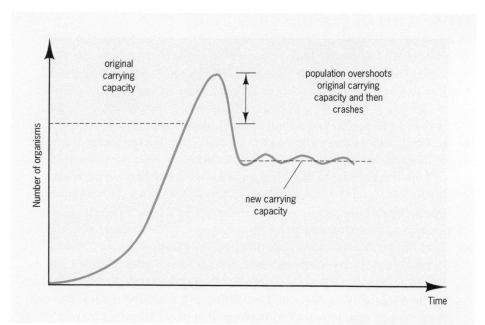

Fig 35.3 Stable, irruptive and cyclic population curves

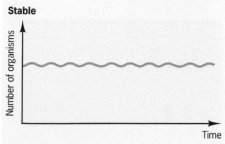

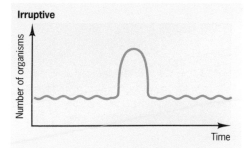

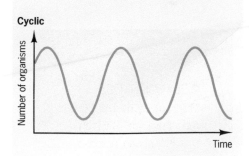

After a few months, the population would start to grow very rapidly and would double at regular intervals. This type of growth is called **exponential growth**, and is illustrated as phase B in the same graph. As more rabbits were born, they would then breed and there would be a huge rabbit population in a couple of years.

At some point, the population growth rate starts to slow down and is eventually prevented from increasing further by various **limiting factors** in the environment – phase C. When food, space and water supply are stretched to their limit, more rabbits would die and they might breed less successfully. (If possible, they would leave the environment.) Because there is less food to go around, they would eventually eat the plants faster than the plants could grow.

Most populations stabilise at some point and the growth curve becomes flat. The number of individuals in the population at this point is the **carrying capacity** of the environment – start of phase D. Sometimes there is a sudden increase in numbers and a population overshoots its carrying capacity (Fig 36.2). The environment cannot provide the resources for this number of organisms and a **population crash** usually follows. The population dies back below the level of the carrying capacity, and then takes time to stabilise. If the overshoot has damaged the environment, the population might stabilise at a different level.

The two growth curves described in Figs 35.1 and 35.2 occur in laboratory situations but you will not find them in real ecosystems. Fig 35.3 shows the three main types of growth curve that occur in natural conditions:

- **Stable**: the population size remains at roughly the same level over a long time. A study of a population of trees in a well-established woodland would produce this sort of curve.
- **Irruptive**: the population size is basically stable, with the occasional dramatic increase followed by a population crash. This type of population growth occurs in algae. A population is mostly stable but, in hot weather, when water is contaminated by phosphate or nitrate fertilisers, there is an **algal bloom** (Fig 35.4).
- **Cyclic**: the population increases and decreases in a regular cycle of growth and die-back.

Fig 35.4 An algal bloom on a dyke

WHAT LIMITS POPULATION SIZE?

In an ecosystem, **abiotic** (physical) factors in the environment can limit population size. When we look at the whole range of abiotic factors that affect population size, we can classify them into three general categories:

- Limiting factors that are *always present*. These include space and light. A cliff that supports a population of kittiwakes, for example, is a fixed size and can accommodate a fixed number of birds (Fig 35.5).

- Limiting factors that *vary predictably over time*. These include changes in temperature and weather patterns in different seasons which lead to population movement or a change in behaviour among different organisms. As winter approaches, birds that live in an ecosystem in the summer might migrate to a warmer place (Fig 35.6). Small mammals such as dormice hibernate.

- Limiting factors that are *unpredictable*. Some unpredictable limiting factors are **density dependent**: they cause a greater effect if there is a high population than if the population is quite small. Sudden changes in the availability of water and nutrients, or a localised natural disaster, for example, can devastate an overcrowded population but leave a widely dispersed population relatively untouched.

- Other factors are **density independent**: they can devastate a population, whatever its size. Fires, floods, severe frosts and other catastrophes, for example, might destroy part or the whole of an ecosystem without warning (Fig 35.7).

Fig 35.5 The size of the ledges is an unchanging limiting factor. The kittiwakes can get nesting sites only by displacing other birds

Fig 35.6 Arctic terns are the world's champion travellers. In their search for food and the right breeding conditions, they range from deep inside the Arctic Circle, over the Atlantic to the pack ice of Antarctica

Fig 35.7 Catastrophes, including those caused by human activity such as oil spills, have a density-independent effect on populations

2 INTERACTIONS BETWEEN ORGANISMS

Every living organism in an ecosystem affects the others. Some of these interactions are **intraspecific**: they take place between organisms of the same species. Others are **interspecific**: they take place between organisms of different species.

> **✓ REMEMBER THIS**
> To remember which type of competition is which, *inter*specific competition is *between* species like *inter*national football matches are *between* nations.

 HOW SCIENCE WORKS

Growth and bacterial populations

Bacteria are among the fastest growing organisms on Earth. A population of *Escherichia coli*, for example, can double in size in only 20 minutes, if the conditions are very favourable. This doubling time is called the **growth rate** of the bacterial population. The doubling time is a fixed characteristic of each type of bacteria, and it can be used to identify unknown cultures.

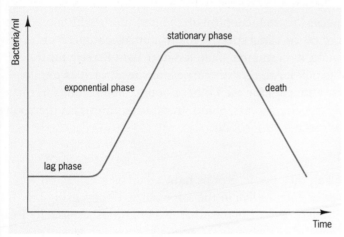

Fig 35.8 A typical bacterial growth curve

A typical growth curve for a population of bacteria is shown in Fig 35.8. There are four basic stages of growth:

The **lag phase** occurs just after some bacteria have been placed in a new container of culture medium. During this period, the bacteria are adjusting to their new surroundings, new food source and new temperature. Their doubling time is relatively slow.

When the **exponential** or **lag phase** begins, the bacteria really get going and multiply very quickly. They have plenty of food and waste has not built up to dangerous levels. They make as many new bacterial cells as they can.

After a period of rapid growth, the population then enters the **stationary phase**. The bacteria run out of room or space, or they have produced toxins that limit their growth. The bacteria are still dividing, but the growth rate is equal to the death rate, and no increase in numbers occurs.

In the **death phase**, the balance shifts again, as more bacteria die than are produced by cell division. For the first time since the bacteria were placed in the new culture, their population shrinks.

Interspecific interactions in a bacterial population

The curve shown in Fig 35.8 describes the growth of a single population of bacteria. When more than one type of bacteria is present in the culture fluid, the two species compete (Fig 35.9). This is an example of an interspecific interaction.

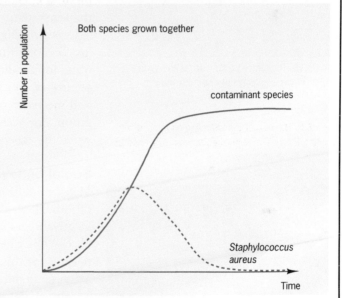

Fig 35.9 Interspecific competition between two bacterial populations

In microbiology laboratories, it is common for cultures to become accidentally contaminated. If a researcher is trying to grow up a culture of *Staphylococcus aureus* to isolate polysaccharides to study their potential as vaccines, it can be disastrous for a second species of bacteria to gain access to the culture. Fig 35.9 shows what can happen when the contaminant is a bacterial species with a faster doubling time. Both species grow equally well at first, but the contaminant species enters the exponential growth phase first, and its numbers increase at a faster rate than the *S. aureus*. The contaminant species may also be more comfortable metabolising different sugars in the medium, and can take more advantage of the food available. After a time, the *S. aureus* population enters a short stationary phase, followed quickly by a decline in population. By the end of the culture time, the researcher returns to the incubator to find a pure culture, not of the *S. aureus* that was expected, but of the rogue contaminant.

INTRASPECIFIC INTERACTIONS

Relationships between organisms of the same species in an ecosystem fall into three main categories:

- reproduction and care of the young;
- social behaviour;
- competition for resources.

Reproduction and care of the young

Organisms of the same species interact at many different stages of sexual reproduction. In some species, the relationship is brief and minimal: a wind-pollinated plant simply releases pollen into the air and accepts pollen from another plant of the same species. In others, the association is complex and long term. For example, swans, like humans, find a mate and stay together for many years, sometimes for life (Fig 35.10). Both parents feed and care for the young and, to some extent, provide companionship and help for each other.

In both examples, the level of interaction achieves the aim of reproduction: producing enough surviving offspring to maintain the population level in the ecosystem.

In many animal species, reproduction involves interactions with more than just one other member of the same species. Many animals compete with each other for a suitable mate: for example, male red deer fight fiercely for the control of groups of females. Animals that are more social, such as termites and meercats, provide joint care for each other's young. Some adults in the group, not necessarily just the parents, guard, transport and protect the young, while the others go off to search for food.

Social behaviour

Many animals are social: they live in groups rather than as a collection of individuals that just happen to live in the same place. Group living has its advantages and disadvantages. Animals that live in groups are usually less likely to be eaten by predators than solitary animals. There are several reasons for this:

- A large number of potential prey in one place can confuse a predator – the predator cannot decide which one to attack.
- Different individuals in a large group of animals can perform different tasks: some can look for food, others care for the young and some can act as look-outs for predators. Large groups therefore spot approaching predators much sooner than animals on their own and so can take more successful evading action.
- A large number of animals are more successful than just a few at defending against an attacking predator. If a predator threatens, elephants group together, forming a circle to protect their young, which are herded into the centre of the group.

Animals that live in groups also stand more chance of hunting or finding food successfully (Fig 35.11, and the How Science Works box opposite). Hyenas, for example, hunt in packs and can bring down a deer that is much larger than one hyena would be prepared to tackle. Other group associations can help to protect animals from the cold as they huddle together.

Fig 35.10 The female swan carries her young, while the male displays to deter intruders

Fig 35.11 There is evidence that bees can communicate the position of good pollen sources by elaborate dancing displays to other bees in the hive.

Costs and benefits of social behaviour

Prairie dogs are rabbit-sized rodents with small ears and short legs (Fig 35.12). They live in large groups in the plains of North America digging a complex of tunnels up to 30 metres long. There are two species of prairie dog, the black-tailed and white-tailed forms, and each community or 'town' may contain up to a thousand animals.

Prairie dogs graze above ground during the day when there are many predators around such as coyotes, badgers and hawks. They have 'sentries' sitting upright outside the burrows keeping a watchful eye for predators. The appearance of a predator provokes a series of whistling barks from the sentries.

A cost–benefit analysis has been carried out on the social behaviour of the prairie dog. An anti-predator response was tested by dragging a stuffed badger through a colony. In the larger groups of prairie dog, the predator was detected sooner. Black-tailed prairie dogs live in larger groups than their white-tailed cousins. Consequently the black-tails can afford to spend significantly less time scanning their surroundings for predators (35 per cent of their average daily time budget compared with 45 per cent in the white-tails). Less time being vigilant means more time to feed.

So why don't white-tailed prairie dogs live in large groups as well? One answer is that living in large groups has costs as well as benefits. Within either species, individuals in large groups fight more often than those in small groups. Also disease-spreading fleas are more common in large groups (3.3 fleas per black-tailed burrow, 0.5 fleas per white-tailed burrow).

The evolution of group behaviour therefore depends on balancing advantages against disadvantages. Other factors may also be involved – black-tailed prairie dogs live in open plains, for example, and, since there are many predators around, it is safer for them to live in larger groups.

Fig 35.12 A prairie dog keeping guard as sentry

Competition for resources

Most of the disadvantages of group living arise because animals have to compete with each other for vital resources. Hyenas may make a bigger kill when they hunt in a pack, but the meat has to fill more stomachs. If food becomes scarce, rivalry within the pack becomes intense and some animals may need to find a new habitat to survive. Animals of the same species also compete for mates, shelter, water and social position.

INTERSPECIFIC INTERACTIONS

A full list of the interactions that can occur between individuals of different species in an ecosystem is given in Table 32.2 on page 518. In this chapter, we will look in detail at four of these:

- predator–prey relationships;
- parasitism and disease;
- mutualism and commensalism;
- competing for resources.

Predator–prey relationships

A basic definition of a predator, including animals that feed on plants, is:

A predator is an organism that eats another organism to obtain energy and nutrients.

The relationship between predator and prey is based on a constant battle. The actual 'catching and eating' part of the battle is only a small part of the story.

 REMEMBER THIS

Symbiosis comes from a Greek word that means 'to live together'. We use it to describe a close physical interaction between two organisms from different species. There are three types of symbiosis:

Parasitism: one of the partners benefits, the other is harmed.

Commensalism: one partner benefits and the other is unaffected.

Mutualism: both partners benefit from the association.

Evolution and natural selection are discussed in Chapter 26.

Fig 35.13 Bats have evolved sophisticated sonar systems to detect flying insects with great accuracy. Some insects, however, have evolved the ability to detect bat sonar signals and respond by closing their wings and dropping out of the sky. This makes it much harder for the bat to catch them and they gain an important survival advantage over other insect species

The more serious conflict involves evolution.

All species evolve by the processes of genetic mutation and natural selection: prey species are constantly evolving and becoming better able to cope with predators; predator species are constantly evolving to out-manoeuvre their prey (Fig 35.13). Sometimes, either the predator or prey loses the long-term battle. If the prey evolves to completely avoid a predator, and that predator has no alternative food supply, the predator species can be driven to extinction. If a predator out-evolves its prey, it will do well in the short term. But, if the prey species becomes extinct, the predator then needs to find an alternative food supply to survive.

In general, predators thrive when they develop adaptations that allow them to eat well and make the most of their food. Herbivores, animals that eat plants, often have specialised mouthparts and digestive system to gain the most nourishment from the plants available to them. Many herbivores are ruminants: their digestive system has four stomachs and they chew their food twice, once after eating it and once after regurgitating it.

Prey species continue to survive because they have evolved defence mechanisms against predators. In animals, these include **camouflage**, **spines** and **barbs** and **chemical deterrents** (Fig 35.14). Many, particularly insects, produce foul tasting or poisonous chemicals and usually advertise this by the colours and patterns on their bodies and wings.

Although plants cannot run away from grazing animals, they are not totally defenceless. Their defences include thorns, tough tissues and thick waxy cuticles. Many plants have also developed **secondary compounds**, chemicals that have no metabolic role but which put off potential grazers. Examples include tannins in oak leaves and cardiac glycosides in foxglove plants.

Fig 35.14 Animals use many different techniques to avoid capture by predators

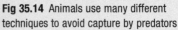
When under threat, a porcupine runs backwards at the enemy with quills raised. An embedded quill can cause a septic wound that would kill a lion

The camouflage of this iguana matches the colours of the trunk and lichens in a Peruvian rainforest

The deadly sting of a scorpion deters predators as well as being useful for hunting food

There is an ecosystem in all of us – the ecology of the human gut

When you were born, your gut was completely sterile. No bacteria lived there. This changed rapidly over the first few days of life and, by the time you were 5 days old, the number of bacteria in your gut already outnumbered the total numbers of cells in your body (Fig 35.15).

This teeming mass of internal bacteria is very important in health and disease. Commensal bacteria, the normal bacteria that live in the gut without causing disease, are actually useful. They help to protect the intestinal epithelium against cell injury by toxins, pathogenic bacteria or parasites. These 'helpful' bacteria help prevent potentially disease – causing bacteria from invading cells and from dividing out of control. They also help the body control its uptake of fat, and they stimulate the continual growth of new blood vessels as the surface of the intestine renews itself.

Despite its importance, we are now only just learning about this fabulously complex ecosystem. In 2005, researchers investigated the gut ecology of three healthy people using DNA fingerprinting technology. Previously, researchers had just cultured bacteria from faecal samples, but the bacteria from faeces only represent the tip of the iceberg. Many of the bacteria in your gut are attached to the villi and the wall of different parts of the intestine – they don't come out in faeces.

By looking at the genetic fingerprint of bacteria found in mucosal samples during colonoscopy from the human colon (the caecum, ascending colon, transverse colon, descending colon, sigmoid colon) and rectum, the researchers built up a fascinating profile of the ecology of the human gut. The DNA was amplified from samples with polymerase chain reaction (PCR) and PCR products were cloned and sequenced. The existence of 245 different bacterial species was identified from the sequences. Of these sequences, 80 per cent were from bacteria that had never been cultivated from faecal samples. Most of the organisms were members of the phyla Firmicutes and Bacteroidetes. Almost all of the bacteria from the Firmicutes phylum were members of the Clostridia class of anaerobic bacteria. Some were related to known butyrate–producing bacteria – butyrate is a compound produced in faeces that is responsible for its foul smell.

Sixty-five different species within the Bacteroidetes phylum were detected, including *Bacteroides thetaiotaomicron*, a species known to be involved in

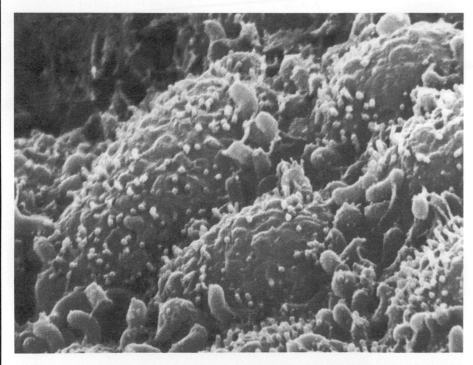

Fig 35.15 Intestinal bacteria. Coloured scanning electron micrograph (SEM) of bacteria (red) on the surface of the duodenum. The duodenum is the first part of the small intestine, the part that receives food from the stomach. A healthy human intestine has a large population of bacteria, most of which are harmless and some of which aid digestion. Cases of food poisoning often involve a bacterial strain that upsets the digestive system. The brown lumps are the intestinal villi, and the white, hair-like structures are microvilli, which absorb nutrients from food.

beneficial functions, including nutrient absorption and epithelial cell maturation and maintenance. There were not many bacteria from the phyla Proteobacteria, Actinobacteria, Fusobacteria and Verrucomicrobia. The Proteobacteria phylum includes the well-known *Escherichia coli*, which is often thought to be a major gut bacterium. However, it is not an anaerobe – it needs oxygen to survive – and it actually makes up less than 0.1% of the bacteria found in the colon.

Overall, the researchers concluded that bacterial diversity within the human colon is greater than expected and most of it was novel – the bacteria there had not been identified before. Interestingly, the sequences from the three different people showed that they had quite different microbial communities in their gut. Subject B had the greatest biodiversity of gut bacteria. The reasons for this probably had a lot to do with diet and lifestyle.

Parasitism and disease

Predators usually kill their prey: parasites have a different strategy because they depend on their prey, the **host** (a different species), staying alive. They live and reproduce using the resources of the host organism that is damaged in the process. This is why many parasites cause disease in humans and other animals (Table 35.1). The more successful parasites cause less damage to their host. Tapeworms (Fig 35.16), for example, can survive in the bovine and human digestive systems for many years, causing relatively minor damage in the affected person.

Table 35.1 Some major parasitic diseases and the parasites that cause them

Disease	Parasite	Transmission and effects
malaria	*Plasmodium* (4 different species of this protoctist parasite cause malaria)	Regular cycles of fever: disease transmitted to people by the *Anopheles* mosquito (see box on page 578).
sleeping sickness	*Trypanosoma* species (also a protoctist parasite)	Transmitted to people by insects such as the tsetse fly which bite infected animals
leishmaniasis (also called kala-azar and oriental sore)	*Leishmania* species (another protoctist parasite)	Transmitted to humans via sandflies
schistosomiasis (used to be called bilharzia)	*Schistosoma* (blood fluke)	Flukes live in water (their intermediate host is a water snail) and infect humans who bathe
tapeworm infection	*Taenia solium* (tapeworm)	Parasite lives in intestine: passed to humans in infected and undercooked meat
toxoplasmosis	*Toxoplasma* species (protoctist parasite)	Causes weakness and fever – passed from mother to baby across placenta. Can be transmitted by domestic cats

Parasitic adaptations

Parasites show very specialised features that fit them to their lifestyle. All parasites show some general adaptations:

- They have ways of getting into the body of the host. Fungi that parasitise plants produce the enzyme **cellulase**, which breaks down plant cell walls. The schistosome (blood fluke), which causes schistosomiasis, can bore through human skin.
- They have structures that enable them to attach to their host. Tapeworms have a ring of hooks called a **scolex** that they insert into the wall of the host's intestine (Fig 35.16).

Schistosomiasis is discussed in Chapter 29.

- They have lost the organ systems that they no longer need. Tapeworms and other gut parasites have, for example, lost their own digestive system because they can absorb already-digested food across their body surface. We call this loss of structure and function **parasitic degeneration**.
- They protect themselves against the internal defences of the host. Trypanosomes, which cause sleeping sickness in humans, produce an outer layer which varies constantly in its structure. Just as the host's body starts to make antibodies to one type of coating, the trypanosome produces a different coat which the antibodies cannot recognise.
- They often have complex life cycles that allow them to infect new hosts (see the How Science Works box overleaf).
- They show great **fecundity**, or capacity for reproduction. Many tapeworm segments break off every day and pass out of the body in the faeces. Each segment can contain hundreds or even thousands of eggs.

QUESTION 1

1 Organisms can interact intraspecifically or interspecifically. What do these two terms mean?

✔ REMEMBER THIS

Legumes, with their root nodules filled with bacteria (see Chapter 34), are an example of a mutualistic association.

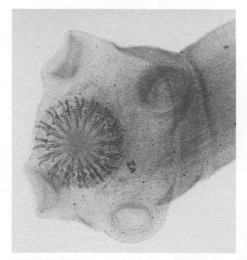

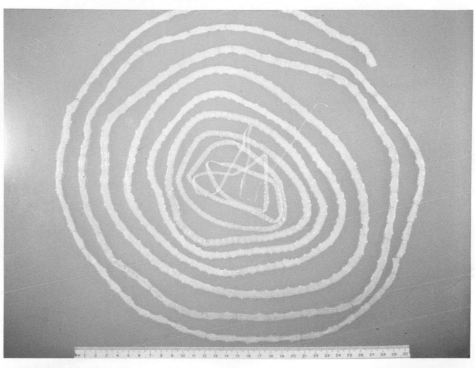

Fig 35.16 Right: the beef tapeworm. The head is at the thin end, attached to many hundreds of segments. They break off and pass out of the host (cattle), to infect another victim. The ruler is 30 cm long. Above: At the head end (the scolex) are four suckers and a ring of hooks that attach the tapeworm to the gut wall of its host

Mutualism and commensalism

The term **mutualism** is used to describe a relationship between two organisms from different species that receive mutual benefit from the association. Neither is harmed and both do better by being together than they would alone. The African bird, the oxpecker, for example, cleans the skin of the rhinoceros, removing all the parasites such as ticks and mites. The bird eats the insects, gaining the advantage of the nutrients.

Two organisms that are **commensal** live together and, while one of the partners benefits, the other does not. The relationship between clownfish and sea anemones is a good example of commensalism (Fig 35.17). The fish hide from predators in the mass of the anemone's stinging tentacles. Although the anemone is unharmed, it does not seem to benefit in any way.

Competing for resources

Species can compete in different ways. Some organisms stop the members of other species from getting near a particular resource. Wrens, for example, defend food plants in their territory. In other situations, two species have access to the resource, but one is better at exploiting it.

Fig 35.17 There are 26 species of clownfish which all associate with sea anemones. This colourful fish hides from predators among the venom-producing tentacles of the sea anemone

The fight against malaria

Malaria is a disease caused by a parasite. Four species of protoctist of the genus *Plasmodium* are responsible for 200 to 300 million cases of malaria every year, worldwide. The disease is common in the tropical and sub-tropical regions of the world – the home of 40 per cent of the world's population. The death rate is appalling, somewhere between one million and five million people each year, many of them children under 5.

Malaria is extremely difficult to control for several reasons. First, it has a complex life cycle (Fig 35.18). Any four of the *Plasmodium* species that cause it can be passed on to human hosts by any one of 60 species of the *Anopheles* mosquito. Draining swamps and marshes and using insecticides to spray mosquito-breeding areas was temporarily successful and reduced the spread of malaria during the 1950s and 1960s. However, not every species of carrier mosquito was wiped out. Many of them have developed resistance to the early insecticides and are now much more difficult to control.

Secondly, the parasite itself is a tricky customer. It produces symptoms that get worse and then better in cycles. Anti-malarial drugs target some but not all stages in the life cycle (Fig 35.19). The sufferer therefore needs to take the drugs over a long period of time, or the parasite 'hides' from their effects. The common anti-malarial drugs have been used so widely that they have put selection pressure on the parasite and it has now developed resistance to many of them.

Although new drugs are being developed all the time (some of the most recent are based on extracts of rainforest plants), workers now concentrate on prevention. Any stagnant water that can be eliminated is removed and mosquito nets in houses are dipped in a safe insecticide such as permethrin. Biological methods of control of the mosquito larvae are being used – fish such as guppies are being bred to eat them. Vegetation around houses is cleared, and trees are being planted in marsh areas to destroy the wet breeding grounds of the mosquito.

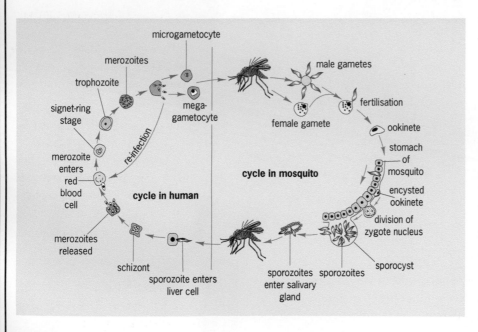

Fig 35.18 The life cycle of the malarial parasite, *Plasmodium vivax*. The *Anopheles* mosquito is the first host; humans are the second host. People with malaria get regular cycles of illness because merozoites are continually released. When the parasite is 'hiding' in the red blood cells, symptoms are not too bad: it is the sudden rush of foreign proteins in the bloodstream that leads to the bouts of fever that are typical of malaria

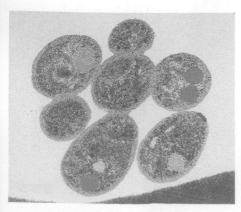

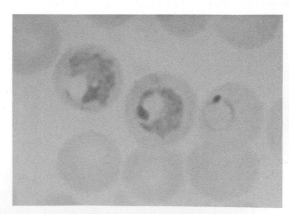

Fig 35.19 The merozoite stage (left) and signet ring stage (right) of *Plasmodium vivax* in human red blood cells

Some species have evolved to **coexist**, sharing the resources that are available in an ecosystem. This is termed **symbiosis**. Some plants, for example, have a mesh of roots close to the surface of the soil to trap water from short showers of rain. Other neighbouring plants have long roots that take advantage of deeper groundwater. Both species are using the same resource, but they are sharing it in a way that allows both to benefit. This is called **resource partitioning** and it also occurs between animal species.

3 A WIDER LOOK AT COMMUNITIES IN AN ECOSYSTEM

We have seen that ecosystems are dynamic systems and that the physical parts of an ecosystem, such as the weather, affect and interact with the organisms that live there. In this section, we look at how an ecosystem forms.

Imagine a completely new physical environment such as the bare rock and ash left behind after a volcano has erupted and then become quiet. After the ground cools and water re-enters the environment, some plants start to grow there. This process is called **colonisation**. After a time, other species of plant establish themselves and eventually, probably after many years, woodland forms. This process is called **succession**.

Succession also describes the more minor changes in the vegetation of an ecosystem that occur after a less catastrophic environmental change. For example, after a small patch of trees in woodland dies following a series of hot, dry summers, the land becomes repopulated by the same sorts of trees very quickly because their seeds are already present in fertile ground.

To distinguish between these two types of situation, we say that the process that results in new vegetation on a bare site is **primary succession**. This takes time, and the final appearance of the community depends on which plant spores and seeds come into the area from neighbouring ecosystems. The process that repopulates a previously well-vegetated area after a minor environmental change is called **secondary succession**.

Strictly, succession is used only to describe the development of and changes in plant communities. In any ecosystem, however, the plant communities help to form the conditions that enable other classes of organism (animals, bacteria, fungi, and so on) to colonise the environment. Once a stable community has formed, we say that a **climax community** is established.

A DEVELOPING ECOSYSTEM

In order to understand the stages that occur in succession, think of an environment consisting just of bare rock (Fig 35.20, overleaf).

Colonisation occurs in many different stages called **seres**. The rocks first provide a home for spiders that feed on insects in the area. Succession begins as mosses and lichens start to grow on the bare rocks. These first plant species are called **pioneer species**. Over many years, perhaps thousands of years, the environment is battered by rain and wind and other climatic events, and the surface of the rock is broken down into smaller fragments. The mosses and lichens help soil to form in cracks and crevices, and eventually a few small broad-leaved plants start to grow. The first ones are what we might think of as weeds – herbaceous plants that can grow in relatively nutrient-poor soil. They have a rapid life cycle, then die back and increase the humus content of the soil. Typical coloniser species are grasses, daisies, dandelions and clover.

REMEMBER THIS

Deflected succession

The process of succession is not always a straightforward progression from colonisation to a climax community. Environmental factors can get in the way, forming what are known as deflected climax communities.

These are very common if there is a fire, whether a natural one started by lightning or one started by humans deliberately burning vegetation. Some plants tolerate fire better than others and survive, whereas fire-sensitive species perish. The result is a fire climax community.

In mountainous areas affected by avalanches, the normal forest climax community can be replaced by a deflected climax community of shrubs that can reach maturity and reproduce in that site in the intervals between avalanches.

REMEMBER THIS

Lichens consist of algae growing inside a fungus. The fungus provides anchorage and protection from drying out. In turn the algae can photosynthesise and give the fungus organic compounds that would otherwise be unavailable from the bare rock.

REMEMBER THIS

Herbaceous plants have no woody tissue, in contrast to shrubs and trees.

Fig 35.20 Succession in a bare site to woodland

Boulders and rocks with mosses and lichens

Grassland with small flowering plants, like daisies

Taller herbaceous plants, like willowherb and foxgloves, grow and cut off light to small ones. Tree seedlings can become established.

Bushes and shrubs, such as hawthorn and bramble, grow. Most non-woody plants die out.

Fast-growing trees, such as birch, grow up, forming dense, low forest.

Larger, slower growing, but stronger oak trees grow above the birch and establish the climax community.

Studying succession

The process of succession normally takes decades or centuries, but there are two common ways of studying the process:

- Clear a patch of ground and watch what happens to the bare soil.

- Study a sand dune system near a beach and observe the changes that occur as you move inland.

1 Clearing a patch of ground

Remove all the plants, fence it off from grazing animals, and observe the changes. The problem is that it takes at least 50 years, but eventually the forest is re-established. There are two conditions that need to be met if the ecosystem is to develop as normal:

- The soil must initially have relatively low humus content.

- There must be no grazing animals that can get on to the plot.

The climax community that develops depends on the climate. The process outlined in the text on this page will not happen, for example, on exposed hillsides where the soil is too thin or the wind too harsh. Other examples of climax vegetation around the world include rainforest, cloud forest, tundra, grassland and coniferous forest.

2 Studying a sand dune system

Sand dunes are useful areas to study because you can observe colonisation and succession

without having to wait 50 years. You can see some of the changes associated with the development of ecosystems as you simply walk inland (Fig 35.21). Near the sea, the dunes are at their youngest and wind tends to pile up the sand. As a result, the profile changes from year to year. Sand is a very difficult medium for plants; water and nutrients drain straight through, and the constant shifting makes it very difficult for the roots to anchor the plant. However some pioneer species, notably marram grass, have a dense root system that binds the sand together. This holds water and humus particles and makes the whole dune more permanent. Once the marram grass has made the environment less hostile, other plants such as ragwort, willow and grasses can take over. As you move inland the sand gets darker because the humus content increases, as does the species diversity.

Fig 35.21 Sand dune succession

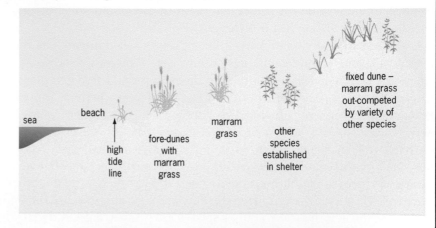

sea

beach

high tide line

fore-dunes with marram grass

marram grass

other species established in shelter

fixed dune – marram grass out-competed by variety of other species

Succession proper is then under way. Succession occurs because the colonisers change the habitat. Once the colonisers have improved the quality of the soil, more species can grow. The **diversity** of herbaceous plants increases greatly as conditions become more favourable. Grasses and then plants with root nodules that have nitrogen-fixing bacteria start to grow. Taller plants cut off the light and so out-compete the shorter ones. The greater diversity of plants attracts more insects that in turn attract birds and small mammals. At this stage, grazing can have a marked effect and prevent any further succession. Many plants, including tree saplings, have their growing points at the top of the stem, and herbivores such as rabbits and sheep prevent any further growth. In contrast grasses grow from the base of the stem, so they thrive despite constant grazing.

ESTABLISHMENT OF THE CLIMAX COMMUNITY

In this phase, small woody plants – shrubs such as hawthorn and bramble – begin to dominate. In turn these are out-competed for light by fast-growing tree species such as birch, which form a low, dense forest. Eventually the large but slow-growing trees – notably the oak – begin to dominate until the **climax community** is established and there is *no further succession*.

REMEMBER THIS

It is a good idea to learn one example of succession in detail, and make sure that you can name some plant *species* at each stage, rather then talking vaguely about 'trees and shrubs', and so on.

4 BIODIVERSITY

The word **biodiversity** is the popular contraction of the term **biological diversity**, which is used generally to refer to the variety of life on Earth. More specifically, we use it to describe the numbers of species present in a particular habitat or to show the range of types of organism found there (Fig 35.22). Take the sea bed, for example. It has a large number of species but it is largely made up of a few basic types: annelid worms, molluscs and echinoderms. A rocky shore may have the same number of species but there is often a greater variety of animals and plants: arthropods, annelid worms, molluscs, echinoderms, cnidarians, fish, birds and seaweeds. We say that the shore has a greater **species diversity**: it has a greater **species richness** (number of different species) and a greater **species disparity** (range of different species types) than the sea bed. One of the most diverse ecosystems in the world is the tropical rainforest (Fig 35.23).

Diversity can be used to indicate the 'biological health' of a particular habitat. A slow increase in plant species diversity is normal for stable habitats such as hedgerows. If a habitat suddenly begins to lose its animal or plant types, ecologists become worried and search for causes (a pollution incident, for example).

You might also have heard of the term **genetic diversity**. Different species are obviously genetically

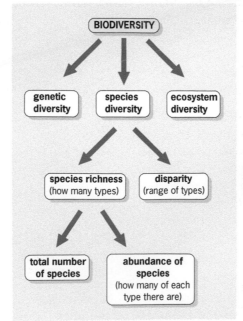

Fig 35.22 What biodiversity means

Fig 35.23 The tropical rainforests of the world encompass a massive diversity of all forms of life

Fig 35.24 Aspirin, the most commonly used drug in the world, was originally extracted from the bark of the willow

different from each other, but there is also genetic variation within a species (no two people look identical, for example).

A third form of biodiversity is **ecosystem diversity**. The environment is an important factor that determines the number and types of species present in an ecosystem. At its simplest, ecosystem diversity is described as the number of different types of habitat in a given area.

WHY BIODIVERSITY IS IMPORTANT

When a species becomes extinct, many other species can be affected: no type of living organism exists independently. No one can really predict the long-term effects of repeated extinctions. Organisms have, of course, always become extinct, because of evolutionary pressures. What concerns us today is the rate at which species are disappearing because of the impact of human populations.

Although the diverse world of animals and plants provides us with a massive resource of food, we actually use only 30 species or so to provide 90 per cent of all the food we eat worldwide. Plants have already given us many useful medicines (Fig 35.24 and the How Science Works box below) and there are undoubtedly more compounds still to be found. In addition, many different animals and plants stabilise soil and so minimise erosion, support fisheries, provide income through tourism and provide beauty and interest.

MEASURING BIODIVERSITY

Estimates for the total number of species that still exist range from 1 million to 50 million. We base our estimates on the knowledge we already have of numbers and types of species found in particular habitats and on the rate at which our knowledge is increasing. Between 1978 and 1987, 367 new vertebrates, 173 new annelids and 7222 new species of insect were identified.

HOW SCIENCE WORKS

Medicines from the rainforest

The weather conditions in the tropical rainforests around the Equator have encouraged the evolution of areas of high plant and animal diversity. The indigenous human populations who live and work in the rainforests need to make a living, and working for developers who sell timber and other products to the developed world is an obvious means of doing this. The consequences for the environment are serious. Some scientists estimate that three species of living things are being made extinct every hour, because the rainforest is being destroyed.

But what difference can a few plants make? Possibly the difference between life and death for many people. Screening plants from the rainforest has shown that they contain many compounds that protect them against predators and parasites. These compounds might also be useful in human medicine (Fig 35.25). Since 1990, at least four compounds have been identified that might provide drugs for use against the virus that causes AIDS (acquired immune deficiency syndrome).

The work of developing medicines needs sophisticated chemical and pharmaceutical techniques, but it cannot be done unless plants that are collected for screening can be

identified accurately and consistently. Speed is also important. If destruction continues at its present rate, the rainforests will have been completely wiped out by the end of the century.

Fig 35.25 Scientists who travel to remote areas of rainforest to search for and identify plants which may be useful in medicine are called ethnobotanists. They work with local traditional healers to narrow down the choice of plants to study. Samples are sent back to laboratories for testing. Research programmes are now being set up to share the profits from successful compounds, so that indigenous people can benefit from the discovery, instead of suffering exploitation

The ecology of human skin

The skin is inhabited by millions of commensal bacteria (Fig 35.26). DNA technology has also recently told us a lot about the populations of different microbes that are found here.

When researchers in the USA analysed the bacteria on the forearms of six healthy subjects, three men and three women, they found some permanent residents of the skin ecosystem, and some that were just passing through. Just over half, 54.4 per cent, of the bacteria identified in the samples were from four genera: *Propionibacteria*, *Corynebacteria*, *Staphylococcus* and *Streptococcus*. These are all known to be regularly found on human skin using culture methods.

The six people in the study all had very different bacterial populations on their skin. In fact, they only had four species of bacteria in common: *Propionibacterium acnes*, *Corynebacterium tuberculostearicum*, *Streptococcus mitis* and *Finegoldia AB109769*. This was quite a surprise but was probably due to different exposure to factors such as temperature, weather, light and cosmetics.

Propionibacterium acnes growing in or near the sebaceous glands of the skin is known to cause the common skin condition acne vulgaris. Mild cases of acne can be treated successful with antibacterial agents such as benzoyl peroxide (e.g. Clearasil). More serious attacks require antibiotic therapy with tetracycline.

Three bacterial species were only found in the male subjects: *Propionibacterium granulosum*, *Corynebacterium singulare*, and *Corynebacterium appendixes*. The sample is too small to draw firm conclusions, but it may be that males and females attract different sets of commensal skin bacteria.

The researchers now plan to look at diseased skin to find out about the bacterial ecology of the skin in diseases like psoriasis and eczema.

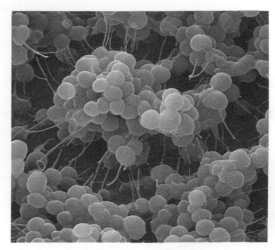

Fig 35.26 *Staphylococcus epidermis* on human skin. This bacterium is one of those responsible for the familiar 'sweaty feet' smell. Cultures of *S. epidermis* smell like a bit like a men's locker room

Why do we need to measure biodiversity?

There is more to measuring biodiversity than just being able to impress your friends by knowing how many species of living things there are on Earth. We need to measure the diversity:

- to compare the diversity of the same habitat over time, to assess, for example, whether pollution is damaging that habitat;
- to compare two different habitats at the same time, to find out which is the more diverse;
- to check whether a new habitat is being colonised by the number of species that it should. This would be important in the case of a polluted river or lake that had been 'cleaned up' and then repopulated with living organisms.

Putting a number on biodiversity – the Simpson diversity index

Being able to put a number on the diversity of a habitat makes the life of an ecologist much easier. A **diversity index** allows us to estimate the variety of living things in a particular area. However, we must make sure that the value we use accurately reflects all aspects of diversity. We could, for example, simply count the number of different species present:

? QUESTION 2

2 The Simpson index was used to calculate the diversity of an oak wood and a conifer plantation. The values obtained were 19.6 for the oak wood and 12.04 for the conifer plantation. Which habitat is the more diverse?

Stream 1 (100 animals)	16 species
Stream 2 (100 animals)	10 species

The conclusion would be that stream 1 was more diverse. But there is a problem: this method gives an idea of species richness, but it does not reflect how many of each species are present. In stream 1, there may be 85 animals of one species and just 1 of each of the others. In stream 2, there may be 10 animals of each species. Ecologists would say that stream 2 was the more diverse habitat, and so our first conclusion would be wrong.

To avoid this problem, we can use the Simpson index. Its formula takes into account that diversity depends on the number of individuals of each species present and the species richness (the number of different species present). The higher the value of the index, the greater the variety of living organisms found in the area.

$$D = \frac{N(N-1)}{\Sigma n(n-1)}$$

where D is the Simpson diversity index, N is the total number of individuals of all species found, n is the total number of organisms of a particular species found, and Σ means 'the sum of'.

EXAMPLE

Q Assume that you have studied a particular habitat and you have recorded the types of 20 animals you have found. The different types are represented by letters of the alphabet, and the record shows five species:

A A A A B B A C C B B A B B A D C C D E

What is the diversity of this habitat?

A Putting the figures into the equation for the Simpson index:

$$D = \frac{20 \times 19}{(7 \times 6) + (6 \times 5) + (4 \times 3) + (2 \times 1) + (1 \times 0)} = \frac{380}{86} = 4.42$$

HOW SCIENCE WORKS

Deforestation, agriculture and species diversity

As well as damaging valuable natural resources, deforestation also reduces biodiversity. The annual destruction of millions of hectares of tropical forests means that thousands of species and varieties of plants and animals are becoming extinct every year. Many of them have never been catalogued scientifically, so we do not even know what we are losing. Some environmental experts estimate that 50 000 separate species are disappearing every year, including insects and animals that live in forests.

Agriculture and other human land use led to loss of about 2 per cent of the world's forests and woodlands between 1980 and 1990. In the developing countries, this was closer to 8 per cent. Intensive farming also has an impact on biodiversity in land and water-based habitats near to fields and farms. Fertilisers, pest-control chemicals, tillage and even crop rotation all reduce the biodiversity of agricultural ecosystems.

How old is my hedge?

As you next walk down a country lane, take a good look at the hedges that you pass. Many of these structures date from the Middle Ages. They were planted by medieval farmers as boundaries to mark off their land and to keep in livestock. They probably planted shrubs such as hawthorn a few feet apart. The gaps at the base of these plants then allowed waves of colonisers (Fig 35.27).

First of all, fast growing, non-woody annual plants such as typical garden weeds would have thrived: chickweed, groundsel and shepherd's purse for example. Then, non-woody perennials such as bluebells, primroses, stinging nettles, dandelions and thistles may have become established. These plants can produce food reserves to survive the winter. Eventually, woody perennials, shrubs and small trees such as elder, holly and even some of the larger forest trees could have invaded the hedge.

Fig 35.27 In theory, it is possible to date a hedge by looking at the range, or diversity of plants that are in it. The more diverse the community of the hedge, the older it is

SUMMARY

By the end of this chapter, you should know and understand the following:

- A **population** grows **exponentially** in ideal conditions. Growth slows down because of **limiting factors** in the environment. This produces an **S-shaped** or **sigmoid growth curve**.
- When the size of the population becomes stable, we say that this is the **carrying capacity** of the environment. Sometimes, population growth exceeds the carrying capacity and a **population crash** follows. The population may then settle at a new carrying capacity.
- Three main types of growth curve occur in nature: **stable**, **irruptive** and **cyclic**.
- Some factors that limit population growth are **density dependent**: their effects are worse in a large, closely packed population. Others are **density independent**: their effects do not depend on the population distribution or size.
- **Intraspecific interactions** occur between organisms of the same species; **interspecific interactions** occur between organisms of different species. You should know some examples of each type.

- The living part of an ecosystem develops by the process of **succession**. If the area has never been **colonised** before, a newly formed volcanic island, for example, the process is called **primary succession**. If the area was previously covered with vegetation, a forest site that suffers a severe fire, for example, **secondary succession** occurs.
- There are three types of **biological diversity**: **species diversity**, **genetic diversity** and **ecosystem diversity**.
- Measuring biodiversity is important because it allows us to monitor a habitat over time, to compare two different habitats and to find out if a new habitat is being colonised by the correct number and type of species. This is vital if we are to assess the impact of our 'clean-up' efforts after environmental tragedies such as oil spills.

 Practice questions and a How Science Works assignment for this chapter are available at www.collinseducation.co.uk/CAS

36

Human activity and the environment

36 HUMAN ACTIVITY AND THE ENVIRONMENT

Your contribution to this planet

The average UK citizen will live for about 83 years and in that time will munch their way through 4.5 cows, 15 pigs, 21 sheep and, wait for it, 1200 chickens. We will also eat 13 345 eggs and drink 15 951 pints of milk. And that's just the animal products. What about the carbohydrates? We will also boil, mash or chip 2327 kg of potatoes and consume 4283 loaves of bread. We will eat 5272 apples and nearly 11 000 carrots.

Nothing influences the planet quite as much as agriculture. It wouldn't be so bad if we ate what we grew locally, but in the UK we import 95 per cent of our fruit. We also like our fruit and vegetables washed, chopped and packaged. The energy costs associated with these demands are huge. Millions and millions of food miles are generated every year, shipping food around the globe from the people who grow it to the affluent counties who want a wide variety of choice all the year round.

Overall, your contribution to this planet will be to eat about 50 tonnes of food (using 8.5 tonnes of food packaging in the process). To show your gratitude, in return you will sit down and produce 3 tonnes of faeces – enough poo to fill a large skip. And if that wasn't enough we will also break wind 15 times a day producing a life total of 35 815 litres of gas. That's a lot of wind to hold in, or blame on someone else. And as for the urine and the vomit …

We will own eight cars in our life, driving a total of 450 000 miles and using 120 000 litres of fuel in the process. We will also own five televisions, eight DVD recorders – the list goes on and on.

To provide our daily milk, the UK has about 2.1 millions cows. Each one can eat up to 100 kg of grass a day, and produce over 250 litres of methane – a major greenhouse gas

Walkers crisps started putting labels on some of their products in 2007 to show the carbon footprint of the product. The label shows how many grams of carbon were produced in the manufacture of the packet of crisps

1 WHY DO HUMANS AFFECT THEIR WORLD SO MUCH?

We humans are a unique species: we live in most of the ecosystems of the world, adapting our habitats to suit us (Fig 36.1), rather than having to adapt to the environment as most other species do. And, unlike other species, our behaviour affects the environment on a global as well as a local level.

The main human activity that affects the world environment is farming. Close behind are industry and mining for fuels such as coal, oil and gas. The more technologically advanced the society, the greater the overall environmental impact of each individual person. So, for example, a child born in the UK into a fairly well-off family may live for 80 years or more. In that time, the amount of resources they use is enormous: think of the furniture, electrical appliances, cars, fuel, food, clothing, etc. used up by someone like this. At the other extreme, a child born in a poor family in India or South America might live only 20 years. They could live in poorly built housing, or be homeless and forced to live on the streets, probably never get quite enough to eat, and would certainly never use a car or any significant amount of fuel.

Fig 36.1 We think of the British countryside as 'natural' but, in fact, it is almost entirely artificially made. Only a few areas in Scotland, Wales and Northern England remain in their 'original' form – all farmland and most forests are the result of human influence on the environment

In this chapter, we take a brief look at some of the complex issues involved in controlling the effect people have on the ecology of the Earth. There is room only to give an overview of some of the problems and potential solutions.

2 FARMING AND ITS EFFECTS ON THE ENVIRONMENT

It is very common to hear environmentalists complain that farming is destroying hedgerows and wildlife habitats, that woodlands and forests are being cut down to make new farmland and that it should all be stopped. In addition, farmers are often portrayed as hard-line businesspeople who put profit before anything else. There is a clear conflict here, but the issue is not as simple as these points of view suggest.

We must consider five basic facts:

- People need to have food to eat: staple foods need to be cheap.
- There are more people on Earth to feed than ever before.
- Farmers need to make a living.
- Farming needs land.
- Some farming techniques are damaging the environment so badly that the capacity of the land to produce food is being reduced.

The key to solving the problem of food production is to develop farming methods that are efficient enough to provide the amount of food that we need in the short term, but that protect the environment to ensure that food production levels can be maintained, or even increased, in the future.

FOOD PRODUCTION ON A GLOBAL SCALE

On a global scale, food production has increased during the past 50 years (Fig 36.2). Until the early 1980s, this increase matched the increase in world population, so that, overall, more people had enough to eat than ever before. There were still, of course, countries and areas where food supply did not meet demand and many millions of people were malnourished or starving.

Since 1984, population growth has increased faster than food production. Part of the reason for this has been the damaging effects of farming practices which produce food in the short term, but lead to long-term environmental damage, particularly soil erosion and desertification (see page 598). Very poor countries where food production has fallen since the mid-1980s have tried to compensate by importing food from Europe and the USA where more food is produced than is needed by the local population. However, this only increases the debt of the poor countries, leading to more poverty and, in the long term, more malnutrition and starvation.

Many scientists believe that soil erosion, water shortages and pollution from excessive use of pesticides and fertilisers will lead to a crisis in food production during the twenty-first century. Only a change in farming techniques might prevent this.

THE CONCEPT OF SUSTAINABLE AGRICULTURE

The problem of providing food for an increasing world population without reducing the capacity of the land to produce food is a difficult one, but it has some solutions. By using farming techniques that are sustainable, rather than successful in the short term only, many areas of the world should be able to continue food production at the right level for the foreseeable future.

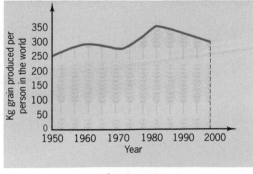

Fig 36.2 The amount of grain produced per person in the world between 1950 and 1999

? QUESTION 1

1 What is the difference between a sustainable resource and an unsustainable one? Think of an example of each.

Moves towards sustainable agriculture would include:

- using new varieties of crop plants, perhaps produced by genetic engineering, which can survive and give good yields in poor soils, dry conditions and without the need for expensive pesticides and fertilisers.
- making use of non-traditional food sources – more plants that have nitrogen-fixing bacteria in their roots (these need no fertiliser), insects (many cultures of the world eat insects and their larvae as protein-rich delicacies) and unusual animals. In Europe, for example, ostrich farming is becoming more widespread.
- reducing the amount of fish caught in the seas to levels at which the sizes of the populations do not continue to decrease. At the same time, the number of fish farms could increase.
- encouraging and rewarding farming methods that do not lead to soil erosion and other environmental problems.
- using biological pest control (Fig 36.3), rather than artificial pesticides which can persist in the environment.
- encouraging farmers in poor countries to grow a selection of crops that can feed local people, rather than cash crops such as tobacco and coffee for export.
- increasing financial and technological aid to poor countries to allow them to set up more sustainable methods of farming and food production.

FARMING IN THE TWENTY-FIRST CENTURY

Monoculture

Until relatively recently, farms were diverse businesses where the farmers would grow several different crops, keep a variety of animals and perhaps have an orchard or two. The produce would be sold in local markets. But in today's supermarket-dominated food industry, farmers are encouraged to grow a single crop such as oilseed rape or wheat, so that the supermarkets have fewer growers to deal with. The word **monoculture** refers to the growing of the same crop in large fields, often year after year. A particular crop will rapidly deplete the soil of particular nutrients at particular depths, and often lead to the accumulation of pests. Without humus to bind it together, soil becomes more powdery and is likely to be eroded by wind and rain, leading to dust storms and the accumulation of silt in rivers, lakes and oceans.

Removal of hedgerows

Hedgerows can be hundreds of years old, and represent an important habitat for many species. Farmers remove hedgerows for a variety of reasons:

- so that they can use large machinery more efficiently;
- to increase the area for growing crops;
- to avoid the need for maintenance (cutting, etc.);
- to prevent them shading the crops;
- to remove what is often seen as a reservoir of crop pests.

The loss of hedgerows has several disadvantages. Hedges, as well as being aesthetically desirable, shelter crops from wind and therefore minimise soil erosion. Without hedgerows, the species diversity of the countryside is lowered as many species of insect, plant, bird and small mammals lose their habitat.

Fig 36.3 Until recently the only biological control for whitefly, a common plague on greenhouse-grown tomatoes (top) was *Encarsia formosa* (centre), a parasitic wasp which eats only adult flies, leaving larvae to develop and so cause a recurring problem. Gardeners and commercial growers can now obtain the beetle *Delphastus pusillus* (bottom) which kills larvae as well as adult flies, so providing much more effective and long-lasting control

Use of fertilisers

One of the key differences between agriculture and a natural ecosystem is that harvesting removes nutrients from the soil. Fertilisers are used to replace these nutrients. There are two types of fertiliser:

- Organic – faeces and urine of animals and humans (sewage sludge or farmyard manure).
- Inorganic – liquid or (usually) pellets containing mineral ions, mainly nitrate, phosphate and potassium.

Both types are added to crops with the same basic aim: to increase crop yield. Both provide mineral ions, but with manure the release of ions is gradual as the manure is decomposed by microbial action. The advantages and disadvantages are given in Table 36.1.

DNA technology, such as genetic engineering, is discussed in Chapter 27.

Table 36.1 Organic v inorganic fertiliser

Type of fertiliser	Advantages	Disadvantages
organic	cheap – farms often generate their own not easily lost by leaching improves soil – better humus levels, water retention, aeration and texture	variable (usually low) nutrient content slow release of nutrients difficult/expensive to store and handle may contain plant or animal pathogens – causing disease may contain metal residues that are passed into food chain requires heavy machinery that can compact soil
inorganic	exact composition known – soil balance can be controlled easy to store and handle can be applied with light machinery – avoids soil compaction	expensive – it's a commercial product most components soluble – rapid leaching into rivers applied in concentrated form – can cause osmotic damage to plants can cause acidification of soil

Balance of food production and conservation

For economic reasons, farmers are under great pressure to:

- concentrate on a small number of crops and grow these in large areas of monoculture;
- remove hedgerows to make more growing space and to make the use of heavy machinery easier;
- use large amounts of chemical fertilisers to increase yield from the depleted soil;
- drain marshy areas and remove woodland;
- use a wide variety of pesticides.

The two activities of producing food and conservation would seem to be mutually exclusive, but there is some room for compromise. There are several strategies that farmers can adopt to minimise damage to the environment.

Some methods of creating more sustainable agriculture

- Use more organic manure, which improves soil structure by providing more humus so retaining more water. As the humus decays the nutrients are released slowly so there is less chance of them being leached away.
- Delay the application of chemical fertilisers until the main growing season so that more is absorbed before it is washed away.
- Leave crop stubble over the winter (i.e. plough later) so that there is less bare soil to blow away.
- Rotate crops – growing a different crop year on year makes better use of minerals available at each soil depth. It also prevents the accumulation of crop-specific pests.
- Set aside – leave areas of wilderness to develop.
- Stop destroying hedgerows.

3 HUMAN ACTIVITIES AND POLLUTION

Pollution occurs when substances are released into the environment in harmful amounts, usually as a direct result of human activity (Fig 36.4). Most pollution results from excessively high concentrations of substances, but there are exceptions.

Some pollutants are harmless substances already found in the natural environment. They cause problems when they are produced in large quantities by human activity; examples include carbon dioxide and organic waste (sewage). Other pollutants produced by human activity are compounds that are known to be toxic, such as pesticides. These often remain in the environment for a long time because there is no living organism to break them down. A third type of pollutant is a chemical that passes all the safety tests, but then has totally unexpected effects, as chlorofluorocarbons have done. Pollutants can affect the air, the sea, waterways and the land.

Fig 36.4
Examples of how we humans pollute our environment.

Fig 36.4a The outlet pipe in the background is discharging raw sewage which is being washed up onto this beach

Fig 36.4b Effluent from the chemical works is polluting the river which supplies it with vital water

Fig 36.4c This landfill site will eventually be levelled and should look more pleasant, but methane production under the surface might still cause problems

Fig 36.4d Pollutants in smoke from industrial plants add to the acid content of rain

AIR POLLUTION: ACID RAIN

To many people, walking through a high Swiss meadow in summer during a shower of rain sounds idyllic. Peace and quiet, fresh mountain air and refreshing rain. But what if the rain has the pH of vinegar or lemon juice (Fig 36.5)?

In Switzerland, Norway, Sweden and other parts of Scandinavia, acid rain falls on some of the most unspoilt natural landscape in the world. It starts off, as far as 1000 kilometres away, as emissions of sulfur and nitrogen oxides released from the more highly industrial areas of Europe. Although there are some natural sources of these gases, by far the largest amount is generated by burning fossil fuels: the petrol used to power road vehicles and the oil, coal and gas used in power stations (Fig 36.6).

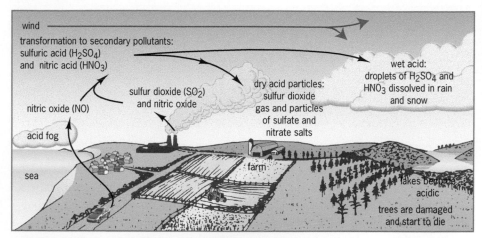

Fig 36.6 As they are carried in the air, oxides of sulfur and nitrogen form secondary pollutants such as nitric acid vapour, droplets of sulfuric acid and particles of nitrates and sulfates. These chemicals reach the surface of the Earth in two forms: wet, as acid rain, snow or cloud vapour; or dry as acidic particles

Normally, rainwater is mildly acidic because of the presence of dissolved carbon dioxide. The pH is normally no lower than 5.6. Any precipitation (rain, hail, sleet or snow) with a pH below this value is defined as acid rain. The cause of acid rain is a reaction between certain gases and the water that makes up clouds. The following gases are mainly responsible:

- Sulfur dioxide, which reacts with water to make sulfuric acid (H_2SO_4). Sulfur dioxide is emitted naturally from volcanoes but the vast majority comes from burning fossil fuels, notably coal, that contain a lot of sulfur. Power stations and metal smelting industries are the major sources of sulfur dioxide, not petrol or diesel engines.
- The oxides of nitrogen, nitrous oxide (N_2O), nitric oxide (NO) and nitrogen dioxide (NO_2). Collectively known as NO_x, these gases combine with water to make nitric acid (HNO_3). NO_x gases are made in engines when nitrogen in the air combines with oxygen during combustion.

Perhaps the most extreme example of acid rain was a 'smog' (smoke/fog) that descended on London for a whole week in 1953. Its pH was measured at 1.6 and was responsible for the deaths of 4000 people who suffered from breathing problems (asthma, bronchitis and emphysema).

The effects of acid rain on the environment are still causing arguments. It is quite difficult to get good evidence to demonstrate a definite link between the occurrence of acid rain and the effects that ecologists and environmental scientists have noticed. For example, many European and North American forests have been damaged during since the 1970s, especially at high altitudes and at the edges of some forests, which have large areas of dying trees (Fig 36.7). In Britain, the pattern of loss of lichen species seems to follow the

Fig 36.5 Unpolluted rain has a pH of about 5.6 because of the carbon dioxide that occurs naturally in the atmosphere: rainwater in most industrialised countries has a pH of between 4 and 4.5. Some cities in the USA are bathed in acid fog with a pH as low as 2.3, which is 1000 times as acidic as normal rain, and the same acidity as lemon juice

Fig 36.7 These trees in the Czech Republic have been killed by acid rain that has resulted from emissions of sulfur dioxide and oxides of nitrogen given off by heavy industry all over Europe

593

distribution of sulfur dioxide pollution. In fact, lichens are so sensitive to this sort of pollution that they are used as pollution indicators. Other possible effects of acid rain are listed in Table 36.2.

Many countries, including Britain, have now recognised acid rain as a serious problem and have been reducing their emissions of sulfur dioxide since the mid-1970s.

Table 36.2 Possible effects of acid rain

Possible effect of acid rain	How it might happen
Fish deaths in Scandinavian lakes and elsewhere	Indirect mercury poisoning: more acidic water is thought to convert inorganic mercury compounds in lake sediments into compounds which are soluble in the fatty tissues of fish
	Indirect aluminium poisoning: in acidic conditions, more aluminium ions are released from the soil and washed into lakes. Fish respond by making excess mucus which clogs their gills, with fatal results
Damage to statues and buildings	Chemical reactions between components of acid rain and rock
Overstimulation of plant growth	Acid rain falling on the ground means excess nitrogen. The plants then use other nutrients faster, reducing soil fertility
Thought to make human respiratory problems, such as asthma, worse	Irritation of surface membranes in lungs and bronchi

AIR POLLUTION: THE OZONE LAYER

The effects of acid rain occur hundreds of kilometres away from the source of the pollutants that cause it. This is bad enough, but other pollutants affect the whole planet. Examples include the effect of greenhouse gases on global warming and ozone depletion.

Ozone, O_3, is a form of oxygen: its molecules contain three oxygen atoms, rather than the two that occur in molecules of atmospheric oxygen. Both gases have been a major feature of the Earth's atmosphere for the past 450 million years. Oxygen is vital to life: it is the fuel for aerobic cell respiration. The role of ozone is less direct: it forms a sort of sunscreen high up in the atmosphere, shielding the biosphere from the harmful effects of ultraviolet light that beams down from the Sun.

Chlorine- and bromine-containing compounds, mainly chlorofluorocarbons (CFCs), that were released into the atmosphere during the twentieth century seem to be causing a measurable thinning of the ozone layer (Fig 36.8). This could cause serious problems in the future.

REMEMBER THIS

Ozone (O_3) is present in the atmosphere as a whole in very small amounts, a few parts per million at the most. However, it is not evenly spread. It occurs as a definite layer in the stratosphere, 15–50 kilometres above the Earth's surface. Low-level ozone occurs at ground level when bright sunlight acts on pollutants produced by heavy traffic – we see this as photochemical smog in large cities in high summer.

REMEMBER THIS

Brands of products that carry the phrase 'CFC-free' may do so as a marketing ploy to persuade the consumer that they are more 'environmentally worthy' than other products. In fact, by law, no aerosols sold in Britain can contain CFCs.

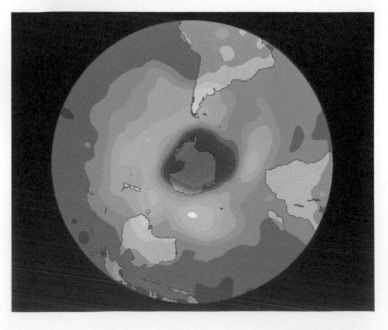

Fig 36.8 In the mid-1980s, scientists discovered something unexpected: sunlight on the Antarctic in spring causing immense destruction of ozone and a huge hole appearing in that part of the ozone layer. This 'ozone hole' has been noticed every year since, and appears to be growing

Originally, CFCs were thought to be 'ideal' chemicals: they were non-flammable, non-toxic, non-corrosive, cheap to make and could be used as coolants in fridges, propellants in aerosol spray cans, and in the manufacture of the packing material Styrofoam. However, in the mid-1970s, scientists showed that CFCs rise slowly into the stratosphere and are then converted by the action of ultraviolet light into chlorine atoms. These accelerate the breakdown of ozone into O_2 and O. CFC molecules can stay in the stratosphere for about 100 years, and each one can destroy hundreds of thousands of molecules of ozone.

CFCs have now been phased out in most countries of the world, but their effects continue. Even if no more CFCs were put into the air from tomorrow, the ozone layer would still take about 100 years to recover from the worst of the damage. Less ozone in the stratosphere allows more of the harmful ultraviolet radiation from sunlight to reach the Earth's surface. Scientists predict that this will lead to a large increase in the number of skin cancers and cataracts (Fig 36.9).

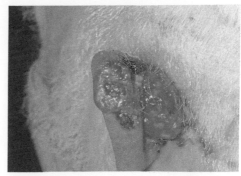

Fig 36.9 In countries such as Australia, New Zealand, South Africa and Chile, which are protected only by a very thin layer of ozone for many months of the year, there have already been many more cases of skin cancer than ever before. Children in Australian schools are required by law to wear hats and blocking creams whenever they go outside

WATER POLLUTION: GETTING INTO HOT WATER

When you think of water pollution, you probably think of chemical effluent (Fig 36.10), outflows of raw sewage or huge oil spills. While these undoubtedly are examples of pollution, probably the most overlooked cause of damage to inland aquatic ecosystems is the warm water released as a by-product of many industrial processes. Fish and water-living invertebrates are killed by warm water, not because they are scalded directly, but because warm water can carry much less dissolved oxygen than cool water and they 'suffocate' (Fig 36.11).

Fig 36.10 This chemical outflow is on its way to the River Mersey

Fig 36.11 As well as causing the death of fish directly, by depriving them of oxygen, warm water can also encourage the growth of parasites which would otherwise not be a problem

WATER POLLUTION AND SEWAGE

Sewage is good stuff. It's a natural product that is an essential part of most ecosystems. Call it what you will, but faeces and urine from any source are a mixture of undigested food, cells from the gut lining, urea, salt and lots of living bacteria and protoctists. When acted upon by decomposers (or saprophytes), sewage releases vital nutrients such as nitrate that are essential for renewed plant (or algal) growth.

Problems with sewage start when there's too much of it to break down. If the entire sewage production of a village or town is pumped into one stream or river, the natural balance is upset.

Contamination of rivers, lakes and seas by sewage causes two main problems. First, it can introduce potentially dangerous pathogens (organisms that can cause disease) into the environment (look back to Fig 36.4a). This is a particular problem if the water is drunk by animals or people. The second problem is that sewage and other organic wastes are decomposed by the action of aerobic bacteria. These use large quantities of oxygen, leaving less for other organisms living in the water. Sewage contamination can lead to **eutrophication** (Fig 36.12).

The term **eutrophic** means over-fertile. A eutrophic river or lake contains more nitrates and phosphates than normal, and abnormal growth, particularly of algae, can occur. Note that the algae themselves do not reduce the oxygen content. They photosynthesise, so actually increase the oxygen levels for a short while. The opposite of eutrophic is **oligotrophic**: this means under-fertile. An oligotrophic river cannot sustain much life.

Biological oxygen demand

It is easy to find out the oxygen requirements of water from a particular site by taking a sample of the water and measuring its oxygen content, usually with an electronic meter. After keeping it sealed and in the dark for five days at 20 °C, the oxygen content is measured again. The rate at which the oxygen has been used up is the **biological oxygen demand (BOD)**. The BOD of unpolluted river water is about 3 mg O_2 per litre (dm^3) of water, per day. Raw sewage uses over 100 times more oxygen in the same time (Table 36.3).

Although pollution by sewage is a major cause of eutrophication, there are others. If large amounts of nitrate- and phosphate-rich fertilisers are put on to farmland, the excess can run off and contaminate local rivers and lakes. Industrial processes which pollute local waterways with large amounts of concentrated sugars or other organic waste products can also cause the BOD of the water to increase significantly.

Find out more about eutrophication in the Assignment for this chapter on the CAS website.

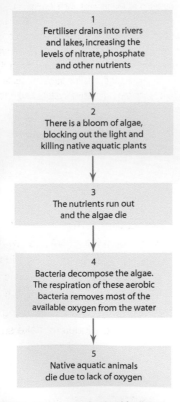

Fig 36.12 A summary of eutrophication

? QUESTION 2

2 Describe how you would measure BOD.

Table 36.3 Typical BOD values per day of different types of water

Water type	BOD value per day
clean river water	3 mg dm^{-3}
water from polluted stream	10 mg dm^{-3}
domestic sewage, untreated	250–350 mg dm^{-3}

CONTAMINATION OF WATER BY OIL AND DETERGENTS

The names *Exxon Valdez*, *Braer* and *Sea Empress* may sound familiar, even if you can't quite remember that they are oil tankers that have famously run aground causing huge oil spills (Fig 36.13). These pollution disasters make headline news but, in 1993, a study by Friends of the Earth estimated that over 1000 times the amount of oil carried by the *Exxon Valdez* (which polluted the coast of Alaska in 1989) is lost in the USA each year – not through national catastrophes, but through the normal operations of washing tankers and releasing the oily water, from pipeline and storage tank leaks and from accidental loss from offshore oil wells.

The effects of oil spills on ocean ecosystems depend on:

- the amount of oil released;
- how far away from the shore the spill happens;
- the weather conditions, water temperature and speed of ocean currents;
- the type of oil released: an area affected by a spill of refined oil takes over three times longer to recover than a similar area that has a crude oil spill.

In the clean-up operation following a large spill, oil is dispersed by detergents. The aim is to spread the oil molecules to limit damage to the local environment. Unfortunately, many detergents are extremely toxic and lead to many sea-bird deaths.

SPOILING THE LAND

People can pollute areas of land directly by dumping toxic waste, by industrial processes such as mining and by simple neglect: leaving non-biodegradable rubbish behind after a picnic at a local beauty spot, for example. However, human activity, particularly farming, is indirectly causing two main problems: **deforestation** and **desertification**. Both have far-reaching global effects.

Deforestation

Deforestation has been practised by humans for centuries. As human population and technology have increased, so has the need to use wood. (The discovery of fire has had an enormous impact.) Many European forests were cleared by various different settlers from about the fourteenth century onwards. Today, the human population is larger than ever before and the process of deforestation has now spread to all parts of the world, particularly the tropical rainforests.

Deforestation is a process that results from the conflict between immediate human need and the necessity to protect the environment from serious and long-term damage. Tropical hardwoods are prized as building materials throughout the world, and other types of wood are used to make the thousands of tonnes of paper that we use every day. Many people in the poor countries of the world depend on wood or charcoal for fuel, and wood, twigs, crop residues and grass are used for fuel in many of the 'better off' countries such as China, Brazil and Egypt. As well as cutting down forests to use the wood they provide, deforestation is also carried on to clear land for farming. People have cleared large areas of forest for plantations of other trees (rubber and palm oil), and for cattle ranches and enormous wheat fields, in order to fulfil the needs of the human population for more raw materials and more food.

36.13 This cormorant was caught in the oil spill that happened at Manobier beach, Wales, in February 1996 after the tanker *Sea Empress* ran aground. Birds covered in oil cannot swim very easily and sometimes tire and drown. Those which end up on the beach try desperately to clean their feathers, ingesting and inhaling large amounts of oil which damages their digestive system and often causes pneumonia

 REMEMBER THIS

Many countries worldwide are now working together to try to reduce deforestation: people are being encouraged to manage forests **sustainably**. This means using different techniques to make sure we have the wood we need without destroying large areas of forest. People are also being persuaded to use **paper substitutes**. There is growing investment in **rainforest communities** to find new medicines from plants and **eco-tourism** is being promoted. If people pay to see wildlife in its natural surroundings, this contributes to the income of an area and makes it less likely that habitats are destroyed.

Medicines from the rainforest are discussed in Chapter 35.

The consequences of deforestation are potentially serious. Locally, clearing the land of trees and plants leads to increased soil erosion (see below). Globally, reducing the biomass of productive trees may contribute to the greenhouse effect as less carbon dioxide is used in photosynthesis. Also, loss of rainforest, one of the most diverse ecosystems of the world, is likely to affect humans and other organisms in ways we cannot predict, although we already know that many of the plants found in rainforests are useful sources of powerful drugs. Balancing the needs of people with those of the environment is difficult.

Desertification

Soil is an important abiotic factor that determines the distribution of organisms in an ecosystem: few plants can grow without the anchorage of soil or the water and minerals it supplies. When the soil is lost, a process called **soil erosion** occurs. This is a worldwide problem.

Soil erosion is a natural process: wind and water movements move surface debris and topsoil from one place to another. In ecosystems that are undisturbed by human activity, healthy plant roots 'bind' the soil, and soil is lost and made at about the same rate. Problems occur when the soil is removed faster than it is formed: this can happen because of farming, deforestation, building, over-grazing by animals, physical damage by off-road vehicles or fire and some kinds of mining.

Farming is probably the worst culprit. Topsoil is eroding on about a third of the world's cropland. Each year, the land must feed 90 million more people with about 25 billion tonnes less topsoil. In Africa, soil erosion is now 20 times faster than it was 30 years ago. The cattle-producing areas are now severely affected by desertification (Fig 36.14). But, there are solutions to soil erosion and desertification, and many of these are starting to be put into practice.

Fig 36.14 Areas of severe desertification occur all over the world. Desertification is defined as the process that reduces the productive potential of hot dry areas by 10 per cent or more. Moderate desertification is a 10–25 per cent drop in productivity: severe desertification is a 25–50 per cent drop. At its most severe, desertification actually creates deserts

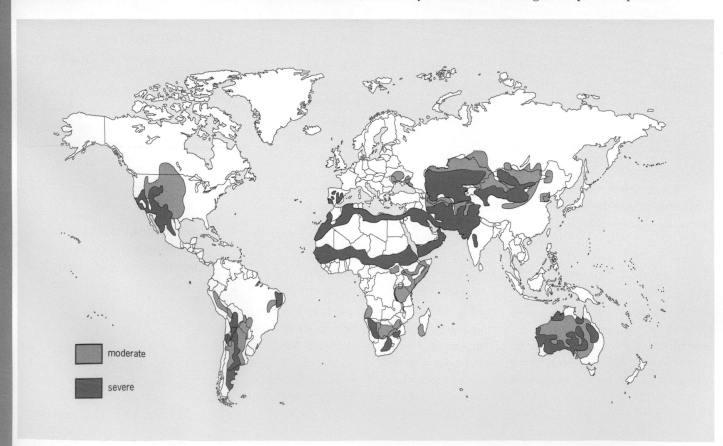

moderate

severe

Waste gone mad

Next time you put something in a bin, think about how much waste you produce. Some of the facts and figures are very scary:

- The UK produces more than 434 million tonnes of waste every year. This rate of rubbish generation would fill the Albert Hall in London in less than 2 hours.
- Every year 3.5 UK households throw away the equivalent of 3½ million double-decker buses (almost 30 million tonnes), a queue of which would stretch from London to Sydney (Australia) and back.
- 70 000 of those double-decker buses just contain babies' nappies. If lined up front to end, the nappy buses would stretch from London to Edinburgh.
- Every year, supermarkets give an estimated 17.5 billion plastic bags away. 17.5 billion seconds ago it was the year 1449.
- 1 litre of oil can pollute 1 million litres of fresh drinking water.
- In 2001 UK households produced the equivalent weight of 245 jumbo jets in waste.

Most experts and many ordinary people are now very worried about waste because it threatens to swamp the planet. We cannot deal with it as fast as we make it. We urgently need to cut down on the amount of waste we produce, and find new ways of managing it.

Most of the waste produced by households currently ends up in landfill sites, with only around 35 per cent of industrial and commercial waste and 12 per cent of household waste recycled or composted. Under the EU Landfill Directive we must dramatically reduce the amount of biodegradable municipal waste sent to landfill.

One policy that experts are trying to put in place is the Best Practicable Environmental Option (BPEO). This identifies waste management options that provide the most environmental benefits, as well as getting rid of the rubbish and not costing too much.

Another policy idea is the waste hierarchy:

- Waste should be prevented or reduced at source – every year each person in the UK produces four times as much packaging waste as his or her luggage allowance on a jumbo jet. How can you cut it down?
- Waste materials should be re-used – how much recycling of glass, plastic, paper and cans do you do?
- Waste materials should be recycled and used as a raw material – do you buy recycled products and materials?
- Waste that cannot be re-used should be used as a substitute for non-renewable energy sources.
- Only waste that cannot be treated in any of the above ways should go to landfill.

Fig 36.15 A landfill site

Reclaiming brownfield sites

The term brownfield is used to describe land that has already been in use, usually for some industrial purpose. The land may have derelict buildings, it may have been left as wasteland and it may or may not be contaminated with chemicals, toxins or other dangerous waste.

Reclaiming brownfield sites is an important step towards sustainable land use. In February 1998, the UK government announced a national target for England that at least 60 per cent of new homes are to be built on brownfield sites by 2008. This is to provide the 3.8 million new households that will be required in England by 2021.

As well as development, brownfield can also be reclaimed for wildlife. This can happen anyway as wastelands become colonised by plants and animals without anyone doing a thing.

However, projects are underway all over the UK to clean up brownfield sites. One project is set among the industrial landscape of Barking Reach where there was once a dumping area for pulverised fuel ash. The 10-hectare Ripple site now supports hundreds of orchids including the southern marsh and common spotted orchid, which flower in the birch woodland floor between

Reclaiming brownfield sites (Cont.)

May and June. The silver birches are next to a new meadow that now has a wide range of native wildflowers and grasses that attract butterflies, hoverflies and bees.

Another project is at Crane Park Island. Once the old Hounslow Gunpowder Mills, this island is now a nature reserve surrounded by the River Crane. It contains woodland, scrub and reed beds, which provide a habitat for the increasingly scarce water vole. There is rich aquatic life in the shallow stretches of the Crane and work is now in progress to turn the tower, a relic of the old mills, into a nature study centre.

Fig 36.16 The Crane park nature reserve survey, UK

4 HUMANS AND THEIR IMPACT ON PLANET EARTH

There are three key questions on global warming:
- Is it happening?
- Is it a bad thing?
- It is man made?

The weight of scientific evidence tells us that the answer is yes to all three. Over the last 100 years, the average surface temperature of the Earth rose 0.6 to 0.9 °C, and the rate of temperature increase is accelerating. The higher temperatures mean that seawater is less dense and so levels rise: 17 cm during the last century. The world's glaciers are melting and Arctic sea ice is shrinking by 2.7 per cent every ten years, also contributing to the rising sea levels.

WHAT CAUSES GLOBAL WARMING?

Carbon dioxide is one of the main greenhouse gases. Together with other gases, it reduces the heat that can escape from the atmosphere (Fig 36.17). This effect is often called the **greenhouse effect**.

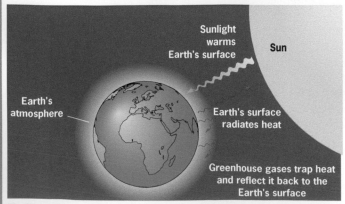

Fig 36.17 The greenhouse effect. Energy from the Sun is either transmitted, reflected or trapped by the atmosphere. Energy radiated back into space is at longer wavelengths (in the infrared region) than the energy arriving from the Sun, and the consequence is that more heat enters than can escape. Greenhouse gases (CO_2, CH_4, nitrous oxides, CFCs), together with water vapour in clouds, contribute to the 'blanketing' effect of the atmosphere

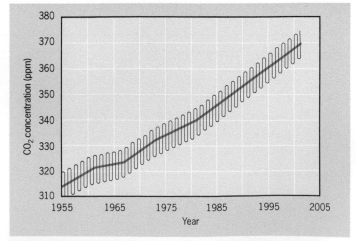

Fig 36.18 The rising level of carbon dioxide in the atmosphere. Data were collected at the Mauna Loa observatory in Hawaii. There is good evidence to suggest that changes in the average temperature at the surface of the Earth are closely linked to changes in the levels of greenhouse gases

Although pollution could be increasing the greenhouse effect, this phenomenon is not caused by pollution: it is a natural effect due to the presence of an atmosphere. In fact, having an atmosphere is one of the reasons that life developed on Earth: it allows surface temperatures to be much more stable than they would otherwise be.

Accurate measurements of atmospheric carbon dioxide concentrations began in the mid-1950s, in Hawaii and at the South Pole. These sites were chosen because they are far away from any major sources of pollution. The curve in Fig 36.18 clearly shows an annual cycle of peaks and troughs corresponding to summer and winter. Superimposed on this is a small but steady rise in overall CO_2 levels. Levels of other greenhouse gases such as methane, nitrogen oxides and chlorofluorocarbons (CFCs) have also increased during the same period (Table 36.4).

Table 36.4 Which gases contribute most to the greenhouse effect? To get some idea of the relative importance of different gases, multiply the abundance by the 'greenhouse factor'. Carbon dioxide does not have the highest factor, but its abundance makes it the most significant greenhouse gas

Gas	% Abundance in atmosphere	Greenhouse factor
nitrogen	78	negligible
oxygen	20	negligible
water vapour	1	0.1
carbon dioxide	0.037	1
methane	0.00018	20
nitrous oxide	0.000031	310
CFCs	0.000000026	3800

EVIDENCE FOR GLOBAL WARMING

How do we know global warming happening and is caused by humans? Could it be a natural variation in climate? Scientists don't think so because of the methods they use to study the Earth's climate, six of which are discussed below.

Weather records

The UK has the longest set of weather records in the world. Measurements began back in 1659 and Fig 36.19 shows the general trend. A computerised line of best fit through the graph shows a definite average temperature increase.

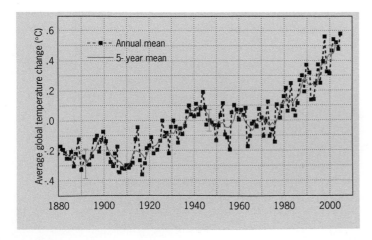

Fig 36.19 Graph of temperature increase

Studying peat bogs

Peat consists of layer upon layer of partly decomposed plant material. Conditions are anaerobic and acidic, so normal bacterial decomposition is inhibited. Consequently, plant material such as leaves and pollen accumulates in layers and gives an accurate record of the plants found in that area, and the time of year they flowered. Pollen grains are very distinctive (Fig 36.20) and can be used to identify the species growing in the area at the time. Cores drawn out of peat bogs can give information going back to the last Ice Age 12,000 years ago, and the exact age of a particular layer can be determined by carbon dating or similar techniques.

This study of pollen, known as **palynology**, shows two trends:

- Many species of plants are flowering earlier.
- The distribution of many cold-intolerant species is creeping towards the poles, suggesting higher average temperatures and an earlier spring.

This evidence is supported by general observations of our natural history. Many of the events that we use to judge the onset of spring – snowdrops appearing, frogs spawning and birds nesting – tend to happen earlier each year.

Fig 36.20 Pollen grains have tough, sculptured outer shells that make them very resistant to rotting. Scanning electron micrograph of pollen grains from a daffodil, a paper birch tree, ragweed, impatiens, mallow, marsh marigold, forsythia, pussy willow and sunflower

Studying tree rings

Dendrochronology, the study of tree ring patterns, can tell us something about climate change during the life of the tree. Warmer and wetter years give thicker rings due to the rapid growth of new xylem fibres. Colder and drier years give narrower rings and denser wood. Overall, thicker rings in recent years are becoming more common in many species.

Glacial ice and ocean sediments

Glacial ice traps tiny samples of Earth's atmosphere, giving scientists a record of greenhouse gases that stretches back more than 650 000 years, and the chemical make-up of the ice provides clues to the average global temperature. From these and other records, scientists have shown that climate changes in the past have been triggered by variations in Earth's orbit, solar variation, volcanic eruptions and greenhouse gases.

Computer modelling

Carefully tested computer models can also help tell us what is driving climate change. By experimenting with the models – removing greenhouse gases emitted by the burning of fossil fuels or changing the intensity of the Sun to see how each influences the climate – it is possible to use the models to explain Earth's current climate and to predict its future climate. So far, the only way scientists can get the models to match the rise in temperature seen over the past century is to include the greenhouse gases that humans have put into the atmosphere. This is strong evidence that humans are responsible for most of the global warming observed during the second half of the twentieth century. Having said that, our climate is one of the most complex systems we can attempt to program into a computer, and there has to be a degree of uncertainty in all calculations and predictions.

Remote sensing

Lots of equipment from space-based satellites to deep-sea thermometers (remote sensors) is used to monitor what is happening to the Earth's climate. Along with the paleoclimate data described above, these sources reveal that the planet has been warming for at least the last 400 years, and possibly the last 1000 years. The monitoring takes into account natural events such as volcanic eruptions but, to date, the overall conclusion is that natural influences such as volcanic eruptions or changes in the Sun's output cannot account for the temperatures changes that have been observed over the last 60 years. Only the increase in greenhouse gases that have come from burning fossil fuels can do that.

WHAT ARE THE LIKELY EFFECTS OF GLOBAL WARMING?

The impact of global warming and climate change is already starting to be felt – it is no longer something that may happen but something that is happening (Fig 36.21). The list of possible consequences is enormous but the main ones are discussed below.

Climate pattern changes

It's not as simple as 'the world warms up and the sea levels rise'. Changing patterns of weather could mean a change in the ocean currents. The UK is kept warmer than other countries of the same latitude by the Gulf Stream. Some scientists predict that the Gulf Stream could be disrupted, making the UK much colder.

In many parts of the world there would be greater drought and more forest fires. We are already seeing changes in rainfall patterns, with heavy downpours causing flooding in many areas of the world. Warmer water in the oceans pumps more energy into tropical storms, making them more intense and potentially more destructive.

Fig 36.21 One of the effects of climate change is the destruction of the ice cap over the Arctic. These photographs clearly show the difference in just 24 years

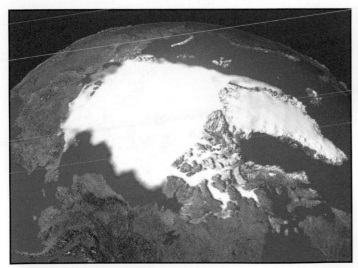

Fig 36.21a Arctic sea ice, 1979

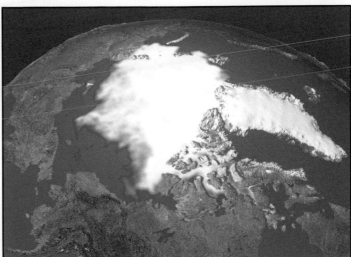

Fig 36.21b Arctic sea ice, 2003

Changing vegetation

Species differ in their tolerance to temperature. In any particular area, the climax community that develops depends on the temperature and other environmental factors. If the temperature changes, different species may become dominant.

Generally speaking, global warming will favour the colder countries, which will be able to grow a wider variety of crops, faster, for more of the year. Conversely, warmer countries may find that previously fertile land will no longer sustain crops – further adding to their problems and the existing inequality of resources. Climate change will also bring about changes in rainfall that will affect farming all over the world. Some plant species face extinction while the dominance of others may spread.

A study published in the journal *Nature* found that at least 279 species of plants and animals are already responding to global warming. Species' geographic ranges have shifted toward the poles at an average rate of four miles per decade and their spring events have shifted earlier by an average of two days per decade.

Different rates of growth and development

Basically, organisms grow and develop faster when the temperature is higher – it's an enzyme thing. Plants photosynthesise faster, assuming that temperature is the limiting factor. Incubation times of eggs get faster, as do all stages of an organism's development and life cycle.

In some species, such as crocodiles, the temperature of egg development determines the sex of the individual. A rise in temperature could lead to all of the hatchlings being of the same sex, which could lead to the extinction of some populations or even whole species.

Different patterns of health and disease

More frequent and more intensive heat waves could result in more heat-related deaths. These conditions could also aggravate local air quality problems. Global warming is expected to increase the potential geographic range and virulence of various tropical diseases as well. Disease-carrying mosquitoes are spreading towards the poles as climate shifts allow them to survive in formerly inhospitable areas. Mosquitoes can carry a variety of pathogens including dengue fever and malaria.

WHAT CAN BE DONE ABOUT GLOBAL WARMING?

Everyone agrees that cutting emissions is a good idea. This means reducing the amount of fossil fuels that we burn: we could make a big difference by improving the efficiency of heating systems and by reusing 'waste' heat from industry. Wherever possible, we should use renewable energy sources such as wind, water and solar power. Although many people object on safety grounds, making more use of nuclear energy would also help to reduce emissions of greenhouse gases.

In addition, we should limit the destruction of forests and use methods of agriculture that do not irreversibly damage the land. Of course, all of this is cancelled out if the human population continues to increase, so efforts should also be made to stabilise the world's population, to prevent further pressure on energy resources.

The United Nations Framework Convention on climate change agreed that industrialised countries would reduce carbon dioxide emissions first, providing developing countries space for development. The Kyoto Protocol (1997) set out that industrialised countries would reduce their emissions by 5.2 per cent compared with 1990 level by 2012. However, not all industrialised countries have put their full backing behind this policy and other strategies are becoming necessary.

Industrialised countries have put forward the idea of creating carbon sinks by planting new forests that will act as traps for carbon dioxide. Whether this would work or not remains to be seen, but the commitment to reduce emission of greenhouse gases during 2008–2012 is being put in jeopardy by the USA and Japan, who propose to use carbon sinks to meet their reduction commitments without curbing emissions from fossil fuels.

Conservation and the role of zoos

The role of zoos has been changing for the last few decades. Early zoos kept animals in barren cages designed to keep them visible at all times, with little or no attempt made to cater for their psychological or medical needs.

In the twenty-first century, the role of zoos includes the following three areas:

- Conservation. Many zoos are part of a worldwide network that collaborates to breed endangered species. **Studbooks** are used to manage captive breeding. The aim is to increase genetic diversity and reduce inbreeding by making sure that closely related individuals do not mate if at all possible. The long-term aim is to create a large, viable, genetically diverse population that can be used to re-stock the wild, if and when conditions are favourable.
- Animal study. Most zoos have research grants, which allow zoologists to improve the care of animals and find ways to increase the breeding of endangered species.
- Education. A zoo environment can teach everything from art to zoology. Zoos work in collaboration with local schools and colleges to provide opportunities to make science education more exciting.

Fig 36.22 Polar bear, Belgrade Zoo, Serbia. On 24 March 1999, NATO launched air strikes against the former Yugoslavia. Some of the animals at Belgrade Zoo were so traumatised that they resorted to self-mutilation and pregnant females aborted their young. The lack of investment has meant that these traumatised animals continue to be housed in small enclosures, the stress of which has led to repetitive behaviour, for example pacing (seen here).

SUMMARY

After reading this chapter, you should know and understand the following:

- Humans influence the environment of the Earth on a *global* as well as a *local* level.
- Farming is the major human activity that causes environmental damage.
- People need to farm to feed the increasing human population but many farming techniques cause long-term damage to the land, lowering its capacity to grow food in the future.
- To avoid this problem, it is important that farmers start to use **sustainable** methods of agriculture.
- Many human activities cause **pollution**. Pollutants can affect the air, the sea and rivers and lakes, and the land.
- Acid rain, use of CFCs and other substances that destroy the ozone layer, and increased production of greenhouse gases all contribute to air pollution.
- Water can be contaminated by organic pollution. This can lead to a sequence of events called **eutrophication**. Heat, oil and other substances can also pollute waterways.
- When forests are cleared to make new farmland, fewer trees are available to photosynthesise, we lose species diversity and the land becomes vulnerable to **erosion**.
- Severe soil erosion can lead to **desertification**. Many sustainable methods of agriculture try to minimise the effects of soil erosion.
- Scientific evidence shows that global warming is occurring due to high concentrations of greenhouse gases in the atmosphere.
- It is likely that global warming is contributing to changes in climate vegetation and even patterns in disease.

 Practice questions and a How Science Works assignment for this chapter are available at www.collinseducation.co.uk/CAS

APPENDIX: USING STATISTICS IN HUMAN BIOLOGY

Statistics is the only useful way of scientifically answering the question 'so what?'. Suppose you counted the number of earthworms in a square metre of soil sample. If you did this just once, you might extract six worms. All that this would really tell you is that some soil samples contain earthworms. If you did the experiment many times, you might be able to draw some conclusions about the density of the earthworm population in that particular field. However, if you did the experiment in two separate fields, you might end up with two very different collections of raw data, and this is where the application of statistics will help you answer the question 'Well, so what?'.

THE NULL HYPOTHESIS AND EXPERIMENTAL DESIGN

In science, progress is made by a process of **conjecture** and **refutation**, which in plain English means thinking up an explanation for an observation ('could this possibly be – ?'), then gathering data, analysing it and coming to the conclusion 'it's highly unlikely' or 'possibly, it could'. A vital point here is that you can never prove anything, but you can devise a **hypothesis**, test it and fail to disprove it, so you gather support for a particular idea.

A hypothesis is a testable idea. There are two types:

- The **experimental hypothesis** states that there will be a significant connection between cause and effect. An example could be 'alcohol slows down an individual's ability to do mental arithmetic'.
- The **null hypothesis** takes the opposite standpoint, such as 'alcohol has no effect on an individual's ability to do mental arithmetic'. A key point here is that **statistical tests only test the null hypothesis.** Statistical tests are needed to tell you whether your results are due to chance or not.

THE MEAN, AND THE STANDARD DEVIATION FROM THE MEAN

These are different types of average values. Averaging is very useful in biology, particularly on ecology field trips: it helps us to make the most of our results from sampling.

If we take the mean of several samples, we can get a more accurate estimate of the total number of organisms in the population that we are studying.

$$\text{Mean} = \frac{\text{total number of animals caught}}{\text{total number of samples taken}}$$

Distribution around the mean

We can work out the mean of a sample, but this does not tell us anything about the *range* of values in our sample. For example, a mean value for neck length in 200 swans might be 47 cm. The swan with the longest neck might have a neck 50 cm long, the one with the shortest might have a neck 44 cm long. Or something really unusual might be happening to the swan population and some swans might have necks only 20 cm long, while others had necks over 60 cm long. The mean itself does not tell us which is the real situation.

Normal distribution

Statistical tests usually involve situations in which there is a normal distribution. Fig A.1 gives an example of a normal distribution. The exact shape of the normal distribution depends on the mean and the **standard deviation** (**SD**), the range of values around the mean. Differences in standard deviation affect the shape of the distribution. All the curves in Fig A.2 represent normal distributions. Although the distribution is symmetrical in all the curves shown, the distribution changes as the standard deviation changes. When the SD increases, the curve becomes flatter; if the SD decreases, the curve is taller.

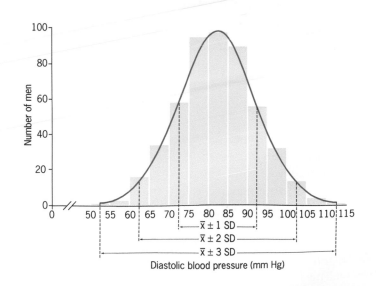

Fig A.1 Normal distribution of blood pressure. The mean blood pressure is 82 mmHg and the standard deviation is 10 mmHg. This means that 68 per cent of the sample have blood pressures between 72 and 92 mmHg and 95 per cent of the sample have blood pressures between 62 and 102 mmHg. It follows that if we sample, there is only a 0.05 per cent (1 in 20 chance) of selecting a male with a blood pressure less than 62 or greater than 102 mmHg

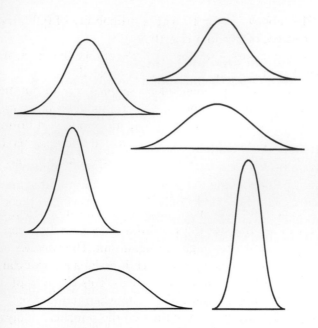

Fig A.2 In a normal distribution:
- Measurements greater than the mean and measurements less than the mean are equally common
- Small deviations from the mean are much more common than large ones
- 68 per cent of all the measurements fall within a range of ± 1 standard deviation from the mean and 95 per cent within ± 2 standard deviations

Standard deviation

We can tell how much individual values deviate away from the mean value by working out the standard deviation:

$$\text{standard deviation} = \sqrt{\frac{\Sigma(x - \bar{x})^2}{n - 1}}$$

where Σ = the sum of, x = measured value, $\bar{x}$ = mean value, and n = number of measurements.

To do this you would:

1 Take each of your measured values and subtract the mean value from each of them.

2 Square each value you obtain in step 1.

3 Add all your squared values together. This gives you the $[\Sigma(x - \bar{x})^2]$ part of the formula.

4 Divide the number you get in step 3 by the number of measurements on which your mean was based minus one. Work out the square root of the number you get. This is your standard deviation from the mean.

5 You can now show your mean value, plus or minus the standard deviation to indicate the range of values in your original sample.

SIGNIFICANCE OF DATA

The *t*-test

The *t*-test is used when we make a range of measurements rather than counting numbers. For example we might wish to compare the thickness of the leaves of a species of plant growing on the north side with those growing on the south

side of a wall. It is only really useful if you have more than about 20 data points – fewer than that, and you might be better off using the Mann–Whitney *U*-test (see later).

Table A.1 shows the results of one such set of measurements.

Table A.1

Thickness of leaves on south side of wall/μm	Thickness of leaves on north side of wall/μm
300	180
270	210
220	220
250	160
210	210
250	190
290	160
190	180
220	200
270	210

The statistical test you can use to find if there is a difference between two means is the *t*-test. The *t*-test compares both the mean and the standard deviation of two populations whose distribution curves overlap (Fig A.3).

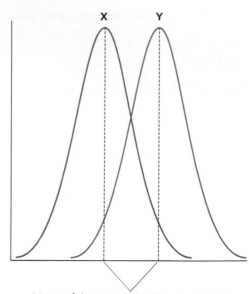

Means of the two overlapping populations

Fig A.3 What the *t*-test measures

(You would not be asked to calculate a *t*-test in an exam so you need not learn the formula. But you may need to use a *t*-test in an investigation.) The *t*-test compares the means and standard deviations of two populations. The *t*-test calculates:

$$\frac{\text{difference in means of two populations}}{\text{variability of the two populations}}$$

The variability of the population is the square of the standard deviation, so the formula for the t-test is

$$t = \frac{|x_1 - x_2|}{\sqrt{\dfrac{\sigma_1^2}{n_1} + \dfrac{\sigma_2^2}{n_2}}}$$

where $t = t$-test value
$|x_1 - x_2| =$ the difference between the mean value of population 1 (x_1) and the mean value of population 2 (x_2)
$\sigma_1^2 =$ the variance (square of the standard deviation) of population 1
$\sigma_2^2 =$ the variance (square of the standard deviation) of population 2
$n_1 =$ the number of measurements in population 1
$n_2 =$ the number of measurements in population 2
   You will see that this formula takes into account both the size of the differences and the size of the sample.
   The null hypothesis is that there is no significant difference between the thickness of leaves on the north and south facing walls.
   First calculate the mean thickness of each population of leaves:

Mean thickness of south facing leaves (x_1) = 247 µm

Mean thickness of north facing leaves (x_2) = 192 µm

$$|x_1 - x_2| \qquad = 247 \text{ µm} - 192 \text{ µm} = 55$$

Now calculate the standard deviation of each of population leaves:

Standard deviation of south facing leaves = 36.2

Standard deviation of north facing leaves = 21.5

$$\sigma_1^2 = (36.2)^2 = 1310.4$$

$$\sigma_2^2 = (21.5)^2 = 462.25$$

$$\frac{\sigma_1^2}{n_1} = \frac{1314.4}{10} = 131.04$$

$$\frac{\sigma_2^2}{n_2} = \frac{462.25}{10} = 46.22$$

$$t = \frac{55}{\sqrt{131.04 + 46.22}} = 4.131$$

In the t-test, the degrees of freedom are given by the sum of the number of measurements in both groups minus 2.
   In this case, the number of degrees of freedom is

$$(10 + 10 - 2) = 18$$

We now compare our values for t with the values of t in a table, available in all data books.

The table value for 10 with a probability of 0.05 and 18 degrees of freedom is 2.101.
   Our calculated value of t is 4.131. Since this is greater than 2.101, the null hypothesis is rejected.
   For the t-test for independent samples you do not have to have the same number of data points in each group. We have to assume that the population follows a normal distribution. The t-test can be performed knowing just the means, standard deviation and number of data points.

The Mann–Whitney U-test
If you have fewer than 20 data points, but more than 6, you can use this test to establish if differences between two groups of data might be significant. The way the Mann–Whitney test works is that it assigns an artificial value to each score in the data fields. This is a way of ironing out the 'patchiness' of data when number of samples is only small. You then do the statistical analysis on the artificial values – called rank scores – and look up the result in a table. It has limitations: it can only be used when the number of values in the data set is between 6 and 20. If it is below 6, you really don't have enough data to analyse, if it is above 20, you ought to use the t-test instead. This test is more approximate than tests done on larger data sets, but it is important because it still provides an answer to the essential question 'Do my results mean anything?'

The Spearman rank correlation
Basically, this is another test where you rank the values you obtain – not unlike the Mann–Whitney test – and then do your stats on the ranked values. You use this test to prove or disprove that there is a correlation between any two observed variables. A typical application would be to analyse the points on a **scattergram** (Fig A.4). Sometimes the data from these things is fairly clear – the scattergram points lie plainly on a straight line, showing you that there is a correlation, either positive or negative, between two sets of data.

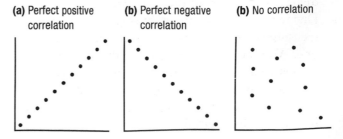

(a) Perfect positive correlation   **(b)** Perfect negative correlation   **(b)** No correlation

Fig A.4 Correlation

The chi-squared test
When we are comparing two or more sets of data, we need to find out if any differences between them are significant; are they different enough to tell us to look for an explanation?

Chi-squared is one statistical test that is used to decide whether results are **significant** or simply due to chance. Basically, the observed results are compared with those that you would expect of they were due to chance. The chi-squared test gives you a result in terms of probability. By convention, if the probability that your results are due to chance is **less than 5 per cent** (or 0.05 as a decimal), you can say that they are significant and so reject the null hypothesis.

The chi-squared test: a case study

A group of scientists carried out an investigation into the preferred resting places of leopards in a game park in Tanzania, to see if there was a difference between males and females. They tracked 200 leopards by sight, by radio collar and by analysis of their droppings. Their findings are shown in Table A.2.

The null hypothesis for this investigation is 'There is no difference in the distribution of male and female leopards at rest'.

Table A.2 The distribution of leopards

		In trees		In rocks		In open grassland	Row total
Males	a	19	b	56	c	25	100
Females	d	51	e	26	f	23	100
Column total		70		82		48	200 (grand total)

The formula for the chi-squared test is

$$\chi^2 = \sum \frac{(O - E)^2}{E}$$

where Σ = the sum of all, O = observed and E = expected.

Having got our observed results from the field studies, how do we work out the expected results? For each category of box, we can work out the expected value E from the formula

$$E = \frac{\text{Row total} \times \text{column total}}{\text{Grand total}}$$

For each box...

$$\text{Box a} = \frac{100 \times 70}{200} = 35$$

$$\text{Box b} = \frac{100 \times 82}{200} = 41$$

$$\text{Box c} = \frac{100 \times 48}{200} = 24$$

$$\text{Box d} = \frac{100 \times 70}{200} = 35$$

$$\text{Box e} = \frac{100 \times 82}{200} = 41$$

$$\text{Box f} = \frac{100 \times 48}{200} = 24$$

Having both the expected and observed values, one of the easiest ways to work out the value is χ^2 is to fill in a table like Table A.3.

Table A.3 Working out the value of χ^2

Box	Observed	Expected	O – E	(O – E)^2	$\frac{(O-E)^2}{E}$
a	19	35	−16	256	7.3
b	56	41	15	225	5.5
c	25	24	1	1	0.04
d	51	35	16	256	7.3
e	26	41	−15	225	5.5
f	23	24	−1	1	0.04
					Total = 25.68

The next step is to work out the degrees of freedom (D of F). This is a measure of the spread of the data; the more categories, the more degrees of freedom. The formula is

D of F = (no. of rows − 1) × (no. of columns − 1)

which in this case is

$$(2 - 1) \times (3 - 1) = 1 \times 2 = 2$$

So we have a value of 25.68 with 2 degrees of freedom.

What next? To get a probability value (the whole point of the test) we look up these values in a table of chi-squared values.

Table A.4 shows that, with two degrees of freedom, the chances of obtaining a χ^2 value of 25.68 is less than 1 per cent. This means there is a less than 1 per cent probability that our results are due to chance (we needed a value of 9.2 for 1 per cent probability and 5.99 for the vital threshold of 5 per cent probability). Values as high as 25 are only found when there is a marked difference in the observed and expected frequencies. From this we can reject the null hypothesis and accept the experimental hypothesis, i.e. there is a difference in the habitats preferred by male and female leopards. Note that we have *gained support* for our theory, but we can *never prove* it to be correct.

Chi-squared in examinations

You will not be asked to work through any examples of χ^2 in examinations – it's far too time consuming and the examiners want to test your biology, not your maths. However, you may be expected to know the basic principles and uses of χ^2.

Table A.4 Chi-squared values

Degrees of freedom	Probability					
	0.50	0.25	0.10	0.05	0.02	0.01
1	0.45	1.32	2.71	3.84	5.41	6.64
2	1.39	2.77	4.61	5.99	7.82	9.21
3	2.37	4.11	6.25	7.82	9.84	11.34
4	3.36	5.39	7.78	9.49	11.67	13.28

ANSWERS TO QUESTIONS

CHAPTER 1

1 **(a)** The full stop is about 0.3 mm = 300 μm.
   (b) 1 million.
2 Dispersal of oil pollution, breakdown of waste plastic.
3 1000.
4 They are all secretory. You would recognise them by their prominent ER and Golgi bodies.
5 Because the ER is a network of thin membranes which would not show up under a light microscope.
6 All its tissues would be destroyed.
7 We could destroy tumours.
8 Because they could digest the organelle/cell that made them.
9 To respire and make ATP/energy for swimming; sperm swim incredible distances and use a vast amount of energy.
10 **(a)** 500, **(b)** 200, **(c)** 5000.
11 The root system, because it receives no light.

CHAPTER 2

1 Ciliated epithelium.
2 Smooth muscle because its contraction is slowest.
3 It fatigues too quickly.
4 The reproductive system.

CHAPTER 3

1 3. No, because there is no dot at the right position for C.
2 Q=25/50=0.5, R=15/50=0.3. R is the least soluble.
3 Sucrose consists of glucose and fructose, both reducing sugars. In the sucrose molecule, the reactive groups of the two component sugars point inwards and so are 'hidden' in the 3D structure. They only become available to react with Benedict's solution when the sucrose has been split apart.
4 The glycosidic link, which is between carbon 1 and carbon 4.
5 It would cause an osmotic imbalance – the cytoplasm would become too concentrated.
6 Amylose because it has more ends to add or remove glucose units from.
7 Because few organisms have the enzymes that can degrade cellulose.
8 Oleic and linoleic acid.
9 Oleic and linoleic. They are both unsaturated.
10 46 x 5 cm = 2.3 metres.
11 TAGCAATGG.
12 Bases (RNA has U instead of T), sugars (RNA has ribose, DNA has deoxyribose), strands (DNA has 2, RNA has 1), life span (DNA is long, RNA is short).

CHAPTER 4

1 1 mm.
2 Water-loving, water-hating. In a phospholipid, the phosphorus end is attracted to water, the lipid end repels water.
3 Because the atoms/molecules in liquids and gases are free to move, but those in solids aren't free to move.
4 To maintain the diffusion gradient.
5 Lung, intestine, leaf, kidney, placenta, root.
6 Because water passes out of the bacteria by osmosis, dehydrating and killing them.
7 You must have a solution to compare it to: hypertonic to what?
8 Cyanide blocks aerobic respiration, so virtually no ATP is available.

CHAPTER 5

1 Protein synthesis.
2 Protein, lipid, nucleic acids.
3 Because they all contain the same food type: starch.
4 Tears, sweat, in the lysosomes of some white cells.
5 It stimulates glycogen production.
6 Peptide bonds.
7 Proteins are denatured. This is irreversible.
8 Because that is the optimum temperature for enzymes.
9 It would be denatured by the acid.
10 There would be no activity at all: the plot would go along the x-axis.

CHAPTER 6

1 The castrated ram has no testes and does not produce testosterone or pheromones; he has nothing with which to turn on the ewes, and they would just ignore him.

2 Fermentation is a vague word, and in biotechnology is taken to mean the metabolics/respiration of microbes. Fermentation can be either aerobic or anaerobic. Glycolysis is the first, anaerobic, process in respiration.

CHAPTER 7

1 The toddler needs energy for rapid growth and development.
2 Benedict's test: add blue Benedict's and heat; will give an orange-red precipitate if sugar is present.
3 Biuret test: will change from blue to lilac if protein is present.
4 The symptoms are those of vitamin A poisoning.
5 The meal is reasonably high in energy, complex carbohydrate and fat. It would be a good basis for a lunchtime meal for an adult man. The brown bread provides some protein, fibre and most of the B vitamins; the cheese and butter provide mainly fat, and vitamin A. It lacks vitamin C, iron and other minerals; it could be improved by the addition of a side salad, or just by adding some salad – tomatoes, green salad leaves and cucumber to the sandwich, or by following it up with a few pieces of fruit.
6 Exercise increases basal metabolic rate and the muscle to fat ratio; both will use up more energy, so a better diet will prevent more fat being added, while exercise will help take existing fat off.

CHAPTER 8

1 They would have to eat smaller, more frequent meals as they would have no stomach to store large meals.
2 About 2–3.
3 Endopeptidase cuts within the chain, exopeptidase removes the end amino acid.
4 **(a)** You would expect rennin production to decrease with age. It does. Young mammals on milk diets need it but older animals do not. **(b)** No, since only mammals produce milk.
5 So they do not self-digest the organs that produced them.
6 The small intestine is a very absorptive region and can take in many of the nutrients produced by bacteria; only a small amount of absorption takes place in the colon, so nutrients produced here will just be lost in the faeces.

CHAPTER 9

1 7.5 litres min^{-1}.
2 One five thousandth.
3 Dead space would be enormous and diver would breathe the same air in and out. Also, the chest muscles are not strong enough to keep expanding the lungs against the water pressure at that depth.

CHAPTER 10

1 **(a)** 6 and **(b)** 15 litres per minute.
2 **(a)** Increase. **(b)** Decrease.
3 120 mmHg = systolic pressure, 70 mmHg = diastolic pressure.
4 **(a)** Increase. **(b)** Decrease.
5 Hb 'steals' oxygen from an oxygen-rich environment and delivers it to tissues that are oxygen-poor.

CHAPTER 11

1 Because they detect a change in internal conditions and then adjust it back to the normal level.
2 The urine of diabetic dogs has high concentrations of sugar. This is very attractive to flies.
3 They are metabolically active and are involved in a rapid exchange of large amounts of materials.
4 Any excess of the daily need would be deaminated and excreted.

CHAPTER 12

1 Humidity changes our ability to lose heat by sweating. Low humidity means a higher upper lethal temperature because we can sweat and lose heat by evaporation more effectively.
2 They must be below, otherwise they would disappear as old skin was shed and replaced by new skin.
3 Piloerection because we have so few hairs.
4 To replace the sodium and chloride ions that they lose constantly by sweating.

CHAPTER 13

1 7.5 litres.
2 14.4 times per hour.
3 Blood pressure would drop, filtration would become ineffective, kidneys would suffer damage and fail.
4 The loop of Henle is shaped like a hairpin, and it behaves in the same way as a countercurrent multiplier.
5 Dark yellow urine indicates dehydration. Brown = trouble!
6 Much more lost in faeces – dehydration.
7 Proximal and distal tubules, ascending limb of loop of Henle.
8 It makes the blood more concentrated, either by adding salt or removing water or both.

CHAPTER 14

1 We are using the lactic anaerobic system. This means that respiration can't be completed and so the intermediate compound (lactate) builds up.

CHAPTER 15

1 A neurone is a single specialised nerve cell, a nerve is a bundle of neurones in connective tissue.
2 **(a)** Active transport mechanism needs specific membrane protein and ATP. **(b)** It moves particles against a diffusion gradient rather than down a diffusion gradient.
3 **(a)** To ensure that a neurone responds to a specific minimum level of stimulation. **(b)** Continuous depolarisation and fatigue of neurone.
4 Myelinated nerves conduct impulses more quickly.
5 Narrow to make diffusion of the neurotransmitter as fast as possible and high resistance to ensure that the action potential can't jump across.
6 Because the transmitter is only made on one side.
7 No action potential.

CHAPTER 16

1 Discussion question, no answer provided.
2 Enclosed in bony verterbral column; protected from shocks by cerebrospinal fluid that acts as cushion; enclosed by the spinal meninges, three tough membranes.
3 Proprioception makes it possible to stand on one leg with your eyes shut but feedback from the eyes helps with balancing, providing further help, so its easier with your eyes open.
4 **Right motor area.**

CHAPTER 17

1 They have to support more weight.
2 The spinal cord.
3 Ligaments connect bones together at joints; ligaments need to be flexible to allow the joined bones to move.
4 A tendon joins a muscle to a bone. As the muscle contracts it exerts a force on the bone, moving it. If the tendon were elastic, the bone would move less than the change in muscle length.
5 **(a)** Smooth, **(b)** smooth, **(c)** skeletal, **(d)** cardiac, **(e)** skeletal.

CHAPTER 18

1 The digestive juices undiluted by food would attack the stomach lining. This could cause an ulcer.

CHAPTER 19

1 Chemoreceptor; exteroreceptor.
2 Exteroreceptors since they respond to stimuli from outside the body.
3 The light-sensitive film.
4 **(a)** The ciliary muscles relax.
   (b) The lens becomes thinner due to tension on the suspensory ligaments. **(c)** The pupil dilates.
5 Night sight (i.e. vision in dim light) needs rods that work properly. Vitamin A is necessary for the formation of retinal in rhodopsin.
6 **(a)** Blue, green and red cones. **(b)** Blue cones only.

CHAPTER 20

1 It would secrete more GnRF, thus stimulating more gonadotrophin release and therefore more oestrogen release.
2 Less energy and resources are wasted; the changes of successful fertilisation are increased.

3 They both contain the same amount of DNA. Both gametes are haploid – they contain only one copy of each chromosome.

4 Day 21.

CHAPTER 21

1 5.

2 No. Growth occurs evenly in the crocodile's body, so the ratio stays the same.

3 (a) The brain is small at birth to allow the baby's head to pass through its mother's pelvis. It then develops rapidly to enable the child to become independent and have a better chance of survival. (b) The reproductive organs do not mature until we are large enough and emotionally mature enough to have children. In our society, cultural pressures delay childbearing until well past the age when females are physiologically ready to have babies.

4 Between the ages of 2 and 3 years.

5 Human tissue.

6 (i) The person might have a condition that prevents them from making it. (ii) The hormone is produced in pulses and the sample was taken between pulses. (iii) The hormone is quickly converted into another form or is broken down as soon as it is released.

7 The number of younger people in the population is growing at an equivalent rate.

CHAPTER 22

1 No: the species certainly goes on but individuals frequently perish.

2 Tall.

CHAPTER 23

1 DNA contains the genetic code, RNA is built up on it, then RNA is used as the instructions for making proteins.

2 $n = 23$.

3 G_1, S and G_2.

4 The amount of DNA would be halved at each division.

5 (a) In ovaries and testes respectively. (b) Because it reduces the chromosome number (by half).

6 In a human cell there are 23 bivalents.

CHAPTER 24

1 (a) UAUGCGAUA. (b) 3.

2 They eat and digest protein, absorb amino acids, transport them in the circulation from the gut to cells.

3 Anabolic.

4 61 different tRNAs carry only 20 different amino acids.

5 Six: all ^{14}N strands, two hybrids.

6 Substitution.

7 With no spindle, the chromosomes double but cannot separate, so a cell with twice the normal number of chromosomes results.

CHAPTER 25

1 50:50 (half Yy and half yy).

2 If you use a plant that is TT, all the offspring of a cross with a Tt or a TT plant will be tall and you wouldn't gain any information.

3 700 000/2500 = 280.

4 In previous centuries, villagers married people from the next village. Because we now have more transport, people move around the world more easily and so people meet and mate with people to whom they are unlikely to be related.

5 XhXh. Female haemophiliacs must be a double recessive whose mother is a female carrier and whose father is a male sufferer. The chance of two such people meeting and marrying is unlikely and male haemophiliacs often don't live long enough to reproduce.

6 One in two chance.

7 No, If the mother is O and the father is A, the child could be only A or O.

8 Yes, one could be aabbcc (white) and the other could be AABBCC (very dark).

9 9 agouti, 3 black, 4 albino.

CHAPTER 26

1 Acquired characteristics don't affect an organism's DNA and so cannot be passed on to the next generation.

2 Insects have short life spans and high reproductive capacity; whales and rhinos have long life spans and only produce a handful of offspring during their lifetime.

3 Peppered moth, *Biston betularia*: selection of melanic form following industrial revolution. Hawks: selection for great visual acuity. Cheetah: selection for high-speed running.

4 (a) Analogous. (b) Homologous.

CHAPTER 27

1 It performs transcription in reverse: it makes DNA from RNA.

2 Because the smaller recognition site is more common in any DNA sequence.

3 An individual has a brain with memories – these memories are not contained in the cell used for cloning. In cloning you may create a new brain but not the memories.

CHAPTER 28

1 We now have vaccines and antibiotics, a better standard of living, less poverty and better sanitation.

2 (a) Most protein digestion occurs in the small intestine, before the pancreatic duct. (b) Pancreas does not produce as much trypsin when it starts to fail because of pancreatitis.

3 Reduce the amount of phenylalanine and increase the amount of tyrosine.

4 Less messy, more convenient to carry around.

5 (a) Endoscopy. (b) X-ray.

CHAPTER 29

1 Infection with *Salmonella typhimurium* does not always cause the disease – some people are carriers.

2 (a) Gloves, masks, sterile instruments. (b) Gloves, mask.

3 There are now effective vaccines for measles, mumps and diphtheria, but none for the common cold.

4 If you don't have the gene to produce the cell receptor for a particular virus, it cannot get into your cells.

5 Antibiotic to kill the bacteria, antitoxin to neutralise the toxin they release when they eat.

6 Bacteria do not divide as quickly at lower temperatures so curve will be lower and flatter.

7 Salmonellosis develops when the bacteria that cause it multiply inside the cells that line the intestine. Other types of food poisoning are caused by bacteria that do not multiply in the body. They simply release toxins, and so cause symptoms more quickly.

8 Generally – a knowledge of the life cycle and mode of transmission of any pathogen is essential in its prevention. The flu virus is transmitted by droplet infection so isolation of patients will limit spread, as will general education such as using tissues when sneezing.

9 Faster global transport – trains, boats etc. Also – a large human population, more overcrowding in cities.

10 Attack from antibodies and cells of the immune system.

11 More people are in the water to get infected in the day.

12 No problems from osmosis.

CHAPTER 30

1 Warm, moist, lots of food.

2 Mucous secretions containing dust and dead cells, which should be swept by cilia to the exterior, remain trapped in the airways to cause congestion and infection.

3 Anticoagulants stop blood clotting so that leech can feed freely. Uses: microsurgery, plastic surgery, dispersal of blood clots, treatment of chronic skin ulcers.

4 Two basophils, a lymphocyte and a monocyte.

5 Not likely to cause an allergic reaction.

6 Some of the donors would have been blood group O, some of the recipients would have been blood group AB, and so no adverse reactions would have occurred.

7 AB blood has A and B antigens on the surface of the red blood cells. These react with antibodies present in the blood of people from the other three groups.

8 Agglutination occurs when red cells stick together due to antibodies making them sticky. Blood clotting is the complex reaction that turns fibrinogen into a fibrin mesh, in which red cells get stuck.

9 Fungi are eukaryotes while antibiotics work by interfering with prokaryote metabolism, and don't affect eukaryotes.

10 If the course is not finished, not all the bacteria will be killed. The most susceptible die first, leaving the more resistant ones to breed back. The behaviour therefore speeds up the evolution of superbugs.

CHAPTER 31

1 They live longer.

2 Ozone hole, people holiday in more exotic locations.

3 Need for stress relief, peer pressure, addictive nature of nicotine.

4 It uses more energy than a muscle at rest.

5 (a) More oxygen and fuel to muscles. (b) Don't have to go to the loo as often.

6 The interplay between control genes and genes that code for the LDL receptor might result in a lower risk of developing high blood cholesterol.

7 Hypertension and smoking.

8 Enzymes digest adhesion proteins that hold normal cells together, making it easier for tumour cells to grow.

9 Doing things that are bad for you is fun. Eating lettuce, drinking mineral water and doing lots of strenuous exercise, generally, isn't fun.

10 Cost:benefit ratio is poor.

CHAPTER 32

1 Temperate deciduous forest.

2 Abiotic: (a), (c), (d). Biotic: (b), (e), (f).

3 Habitat is where an organism lives, the term ecological niche also describes what it does and how it relates to its physical environment and to other organisms.

CHAPTER 33

1 We still need to maintain body temperature, breathe, keep our hearts beating and keep cellular processes going. All this requires energy.

2 False. Some heterotrophs eat other heterotrophs (e.g. lions eat gazelles).

3 The transfer of energy from one organism to another.

4 **Class exercise**

5 Provided – water, carbon dioxide; taken away – oxygen, glucose/sugar.

6 Oxidation is the loss of electrons, reduction is gain (OILRIG).

7 When a molecule is reduced it gains energy.

8 More gas can be exchanged per unit time. Thin-walled, moist, good vascular network.

9 Organic 'fuel' is oxidised to release energy and do useful work, and it produces heat as waste.

10 It would take too long to break glucose down. ATP releases energy when one phosphate group is removed.

11 The first uses a substrate as phosphate source, the second uses free phosphate and is powered by energy gained from redox reactions in the electron transport chain.

12 An extra electron.

13 It goes into the blood, is taken to the lungs and is breathed out.

14 Because each glucose provides two pyruvates and therefore two acetyl-CoAs.

15 These more metabolically active cells need more ATP to power their processes and therefore need more mitochondria (the sites of ATP synthesis in the cell).

16 Active transport requires energy.

17 (a) Each one of the 20 amino acids has a different number of carbon atoms. In respiration, the reactions which break each one down produce different volumes of CO_2 and use different volumes of O_2. (b) CO_2 is produced but no O_2 is used: any value divided by 0 is infinity.

18 It prevents them being eaten.

19 50 units.

20 The end-products are different (though the processes are similar).

CHAPTER 34

1 Nitrogen fixing bacteria in the root nodules of clover would increase the amount of nitrogen compounds in the soil. When the clover was ploughed in at the end of the year, these compounds would decompose into nitrates and would increase crop fertility. The advantage of this from an environmental point of view is that the farmer would need to use fewer artificial nitrogen fertilisers.

CHAPTER 35

1 Intra: within a species. Inter: between different species.

2 The oak wood.

CHAPTER 36

1 Sustainable: used but continually replaced, e.g. solar power, hydroelectric power. Unsustainable: when used it is gone forever, e.g. coal deposits, oil, gas.

2 Take two water samples A and B. Record the oxygen concentration of A. Incubate B in the dark at 20 °C for five days. Then measure the oxygen concentration of B. The BOD is B – A.

PHOTOGRAPH CREDITS

The publishers would like to thank the following for permission to reproduce photographs (T = Top, B = Bottom, C = Centre, L= Left, R = Right):

A-Z Botanical Collection/M Nimmo, 26.6; **Allsport**/T Duffy, 3.24, M Powell, 7.12, M Steele, 14.1, S Forster, 14.4, 33.35; **Aquarius Library**, 27.1; **Ardea**/S Gooders, 32.2L, D&K Urry, 35.6; **F G Banting Papers, Thomas Fisher Rare Book Library, University of Toronto**, 11.4; **Biophoto Associates**, 1.15a, 1.16a, 1.18, 1.20, 1.29, 1.32, 1.33, 2.4, 4.18b, 4.18c, 4.22, 7.1a, 7.1b, 7.8, 8.2L, 8.3L, 8.4L, 8.5L, 8.6L, 11.7d, 11.9, 15.13b, 17.3b, 19.2, 21.15a, 23.1a, 23.6, 23.10, 23.11, Table 23.3, 23.15b, 28.3, 29.22e, 30.5, 33.3B, 33.7, 33.18, 33.42a, 33.52, 35.16L, 35.19R; **John Birdsall Social Issues Photo Library**, p.318, 12.10, 21.1R; © **Prof W Bishop, M.D & University of Iowa. All rights reserved**, 8.2R, 8.5R; **Michael Bliss, 'The Discovery of Insulin' (Paul Harris Publishing, Edinburgh, 1983)**, p.174; **Chris Bonington Picture Library**/D Scott, 9.10a; **Bridgeman Art Library, London/Prado, Madrid, 'The Three Graces' by Rubens**, 3.21a; **Bridgeman Art Library, London/Prado, Madrid/Index, 'Adam and Eve' by Titian**, p.392; **Courtesy of Centres for Disease Control and Prevention**, 29.8; **Cephas**/M Schilder, 6.11T, TOP/C Adam, 6.11B, 33.49TR, 33.49BL, StockFood, 33.49TL, 33.49BR; **Martyn Chillmaid**, 9.7; **Cleveland Museum of Natural History, Ohio**, 22.3T, 26.23; **Bruce Coleman Ltd**, 19.13R, R Williams, 19.13L, W Layer, 21.1L, K Taylor, 24.9TL, 24.15b, 26.14b, 26.19, 35.11, 35.13, A Purcell, 24.9TR, J Grayson, 24.9BL, F Labhardt, 24.9BR, A Bacchella, 26.9, S Krasemann, 26.17L, P Davey, 26.22, Dr F Sauer, 33.8L, D Davies, 33.8R, J Burton, 35.17; **Collections/** A Sieveking, p.298, G Howard, 21.11a; **College of Veterinary Medicine, Texas A&M University**, p.416; **Corbis/epa/**Juan Manuel Villasenor, 21.7; **Corbis-Bettmann/UPI**, 29.2; **Brian Culcheth**, 32.3; **H Damasio, T Grabowski, R Frank, A M Galaburda, A R Damasio, 'The Return of Phineas Gage: Clues about the brain from a famous patient' (Science, 264: 1102-1105, 1994), Dept of Neurology & Image Analysis Facility, University of Iowa**, p.242; **Environmental Images**, 36.4b, 36.4c, 36.10, 36.11R, 36.13; **flpa-images.co.uk/**P Peacock, 6.4, M J Thomas, 25.7, R Wilmshurst, 26.4T, 35.4, S Hosking, 26.4B, D Hosking, 26.16, T Whittaker, 26.17R, Panda, 33.15, P Perry, 35.10, M Gore, 35.14BR, N Cattlin, 33.26, 33.27, 34.6, 36.1, 36.3; **Vivien Fifield**, 26.3; **Fresh Food Images/** Maximilian, p.130; **GSF Photo Library**, 26.6L; **Getty Images**, 7.16, p.268, Stone/B Thomas, 14.10, p.568, Riser/J Riley, 18.1, Stone/ E Honowitz, 22.5, Hulton Archive/J J Lambe, 29.20, Riser/T Macpherson, 31.2; **Paul Glendell**, 36.11L; **GlucoWatch ® Biographer (registered trademark of Cygnus Inc, Redwood City, California, USA)**, 6.20; **www.goldenrice.com**, 6.8, 6.9, 6.10; **Green Moon Ltd**, 12.3; **Sally & Richard Greenhill Photo Library**, 21.2, 21.3; **HarperCollins Publishers**, 33.46; **Istockphoto**, 7.11, 12.4, p.438, p.588T, 36.15, S Rohrer, 7.13, N Silva, 12.4, M Gascho, 12.14a, A Pustovaya, 14.5, J Rihak, 19.1, L Svara, 20.20, Z Kostic, p.356, Y Chamberlain, 25.1, P Rosado, 27.4; © **Jupiterimages 2008**, 16.12; **Courtesy Kobal Collection/20th Century Fox/Paramount**, 12.5; **Courtesy Kobal Collection/ Dreamworks/Warner Bros**, 22.1; **Andrew Lambert**, 3.4, 3.10, 3.13, 3.16, p.74, 7.5, 13.2, 14.8, 31.23; **Lommond Mountain Rescue Team/** G Baird, 12.11; **London Borough of Richmond upon Thames**, 36.16; **MPL International/**R Adshead, 33.14; **Microscopix/**A Syred, 17.2b; **www.monsato.com**, 6.7; **Used with permission of the Muscular Dystryophy Association of the United States**, 25.15; **NASA**, p.258; **NHM Photo Library**, 16.8a, 26.1; **NHPA/**M Stelzner, 6.3, B Wood, 12.6; **Natural Visions/**Heather Angel, 32.6d, 32.11B, 33.17, 33.47, 34.24, B Rogers, 32.11T, I Tait, 32.14; **Nature Picture Library/**D Wechsler, 35.25; **Nature and Science AG/**F-L Vaduz, 13.7a; **Novo Nordisk A/S**, 11.5, 11.6; **Oxford Scientific Films (OSF)/**D Broomhall, 10.23B, H Pooley, 12.1, M Hill, 32.1, T de Roy, 32.6a, S Osolinski, 32.6b, R Redfern, 32.9, D H Thompson, 32.12, D Macdonald, 32.13, G I Bernard, 33.1, W Shattil & B Rozinski, 35.12; **PA Photos/**Abaca/EMPICS Entertainment, 3.21b, Sutton, 14.7, AP, p.280; **Panos Pictures/**T Page, 7.6, P Harrison, 28.1, H Davis, 29.24, J Hammond, 29.28; **Dr C Paterson, Ninewell's Hospital, Dundee**, 3.39; **www.phil.cdc.gov**, 29.11, 29.13; **Photofusion/**M Murray, 20.1; **Photoshot/**Woodfall Wild Images/B Gibbons, 35.27; **Planet Earth/** E Darack Photography, 26.14a, M Snyderman, 26.14cL, J Jackson, 26.14cR,

T Dressler, 33.3TR, Purdy & Matthews, 33.3C, M Welby, 35.5, A Bartschi, 35.14T, N Garbutt, 35.14BL; **Punch Limited**, 28.2; **Purdue News Service/** D Umbeger, p.24; **Redferns/**Michael Och's Archive, p.374; **Rex Features Ltd**, 7.4, p.186, 13.1; **Ronald Grant Archive**, 24.15a; **Roselin Institute, Edinburgh**, 23.1b, 23.1c, 27.12; **Royal Holloway College, University of London, EM Unit**, 1.22, 1.30a; **Science Photo Library (SPL)**, 1.10, 1.24, 3.20, 10.3c, 10.12, 15.11, 21.15b, 23.5, 25.2, 29.12, 29.22b, 29.27, p.470, SPL/Dr Y Nikas, p.2, 20.16, 27.2, 27.22, SPL/ Dr G Murti, 1.2, 24.16T, 27.11, SPL/D Scharf, 1.4a, 1.4b, 1.6, 23.3, 31.28, 33.50, 35.26, SPL/Moredun Animal Health Ltd, 1.7a, 35.19L, SPL/Lawrence Livermore Laboratory, 1.11, SPL/D Fawcett, 1.13a, SPL/Prof Gimenez-Martin, 1.14, 23.9a, 23.9b, 23.9c, 23.9d, 23.9e, SPL/Dr J Burgess, 1.26, SPL/J Berger, 1.27, SPL/D Phillips, 2.1, SPL/John Radcliff Hospital, 2.3a, SPL/BSIP Leca, 2.3b, SPL/D Van Ravenswaay, p.32, SPL/C D Winters, 3.6, SPL/S Gschmeissner, 3.19, SPL/K Eward/Biografx, 3.32, SPL/Professor K Seddon & Dr T Evans, Queen's University, Belfast, 3.35, SPL/B Nelson/Custom Medical Stock Photo, p.58, SPL/GJLP, 4.6, 16.1b, SPL/O Burriel, 4.17, 7.3a, 7.3b, 7.14, SPL/J C Revy, 5.2, 14.2, 27.14, 27.20, 30.11b, 33.33, SPL/W Baumeister, 5.8, SPL/A Bartel, 5.15, 22.3CR, SPL/J Mason, p.88, SPL/D Leah, 6.1, SPL/F Sauze, 6.14, SPL/Saturn Stills, 6.19, 11.2, 28.9, 30.13, SPL/Centre Jean Perrin, 6.24, SPL/H Young, 6.25T, 31.5, SPL/J King-Holmes/D Mercer, p.112, SPL/M Clarke, 7.7, 20.19, SPL/C Pedrazzini, 7.10, SPL/CNRI, 8.3R, 8.4R, 15.10b, 22.2, p.342B, 23.7, 29.19, 16.3b, SPL/D M Martin, M.D, 8.6R, SPL/D Parker, 8.10, 12.12, 35.23, SPL/A Tsiaras, p.144, 20.17, SPL/M Dohrn, Royal College of Surgeons, 9.2b, 29.6, 33.30, SPL/A & H Frieder Michler, 9.6a, SPL/BSIP, Edwige, 9.8, SPL/Prof P Motta/Dept of Anatomy/University "La Sapienza", Rome, 9.9, 13.7b, 17.2c, 19.8, 20.4b, SPL/Biophoto Associates, 10.3b, 15.3b, 31.17, SPL/G Tompkinson, 10.4a, 19.7, SPL/A Crump, TDR, WHO, 10.15, SPL/E Reschke, P Arnold Inc, 10.16, SPL/E Grave, 10.20a, 29.22a, 29.22c, 35.16R, SPL/SIU, 11.7a, SPL/Dr R Clark & M R Goff, 12.2, SPL/Dr P Marazzi, 12.17, 21.15c, 26.20, 30.1a, 30.1b, 31.24, 36.9, SPL/ H Morgan, p.200, 16.10a, 16.10b, 16.13, 20.15, 27.21a, SPL/M Chillmaid, 13.14, SPL/J King-Holmes, p.212, 27.21b, p.486, SPL/M Kage, 15.3a, 19.3a, SPL/Eye of Science, 15.13a, 29.5, 29.22d, SPL/D Lovegrove, 15.18, SPL/ M Delvin, 16.1a, SPL/G Watson, 16.3a, SPL/P Plailly, Eurelios, 17.8, 27.13, SPL/P Menzel, 17.12, 22.3BR, 27.15b, SPL/ Dr H C Robinson, 18.2, 30.6, SPL/J Lepore, 18.8, SPL/P Jude, 19.5, SPL/A McClenaghan, 19.15b, SPL/ Prof P M Motta, G Macchiavelli, S A Nottola, 20.8b, SPL/S Terry, 20.11, SPL/Edelmann, 22.3CL, 27.3, SPL/J Holmes/Celltech Ltd, 22.3BL, 27.10, SPL/S Fraser/Royal Victoria Infirmary, Newcastle upon Tyne, p.342T, SPL/Profs P M Motta & J Van Blerkom, 23.2, SPL/M Sklar, 23.4, SPL/ H Young, 23.15a, 28.5, SPL/Omikron, 24.16B, SPL/A Hart Davis, 25.16, SPL/National Library of Medicine, 26.2, SPL/T Craddock, 26.10, 36.5, SPL/R Sutherland, 27.5, SPL/Dept of Clinical Cytogenetics, Addenbrookes Hospital, 27.6, SPL/K Guldbrandsen, 27.15a, SPL/P Plailly, 27.19, 31.26, SPL/Deep Light Productions, 28.4, SPL/Dr L Stannard, UCT, 28.10, SPL/ P Saada/Eurelios, 28.11, SPL/B Dowsett, p.448, 33.43, SPL/H Pincis, 29.3, SPL/M Meadows, P Arnold Inc, 29.4, SPL/NIBSC, 29.7, 30.4, SPL/S Ford, 29.17, 29.26, 33.51, SPL/L Georgia, 29.18, SPL/A Dex, Publiphoto Diffusion, 29.21, SPL/Dr A Brody, 30.7, SPL/Dr K Lounatmaa, 30.18, SPL/J Varney, 31.15, SPL/J Ashton, 31.19, SPL/J Stevenson, 31.22, SPL/ P Jerrican, 31.25a, SPL/Breast Screening Unit, Kings College Hospital, London, 31.25b, SPL/J Durham, 31.27, SPL/ Tom van Sant/Geosphere Project, Santa Monica, p.520, SPL/P M Andrews, 33.2, SPL/J Howard, 33.3TL, SPL/A Syred, 33.16, SPL/© Health & Safety Laboratory, 33.31, SPL/Dr K Porter, 33.42b, SPL/S Fraser, 35.7, 36.4a, 36.7, SPL/Biomedical Imaging Unit, Southampton General Hospital, 35.15, SPL/Y Hamel/Publiphoto Diffusion, 36.4d, SPL/S Camazine, 36.20, SPL/NASA, 36.21a, 36.21b, SPL/R Brook, 36.22; **Science and Society Picture Library/Science Museum**, 1.1, 9.12; **Kath Senior**, p.588B; **Shout Pictures**, p.156, 31.4; **Still Pictures/**N Cobbing, 6.6, E Parker, 6.17, J Schytte, 6.18, M Edwards, 13.4, p.224, N Dickinson, 32.4, F Bavendam, 32.6c, J Isaac, 33.29; **C Thompson**, 13.5; **John Walmsley Photography**, 32.2R; **The Wellcome Trust**, 7.15, 12.14b, 24.13, 28.6, 30.8, 31.7; **Robin Westlake, US National Marine Fisheries Service, Southwest Fisheries Science** Center, p.560; **Wikipedia/**GDFL, 20.6, I Giel, 21.14, J Pfeifer, p.332; **Photo courtesy of Eyam Parish Church/Mike Williams**, 29.1

Cover: Science Photo Library/Eye of Science

Chapter contents photo: Istockphoto/S Kaulitzki

INDEX